Nanobiotechnology
Human Health and the Environment

Nanobiotechnology
Human Health and the Environment

Edited by
Alok Dhawan
Sanjay Singh
Ashutosh Kumar
Rishi Shanker

Boca Raton   London   New York

CRC Press is an imprint of the
Taylor & Francis Group, an **informa** business

CRC Press
Taylor & Francis Group
6000 Broken Sound Parkway NW, Suite 300
Boca Raton, FL 33487-2742

© 2018 by Taylor & Francis Group, LLC
CRC Press is an imprint of Taylor & Francis Group, an Informa business

No claim to original U.S. Government works

Printed on acid-free paper

International Standard Book Number-13: 978-1-4987-2142-4 (Hardback)

This book contains information obtained from authentic and highly regarded sources. Reasonable efforts have been made to publish reliable data and information, but the author and publisher cannot assume responsibility for the validity of all materials or the consequences of their use. The authors and publishers have attempted to trace the copyright holders of all material reproduced in this publication and apologize to copyright holders if permission to publish in this form has not been obtained. If any copyright material has not been acknowledged please write and let us know so we may rectify in any future reprint.

Except as permitted under U.S. Copyright Law, no part of this book may be reprinted, reproduced, transmitted, or utilized in any form by any electronic, mechanical, or other means, now known or hereafter invented, including photocopying, microfilming, and recording, or in any information storage or retrieval system, without written permission from the publishers.

For permission to photocopy or use material electronically from this work, please access www.copyright.com (http://www.copyright.com/) or contact the Copyright Clearance Center, Inc. (CCC), 222 Rosewood Drive, Danvers, MA 01923, 978-750-8400. CCC is a not-for-profit organization that provides licenses and registration for a variety of users. For organizations that have been granted a photocopy license by the CCC, a separate system of payment has been arranged.

Trademark Notice: Product or corporate names may be trademarks or registered trademarks, and are used only for identification and explanation without intent to infringe.

Library of Congress Cataloging-in-Publication Data

Names: Dhawan, Alok, editor. | Singh, Sanjay, 1980- editor. | Kumar, Ashutosh, 1984 March 26- editor. | Shanker, Rishi, 1960- editor.
Title: Nanobiotechnology : human health and the environment / [edited by] Alok Dhawan, Sanjay Singh, Ashutosh Kumar, and Rishi Shanker.
Other titles: Nanobiotechnology (Dhawan)
Description: Boca Raton : Taylor & Francis, 2018. | Includes bibliographical references and index.
Identifiers: LCCN 2017056308| ISBN 9781498721424 (hardback : alk. paper) | ISBN 9781351031585 (ebook)
Subjects: | MESH: Nanotechnology--methods | Nanomedicine | Nanostructures
Classification: LCC R857.N34 | NLM QT 36.5 | DDC 610.285--dc23
LC record available at https://lccn.loc.gov/2017056308

Visit the Taylor & Francis Web site at
http://www.taylorandfrancis.com

and the CRC Press Web site at
http://www.crcpress.com

Contents

Preface vii
Acknowledgments ix
Editors xi
Contributors xv

1. **Contemporary Developments in Nanobiotechnology: Applications, Toxicity, Sustainability, and Future Perspective** 1
   Anubhav Kaphle, Navya Nagaraju, and Hemant Kumar Daima

2. **Nanotheranostics: Implications of Nanotechnology in Simultaneous Diagnosis and Therapy** 35
   Brahmeshwar Mishra and Ravi R. Patel

3. **Nanodevices for Early Diagnosis of Cardiovascular Disease: Advances, Challenges, and Way Ahead** 69
   Alok Pandya and Madhuri Bollapalli

4. **Emerging Trends in Nanotechnology for Diagnosis and Therapy of Lung Cancer** 105
   Nanda Rohra, Manish Gore, Sathish Dyawanapelly, Mahesh Tambe, Ankit Gautam, Meghna Suvarna, Ratnesh Jain, and Prajakta Dandekar

5. **Surface Modification of Nanomaterials for Biomedical Applications: Strategies and Recent Advances** 171
   Ragini Singh and Sanjay Singh

6. **Nanoparticle Contrast Agents for Medical Imaging** 219
   Rabee Cheheltani, Johoon Kim, Pratap C. Naha, and David P. Cormode

7. **Recapitulating Tumor Extracellular Matrix: Design Criteria for Developing Three-Dimensional Tumor Models** 251
   Dhaval Kedaria and Rajesh Vasita

8. **Understanding the Interaction of Nanomaterials with Living Systems: Tissue Engineering** 279
   Shashi Singh

9. **Nanoparticles and the Aquatic Environment: Application, Impact and Fate** 299
   Violet A. Senapati and Ashutosh Kumar

10. **Iron Nanoparticles for Contaminated Site Remediation and Environmental Preservation** 323
    Adam Truskewycz, Sayali Patil, Andrew Ball, and Ravi Shukla

11. **Solubility of Nanoparticles and Their Relevance in Nanotoxicity Studies** 375
    Archini Paruthi and Superb K. Misra

12. **Nanotechnology in Functional Foods and Their Packaging** 389
    Satnam Singh, Shivendu Ranjan, Nandita Dasgupta, and Chidambaram Ramalingam

13. **Use of Nanotechnology as an Antimicrobial Tool in the Food Sector** 413
    María Ruiz-Rico, Édgar Pérez-Esteve, and José M. Barat

14. **Interface Considerations in the Modeling of Hierarchical Biological Structures** 453
    Parvez Alam

Index 463

Preface

The rapid advancements in nanoscience of the past two decades have firmly enthroned nanotechnology as the technology of the new millennium. Nanomaterials are novel materials due to their unique physicochemical and optoelectronic properties. The small size, in the range of 1–100 nm, has the potential for almost unlimited applications. This enabling technology holds the key to several recent innovations in diverse sectors, including pharmaceutical, electronic, energy, textile, environmental, agricultural and consumer products.

This book, *Nanobiotechnology: Human Health and the Environment*, intends to present a comprehensive overview of the current progress at the intersection of nanotechnology and biology. This book addresses various aspects of nanomaterials research and development that are closely related to human and environmental health. This book contains 14 chapters primarily focused on areas like bio-inspired/bio-mimetic, bio-imaging, disease treatment, diagnostics, food products, and the environment.

Chapter 1 introduces the reader to contemporary developments in nanobiotechnology with special reference to toxicity, sustainability, and future perspective. A few chapters focus on nanotheranostics and nanodevices for early diagnosis and therapy of diseases like cardiovascular disease and lung cancer. Specific chapters related to the surface modification of nanomaterials, contrast agents for medical imaging, and detection of cancer cells provide useful information for development of novel materials for biomedical applications.

The impact and fate of nanomaterials in the environment has received attention in this book, as development of safe nanoproducts is an intrinsic component of nanobiotechnology. This book contains chapters related to the application of nanoparticles in bioremediation, solubility of nanoparticles, and their relevance to nanotoxicity studies. Additionally, Chapter 12, "Nanotechnology in Functional Foods and Their Packaging," is intended to add value to this book in the evolving dimension of food and nanotechnology.

This book aims to address the impact of nanomaterials on human and environmental health as well as to improve our understanding of these novel materials for sustainable mega-exploitation of the nano dimension.

Acknowledgments

The editors wish to acknowledge with thanks the unstinting support and contributions of all the authors in this book. Generous funding from the Council of Scientific and Industrial Research, India, under its network projects NWP34, NWP 35, INDEPTH (BSC0111), and NanoSHE (BSC 0112), Department of Biotechnology for the project NanoToF (grant number -BT/PR10414/PFN/20/961/2014), and Gujarat Institute of Chemical Technology, India, for CENTRA (Centre for Nanotechnology and Research Application) & Risk assessment project. Funding received from Department of Science and Technology, Science and Engineering Research Board (SERB), for the project "Nanosensors for the Detection of Food Adulterants and Contaminants" (Grant number-EMR/2016/005286) and "Multi target antioxidant nanoconstruct for prostate cancer treatment" (Grant number-SB/YS/LS-159/2014) is gratefully acknowledged. The research fellowship provided under Endeavour Research Fellowship by the government of Australia and YY Memorial Research Grant by Union of International Cancer Control, Japan, are also acknowledged.

Editors

Professor Alok Dhawan is currently director, CSIR-Indian Institute of Toxicology Research, Lucknow. He also served as founding director, Institute of Life Sciences, and dean, Planning and Development, Ahmedabad University, Gujarat. Before joining as director, CSIR-IITR, he also worked in different scientific positions, like scientist, C; senior principal scientist; principal scientist; and so on. He obtained his PhD in biochemistry from University of Lucknow, India, in 1991. He was awarded a DSc (Honorary) from University of Bradford, U.K., 2017. He was a visiting scholar, Michigan State University, USA, and a BOYSCAST Fellow, University of Surrey, Wales & Bradford, UK. Professor Dhawan started the area of nanomaterial toxicology in India and published a guidance document on the safe use of nanomaterials. His group elucidated the mechanism of toxicity of metal oxide nanoparticles in human and bacterial cells. His work has been widely cited. He set up a state-of-the-art nanomaterial toxicology facility at CSIR-IITR as well as at the Institute of Life Sciences.

Professor Dhawan has won several honors and awards, including the INSA Young Scientist Medal in 1994; the CSIR Young Scientist Award in 1999; the Shakuntala Amir Chand Prize of ICMR in 2002; and the Vigyan Ratna by the Council of Science and Technology, UP, in 2011. His work in the area of nanomaterial toxicology has won him international accolades as well, and he was awarded two Indo-UK projects under the prestigious UK-IERI program. He was also awarded two European Union Projects under the FP7 and New INDIGO programs.

He founded the Indian Nanoscience Society in 2007. In recognition of his work, he has been elected fellow, Royal Society of Chemistry, UK; fellow, The National Academy of Sciences, India; fellow, The Academy of Toxicological Sciences, USA; fellow, The Academy of Environmental Biology; fellow, Academy of Science for Animal Welfare; fellow, Society of Toxicology (India); founder fellow, Indian Nanoscience Society; fellow, Gujarat Science Academy; vice president, Environmental Mutagen Society of India (2006–2007); member, National Academy of Medical Sciences; member, United Kingdom Environmental Mutagen Society, UK; and member, Asian Association of Environmental Mutagen Societies, Japan.

He has more than 125 publications to his credit in peer-reviewed international journals, 18 reviews/book chapters, four patents, and two copyrights, and has edited two books. He is the editor-in-chief of the *Journal of Translational Toxicology* published by American Scientific Publishers, USA, and serves as a member of the editorial boards of *Mutagenesis*, *Nanotoxicology*, *Mutation Research—Reviews*, and other journals of repute.

Dr. Sanjay Singh is an associate professor in the Division of Biological & Life Sciences, Ahmedabad University, India. He obtained his MSc from University of Allahabad (2004), and his PhD from CSIR-National Chemical Laboratory India (2008). Prior to joining Ahmedabad University as an assistant professor (2012), he worked as a postdoctoral research fellow (2008–2012) at the University of Central Florida (Orlando, USA) and Pennsylvania State University (Hershey, USA).

Dr. Singh's research area includes the synthesis of catalytic nanomaterials exhibiting biological enzyme-like properties (nanozymes), which are exploited for the development of biosensors and detection of cancer cells. His research group works to develop novel nanoprobes for imaging of cancerous cells using chemiluminescence, absorbance, and photoluminescence approaches that would yield improved diagnoses and greater information about the underlying pathophysiology of human diseases. He also works to develop nanoliposomes encapsulating multiple pharmacological agents for targeted damage to cancer cells.

Dr. Singh has received several honors and awards, such as the Endeavour Research Fellowship by the government of Australia (2015); Research Excellence Award by Ahmedabad University, India (2016); Yamagiwa-Yoshida (YY) Memorial Award by Union of International Cancer Control (2016); and the International Association of Advanced Materials Scientist Medal (2017). He has served as visiting professor at RMIT University, Australia, and University of Pennsylvania, USA.

Dr. Singh has authored more than 50 papers in international journals and has had 4 book chapters published in the field of nanoscience and nanotechnology. He is serving as an associate editor for the journal *3Biotech*, published by Springer, and an editorial board member of the *American Journal of Nanoscience and Nanotechnology* and the *International Journal of Medical Sciences*. Dr. Singh also serves as consulting editor for *Research and Reports in Transdermal Drug Delivery*, published by Dove Press.

Dr. Ashutosh Kumar is currently working as an assistant professor in the Division of Biological and Life Sciences, School of Arts and Sciences, Ahmedabad University, Gujarat, India. He obtained his master's degree in applied microbiology from Vellore Institute of Technology, Vellore, in 2008 and worked at CSIR-Indian Institute of Toxicology Research, Lucknow, for his doctorate degree in biotechnology.

Dr. Kumar's research group works majorly in the area of nanomedicines for cancer and arthritis, nano-based drug and gene delivery, DNA biochips for the detection of pathogens, nanoemulsions for food applications, environmental nanotechnology, nanotoxicology, and genetic toxicology. Dr. Kumar established several new methods for nanomaterial safety assessment in India. He developed a novel method for the detection of the uptake of nanoparticles in live bacteria for several generations. His work was focused on deciphering the mechanism of metal oxide nanoparticle–induced genotoxicity and apoptosis in the prokaryotic, eukaryotic, and *in vivo* models. In the area of environmental biology, he was involved in understanding the fate, impact, and biomagnification of nanomaterials in aquatic ecosystems using the paramecium, daphnia, and zebra fish as a model organisms. The work conducted by him is on the frontiers of the nanomaterial safety area worldwide. His studies will have far-reaching applications in predicting the adverse health effects of nanoparticles and safe use of nanomaterials to protect human and environmental health.

He has published more than 40 research papers and 12 book chapters in internationally reputed peer-reviewed journals. He has received several national and international awards, including the INSA Medal for Young Scientist 2014 and NASI-Young Scientist Platinum Jubilee Award (2015) in the area of Health Sciences for his scientific contributions.

Professor Rishi Shanker currently serves as guest faculty at Biotechnology Park, Lucknow, India, and as a consultant to the biotechnology industry. He has served as professor and associate dean at the School of Arts & Sciences, Ahmedabad University, Gujarat, India (2014–2016). Prior to joining Ahmedabad University, he served as chief scientist and area coordinator—Environmental & Nanomaterial Toxicology Groups at CSIR-Indian Institute of Toxicology Research, Lucknow (2001–2013). He also served as principal scientist at CSIR-National Environmental Engineering Research Institute, Nagpur. He holds a master's in biochemistry from University of Lucknow and a PhD in environmental microbiology and toxicology from CSIR-IITR and CSJM University (1985). Dr. Shanker's postdoctoral research addressed methanogen microbiology, deep subsurface microbiology, and protein engineering at University of London and Pennsylvania State University, USA (1987–1990).

Dr. Shanker's research contributions range from genetically engineered bacteria for bioremediation and molecular probes for pathogen detection to alternate models in toxicity assessment of chemicals and engineered nanomaterials. He has successfully steered over 34 national and international research projects, including Indo-US program: Common Agenda for Environment, Indo-Swiss Program in Biotechnology, Indo-German, EU FP7, EU New Indigo & Inno Indigo program, and Unilever. He has mentored 11 PhD students and has more than 85 publications and 20 reviews/book chapters to his credit in the aforementioned areas.

His work on pathogen detection and water quality received recognition in form of Vigyan Ratna, conferred by the government of Uttar Pradesh, India. He was awarded the Visiting Research Fellowship of Society for General Microbiology, UK, to work at the Anaerobic Microbiology Laboratory, Queen Mary College, University of London. He has served as a visiting scientist and visiting professor at the University of Washington, Seattle, USA; Pasteur Institute & CEA, France; and Pohang University of Science & Technology, Republic of Korea.

Contributors

Parvez Alam
School of Engineering
Institute for Materials and Processes
University of Edinburgh
Edinburgh, United Kingdom

Andrew Ball
School of Science
and
Centre for Environmental Sustainability and Remediation
RMIT University
Bundoora, Australia

José M. Barat
Departamento de Tecnología de Alimentos
Universitat Politècnica de València
Valencia, Spain

Madhuri Bollapalli
Division of Biological & Life Sciences
Ahmedabad University
Ahmedabad, India

Rabee Cheheltani
Department of Radiology
University of Pennsylvania
Philadelphia, Pennsylvania

David P. Cormode
Department of Radiology
and
Department of Bioengineering
and
Department of Cardiology
University of Pennsylvania
Philadelphia, Pennsylvania

Hemant Kumar Daima
Department of Biotechnology
Siddaganga Institute of Technology
Tumkur, India

and

Amity Institute of Biotechnology
Amity University Rajasthan
Jaipur, India

Prajakta Dandekar
Department of Pharmaceutical Sciences and Technology
Institute of Chemical Technology
Matunga, India

Nandita Dasgupta
Computational Modelling and Nanoscale Processing Unit
Indian Institute of Food Processing Technology
Thanjavur, India

Sathish Dyawanapelly
Department of Pharmaceutical Sciences and Technology
Institute of Chemical Technology
Matunga, India

Ankit Gautam
Department of Pharmaceutical Sciences and Technology
Institute of Chemical Technology
Matunga, India

Manish Gore
Department of Pharmaceutical Sciences and Technology
Institute of Chemical Technology
Matunga, India

Ratnesh Jain
Department of Chemical Engineering
Institute of Chemical Technology
Matunga, India

Anubhav Kaphle
Department of Biotechnology
Siddaganga Institute of Technology
Tumkur, India

and

Göttingen Graduate School for Neurosciences, Biophysics and Molecular Biosciences (GGNB)
International Max Planck Research School for Molecular Biology
Georg-August-Universität Göttingen
Göttingen, Germany

Dhaval Kedaria
School of Life Sciences
Central University of Gujarat
Gandhinagar, India

Johoon Kim
Department of Bioengineering
University of Pennsylvania
Philadelphia, Pennsylvania

Ashutosh Kumar
Division of Biological & Life Sciences
Ahmedabad University
Ahmedabad, India

Brahmeshwar Mishra
Department of Pharmaceutical Engineering & Technology
Indian Institute of Technology
Banaras Hindu University
Varanasi, India

Superb K. Misra
Materials Science and Engineering
Indian Institute of Technology
Gandhinagar, India

Navya Nagaraju
Department of Biotechnology
Siddaganga Institute of Technology
Tumkur, India

Pratap C. Naha
Department of Radiology
University of Pennsylvania
Philadelphia, Pennsylvania

Alok Pandya
Department of Chemistry
University and Institute of Advanced Research
Gandhinagar, India

Archini Paruthi
Materials Science and Engineering
Indian Institute of Technology
Gandhinagar, India

Ravi R. Patel
Department of Pharmaceutical Engineering & Technology
Indian Institute of Technology
Banaras Hindu University
Varanasi, India

Sayali Patil
Department of Environmental Science
Savitribai Phule Pune University
Pune, India

Édgar Pérez-Esteve
Departamento de Tecnología de Alimentos
Universitat Politècnica de València
Valencia, Spain

Chidambaram Ramalingam
Nano-food Research Group
Instrumental and Food Analysis Laboratory
Industrial Biotechnology Division
School of Bio Sciences and Technology
VIT University
Vellore, India

Shivendu Ranjan
Computational Modelling and Nanoscale Processing Unit
Indian Institute of Food Processing Technology
Thanjavur, India

Nanda Rohra
Department of Chemical Engineering
Institute of Chemical Technology
Matunga, India

María Ruiz-Rico
Departamento de Tecnología de Alimentos
Universitat Politècnica de València
Valencia, Spain

Violet A. Senapati
Division of Biological & Life Sciences
Ahmedabad University
Ahmedabad, India

Ravi Shukla
School of Science
RMIT University
Bundoora, Australia

and

Nanobiotechnology Research Laboratory
and
Centre for Advanced Materials and Industrial Chemistry
RMIT University
Melbourne, Australia

Ragini Singh
Division of Biological and Life Sciences
Ahmedabad University Central Campus
Ahmedabad, India

Sanjay Singh
Division of Biological and Life Sciences
Ahmedabad University Central Campus
Ahmedabad, India

Satnam Singh
Nanyang Technical University
Nanyang Ave
Singapore

Shashi Singh
Centre for Cellular and Molecular Biology
Hyderabad, India

Meghna Suvarna
Department of Chemical Engineering
Institute of Chemical Technology
Matunga, India

Mahesh Tambe
Department of Chemical Engineering
Institute of Chemical Technology
Matunga, India

Adam Truskewycz
School of Science
and
Centre for Environmental Sustainability and Remediation
RMIT University
Bundoora, Australia

Rajesh Vasita
School of Life Sciences
Central University of Gujarat
Gandhinagar, India

1

Contemporary Developments in Nanobiotechnology: Applications, Toxicity, Sustainability, and Future Perspective

Anubhav Kaphle, Navya Nagaraju, and Hemant Kumar Daima

CONTENTS

1.1 Introduction 1
1.2 Types of Nanomaterials and Their Applications 5
    1.2.1 Gold (Au)-Based Nanomaterials 7
    1.2.2 Iron-Oxide Nanoparticles 9
    1.2.3 Quantum Dots 9
    1.2.4 Carbon-Based Materials 10
1.3 Physicochemical Properties of Nanomaterials 13
1.4 Exposure State and Routes of Uptake of Nanomaterials 16
1.5 Interaction with Biological Systems, Cellular Uptake, and Distribution 17
1.6 Nanotoxicity and Possible Molecular Mechanisms 20
1.7 Assessing Nanotoxicity: Characterization Steps, Assay Methods, and Reliability ... 21
1.8 Conclusion, Sustainability, and Future Outlook 24
Competing Interests 25
Acknowledgments 25
References 25

1.1 Introduction

Nanomaterials are engineered miniscule constructs of elements with at least one dimension in the 100-nm range that exhibit properties that are different from their bulk counterparts. The National Nanotechnology Initiative (NNI) defines nanotechnology as the research and developmental efforts at the atomic or molecular level to create structures and systems applicable in diverse aspects (http://www.nano.gov/nanotech-101/what/definition) (Drexler and Peterson 1989, Whitesides 2003, Balzani 2005). Though the modern era, discoveries in the field of nanotechnology have been crucial and have set milestones in the human ability to manipulate matter at the atomic scale, there is a long history of how mankind puzzled out its peculiar properties yet lacked the knowledge and techniques to explain them. One important example of a premodern object is the fourth-century Lycurgus Cup—a dichroic glass now at the British Museum—which was contaminated with gold and silver nanoparticles, causing it to look either opaque green or translucent red depending on whether light was shining from outside or inside, respectively, as shown in Figure 1.1a and b (Freestone et al. 2007). Similarly, as illustrated in Figure 1.1c, the

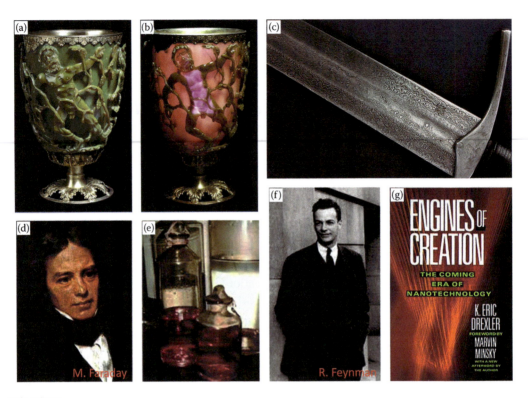

FIGURE 1.1
Lycurgus cup glowing in different colors depending on the shining of light (a and b) (Reprinted with permission from Freestone et al. 2007. *Gold Bulletin* 40(4):270–277.); seventeenth-century "Damascus" saber blades encompassing carbon nanotubes and cementite nanowires (c) (Courtesy of https://blogs.systweak.com/2017/04/ancient-tech-that-we-still-cant-figure-out/.); Michael Faraday and a display of his colloidal solutions of gold at the Royal Institution (d and e) (Courtesy of https://commons.wikimedia.org/wiki/File:Faraday.jpg and http://slideplayer.com/slide/6247959/21/images/43/AuCl3+Au+Michael+Faraday+1791+−+1867+P+CS2.jpg.); photograph of Richard Feynman (f) (Courtesy of http://ffden-2.phys.uaf.edu/211_fall2002.web.dir/David_Boniface-Jones/Page1.htm.); and hardcover edition of book engines of creation (g) by Eric Drexler (Courtesy of https://images-na.ssl-images-amazon.com/images/I/517nqcDDGEL._SX321_BO1,204,203,200_.jpg.).

seventeenth-century "Damascus" saber blades contained carbon nanotubes and cementite nanowires—an ultrahigh-carbon steel formulation that provided them with strength and resilience (Reibold et al. 2006). The incorporation of nanomaterials in these objects has been attributed to the use of high heat during casting (http://www.nano.gov/timeline).

Nanomaterials also had an important role in traditional medical formulations. Metal colloids were a part of antique medical formulations in various civilizations. Scientists in India and China succeeded in preparing gold colloids as early as the fourth and fifth centuries BC and used it for decorative and therapeutic purposes (Sarkar and Chaudhary 2010, Paul and Chugh 2011). Gold colloids were also studied and employed in many parts of Europe during the middle ages. In particular, Paracelsus applied colloidal gold in treating mental disorders and syphilis (Dykman and Khlebtsov 2012). Years later, in 1618, philosopher and doctor of medicine Francisco Antonii published a first-of-its-kind book on the preparation and therapeutic applications of colloidal gold, which is still preserved (Antonii 1618). In the year 1857, the first scientific publication on colloidal gold was authored by Michael Faraday (Figure 1.1d), wherein he attributed the ruby color of the solution to it being "divided," not the dissolved state of elemental gold, as shown in Figure 1.1e (Boisselier and Astruc 2009). However, it was not until the famous 1959 talk by

Richard Feynman (Figure 1.1f) at Caltech, USA (Feynman 1960), that scientists and engineers were inspired to look down into the entities that compose material and manipulate them bit by bit to produce things as desired. This vision was popularized by Eric Drexler in the 1980s through his particular book *Engines of Creation: The Coming Era of Nanotechnology* (the cover of the book is shown in Figure 1.1g) (Drexler and Minsky 1990), which has inspired many generations of researchers to pursue their careers in the field.

Furthermore, the invention of the scanning tunneling microscope in 1981 (Binnig and Rohrer 1983) provided researchers with a modern tool to probe into molecular systems and design as desired. The invention of characterization and manipulation tools and the discovery of many different nanostructures like fullerenes in 1985 (Kroto et al. 1985) caught the interest of researchers from various disciplines in developing materials and systems for widespread applications. We can now work at the nanoscale level, build materials atom by atom, and impose desired characteristics for numerous applications in almost every area, such as composite materials development, electronics, nano-electro-mechanical-systems (NEMS), biomedical technologies, renewable energy solutions, environmental remediation, and so on (Katz and Willner 2004, Li et al. 2007, Baruah and Dutta 2009, Gong et al. 2009, Yang et al. 2009a, Daima 2013, Daima et al. 2013, Liu et al. 2013b, Sharma et al. 2014, Chen et al. 2015a, Daima and Bansal 2015, Dubey et al. 2015, Shankar et al. 2015, Navya and Daima 2016). From this perspective, various nanomaterials or systems have been developed to cater to needs in numerous areas. In the context of biomedical applications, nanobiotechnology is an extended term that can be defined as the manipulation of materials at molecular levels with aims to produce nontoxic bioactive devices that have specificity toward the desired tissue type and location. Nanobiotechnology can find its niche in a multitude of applications, wherein nanomaterials provide great advantages in developing small detection and diagnostic systems given their small size, thereby allowing us to detect at the single cell or molecule level (Fortina et al. 2005, Jain 2007). For example, devices that can selectively detect and concentrate sparse cells for identification of rare cell types, cancer subtypes, or even fetus cells from the mother's blood to check for genotypic differences in them are of great use in biomedical science (Tao et al. 2013, Lamish and Klein 2015). These methods would greatly enhance our understanding of the biology of these cells. Also, nanomaterials can greatly enhance imaging of cells. Advanced florescence microscopy techniques have achieved high resolutions up to ~10 nm; however, conventional staining using large affinity bodies doesn't provide accurate resolution for super-resolution imaging (Ishikawa-Ankerhold et al. 2012). This is where the importance of small-sized nanoparticles with higher affinities toward particular molecules comes in. Nanomaterials also contribute to the development of protein arrays, allowing multiplexing of biochemical studies on proteins such as protein identification and protein–protein interactions (Liu et al. 2013a; Jain 2007, Li et al. 2017). Additionally, nanomaterials have been studied as probable therapeutic agents for cardiac disorders (Behera et al. 2015, Vaidya and Gupta 2015), in dental applications, and in cosmetics (Besinis et al. 2015, Kwon et al. 2015). According to a Business Communications Company (BCC) research report, the global market for nanomaterials used in biomedical, pharmaceutical, and cosmetic applications increased from $170.17 million in 2006 to around $204.6 million in 2007. In 2012, it reached nearly $684.4 million, with an expected compound annual growth rate of 27.3% (http://www.bccresearch.com/market-research/nanotechnology/nanostructured-materials-markets-nan017d.html).

Sitharaman (2011) categorized the basic types of nanomaterials into nine groups based on their use, mostly in a biological setting, though most of them are used in other sectors as well. They have classified nanobiomaterials as (i) metallic nanomaterials, (ii) ceramic nanomaterials, (iii) semiconductor-based nanomaterials, (iv) organic/carbon-based nanomaterials, (v) organic-inorganic hybrid nanomaterials, (vi) silica-based nanomaterials, (vii) polymeric nanomaterials and nanoconjugates, (viii) biological nanomaterials, and, finally, (ix) biologically directed/self-assembled nanobiomaterials. The peculiar behavior of nanomaterials compared to their bulk counterparts comes from the quantum effects that govern physics at that scale. Also, the increased surface-to-volume ratio confers high reactivity and can be used to conjugate probes or payloads (Biju 2014) as well as to improve catalysis in most chemical reactions (Burda et al. 2005). In biomedical applications, the high aspect ratio contributes to tunable properties that can be used in selective separations of drugs, gene delivery, sensing, and many more (Bauer et al. 2004, Salata 2004, Daima et al. 2012, Harrison and Atala 2007, Daima 2013, Goenka et al. 2014, Zhang et al. 2014, Sanghavi et al. 2015, Daima and Navya 2016). There is an immense potential for the use of nanomaterials in solving biomedical problems given the comparable size of materials with biological entities, and it opens up new possibilities to engineer materials compatible with biological systems (Whitesides 2003, Niemeyer and Mirkin 2004, Whitesides 2005, Mirkin and Niemeyer 2007, Suh et al. 2009, Daima and Navya 2016).

According to Lux Research, the nano-based product market is predicted to be around $4.4 trillion by 2018 (https://portal.luxresearchinc.com/research/report_excerpt/16215), suggesting an inevitable increment in human use and thus exposure to nanomaterials in subsequent years. Wiesner et al. (2006) argued that the production, use, and disposal of nanomaterials will soon start showing up in water, air, soil, and the environment. The environmental impact, health consequences, and life cycles of these nanomaterials are prior concerns to any regulatory body in developing guidelines in their uses. Nanomaterials used for food and biomedical applications demand a great deal of attention in particular, as they directly relate to public health. Toxicity associated with any nanomaterials is now a hot topic of study and debate (Navya and Daima 2016). Numerous studies have been carried out to assess "nanotoxicity" in organisms as well as the impact of the long life of residual nanomaterials on the ecosystem. However, a clear-cut consensus is lacking, and many of the studies contradict each other in terms of apparent toxicity (Alkilany and Murphy 2010). Therefore, an assessment of the benefits over risks of the use of nanomaterials is currently needed and will eventually set the directions for their further uses and developments (Lyass et al. 2000, Gokhale et al. 2002, Yoshida et al. 2003, Yamamoto et al. 2004, Garnett and Kallinteri 2006, Fischer and Chan 2007, Klaine et al. 2008, Lewinski et al. 2008, Murphy et al. 2008, Maynard et al. 2010, Soenen et al. 2011, Walkey and Chan 2012, Chen et al. 2015b, Daima and Navya 2016). In recent times, the concept of "sustainable nanotechnology" has grown rapidly in the literature, which calls for a comprehensive evaluation of the risks and impacts of nanotechnology and allied sciences at an early stage of the nano-enabled product life cycle. The triple-bottom-line (TBL) approach is one of the methods to frame sustainable nanotechnology, and it involves the ecological, economic, and societal "stakes" that contribute to the inclusive sustainability of a nano-supported product. Safeguard of the ecosystem is the mandate of regulators due to uncertainty and probable risk to humans and the environment. Therefore, sustainability may be observed as employing resources and evolving products and processes in a way that ensures and promotes a legacy of economic viability, social equity, and environmental accountability for existing and forthcoming generations. Nano-enabled products are hindered by data gaps and knowledge limitations in evaluating their risks and impacts. As discussed previously, nanotechnology is widely seen as having enormous prospects to bring about positive changes in many areas of research and application, and this is attracting growing investments globally from industries and governments. At the same time, it is predictable that its application may raise different challenges in the safety, regulatory, and ethical domains that will require societal debate (Brundtland Commission 1987, Helland and Kastenholz 2008, Subramanian et al. 2014). Therefore, in this chapter, we focus on different facets of sustainable nanotechnology development by discussing a series of important nanomaterials designed for biomedical applications, the importance of various physicochemical properties with suitable examples employed for biomedicine along with the interaction of nanomaterials with biological entities at the nano–bio interface, possible toxicological risks associated with engineered nanomaterial constructs after their exposure, and a future outlook toward safe and reliable utility of nanomaterials. Furthermore, we have documented the progress that has been made in this field and highlighted important achievements and gaps at the theoretical and applied levels.

1.2 Types of Nanomaterials and Their Applications

As discussed earlier, nanomaterials can be of different types, compositions, shapes, and sizes. In general, nanomaterials are classified into nine major groups based on their chemical nature and use, mostly surveyed for applications in the biological realm. Interestingly, nanomaterials are expected to cross biological obstacles, including blood-brain barriers (BBBs), gaining entry to the body; subsequently, the nano size of materials may govern their kinetics, absorption, distribution, metabolism, and excretion, which would not be imaginable otherwise with bulk material of similar composition (Whitesides 2003, Whitesides 2005, Longmire et al. 2008). Table 1.1 lists categories of numerous nanomaterials along with some examples to give an overview.

The ability to synthesize nanomaterials from a wide range of chemical species as well as to tune their physicochemical properties has shown promise in their feasible employment in

TABLE 1.1

An Overview of Major Categories of Basic Types of Nanomaterials Established on Their Usage

S. No.	Category	Examples
i	Metal-based nanomaterials	Nanomaterials based on gold, silver, iron oxide, gadolinium, etc.
ii	Ceramic nanomaterials	Nanomaterials based on alumina, zirconia, hydroxyapatite tricalcium phosphate, silicon nitride, etc.
iii	Semiconductor-based nanomaterials (quantum dots)	Nanomaterials based on cadmium sulfide, cadmium telluride, cadmium selenide, etc.
iv	Organic/carbon-based nanomaterials	Carbon nanotubes, fullerenes, etc.
v	Organic-inorganic hybrid nanomaterials	SiO_2/PAPBA (poly(3-aminophenylboronic acid), Ag_2S/PVA (polyvinylalcohol), CuS/PVA, Gold/polyethylene hybrid, etc.
vi	Silica-based nanomaterials	Monoliths, rod-like particles, fibers, hollow or solid nanospheres, silica nanotubes, and mesoporous silica nanoparticles (MSNs)
vii	Polymeric nanomaterials and nanoconjugates	Poly(lactic acid), poly(glycolic acid), poly(ε-caprolactone), poly(hydroxyl butyrate), proteins (collagen, silk), polysaccharides (starch, chitin/chitosan, hyaluronic acid, alginate), etc.
viii	Biological nanomaterials	Lipoprotein- or peptide-based nanomaterials
ix	Biologically directed/self-assembled materials	Virus-derived nanomaterials (Cowpea mosaic virus or CPMV)

Source: Sitharaman, B. 2011. *Nanobiomaterials Handbook.* Boca Raton, Florida: CRC Press.

numerous applications. Nanoparticles have been shown to behave as potent antimicrobial agents, which is important in the time of increasing microbial drug resistance cases (Daima et al. 2011, Huh and Kwon 2011, Daima et al. 2013, Daima et al. 2014b, Daima and Bansal 2015); to mimic enzymes, thereby overcoming their limitations such as an inhibitor effect or temperature susceptibility (Daima, Selvakannan et al. 2014a, Gao et al. 2007, Wei and Wang 2008, Sharma et al. 2014); and so on. Generally, metal-based nanomaterials are widely used for medical imaging due to their high electron density. Also, they can be used as carriers for drug molecules, biologics, and probe molecules (Lehner et al. 2013, Selvakannan et al. 2013), as well as in photothermal therapy against cancer (Soenen et al. 2011). Metal nanoparticles come in a variety of shapes and sizes, as shown in Figure 1.2a, wherein quantum dots (QDs) are used for fluorescence based bio-imaging (Chan et al. 2002, Sitharaman 2011), and polymeric nanomaterials and nanoconjugates are considered biocompatible (Sitharaman 2011), thus serving as intracellular delivery vehicles for various probes and therapeutics. Carbon-based nanomaterials are used for cell tracking, photoacoustics, pressure sensing, gene and protein microarray, and so on (Sitharaman 2011).

In the next section, we take some examples of selected types of nanomaterials that have shown promise for diverse biomedical applications, their properties that confer potential for use, and advances in the area for better understanding of the field of nanobiotechnology/nanomedicine.

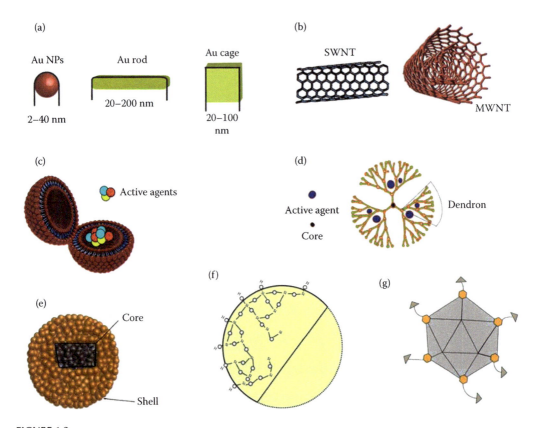

FIGURE 1.2
Schematic representation of some important nanomaterials. (a) metallic nanomaterials, (b) single and multiwalled carbon nanotubes, (c) polymeric nanomaterials, (d) dendrimers, (e) semiconductor-based nanomaterials, (f) silica-based nanomaterials, and (g) virus-derived nanomaterials.

1.2.1 Gold (Au)-Based Nanomaterials

Au colloids have been used in China, India, and the Middle East for therapeutic and decorative purposes since antiquity (Boisselier and Astruc 2009, Dykman and Khlebtsov 2012). Due to its long history of usage, gold is considered biocompatible, and no apparent toxicity has been concluded (Connor et al. 2005), although there have been some toxicity reports in recent times (Chen et al. 2009c, Cho et al. 2009). Au nanomaterials are desirable due to their easy synthesis, high stability, easy surface functionalization, unique optical properties, and potential biomedical applications (Murphy et al. 2008, Daima et al. 2011, Daima 2013, Daima et al. 2013, Selvakannan et al. 2013, Daima et al. 2014a, Sharma et al. 2014, Dubey et al. 2015, Shankar et al. 2015).

The unique optical properties of Au-based nanomaterials come from their localized surface plasmon resonance effect (LSPRE). Surface plasmon refers to the coherent oscillation of conduction electrons on the metal surface when irradiated by a suitable wavelength of light, as shown in Figure 1.3. The absorption and scattering of light by the particles is due to this phenomenon (Stewart et al. 2008, Cao et al. 2014). The light scattered by Au nanoparticles falls within the visible range of the electromagnetic spectrum, thereby allowing for noninvasive imaging and tracking in the cells or whole body (Murphy et al. 2008, Stewart et al. 2008). Au-based nanomaterials can have different shapes and sizes, as mentioned earlier. Anisotropic materials are pretty much focused nowadays, due to their manipulable properties. For example, nanorods have two plasmon bands, one transverse and the other longitudinal, as shown in Figure 1.3. The plasmon band position can be tuned by controlling dimensions to make the particles absorb in the near-infrared (NIR) spectrum range of ~800–1200 nm, which is considered "the biological water window," where light can penetrate deep into tissues, water does not absorb, and background fluorescence is low, thus providing opportunities for deep-tissue imaging and medical "tagging" (Alkilany and Murphy 2010).

Furthermore, the anisotropic form of the topology helps in inculcating many noble properties, particularly in the case of Au nanoparticles. Au nanoparticles can be synthesized as spherical single particles, nanorods of a high aspect ratio, or nanocages with a hollow interior and porous walls (Soenen et al. 2011). Figure 1.4 displays the change in the aspect ratio of Au nanoparticles,

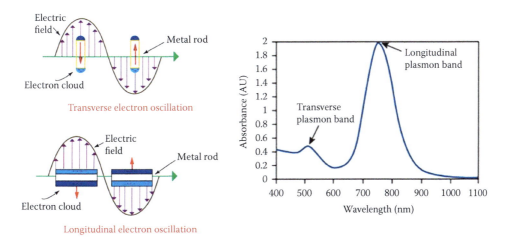

FIGURE 1.3
Pictorial representation of different plasmon bands for Au nanorods. (Reprinted from *Sensors and Actuators B: Chemical*, 195, Cao, J. et al. Gold nanorod-based localized surface plasmon resonancebiosensors: A review, 332–351, Copyright (2014), with permission from Elsevier. doi: http://dx.doi.org/10.1016/j.snb.2014.01.056.)

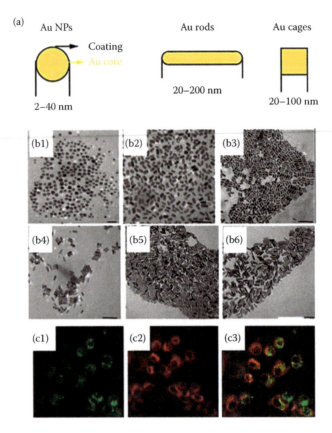

FIGURE 1.4
Illustration of different topology of gold (Au) nanoparticles and colors emitted by them during cell imaging. (Reprinted from *Nano Today*, 6(5), Soenen, S. J. et al. Cellular toxicity of inorganic nanoparticles: Common aspects and guidelines for improved nanotoxicity evaluation, 446–465, Copyright (2011), with permission from Elsevier.)

which leads to a change in their plasmon band, resulting in different-colored light emission that can be used for differential biomedical imaging for "color-blind" cell types or if stains are not significant in imparting colors, as shown in Figure 1.4c1–c3. Moreover, they can be used for deep tissue imaging because of the tunability of Au nanoparticles to make them absorb at NIR range. Interestingly, due to their unique physical and chemical properties as well as an easy synthesis process and stability, Au nanoparticles look prominent in many areas of biomedicine.

Besides medical imaging and tagging, Au nanomaterials can be used for diagnosis, drug delivery, and photothermal effects in cancer therapy (Murphy et al. 2008). The pie chart in Figure 1.5 demonstrates a schematic representation of possible uses of Au nanoparticles for various biomedical areas. The majority of applications include sensing, as shown in Figure 1.5. Within the sensing application of Au nanoparticles, surface-enhanced Raman spectroscopy (SERS) is very promising. The surface plasmon of the excited Au nanoparticles creates an electrical field around it, thereby enhancing the Raman scattering cross-section of nearby molecules. This can facilitate, in theory, single molecule detection and identification. For example, Senapati et al. have synthesized tryptophan-protected popcorn-shaped Au nanomaterials as SERS probes to detect Hg(II) ions with a 5-ppb limit of detection from an aqueous solution in the presence of competing analytes, implying high sensitivity for the construct (Senapati et al. 2011).

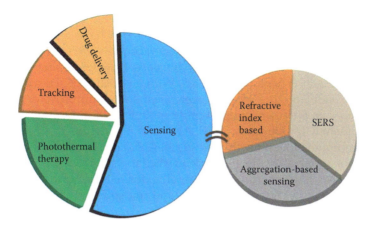

FIGURE 1.5
Pie chart depicting possible areas of biomedical applications of Au-based nanomaterials. (Reprinted with permission from Murphy, C. J. et al. 2008. Gold nanoparticles in biology: Beyond toxicity to cellular imaging. *Accounts of Chemical Research* 41(12), 1721–1730. Copyright 2008, American Chemical Society.)

1.2.2 Iron-Oxide Nanoparticles

Iron oxide in bulk is ferromagnetic and is influenced by any externally applied magnetic field. The two main forms of iron oxide are magnetite (Fe_3O_4) and its oxidized form, maghemite (γ-Fe_2O_3), which are preferred in biology and medicine due to their biocompatible nature. The problem associated with bulk iron oxide is the residual magnetic effect left over after the external field has been turned away, which poses biomedical issues. However, iron oxide is easily degradable and therefore useful for many *in vivo* applications (Jain et al. 2005, Laurent et al. 2008, Hu et al. 2009, Mahmoudi 2011c,d). Thus, small iron-oxide nanoparticles (IONPs), whose size falls into the superparamagnetic range and which are susceptible to a magnetic field, but without any residual field after its removal (Soenen et al. 2011), are preferred. These can be used as magnetic resonance imaging (MRI) contrast agents, as well as in hyperthermia killing of cancer cells. In hyperthermia killing, INOPs are "vibrated" in an external field, which causes dissipation of thermal energy due to friction, thereby killing nearby cells. MagForce Technologies, Berlin, Germany, has developed similar working technology for cancer treatment. Similarly, as shown by Gao et al. (2007), IONPs can mimic peroxidase activity that can be used for glucose detection in diagnostic kits as a rapid and robust method of detection. This feature also overcomes the intrinsic susceptibility of enzymes to inhibitors, temperature, pH, and so on (Wei and Wang 2008).

1.2.3 Quantum Dots

QDs are semiconductor crystals constituting elements from group II to IV or III to V with physical dimensions smaller than the exciton Bohr radius (Chan et al. 2002). QDs can be made by numerous methods, including colloidal synthesis, plasma synthesis, or mechanical fabrication, and they tightly confine either electrons or electron holes in all three spatial dimensions. QDs range in size from 2 to 10 nm in diameter. QDs exhibit high light stability and wide and continuous absorption spectra with narrow emission spectra, thus emitting a precise wavelength of light (Bruchez et al. 1998). They also have high stability against photo bleaching caused by prolonged exposure to light (William et al. 2006).

QDs in biomedical applications are mainly used in bio-imaging by using them as label-tags for precisely locating markers in cells or systems (Michalet et al. 2005). QDs can be tuned to emit various wavelengths of light depending on their size and composition and can be used flexibly for many assays (Alivisatos 1996, Chan and Nie 1998, Gerion et al. 2001, Gao et al. 2005, Michalet et al. 2005, William et al. 2006, Deerinck 2008). A QD consists of a semiconductor core with an overcoated shell to improve optical properties. They can be coated with polymers or hydrophilic moieties to improve their solubility in the aqueous environment (Ghasemi et al. 2009). Also, passivizing them by coating them with nonantigenic compounds like polyethylene glycol (PEG) can prevent their entrapment by the immune system or uptake by the reticuloendothelial system (RES), thus increasing their half-life in the system and their efficiency in tagging desired markers (Michalet et al. 2005). QDs made up of a cadmium selenide (CdSe) core and zinc sulfide (ZnS) shell (CdSe/ZnS) are presently most commonly used as materials for secondary antibody conjugates (Deerinck 2008). They have also been used as florescent labels for fluorescence *in situ* hybridization (FISH) of biotinylated DNA to the human lymphocyte metaphase chromosome (Xiao and Barker 2004). Similarly, Bruchenz et al. demonstrated dual-color imaging of mouse fibroblast nuclei and actin filaments using silica-coated CdSe/CdS QDs (Bruchez et al. 1998).

1.2.4 Carbon-Based Materials

Carbon-based materials include single- or multiwalled carbon nanotubes (CNTs), fullerenes, carbon dots, nanodiamonds, and graphene (Krueger 2010). They have been highlighted as highly attractive materials due to their versatile properties that can be exploited for multifunctional applications. They possess excellent mechanical strength, electrical and thermal conductivity, and optical properties. These properties are exploited in a variety of applications, such as high resilient nanocomposites, electronics, scaffold material for tissue engineering, and sensing (Harrison and Atala 2007).

CNTs have been used for their high strength in mechanical applications, high conductivity, and optical properties. The chemical structure of CNTs can be modified to attach moieties of important functional groups, such as polymers that can be very useful for loading payloads such as drugs or any sensing agents (Bianco et al. 2005). CNTs also possess good optical transition in the near-infrared region, which is very desirable in biological applications, as it can penetrate deep into tissues wherein light intensity loss through absorption and excitation is minimal (Cherukuri et al. 2004). Likewise, CNTs can be used as very good Raman scattering agents because of their symmetrical carbon bonds (Dresselhaus et al. 2002). Therefore, they can be used as tissue or cell probes by exploiting their Raman signatures. Liu et al. have monitored CNTs in various tissues after delivering them intravenously using Raman spectroscopy (Liu et al. 2008). In addition to the above-discussed applications, interestingly, the diameter of CNTs plays a fundamental role in their biological action, which has been confirmed by their antibacterial effects, as depicted in Figure 1.6a and b, wherein the main CNT-toxicity mechanism was found to be cell membrane damage by direct contact of CNTs and bacterial cells. Scanning electron microscopic observations exposed severe cellular damage to *Escherichia coli* bacteria after their incubation with multiwalled carbon nanotube (MWNT) and single-walled carbon nanotube (SWNT) for 1 hour. Moreover, release of DNA into the solution further supported the severe destruction by the expression of genes related to cell damage, whereas the molecular-level study of DNA microarrays and the facts of gene expression confirm that in the existence of both MWNTs and SWNTs, *E. coli* cells express higher levels of stress-related gene products associated to cell membrane damage and oxidative stress, albeit with

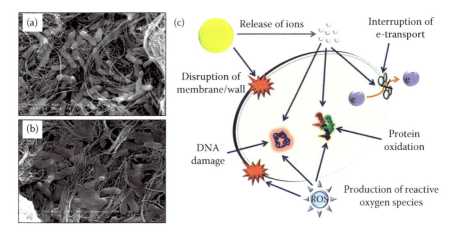

FIGURE 1.6
Scanning electron microscopic observations of *E. coli* cells treated with MWNTs (a) and SWNTs (b). The scale bars in both images represent 2 μm. (Adapted from Kang, S. et al. 2008a. *Langmuir* 24(13):6409–6413. doi: 10.1021/la800951v.) Illustration of possible mechanisms by which various nanomaterials exhibit antibacterial activities. (Adapted from Daima, H. K., and V. Bansal. 2015. *Nanotechnology in Diagnosis, Treatment and Prophylaxis of Infectious Diseases*, edited by Mahendra R. and K. Kon, 151–166. Boston: Academic Press.)

the quantity and degree of expression being noticeably greater in the existence of SWNTs (Kang et al. 2008a). The schematic representation in Figure 1.6c shows various possible mechanisms by which a range of nanomaterials may exhibit antibacterial activities.

Like CNTs, graphene is another carbon-based material that can provide a large surface area as well as modifiable functional moieties on the surface to load active molecules for therapy or diagnostic applications. The oxidized form of graphene, graphene oxide (GO), has been used in composite with copper oxide nanoparticles for effective sensing of glucose molecules and hydrogen peroxide (Liu et al. 2013b). Moreover, GO has been used very precisely for signal detection of pathogens based on conjugated biomarkers (Jung et al. 2010). Carbon dots are yet another group of carbon-based materials that have a luminous nature with a wide tunable absorption spectrum and narrow emission spectra (Yang et al. 2009b). Carbon dots have been used to deliver siRNA and DNA for successful silencing of genes and successful transfection of cells, respectively (Wang et al. 2014). Due to their low synthesis cost and amenable characteristics, carbon dots are extensively used for bioimaging. Also, because of their organic nature, carbon dots have been found to pose less toxicity as compared to QDs, as QDs suffer from in solution leaching of heavy metals that are detrimental to whole biosystems.

In addition to the above-discussed, development of protein chips, cardiac therapy, and dental applications are emerging applications of nanomaterials. Protein chips play major roles in functional studies of proteins such as protein–protein interaction, catalysis, identification, and detection (Chen et al. 2009a). One major consideration for the development of chips is the adsorption of the recognition element to the surface of a substrate. This process of immobilization is critical in the production of chips and should be achieved with minimal to nil structural changes that can retain maximum activity. Nanomaterials and nanostructured films offer advantageous properties that can be tuned to minimize any structural defects and maximize interactions with proteins to maximize activity recognition and minimize "leaching" (Bhakta et al. 2015). Also, nanomaterials can be used to increase signal amplification in protein microarray assays, which have always been a bottleneck where traditionally expensive and unstable enzymes with additional biochemical reaction usages limit practical applications. Liu et al. produced mulberry-structure ZnO nanoparticles with low autoflorescence and significant fluorescence amplification capability with detection up to as low as 1 pg mL^{-1} with a wide dynamic range (Liu et al. 2014). This provides an enzyme-free system to amplify an array signal. Wu et al. developed a simple and cheap route for the production of luminescent dye-doped silica nanoparticles with their efficient application as a probe for trace amounts of human immunoglobulin E (IgE) on a reverse-phase protein microarray (Wu et al. 2008).

Heart diseases are a major lifestyle-associated chronic type of condition prevalent in major parts of the world. Myocardial infarction (MI) alone accounts for 1 in 6 of the total deaths in the United States (Sidney et al. 2013). Similarly, ischemic heart diseases are also among the major causes of death (Zhu et al. 2015). Early response and treatment strategies might help in preventing major fatality. For example, injectable biomaterials designed to treat MI in early stages are administered through intramyocardial, intracoronary, or intravenous routes (Nguyen et al. 2015) and help decrease the risk of heart failure. However, this is limited by poor absorption and availability of active agents in the tissue. Nanomaterials are prominent options for effective and targeted drug delivery into the cardiac tissues. Also, they provide advantages for minimally invasive delivery, as they may be administered by intravenous (IV) injection with active cardiac region targeting. Injectable, self-assembling peptides are one class of nanofibrous materials that have been studied for the treatment of MI. These compounds are relatively smaller and can be assembled into a nanofibrous, hydrogel network under physiological conditions with sizes of around 5 nm in diameter (Davis et al. 2005, Davis et al. 2006, Lin et al. 2010). Using RAD16-II as a carrier for the delivery of vascular endothelial growth factor (VEGF) has been studied in rodent and porcine models (Lin et al. 2012). Similarly, a synthetic polymer composed of ureido-pyroimidinone and polyethylene glycol was used as a construct to deliver hepatocyte growth factor (HGF) and insulin-like growth factor (IGF-1) using a catheter injection strategy (Bastings et al. 2014). Also, another hybrid system composed of a methacrylated gelatin hydrogel and polyethylenimine-functionalized graphene oxide nanosheets was developed for gene delivery of VEGF-165 (Paul et al. 2014). Moreover, researchers are attempting to use nanomaterials in conjunction with stem cells for the treatment of ischemic heart disease. They are using nanomaterial constructs made of metal nanoparticles to perform genome engineering, retention, and stabilization of stem cells in the heart (Zhu et al. 2015).

Nanomaterials use in dental practice involve application as dental restorative material, endodontic retro-fill cements, and dental implants. Most dental treatments become a concern when pathological organisms colonize the dentine and enamel, gaps between dentine and enamel and dental restorations, prosthetic materials, and neighboring soft tissue (Rao et al. 2011). Bacteria, in particular *Streptococcus mutans* and *S. lactobacilli*, produce acids that can cause extensive dental caries and damage of tissues. It is a real concern when fillings get contaminated with these pathogens. Therefore, the fillings can be treated first with antimicrobials before application. As discussed, metal nanoparticles such as silver particles have an antimicrobial effect with minimal toxicity toward humans, and they can be formulated as coating materials or filling materials. Alginate nanomaterials can also be used as antimicrobial agents (Ahn et al. 2009). However, the addition of these materials should not affect the mechanical properties or hardening behavior of the fillings. Incorporation of silver nanoparticles into bonding adhesives was tried by Moszner et al., and they were successful in maintaining both physical stability and antimicrobial properties (Moszner and Salz 2007). Similarly, Mahmoud and his team managed to synthesize calcium fluoride/fluorinated hydroxyapatite (CF/FHAp) nanocrystals that can be used as osteoconductive dental fillers or implants with antimicrobial activity that can prevent dental caries due to the release of fluorine ions (Azami et al. 2011).

Furthermore, nanomaterial applications beyond biomedical areas include ceria (cerium oxide nanoparticles), which is used as an important diesel fuel catalyst (Hardas et al. 2010); fullerene nanotubes used in tires, tennis rackets, and video screens; fullerene cages used in cosmetic products; silica nanomaterials used as solid lubricants in machinery; metal nanoparticles, which also find a role in underground water remediation and purification; and protein-based nanomaterials, which are used in detergents and soaps as immobilized agents trapped inside a matrix (Wiesner et al. 2006). Given so many potential applications of nanomaterials, it is true that, in the near future, we are going to witness a huge rise in their use and consumption that spans all current aspects of human needs. As a result, sensibly primed nanomaterials are in widespread demand for biomolecular detection and diagnostics, therapeutics, drug and gene delivery, fluorescent labeling, tissue engineering, biochemical sensing, and other pharmaceutical applications, as discussed. Interestingly, to fulfill this demand, contemporary progress in nanosciences and nanotechnology has given us the ability to rationally design assorted nanomaterials and manipulate their chemical, physical, and prospective biological properties (Daima and Navya 2016). Therefore, in the next section of this chapter, we aim to reconnoiter current knowledge at the nano-scale level to engineer various physicochemical features of materials for their biomedical applications.

1.3 Physicochemical Properties of Nanomaterials

As discussed in earlier sections, nanomaterials present unique physicochemical properties, which can be applied for various biomedical applications. The various physical and chemical properties of nanomaterials may independently or cooperatively affect the nano–bio interfacial interactions, nanomaterial adhesion on cells, and their uptake or penetration inside cells, which eventually translates into the biocompatibility or biotoxicity of these nanomaterials leading toward therapeutic or adversative effects (Nel et al. 2006, Nel et al. 2009, Daima and Bansal 2015, Daima and Navya 2016). Therefore, it is vital to advance deeper insights into physicochemical properties of nanomaterials and their biological aspects to prepare improved nanomaterials for future biomedical applications.

As illustrated in Figure 1.7, various physicochemical properties of nanomaterials such as their size, shape, composition, surface charge, surface corona, or aggregation may influence their biomedical applications. It has been recognized that the size of a nanomaterial is principally dominant, while other properties are controlled because size reduction of a nanomaterial provides an opportunity for improved uptake and probabilities to interact with biological tissues to a greater extent (Nel et al. 2006). To confirm this, systematic assessments of size-dependent biological profile and biodistribution using a variety of nanomaterials have been evaluated. For example, in a recent study, three monodisperse drug-silica nanoconjugates of 20, 50, and 200 nm were prepared, and it was proved that the 50-nm-size conjugate system had the highest retention over time in cancer tissue, leading to deeper penetration and effective internalization within the cells, along with slower clearance (Tang et al. 2014).

Furthermore, right-sized nanomaterials functionalized with antibodies can regulate the process of membrane receptor internalization, which may lead to a downregulated cellular expression level, which is essential for basic cell functioning (Jiang et al. 2008). Like the

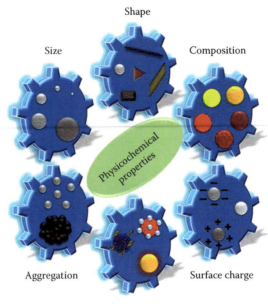

FIGURE 1.7
Graphic illustration of major physicochemical properties of nanomaterials that influence their biomedical applications.

size of nanomaterial, the composition or type of material and its morphology (shape) are key characteristics at the nano level. Along with the size of the nanomaterial, the shape of the nanomaterial also has a substantial influence on cellular uptake; although the size of a material is a key element that is linked with the surface area for a specific mass-dosage. Nevertheless, the contribution of the shape of the nanomaterials will be noteworthy to the overall surface area. This can be understood by a simple example: an octagon-shaped nanomaterial will have a dissimilar surface area compared to a spherical nanomaterial of equivalent size. Moreover, since surface atoms at the nano level have an inclination to hold unsatisfied high energy bonds, the higher catalytic activity of nanomaterials with greater surface areas enhances their reactivity, leading to effective interaction between nanomaterials and the biomolecules of cells, triggering direct cellular destruction and promoting oxidative stress (Mulvaney 1996, Narayanan and El-Sayed 2004a,b, Burda et al. 2005, Bansal et al. 2010, Mahmoudi et al. 2011a,b). Additionally, the importance of the shape of nanomaterials has been demonstrated in terms of their uptake by biological systems, where spherical nanoparticles display a higher uptake over other shaped nanoparticles. Furthermore, in the case of nanorods, internalization has been dependent on their dimensions, and high-aspect ratio rods show considerably faster internalization than low-aspect ratio nanorods (Mahmoudi et al. 2011a). Similarly, in the recent past, Pal and co-workers have demonstrated that nano size and the existence of the (111) lattice plane combine to inspire antibacterial potential against *E. coli*, and silver nanoparticles commenced shape-dependent interaction with bacterial cells (Pal et al. 2007). In one of our studies, we have demonstrated the antimicrobial and *in-vitro* peroxidase-like behavior of gold and silver nanoparticles, which was composition reliant. Through this study, we have established that composition is another important characteristic of nanomaterials, along with other physicochemical properties, and it has considerable impact on nanomaterial antimicrobial potential and *in-vitro* peroxidase-like behavior (Daima et al. 2011, Daima et al. 2014a, Daima and Navya 2016).

In addition to above discussion, it is essential to recognize that the number of surface molecules increases exponentially when the size of the nanomaterial falls below 100 nm. The total number of atoms/molecules that are present on the surface of a nanomaterial defines its reactivity and biological action. Moreover, the size of the nanomaterial and number of surface-expressed molecules display an inverse relationship; therefore, an increment in the surface area at the nanoscale level will allow a greater population of the material's atoms/molecules to be displayed on the surface rather than the interior. For example, in a material of 10-nm size, about 20% of its molecules will be expressed on the surface; whereas nanomaterials with 3-nm size will have 50% intensification in the surface-expressed molecules. The amplified surface-to-volume ratio reveals the significance of the chemical and biological activities of the nanomaterial since it can be positive and desirable or negative and undesirable. Besides, the size reduction of the material under the nanometer range may create irregular crystal planes, increasing structural defects, which can disrupt electronic configuration and alter the electronic properties of the material. All such changes can make specific surface groups that may function as reactive sites, like hydrophilic or hydrophobic, catalytically active or passive sites, and so on depending upon the composition of the material (Oberdörster et al. 2005, Nel et al. 2006, Krug and Wick 2011).

In addition to the above discussed, the aggregation phenomenon of nanomaterials has frequently been considered trivial for many biological and medical applications and often overlooked. Nanomaterials prepared in solvents are typically unstable due to their higher free surface energy resulting from their nano size, and such nanomaterials have an inclination toward aggregation, which may limit their biomedical potential. Therefore, the stability of nanomaterials need to be of the utmost concern before using them for biological applications, and it depends on various parameters (Kallay and Zalac 2002). Control over aggregation in the case of metallic nanomaterials in a solution can be achieved by the addition of shielding or protecting groups, which provides surface coatings/functionalization to metal nanoparticles (Bonnemann and Richards 2001, Richards and Bönnemann 2005). Such tailored surfaces on materials may produce interesting opportunities for nanomaterials in a highly controlled fashion for biomedical applications. In this context, various nanomaterials like silica nanoparticles, magnetic iron oxide nanoparticles, zinc oxide, carbon nanotubes, gold or silver nanoparticles, and many more have been functionalized by small molecules, ligands, polymers, or biomolecules to control their aggregation for biomedical applications. Moreover, recent investigations strongly recommend that control of nanomaterial surfaces may play a considerable part in governing their biological activities due to surface functionalization/corona (Daniel and Astruc 2004, Laurent et al. 2008, Boisselier and Astruc 2009, Mout et al. 2012, Daima 2013, Daima et al. 2013,, Daima et al. 2014a,b, Sharma et al. 2014, Dubey et al. 2015, Daima and Navya 2016).

On the contrary, it is interesting that the aggregation-based immunoassay technique has been developed by employing gold nanoparticles for antiprotein A, which is extremely sensitive and specific (Thanh and Rosenzweig 2002). A similar concept has been used for preparation of DNA-functionalized gold nanoparticles, which opens up new opportunities for accurate, speedy, easy, and reliable genetic diagnosis (Sato et al. 2003). In an aqueous milieu, in conjunction with surface functionalization/corona, the surface charge of nanomaterials is a distinguishing physicochemical property to control nanomaterial aggregation and to dictate its biological behavior under different experimental setups. Surface charge plays a leading role in initial electrostatic interaction at the nano–bio interface, and nanomaterials bearing positive surface charges are recognized to produce toxicity in living organisms.

Generally, cell membranes have a negative charge and, therefore, it is believed that materials with a negative surface charge may internalize slowly compared to their positively surface charged materials. In this context, it has also been testified that a negative charge on DNA is more likely to intermingle with cationic surface charged nanomaterials, which triggers genotoxicity. On the other hand, if a nanomaterial surface charge is similar to the charge of cell membranes, it will induce repulsion and avoid nanomaterial–cell contact (Ozay et al. 2010, Wiesner et al. 2006).

As discussed, the inherent physicochemical characteristics of nanomaterials have a notable impact on their potential usages. In addition to their inherent characteristics, the purity of a nanomaterial, surface smoothness/roughness/defects, hydrophobicity or hydrophilicity, electron transfer capability, oxidizability in physiological conditions, and effects of counter ions are additional necessary parameters for nano-objects that need to be considered while engineering nanomaterials for biological or medical applications (Nel et al. 2006, Aillon et al. 2009, Gatoo et al. 2014, Manshian et al. 2014, Daima and Bansal 2015, Daima and Navya 2016). From the discussion, it is clear that nanomaterials should no longer be observed as simple carriers only but essentially be engineered at the nano-scale level for delivery, diagnosis, and therapeutic applications.

1.4 Exposure State and Routes of Uptake of Nanomaterials

It is certain that the growing developments in the field of nanotechnology will cause an increased production of nanomaterials that can leak into the environment and then persist for longer periods. Figure 1.8 shows the sources that can contribute a high amount of nanomaterials in the environment, their exposure, and possible methods of removal. The stability and durability of nanoparticles in the environment depends on their type, their transportation, and, importantly, the charge and surface coats that confer their stability. Sharma (2013) writes that high-molecular-weight compounds like humic and fulvic acids provide stability, up to months, to silver nanoparticles and thus are likely to be transported to various locations and have ecological risk associated with them. Also, as we will see later, the effect of surface properties will likely cause selective uptake and distribution of particles in the physiological system. Hence, it is necessary to consider that, while establishing threshold parameters for apparent toxicity and exposure concentration of nanoparticles, the intrinsic properties of surface and types of nanomaterials "leaked" out have to be properly analyzed. Because nanoparticles are too small to bar their entry into organisms, easy access into the body is seen in most cases. They are easily taken up by cells and transcytosed through epithelial and endothelial cells into the blood and lymph vessels and could possibly be translocated into sensitive sites like the heart, liver, lungs, and brain (Oberdörster et al. 2005). The entry points of nanomaterials into the body can be summarized in six routes: intravenous, dermal, subcutaneous, inhalation, intraperitoneal, and oral (Fischer and Chan 2007) as vulnerable sites of entry.

One important question that needs to be addressed while defining the risk associated with nanomaterials is: "How much are we exposed to them?" Anything beyond a certain threshold is toxic; thus, a proper definition of the environmental amount that poses a risk to human health as well as disrupting the ecosystem balance has to be established. Workplace exposure is the main source of high dosage of nanomaterials. There have been very few studies regarding the potential risk of "nano-overdose," but many companies derived material

Contemporary Developments in Nanobiotechnology 17

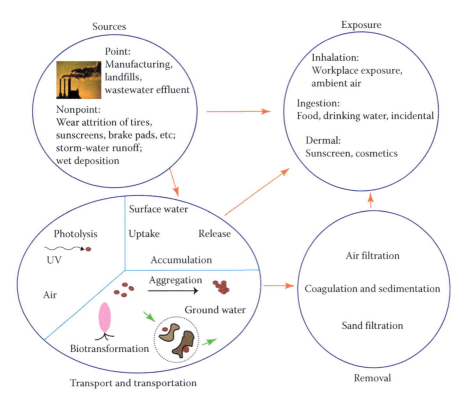

FIGURE 1.8
Various sources of nanomaterials contributing to growing environmental amounts along with exposure routes and removal methods. (Reprinted with permission from Wiesner, M. R. et al. 2006. Assessing the risks of manufactured nanomaterials. *Environmental Science and Technology* 40(14), 4336–4345. Copyright 2006, American Chemical Society.)

safety data sheets (MSDSs) from bulk particle data, which is very much unrelated in those circumstances (Colvin 2003). Thus, it requires strong regulation for the safety of workers at "nano production plants." As suggested by Dawidczyk et al. (2014), the no observable effect level (NOEL), which is the maximum dose with no observable adverse effects, needs to be established from *in vivo* studies. Also, a reference quantity, which is the daily exposure that doesn't have likely effects seen during an individual lifetime, has to be properly assessed to set up any threshold exposure concentration that may have an adverse effect on organisms.

1.5 Interaction with Biological Systems, Cellular Uptake, and Distribution

Two sciences—nanosciences and biology—merge at the interface between nanomaterials and biological systems and will define the probable fate of nanomaterials in the system as well as any responses that will emerge downstream of the process, like improved therapeutics, signaling, uptake, or any associated apparent toxicity. Colloidal forces and biophysicochemical interactions govern interfaces and thus the subsequent formation of protein corona, wrapping of particles, and immunological responses that define either their

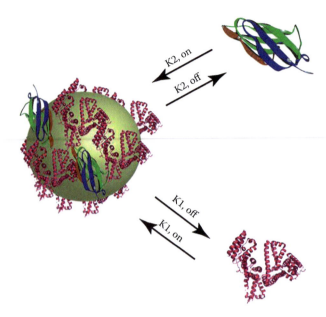

FIGURE 1.9
Figure depicting a dynamic exchange process of molecules on the surface of nanomaterials along with equilibrium constants. (Adapted from Lynch, I., and K. A. Dawson. 2008. *Nano Today* 3(1):40–47.)

biocompatibility or bio-adverse aspects (Cedervall et al. 2007b, Lynch et al. 2007, Lundqvist et al. 2008, Lynch et al. 2009, Nel et al. 2009, Suh et al. 2009, Walczyk et al. 2010, Monopoli et al. 2011, Dell'Orco et al. 2012, Walkey and Chan 2012). Figure 1.9 shows the dynamics of different proteins associated with the nanoparticle. There is a difference in the affinity of proteins for a given surface of nanomaterials, which depends on the type of interactions evoked, such as polar or hydrophobic interactions, thus creating a wide range of residence times on the surface. This essentially means we expect to have a range of equilibrium constants (one for each protein type) that highlights the diverse binding mechanism present. As represented with K_1 and K_2 in Figure 1.9, the composition of the protein corona depends on the abundance and types of biomolecules present in the vicinity as well as the kinetic rate of binding (Lynch and Dawson 2008).

Whenever nanomaterials are administered inside a biological system, their identity changes. It is known that protein molecules as well as other molecules (minute traces of lipids bound to particles have been reported [Rahman et al. 2013]) in body fluids bind to nanomaterials in a very dynamic way. First, abundant protein molecules of relatively low affinity bind to the particles, which will be replaced by molecules of high affinities and low concentration (Vroman's effect) over time (Rahman et al. 2013). The binding of molecules depends on the concentration of molecules present as well as the relative affinity of the molecules, ultimately introducing a different "biological identity" to a bare nanoparticle. It has been reported that building of the protein corona caused the loss of specificity in targeting for transferrin-coated nanomaterials (Salvati et al. 2013). Thus, it is this identity that needs to be considered for any assessment of nanomaterial effects, toxicity, kinetics, circulation, lifetime and clearance, signaling, responses, and subcellular localization inside a physiological system (Lynch and Dawson 2008, Rahman et al. 2013).

The formation of the protein corona depends on many physicochemical factors—size, shape, and surface properties (charge and coatings) of nanomaterials are the important

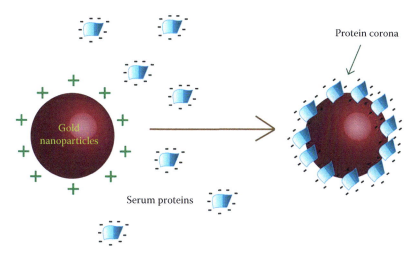

FIGURE 1.10
Pictographic representation showing how the net surface charge on gold nanoparticles changes with the binding of serum protein. (With kind permission from Springer Science+Business Media: *Journal of Nanoparticle Research*, Toxicity and cellular uptake of gold nanoparticles: What we have learned so far?, 12, 2010, 2313–2333, Alkilany, A. M., and C. J. Murphy.)

ones, as well the local environment parameters (temperature, localization, fluids) (Rahman et al. 2013). The protein corona differs with each type of nanomaterial, and it cannot be predicted for a given material (Alkilany et al. 2009). The building of a protein corona on a nanomaterial confers on it properties different from its native form. For example, if a nanoparticle has positive surface charge (cationic), upon introduction to proteins with overall negative charge, the formed nanomaterial corona adopts a negative charge, thus completely losing its identity, as shown in Figure 1.10 (Alkilany and Murphy 2010).

Cedervall et al. (2007a) noted that copolymer nanoparticles, upon introduction to the blood serum, bind to human serum albumin, which are highly abundant serum proteins but are subsequently replaced by proteins with high-affinity and slower-exchanging apolipoproteins like AII, AIV, and E. Many reports are available that suggest the same or a similar group of serum proteins binding to a nanomaterial's core. Apolipoproteins are involved in the cholesterol metabolism pathway. As apolipoprotein E is known to be involved in the trafficking to the brain (Lynch and Dawson 2008), this suggests that nanomaterials can exploit the protein corona to localize in systemic parts of the body. Sometimes, the protein corona increases the aggregation state of nanomaterials, and sometimes they can cause protein fibrillation under certain conditions (Lynch and Dawson 2008). Thus, the "nature" of the corona is important to assess systemic distribution and localization of nanoparticles.

As discussed, while talking about biodistribution and localization, it is important to consider properties like size, shape, surface properties (amount of hydrophobicity and charge [Yu et al. 2013]), and the protein corona. As studies have exhibited, objects less than 12 nm can cross the blood-brain barrier, and objects less than 30 nm can be endocytosed (Alkilany and Murphy 2010). In many cases, it is reported that endocytosis of nanomaterials is through receptor-mediated endocytosis (RME), while other pathways like clathrin-mediated endocytosis or caveolae-mediated endocytosis uptake could also be present (Chithrani and Chan 2007). It is also reported that preferential size- and shape-dependent uptake were favored for nanomaterials with a 50-nm size as the most optimal for uptake, and spherical-shaped nanoparticles were favored more than anisotropic particles, which can be explained by the "wrapping-time effect." This explanation says that the time required for the uptake of anisotropic nanoparticles like gold nanorods will be greater when compared to spherical shapes, as the membrane has to spread out more to engulf those particles (Chithrani et al. 2006). Also, it is to be noted that cellular response via receptor activation or cross-linking is also size dependent and can influence cellular signaling, as mentioned by Jiang and his research team (Jiang et al. 2008). As mentioned earlier, the corona also tends to increase the effective size of a nanoparticle. Hence, for any study on nanomaterial effective physicochemical properties as well as physiological tendencies, all these parameters need to be taken into account.

Only a few reports are available in the literature about the possible mechanism of interaction with the bio-interface (see Nel et al. 2009 for some details about it), and proper assessment of the probable fate of nanomaterials in the body is still murky. It is still unclear how a particle behaves when surrounded by the corona, what the distribution will be, and what responses are incurred. Studies conducted to assess these questions provide faint hints about the possible effects. There is still a need for a governing principle or at least rightful predictions about the fate of nanomaterials when certain administration routes are chosen. This could possibly help in the development of possible therapeutic nanosystems that can resolve the issues of toxicity and improve the pharmacokinetics of drug molecules bound, improve delivery of payloads, facilitate active targeting of, say, cancer or Alzheimer's while minimizing any adverse side effects as well as being quite versatile to translate into a clinical application—which is a must, as Juliano pointed out in his comment (Juliano 2013).

1.6 Nanotoxicity and Possible Molecular Mechanisms

Nanomaterials, for example in drug delivery, aren't just mere carriers for a drug or any therapeutics bound to them, they are much more dependent on the inherent properties they possess to incur any responses in the body (Jiang et al. 2008). Elemental gold, generally thought of as nontoxic, doesn't translate into a "nontoxic" entity at the nano level as suggested by some studies (Chen et al. 2009c, Cho et al. 2009). However, in another scenario, some groups have not found any toxicity associated with them (Connor et al. 2005). Similarly, in a study (Alkilany et al. 2009), it was shown that bare gold nanoparticles didn't induce toxicity, but the desorption of surface-coating molecules in this case was due to cetyltrimethylammonium bromide (CTAB), which was responsible for the toxicity observed. This study implies that some of the reducing molecules used for the synthesis of nanomaterials have a tendency to desorb from the nanoparticle shell surface, distribute around the body, and cause toxicity. In contrast to it, surprisingly, PEG-coated gold nanoparticles, (PEGylation is used to evade any immune response or make material biocompatible) caused acute inflammation and apoptosis, as shown by Cho et al. (2009). Thus, these ranges of studies don't provide a general conclusion to assess the associated toxicity, but do highlight the need for a wide range of experimentation to assess it in *in vivo* conditions to exactly know how the systemic effects are initiated by the particles.

One more concerning risk of nanomaterials is their residual environmental amount and their possible entry into the trophic level through food chains. In one study, it was shown that cadmium sellenide quantum dots amplified in a protozoan that fed bacteria in an experimental food chain, a result one cannot ignore (Werlin et al. 2011). The risk associated with cadmium quantum dots is that the cadmium can leach out from the particle surface and move into the body fluids as ions and cause severe toxicity (Soenen et al. 2011). It was also shown that these quantum dots interact with mitochondria and cause the production of reactive oxygen species (ROS), which subsequently damage DNA and also cause apoptosis. Similarly, Hong et al. (2014) showed that ceria (CeO_2, used as a diesel fuel catalyst) was distributed in grown cucumber tissues through the interaction with leaves in the same amount as would be from the root uptake, suggesting multiple uptake routes of materials by plants. These findings show the possibility of accumulation of particles in the environment, resulting in toxicity in the same way heavy metals accumulated in sea fish cause poisoning to consumers.

There have been many studies on the toxicity caused by a wide range of nanomaterials, besides frequently used particles. Carbon nanotubes have been shown to stimulate immunological responses by activating the complement system of the immune system (Salvador-Morales et al. 2006). In one review, it was mentioned that, generally, cationic nanomaterials (surface charge being positive) induced the production of cytokines—immune-stimulatory molecules. Furthermore, it has been stated that nanoparticles are capable of inducing even adaptive immunity that causes increased production of antibodies against them (Dobrovolskaia and McNeil 2007). Zinc oxide nanomaterials generally used in paints, coatings, and cosmetics cause autophagic cell and mitochondrial damage, as illustrated by Yu and co-workers (Yu et al. 2013) in an *in vitro* setting. A similar study by Ye et al. (2010) demonstrated toxicity caused by nano-SiO_2 through oxidative stress. There have been many *in vitro* as well as *in vivo* assessments of possible toxicity associated with different nanoparticles. We have provided a summary of just a few select studies. However, in some of the cases explained here, the researchers did not probe for any corona developments, its nature, or its subsequent influence on the observed toxicity (the corona can increase or decrease observed toxicity).

As hinted above, the mechanism of toxicity caused by nanomaterials is the production of reactive oxygen species in most of the cases, as illustrated in Figure 1.11. Also, the accumulation of nanomaterials in endolysosomes and their delayed removal cause ions "overload" in the cell, which can cross the LD_{50} (lethal dose at which 50% of cells or animals die) limit level (Soenen et al. 2011). For nanoparticles of cationic nature, it has been shown that they cause cell membrane disruption, thereby causing an imbalance in ion concentration inside and outside cells and osmotic shock, causing cell death (Chen et al. 2009b). For smaller nanoparticles of size <10 nm, it has been revealed that they can bind to the major grooves of DNA and induce epigenetic effects (Soenen et al. 2011). Tang et al. showed that the mechanism of associated toxicity was due to oxidative stress and DNA damage as supported by measuring differential gene expression profiles for stress response and DNA repair genes (Tang et al. 2015).

1.7 Assessing Nanotoxicity: Characterization Steps, Assay Methods, and Reliability

The schematic illustrates (Figure 1.12) steps of characterization that need to be established to assess any biological response, especially when the particles are used for biomedical applications. As mentioned earlier, "the biological identity" of a particle is what is taken into account by cells or systems to respond to or start any cascade of events; the methods to determine it need a full standardization. The bias in various toxicological studies related

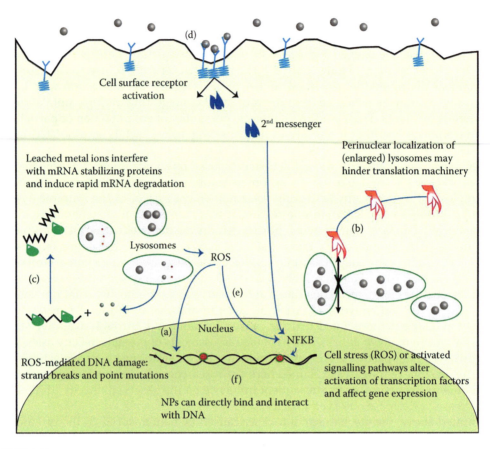

FIGURE 1.11
Schematics showing possible mechanisms of ROS generation by nanoparticles. (Reprinted from *Nano Today*, 6(5), Soenen, S. J. et al. Cellular toxicity of inorganic nanoparticles: Common aspects and guidelines for improved nanotoxicity evaluation, 446–465, Copyright (2011), with permission from Elsevier.)

to nanoparticles emerges because of variations of lab settings and conditions, parameters, toxicity assays used, cell lines, and nanoparticle chemical and/or physical properties (Alkilany and Murphy 2010) and indeed make the findings insignificant or of less value.

Marquis and co-workers (Marquis et al. 2009) suggested some methods and techniques for uptake, distribution, and nanotoxicity assessment. *In vitro* uptake and localization can be characterized by using a transmission electron microscope (TEM) (Li et al. 2008), electron-dispersive x-ray analysis (EDS) (Asharani et al. 2008b), and electron energy loss spectroscopy (EELS) used in conjunction with TEM (Porter et al. 2007). For elemental analysis, inductively coupled plasma atomic emission spectroscopy (ICP-AES) can be used as a powerful and precise instrument to determine elemental concentration of particles inside cells (Szpunar 2000). However, this technique does have the limitation that it cannot determine whether nanoparticles are internalized into the cell or are externally associated with the cell. Similarly, florescence spectroscopy can be used qualitatively or quantitatively for the assessment of the uptake in the cell (Bagalkot et al. 2007, Hu et al. 2007).

For *in vitro* toxicity, studies generally use cytotoxicity assays for any chemical using certain animal cell lines, and are also being used for nanoparticle toxicity assessment. These include

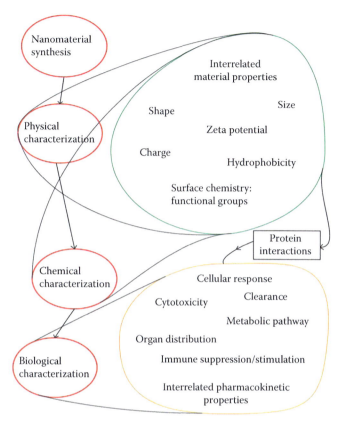

FIGURE 1.12
Fischer and Chang illustrated the steps to characterize nanomaterials. (Reprinted from *Current Opinion in Biotechnology*, 18, Fischer, H. C., and W. C. W. Chan, Nanotoxicity: The growing need for in vivo study, 565–571, Copyright (2007), with permission from Elsevier.)

proliferative assays like the 3-(4,5-dimethylthiazol-2-Yl)-2,5-diphenyltetrazolium bromide (MTT) assay (Mosmann 1983) or necrosis assay (Sayes et al. 2004) that uses dyes to check for viability of cells or an apoptosis assay like DNA laddering assay or annexin-V (Pan et al. 2007), Comet assay (Jin et al. 2007, Omidkhoda et al. 2007), or terminal deoxynucleotidyl transferase dUTP-mediated nick end labeling (TUNEL) assay (Bhol and Schechter 2005) that detects any morphological changes induced by nanoparticles. Similarly, oxidative stress induced by nanomaterials is measured by using various oxidative stress probes to examine the level of reactive oxygen species being generated from the interaction (Lee et al. 2007, Kang et al. 2008b).

Biodistribution and localization *in vivo* can be assessed by ICP-AES or inductively coupled mass spectroscopy (IC-MS) as mentioned earlier for *in vitro* assessment. For *in vivo* toxicity studies, hematological screening and serum analysis (for example, identifying and quantifying cytokines that cause inflammation) for possible biomarkers that correlate to nanoparticle interactions in the body. Also, histopathological tests can be conducted to examine any tissue damage or morphological changes induced using TEM as a viewing tool.

It is to be understood, however, that though *in vitro* methods give us a rapid examination of any toxicity associated with nanoparticles as well as preventing the sacrifice of huge number of model animals for studies, there are some limitations inherent to them. Factors like pharmacokinetics of particles and possible tissue and systemic responses are important when the data are interpreted in the *in vivo* realm. The difference in the exposure time, circulating time, clearance time, and effective concentration when particles reach certain parts of the body differ from what will be used *in vitro* and hence are to be noted when the results are interpreted for the assessment of plausible *in vivo* toxicity. Also, sometimes the *in vitro* screening methods give us false results due to the influence of particles under examination. *In vitro* techniques normally use fluorescence or colorimetric methods to determine the amount of toxicity incurred, but sometimes the color gradient being built up will be interfered with the absorption of light by the particles themselves in the visible range, as reported by Asharani and team (Asharani et al. 2008a). Also, florescence probes or dyes used will be adsorbed on the nanoparticle surface and will quench the produced color intensity and lead to aberrant results.

There have been some innovations to develop novel techniques for an effective, robust, specific assessment of interaction of nanomaterials with the biological system and subsequently the toxicity, if associated. High-throughput methods like OMICS technology are feasible for *in vitro* as well as *in vivo* studies. Metabolomics, especially, can be applied to study the markers that are relevant to the type of effects induced. Most commonly employed methods for metabolomics involve liquid chromatography/mass spectrometry (LC/MS) techniques and nuclear magnetic resonance (NMR) spectroscopy. Other analytical techniques include gas chromatography coupled to mass spectrometry (GC/MS), thin-layer chromatography coupled to mass spectrometry (TLC/MS), matrix-assisted laser desorption/ionization mass spectrometry (MALDI/MS), and Fourier transform mass spectrometry (FTMS) (Schnackenberg et al. 2012). Transcriptomics can also be used to count the differentially expressed total transcripts of genes related to various pathways like signaling, stress response, and DNA damage control pathways to assess the mechanism of toxicity as used in the study in reference (Tang et al. 2015). Similarly, mass spectroscopy-based proteomics can be applied to study the nature of the protein corona associated with nanoparticles *in vivo* and thus facilitate the study of cellular signaling and biological behavior (Lai et al. 2012).

1.8 Conclusion, Sustainability, and Future Outlook

There has been a rapid expansion in the field of nanotechnology since its first methodological inception by Richard Feynman. Novel works related to precise manipulation of materials to cater to needs in different areas such as energy storage and management, high-strength composites, fluorescence bio-imaging, drug delivery, internal implants, and tissue engineering are on the rise. There have also been substantive investments from both the government and private sectors. With such high interest and investments in nanotechnology, we, indeed in the near future, will be using or consuming nanoproducts in various regards. However, with this outburst, it is certain that until a proper monitored use is mandated, the biological ecosystem is going to interface with these synthetic constructs at large scales. This rising amount of nanoparticles in the environment surrounding us should be seen as a matter of concern, as the fate and clearance time of these nanoparticles are still obscure. These "foreign-by-nature" particles can evoke many unknown responses in organisms. Also to be noted is that a few reports of toxicity in some research papers can create bias—a situation where hype overtakes rationality, which can spread fear among the public. Things of these sorts may put pressure on guideline setters such as the WHO or FDA in a way that can be solved only through properly phased unbiased high-scale research into the systemic characterization of these particles' effects in living matter. There is a dire need to conduct *in vivo* studies for the assessment of toxicity related to nanoparticles, as *in vitro* studies lack fundamental aspects of systemic architecture like pharmacokinetics or responses. There are huge prospects for improvising virtually everything using nanotechnology. We can create high-energy-efficient devices; rapid therapeutics and diagnostics; high-strength materials; and shape-shifting, self-healing materials. However, sustainability in the use of nanomaterials demands carefully discussed terms of usage and more scrutiny.

Competing Interests

The authors declare that they have no competing interests.

Acknowledgments

HKD gratefully acknowledges the Department of Science & Technology (DST), Government of India, for ITS Grant (Grant No. SB/ITS-Y/0988/2014-15). HKD and NPN also acknowledge in-house financial support from SIT, Sree Siddaganga Education Society, Tumkur, Karnataka, India, toward establishing a research laboratory at the Department of Biotechnology. AK gratefully acknowledges a scholarship supported by the International Max Planck Research School to pursue his MSc/PhD in molecular biology at the University of Göttingen, Germany.

References

Ahn, S.-J., S.-J. Lee, J.-K. Kook, and B.-S. Lim. 2009. Experimental antimicrobial orthodontic adhesives using nanofillers and silver nanoparticles. *Dental Materials* 25(2):206–213.

Aillon, K. L., Y. Xie, N. El-Gendy, C. J. Berkland, and M. Laird Forrest. 2009. Effects of nanomaterial physicochemical properties on *in vivo* toxicity. *Advanced Drug Delivery Reviews* 61(6):457–466.

Alivisatos, A. P. 1996. Semiconductor clusters, nanocrystals, and quantum dots. *Science* 271(5251):933–937.

Alkilany, A. M., and C. J. Murphy. 2010. Toxicity and cellular uptake of gold nanoparticles: What we have learned so far? *Journal of Nanoparticle Research* 12(7):2313–2333.

Alkilany, A. M., P. K. Nagaria, C. R. Hexel, T. J. Shaw, C. J. Murphy, and M. D. Wyatt. 2009. Cellular uptake and cytotoxicity of gold nanorods: Molecular origin of cytotoxicity and surface effects. *Small* 5(6):701–708.

Antonii, F. 1618. Panacea aurea-auro potabile. *Hamburg: Ex Bibliopolio Frobeniano* 205.

Asharani, P. V., G. Low Kah Mun, M. P. Hande, and S. Valiyaveettil. 2008a. Cytotoxicity and genotoxicity of silver nanoparticles in human cells. *ACS Nano* 3(2):279–290.

Asharani, P. V., Y. L. Wu, Z. Gong, and S. Valiyaveettil. 2008b. Toxicity of silver nanoparticles in zebrafish models. *Nanotechnology* 19(25):255102.

Azami, M., S. Jalilifiroozinezhad, M. Mozafari, and M. Rabiee. 2011. Synthesis and solubility of calcium fluoride/hydroxy-fluorapatite nanocrystals for dental applications. *Ceramics International* 37(6):2007–2014.

Bagalkot, V., L. Zhang, E. Levy-Nissenbaum, S. Jon, P. W. Kantoff, R. Langer, and O. C. Farokhzad. 2007. Quantum dot-aptamer conjugates for synchronous cancer imaging, therapy, and sensing of drug delivery based on bi-fluorescence resonance energy transfer. *Nano Letters* 7(10):3065–3070.

Balzani, V. 2005. Nanoscience and nanotechnology: A personal view of a chemist. *Small* 1(3):278–283. doi: 10.1002/smll.200400010.

Bansal, V., V. Li, A. P. O'Mullane, and S. K. Bhargava. 2010. Shape dependent electrocatalytic behaviour of silver nanoparticles. *CrystEngComm* 12(12):4280–4286.

Baruah, S., and J. Dutta. 2009. Nanotechnology applications in pollution sensing and degradation in agriculture: A review. *Environmental Chemistry Letters* 7(3):191–204.

Bastings, M., S. Koudstaal, R. E. Kieltyka, Y. Nakano, A. C. H. Pape, D. A. M. Feyen, F. J. van Slochteren, P. A. Doevendans, J. P. G. Sluijter, and E. W. Meijer. 2014. A fast pH-switchable and self-healing supramolecular hydrogel carrier for guided, local catheter injection in the infarcted myocardium. *Advanced Healthcare Materials* 3(1):70–78.

Bauer, L. A., N. S. Birenbaum, and G. J. Meyer. 2004. Biological applications of high aspect ratio nanoparticles. *Journal of Materials Chemistry* 14(4):517–526.

Behera, S. S., K. Pramanik, and M. K Nayak. 2015. Recent advancement in the treatment of cardiovascular diseases: Conventional therapy to nanotechnology. *Current Pharmaceutical Design* 21(30):4479–4497.

Besinis, A., T. De Peralta, C. J. Tredwin, and R. D. Handy. 2015. Review of nanomaterials in dentistry: Interactions with the oral microenvironment, clinical applications, hazards, and benefits. *ACS Nano* 9(3):2255–2289.

Bhakta, S. A., E. Evans, T. E. Benavidez, and C. D. Garcia. 2015. Protein adsorption onto nanomaterials for the development of biosensors and analytical devices: A review. *Analytica Chimica Acta* 872:7–25.

Bhol, K. C., and P. J. Schechter. 2005. Topical nanocrystalline silver cream suppresses inflammatory cytokines and induces apoptosis of inflammatory cells in a murine model of allergic contact dermatitis. *British Journal of Dermatology* 152(6):1235–1242.

Bianco, A., K. Kostarelos, C. D. Partidos, and M. Prato. 2005. Biomedical applications of functionalised carbon nanotubes. *Chemical Communications* (5):571–577.

Biju, V. 2014. Chemical modifications and bioconjugate reactions of nanomaterials for sensing, imaging, drug delivery and therapy. *Chemical Society Reviews* 43(3):744–764.

Binnig, G., and H. Rohrer. 1983. Scanning tunneling microscopy. *Surface Science* 126(1):236–244.

Boisselier, E., and D. Astruc. 2009. Gold nanoparticles in nanomedicine: Preparations, imaging, diagnostics, therapies and toxicity. *Chemical Society Reviews* 38(6):1759–1782. doi: 10.1039/B806051G.

Bonnemann, H., and R. M. Richards. 2001. Nanoscopic metal particles—Synthetic methods and potential applications. *European Journal of Inorganic Chemistry* 2001(10):2455–2480.

Bruchez, M., M. Moronne, P. Gin, S. Weiss, and A. Paul Alivisatos. 1998. Semiconductor nanocrystals as fluorescent biological labels. *Science* 281(5385):2013–2016.

Brundtland Commission. 1987. *Our Common Future*. Oxford: Oxford University Press.

Burda, C., X. Chen, R. Narayanan, and M. A. El-Sayed. 2005. Chemistry and properties of nanocrystals of different shapes. *Chemical Reviews* 105(4):1025–1102.

Cao, J., T. Sun, and K. T. V. Grattan. 2014. Gold nanorod-based localized surface plasmon resonance biosensors: A review. *Sensors and Actuators B: Chemical* 195:332–351. doi: http://dx.doi.org/10.1016/j.snb.2014.01.056.

Cedervall, T., I. Lynch, M. Foy, T. Berggård, S. C. Donnelly, G. Cagney, S. Linse, and K. A. Dawson. 2007a. Detailed identification of plasma proteins adsorbed on copolymer nanoparticles. *Angewandte Chemie International Edition* 46(30):5754–5756.

Cedervall, T., I. Lynch, S. Lindman, T. Berggård, E. Thulin, H. Nilsson, K. A. Dawson, and S. Linse. 2007b. Understanding the nanoparticle–protein corona using methods to quantify exchange rates and affinities of proteins for nanoparticles. *Proceedings of the National Academy of Sciences* 104(7):2050–2055. doi: 10.1073/pnas.0608582104.

Chan, W. C. W., D. J. Maxwell, X. Gao, R. E. Bailey, M. Han, and S. Nie. 2002. Luminescent quantum dots for multiplexed biological detection and imaging. *Current Opinion in Biotechnology* 13(1):40–46.

Chan, W. C. W., and S. Nie. 1998. Quantum dot bioconjugates for ultrasensitive nonisotopic detection. *Science* 281(5385):2016–2018. doi: 10.1126/science.281.5385.2016.

Chen, H., C. Jiang, C. Yu, S. Zhang, B. Liu, and J. Kong. 2009a. Protein chips and nanomaterials for application in tumor marker immunoassays. *Biosensors and Bioelectronics* 24(12):3399–3411.

Chen, J., J. A. Hessler, K. Putchakayala, B. K. Panama, D. P. Khan, S. Hong, D. G. Mullen, S. C. DiMaggio, A. Som, and G. N. Tew. 2009b. Cationic nanoparticles induce nanoscale disruption in living cell plasma membranes. *The Journal of Physical Chemistry B* 113(32):11179–11185.

Chen, J., Z. Jiang, J. D. Ackerman, M. Yazdani, S. Hou, S. R. Nugen, and V. M. Rotello. 2015a. Electrochemical nanoparticle-enzyme sensors for screening bacterial contamination in drinking water. *Analyst* 140(15):4991–4996.

Chen, L. Q., L. Fang, J. Ling, C. Z. Ding, B. Kang, and C. Z. Huang. 2015b. Nanotoxicity of silver nanoparticles to red blood cells: Size dependent adsorption, uptake, and hemolytic activity. *Chemical Research in Toxicology* 28(3):501–509.

Chen, Y.-S., Y.-C. Hung, I. Liau, and G. Steve Huang. 2009c. Assessment of the *in vivo* toxicity of gold nanoparticles. *Nanoscale Research Letters* 4(8):858–864.

Cherukuri, P., S. M. Bachilo, S. H. Litovsky, and R. Bruce Weisman. 2004. Near-infrared fluorescence microscopy of single-walled carbon nanotubes in phagocytic cells. *Journal of the American Chemical Society* 126(48):15638–15639.

Chithrani, B. D., and W. C. W. Chan. 2007. Elucidating the mechanism of cellular uptake and removal of protein-coated gold nanoparticles of different sizes and shapes. *Nano Letters* 7(6):1542–1550.

Chithrani, B. D., A. A. Ghazani, and W. C. W. Chan. 2006. Determining the size and shape dependence of gold nanoparticle uptake into mammalian cells. *Nano Letters* 6(4):662–668.

Cho, W.-S., M. Cho, J. Jeong, M. Choi, H.-Y. Cho, B. S. Han, S. H. Kim, H. O. Kim, Y. T. Lim, and B. H. Chung. 2009. Acute toxicity and pharmacokinetics of 13 nm-sized PEG-coated gold nanoparticles. *Toxicology and Applied Pharmacology* 236(1):16–24.

Colvin, V. L. 2003. The potential environmental impact of engineered nanomaterials. *Nature Biotechnology* 21(10):1166–1170.

Connor, E. E., J. Mwamuka, A. Gole, C. J. Murphy, and M. D. Wyatt. 2005. Gold nanoparticles are taken up by human cells but do not cause acute cytotoxicity. *Small* 1(3):325–327.

Daima, H. K. 2013. *Towards fine-tuning the surface corona of inorganic and organic nanomaterials to control their properties at nano-bio interface*. PhD, School of Applied Sciences RMIT, Melbourne, Australia. 1–236.

Daima, H. K., and V. Bansal. 2015. Chapter 10—Influence of physicochemical properties of nanomaterials on their antibacterial applications. In *Nanotechnology in Diagnosis, Treatment and Prophylaxis of Infectious Diseases*, edited by Mahendra R. and K. Kon, 151–166. Boston: Academic Press.

Daima, H. K., and P. N. Navya. 2016. Rational engineering of physicochemical properties of nanomaterials for biomedical applications with nanotoxicological perspectives. *Nano Convergence* 3(1):1–14.

Daima, H. K., P. R. Selvakannan, S. K. Bhargava, S. K. Shastry, V. Bansal. 2014a. Amino acids-conjugated gold, silver and their alloy nanoparticles: Role of surface chemistry and metal composition on peroxidase like activity. Paper read at *Technical Proceedings of Nanotech 2014 TechConnect World Conference and Expo*. Vol. 2, 275–278. Washington, USA. ISBN: 978-1-4822-5827-1.

Daima, H. K., P. Selvakannan, Z. Homan, S. K. Bhargava, and V. Bansal. 2011. Tyrosine mediated gold, silver and their alloy nanoparticles synthesis: Antibacterial activity toward gram positive and gram negative bacterial strains. *Paper Read at 2011 International Conference on Nanoscience, Technology and Societal Implications, NSTSI11*, Bhubaneswar, Orissa, India. 1-6. IEEE Press, ISBN: 9781457720352.

Daima, H. K., P. R. Selvakannan, A. E. Kandjani, R. Shukla, S. K. Bhargava, and V. Bansal. 2014b. Synergistic influence of polyoxometalate surface corona towards enhancing the antibacterial performance of tyrosine-capped Ag nanoparticles. *Nanoscale* 6(2):758–765. doi: 10.1039/C3NR03806H.

Daima, H. K., P. R. Selvakannan, R. Shukla, S. K. Bhargava, and V. Bansal. 2013. Fine-tuning the antimicrobial profile of biocompatible gold nanoparticles by sequential surface functionalization using polyoxometalates and lysine. *PloS One* 8(10):1–14.

Daima, H. K., S. Shankar, P. R. Selvakannan, S. K. Bhargava, and V. Bansal. 2012. Self Assembled Nanostructures of Triblock Co-polymer and Plasmid DNA for Gene Delivery, NSTI-Nanotech 3:178–181.

Daniel, M. C., and D. Astruc. 2004. Gold nanoparticles: Assembly, supramolecular chemistry, quantum-size-related properties, and applications toward biology, catalysis, and nanotechnology. *Chemical Reviews* 104(1):293–346.

Davis, M. E., P. C. H. Hsieh, T. Takahashi, Q. Song, S. Zhang, R. D. Kamm, A. J. Grodzinsky, P. Anversa, and R. T. Lee. 2006. Local myocardial insulin-like growth factor 1 (IGF-1) delivery with biotinylated peptide nanofibers improves cell therapy for myocardial infarction. *Proceedings of the National Academy of Sciences* 103(21):8155–8160.

Davis, M. E., J. P. Michael Motion, D. A. Narmoneva, T. Takahashi, D. Hakuno, R. D. Kamm, S. Zhang, and R. T. Lee. 2005. Injectable self-assembling peptide nanofibers create intramyocardial microenvironments for endothelial cells. *Circulation* 111(4):442–450.

Dawidczyk, C. M., C. Kim, J. H. Park, L. M. Russell, K. H. Lee, M. G. Pomper, and P. C. Searson. 2014. State-of-the-art in design rules for drug delivery platforms: Lessons learned from FDA-approved nanomedicines. *Journal of Controlled Release* 187:133–144.

Deerinck, T. J. 2008. The application of fluorescent quantum dots to confocal, multiphoton, and electron microscopic imaging. *Toxicologic Pathology* 36(1):112–116.

Dell'Orco, D., M. Lundqvist, T. Cedervall, and S. Linse. 2012. Delivery success rate of engineered nanoparticles in the presence of the protein corona: A systems-level screening. *Nanomedicine: Nanotechnology, Biology and Medicine* 8(8):1271–1281.

Dobrovolskaia, M. A., and S. E. McNeil. 2007. Immunological properties of engineered nanomaterials. *Nature Nanotechnology* 2(8):469–478.

Dresselhaus, M. S., G. Dresselhaus, A. Jorio, A. G. Souza Filho, and R. Saito. 2002. Raman spectroscopy on isolated single wall carbon nanotubes. *Carbon* 40(12):2043–2061.

Drexler, K. E., and M. Minsky. 1990. *Engines of Creation: The Coming Era of Nanotechnology*. London: Fourth Estate.

Drexler, K. E., and C. Peterson. 1989. Nanotechnology and enabling technologies. *Foresight Briefing* 2. http://www.foresight.org/Updates/Briefing2.html.

Dubey, K., B. G. Anand, R. Badhwar, G. Bagler, P. N. Navya, H. K. Daima, and K. Kar. 2015. Tyrosine- and tryptophan-coated gold nanoparticles inhibit amyloid aggregation of insulin. *Amino Acids* 47(12):2551–2560. doi: 10.1007/s00726-015-2046-6.

Dykman, L., and N. Khlebtsov. 2012. Gold nanoparticles in biomedical applications: Recent advances and perspectives. *Chemical Society Reviews* 41(6):2256–2282.

Feynman, R. P. 1960. There's plenty of room at the bottom. *Engineering and Science* 23(5):22–36.

Fischer, H. C., and W. C. W. Chan. 2007. Nanotoxicity: The growing need for *in vivo* study. *Current Opinion in Biotechnology* 18(6):565–571.

Fortina, P., L. J. Kricka, S. Surrey, and P. Grodzinski. 2005. Nanobiotechnology: The promise and reality of new approaches to molecular recognition. *Trends in Biotechnology* 23(4):168–173.

Freestone, I., N. Meeks, M. Sax, and C. Higgitt. 2007. The Lycurgus cup—a Roman nanotechnology. *Gold Bulletin* 40(4):270–277.

Gao, L., J. Zhuang, L. Nie, J. Zhang, Y. Zhang, N. Gu, T. Wang, J. Feng, D. Yang, and S. Perrett. 2007. Intrinsic peroxidase-like activity of ferromagnetic nanoparticles. *Nature Nanotechnology* 2(9):577–583.

Gao, X., L. Yang, J. A. Petros, F. F. Marshall, J. W. Simons, and S. Nie. 2005. *In vivo* molecular and cellular imaging with quantum dots. *Current Opinion in Biotechnology* 16(1):63–72. doi: http://dx.doi.org/10.1016/j.copbio.2004.11.003.

Garnett, M. C., and P. Kallinteri. 2006. Nanomedicines and nanotoxicology: Some physiological principles. *Occupational Medicine* 56(5):307–311. doi: 10.1093/occmed/kql052.

Gatoo, M. A., S. Naseem, M. Y. Arfat, A. M. Dar, K. Qasim, and S. Zubair. 2014. Physicochemical properties of nanomaterials: Implication in associated toxic manifestations. *BioMed Research International* 2014:8. doi: 10.1155/2014/498420.

Gerion, D., F. Pinaud, S. C. Williams, W. J. Parak, D. Zanchet, S. Weiss, and A. P. Alivisatos. 2001. Synthesis and properties of biocompatible water-soluble silica-coated CdSe/ZnS semiconductor quantum dots. *Journal of Physical Chemistry B* 105(37):8861–8871.

Ghasemi, Y., P. Peymani, and S. Afifi. 2009. Quantum dot: Magic nanoparticle for imaging, detection and targeting. *Acta Bio Medica Atenei Parmensis* 80(2):156–165.

Goenka, S., V. Sant, and S. Sant. 2014. Graphene-based nanomaterials for drug delivery and tissue engineering. *Journal of Controlled Release* 173:75–88.

Gokhale, P. C., C. Zhang, J. T. Newsome, J. Pei, I. Ahmad, A. Rahman, A. Dritschilo, and U. N. Kasid. 2002. Pharmacokinetics, toxicity, and efficacy of ends-modified raf antisense oligodeoxyribonucleotide encapsulated in a novel cationic liposome. *Clinical Cancer Research* 8(11):3611–3621.

Gong, J., L. Wang, and L. Zhang. 2009. Electrochemical biosensing of methyl parathion pesticide based on acetylcholinesterase immobilized onto Au-polypyrrole interlaced network-like nanocomposite. *Biosensors and Bioelectronics* 24(7):2285–2288.

Hardas, S. S., D. Allan Butterfield, R. Sultana, M. T. Tseng, M. Dan, R. L. Florence, J. M. Unrine, U. M. Graham, P. Wu, and E. A. Grulke. 2010. Brain distribution and toxicological evaluation of a systemically delivered engineered nanoscale ceria. *Toxicological Sciences* 116(2):562–576.

Harrison, B. S., and A. Atala. 2007. Carbon nanotube applications for tissue engineering. *Biomaterials* 28(2):344–353.

Helland, A., and H. Kastenholz. 2008. Development of nanotechnology in light of sustainability. *Journal of Cleaner Production* 16(8–9):885–888. doi: http://dx.doi.org/10.1016/j.jclepro.2007.04.006.

Hong, J., J. R. Peralta-Videa, C. Rico, S. Sahi, M. N. Viveros, J. Bartonjo, L. Zhao, and J. L. Gardea-Torresdey. 2014. Evidence of translocation and physiological impacts of foliar applied CeO_2 nanoparticles on cucumber (*Cucumis sativus*) plants. *Environmental Science & Technology* 48(8):4376–4385.

Hu, X., S. Cook, P. Wang, and H. M. Hwang. 2009. *In vitro* evaluation of cytotoxicity of engineered metal oxide nanoparticles. *Science of the Total Environment* 407(8):3070–3072.

Hu, Y., J. Xie, Y. W. Tong, and C.-H. Wang. 2007. Effect of PEG conformation and particle size on the cellular uptake efficiency of nanoparticles with the HepG2 cells. *Journal of Controlled Release* 118(1):7–17.

Huh, A. J., and Y. J. Kwon. 2011. "Nanoantibiotics": A new paradigm for treating infectious diseases using nanomaterials in the antibiotics resistant era. *Journal of Controlled Release* 156(2):128–145. doi: http://dx.doi.org/10.1016/j.jconrel.2011.07.002.

Ishikawa-Ankerhold, H. C., R. Ankerhold, and G. P. C. Drummen. 2012. Advanced fluorescence microscopy techniques—Frap, flip, flap, fret and flim. *Molecules* 17(4):4047–4132.

Jain, K. K. 2007. Applications of nanobiotechnology in clinical diagnostics. *Clinical Chemistry* 53(11):2002–2009.

Jain, T. K., M. A. Morales, S. K. Sahoo, D. L. Leslie-Pelecky, and V. Labhasetwar. 2005. Iron oxide nanoparticles for sustained delivery of anticancer agents. *Molecular Pharmaceutics* 2(3):194–205.

Jiang, W., B. Y. S. Kim, J. T. Rutka, and W. C. W. Chan. 2008. Nanoparticle-mediated cellular response is size-dependent. *Nature Nanotechnology* 3(3):145–150.

Jin, Y., S. Kannan, M. Wu, and J. X. Zhao. 2007. Toxicity of luminescent silica nanoparticles to living cells. *Chemical Research in Toxicology* 20(8):1126–1133.

Juliano, R. 2013. Nanomedicine: Is the wave cresting? *Nature Reviews Drug Discovery* 12(3):171–172.

Jung, J. H., D. S. Cheon, F. Liu, K. B. Lee, and T. S. Seo. 2010. A graphene oxide based immuno-biosensor for pathogen detection. *Angewandte Chemie International Edition* 49(33):5708–5711.

Kallay, N., and S. Zalac. 2002. Stability of nanodispersions: A model for kinetics of aggregation of nanoparticles. *Journal of Colloid and Interface Science* 253(1):70–76.

Kang, S., M. Herzberg, D. F. Rodrigues, and M. Elimelech. 2008a. Antibacterial effects of carbon nanotubes: Size does matter! *Langmuir* 24(13):6409–6413. doi: 10.1021/la800951v.

Kang, S. J., B. M. Kim, Y. J. Lee, and H. W. Chung. 2008b. Titanium dioxide nanoparticles trigger p53-mediated damage response in peripheral blood lymphocytes. *Environmental and Molecular Mutagenesis* 49(5):399–405.

Katz, E., and I. Willner. 2004. Integrated nanoparticle–Biomolecule hybrid systems: Synthesis, properties, and applications. *Angewandte Chemie International Edition* 43(45):6042–6108. doi: 10.1002/anie.200400651.

Klaine, S. J., P. J. J. Alvarez, G. E. Batley, T. F. Fernandes, R. D. Handy, D. Y. Lyon, S. Mahendra, M. J. McLaughlin, and J. R. Lead. 2008. Nanomaterials in the environment: Behavior, fate, bioavailability, and effects. *Environmental Toxicology and Chemistry* 27(9):1825–1851. doi: 10.1897/08-090.1.

Kroto, H. W., J. R. Heath, S. C. O'Brien, R. F. Curl, and R. E. Smalley. 1985. C 60: Buckminsterfullerene. *Nature* 318(6042):162–163.

Krueger, A. 2010. *Carbon Materials and Nanotechnology*. Germany: Wiley-VCH Verlag GmbH & Co. KGaA.

Krug, H. F., and P. Wick. 2011. Nanotoxicology: An interdisciplinary challenge. *Angewandte Chemie International Edition* 50(6):1260–1278. doi: 10.1002/anie.201001037.

Kwon, T.-Y., D. S. Oh, and R. Narayanan. 2015. Nanomaterials for medical and dental applications. *Journal of Nanomaterials* 2015:1–2. doi:10.1155/2015/707683.

Lai, Z. W., Y. Yan, F. Caruso, and E. C. Nice. 2012. Emerging techniques in proteomics for probing nano–bio interactions. *ACS Nano* 6(12):10438–10448.

Lamish, A., and O. Klein. 2015. Magnetic separation of rare cells. Patent number: US 2017/0121669 A1 2017.

Laurent, S., D. Forge, M. Port, A. Roch, C. Robic, L. Vander Elst, and R. N. Muller. 2008. Magnetic iron oxide nanoparticles: Synthesis, stabilization, vectorization, physicochemical characterizations and biological applications. *Chemical Reviews* 108(6):2064–2110. doi: 10.1021/cr068445e.

Lee, K. J., P. D. Nallathamby, L. M. Browning, C. J. Osgood, and X.-H. N. Xu. 2007. In vivo imaging of transport and biocompatibility of single silver nanoparticles in early development of zebrafish embryos. *ACS Nano* 1(2):133–143.

Lehner, R., X. Wang, S. Marsch, and P. Hunziker. 2013. Intelligent nanomaterials for medicine: Carrier platforms and targeting strategies in the context of clinical application. *Nanomedicine: Nanotechnology, Biology and Medicine* 9(6):742–757.

Lewinski, N., V. Colvin, and R. Drezek. 2008. Cytotoxicity of nanoparticles. *Small* 4(1):26–49. doi: 10.1002/smll.200700595.

Li, J. J., L. Zou, D. Hartono, C.-N. Ong, B.-H. Bay, and L.-Y. Lanry Yung. 2008. Gold nanoparticles induce oxidative damage in lung fibroblasts *in vitro*. *Advanced Materials* 20(1):138–142.

Li, M., H. X. Tang, and M. L. Roukes. 2007. Ultra-sensitive NEMS-based cantilevers for sensing, scanned probe and very high-frequency applications. *Nature Nanotechnology* 2(2):114–120.

Li, Z., Z. Li, Q. Niu, H. Li, M. Vuki, and D. Xu. 2017. Visual microarray detection for human IgE based on silver nanoparticles. *Sensors and Actuators B: Chemical* 239:45–51.

Lin, Y.-D., C.-Y. Luo, Y.-N. Hu, M.-L. Yeh, Y.-C. Hsueh, M.-Y. Chang, D.-C. Tsai, J.-N. Wang, M.-J. Tang, and E. I. H. Wei. 2012. Instructive nanofiber scaffolds with VEGF create a microenvironment for arteriogenesis and cardiac repair. *Science Translational Medicine* 4(146):146ra109.

Lin, Y.-D., M.-L. Yeh, Y.-J. Yang, D.-C. Tsai, T.-Y. Chu, Y.-Y. Shih, M.-Y. Chang, Y.-W. Liu, A. C. L. Tang, and T.-Y. Chen. 2010. Intramyocardial peptide nanofiber injection improves postinfarction ventricular remodeling and efficacy of bone marrow cell therapy in pigs. *Circulation* 122(11 suppl. 1):S132–S141.

Liu, H., Y. Wang, H. Li, Z. Wang, and D. Xu. 2013a. Luminescent rhodamine B doped core–shell silica nanoparticle labels for protein microarray detection. *Dyes and Pigments* 98(1):119–124.

Liu, M., R. Liu, and W. Chen. 2013b. Graphene wrapped Cu_2O nanocubes: Non-enzymatic electrochemical sensors for the detection of glucose and hydrogen peroxide with enhanced stability. *Biosensors and Bioelectronics* 45:206–212.

Liu, Y., W. Hu, Z. Lu, and C. M. Li. 2014. ZnO nanomulberry and its significant nonenzymatic signal enhancement for protein microarray. *ACS Applied Materials & Interfaces* 6(10):7728–7734.

Liu, Z., C. Davis, W. Cai, L. He, X. Chen, and H. Dai. 2008. Circulation and long-term fate of functionalized, biocompatible single-walled carbon nanotubes in mice probed by Raman spectroscopy. *Proceedings of the National Academy of Sciences* 105(5):1410–1415.

Longmire, M., P. L. Choyke, and H. Kobayashi. 2008. Clearance properties of nano-sized particles and molecules as imaging agents: Considerations and caveats. *Nanomedicine* 3(5):703–717. doi: 10.2217/17435889.3.5.703.

Lundqvist, M., J. Stigler, G. Elia, I. Lynch, T. Cedervall, and K. A. Dawson. 2008. Nanoparticle size and surface properties determine the protein corona with possible implications for biological impacts. *Proceedings of the National Academy of Sciences* 105(38):14265–14270. doi: 10.1073/pnas.0805135105.

Lyass, O., B. Uziely, R. Ben-Yosef, D. Tzemach, N. I. Heshing, M. Lotem, G. Brufman, and A. Gabizon. 2000. Correlation of toxicity with pharmacokinetics of pegylated liposomal doxorubicin (Doxil) in metastatic breast carcinoma. *Cancer* 89(5):1037–1047.

Lynch, I., T. Cedervall, M. Lundqvist, C. Cabaleiro-Lago, S. Linse, and K. A. Dawson. 2007. The nanoparticle-protein complex as a biological entity; a complex fluids and surface science challenge for the 21st century. *Advances in Colloid and Interface Science* 134–135(0):167–174.

Lynch, I., and K. A. Dawson. 2008. Protein-nanoparticle interactions. *Nano Today* 3(1):40–47.

Lynch, I., A. Salvati, and K. A. Dawson. 2009. Protein-nanoparticle interactions: What does the cell see? *Nat Nano* 4(9):546–547.

Mahmoudi, M., K. Azadmanesh, M. A. Shokrgozar, W. S. Journeay, and S. Laurent. 2011a. Effect of nanoparticles on the cell life cycle. *Chemical Reviews* 111(5):3407–3432.

Mahmoudi, M., I. Lynch, M. R. Ejtehadi, M. P. Monopoli, F. B. Bombelli, and S. Laurent. 2011b. Protein-nanoparticle interactions: Opportunities and challenges. *Chemical Reviews* 111(9):5610–5637.

Mahmoudi, M., M. A. Sahraian, M. A. Shokrgozar, and S. Laurent. 2011c. Superparamagnetic iron oxide nanoparticles: Promises for diagnosis and treatment of multiple sclerosis. *ACS Chemical Neuroscience* 2(3):118–140. doi: 10.1021/cn100100e.

Mahmoudi, M., V. Serpooshan, and S. Laurent. 2011d. Engineered nanoparticles for biomolecular imaging. *Nanoscale* 3(8):3007–3026.

Manshian, B. B., D. F. Moyano, N. Corthout, S. Munck, U. Himmelreich, V. M. Rotello, and S. J. Soenen. 2014. High-content imaging and gene expression analysis to study cell–nanomaterial interactions: The effect of surface hydrophobicity. *Biomaterials* 35(37):9941–9950. doi: 10.1016/j.biomaterials.2014.08.031.

Marquis, B. J., S. A. Love, K. L. Braun, and C. L. Haynes. 2009. Analytical methods to assess nanoparticle toxicity. *Analyst* 134(3):425–439.

Maynard, A. D., D. B. Warheit, and M. A. Philbert. 2010. The new toxicology of sophisticated materials: Nanotoxicology and beyond. *Toxicological Sciences* 120(Suppl 1):S109–S129.

Michalet, X., F. F. Pinaud, L. A. Bentolila, J. M. Tsay, S. J. J. L. Doose, J. J. Li, G. Sundaresan, A. M. Wu, S. S. Gambhir, and S. Weiss. 2005. Quantum dots for live cells, *in vivo* imaging, and diagnostics. *Science* 307(5709):538–544.

Mirkin, C. A., and C. M. Niemeyer, eds. 2007. *Nanobiotechnology II: More Concepts and Applications.* Wiley-VCH Verlag GmbH & Co. KGaA. ISBN: 978-3-527-31673-1.

Monopoli, M. P., D. Walczyk, A. Campbell, G. Elia, I. Lynch, F. B. Bombelli, and K. A. Dawson. 2011. Physical-chemical aspects of protein corona: Relevance to *in vitro* and *in vivo* biological impacts of nanoparticles. *Journal of the American Chemical Society* 133(8):2525–2534. doi: 10.1021/ja107583h.

Mosmann, T. 1983. Rapid colorimetric assay for cellular growth and survival: Application to proliferation and cytotoxicity assays. *Journal of Immunological Methods* 65(1):55–63.

Moszner, N., and U. Salz. 2007. Recent developments of new components for dental adhesives and composites. *Macromolecular Materials and Engineering* 292(3):245–271.

Mout, R., D. F. Moyano, S. Rana, and V. M. Rotello. 2012. Surface functionalization of nanoparticles for nanomedicine. *Chemical Society Reviews* 41(7):2539–2544.

Mulvaney, P. 1996. Surface plasmon spectroscopy of nanosized metal particles. *Langmuir* 12(3):788–800. doi: 10.1021/la9502711.

Murphy, C. J., A. M. Gole, J. W. Stone, P. N. Sisco, A. M. Alkilany, E. C. Goldsmith, and S. C. Baxter. 2008. Gold nanoparticles in biology: Beyond toxicity to cellular imaging. *Accounts of Chemical Research* 41(12):1721–1730.

Narayanan, R., and M. A. El-Sayed. 2004a. Changing catalytic activity during colloidal platinum nanocatalysis due to shape changes: Electron-transfer reaction. *Journal of the American Chemical Society* 126(23):7194–7195.

Narayanan, R., and M. A. El-Sayed. 2004b. Shape-dependent catalytic activity of platinum nanoparticles in colloidal solution. *Nano Letters* 4(7):1343–1348.

Navya, P. N., and H. K. Daima. 2016. Rational engineering of physicochemical properties of nanomaterials for biomedical applications with nanotoxicological perspectives. *Nano Convergence* 3(1):1.

Nel, A., T. Xia, L. Mädler, and N. Li. 2006. Toxic potential of materials at the nanolevel. *Science* 311(5761):622–627. doi: 10.1126/science.1114397.

Nel, A. E., L. Mädler, D. Velegol, T. Xia, E. M. V. Hoek, P. Somasundaran, F. Klaessig, V. Castranova, and M. Thompson. 2009. Understanding biophysicochemical interactions at the nano–bio interface. *Nature Materials* 8(7):543–557.

Nguyen, M. M., N. C. Gianneschi, and K. L. Christman. 2015. Developing injectable nanomaterials to repair the heart. *Current Opinion in Biotechnology* 34:225–231.

Niemeyer, C. M., and C. A. Mirkin, eds. 2004. *Nanobiotechnology: Concept, Application and Perspectives*. Wiley-VCH Verlag GmbH & Co. KGaA. ISBN: 978-3-527-30658-9.

Oberdörster, G., E. Oberdörster, and J. Oberdörster. 2005. Nanotoxicology: An emerging discipline evolving from studies of ultrafine particles. *Environmental Health Perspectives* 113(7):823–839.

Omidkhoda, A., H. Mozdarani, A. Movasaghpoor, and A. A. P. Fatholah. 2007. Study of apoptosis in labeled mesenchymal stem cells with superparamagnetic iron oxide using neutral comet assay. *Toxicology in Vitro* 21(6):1191–1196.

Ozay, O., A. Akcali, M. T. Otkun, C. Silan, N. Aktas, and N. Sahiner. 2010. P(4-VP) based nanoparticles and composites with dual action as antimicrobial materials. *Colloids and Surfaces B: Biointerfaces* 79(2):460–466.

Pal, S., Y. K. Tak, and J. M. Song. 2007. Does the antibacterial activity of silver nanoparticles depend on the shape of the nanoparticle? A study of the gram-negative bacterium *Escherichia coli*. *Applied and Environmental Microbiology* 73(6):1712–1720.

Pan, Y., S. Neuss, A. Leifert, M. Fischler, F. Wen, U. Simon, G. Schmid, W. Brandau, and W. Jahnen-Dechent. 2007. Size-dependent cytotoxicity of gold nanoparticles. *Small* 3(11):1941–1949.

Paul, A., A. Hasan, H. A. Kindi, A. K. Gaharwar, V. T. S. Rao, M. Nikkhah, S. R. Shin, D. Krafft, M. R. Dokmeci, and D. Shum-Tim. 2014. Injectable graphene oxide/hydrogel-based angiogenic gene delivery system for vasculogenesis and cardiac repair. *ACS Nano* 8(8):8050–8062.

Paul, S., and A. Chugh. 2011. Assessing the role of ayurvedic Bhasmas as ethno-nanomedicine in the metal based na.nomedicine patent regime. *Journal of Intellectual Property Rights* 16:509–515.

Porter, A. E., M. Gass, K. Muller, J. N. Skepper, P. A. Midgley, and M. Welland. 2007. Direct imaging of single-walled carbon nanotubes in cells. *Nature Nanotechnology* 2(11):713–717.

Rahman, M., S. Laurent, N. Tawil, L. Yahia, and M. Mahmoudi. 2013. *Protein-Nanoparticle Interactions*. Berlin Heidelberg: Springer-Verlag.

Rao, C. V. S., P. Pranau Vanajasan, and V. S. Chandana. 2011. Scope of biomaterials in conservative dentistry and endodontics. *Trends Biomater Artif Organs* 25:75–78.

Reibold, M., P. Paufler, A. A. Levin, W. Kochmann, N. Pätzke, and D. C. Meyer. 2006. Materials: Carbon nanotubes in an ancient Damascus sabre. *Nature* 444(7117):286–286.

Richards, R., and H. Bönnemann. 2005. Synthetic approaches to metallic nanomaterials. In *Nanofabrication towards Biomedical Applications*, Kumar, C. S. S. R., Hormes J., and Leuschner C., eds., 1–32. Wiley-VCH Verlag GmbH & Co. KGaA.

Salata, O. V. 2004. Applications of nanoparticles in biology and medicine. *Journal of Nanobiotechnology* 2(1):3.

Salvador-Morales, C., E. Flahaut, E. Sim, J. Sloan, M. L. H. Green, and R. B. Sim. 2006. Complement activation and protein adsorption by carbon nanotubes. *Molecular Immunology* 43(3):193–201.

Salvati, A., A. S. Pitek, M. P. Monopoli, K. Prapainop, F. B. Bombelli, D. R. Hristov, P. M. Kelly, C. Åberg, E. Mahon, and K. A. Dawson. 2013. Transferrin-functionalized nanoparticles lose their targeting capabilities when a biomolecule corona adsorbs on the surface. *Nature Nanotechnology* 8(2):137–143.

Sanghavi, B. J., O. S. Wolfbeis, T. Hirsch, and N. S. Swami. 2015. Nanomaterial-based electrochemical sensing of neurological drugs and neurotransmitters. *Microchimica Acta* 182(1-2):1–41.

Sarkar, P. K., and A. K. Chaudhary. 2010. Ayurvedic Bhasma: The most ancient application of nanomedicine. *Journal of Scientific and Industrial Research* 69(12):901.

Sato, K., K. Hosokawa, and M. Maeda. 2003. Rapid aggregation of gold nanoparticles induced by non-cross-linking DNA hybridization. *Journal of the American Chemical Society* 125(27):8102–8103. doi: 10.1021/ja034876s.

Sayes, C. M., J. D. Fortner, W. Guo, D. Lyon, A. M. Boyd, K. D. Ausman, Y. J. Tao, B. Sitharaman, L. J. Wilson, and J. B. Hughes. 2004. The differential cytotoxicity of water-soluble fullerenes. *Nano Letters* 4(10):1881–1887.

Schnackenberg, L. K., J. Sun, and R. D. Beger. 2012. Metabolomics techniques in nanotoxicology studies. In *Nanotoxicity*, Joshua, R. ed., 141–156. Springer.

Selvakannan, P. R., R. Ramanathan, B. J. Plowman, Y. M. Sabri, H. K. Daima, A. P. O'Mullane, V. Bansal, and S. K. Bhargava. 2013. Probing the effect of charge transfer enhancement in off resonance mode SERS via conjugation of the probe dye between silver nanoparticles and metal substrates. *Physical Chemistry Chemical Physics* 15(31):12920–12929. doi: 10.1039/C3CP51646F.

Senapati, T., D. Senapati, A. K. Singh, Z. Fan, R. Kanchanapally, and P. C. Ray. 2011. Highly selective SERS probe for Hg (II) detection using tryptophan-protected popcorn shaped gold nanoparticles. *Chemical Communications* 47(37):10326–10328.

Shankar, S., S. K. Soni, H. K. Daima, P. R. Selvakannan, J. M. Khire, S. K. Bhargava, and V. Bansal. 2015. Charge-switchable gold nanoparticles for enhanced enzymatic thermostability. *Physical Chemistry Chemical Physics* 17(33):21517–21524. doi: 10.1039/C5CP03021H.

Sharma, T. K., R. Ramanathan, P. Weerathunge, M. Mohammadtaheri, H. K. Daima, R. Shukla, and V. Bansal. 2014. Aptamer-mediated "turn-off/turn-on" nanozyme activity of gold nanoparticles for kanamycin detection. *Chemical Communications* 50(100):15856–15859.

Sharma, V. K. 2013. Stability and toxicity of silver nanoparticles in aquatic environment: A review. *Sustainable Nanotechnology and the Environment: Advances and Achievements* 1124:165–179.

Sidney, S., W. D. Rosamond, V. J. Howard, and R. V. Luepker. 2013. The "heart disease and stroke statistics—2013 update" and the need for a national cardiovascular surveillance system. *Circulation* 127(1):21–23.

Sitharaman, B. 2011. *Nanobiomaterials Handbook*. Boca Raton, Florida: CRC Press.

Soenen, S. J., P. Rivera-Gil, J.-M. Montenegro, W. J. Parak, S. C. De Smedt, and K. Braeckmans. 2011. Cellular toxicity of inorganic nanoparticles: Common aspects and guidelines for improved nanotoxicity evaluation. *Nano Today* 6(5):446–465.

Stewart, M. E., C. R. Anderton, L. B. Thompson, J. Maria, S. K. Gray, J. A. Rogers, and R. G. Nuzzo. 2008. Nanostructured plasmonic sensors. *Chemical Reviews* 108(2):494–521.

Subramanian, V., E. Semenzin, D. Hristozov, A. Marcomini, and I. Linkov. 2014. Sustainable nanotechnology: Defining, measuring and teaching. *Nano Today* 9(1):6–9. doi: 10.1016/j.nantod.2014.01.001.

Suh, W. H., K. S. Suslick, G. D. Stucky, and Y.-H. Suh. 2009. Nanotechnology, nanotoxicology, and neuroscience. *Progress in Neurobiology* 87(3):133–170.

Szpunar, J. 2000. Bio-inorganic speciation analysis by hyphenated techniques. *Analyst* 125(5):963–988.

Tang, L., X. Yang, Q. Yin, K. Cai, H. Wang, I. Chaudhury, C. Yao et al. 2014. Investigating the optimal size of anticancer nanomedicine. *Proceedings of the National Academy of Sciences* 111(43):15344–9. doi: 10.1073/pnas.1411499111.

Tang, S., Y. Wu, C. N. Ryan, S. Yu, G. Qin, D. S. Edwards, and G. D. Mayer. 2015. Distinct expression profiles of stress defense and DNA repair genes in Daphnia pulex exposed to cadmium, zinc, and quantum dots. *Chemosphere* 120:92–99.

Tao, Y., Y. Lin, Z. Huang, J. Ren, and X. Qu. 2013. Incorporating graphene oxide and gold nanoclusters: A synergistic catalyst with surprisingly high peroxidase-like activity over a broad pH range and its application for cancer cell detection. *Advanced Materials* 25(18):2594–2599.

Thanh, N. T. K., and Z. Rosenzweig. 2002. Development of an aggregation-based immunoassay for antiprotein a using gold nanoparticles. *Analytical Chemistry* 74(7):1624–1628. doi: 10.1021/ac011127p.

Vaidya, B., and V. Gupta. 2015. Novel therapeutic strategies for cardiovascular diseases treatment: From molecular level to nanotechnology. *Current Pharmaceutical Design* 21(30):4367.

Walczyk, D., F. B. Bombelli, M. P. Monopoli, I. Lynch, and K. A. Dawson. 2010. What the cell sees in bionanoscience. *Journal of the American Chemical Society* 132(16):5761–5768. doi: 10.1021/ja910675v.

Walkey, C. D., and W. C. W. Chan. 2012. Understanding and controlling the interaction of nanomaterials with proteins in a physiological environment. *Chemical Society Reviews* 41(7):2780–2799.

Wang, L., X. Wang, A. Bhirde, J. Cao, Y. Zeng, X. Huang, Y. Sun, G. Liu, and X. Chen. 2014. Carbon-dot-based two-photon visible nanocarriers for safe and highly efficient delivery of siRNA and DNA. *Advanced Healthcare Materials* 3(8):1203–1209.

Wei, H., and E. Wang. 2008. Fe_3O_4 magnetic nanoparticles as peroxidase mimetics and their applications in H_2O_2 and glucose detection. *Analytical Chemistry* 80(6):2250–2254.

Werlin, R., J. H. Priester, R. E. Mielke, S. Krämer, S. Jackson, P. K. Stoimenov, G. D. Stucky, G. N. Cherr, E. Orias, and P. A. Holden. 2011. Biomagnification of cadmium selenide quantum dots in a simple experimental microbial food chain. *Nature Nanotechnology* 6(1):65–71.

Whitesides, G. M. 2003. The "right" size in nanobiotechnology. *Nat Biotech* 21(10):1161–1165.

Whitesides, G. M. 2005. Nanoscience, nanotechnology, and chemistry. *Small* 1(2):172–179.

Wiesner, M. R., G. V. Lowry, P. Alvarez, D. Dionysiou, and P. Biswas. 2006. Assessing the risks of manufactured nanomaterials. *Environmental Science and Technology* 40(14):4336–4345.

William, W. Y., E. Chang, R. Drezek, and V. L. Colvin. 2006. Water-soluble quantum dots for biomedical applications. *Biochemical and Biophysical Research Communications* 348(3):781–786.

Wu, H., Q. Huo, S. Varnum, J. Wang, G. Liu, Z. Nie, J. Liu, and Y. Lin. 2008. Dye-doped silica nanoparticle labels/protein microarray for detection of protein biomarkers. *Analyst* 133(11):1550–1555.

Xiao, Y., and P. E. Barker. 2004. Semiconductor nanocrystal probes for human metaphase chromosomes. *Nucleic Acids Research* 32(3):e28–e28.

Yamamoto, A., R. Honma, M. Sumita, and T. Hanawa. 2004. Cytotoxicity evaluation of ceramic particles of different sizes and shapes. *Journal of Biomedical Materials Research Part A* 68A(2):244–256. doi: 10.1002/jbm.a.20020.

Yang, L., X. Ren, F. Tang, and L. Zhang. 2009a. A practical glucose biosensor based on Fe 3O 4 nanoparticles and chitosan/nafion composite film. *Biosensors and Bioelectronics* 25(4):889–895.

Yang, S.-T., X. Wang, H. Wang, F. Lu, P. G. Luo, L. Cao, M. J. Meziani, J.-H. Liu, Y. Liu, and M. Chen. 2009b. Carbon dots as nontoxic and high-performance fluorescence imaging agents. *The Journal of Physical Chemistry C* 113(42):18110–18114.

Ye, Y., J. Liu, J. Xu, L. Sun, M. Chen, and M. Lan. 2010. Nano-SiO_2 induces apoptosis via activation of p53 and Bax mediated by oxidative stress in human hepatic cell line. *Toxicology in Vitro* 24(3):751–758.

Yoshida, K., M. Morita, and H. Mishina. 2003. Cytotoxicity of metal and ceramic particles in different sizes. *JSME International Journal, Series C: Mechanical Systems, Machine Elements and Manufacturing* 46(4):1284–1289.

Yu, K.-N., T.-J. Yoon, A. Minai-Tehrani, J.-E. Kim, S. J. Park, M. S. Jeong, S.-W. Ha, J.-K. Lee, J. S. Kim, and M.-H. Cho. 2013. Zinc oxide nanoparticle induced autophagic cell death and mitochondrial damage via reactive oxygen species generation. *Toxicology in Vitro* 27(4):1187–1195.

Zhang, L., J. Wang, and Y. Tian. 2014. Electrochemical in-vivo sensors using nanomaterials made from carbon species, noble metals, or semiconductors. *Microchimica Acta* 181(13-14):1471–1484.

Zhu, K., J. Li, Y. Wang, H. Lai, and C. Wang. 2015. Nanoparticles-assisted stem cell therapy for ischemic heart disease. *Stem Cells International* 2016:1–9.

2
Nanotheranostics: Implications of Nanotechnology in Simultaneous Diagnosis and Therapy

Brahmeshwar Mishra and Ravi R. Patel

CONTENTS

2.1 Introduction 35
2.2 Different Types of Theranostic Nanomedicines 37
    2.2.1 Polymeric Nanoparticles 37
    2.2.2 Polymer Drug Conjugates 39
    2.2.3 Dendrimers 41
    2.2.4 Gold Nanoparticles 43
    2.2.5 Magnetic Nanoparticles 46
    2.2.6 Silica Nanoparticles 48
    2.2.7 Solid Lipid Nanoparticles 48
    2.2.8 Liposomes 49
    2.2.9 Polymeric Micelles 51
    2.2.10 Nanoemulsion 53
    2.2.11 Carbon-Based Nanostructures 55
2.3 Conclusion and Future Perspective 57
References 58

2.1 Introduction

Nanocarriers, in their various forms, have been extensively probed for delivery of numerous therapeutic molecules, including drugs and imaging agents. Recent advancements in the nanotechnological world have drawn considerable attention among researchers to explore the new horizons of nanocarriers in the diagnostic and therapeutic arenas. Nanotheranostics are integrated nanostructured vehicles composed of therapeutic molecules and imaging probes in a single carrier for simultaneous application of treatment and diagnosis of diseases. Imaging-guided nanomedicines are able to detect and treat various menacing diseases including cancer at their earliest stages, when they are easy to cure. Moreover, prolonged, controlled, and targeted co-delivery of the imaging agent and therapeutic entity can be easily achieved by using multifaceted nanocarriers in order to achieve a superior theranostic effect without any serious adverse effects. As a result, miscellaneous nanoscopic vectors, including polymeric nanoparticles, liposomes, micelles, dendrimers, nanoemulsion, polymeric conjugates, carbon-based nanomaterials, metal and inorganic nanoparticles, and so on have been engineered and show excellent potential and promise for better theranostic effects. Numerous therapeutic substances, such as drugs, genes, proteins and peptides, and so on, can be delivered in a smarter way with the aid of nanotheranostics for safe and precise targeting of diseased tissue to achieve higher efficacy. Additionally, various imaging modalities, such as optical imaging, magnetic resonance imaging (MRI), nuclear imaging, ultrasound imaging, photoacoustic imaging, and computer tomography (CT), can be conferred on nanotheranostics for making a contrast between the diseased site and normal tissue at the cellular and molecular levels (Figure 2.1). Hence, next-generation theranostics nanomedicine is promising in terms of creating effective treatment with a favorable prognosis at a lower cost without compromising the quality of life of the patients. The current chapter thus presents an elaborated view related to the various nanotheranostics delivery systems, such as polymeric nanoparticles, liposomes, micelles, dendrimers, polymeric conjugates, carbon nanotubes, metal and inorganic nanoparticles, and so on, developed so far for theranostic application in the treatment of various diseases, along with their recent advancements.

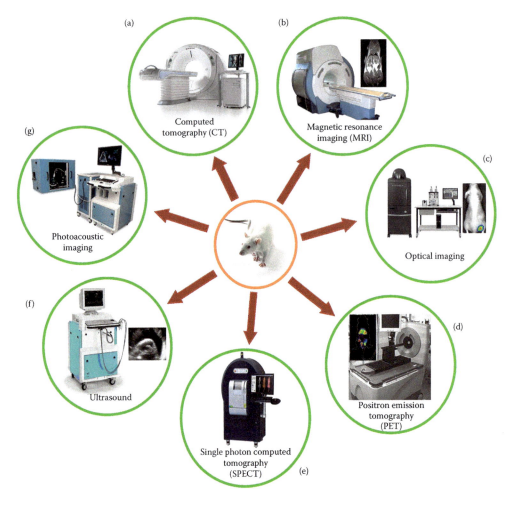

FIGURE 2.1
Various molecular imaging techniques employed for the theranostic applications. (a) Computed tomography (CT) (b) Magnetic resonance imaging (MRI), (c) Optical imaging, (d) Positron emission tomography (PET) (e) Single-photon emission-computed tomography (SPECT), (f) Ultrasound imaging, (g) Photoacoustic imaging.

2.2 Different Types of Theranostic Nanomedicines

2.2.1 Polymeric Nanoparticles

A wide spectrum of polymeric nanoparticles made from biodegradable polymers has been explored for the delivery of therapeutic substances. Some obvious characteristics, such as biocompatibility, biodegradability, stability in the biological milieu, nontoxicity, and controlled release capability, make them highly attractive for targeted co-delivery of therapeutic and imaging agents (Mu and Feng 2003, Muthu and Singh 2009, Luk et al. 2012). To make them highly biocompatible, in most cases, biodegradable polymers or co-polymeric blocks are used, which also helps in avoiding phagocytic uptake and thereby prolongs the systemic circulation. Moreover, they also exhibit remarkable stability during storage as compared to the vesicular system, that is, liposomes and micelles (Hu et al. 2014). Among all biodegradable polymers, poly(D,L-lactide-co-glycolide) (PLGA), poly(D,L-lactide-co-glycolide)-polyethylene glycol (PLGA-PEG), and poly(lactic acid)-D-α-tocopheryl polyethylene glycol 1000 succinate (PLA-TPGS), are commonly employed for theranostic application in the form of polymeric nanoparticles. They are easily cleared from the body through metabolic pathways after hydrolysis into their monomers. They are mainly prepared by well-reported synthetic methods, including the nanoprecipitation method, salting out, the solvent evaporation method, the emulsification method, the direct polymerization method, and so on (Moghimi et al. 2001, Soppimath et al. 2001, Mok and Park 2012, Svenson 2013, Vijayakumar et al. 2013).

A wide range of polymeric nanoparticles in their various sizes and shapes has been engineered for simultaneous delivery of drugs and imaging agents to make theranostic systems. Numerous functional groups have also been attached on the surface of the polymeric nanoparticles to confer targeting potential on them. Pan et al. developed folate-anchored PLGA- and D-alpha-tocopheryl polyethylene glycol succinic acid (TPGS-COOH)-based polymeric nanoparticles for targeted delivery of doxorubicin, along with quantum dots (QDs) for simultaneous imaging and treatment in breast cancer (Pan et al. 2010). By modulating the ratio of both polymer mixtures, they obtained high encapsulation efficiency and high cellular adhesion with controlled drug release. During an *in-vitro* cytotoxicity study, folate-anchored nanoparticles showed enhanced uptake in MCF-7 cells as compared to NIH-3T3 cells, confirming the targeting potential of nanotheranostics. Likewise, folate-conjugated PLA-TPGS and TPGS-COOH mixed polymeric nanoparticles were prepared by co-encapsulating docetaxel and quantum dots, which showed higher uptake and thereby cytotoxicity in folate overexpressed MCF-7 cells as compared to NIH-3T3 fibroblast cells. The higher internalization of folate modified folate-decorated quantum dot-loaded polymeric nanoparticles proved the potential of targeting the ligand-conjugated TPGS molecule for theranostic application with different polymers. Thus, multimodality can be easily conferred on theranostic nanomedicines by decorating their surfaces with different homing carriers/ligands. Recently, PLA-TPGS–based nanoparticles have been developed by encapsulating supermagnetic iron oxide and quantum dots in order to bestow multimodality via MRI and fluorescence imaging (Tan et al. 2011). Researchers were also able to target cancer cells with controlled imaging, which ultimately helps in reduction of cell toxicity due to the high exposure of contrasting agent. These showed sharp and clear detection with good depth of penetration in various organs such as liver, kidney, tumor, brain, and so on in a xenograft model, indicating the possibility of individual imaging during the treatment. Hence, by utilizing the above concept, it is possible to develop advanced multimodal theranostic nanoparticles by encapsulating and conjugating the homing carrier.

Some researchers have utilized the copolymerization potential of several polymers, that is, poly(d-,l-lactic acid), in order to form theranostic nanoparticles carrying hydrophobic drugs with high payloads. Lu et al. have engineered theranostic nanoparticles by the self-assembling of poly(N-vinylimidazole-co-N-vinylpyrrolidone)-g-poly(d-, l-lactide) grafted methoxy functionalized polyethylene glycol-poly(lactic acid) (PEG-PLA) di-block copolymer for the delivery of doxorubicin to cancerous tissues (Lu et al. 2011). Single-photon emission computed tomography confirmed the intratumoral localization of nanoparticles owing to their tagging with ^{123}I. Feng et al. developed theranostic polymeric nanoparticles by electrostatic interaction between positively charged fluorophore-conjugated polymer and negatively charged doxorubicin-coupled poly(glutamic acid) (Feng et al. 2010). Doxorubicin was released by hydrolysis of poly(glutamic acid) after cellular uptake, which in turn activated the fluorescent polymer. This activated fluorescence can be easily detected by fluorescence microscopy in order to visualize the drug release dynamically from the polymeric nanoparticles at the molecular level in the targeted tumor cells during *in vivo* studies. Magnetic nanoparticles, that is, paramagnetic gadolinium nanoparticle-incorporated polymeric nanoparticles, have also been developed for theranostic application. Rowe et al. fabricated copolymers of poly-(n-isopropylacrylamide)-co-poly(n-acryloxysuccinimide)-co-poly(fluorescein O-methacrylate)-conjugated gadolinium nanoparticles for the targeted treatment of cancer, which can also be visualized under MRI (Rowe et al. 2009).

Therapeutic agents can be encapsulated or covalently conjugated on the surface of polymeric nanoparticles. Lammers et al. engineered doxorubicin- and gemcitabine-encapsulated, image-guided, and passively targeted N-(2-hydroxypropyl)methacrylamide copolymer-based polymeric nanoparticles for theranostic application (Lammers et al. 2008). Doxorubicin and gemcitabine were conjugated on the surface of the polymeric nanoparticles via a lysosomally degradable spacer for targeted delivery of combined anticancer agents in order to improve the therapeutic efficacy *in vivo*. Here, gemcitabine acted as radiosensitizer for radiotherapy.

Photosensitizer-encapsulated polymeric nanoparticles have also been explored for theranostic applications. McCarthy et al. encapsulated meso-tetraphenylporpholactol into PLGA nanoparticles for theranostic application (McCarthy et al. 2005) in which photosensitizers were not able to produce fluorescence until they remained encapsulated. However, photosensitizers were effectively released from the nanoparticles after their cellular uptake and showed marked fluorescence inside the cell, indicating improvement in the photosensitizer-mediated cytotoxicity, which was further confirmed during *in-vivo* studies in a prostate cancer mouse model. However, the visible light–absorbing property of the photosensitizer and lower tissue penetration of the visible light hinder their further exploration for the theranostic applications. Hence, near-infrared (NIR) fluorescence dyes have been incorporated for exploiting the imaging potential of natural polymer-based nanoparticles.

Natural polymers such as chitosan have been employed for the development of theranostic nanomedicines. Paclitaxel-encapsulated chitosan nanoparticles with Cy5.5 have been prepared for the simultaneous treatment and live imaging of cancer cells. Hydrophobic paclitaxel was encapsulated by modifying the native chitosan with 5β-cholanic acid. Enhanced anticancer activity owing to nanoparticles was visualized by real-time imaging, which revealed that longer systemic circulation and higher penetration in the leaky tumor vasculature were the key mechanisms for the superior antitumor activity (Na et al. 2011). Likewise, Kim et al. encapsulated Cy5.5 and paclitaxel inside the tumor-homing multifunctional, glycol chitosan-based polymeric nanoparticles for passive targeting of cancer cells (Kim et al. 2010a,b). In another instance, poly(amino acid) has also been investigated as a carrier for the design of tunable theranostic polymeric nanoparticles. Santra and Perez have designed poly(amino acid)-based polymeric nanoparticles by using N-alkylated amino acid–based monomers (Santra and Perez 2011). The therapeutic agent taxol and imaging agent DiI dye were encapsulated in the hydrophobic nanocavities for theranostic application in the treatment of cancer.

2.2.2 Polymer Drug Conjugates

The covalent conjugation of drugs with polymeric materials at the nano scale has gained considerable recognition in the field of drug delivery, which has caused researchers to explore the potential of polymer-drug conjugates for theranostic application. A wide range of polymer-drug/imaging agent-conjugates has been prepared in the past (Wang et al. 2014). They are generally prepared by covalent interactions between the functional groups of polymers and drug/imaging agents by employing different chemical pathways (Lammers and Ulbrich 2010). Different types of polymers, including natural, proteins or peptides, and synthetic, have been used for the design of polymer conjugates.

Natural polymers have been employed for the delivery of drug and imaging agents due to their diversified characteristics. Their biodegradability, biocompatibility, and nontoxicity have captured marked attention for drug delivery purposes (Wang et al. 2014). Numerous polymeric conjugates with drug and imaging agents have been prepared by employing polysaccharides, that is, hyaluronic acid (HA), dextran, chitosan, and alginate. HA-based conjugates are used for preferential targeting to CD44 cells, which are found overexpressed in some cancers (Zoller 1995, Surace et al. 2009). HA conjugates are also used for treatment of type 2 diabetes after covalently conjugating with exendin-4 (Kong et al. 2010). Photodynamic therapy-assisted cancer treatment is also possible by linking HA conjugates with photosensitizers (Li et al. 2010). Selective tumor accumulation with an intense imaging signal was obtained with near infrared dye–conjugated HA in animals (Choi et al. 2010). Chitosan in their various conjugations can be used for a variety of diseases, such as anti-*Streptococcus suis* infection (Xu et al. 2010), cancer (Park et al. 2011), chronic renal diseases (He et al. 2012), and so on. Recently, sialyl Lewis X–chitosan conjugates were prepared for anti-inflammatory action (Hanand and Li 2011). Likewise, galactose functionalized dextran was conjugated with Cy5.5 dye and ^{99m}Tc for NIR fluorescence and single-photon emission computed tomography (SPECT) imaging, which could be used for theranostic application after oral administration in ulcerative colitis–like diseases (Era et al. 2005). Additionally, normal dextran can also be tethered to ^{99m}Tc for lymphoscintigraphical and angiocardiographical imaging (Matsunaga et al. 2005). Alginate-based conjugates have also been tried for theranostic purposes; however, their high molecular weight limit their widespread application as drug and imaging agent carriers (Wang et al. 2014).

Interestingly, naturally occurring proteins and peptide molecules are also serving as choices of polymeric carriers for theranostic application in the form of protein conjugates (Wang et al. 2014). Doxorubicin-elastin conjugates have shown enhanced systemic circulation with high cellular uptake. They have the property of self-assembling into nanoparticles upon contact with an aqueous solution (Dreher et al. 2003). In another study, novel chimeric polypeptide-based conjugation was prepared with doxorubicin, which exhibited complete tumor regression with a fourfold enhanced maximum tolerated dose compared to that of free doxorubicin upon a single administration. It also has the self-assembling characteristic to form nanostructures (MacKay et al. 2009). Here, doxorubicin itself acts as an imaging agent owing to its intrinsic fluorescent property, which can be of immense importance for measuring the extent of drug distribution in different organs during *in-vivo* studies (Mohanand and Rapoport 2010). Polyglutamic acid (PGA) has also been widely studied among researchers for theranostic application. It is commonly used by conjugating with gadolinium (Gd) chelate for molecular imaging. The formed conjugate is biodegradable and can easily clear from the body (Lu et al. 2003, Wen et al. 2004). It has been shown that the angiogenic biomarker integrin αvβ3 could easily be detected by functionalizing the PGA-Gd conjugate with cyclic Arg-Gly-Asp-D-Phe-Lys [c(RGDfK)] peptide under T1-weighted MRI (Ke et al. 2007). Vaidya et al. demonstrated the potential of PGA-photosentisizer/Gd conjugates for theranostic purposes in xenograft tumors by employing two different techniques, such as photodynamic therapy and MRI (Vaidya et al. 2006). The sharp contrast between normal organs and tumors can be achieved by contrast-enhanced MRI on account of the enhanced permeation and retention (EPR) effect in the tumor cells, which enforces the preferential accumulation of PGA conjugates in the tumor region through tumor vasculature (Tanand and Lu 2011). Some other peptides, such as poly(aspartic acid), polylysine (Hudecz 1995, Atmaja et al. 2010), and various chimeric polypeptides, have also been utilized in the past for developing nanoscopic drug-imaging agent conjugates (McDaniel et al. 2010, Liu et al. 2010, Wang et al. 2014).

Despite several advantages of natural polymers and peptides, some of the properties, such as simple structure, faster degradation, being prone to hydrolysis and proteolysis, and so on, demand the use of biocompatible, synthetic polymer as a theranostic carrier. Among all synthetic polymers, N-(2-hydroxypropyl) methacrylamide (HPMA) has been a widely used copolymer for theranostic application owing to its obvious advantages such as being nontoxic, biocompatible, nonimmunogenic, and stable in biological media, which makes it a choice for *in-vivo* application (Kopecek and Kopeckova 2010, Lammers and Ulbrich 2010, Nakamura et al. 2014). HPMA copolymer-based conjugates are functionalized with therapeutic and imaging agents by means of chemical conjugation and copolymerization to make them theranostics (Lammers and Ulbrich 2010, Allmeroth et al. 2013). Some HPMA-based conjugates are currently in different phases of clinical trials, indicating clinical translation. The Phase I clinical trial of HPMA-based conjugate (i.e., Doxorubicin-HPMA-Gly-Phe-Leu-Gly conjugates) was initiated in 1994 and was developed by Vasey et al. They have developed Doxorubicin-HPMA-Gly-Phe-Leu-Gly conjugates (PK1) by covalent interactions through peptidyl linker, which was further tagged with I-131 to act as a theranostic for passive targeting to breast cancer cells. The developed conjugate is stable in systemic circulation, while after cellular uptake, cathepsin B, the lysosomal enzyme, cleaves the peptidyl linker of the HPMA-based conjugate and releases the drug. By forming the conjugate, the maximum tolerated dose is increased up to fivefold that of the pure drug, indicating the importance of the conjugate (Vasey et al. 1999, Wang et al. 2014).

Borgman et al. designed Arg-Gly-Asp (RGD)-HPMA-Indium-111 (In-111)-conjugated nanotheranostics by the copolymerization method for targeted delivery of radiotherapeutics to alpha(v)beta(3) integrin-expressed solid tumors, which was confirmed by scintigraphy (Borgman et al. 2008). Gadolinium is also used as an effective contrast agent for MRI. Gadolinium-tagged HPMA conjugates were also prepared for targeting integrin αvβ3 by functionalizing with c(RGDfK) peptide, which has shown increased blood circulation time with active targeting, indicating the potential of conjugates for tumor therapy monitoring (Zarabi et al. 2009). Very recently, Yuan et al. developed Cu-64 (intrinsic theranostic agent)-conjugated HPMA copolymers for active targeting of tumor angiogenesis. They have utilized RGD as a targeting ligand. The targeting potential was quantitatively confirmed by measuring the Cu-64 radioactivity with the aid of positron emission tomography (PET), which showed significantly higher tumor localization of RGD-bound nanotheranostics as compared to a bare one upon intravenous administration in human prostate cancer cells bearing mice xenografts. Further,

HPMA conjugates exhibited an enhanced pharmacokinetic profile of Cu-64; approximately onefold in the tumor region (Yuan et al. 2013). Likewise, RGD4C-conjugated HPMA copolymer was radiolabeled with ^{99m}Tc, which indicated retention of conjugated structure for a prolonged time during scintigraphic imaging (Mitra et al. 2005, Wang et al. 2014).

2.2.3 Dendrimers

Dendrimers are polymer-based nanosized structures composed of highly branched polymers in a spherical manner. Nanomedicine-based dendrimers are of 10–100 nm in size and can be used for theranostic purposes (Fahmy et al. 2007). The basic architecture of dendrimers involves the repeated, covalently linked branching surrounding the core to form a three-dimensional geometrical pattern. Based on the number of layered structures, dendrimers are classified into different generations. As the layered structure increases, higher-generation dendrimers are obtained (Kaminskas et al. 2012). The functional properties and activities of the dendrimers are a result of their surface functional groups from the hyperbranched structures. During synthesis, branching units are added repetitively to the core to form dendrimers. They can be synthesized mainly by two strategies: divergent synthesis (i.e., synthesis begins from functional core to periphery) and convergent synthesis (i.e., synthesis begins from periphery to reactive core) (Oliveira et al. 2010). Poly(amidoamine) (PAMAM)- and polypropylenimine (PPI)-based dendrimers are synthesized by the divergent method, whereas poly(ether-imide)s dendrimers are synthesized by the convergent method (Esfand and Tomalia 2001, Leu et al. 2001). Dendrimers generally possess a positive surface charge, which might result in severe toxicity by perturbing the cellular membrane. The size, shape, molecular weight, surface chemistry, and chemical composition of dendrimers can be effectively harnessed by controlling the degree of polymerization. These properties confer monodispersity, longer blood circulation potential, and tunable biodistribution capability to the dendrimers, along with controlled drug release characteristics (Li et al. 2007).

Hydrophilic as well as hydrophobic therapeutic molecules can be loaded either by encapsulation inside the dendrimers or through covalent conjugation with the surface. In the case of covalent-conjugation, therapeutic agents are conjugated through different reversible labile linkers, such as peptide, ester, disulfide, and so on, in order to improve drug loading and control drug release (Lazniewska et al. 2012). Other drug-loading methods through noncovalent bonding, such as hydrogen bonding, and ionic interactions have also been employed for encapsulation of therapeutic agents without any chemical synthesis. However, in noncovalent conjugation, a burst release of the encapsulated drug from the dendrimers was observed during targeted delivery before reaching the targeted site and resulting in limited therapeutic benefits (Bosman et al. 1999, Frechet 1994). This can be prevented by spherical-shaped, higher-generation dendrimers, which are able to incorporate the therapeutic agents inside their various cavities and branches and thereby limit rapid release of encapsulated agents. Higher-generation dendrimers have the capability to incorporate therapeutic agents and imaging agents with precise structural architecture, which renders them attractive platform for theranostic nanomedicines. Specially, fifth-generation dendrimers with comparatively higher hydrophobicity are used for the theranostic application owing to their enhanced drug stability (Jansen et al. 1994, Li et al. 2007). Many anticancer drugs, such as doxorubicin, methotrexate, paclitaxel, and 5-fluorouracil, have been encapsulated in dendrimers through noncovalent conjugation (Ooya et al. 2004, Dhanikula and Hildgen 2007, Davis et al. 2008, Ren et al. 2010). Moreover, the presence of large numbers of surface functional groups of dendrimers allows the attaching of various targeting agents and surface modifiers.

Among the array of dendrimers, PAMAM- and PPI-based dendrimers have been widely employed to serve the emerging interest in various multimodal imaging. They do not have any intrinsic property to render themselves self-luminous under various imaging modalities such as optical imaging, PET, SPECT, and MRI. In order to confer imaging modalities on the dendrimers, numerous imaging agents, including NIR fluorescence dies, chelating agents, that is, $^{64}Cu^{2+}$, $^{111}In^{3+}$, Gd^{3+}, and so on, have been conjugated to the surface of theranostic dendrimers. Additionally, targeted drug delivery through dendrimers can be possible either by passive targeting via the EPR effect or active targeting by functionalizing with different targeting ligands (Lo et al. 2010). Several homing ligands such as antibodies, peptides, and aptamers have been conjugated on the surface of dendrimers for active targeting of tumor cells (Patri et al. 2004, Shukla et al. 2005, Wu et al. 2011). To this end, Saad et al. encapsulated a therapeutic agent, that is, paclitaxel, and diagnostic agent (Cy5.5) into theranostic dendrimers, which were further conjugated with luteinizing hormone-releasing hormone peptide as a homing carrier for tumor targeted drug delivery (Saad et al. 2008). A significant improvement in the cellular uptake and thereby cytotoxicity was observed for the targeted theranostic dendrimers compared to nontargeted ones. Additionally, tumor-selective paclitaxel accumulation and reduced normal cell distribution of the targeted theranostic dendrimers suggested the specificity of the targeted dendrimers, which might have reduced the side effects to healthy organs. Further, imaging using the fluorescent probe, Cy5.5, proved the tumor-specific theranostic potential of dendrimers for the effective treatment of cancer. Similarly, Grunwald et al. developed fifth-generation poly(amidoamine) dendrimers for incorporating the sodium-iodide symporter (hNIS) gene along with a replication-deficient adenovirus vector and imaging agent, that is, I-123, for the treatment of liver cancer (Grünwald et al. 2013). They have utilized scintigraphy for visualizing the transduction efficacy of the hNIS gene via I-123. *In-vitro* studies showed significant enhancement in the transduction efficacy along with protection from neutralizing antibodies. Subsequently, *in-vivo* studies in the liver cancer xenograft mice model demonstrated significant reduction in hepatic transgene expression following the intravenous administration of hNIS containing replication-deficient adenovirus–encapsulated dendrimers, which was further simultaneously visualized by I-123 scintigraphy. Interestingly, some researchers have incorporated different imaging agents for bimodal imaging. To this end, Koyama et al. developed PAMAM-based sixth-generation dendrimers by incorporating Cy5.5 and Gd^{3+} for bimodal NIR optical imaging and MRI (Koyama et al. 2007).

Taratula et al. have delivered phthalocyanines by incorporation inside the dendrimers to utilize their theranostic potential in the treatment of cancer (Taratula et al. 2013). Phthalocyanines have been encapsulated with high payloads inside fourth-generation poly-propylenimine dendrimers by modifying through a hydrophobic linker. Further, dendrimers were surface functionalized by PEG and luteinizing hormone-releasing hormone peptide for targeted drug delivery with improved biocompatibility. The intrinsic fluorescence characteristic of phthalocyanines has conferred theranostic potential on dendrimers, which was further confirmed by visualizing *in-vitro* intracellular localization and *in-vivo* tissue distribution through fluorescence imaging. Additionally, the NIR absorption property of phthalocyanines was also utilized for photodynamic therapy through target-oriented dendrimers. The results of various *in-vitro* and *in-vivo* studies have suggested that targeting agent–functionalized dendrimers have the capability to produce higher cytotoxicity with significant photodynamic therapy through enhanced tumor cell accumulation and cellular internalization. To this end, dendrimers can be effectively employed as NIR theranostic agents for the management of serious diseases, which is further supported by large number of successful preclinical studies. Recently, Al-Jamal et al. developed polylysine-based sixth-generation theranostic dendrimers for improving the efficacy of doxorubicin in lung cancer. An *in-vivo* study in a lung cancer xenograft mouse model demonstrated significant enhancement in the anticancer efficacy of doxorubicin (Al-Jamal et al. 2013). Further, live whole-animal fluorescence imaging showed the tumor-specific accumulation of dendrimers as compared to a free drug, proving the theranostic potential of dendrimers. The ability to carry various biologically active molecules in a well-defined and controlled manner makes dendrimers stand out as a promising theranostic platform.

2.2.4 Gold Nanoparticles

Gold nanoparticles (GNPs) represent a versatile class of nanocarrier system, having a core size of 1.5–10 nm, which can be used for multifunctional theranostic application (Link and El-Sayed 2002, Daniel and Astruc 2004). The unique surface plasmon resonance (SPR) along with larger surface area bestows space for the effective encapsulation of therapeutic and imaging agents in order to act as an advanced theranostic medicine. In addition, well-established strategies for surface functionalization with the aid of homing ligands can serve as template for active targeting to diseased organs (Connor et al. 2005, Kumar et al. 2013). GNPs can be traced by optical scanning with an electromagnetic wave, owing to their ability to produce intense absorption and scattering signals. Moreover, their absorption and scattering intensity can be easily adjusted by tuning their size and morphological characteristics, making them favorable for *in-vivo* imaging in the NIR region (Xie et al. 2010b, Xia et al. 2011).

GNPs offer numerous advantages such as tunable particle size, monodispersity, diagnostic ability via light scattering, superior biocompatibility, ease of surface functionalization through gold-thiol (Au-S) bonding, easy fabrication, and lower toxicity (Han et al. 2006). Various shapes of GNPs such as nanorod, nanoshell, and nanocage with different thermal and optical properties have been prepared in the past as prospective theranostic agents (Xie et al. 2010b). Generally, GNPs are engineered by chemical reaction with hydrogen tetrachloroaurate. Numerous therapeutic agents (i.e., carboxylated paclitaxel) and targeting as well as imaging ligands (i.e., tumor necrosis factor) have been conjugated to the surface of GNPs by means of gold-thiol bonding for theranostic applications (Paciotti et al. 2004, Gibson et al. 2007). Moreover, strong covalent interaction generated between the gold and thiol groups of sulfur also confers stability on the GNPs (Hakkinen 2012). Therapeutic agents can be loaded mainly two ways: noncovalent bonding through electrostatic interactions and covalent conjugation through chemical bonding. Nevertheless, other materials such as PEG are used to coat the surface of GNPs to improve their biological activity by improving hydrophilicity as well as systemic circulation time (Niidome et al. 2006). After administration in the body, GNPs release the therapeutic agent in the cytoplasm of the cell through the exchange of abundantly present cytosolic glutathione (~10 mM) (Xie et al. 2010b).

Additionally, GNPs are not only useful for sensing purposes, they can also treat malignancy by inducing a photothermal effect upon laser irradiation, owing to their SPR property, which makes them serve as energy transducers. This property of GNPs is gaining more interest among researchers compared to conventional treatment paradigms due to significant reduction in normal cell toxicity by confining the illuminating area, thus avoiding serious side effects related to chemotherapeutic agents. Chen et al. demonstrated a significant increase in tumor surface temperature, up to 54°C within 2 min, by photothermal therapy of

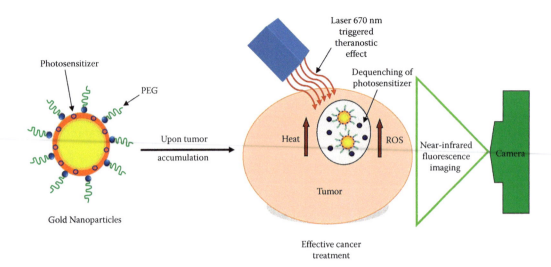

FIGURE 2.2
Schematic representation of photothermal therapy by using photosensitizer-incorporated gold nanoparticles.

PEG-coated gold nanocages under the exposure of NIR light (Chen et al. 2010a,b). They can be used for photoacoustic imaging despite their poor intrinsic fluorescence-emitting attribute (Figure 2.2). Several advanced techniques, such as surface-enhanced Raman spectroscopy, x-ray, SPECT, PET, MRI, and CT can be used for diagnostic imaging of GNPs, owing to their higher x-ray absorption coefficient as well as their having a higher atomic number than iodine, making them a contrasting agent (Melancon et al. 2008, Xu et al. 2008, Xie et al. 2010a, Van Schooneveld et al. 2010, Bhatnagar et al. 2013, Qin et al. 2013).

GNPs can be targeted to tumor sites in two main ways; passive targeting and active targeting. In passive targeting, GNPs can easily be delivered to tumor tissue via the EPR effect through leaky blood vessels in the tumor, due to their smaller size (Li et al. 2009). Various targeting ligands or carriers such as antibodies, peptides, ligands, and antibody fragments have been attached to GNPs for active targeting of tumor cells (Qian et al. 2008, Lukianova-Hleb et al. 2011, Chen et al. 2012, Kumar et al. 2012a,b).

Paclitaxel, PEG, rhodamine B-linked β-cyclodextrin, and biotin receptor-functionalized GNPs have been successfully developed by Heo et al. for theranostic purposes in the treatment of cancer (Heo et al. 2012). Paclitaxel was conjugated with GNPs with the aid of β cyclodextrin-based inclusion complex. The researchers showed that significant improvement in anticancer activity and affinity against various cancer cells, including HeLa, MG63, and A549, compared to that of normal cells was achieved with multiple ligand-modified GNPs, demonstrating the use of GNPs for versatile theranostic platforms. In another work, Kim et al. developed aptamer-conjugated, doxorubicin-loaded GNPs for the theranostic treatment of prostate cancer with the aid of molecular CT imaging (Kim et al. 2010a,b). The surface of the GNPs was functionalized by bioconjugating with prostate-specific membrane antigen RNA aptamer. The cell cytotoxicity study, along with CT imaging, proved the targeted and theranostic potential of GNPs through significant improvement in anticancer activity against LNCaP prostate cancer cells as compared to PC3 cells. Likewise, Chen et al. engineered doxorubicin-encapsulated smart theranostic GNPs. Gold-thiol bonding via peptide linker Cys-Pro-Leu-Gly-Leu-Ala-Gly-Gly was utilized for surface binding of doxorubicin (Chen et al. 2013). *In-vivo* studies in tumor-bearing mice showed that intravenous administration of GNPs results in enhanced anticancer activity by rapid release of doxorubicin in the tumor cells via intracellular glutathione and selective protease cleavage. Further, tumor tissue localization of GNPs was simultaneously visualized through fluorescent imaging. The prolific results demonstrated the feasibility of stimuli-responsive drug release of smart, multifunctional GNPs in the treatment of cancer as theranostic nanomedicine.

Some researchers have also used GNPs as radiotherapy sensitizers in the treatment of cancer radiotherapy to reduce morbidity. Surface-modified GNPs can accumulate in the selective target tissue and are able to sensitize cancer cells during the exposure of radiation without affecting normal cells. Huang et al. employed radiation and photothermal therapy for the treatment of cancer by using folic acid-conjugated silica functionalized rod shaped GNPs (Huang et al. 2011). Dual-mode x-ray and CT imaging proved the potential of multifunctional GNPs for tumor-selective efficacy in gastric cancer without affecting neighboring cells. GNPs have also been developed for indirect utilization as a theranostic agent by generating a photothermal vapor-based plasmonic nanobubble. They are generated optically and are able to damage targeted cells in a controlled, specific, and speedy mechanical manner. Wagner et al. demonstrated the plasmonic nanobubble-based indirect potential of GNPs as a cellular theranostic platform through *in-vivo* study in zebrafish hosting prostate cancer xenografts (Wagner et al. 2010).

Matrix metalloprotease (MMP)-based novel probes were also used for simultaneous bioimaging and photothermal therapy of carcinoid tumors using GNPs. The rationale behind the use of MMP as a diagnostic agent is its involvement in cancer metabolism. By using this approach, Yi et al. designed MMP-conjugated gold nanorods for treatment and diagnosis of cancer (Yi et al. 2010). The surface of the GNPs was modified by conjugating with fluorescent Cy5.5 along with an MMP-degradable peptide linker for detecting MMP (Bremer et al. 2001). Upon photothermal exposure, GNPs were able to damage cancer cells by increasing the temperature to 60°C. Moreover, the progress and stage of cancer could be monitored by measuring the recovery of fluorescence from quenched Cy5.5. Core-shell structures with two different materials such as gold and silver have been also employed for theranostic purposes. Hu et al. developed GNPs surrounded by silver shells for the photothermal therapy of malignant cells (Hu et al. 2009). These structures were able to destroy cancer cells with half the laser power compared to GNPs, proving their efficacy for the theranostic purposes in cancer. Likewise, nanocubes developed by Wu et al. also showed promising results in terms of photo-killing efficiency as well as cellular imaging in the treatment of human liver cancer cells (QGY) and human embryo kidney cells (293T) by the common single-photon excitation method (Wu et al. 2009).

The *in-vivo* use of spherical-shaped GNPs is limited by their shape, which curtails their imaging by reducing the penetration depth of the electromagnetic wave. In order to improve the detection limit, various nonspherical or hollow-shaped GNPs, that is, cubes, rods, cages, wires, and so on, have been developed that can be detected by modulating their SPR band from the visible range to the NIR range. Xia et al. designed gold nanocages for theranostic application by producing the NIR photothermal effect (Xia et al. 2011). Similarly, Moon et al. developed gold nanocages by filling their cores with the phase-convertible, 1-tetradecanol (melting point 38–39°C) for controlled release application. Upon application of focused high-intensity ultrasound or NIR laser, the temperature of the medium increased and therapeutic molecules were released in a controlled manner from gold nanocages in the diseased cells. These nanocages can be used for the site-specific controlled release of very potent drugs in order to reduce the emergence of adverse effects as well as for diagnostic purposes (Moon et al. 2011).

2.2.5 Magnetic Nanoparticles

Magnetic nanoparticles, also known as iron-oxide nanoparticles, are one type of nanocrystals made from hematite or magnetite (Gupta and Gupta 2005). Magnetic nanoparticles have several advantages, such as excellent biocompatibility, cost-effectiveness, and superior magnetic properties, that make them the choice of contrasting nanostructured material for the theranostic bioapplication using MRI (Huang et al. 2013, Li et al. 2013, Yen et al. 2013). Superparamagnetic iron oxide nanoparticles, mainly made from magnetite and maghemite, are mainly used for theranostic purposes. Simple magnetic nanoparticles are unstable due to their high surface activity, which results in aggregation in the colloidal dispersion. The stability to magnetic nanoparticles can be bestowed by passivating the surface with hydrophilic polymers such as dextran, PEG, oleic acid, polyaniline, and polyvinylpyrrolidone (Mornet et al. 2004, Laurent et al. 2008). Among all these, dextran and its derivatives have been widely exploited for extensive studies. Additionally, the polymeric shell also facilitates effective drug encapsulation with surface covalent conjugation and also retards the burst release of therapeutic agents (Wang et al. 2012).

Santra et al. covered magnetic nanoparticles with poly(acrylic acid) (PAA) for effective encapsulation of therapeutic agents, that is, a taxol and imaging agent (NIR dye). Further, the surface of the magnetic nanoparticles was functionalized by folic acid in order to target the folate receptor overexpressed cancer cells (Santra et al. 2009). The theranostic potential of magnetic nanoparticles was visualized by fluorescence imaging and MRI. Rastogi et al. modulated the surface of magnetic nanoparticles with three different polymers, N-isopropylacrylamide (NIPAAM), acrylic acid, and PEG (Rastogi et al. 2011). Different functionalities such as thermoresponsive drug release, tuning of critical temperature, longer circulation, and improved conjugation were conferred on the magnetic nanoparticles. Moreover, stealth coating of methacrylate PEG also bestowed the reactive hydroxyl group for the conjugation of folic acid. NIPAAM has controlled doxorubicin release by its thermoresponsive characteristics, which results in nearly 2.5 times higher release as compared to normal conditions for 48 hr.

Some researchers have also exploited multifunctional amphiphilic polymers for surface modulation of magnetic nanoparticles instead of different polymers. Pluronic F 127 was coated on magnetic nanoparticles with the help of β-cyclodextrin to increase the encapsulation of curcumin in the treatment of cancer. Optimized nanocarriers have demonstrated superior anticancer activity by producing a hyperthermia effect along with contrast properties for imaging (Yallapu et al. 2011). Likewise, study has also been conducted by Foy et al. to prolong the contrasting characteristic during MRI along with high payload and sustained release of anticancer agent (Foy et al. 2010). Five different NIR dyes have also been investigated for visualization of tumor localization with complete distribution profile in the absence of a magnetic field. An *in-vivo* study in mice demonstrated slow tumor accumulation of the developed magnetic nanoparticles upon intravenous administration, which needs 48 hr to reach the highest concentration and was detected for the next 11 days producing anticancer activity. Lim et al. designed a smart magnetic nanocarrier modified with herceptin receptor for the targeted delivery of pH-sensitive drugs such as doxorubicin, along with molecular imaging in the treatment of cancer (Lim et al. 2011) (Figure 2.3). The α-pyrenyl-ω-carboxyl PEG was employed as a matrix to encapsulate doxorubicin and magnetic nanoparticles by the emulsification method. The PEG matrix enabled the pH-dependent release of doxorubicin in cancer cells and showed significant reduction in tumor growth. Further, targeting the herceptin-overexpressed mouse fibroblast cell line was visualized under MRI, proving the theranostic potential of magnetic nanoparticles. Similarly, Chertok et al. developed β-galactosidase encapsulated in heparin-covered

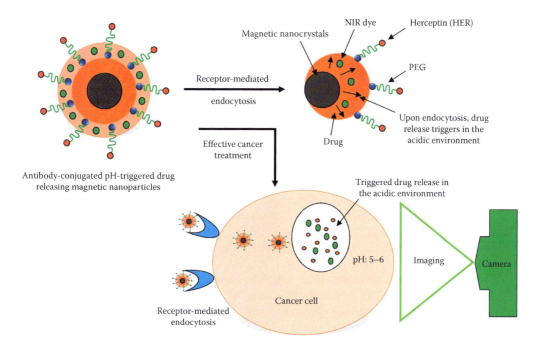

FIGURE 2.3
Schematic representation of antibody-conjugated pH-triggered drug-releasing magnetic nanoparticles for theranostic application in cancer by visualizing through MRI.

magnetic nanoparticles by electrostatic interaction in the treatment of brain cancer (Chertok et al. 2011). *In-vivo* studies in the rat glioma model exhibited a therapeutically active amount of β-galactosidase level in a brain tumor, even though there was a 10- to 50-fold reduction in the administration dose. To this end, the present theranostic platform can be extended for a wide range of protein molecules for various diseases.

Co-delivery of multiple drugs by single theranostic magnetic nanoparticles has also been exploited by several scientists. Singh et al. engineered PLGA-magnetic nanoparticles containing composites for multiple drug loading, that is, cisplatin, paclitaxel, and rapamycin (Singh et al. 2011). During *in-vivo* studies in rats, magnetic nanocomposites demonstrated higher contrasting capability compared to conventional contrasting probes, with the potential of co-delivery of multiple drugs at tumor sites. Additionally, targeting with magnetic nanocomposites can also be achieved by functionalizing with targeting carriers for superior activity.

Several nonpolymeric materials have also been explored for surface modification of magnetic nanoparticles with therapeutic as well as targeting ligands. Das et al. designed biocompatible and biodegradable N-phosphonomethyl iminodiacetic acid-coated magnetic nanoparticles for the coupling of various therapeutic agents, that is, methotrexate, rhodamine B isothiocyanate, and folic acid, in order to obtain a multifunctional nanostructure (Das et al. 2009). They were able to deliver the therapeutic agent in the acidic microenvironment of the tumor cells by pH-labile ester linkage. Simultaneous targeting, magnetically guided drug delivery, and imaging proved the multifaceted advantages of the theranostic magnetic nanoparticles. Likewise, Mitra et al. fabricated lectin-coated magnetic nanoparticles for effective active targeted delivery of paclitaxel to tumor cells. Lectin was functionalized on the surface of magnetic nanoparticles with the aid of chemical conjugation by ethyl(dimethylaminopropyl) carbodiimide/N-hydroxysuccinimide (Mitra et al. 2012). *In-vivo* studies demonstrated enhanced anticancer activity along with improved bioavailability and longer blood circulation, suggesting the higher theranostic potential of the developed modified magnetic nanocarriers compared to that of unmodified ones.

2.2.6 Silica Nanoparticles

Silica nanoparticles (SNs) have emerged as tunable nanomedicines due to their enhanced surface area and size-dependent properties. SNs show similar physical properties at the nanoscale level as those of silica materials. However, the well-defined siloxane structural chemistry confers a significant surface active property to SNs, which can easily be modulated for delivering therapeutic and imaging agents for theranostic utilization. Various modified SNs have been engineered for theranostic applications (Wang et al. 2012). He et al. designed bifunctional phosphate-terminated SNs for simultaneous imaging and treatment by loading a photosensitizer (i.e., methylene blue) (He et al. 2009). Upon exposure to light of 635 nm, SNs have induced photodynamic damage to HeLa cells by producing singlet oxygen. Further, site-specific damage to cancer cells was visualized by measuring the emitted NIR luminescence from methylene blue–encapsulated SNs. Likewise, in order to utilize the potential of the high surface area of SNs as an advanced theranostic platform, mesoporous SNs have been engineered. They furnish a sufficiently large surface area for modulating their site-specific drug-delivering potential via targeting ligands. To this end, Zhang et al. have developed hematoporphyrin, a photosensitizer-encapsulated, multifunctional, mesoporous SN with a dye-doped silica core for theranostic application through the photo-oxidation mechanism (Zhang et al. 2009). Similarly, Cheng et al. designed cRGDyK peptide-functionalized mesoporous SNs for theranostic application in the treatment of cancer. *In-vitro* studies revealed the enhancement of therapeutic efficacy by targeting of $\alpha v \beta 3$ integrins of human glioblastoma cells with the least collateral damage (Cheng et al. 2010).

SNs are biocompatible and safe for *in-vivo* administration. The porous architecture of mesoporous SNs provides space for residing therapeutic and diagnostic agents. Additionally, numerous functionalizing agents have been attached on the outer surface of SNs to impart multifunctionalities to widen their applicability. Lu et al. investigated biocompatibility, biodistribution, and therapeutic efficacy of folate receptor-anchored mesoporous SNs in the treatment of cancer (Lu et al. 2010). An *in-vivo* study in mice containing human cancer xenograft demonstrated significant reduction in tumor growth and tumor-specific localization with high tolerability and safety. Further, accumulation of SNs in tumors was imaged by fluorescence microscopy and mass spectroscopy, confirming the site-specific delivery of therapeutic agents with high therapeutic efficacy. In another instance, Lee et al. developed magnetite nanocrystal-decorated, versatile, and uniform mesoporous SNs for extending diagnostic capability, from fluorescence imaging to MRI (Lee et al. 2009). Recently, Meng et al. designed dual therapeutic agent (i.e., doxorubicin and P-glycoprotein small interfering RNA for the siRNA [siRNA])-loaded, polyethyleneimine-anchored mesoporous SNs to enhance the treatment efficacy in multiple drug-resistant cancer (Meng et al. 2010). Here, siRNAs have the capability to silence the expression of efflux transporter and thereby reduce drug resistance. Further, doxorubicin release and its distribution in the tumor cells can be visualized by fluorescence imaging, suggesting the targeted theranostic potential of SNs for the management of cancer.

2.2.7 Solid Lipid Nanoparticles

Solid lipid nanoparticles (SLNs) have attracted plenty of attention owing to their remarkable effectiveness, versatility, and safety in terms of biocompatibility compared to conventional nanocarriers. SLNs are composed of a surfactant stabilized lipid matrix in which therapeutic drug molecules are dispersed or entrapped (Attama et al. 2012, Mendes et al. 2013). SLNs capitalize on the potential of two prominent therapeutic carrier systems: polymeric nanoparticles and lipidic emulsion, at the same time ameliorating the commonly persisting drawbacks associated with polymeric carriers, that is, temporal and *in-vivo* stability (Mendes et al. 2013). SLNs are made from solid, biocompatible lipids with the aid of commonly used preparation methods such as high-speed homogenization, solvent evaporation, nanoprecipitation, and so on. SLNs emerged as worthwhile tools as a result of their high physical and chemical stability, lower toxicity, controlled release and easy tailoring potential, commercial viability, and lower cost. Moreover, their high encapsulation potential and ability to cross natural biological barriers in the human body keep them at the forefront in the rapidly developing field of nanotechnology (Müller et al. 2000, Mehnert and Mader 2001, Mukherjee et al. 2009, Attama et al. 2012, Chalikwar et al. 2012).

As a result of their multifaceted advantages, many researchers have utilized their potential in theranostic applications (Muller et al. 1997, Wissing et al. 2004). Recently, Shuhendler et al. designed NIR light-emitting SLNs for live imaging of a tumor-targeted drug delivery system (Shuhendler et al. 2012). NIR quantum dots along with $\alpha(v)\beta(3)$ integrin-specific ligand, that is, cyclic Arg-Gly-Asp (cRGD), were utilized for theranostic purposes. Pharmacokinetic studies and biodistribution studies followed by optical imaging suggested the liver-, spleen-, and kidney-specific distribution of cRGD-labeled SLNs upon intravenous administration. Moreover, longer blood circulation along with tumor cell-specific targeting are enough to prove the theranostic potential of SLNs for cell-specific treatment and imaging to cancerous organs.

Likewise, Bae et al. developed paclitaxel and siRNA (Bcl-2 targeting)-encapsulated SLNs labeled with quantum dots to confer multimodality of treatment and imaging to low-density lipoprotein-mimetic SLNs (Bae et al. 2013). The SLNs were of a core-shell architecture in which paclitaxel and quantum dots resided in the core of the SLNs, whereas siRNA molecules were bonded to the external surface of the SLNs through electrostatic interactions. Synergistic anticancer activity against human lung carcinoma cells was offered by SLNs, suggesting the simultaneous delivery potential of SLNs for both paclitaxel and siRNA to cancer cells. Further, intracellular localization of SLNs into cancer cells was visualized by measuring the quantum dot-emitted fluorescence. Simultaneous optical tracing and treatment potential confirmed the utilization of SLNs for multimodal theranostic application.

2.2.8 Liposomes

Liposomal drug delivery systems have been drawing tremendous attention as compared to other nanotechnology-based delivery systems, as suggested by their long history of preclinical and clinical studies. Liposomes serve as pioneering nanomedicine for targeting as well as theranostic application by incorporating multiple targeting and imaging agents with high payloads, as evident from current approved clinical use. Liposomes have been evidenced as a versatile delivery vehicle for improving bioavailability, drug efficacy, targeting potential, and safety profile. They possess unique characteristics of carrying and delivering both hydrophobic as well as hydrophilic, multiple therapeutic molecules including biologics and macromolecules, that is, DNA, siRNA, proteins, and peptides. Hence, theranostic liposomes can provide a wide scope for personalized medicine (Sharma and Sharma 1997).

Liposomes are spherical, bilayer, colloidal vesicular systems enclosing an inner aqueous core. They are composed of amphiphilic lipid molecules, both natural and synthetic lipid, in which hydrophobic drugs are embedded within the lipid bilayer (Lasic 1998, Pinto-Alphandary et al. 2003). The lipid bilayer is analogous to a biological membrane model that provides superior biocompatibility and allows interacting closely with a host cell. The lipidic stratum also protects drugs by shielding the from the extraneous environment and allows easy surface modification for prolonged blood circulation as well as targeting cells using different homing carriers. Moreover, the use of natural lipids makes liposome biodegradable, nontoxic, and immune-inert, resulting in an ideal vehicle for therapeutic agent delivery application, including therapeutic and imaging agents (Voinea and Simionescu 2002, Torchilin 2005, Grange et al. 2010). Several diagnostic and imaging agents, including quantum dots, gold nanoparticles, and magnetic nanoparticles, have been encapsulated within liposomes to explore their theranostic potential (Papahadjopoulos et al. 1991, Al-Jamal et al. 2008, Torchilin 2005, Al-Jamal and Kostarelos 2011, Nie et al. 2012, Leung and Romanowski 2012). Nevertheless, the lack of structural integrity, poor drug loading, burst release of encapsulated drugs, poor shelf life due to physical and chemical instability of vesicles, leakage of hydrophilic drugs during storage, batch-to-batch reproducibility, and complicated multistep manufacturing process hamper their widespread use (Sharma and Sharma 1997, Abra et al. 2002, Gabizon et al. 2012).

Liposomes for theranostic application are mainly prepared by solvent dispersion, mechanical dispersion, and the detergent removal method (Sharma and Sharma 1997, Voinea and Simionescu 2002, Grange et al. 2010). Liposomes are mainly classified into two types based on size and number of bilayers: unilamellar and multilamellar vesicles. There are two ways to incorporate drugs into liposomes: active and passive loading. Liposomes with a diameter of 50–200 nm with neutral to negative charge are employed for image-guided drug delivery. Various homing ligands and molecular biomarkers have been decorated on the lipidic bilayer in order to explore them as advanced theranostic liposomes for facilitating active targeting (Elbayoumi and Torchilin 2010, Akbarzadeh et al. 2013). Only folic acid ligands have been successfully employed for theranostic application (Muthu et al. 2012). However, other targeting templates such as mannose receptor, sigma receptor, and cannabinoid receptors can also be used to explore their targeting potential for improving theranostic application. Additionally, longer blood circulation characteristics for preventing opsonization can be conferred on liposomes through steric stabilization via PEG conjugation, which also improves their *in-vivo* serum stability (Maeda et al. 2001). The diagnostic potential of liposomes has been explored with the aid of various imaging agents, including optical imaging, SPECT, PET, MRI, and radioisotopes (Boerman et al. 2000, Torchilin 1996, Gore et al. 1986, Caride 1985). Liposomes have also been employed for successful loading of inorganic nanoparticles, nanocrystals, and macromolecules for conferring multimodality.

The role of imaging agent in theranostic liposomes mainly depends on its type of incorporation and the selection of bioimaging modality. When imaging agents are conjugated on the outer surface of liposomes, then the visualization of the imaging agent gives an idea of the biodistribution of the liposomal in the body, and drug delivery at the target site can thereby be visualized. These insights can help in the development of personalized medicine with reduced adverse effects. However, compatibility between therapeutic agent and imaging agent is a prime requirement for the utilization of liposomes for theranostic purposes. To make them compatible and nonreactive with each other, they should be incorporated into separate compartments of the liposomes.

Liposomes have demonstrated great versatility for multimodal theranostic applications. D-alpha-tocopheryl polyethylene glycol succinate (TPGS)-modified theranostic liposomes have been successfully utilized for active and passive targeting through the EPR effect.

Muthu et al. developed doxorubicin and quantum dot-encapsulated theranostic liposomes for the treatment of breast cancer (Muthu et al. 2012). Further, liposomes were functionalized by TPGS and folic acid for targeted delivery to folate receptor-overexpressed MCF-7 breast cancer cells. *In-vitro* cell cytotoxicity studies showed significant improvement in cellular uptake and anticancer activity by incubating with targeted theranostic liposomes compared to that of bare liposomes. Likewise, Wen et al. designed theranostic liposomes containing apomorphine and quantum dots for enhancing brain targeting as well as eliminating liver uptake (Wen et al. 2012). An *in-vivo* bioimaging study and *in vitro* cellular uptake study demonstrated significant improvement in brain uptake by receptor-mediated endocytosis (both clathrin- and caveola-mediated)-forming theranostic liposomes as compared to free quantum dots. Additionally, fluorescence as a result of quantum dot-encapsulated liposomes was significantly decreased in the liver and heart. In the case of the brain, stronger fluorescence corresponding to quantum dots was retained up to 1 hr compared to that of free quantum dots, proving the enhanced bioimaging potential of quantum dots by incorporating them into theranostic liposomes.

Recently, the Wen research group has extended the theranostic potential of liposomes by incorporating camptothecin, irinotecan, and quantum dots for simultaneous drug delivery and bioimaging (Wen et al. 2013). Different types of liposomes such as cationic, deformable, and PEGylated liposomes have been engineered for assessing anticancer activity through cell cytotoxicity and cell migration assay. *In-vitro* and *in-vivo* studies suggested that cationic liposomes were able to interact with solid tumors with improved efficacy, which was further confirmed by measuring the fluorescence intensity for a longer period of time (24 hr). Further, the multimodal theranostic potential of liposomes can be extended to numerous anticancer agents to prolong their *in-vivo* activity.

Magnetic liposomes serve as one of the multifunctioning theranostic platforms for simultaneous imaging and controlling drug release by employing a magnetic field for the treatment of cancer. In addition to drug release controlling and imaging potential, they also offer biocompatibility and an inherent hyperthermic effect for the management of cancer. The drug release from magnetic liposomes can be modulated by functionalizing with a targeting ligand, which can trigger the drug release by modulating the magnetic field. Further, the hyperthermic effect of magnetic liposome produced by altering the magnetic field also offers a synergistic effect for killing cancer cells. Chen et al. engineered maghemite-embedded theranostic magnetoliposomes for heat-triggered drug release as well as imaging under the magnetic field generated by MRI (Chen et al. 2010a,b). Similarly, Kenny et al. delivered anticancer siRNA with the aid of theranostic magnetoliposomes to enhance tumor reductive action (Kenny et al. 2011). Li et al. designed multimodal theranostic liposomes for controlled and targeted delivery of doxorubicin as well as rhenium 186/rhenium 188, along with *in-vivo* imaging via radionuclide technetium 99 m (Li et al. 2012). Further, imaging and therapeutic agents could be incorporated during and postprocessing via lipid conjugates, proving the versatility of theranostic liposomes for successful clinical translation.

2.2.9 Polymeric Micelles

Polymeric micelles are colloidal structures with a size of less than 100 nm (Mahmud et al. 2007). They are formed by self-assembling of polymeric amphiphilic materials, which results in the formation of a hydrophobic micellar core and hydrophilic corona. Polymeric micelles have emerged as an attractive theranostic nanocarrier system for simultaneous delivery of theranostic agents and imaging agents due to their higher safety and compatibility in the treatment of various diseases, including cancer. Hence, hydrophobic therapeutic agents can be easily delivered through the parenteral route by incorporation inside micelles (Kataoka et al. 2001, Sawant et al. 2014). The outer envelope is generally composed of hydrophilic polymers, that is, PEG, HPMA, and so on, which bestows biocompatibility, a template for surface functionalization and the stealth property. Moreover, the advantages such as uniform size structures, ease of preparation, higher encapsulation of hydrophobic therapeutic agents, multifunctionalities, stability, easy scale-up, successful clinical translation, and so on put polymeric micelles at the forefront of the emerging theranostic platform of nanomedicine. Two main methods are used for the preparation of micelles: the direct dissolution method and the organic solvent method (Kumar et al. 2012a,b).

A hydrophobic core provides the space for incorporating therapeutic and imaging agents, whereas the hydrophilic shell provides a template for conjugation of targeting ligands (Torchilin et al. 2003, Mi et al. 2011, Kumar et al. 2012a,b, Liu et al. 2012). As they are lacking a hydrophilic core, hydrophobic therapeutic and imaging agents must be conjugated with amphiphilic polymeric material prior to the formation of micelles. The stability of polymeric micelles depends on the attractive force between the therapeutic agent and carrier polymer, their critical micelle concentration, and the crosslinking of the polymeric matrix (Vriezema et al. 2005). However, burst release or premature drug release is observed for micelles before reaching the targeted tissue, which may be due to the exchange between physiological components of the cell membrane, and that can be improved by optimizing the crosslinking potential (Moghimi et al. 2004). Theranostic polymeric micelles can be effectively used for the treatment of cancer owing to their smaller size, which helps enhance permeation inside solid tumors via the EPR effect. Additionally, their ability to bypass the reticuloendothelial system also improves the efficacy of therapeutic agents with reduced doses (Matsumura 2011).

Polymeric micelles have been successfully developed for the treatment of cancer. Among them, some are presently in clinical trials, that is, Genexol-PM™, for evaluating safety and efficacy in solid tumors, metastatic breast cancer, and non-small-cell lung cancer (Kim et al. 2001, Kim et al. 2007, Lee et al. 2008). Various TPGS-based theranostic micelles have been engineered for targeted delivery of therapeutic agents (Kutty and Feng 2013, Zhao et al. 2013). Chandrasekharan et al. developed TPGS-based micelles by incorporating superparamagnetic iron oxide nanoparticles for improving thermotherapy and MRI in the treatment of cancer (Chandrasekharan et al. 2011). The prepared micelles were stable with narrow particle size distribution. *In-vitro* cytotoxicity study showed enhancement in cellular uptake and cytotoxicity by incubating micelles with the MCF-7 breast cancer cell line, followed by hyperthermic treatment. Further, *in-vivo* imaging in mice demonstrated significant enhancement in tumor accumulation by forming TPGS micelles compared to marketed Resovist® and Pluronic-F127 micelles. Recently, Kim et al. developed gold shell–coated, doxorubicin-hyaluronic acid–conjugated, self-assembled micelles for theranostic application (Kim et al. 2012). Doxorubicin was encapsulated by amide conjugation between the amine group of doxorubicin and the carboxylic group of hyaluronic acid. The gold shell was enveloped using the thermal vapor deposition method on the surface of the micelles. An *in-vitro* cytotoxicity study showed enhancement in the cytotoxic activity of doxorubicin by forming micelles. Further, the gold coat might have contributed to prolonging the residence time of the micelles in the tumor region. Hence, gold-coated micelles can be employed as a theranostic platform for successful clinical translation.

Photosensitizer-encapsulated micellar delivery enables the tracking of their distribution and accumulation in the target organ due to their intrinsic fluorescence property. Peng and coworkers have engineered IR-780-incorporated multifaceted PEG-polycaprolactone diblock copolymer-based micelles for the treatment of colorectal cancer. An *in-vivo* study on HCT-116 colorectal carcinoma-bearing BALB/c athymic nude mice showed significant improvement in anticancer efficacy with enhanced intratumor localization. Similarly, Chlorin e6 (Ce6) has also been employed as a photosensitizer for theranostic application (Peng et al. 2011). Quantum dots have also been exploited for theranostic application as potential imaging agents. Generally, CdSe- and CdTe-based semiconductor materials are used as QDs for imaging applications. They show stronger and stable fluorescence for prolonged imaging of live cellular structures (Resch-Genger et al. 2008). To this end, Song et al. developed paclitaxel and CdTe QD-loaded antibody-conjugated pluronic triblock copolymer-based micelles (i.e., hypoxia inducible factor-1α) for targeted therapy of stomach cancer cells, which was further visualized by fluorescence microscopy (Song et al. 2010).

Guthi et al. developed lung cancer-targeting peptide-functionalized magnetic nanoparticles and doxorubicin-encapsulated polymeric micelles for targeted delivery in the treatment of lung cancer. Maleimide-terminated PEG-co-poly(d-, l-lactic acid) (Mal-PEG-PLA) and methoxy-terminated PEG-co-poly(d-, l-lactic acid)-based amphiphilic block copolymers were utilized for the formation of the micellar core. Enhanced targeting potential with higher cellular uptake in αvβ6-expressing lung cancer cells suggested the enhancement in efficacy of doxorubicin by forming micelles, which was further visualized by ultrasensitive MRI in order to confirm tumor accumulation (Guthi et al. 2009). Multiple therapeutic agents have also been encapsulated in micelles for theranostic application. In a recent instance, three hydrophobic therapeutic agents, paclitaxel, rapamycin, and 17-allylamino-17-demethoxygelda-namycin, were incorporated in PEG-PLA micelles for the treatment of colon cancer. Intravenous administration in an LS180 human colon xenograft model showed significant reduction in tumor volume with improved tumor-to-muscle ratio upon administration of PEG-PLA micelles. Subsequently, carbocyanine dye-incorporated PEG-block-poly-ε-caprolactone (PEG-PCL) micelle administration followed by optical imaging demonstrated a higher NIR optical signal from cancer cells, confirming the theranostic potential of PEG-PLA/PEG-PCL micelles with a high apoptosis index for cancer cells (Cho and Kwon 2011). PEG-b-poly(glutamic acid)-based micelles have also been developed for cancer theranostics by incorporating a therapeutic agent, (1,2-diaminocyclohexane)platinum(II), and an imaging agent, gadolinium (Gd)-diethylenetriaminepentaacetic acid (Kaida et al. 2010).

2.2.10 Nanoemulsion

Nanoemulsions are kinetically stable, oil-in-water emulsions with a droplet size typically in the range of 100–500 nm (McClements 2012). They usually contain a higher amount of oil along with one or a combination of surfactants for stabilizing the dispersed phase. They can also be stabilized by steric effects. They have been widely employed in the pharmaceutical field for enhancing solubility and thereby bioavailability of therapeutic agents. Multiple imaging modalities can be incorporated in the nanoemulsion without compromising drug loading and controlled release potential in order to confer theranostic capability on them (Sarker 2005, Rajpoot et al. 2011, Mitri et al. 2012, Muller et al. 2012). Nanoemulsions are generally prepared by different methods such as high-pressure homogenization, the low-energy emulsification method (i.e., phase inversion temperature and concentration method), microfluidization, and sonication. Among these, the sonication and microfluidization methods are generally used for the preparation of theranostic nanoemulsion due to their ease of preparation and capability of formation of uniform smaller droplets both at a small scale and industrial scale (Constantinides et al. 2008, Maali and Mosavian 2013). Therapeutic and imaging agents can be easily encapsulated with high payloads by two main methods: direct encapsulation and covalent conjugation. A lower amount of surfactant (i.e., ionic or nonionic) is generally needed for stabilization of nanoemulsions. They are resistant to creaming and sedimentation owing to their smaller size, which also makes them translucent (McClements 2012). They might be degraded by Ostwald ripening through particle growth by molecular diffusion. They can be delivered in the form of capsule and gel with high drug encapsulation. Moreover, their higher stability and easy scalability make them a promising theranostic nanocarrier system for successful industrial acceptance and thereby clinical translation.

Numerous imaging techniques, such as ultra-sound, MRI, optical imaging, and photoacoustic imaging, have been employed for visualizing nanoemulsion-mediated treatment therapy. Nanoemulsions have been visualized by utilizing the perfluorocarbons as ultrasound imaging and MRI contrast agents. Perfluorocarbon-incorporated nanoemulsions have attracted ample recognition as a theranostic carrier system owing to their ease of manufacturing, scalability, prolonged release potential, and ability to confer multimodalities. Different perfluorocarbons such as perfluoropolyether, perfluorooctylbromide, perfluoropentane, perfluorodecaline, peruoro-15-crown-5 ether, and so on have been incorporated for the development of theranostic nanoemulsions. They are generally used for the imaging of different organs, that is, the liver and tumors, as well as cellular tracking by employing MRI (Longmaid et al. 1985, Janjic and Ahrens 2009). Ultrasound-mediated imaging of nanoemulsions has been evaluated in some preclinical models (Main et al. 2009). Additionally, perfluorocarbon-incorporated nanoemulsions can also be tagged with fluorescent and NIR dyes for dual-mode imaging. To this end, Akers et al. have visualized nanoemulsions by two different techniques, optical imaging and photoacoustic imaging (Akers et al. 2011). Figure 2.4 shows the *in-vivo* imaging potential of perfluorocarbon-incorporated nanoemulsions by using ultrasound modality.

Recently, Janjic et al. incorporated a nonsteroidal anti-inflammatory agent, celecoxib, inside the hydrocarbon core of a theranostic nanoemulsion. NIR and ^{19}F MR signals generated by nanoemulsions have been used for imaging of drug release in the cytoplasm (Janjic et al. 2008). In another study, Pan et al. have developed perfluorocarbon-incorporated theranostic nanoemulsions named "nanobees" for tumor-specific delivery of cytolytic peptide in order to produce a significant anticancer effect (Pan et al. 2011). Further, the bee venom–derived peptide melittin has been encapsulated in a nanoemulsion and showed superior anticancer activity in a mouse model along with ^{19}F MR images, confirming the theranostic potential of nanobees.

Drug release from theranostic nanoemulsions can be controlled by employing low-molecular-weight and low-boiling-point perfluorocarbon, which promotes drug release upon exposure of ultrasound. Hence, perfluorocarbon act as a dual-functioning, versatile agent for simultaneous imaging and controlling drug release (Rapoport et al. 2011). Rapoport et al. developed paclitaxel-incorporated perfluoropentane nanoemulsion for ultrasound and MRI-mediated imaging in a mouse tumor model (Rapoport et al. 2004). In another instance, they utilized low-boiling-point dodecafluoropentane for development of ultrasound-responsive theranostic nanoemulsions, which convert into microbubbles from the nanodroplets under exposure to ultrasound for controlling drug release (Rapoport 2012).

Nanoemulsions can be targeted to specific organs or tumors via passive targeting by virtue of their lower droplet size (150–200 nm), which favors their selective accumulation into the inflamed tissue through the EPR effect. Perfluorocarbon-incorporated nanoemulsions have been widely explored for theranostic applications to inflamed tissue by passive targeting (Kadayakkara et al. 2010, Weise et al. 2011, Stoll et al. 2012, Balducci et al. 2013). Recently,

Nanotheranostics

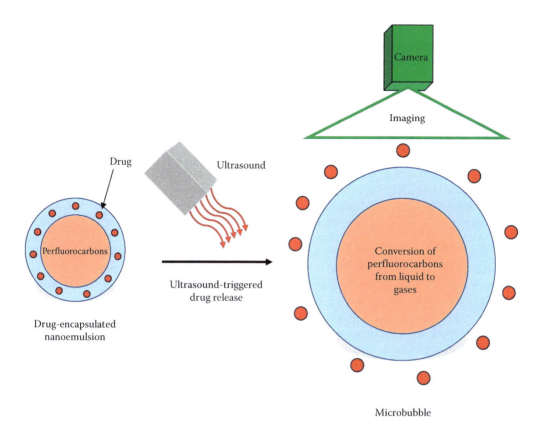

FIGURE 2.4
Schematic representation of drug-incorporated perfluorocarbons nanoemulsion for theranostic application upon exposure of ultrasound.

antigens have been delivered to dendritic cells by means of theranostic nanoemulsion in order to boost the immune response (Dewitte et al. 2013). Nanoemulsions have also been utilized for active targeting by functionalizing the surface of the emulsion droplet through covalent conjugation. Lanza et al. have modulated the surface of doxorubicin and taxol-encapsulated perfluorooctyl bromide nanoemulsions with tissue factor–specific antibodies for active targeting of smooth muscle cells (Lanza 2002, Caruthers et al. 2009).

Some researchers have also developed nonfluorinated nanoemulsions for theranostic application. In a recent study, Gianella et al. designed an iron oxide nanocrystal, fluorescent dye Cy7, and prednisolone acetate valerate-incorporated nanoemulsion for the theranostic treatment of cancer. Further, the surface of the nanoemulsion was modulated by $\alpha_v\beta_3$-specific RGD peptides for targeting of angiogenesis in a mouse tumor model. Dual-imaging modalities, that is, MRI and NIRF imaging, were utilized for the visualizing the nanoemulsion distribution as well as tumor growth inhibition. Moreover, enhancement in the anticancer efficacy of the nanoemulsion was confirmed by means of molecular and histopathological analysis of the excised tissue (Gianella et al. 2011).

2.2.11 Carbon-Based Nanostructures

Different carbon-based nanostructures (CBNs) such as carbon dots (nanoclusters), nanodiamonds, zero-dimensional fullerene, one-dimensional carbon nanotubes, and two-dimensional graphene have been explored for theranostic application due to their exceptional physical and chemical characteristics. CBNs are classified based on their bonding structures, that is, sp2 or sp3 (Baughman et al. 2002, Liu and Liang 2012). Among them, carbon dots and nanodiamonds have intrinsic self-laminating characteristics to give fluorescence emission with the aid of their optoelectronic properties. Carbon nanotubes (CNTs) have been widely employed for theranostic purposes owing to their tunable properties and higher surface area, which confers on them characteristic mechanical, thermal, electrical, and biological attributes in order to encapsulate therapeutic and imaging agents (McDevitt et al. 2007, Porter et al. 2007, Fubini et al. 2010). CNTs are mainly classified into two types, single-walled CNTs and multiwalled CNTs, based on the number of graphene sheet layer surrounding the cylindrical layer to furnish stability (Boncel et al. 2011, Shen et al. 2012, Yang et al. 2013). Different methods, such as the laser method, chemical vapor deposition method, discharge method, and ball milling method, have been employed for the manufacturing of CNTs. Numerous synthetic methods are used for the functionalization of CNTs by virtue of their higher surface area, which provides a template for surface functionalization and makes them multifaceted, biocompatible, and targetable nanomedicines. Hence, CNTs can be designed with strong absorbance in the NIR region for theranostic applications by Raman and photoacoustic imaging as well as photothermal therapy (PTT) (Choi et al. 2012, Liu and Liang 2012).

Robinson et al. designed PEGylated phospholipid modulated single-walled CNTs for theranostics application in the treatment of cancer through PTT and NIR imaging (Robinson et al. 2010). They acted as dual-functioning agents due to their inherent photoluminescent and photothermal behavior. CNTs showed *in-vivo* imaging ability by absorbing in the NIR region, which in turn emitted photoluminescence in the range of 1.0–1.4 μm. Moreover, laser irradiation (strong NIR absorbance) could have also resulted in the generation of heat for photothermal therapy, which resulted in significant tumor cell apoptosis by raising the temperature of the tumor around 60°C within 5 min. Tumor-specific distribution of CNTs was confirmed by Raman imaging with high spatial resolution during *ex-vivo* studies. Further, photothermal treatment-mediated cell cytotoxicity along with enhanced cellular uptake was confirmed by fluorescence imaging after intravenous administration in mice. The significant reduction in side effects with improved anticancer efficacy at 10-fold lower doses of CNTs and irradiation power confirmed the intrinsic theranostic potential of CNTs for the treatment of cancer.

Similarly, Das et al. fabricated different four functional agents, methotrexate (therapeutic agent), Technitium-99m (radionucleide), Alexa-fluor (fluorochrome), and folic acid (targeting agent), incorporated into multiwalled CNTs for the treatment of breast and lung cancer (Das et al. 2013). Folic acid-modulated CNTs showed cancer cell selective higher uptake via endocytosis in the folate receptor-overexpressed MCF-7 breast cancer cell and A549 lung cancer cells. Further, *in-vivo* study in cancer cell-xenografted mice confirmed the tumor-specific localization of the targeted multiwalled CNTs along with the prolonged release of methotrexate. Additionally, diagnostic ability by optical imaging and radiotracing due to fluorochrome and radionucleide, respectively, established the multimodal theranostic potential of CNTs in the management of cancer treatment.

Graphene has also been employed as theranostic nanostructures owing to its capability of optical absorption in the NIR region, which ultimately results in photothermal transduction (Moon et al. 2009). To this end, Yang et al. demonstrated its enhanced uptake and thereby anticancer efficacy by developing nanographene, as compared to CNTs, via NIR-laser mediated photothermal therapy in tumor-xenografted mouse models (Yang et al. 2010). Further, PEGylation of nanographene showed passive targeting of cancer cells along with escape from the reticuloendothelial system upon intravenous administration. Such a developed system has also shown remarkable safety in animals as revealed by histology and blood panel analysis after treatment. Recently, Qin et al. have designed a theranostic nanohybrid by using graphene oxide as a platform for diagnosis and combinatorial cancer treatment (Qin et al. 2014). They have conjugated doxorubicin via matrix metalloproteinase-2-cleavable peptide linkage to poly(ethylenimine)-co-poly(ethylene glycol)-modulated graphene oxide. The matrix metalloproteinase-2 enzyme available in cancerous cells cleaved the peptide linkage and released the doxorubicin to produce cytotoxic activity. Further, the released doxorubicin was simultaneously visualized by fluorescence imaging due to its intrinsic fluorescence property. These positive results suggest that graphene can also be utilized as a promising platform for tumor treatment by simultaneous diagnosis and targeted chemotherapy. Interestingly, the higher efficacy and safety of CBNs during preclinical studies are enough to indicate the suitability of CBNs for clinical translation as efficient theranostic nanomedicines for upcoming decades.

2.3 Conclusion and Future Perspective

Nanotheranostics have begun at a phenomenal pace in the world of nanotechnology with an intention to bestow the dual functionality of simultaneous diagnosis and treatment on nanomedicines in a single vehicle. During the last decade, numerous strategies have been employed for advanced, versatile theranostics application, with improved safety and efficacy to cure diseases at the earliest stage. As a result, multifarious nanocarriers, including polymeric nanoparticles, liposomes, micelles, dendrimers, polymeric conjugates, carbon nanotubes, metal and inorganic nanoparticles, and so on, have been engineered and showed excellent potential and promise for better theranostic effects. Theranostic nanoparticles can deliver therapeutic and imaging agents in a controlled manner at the targeted site to diagnose and treat various dangerous diseases at the cellular as well as the molecular level. The theranostic ability of nanomedicines can easily be tailored by means of various biological homing carriers for targeting as well as prolonging their systemic residence by evading the host immune system. Multifunctionality can also be conferred on nanotheranostics for spatiotemporal release of drug and imaging agents for synergistic therapy, oral delivery, combinatory therapy, gene co-delivery, multimodality therapies, and delivery across natural tissue barriers to avoid multidrug resistance.

Numerous drawbacks such as their nonbiodegradable nature, high cost, toxicity, stability, and complexity of carriers hinder the clinical setting of nanotheranostics, hence necessitate the further exploration for developing personalized medicines. Theranostic nanomedicines have mostly been evaluated through *in-vitro* studies for their theranostic potential. Even though early *in-vitro* studies of nanotheranostics are enough to evince the potential of delivery systems as proof of concept, still, detailed *in-vivo* studies at the molecular level are needed in order to provide the collective influence of various chemical, pharmaceutical, and physiological factors on their theranostic efficacy. These *in-vivo* studies will cross the barrier for the successful translation of *in-vivo* applications to preclinical and clinical levels. Additionally, various physicochemical properties, including particle size, shape, and targeting, as well as imaging agents, should be rugged enough to generate reproducible effects during *in-vivo* studies. Most theranostic nanomedicines have been explored for the treatment of cancer and have also exhibited superiority over other systems. Future research should focus on the area of infectious diseases in which the potential of theranostic nanomedicines could be explored for cellular delivery and imaging of multiple drugs or genes in a single vehicle (Muthu et al. 2014).

The safety of nanotheranostics is still a major concern and needs to be sufficiently explored. Some of the metal- and carbon-based nanomaterials should be rigorously studied in order to establish their biodistribution, safety, and toxicity potential. In addition, newer biocompatible and biodegradable materials should be explored for developing highly efficacious theranostic nanocarriers with high safety potential. Moreover, molecular biomarker-targeted nanotheranostics need to be developed to improve the precision in the treatment of various dreadful diseases. Further, future research needs to explore the existing clinical platform of nanocarriers in order to develop tunable, multifaceted nanotheranostics in a shorter time duration. Cytotoxicity, genotoxicity, and immunotoxicity of nanotheranostics should be established in order to employ them during routine health problems and pave the way to personalized and predictive medicines. Imaging modalities also need to be technically developed to overcome the problems related to resolution and background noise due to photoleaching in order to visualize the theranostic effect up to the cellular level. Moreover, development of easily functionalizable fluorescent dyes is needed for generating reproducible images with high resolution during biomedical imaging.

Earlier detection, treatment, and assessment of therapy throughout the treatment regimen for various fatal diseases including cancer will bring nanotheranostics to the forefront for their clinical utilization and pledge to provide personalized therapy for achieving maximum efficacy without compromising on safety issues. It is likely that the pharmaceutical, clinical, and commercial impact of nanotheranostics will be fully explored in the upcoming decade and the therapeutic potential will be further expanded to various indications. If the present need for future research in the area of nanotheranostics will be critically exploited, the ultimate destination of their clinical realization to provide personalized therapy and cures is not so far away.

References

Abra, R. M., Bankert, R. B., Chen, F. et al. 2002. The next generation of liposome delivery systems: Recent experience with tumor-targeted, sterically-stabilized immunoliposomes and active-loading gradients. *J Liposome Res* 12:1–3.

Akbarzadeh, A., Rezaei-sadabady, R., Davaran, S. et al. 2013. Liposome: Classification, preparation, and applications. *Nanoscale Res Lett* 8:102–11.

Akers, W. J., Kim, C., Berezin, M. et al. 2011. Noninvasive photoacoustic and fluorescence sentinel lymph node identification using dye-loaded perfluorocarbon nanoparticles. *ACS Nano* 5:173–82.

Al-Jamal, K. T., Al-Jamal, W. T., Wang, J. T. et al. 2013. Cationic poly-l-lysine dendrimer complexes doxorubicin and delays tumor growth *in vitro* and *in vivo*. *ACS Nano* 7:1905–17.

Al-Jamal, W. T., Al-Jamal, K. T., Tian, B. et al. 2008. Lipid-quantum dot bilayer vesicles enhance tumor cell uptake and retention *in vitro* and *in vivo*. *ACS Nano* 2:408–18.

Al-Jamal, W. T. and Kostarelos, K. 2011. Liposomes: From a clinically established drug delivery system to a nanoparticle platform for theranostic nanomedicine. *Acc Chem Res* 44:1094–104.

Allmeroth, M., Moderegger, D., Gündel, D. et al. 2013. PEGylation of HPMA-based block copolymers enhances tumor accumulation *in vivo*: A quantitative study using radiolabeling and positron emission tomography. *J Control Release* 172:77–85.

Atmaja, B., Lui, B. H., Hu, Y. H. et al. 2010. Targeting of cancer cells using quantum dot-polypeptide hybrid assemblies that function as molecular imaging agents and carrier systems. *Adv Funct Mater* 20:4091–7.

Attama, A. A., Momoh, M. A., and P. F. Builders. 2012. Lipid nanoparticulate drug delivery systems: A revolution in dosage form design and development. In *Recent Advances in Novel Drug Carrier Systems*, ed. A. D. Sezer, Chapter 5. InTech, Europe.

Bae, K. H., Lee, J. Y., Lee, S. H. et al. 2013. Optically traceable solid lipid nanoparticles loaded with siRNA and paclitaxel for synergistic chemotherapy with *in situ* imaging. *Adv Healthcare Mater* 2:576–84.

Balducci, A., Wen, Y., Zhang, Y. et al. 2013. A novel probe for the non-invasive detection of tumor-associated inflammation. *Oncoimmunology* 2:e23034.

Baughman, R. H., Zakhidov, A. A., and W. A. De Heer. 2002. Carbon nanotubes—The route toward applications. *Science* 297:787–92.

Bhatnagar, P., Li, Z., Choi, Y. et al. 2013. Imaging of genetically engineered T cells by PET using gold nanoparticles complexed to Copper-64. *Integr Biol (Camb)* 5:231–8.

Boerman, O. C., Laverman, P., Oyen, W. J. et al. 2000. Radiolabeled liposomes for scintigraphic imaging. *Prog Lipid Res* 39:461–75.

Boncel, S., Müller, K. H., Skepper, J. N. et al. 2011. Tunable chemistry and morphology of multi-wall carbon nanotubes as a route to non-toxic, theranostic systems. *Biomaterials* 32:7677–86.

Borgman, M. P., Coleman, T., Kolhatkar, R. B. et al. 2008. Tumor-targeted HPMA co-polymer-(RGDfk)-(CHX-A″-DTPA) conjugates show increased kidney accumulation. *J Control Release* 132:193–9.

Bosman, A. W., Janssen, H. M., and E. W. Meijer. 1999. About dendrimers: Structure, physical properties, and applications. *Chem Rev* 99:1665–88.

Bremer, C., Tung, C. H., and R. Weissleder. 2001. *In vivo* molecular target assessment of matrix metalloproteinase inhibition. *Nat Med* 7:743–8.

Caride, V. J. 1985. Liposomes as carriers of imaging agents. *Crit Rev Ther Drug Carrier Syst* 1:121–53.

Caruthers, S. D., Cyrus, T., Winter, P. M. et al. 2009. Anti-angiogenic perfluorocarbon nanoparticles for diagnosis and treatment of atherosclerosis. *Wiley Interdiscip Rev Nanomed Nanobiotechnol* 1:311–23.

Chalikwar, S. S., Belgamwar, V. S., Talele, V. R. et al. 2012. Formulation and evaluation of nimodipine-loaded solid lipid nanoparticles delivered via lymphatic transport system. *Colloids Surf B: Biointerfaces* 97:109–16.

Chandrasekharan, P., Maity, D., Yong, C. X. et al. 2011. Vitamin E (D-alpha-tocopheryl-co-poly(ethylene glycol) 1000 succinate) micelles-superparamagnetic iron oxide nanoparticles for enhanced thermotherapy and MRI. *Biomaterials* 32:5663–72.

Chen, H., Li, S., Li, B. et al. 2012. Folate-modified gold nanoclusters as near-infrared fluorescent probes for tumor imaging and therapy. *Nanoscale* 4:6050–64.

Chen, J. Y., Glaus, C., Laforest, R. et al. 2010a. Gold nanocages as photothermal transducers for cancer treatment. *Small* 6:811–7.

Chen, W. H., Xu, X. D., Jia, H. Z. et al. 2013. Therapeutic nanomedicine based on dual-intelligent functionalized gold nanoparticles for cancer imaging and therapy *in vivo*. *Biomaterials* 34:8798–807.

Chen, Y., Bose, A., and G. D. Bothun. 2010b. Controlled release from bilayer-decorated magnetoliposomes via electromagnetic heating. *ACS Nano* 4:3215–21.

Cheng, S. H., Lee, C. H., Chen, M. C. et al. 2010. Tri-functionalization of mesoporous silica nanoparticles for comprehensive cancer theranostics - The trio of imaging, targeting and therapy. *J Mater Chem* 20:6149–57.

Chertok, B., David, A. E., and V. C. Yang. 2011. Magnetically-enabled and MR-monitored selective brain tumor protein delivery in rats via magnetic nanocarriers. *Biomaterials* 32:6245–53.

Cho, H. and G. S. Kwon. 2011. Polymeric micelles for neoadjuvant cancer therapy and tumor-primed optical imaging. *ACS Nano* 5:8721–9.

Choi, K. Y., Chung, H., Min, K. H. et al. 2010. Self-assembled hyaluronic acid nanoparticles for active tumor targeting. *Biomaterials* 31:106–14.

Choi, K. Y., Liu, G., Lee, S. et al. 2012. Theranostic nanoplatforms for simultaneous cancer imaging and therapy: Current approaches and future perspectives. *Nanoscale* 4:330–42.

Connor, E. E., Mwamuka, J., Gole, A. et al. 2005. Gold nanoparticles are taken up by human cells but do not cause acute cytotoxicity. *Small* 1:325–7.
Constantinides, P. P., Chaubal, M. V., and R. Shorr. 2008. Advances in lipid nanodispersions for parenteral drug delivery and targeting. *Adv Drug Deliv Rev* 60:757–67.
Daniel, M. C. and D. Astruc. 2004. Gold nanoparticles: Assembly, supramolecular chemistry, quantum-size-related properties, and applications toward biology, catalysis, and nanotechnology. *Chem Rev* 104:293–346.
Das, M., Datir, S. R., Singh, R. P. et al. 2013. Augmented anticancer activity of a targeted, intracellularly activatable, theranostic nanomedicine based a fluorescent and radiolabeled, methotrxate-folic acid-multiwalled carbon nanotube conjugate. *Mol Pharm* 10:2543–57.
Das, M., Mishra, D., Dhak, P. et al. 2009. Biofunctionalized, phosphonate-grafted, ultrasmall iron oxide nanoparticles for combined targeted cancer therapy and multimodal imaging. *Small* 5:2883–93.
Davis, S. C., Pogue, B. W., Springett, R. et al. 2008. Magnetic resonance-coupled fluorescence tomography scanner for molecular imaging of tissue. *Rev Sci Instrum* 79:064302.
Dewitte, H., Geers, B., Liang, S. et al. 2013. Design and evaluation of theranostic perfluorocarbon particles for simultaneous antigen-loading and (19)F-Mri tracking of dendritic cells. *J Control Release* 169:141–9.
Dhanikula, R. S. and P. Hildgen. 2007. Influence of molecular architecture of polyether-co-polyester dendrimers on the encapsulation and release of methotrexate. *Biomaterials* 28:3140–52.
Dreher, M. R., Raucher, D., Balu, N. et al. 2003. Evaluation of an elastin-like polypeptide-doxorubicin conjugate for cancer therapy. *J Control Release* 91:31–43.
Elbayoumi, T. A. and V. P. Torchilin. 2010. Current trends in liposome research. *Methods Mol Biol* 605:1–27.
Era, D. R., Hall, D. J., Hoh, C. K. et al. 2005. Cy5.5-DTPA-galactosyl-dextran: A fluorescent probe for *in vivo* measurement of receptor biochemistry. *Nucl Med Biol* 32:687–93.
Esfand, R. and D. A. Tomalia. 2001. Poly(amidoamine) (PAMAM) dendrimers: From biomimicry to drug delivery and biomedical applications. *Drug Discov Today* 6:427–36.
Fahmy, T. M., Fong, P. M., Park, J. et al. 2007. Nanosystems for simultaneous imaging and drug delivery to T cells. *AAPS Journal* 9:E171–80.
Feng, X. L., Lv, F. T., Liu, L. B. et al. 2010. Conjugated polymer nanoparticles for drug delivery and imaging. *ACS Appl Mater Interfaces* 2:2429–35.
Foy, S. P., Manthe, R. L., Foy, S. T. et al. 2010. Optical imaging and magnetic field targeting of magnetic nanoparticles in tumors. *ACS Nano* 4:5217–24.
Frechet, J. M. 1994. Functional polymers and dendrimers—Reactivity, molecular architecture, and interfacial energy. *Science* 263:1710–5.
Fubini, B., Ghiazza, M., and I. Fenoglio. 2010. Physio-chemical features of engineered nanoparticles relevant to their toxicity. *Nanotoxicology* 4:347–63.
Gabizon, A., Amitay, Y., Tzemach, D. et al. 2012. Therapeutic efficacy of a lipid-based prodrug of mitomycin C in PEGylated liposomes: Studies with human gastro-entero-pancreatic ectopic tumor models. *J Control Release* 160:245–53.
Gianella, A., Jarzyna, P. A., Mani, V. et al. 2011. Multifunctional nanoemulsion platform for imaging guided therapy evaluated in experimental cancer. *ACS Nano* 5:4422–33.
Gibson, J. D., Khanal, B. P., and E. R. Zubarev. 2007. Paclitaxel-functionalized gold nanoparticles. *J Am Chem Soc* 129:11653–61.
Gore, J. C., Sostman, H. D., and V. J. Caride. 1986. Liposomes for paramagnetic contrast enhancement in n.m.r. imaging. *J Microencapsul* 3:251–64.
Grange, C., Geninatti-Crich, S., Esposito, G. et al. 2010. Combined delivery and magnetic resonance imaging of neural cell adhesion molecule-targeted doxorubicin-containing liposomes in experimentally induced Kaposi's sarcoma. *Cancer Res* 70:2180–90.
Grünwald, G. K., Vetter, A., Klutz, K. et al. 2013. Systemic image-guided liver cancer radiovirotherapy using dendrimer-coated adenovirus encoding the sodium iodide symporter as theranostic gene. *J Nucl Med* 54:1450–7.

Gupta, A. K. and M. Gupta. 2005. Synthesis and surface engineering of iron oxide nano-particles for biomedical applications. *Biomaterials* 26:3995–4021.

Guthi, J. S., Yang, S. G., Huang, G. et al. 2009. MRI-visible micellar nanomedicine for targeted drug delivery to lung cancer cells. *Mol Pharm* 7:32–40.

Hakkinen, H. 2012. The gold-sulfur interface at the nanoscale. *Nat Chem* 4:443–55.

Han, G., Martin, C. T., Rotello, V. M. et al. 2006. Stability of gold nanoparticle-bound DNA toward biological, physical, and chemical agents. *Chem Biol Drug Des* 67:78–82.

Hanand, J. and X. B. Li. 2011. Chemoenzymatic syntheses of sialyl Lewis X-chitosan conjugate as potential anti-inflammatory agent. *Carbohydr Polym* 83:137–43.

He, X., Wu, X., Wang, K. et al. 2009. Methylene blue-encapsulated phosphonate-terminated silica nanoparticles for simultaneous *in vivo* imaging and photodynamic therapy. *Biomaterials* 30:5601–9.

He, X. K., Yuan, Z. X., Wu, X. J. et al. 2012. Low molecular weight hydroxyethyl chitosan-prednisolone conjugate for renal targeting therapy: Synthesis, characterization and *in vivo* studies. *Theranostics* 2:1054–63.

Heo, D. N., Yang, D. H., Moon, H. J. et al. 2012. Gold nanoparticles surface-functionalized with paclitaxel drug and biotin receptor as theranostic agents for cancer therapy. *Biomaterials* 33:856–66.

Hu, C. M., Fang, R. H., Luk, B. T. et al. 2014. Polymeric nanotherapeutics: Clinical development and advances in stealth functionalization strategies. *Nanoscale* 6:65–75.

Hu, K. W., Liu, T. M., Chung, K. Y. et al. 2009. Efficient near-IR hyperthermia and intense nonlinear optical imaging contrast on the gold nanorod-in-shell nanostructures. *J Am Chem Soc* 131:14186–7.

Huang, G., Chen, H., Dong, Y. et al. 2013. Superparamagnetic iron oxide nanoparticles: Amplifying ROS stress to improve anticancer drug efficacy. *Theranostics* 3:116–26.

Huang, P., Bao, L., Zhang, C. et al. 2011. Folic acid-conjugated silica-modified gold nanorods for x-ray/CT imaging-guided dual-mode radiation and photo-thermal therapy. *Biomaterials* 32:9796–809.

Hudecz, F. 1995. Design of synthetic branched-chain polypeptides as carriers for bioactive molecules. *Anti-Cancer Drugs* 6:171–93.

Janjic, J. M. and E. T. Ahrens. 2009. Fluorine-containing nanoemulsions for MRI cell tracking. *Wiley Interdiscip Rev Nanomed Nanobiotechnol* 1:492–501.

Janjic, J. M., Srinivas, M., Kadayakkara, D. K. et al. 2008. Self-delivering nanoemulsions for dual fluorine-19 MRI and fluorescence detection. *J Am Chem Soc* 130:2832–41.

Jansen, J. F., de Brabander-van den Berg, E. M., and E. W. Meijer. 1994. Encapsulation of guest molecules into a dendritic box. *Science* 266:1226–9.

Kadayakkara, D. K., Beatty, P. L., Turner, M. S. et al. 2010. Inflammation driven by overexpression of the hypoglycosylated abnormal mucin 1 (MUC1) links inflammatory bowel disease and pancreatitis. *Pancreas* 39:510–5.

Kaida, S., Cabral, H., Kumagai, M. et al. 2010. Visible drug delivery by supramolecular nanocarriers directing to single-platformed diagnosis and therapy of pancreatic tumor model. *Cancer Res* 70:7031–41.

Kaminskas, L. M., Mcleod, V. M., Porter, C. J. et al. 2012. Association of chemotherapeutic drugs with dendrimer nanocarriers: An assessment of the merits of covalent conjugation compared to noncovalent encapsulation. *Mol Pharm* 9:355–73.

Kataoka, K., Harada, A., and Y. Nagasaki. 2001. Block copolymer micelles for drug delivery: Design, characterization and biological significance. *Adv Drug Deliv Rev* 47:113–31.

Ke, T. Y., Jeong, E. K., Wang, X. L. et al. 2007. RGD targeted poly(L-glutamic acid)-cystamine-(Gd-DO3A) conjugate for detecting angiogenesis biomarker alpha(v)beta(3) integrin with MRT1 mapping. *Int J Nanomedicine* 2:191–9.

Kenny, G. D., Kamaly, N., Kalber, T. L. et al. 2011. Novel multifunctional nanoparticle mediates siRNA tumour delivery, visualisation and therapeutic tumour reduction *in vivo*. *J Control Release* 149:111–6.

Kim, D., Jeong, Y. Y., and S. Jon. 2010a. A drug-loaded aptamer-gold nanoparticle bioconjugate for combined CT imaging and therapy of prostate cancer. *ACS Nano* 4:3689–96.

Kim, D. W., Kim, S. Y., Kim, H. K. et al. 2007. Multicenter phase II trial of Genexol-PM, a novel Cremophor-free, polymeric micelle formulation of paclitaxel, with cisplatin in patients with advanced non-small-cell lung cancer. *Ann Oncol* 108:2009–14.

Kim, K., Kim, J. H., Park, H. et al. 2010b. Tumor-homing multifunctional nanoparticles for cancer theragnosis: Simultaneous diagnosis, drug delivery, and therapeutic monitoring. *J. Control Release* 146:219–27.

Kim, K. S., Park, S. J., Lee, M. Y. et al. 2012. Gold half-shell coated hyaluronic ac-id-doxorubicin conjugate micelles for theranostic applications. *Macromol Res* 20:277–82.

Kim, S. C., Kim, D. W., Shim, Y. H. et al. 2001. In vivo evaluation of polymeric micellar paclitaxel formulation: Toxicity and efficacy. *J Control Release* 72:191–202.

Kong, J. H., Oh, E. J., Chae, S. Y. et al. 2010. Long acting hyaluronate-exendin 4 conjugate for the treatment of type 2 diabetes. *Biomaterials* 31:4121–8.

Kopecek, J. and P. Kopeckova. 2010. HPMA copolymers: Origins, early developments, present, and future. *Adv Drug Del Rev* 62:122–49.

Koyama, Y., Talanov, V. S., Bernardo, M. et al. 2007. A dendrimer-based nanosized contrast agent dual-labeled for magnetic resonance and optical fluorescence imaging to localize the sentinel lymph node in mice. *J Magn Reson Imaging* 25:866–71.

Kumar, A., Ma, H., Zhang, X. et al. 2012a. Gold nanoparticles functionalized with therapeutic and targeted peptides for cancer treatment. *Biomaterials* 33:1180–9.

Kumar, R., Korideck, H., Ngwa, W. et al. 2013. Third generation gold nanoplatform optimized for radiation therapy. *Transl Cancer Res* 2: doi: 10.3978/j.issn.2218-676X.2013.07.02.

Kumar, R., Kulkarni, A., Nagesha, D. K. et al. 2012b. In vitro evaluation of theranostic polymeric micelles for imaging and drug delivery in cancer. *Theranostics* 2:714–22.

Kutty, R. V. and S. S. Feng. 2013. Cetuximab conjugated vitamin E TPGS micelles for targeted delivery of docetaxel for treatment of triple negative breast cancers. *Biomaterials* 34:10160–71.

Lammers, T., Subr, V., Peschke, P. et al. 2008. Image-guided and passively tumour-targeted polymeric nanomedicines for radiochemotherapy. *Br J Cancer* 99:900–10.

Lammers, T. and K. Ulbrich. 2010. HPMA copolymers: 30 years of advances. *Adv Drug Del Rev* 62:119–21.

Lanza, G. M. 2002. Targeted antiproliferative drug delivery to vascular smooth muscle cells with a magnetic resonance imaging nanoparticle contrast agent: Implications for rational therapy of restenosis. *Circulation* 106:2842–7.

Lasic, D. D. 1998. Novel applications of liposomes. *Trends Biotechnol* 16:307–21.

Laurent, S., Forge, D., Port, M. et al. 2008. Magnetic iron oxide nanoparticles: Synthesis, stabilization, vectorization, physicochemical characterizations, and biological applications. *Chem Rev* 108:2064–110.

Lazniewska, J., Milowska, K., and T. Gabryelak. 2012. Dendrimers—Revolutionary drugs for infectious diseases. *Wiley Interdiscip Rev Nanomed Nanobiotechnol* 4:469–91.

Lee, J. E., Lee, N., Kim, H. et al. 2009. Uniform mesoporous dye-doped silica nanoparticles decorated with multiple magnetite nanocrystals for simultaneous enhanced magnetic resonance imaging, fluorescence imaging, and drug delivery. *J Am Chem Soc* 132:552–7.

Lee, K. S., Chung, H. C., Im, S. A. et al. 2008. Multicenter phase II trial of Genexol-PM, a cremophor-free, polymeric micelle formulation of paclitaxel, in patients with metastatic breast cancer. *Breast Cancer Res Treat* 108:241–50.

Leu, C. M., Shu, C. F., Teng, C. F. et al. 2001. Dendritic poly(ether-imide)s: Synthesis, characterization, and modification. *Polymer* 42:2339–48.

Leung, S. J. and M. Romanowski. 2012. Light-activated content release from liposomes. *Theranostics* 2:1020–36.

Li, F., Bae, B. C., and K. Na. 2010. Acetylated hyaluronic acid/photosensitizer conjugate for the preparation of nanogels with controllable phototoxicity: Synthesis, characterization, autophotoquenching properties, and in vitro phototoxicity against HeLa cells. *Bioconjug Chem* 21:1312–20.

Li, L., Jiang, W., Luo, K. et al. 2013. Superparamagnetic iron oxide nanoparticles as MRI contrast agents for non-invasive stem cell labeling and tracking. *Theranostics* 3:595–615.

Li, M. L., Wang, J. C., Schwartz, J. A. et al. 2009. *In-vivo* photoacoustic microscopy of nanoshell extravasation from solid tumor vasculature. *J Biomed Opt* 14:010507.

Li, S., Goins, B., Zhang, L. et al. 2012. Novel multifunctional theranostic liposome drug delivery system: Construction, characterization, and multimodality MR, near-infrared fluorescent, and nuclear imaging. *Bioconjug Chem* 23:1322–32.

Li, Y., Cheng, Y., and T. Xu. 2007. Design, synthesis and potent pharmaceutical applications of glycodendrimers; a mini review. *Curr Drug Discov Technol* 4:246–54.

Lim, E. K., Huh, Y. M., Yang, J. et al. 2011. pH-triggered drug-releasing magnetic nanoparticles for cancer therapy guided by molecular imaging by MRI. *Adv Mater* 23:2436–42.

Link, S. and M. A. El-Sayed. 2002. Shape and size dependence of radiative, non-radiative and photothermal properties of gold nanocrystals. *Int Rev In Physical Chemistry* 19:409–53.

Liu, L., Yong, K. T., Roy, I. et al. 2012. Bioconjugated pluronic triblock-copolymer micelle-encapsulated quantum dots for targeted imaging of cancer: *In vitro* and *in vivo* studies. *Theranostics* 2:705–13.

Liu, W. G., MacKay, J. A., Dreher, M. R. et al. 2010. Injectable intratumoral depot of thermally responsive polypeptide-radionuclide conjugates delays tumor progression in a mouse model. *J Control Release* 144:2–9.

Liu, Z. and X. J. Liang. 2012. Nano-carbons as theranostics. *Theranostics* 2:235–7.

Lo, S. T., Stern, S., Clogston, J. D. et al. 2010. Biological assessment of triazine dendrimer: Toxicological profiles, solution behavior, biodistribution, drug release and efficacy in a pegylated, paclitaxel construct. *Mol Pharm* 7:993–1006.

Longmaid, H. E., Adams, D. F., Neirinckx, R. D. et al. 1985. In vivo 19F NMR imaging of liver, tumor, and abscess in rats. preliminary results. *Invest Radiol* 20:141–5.

Lu, J., Liong, M., Li, Z. et al. 2010. Biocompatibility, biodistribution, and drug-delivery efficiency of mesoporous silica nanoparticles for cancer therapy in animals. *Small* 6:1794–805.

Lu, P. L., Chen, Y. C., Ou, T. W. et al. 2011. Multifunctional hollow nanoparticles based on graft-diblock copolymers for doxorubicin delivery. *Biomaterials* 32:2213–21.

Lu, Z. R., Wang, X. H., Parker, D. L. et al. 2003. Poly(L-glutamic acid) Gd(III)-DOTA conjugate with a degradable spacer for magnetic resonance imaging. *Bioconjug Chem* 14:715–9.

Luk, B. T., Fang, R. H., and L. Zhang. 2012. Lipid- and polymer-based nanostructures for cancer theranostics. *Theranostics* 2:1117–26.

Lukianova-Hleb, E. Y., Oginsky, A. O., Samaniego, A. P. et al. 2011. Tunable plasmonic nanoprobes for theranostics of prostate cancer. *Theranostics* 1:3–17.

Maali, A. and M. T. H. Mosavian. 2013. Preparation application of nanoemulsions in the last decade (2000–2010). *J Dispers Sci Tech* 34:92–105.

MacKay, J. A., Chen, M. N., McDanie, J. R. et al. 2009. Self-assembling chimeric polypeptide-doxorubicin conjugate nanoparticles that abolish tumours after a single injection. *Nat Mater* 8:993–9.

Maeda, H., Sawa, T., and T. Konno. 2001. Mechanism of tumor-targeted delivery of macromolecular drugs, including the EPR effect in solid tumor and clinical overview of the prototype polymeric drugs ManCs. *J Control Release* 74:47–61.

Mahmud, A., Xiong, X. B., Aliabadi, H. M. et al. 2007. Polymeric micelles for drug targeting. *J Drug Target* 15:553–84.

Main, M. L., Goldman, J. H., and P. A. Grayburn. 2009. Ultrasound contrast agents: Balancing safety versus efficacy. *Expert Opin Drug Saf* 8:49–56.

Matsumura, Y. 2011. Preclinical and clinical studies of NK012, an SN-38-incorporating polymeric micelles, which is designed based on EPR effect. *Adv Drug Deliv Rev* 63:184–92.

Matsunaga, K., Hara, K., Imamura, T. et al. 2005. Technetium labeling of dextran incorporating cysteamine as a ligand. *Nucl Med Biol* 32:279–85.

McCarthy, J. R., Perez, J. M., Bruckner, C. et al. 2005. Polymeric nanoparticle preparation that eradicates tumors. *Nano Letter* 5:2552–56.

McClements, D. J. 2012. Nanoemulsions versus microemulsions: Terminology, differences, and similarities. *Soft Mat* 8:1719–29.

McDaniel, J. R., Callahan, D. J., and A. Chilkoti. 2010. Drug delivery to solid tumors by elastin-like polypeptides. *Adv Drug Deliv Rev* 62:1456–67.

McDevitt, M. R., Chattopadhyay, D., Kappel, B. J. et al. 2007. Tumor targeting with antibody-functionalized, radiolabeled carbon nanotubes. *J Nuclear Med* 48:1180–9.

Mehnert, W. and K. Mader. 2001. Solid lipid nanoparticles: Production, characterization and applications. *Adv Drug Deliv Rev* 47:165–96.

Melancon, M. P., Lu, W., Yang, Z. et al. 2008. *In vitro* and *in vivo* targeting of hollow gold nanoshells directed at epidermal growth factor receptor for photothermal ablation therapy. *Mol Cancer Ther* 7:1730–9.

Mendes, A. I., Silva, A. C., Catita, J. A. M. et al. 2013. Miconazole-loaded nanostructured lipid carriers (NLC) for local delivery to the oral mucosa: Improving antifungal activity. *Colloids Surf B: Biointerfaces* 111:755–63.

Meng, H., Liong, M., Xia, T. et al. 2010. Engineered design of mesoporous silica nanoparticles to deliver doxorubicin and P-glycoprotein siRNA to overcome drug resistance in a cancer cell line. *ACS Nano* 4:4539–50.

Mi, Y., Liu, Y., and S. S. Feng. 2011. Formulation of docetaxel by folic acid-conjugated d-α-tocopheryl polyethylene glycol succinate 2000 (Vitamin E TPGS(2k)) micelles for targeted and synergistic chemotherapy. *Biomaterials* 32:4058–66.

Mitra, A., Mulholland, J., Nan, A. et al. 2005. Targeting tumor angiogenic vasculature using polymer-RGD conjugates. *J Control Release* 102:191–201.

Mitra, R. N., Doshi, M., Zhang, X. et al. 2012. An activatable multimodal/multi-functional nanoprobe for direct imaging of intracellular drug delivery. *Biomaterials* 33:1500–8.

Mitri, K., Vauthier, C., Huang, N. et al. 2012. Scale-up of nanoemulsion produced by emulsification and solvent diffusion. *J Pharm Sci* 101:4240–7.

Moghimi, S. M., Hunter, A. C., and J. C. Murray. 2001. Long-circulating and target-specific nanoparticles: Theory to practice. *Pharmacol Rev* 53:283–318.

Moghimi, S. M., Hunter, A. C., Murray, J. C. et al. 2004. Cellular distribution of nonionic micelles. *Science* 303:626–8.

Mohanand, P. and N. Rapoport. 2010. Doxorubicin as a molecular nanotheranostic agent: Effect of doxorubicin encapsulation in micelles or nanoemulsions on the ultrasound-mediated intracellular delivery and nuclear trafficking. *Mol Pharm* 7:1959–73.

Mok, H. and T. G. Park. 2012. Hybrid polymeric nanomaterials for siRNA delivery and imaging. *Macromol Biosci* 12:40–8.

Moon, G. D., Choi, S. W., Cai, X. et al. 2011. A new theranostic system based on gold nanocages and phase-change materials with unique features for photoacoustic imaging and controlled release. *J Am Chem Soc* 133:4762–5.

Moon, H. K., Lee, S. H., and H. C. Choi. 2009. *In vivo* near-infrared mediated tumor destruction by photothermal effect of carbon nanotubes. *ACS Nano.* 3:3707–13.

Mornet, S., Vasseur, S., Fabien, G. et al. 2004. Magnetic nanoparticle design for medical diagnosis and therapy. *J Mater Chem* 14:2161–75.

Mu, L. and S. S. Feng. 2003. PLGA/TPGS nanoparticles for controlled release of paclitaxel: Effects of the emulsifier and drug loading ratio. *Pharm Res* 20:1864–72.

Mukherjee, S., Ray, S., and R. S. Thakur. 2009. Solid lipid nanoparticles: A modern formulation approach in drug delivery system. *Indian J Pharm Sci* 71:349–58.

Muller, R. H., Harden, D., and C. M. Keck. 2012. Development of industrially feasible concentrated 30% and 40% nanoemulsions for intravenous drug delivery. *Drug Dev Ind Pharm* 38:420–30.

Muller, R. H., Rühl, D., Runge, S. et al. 1997. Cytotoxicity of solid lipid nanoparticles as a function of the lipid matrix and the surfactant. *Pharm Res* 14:458–62.

Muthu, M. S., Kulkarni, S. A., Raju, A. et al. 2012. Theranostic liposomes of TPGS coating for targeted co-delivery of docetaxel and quantum dots. *Biomaterials* 33:3494–501.

Muthu, M. S., Leong, D. T., Mei, L. et al. 2014. Nanotheranostics—Application and further development of nanomedicine strategies for advanced theranostics. *Theranostics* 4:660–77.

Muthu, M. S. and S. Singh. 2009. Targeted nanomedicines: Effective treatment modalities for cancer, AIDS and brain disorders. *Nanomedicine (Lond)* 4:105–18.

Müller, R. H., Mäder, K., and S. Gohla. 2000. Solid lipid nanoparticles (SLN) for controlled drug delivery—A review of the state of the art. *Eur J Pharm Biopharm* 50:161–77.

Na, J. H., Koo, H., Lee, S. et al. 2011. Real-time and non-invasive optical imaging of tumor-targeting glycol chitosan nanoparticles in various tumor models. *Biomaterials* 32:5252–61.

Nakamura, H., Etrych, T., Chytil, P. et al. 2014. Two step mechanisms of tumor selective delivery of N-(2-hydroxypropyl) methacrylamide copolymer conjugated with pirarubicin via an acid-cleavable linkage. *J Control Release* 174:81–7.

Nie, Y., Ji, L., Ding, H. et al. 2012. Cholesterol derivatives based charged liposomes for doxorubicin delivery: Preparation, *in vitro* and *in vivo* characterization. *Theranostics* 2:1092–103.

Niidome, T., Yamagata, M., Okamoto, Y. et al. 2006. PEG-modified gold nanorods with a stealth character for *in vivo* applications. *J Control Release* 114:343–7.

Oliveira, J. M., Salgado, A. J., Sousa, N. et al. 2010. Dendrimers and derivatives as a potential therapeutic tool in regenerative medicine strategies: A review. *Prog Polym Sci* 35:1163–94.

Ooya, T., Lee, J., and K. Park. 2004. Hydrotropic dendrimers of generations 4 and 5: Synthesis, characterization, and hydrotropic solubilization of paclitaxel. *Bioconjug Chem* 15:1221–9.

Paciotti, G. F., Myer, L., Weinreich, D. et al. 2004. Colloidal gold: A novel nanoparticle vector for tumor directed drug delivery. *Drug Deliv* 11:169–83.

Pan, H., Soman, N. R., and P. H. Schlesinger. 2011. Cytolytic peptide nanoparticles ("nanoBees") for cancer therapy. *Wiley Interdiscip Rev Nanomed Nanobiotechnol* 3:318–27.

Pan, J., Liu, Y., and S. S. Feng. 2010. Multifunctional nanoparticles of biodegradable copolymer blend for cancer diagnosis and treatment. *Nanomedicine (Lond)* 5:347–60.

Papahadjopoulos, D., Allen, T. M., Gabizon, A. et al. 1991. Sterically stabilized liposomes—Improvements in pharmacokinetics and antitumor therapeutic efficacy. *Proc Natl Acad Sci USA* 88:11460–4.

Park, S. Y., Baik, H. J., Oh, Y. T. et al. 2011. A smart polysaccharide/drug conjugate for photodynamic therapy. *Angew Chem Int Ed Engl* 50:1644–7.

Patri, A. K., Myc, A., Beals, J. et al. 2004. Synthesis and *in vitro* testing of J591 antibody-dendrimer conjugates for targeted prostate cancer therapy. *Bioconjug Chem* 15:1174–81.

Peng, C. L., Shih, Y. H., Lee, P. C. et al. 2011. Multimodal image-guided photothermal therapy mediated by Re-188-labeled micelles containing a cyanine-type photosensitizer. *ACS Nano* 5:5594–607.

Pinto-Alphandary, H., Andremont, A., and P. Couvreur. 2003. Targeted delivery of antibiotics using liposomes and nanoparticles: Research and applications. *Int J Antimicrob Agents* 13:155–68.

Porter, A. E., Gass, M., Muller, K. et al. 2007. Direct imaging of single-walled carbon nanotubes in cells. *Nat Nanotechnol* 2:713–7.

Qian, X., Peng, X. H. Ansari, D. O. et al. 2008. *In vivo* tumor targeting and spectroscopic detection with surface-enhanced raman nanoparticle tags. *Nat Biotechnol* 26:83–90.

Qin, H., Zhou, T., Yang, S. et al. 2013. Gadolinium(iii)-gold nanorods for MRI and photoacoustic imaging dual-modality detection of macrophages in atherosclerotic inflammation. *Nanomedicine* 8:1611–24.

Qin, S. Y., Feng, J., Rong, L. et al. 2014. Theranostic GO-based nanohybrid for tumor induced imaging and potential combinational tumor therapy. *Small* 10:599–608.

Rajpoot, P., Pathak, K., and V. Bali. 2011. Therapeutic applications of nanoemulsion based drug delivery systems: A review of patents in last two decades. *Recent Pat Drug Deliv Formul* 5:163–72.

Rapoport, N. 2012. Phase-shift, stimuli-responsive perfluorocarbon nanodroplets for drug delivery to cancer. *Wiley Interdiscip Rev Nanomed Nanobiotechnol* 4:492–510.

Rapoport, N., Nam, K. H., Gupta, R. et al. 2011. Ultrasound-mediated tumor imaging and nanotherapy using drug loaded, block copolymer stabilized perfluorocarbon nanoemulsions. *J Control Release* 153:4–15.

Rapoport, N. Y., Christensen, D. A., Fain, H. D. et al. 2004. Ultrasound-triggered drug targeting of tumors *in vitro* and *in vivo*. *Ultrasonics* 42:943–50.

Rastogi, R., Gulati, N., Kotnala, R. K. et al. 2011. Evaluation of folate conjugated PEGylated thermosensitive magnetic nanocomposites for tumor imaging and therapy. *Colloids Surf B Biointerfaces* 82:160–7.

Ren, Y., Kang, C. S., Yuan, X. B. et al. 2010. Co-delivery of as-mir-21 and 5-FU by poly(amidoamine) dendrimer attenuates human glioma cell growth *in vitro*. *J Biomater Sci Polym Ed* 21:303–14.

Resch-Genger, U., Grabolle, M., Cavaliere-Jaricot, S. et al. 2008. Quantum dots versus organic dyes as fluorescent labels. *Nat Methods* 5:763–75.

Robinson, J. T., Welsher, K., Tabakman, S. M. et al. 2010. High performance *in vivo* near-IR (>1 μm) imaging and photothermal cancer therapy with carbon nano-tubes. *Nano Res* 3:779–93.

Rowe, M. D., Thamm, D. H., Kraft, S. L. et al. 2009. Polymer-modified gadolinium metal-organic framework nanoparticles used as multifunctional nanomedicines for the targeted imaging and treatment of cancer. *Biomacromolecules* 10:983–93.

Saad, M., Garbuzenko, O. B., Ber, E. et al. 2008. Receptor targeted polymers, dendrimers, liposomes: Which nanocarrier is the most efficient for tumor-specific treatment and imaging? *J Control Release* 130:107–14.

Santra, S., Kaittanis, C., Grimm, J. et al. 2009. Drug/dye-loaded, multifunctional iron oxide nanoparticles for combined targeted cancer therapy and dual optical/magnetic resonance imaging. *Small* 5:1862–8.

Santra, S. and J. M. Perez. 2011. Selective N-alkylation of β-alanine facilitates the synthesis of a poly(amino acid)-based theranostic nano agent. *Biomacromolecules* 12:3917–27.

Sarker, D. K. 2005. Engineering of nanoemulsions for drug delivery. *Curr Drug Deliv* 2:297–310.

Sawant, R. R., Jhaveri, A. M., Koshkaryev, A. et al. 2014. Targeted transferrin-modified polymeric micelles: Enhanced efficacy *in vitro* and *in vivo* in ovarian carcinoma. *Mol Pharm* 11:375–81.

Sharma, A. and U. S. Sharma. 1997. Liposomes in drug delivery: Progress and limitations. *Int J Pharm* 154:123–40.

Shen, H., Zhang, L., Liu, M. et al. 2012. Biomedical applications of graphene. *Theranostics* 2:283–94.

Shuhendler, A. J., Prasad, P., Leung, M. et al. 2012. A novel solid lipid nanoparticle formulation for active targeting to tumor α(v) β(3) integrin receptors reveals cyclic RGD as a double-edged sword. *Adv Healthcare Mater* 1:600–8.

Shukla, R., Thomas, T. P., Peters, J. et al. 2005. Tumor angiogenic vasculature targeting with PAMAM dendrimer-RGD conjugates. *Chem Commun (Camb)* (46):5739–41.

Singh, A., Dilnawaz, F., Mewar, S. et al. 2011. Composite polymeric magnetic nanoparticles for co-delivery of hydrophobic and hydrophilic anticancer drugs and MRI imaging for cancer therapy. *ACS Appl Mater Interfaces* 3:842–56.

Song, H., He, R., Wang, K. et al. 2010. Anti-HIF-1 alpha antibody-conjugated pluronic triblock copolymers encapsulated with paclitaxel for tumor targeting therapy. *Biomaterials* 31:2302–12.

Soppimath, K. S., Aminabhavi, T. M., Kulkarni, A. R. et al. 2001. Biodegradable polymeric nanoparticles as drug delivery devices. *J Control Release* 70:1–20.

Stoll, G., Basse-Lusebrink, T., Weise, G. et al. 2012. Visualization of inflammation using (19) F-magnetic resonance imaging and perfluorocarbons. *Wiley Interdiscip Rev Nanomed Nanobiotechnol* 4:438–47.

Surace, C., Arpicco, S., and A. Dufay-Wojcicki et al.. 2009. Lipoplexes targeting the CD44 hyaluronic acid receptor for efficient transfection of breast cancer cells. *Mol Pharm* 6:1062–73.

Svenson, S. 2013. Theranostics: Are we there yet? *Mol Pharm* 10:848–56.

Tan, Y. F., Chandrasekharan, P., Maity, D. et al. 2011. Multimodal tumor imaging by iron oxides and quantum dots formulated in poly (lactic acid)-d-alpha-tocopheryl polyethylene glycol 1000 succinate nanoparticles. *Biomaterials* 32:2969–78.

Tanand, M. and Z. R. Lu. 2011. Integrin targeted MR imaging. *Theranostics* 1:83–101.

Taratula, O., Schumann, C., Naleway, M. A. et al. 2013. A multifunctional theranostic platform based on phthalocyanine-loaded dendrimer for image guided drug delivery and photodynamic therapy. *Mol Pharm* 10:3946–58.

Torchilin, V. P. 1996. Liposomes as delivery agents for medical imaging. *Mol Med Today* 2:242–9.

Torchilin, V. P. 2005. Recent advances with liposomes as pharmaceutical carriers. *Nat Rev Drug Discov* 4:145–60.

Torchilin, V. P., Lukyanov, A. N., Gao, Z. et al. 2003. Immunomicelles: Targeted pharma-ceutical carriers for poorly soluble drugs. *Proc Natl Acad Sci USA* 100:6039–44.

Vaidya, A., Sun, Y., Ke, T. et al. 2006. Contrast enhanced MRI-guided photodynamic therapy for site-specific cancer treatment. *Magn Reson Med* 56:761–7.

Van Schooneveld, M. M., Cormode, D. P., Koole, R. et al. 2010. A fluorescent, paramagnetic and PEGylated gold/silica nanoparticle for MRI, CT and fluorescence imaging. *Contrast Media Mol Imaging* 5:231–6.

Vasey, P. A., Kaye, S. B., Morrision, R. et al. 1999. Phase I clinical and pharmacokinetic study of PK1 [N-(2-hydroxypropyl) methacrylamide copolymer doxorubicin]: First member of a new class of chemotherapeutic agents-drug-polymer conjugates. Cancer research campaign phase I/II committee. *Clin Cancer Res* 5:83–94.

Vijayakumar, M. R., Muthu, M. S., and S. Singh. 2013. Copolymers of poly(lactic acid) and D-α-tocopheryl polyethylene glycol 1000 succinate-based nanomedicines: Versatile multifunctional platforms for cancer diagnosis and therapy. *Expert Opin Drug Deliv* 10:529–43.

Voinea, M. and M. Simionescu. 2002. Designing of "intelligent" liposomes for efficient delivery of drugs. *J Cell Mol Med* 6:465–74.

Vriezema, D. M., Comellas, A. M., Elemans, J. A. et al. 2005. Self-assembled nanoreactors. *Chem Rev* 105:1445–89.

Wagner, D. S., Delk, N. A., Lukianova-Hleb, E. Y. et al. 2010. The *in vivo* performance of plasmonic nano-bubbles as cell theranostic agents in zebrafish hosting prostate cancer xenografts. *Biomaterials* 31:7567–74.

Wang, L. S., Chuang, M. C., and J. A. Ho. 2012. Nanotheranostics—A review of recent publications. *Int J Nanomedicine* 7:4679–95.

Wang, Z., Niu, G., and X. Chen. 2014. Polymeric materials for theranostic applications. *Pharm Res* 31:1358–76.

Weise, G., Basse-Luesebrink, T. C., Wessig, C. et al. 2011. *In vivo* imaging of inflammation in the peripheral nervous system by (19)F MRI. *Exp Neurol* 229:494–501.

Wen, C. J., Sung, C. T., Aljuffali, I. A. et al. 2013. Nanocomposite liposomes containing quantum dots and anticancer drugs for bioimaging and therapeutic delivery: A comparison of cationic, PEGylated and deformable liposomes. *Nanotechnology* 24:325101.

Wen, C. J., Zhang, L. W., Al-Suwayeh, S. A. et al. 2012. Theranostic liposomes loaded with quantum dots and apomorphine for brain targeting and bioimaging. *Int J Nanomedicine* 7:1599–611.

Wen, X. X., Jackson, E. F., Price, R. E. et al. 2004. Synthesis and characterization of poly(L-glutamic acid) gadolinium chelate: A new b iodegradable MRI contrast agent. *Bioconjug Chem* 15:1408–15.

Wissing, S. A., Kayser, O., and R. H. Müller. 2004. Solid lipid nanoparticles for parenteral drug delivery. *Adv Drug Del Rev* 56:1257–72.

Wu, X., Ding, B., Gao, J. et al. 2011. Second-generation aptamer-conjugated psMa-targeted delivery system for prostate cancer therapy. *Int J Nanomed* 6:1747–56.

Wu, X., Ming, T., Wang, X. et al. 2009. High-photoluminescence-yield gold nanocubes: For cell imaging and photothermal therapy. *ACS Nano* 4:113–20.

Xia, Y., Li, W., Cobley, C. M. et al. 2011a. Gold nanocages: From synthesis to theranostic applications. *Acc Chem Res* 44:914–24.

Xie, H., Wang, Z. J., and Bao, A. et al. 2010a. *In vivo* PET imaging and biodistribution of radiolabeled gold nanoshells in rats with tumor xenografts. *Int J Pharm* 395:324–30.

Xie, J., Lee, S., and X. Chen. 2010b. Nanoparticle-based theranostic agents. *Adv Drug Deliv Rev* 62:1064–79.

Xu, C., Tung, G. A., and S. Sun. 2008. Size and concentration effect of gold nanoparticles on x-ray attenuation as measured on computed tomography. *Chem Mater* 20:4167–9.

Xu, Y. Z., Fan, H. J., Lu, C. P. et al. 2010. Synthesis of galabiose-chitosan conjugate as potent inhibitor of *Streptococcus suis* adhesion. *Biomacromolecules* 11:1701–4.

Yallapu, M. M., Othman, S. F., Curtis, E. T. et al. 2011. Multi-functional magnetic nanoparticles for magnetic resonance imaging and cancer therapy. *Biomaterials* 32:1890–905.

Yang, K., Feng, L., Shi, X. et al. 2013. Nano-graphene in biomedicine: Theranostic applications. *Chem Soc Rev* 42:530–47.

Yang, K., Zhang, S., Zhang, G. et al. 2010. Graphene in mice: Ultrahigh *in vivo* tumor uptake and efficient photothermal therapy. *Nano Letter* 10:3318–23.

Yen, S. K., Padmanabhan, P., and S. T. Selvan. 2013. Multifunctional iron oxide nanoparticles for diagnostics, therapy and macromolecule delivery. *Theranostics* 3:986–1003.

Yi, D. K., Sun, I. C. Ryu, J. H. et al. 2010. Matrix metalloproteinase sensitive gold nanorod for simultaneous bioimaging and photothermal therapy of cancer. *Bioconjug Chem* 21:2173–7.

Yuan, J., Zhang, H., Kaur, H. et al. 2013. Synthesis and characterization of theranostic poly(HPMA)-c(RGDyK)-DOTA-64Cu copolymer targeting tumor angiogenesis: Tumor localization visualized by positron emission tomography. *Mol Imaging* 12:1–10.

Zarabi, B., Borgman, M. P., Zhuo, J. C. et al. 2009. Noninvasive monitoring of HPMA copolymer-RGDfK conjugates by magnetic resonance imaging. *Pharm Res* 26:1121–9.

Zhang, R., Wu, C., Tong, L. et al. 2009. Multifunctional core-shell nanoparticles as highly efficient imaging and photosensitizing agents. *Langmuir* 25:10153–8.

Zhao, J., Mi, Y., and S. S. Feng. 2013. Targeted co-delivery of docetaxel and siPlk1 by her-ceptin-conjugated vitamin E TPGS based immunomicelles. *Biomaterials* 34:3411–21.

Zoller, M. 1995. CD44-physiological expression of distinct isoforms as evidence for organ-specific metastasis formation. *J Mol Med* 73:425–38.

3
Nanodevices for Early Diagnosis of Cardiovascular Disease: Advances, Challenges, and Way Ahead

Alok Pandya and Madhuri Bollapalli

CONTENTS

3.1 Understanding Cardiovascular Disease 70
    3.1.1 What Are Cardiovascular Diseases? 70
3.2 What Are the Risk Factors for Cardiovascular Disease? 72
3.3 Cardiovascular Disease in the Developing World 73
    3.3.1 Why Is the Rate of CVD Increasing in the Developing World? 75
3.4 Current Tools for Diagnosis of CVDs 76
    3.4.1 Electrocardiography 77
    3.4.2 Imaging Techniques 77
    3.4.3 Echocardiography 78
    3.4.4 Coronary Angiography 78
    3.4.5 Immunoassay Technique 78
3.5 Biomarkers and Their Role in Cardiovascular Risk 80
3.6 Detection of Cardiac Biomarkers 82
    3.6.1 Enzyme-Linked Immune Sorbent Assay 82
    3.6.2 Chemiluminescence-Based Immunoassays 83
    3.6.3 Fluorescence-Based Immunoassays 84
    3.6.4 Electrical Detection-Based Assay 85
    3.6.5 Surface Plasmon Resonance-Based Detection 88
    3.6.6 Colorimetric Detection-Based Assay 90
3.7 Point-of-Care Assay and Technology in the Market 90
    3.7.1 Point-of-Care Devices Currently Available on the Market 93
    3.7.2 Point of Care under Development in the Laboratory 95
    3.7.3 Microfluidic-Based Immunoassay and Chip-Based Devices 96
    3.7.4 Paper-Based Devices 96
    3.7.5 Aptamer 96
3.8 Challenges and Way Ahead 97
Acknowledgments 98
References 99
Suggested Further Reading 103

3.1 Understanding Cardiovascular Disease

Cardiovascular disease is a term that refers to more than one disease of the circulatory system, including the heart and blood vessels. CVD has threatened human lives for decades and is still the leading cause of death globally. The incidence of cardiovascular disease is increasing because of the increase in major risk factors such as diabetes, obesity, rise in age of the population, and so on, thereby posing a challenge for developing efficient treatment (Braunwald, 1997). To begin with, we need to understand the cardiovascular system, also called the *circulatory system*, which moves blood throughout the human body. It is composed of the heart, arteries, veins, and capillaries. It transports oxygenated blood from the lungs and heart throughout the whole body through the arteries. Blood is pumped out from the heart, transported through arteries, and finally reaches the capillaries in all tissues and organs After delivering oxygen and nutrients and having collecting waste products, blood is brought back to the right chambers of the heart through a system of veins. During circulation through the liver, waste products are removed. Cardiovascular disease is a kind of illness involving the blood vessels, including veins, arteries and capillaries, or the heart, or all (Dahlöf, 2010). Thus, cardiovascular disease is termed diseases of the heart and blood vessels. These are usually acute events and mainly caused by a blockage that prevents blood flow to the heart or brain, essentially due to a buildup of fatty deposits (waxy substances) on the inner walls of the blood vessels that supply blood to the heart or brain. Moreover, strokes could occur because of bleeding from a blood vessel in the brain or from a blood clot. The term CVD commonly includes diseases such as coronary heart disease, heart failure, congenital heart disease, cardiomyopathy, peripheral vascular disease, and stroke (Lilly and Braunwald, 2012; Labarthe and Dunbar, 2012), which will be discussed further.

3.1.1 What Are Cardiovascular Diseases?

Cardiovascular diseases are a group of disorders of the heart and blood vessels. The human body, indicating the particular body area generally associated with the various cardiovascular diseases, is illustrated in Figure 3.1 (Levine, 2016). The types of CVD are described below:

1. *Coronary heart disease*—Also known as ischemic heart disease, this arises in two major forms: a heart attack, known as acute myocardial infarction, and angina. *Acute myocardial infarction* (AMI) is caused when the blood supply to the heart is blocked completely, often causing damage to the heart muscle and its function. *Angina* is a chronic condition where short episodes of chest pain occur periodically, caused by a temporary shortage of blood supply to the heart.
2. *Cerebrovascular disease*—Stroke is a type of cerebrovascular disease and occurs when an artery supplying blood to the brain either suddenly becomes blocked or begins to bleed. This may result in part of the brain dying, leading to sudden impairment of one or more capacities, such as speaking, thinking, and/or movement.
3. *Peripheral arterial disease*—This term refers to disease of the large arteries that supply blood to the peripheries and can be caused by blockage of arteries due to cholesterol or fatty substances, or by widening of the arteries such as the aorta, which in severe

Nanodevices for Early Diagnosis of Cardiovascular Disease

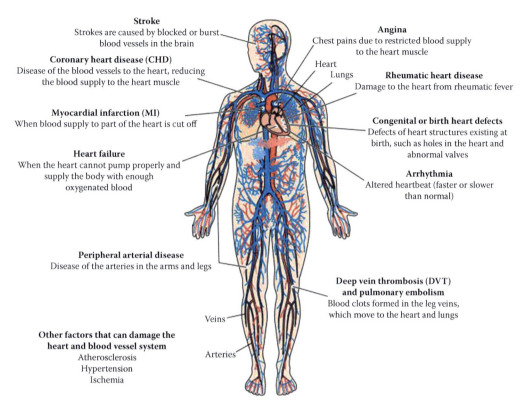

FIGURE 3.1
Outline of the human body indicating the particular body area generally associated with various cardiovascular diseases. (Adapted from http://www.open.edu/openlearn/ocw/mod/oucontent/view.php?printable=1&id=2613.)

cases can lead to rupture of the arterial wall. It mainly occurs in blood vessels supplying the arms and legs.

4. *Rheumatic heart disease (RHD)*—Rheumatic heart disease is a chronic heart condition caused by rheumatic fever, which is caused by a preceding group A streptococcal (strep) infection. It causes damage to the heart muscle and heart valves from rheumatic fever, caused by streptococcal bacteria. RHD is the most common acquired heart disease in children in many countries of the world, especially in developing countries.

5. *Congenital heart disease*—It is an abnormality of the heart structure or any defect of the heart or central blood vessels present at birth. It includes abnormalities of the heart or heart valves, such as a hole between chambers of the heart, narrowing of major blood vessels, or combinations of disorders. These defects can include the walls of the heart, the valves of the heart, and the arteries and veins near the heart. Congenital heart defects alter the normal flow of blood through the heart.

6. *Deep vein thrombosis and pulmonary embolism*—A pulmonary embolism (PE) usually happens when a blood clot called a deep vein thrombosis (DVT), often in the leg, travels to the lungs and blocks a blood vessel, which leads to low oxygen levels in the blood. Deep vein thrombosis is a serious condition that happens when a blood

clot is formed in a vein located deep inside the body. Deep vein blood clots generally form in the thigh or lower leg, but may also develop in other areas of the body. A major and life-threatening complication of DVT is pulmonary embolism. A PE occurs when a blood clot breaks off, travels through the blood stream, and lodges in the lung.

3.2 What Are the Risk Factors for Cardiovascular Disease?

There are many risk factors associated with coronary heart disease and stroke. Nonmodifiable and modifiable risk factors can increase the probability of developing CVD. "Nonmodifiable" risk factors are those that cannot be changed. They include:

1. Age—Risk increases as one gets older.
2. Gender—Before the age of 60, men are at greater risk for CVD than women.
3. Family history—Risk of developing CVD may increase if close blood relatives experienced early heart disease.

Modifiable risk factors are those that can be prevented, treated, and controlled. They include:

1. *Hypertension (high blood pressure)*—High blood pressure is harmful to the arteries and increases the risk of heart attack, heart failure, and stroke. The condition tends to run in families, but is also influenced by lifestyle. To prevent blood pressure from rising, it is important to achieve and maintain a healthy body weight, keep alcohol intake moderate, reduce salt intake, manage and reduce stress, and be physically active. If these measures fail, there are drugs that are effective in reducing elevated blood pressure. It is the number one major risk factor for CVD by far. If hypertension is poorly controlled, the artery walls may become damaged, which raises the risk of developing a blood clot.
2. *Radiation therapy*—Scientists from the Karolinska Institute, Sweden, reported that radiation therapy can increase the risk of cardiovascular disease later in life (a new study published in *The New England Journal of Medicine*).
3. *Smoking*—Habitual smoking can narrow the blood vessels, especially the coronary arteries.
4. *Lack of sleep*—People who sleep less than 7.5 hours each day have a higher risk of developing cardiovascular disease, as reported by researchers from Jichi Medical University, Tochigi, Japan.
5. *Hyperlipidemia (high blood cholesterol)*—In increased cholesterol, there is a higher chance of narrowing of the blood vessels and blood clots. If body accumulates too much cholesterol, it can become deposited in the walls of arteries, which become damaged and may become blocked. This could result in a heart attack.
6. *Having a partner with cancer*—A person whose partner has cancer has a nearly 30% higher risk of developing stroke or coronary heart disease, which was investigated at the Centre for Primary Healthcare Research in Malmö, Sweden, and revealed in the journal *Circulation*.

7. *Diabetes*—People with both type 1 and 2 diabetes are at a much higher risk of CVD. High blood sugar levels can harm the arteries. People with type 2 diabetes are often overweight or obese, and this increases risk factors for cardiovascular disease. People with diabetes are two to four times more likely to die from heart disease than nondiabetics. Experts say that blood glucose control measurements can help predict a diabetes patient's cardiovascular disease risk.
8. *Unhealthy eating*—A diet that is high in fat combined with carbohydrates (mainly of fast foods) can accelerate the accumulation of fatty deposits inside the arteries, which raises the risk of obesity, hypertension, and hyperlipidemia. A diet lacking sufficient amounts of fruit, vegetables, fiber, whole grains, and essential nutrients is not good for cardiovascular health. The World Health Organization recommends that adults should consume less than 2 g of sodium per day, and a new study finds that sodium intake above this recommendation accounts for almost 1 in 10 cardiovascular deaths globally each year.
9. *Physical inactivity*—People who lead predominantly sedentary lives tend to have higher blood pressure, more stress hormones, and higher blood cholesterol levels, and are more likely to be overweight. These are all risk factors for cardiovascular diseases.
10. *Excessive alcohol consumption*—Consumption of excess alcohol is harmful to the heart and other organs. It can directly damage the heart muscle and cause irregular beating of the heart. Alcohol also contributes to weight gain, high triglycerides, high blood pressure, strokes, and so on.
11. *Stress*—Hormones associated with (mental) stress, such as cortisone, raise blood sugar levels. Stress can exacerbate symptoms in people with pre-existing heart disease, and can cause high blood pressure.
12. *Air pollution*—Belgian researchers reported in *The Lancet* that air pollution causes about the same number of heart attacks as other individual risk factors. The investigators assessed 36 separate studies that focused on air pollution.
13. *COPD and reduced lung function*—A study presented at the European Respiratory Society's Annual Congress in Amsterdam showed that people with chronic obstructive pulmonary disease (COPD) have a significant risk of developing cardiovascular disease. The researchers, from the Sunderby Hospital in Sweden, added that patients with reduced lung function are also at higher risk.
14. *The age of first menstruation*—Women who start menstruating early are more likely to become obese and have cardiovascular disease risk factors, researchers reported in the *Journal of Clinical Endocrinology & Metabolism*.
15. Having more than one risk factor suggests that the overall risk of CVD is much higher.

3.3 Cardiovascular Disease in the Developing World

Over the past two decade, the pervasiveness of CVD has made it the leading cause of death in both sexes and all races and ethnicities across the world today (Gaziano, 2005). In particular, CVD is a greater burden in developing countries than in developed countries (Figure 3.2). Therefore, the United Nations General Assembly (UNGA) has recognized CVD

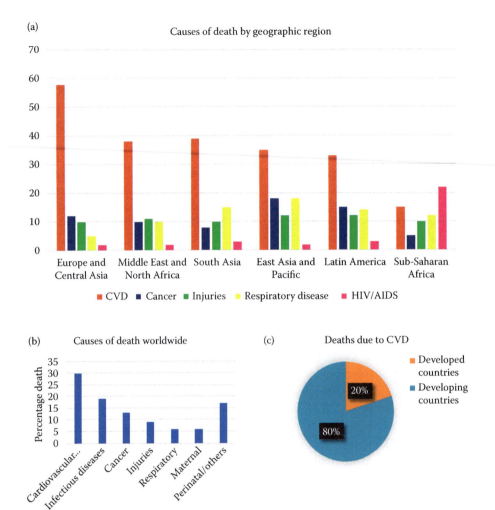

FIGURE 3.2
(a) Causes of death in the developing world, (b) Causes of death throughout the world, and (c) Comparison of CVD deaths in developed vs. developing countries. (Adapted from Gaziano, T. A. et al. 2005. *Circulation*, 112, 3547–3553.)

and other chronic conditions as "one of the major challenges for development in the twenty-first century" (Oliveira et al., 2015).

Among CVDs, atherosclerotic and hypertensive diseases have been projected as the leading cause of death for the year 2020 (Oliveira et al., 2015). As discussed earlier, CVD is the number-one cause of death among all races and ethnicities, and occurs almost equally in men and women as of today. An estimated 33% of all global deaths (20 million people) were due to CVD in 2014, and roughly over 80% of these take place in low- and middle-income countries (Oliveira et al., 2015). The high burden of premature deaths due to CVD is a great cause of concern in developing countries. For example, the proportion of CVD mortalities in 1995 occurring below the age of 70 years was 30% in developed countries compared with 48.9% in developing countries (Reddy and Yusuf, 1998). Many recent reports (2010–2012), like "Race Against Time," have projected a devastating impact of CVD, specifically on the population 35–64 years in age in low- and middle-income countries, and this is projected to increase by the year 2030. Recent data show that CVD places a significant burden on national economies. For example, in China, annual direct costs accrued from CVD are estimated at more than $40 billion (4% of gross national income) (Srinath Reddy, 1999).

3.3.1 Why Is the Rate of CVD Increasing in the Developing World?

The burden of CVD in developing countries has risen enormously in the last 30 years, and is likely to continue to rise, which creates a large healthcare and economic burden with potentially upsetting consequences. This increase in CVD burden has been driven by two important factors: urbanization (modernization) and the modern epidemiological transition.

The World Heart Federation (WHF) argues that modern urbanization in the developing world is mainly responsible for the rise in CVD. According to its report, only 10% of the population lived in cities in 1900, whereas 50% of the world's population lives in an urban setting today, and it is estimated that this proportion will reach 75% by 2050. Urban modernization leads to insanitary lifestyles through larger exposure to advertisements for tobacco and alcohol and greater access to high salt/fat content in food (Stewart, 2011). This modernization also leads to less physical activity and unhealthier diets. The lack of physical activity is cultivated by public transportation and lack of green space in modern cities. It further encourages a move away from traditional cooking to ready-prepared and processed food high in calories, salt, and fatty content. Unhealthy lifestyles often begin early in life, and the WHF estimates that there are currently 35 million overweight children living in developing countries (Stewart, 2011). Recent projections from the WHO suggest a reversal of socioeconomic gradients; tobacco use will account for 12.3% of global deaths by the year 2020, and a large component of this will be in the form of CVD deaths. In India alone, the attributable toll of tobacco is projected to rise from 1.4% in 1990 to 13.3% in 2020 (Vollset, 2006). Smoking is the direct reason for 1/10 of CVDs worldwide. The risk of nonfatal MI increases by 5.6% for every daily cigarette smoked (Smith et al., 2012), and doubles for chewing tobacco. Indeed, smoking bans have decreased the rates of heart attacks. Quitting smoking has been demonstrated to have a multitude of positive effects. There was a higher prevalence of CVD in the urban population when compared to their rural counterparts. Additionally, it was also noted that there was a rise in CVD due to higher body mass index, blood pressure, fasting lipids (total cholesterol, ratio of cholesterol to HDL cholesterol, triglycerides), and prevalence of diabetes (World Health Organization).

The modern epidemiological transition is the second reason for the rise in CVD mortality in the developing world. In the late twentieth century, most developing countries witnessed a rise in life expectancy in relation to efforts made to combat perinatal, nutritional, and infectious disorders. As a result, more children are living to adulthood, thus resulting in higher numbers of middle-aged and older adults. This decline in deaths from infectious causes and rise in deaths from chronic diseases is referred to as the "modern epidemiological transition" (Reddy, 1993). As the number of deaths from infectious, perinatal, and nutritional disorders declined, the number of deaths from NCDs (chiefly CVD) increased. Current estimates from epidemiologic studies from various parts of the country indicate the prevalence of coronary heart disease (CHD) is between 7% and 20% in urban and 2% and 7% in rural populations. CVD mortality is expected to increase based on both the increasing age of populations as well as urbanization and the subsequent rise in risk factor exposure in the developing countries. The demographic age shift in developing societies alone will result in an anticipated rise in CVD mortality. Because urbanization contributes to increased risk factor exposure, the rise in CVD mortality may even be larger than these estimates based solely on demographic shifts.

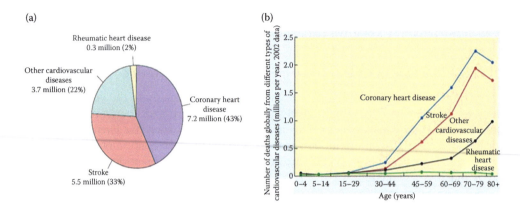

FIGURE 3.3
(a) Pie chart showing the proportion of global deaths from various cardiovascular diseases in 2002; (b) The number of deaths per year throughout the world for different types of cardiovascular diseases subdivided by age into different groups. As you can see, in early adulthood, coronary heart disease and stroke become the common causes of cardiovascular deaths for large numbers of people. (Adapted from http://www.open.edu/openlearn/ocw/mod/oucontent/view.php?printable=1&id=2613.)

To conclude, globally, cardiovascular diseases are the main cause of premature death (before the age of 75) in the UK, across Europe, and in the USA—and across many parts of the world (Figure 3.3). One-third of global deaths (16.7 million people) in 2002 were from cardiovascular diseases (WHO). Coronary heart disease is the single largest cause of death in the developed countries and is one of the leading causes of disease burden in developing countries. In 2001, there were 7.2 million deaths due to CHD worldwide, which occurred in the low- and middle-income countries. We, therefore, will further focus on coronary heart disease and the prevention and effective diagnostic technology of CHD.

Coronary heart disease, also known as coronary artery disease, is that in which a waxy substance called plaque builds up inside the coronary arteries (Ahmad, 2005). These arteries supply oxygen-rich blood to the heart muscle. When plaque builds up in the arteries, the condition is called atherosclerosis. The buildup of plaque occurs over many years. Atherosclerosis is a specific disease of the intima (inner layer) of the arteries, not necessarily related to aging, and is responsible for the most prevalent serious disease of the heart, coronary-artery disease. It typically affects the aorta and the arteries supplying blood to the heart, the brain, and the lower extremities. It should not be confused with arteriosclerosis, commonly known as hardening of the arteries, a degenerative process affecting the arteries that is a part of aging. It affects the arterial intima as well as the media (middle layer), producing abnormalities that generally do not have adverse effects on the flow of blood in the arteries. Today, doctors diagnose CHD based on a person's medical and family histories, risk factors for CHD, physical examination, and the results from tests and procedures. No single test can diagnose CHD. If a physician suspects CHD, he or she may recommend one or more of the tests that will be discussed further.

3.4 Current Tools for Diagnosis of CVDs

People have long been alarmed about heart disease due to unexpected and sudden death after chest pain on the left side. Many scientists have found methods of instrumental diagnosis of heart disease. Only electrophysics achievements made possible the creation of pioneer methods of heart function recording. The current diagnosis of CVD is dominated by either expensive imaging techniques or risky invasive techniques. Sophisticated imaging techniques like MRI or ultrafast CT require expensive equipment and highly skilled staff. Some affordable, user-friendly, and less invasive diagnostic approaches have been developed as alternatives, but these come with their own limitations. A more clinical diagnostic approach is presented by scoring methods like the Framingham study or the Prospective Cardiovascular Münster Study (PROCAM), which considers several simple risk factors and classical symptoms of myocardial ischemia. Current methods demand medical professionals to focus their resources on people complaining of chest pains reporting at emergency department as having a potential AMI (Cai, 2005). This leads to situations where patients with milder CVDs are needlessly admitted and tested for possible heart attack. Driven by need, much recent research has focused on identifying individuals most at risk of primary cardiovascular events at the point of care (Cai, 2005). This section will continue with some of the currently used diagnostic techniques for cardiovascular diseases. It will highlight the advantages and challenges associated with the current methods.

3.4.1 Electrocardiography

Electrocardiography (ECG) is an objective method for detecting potential differences in the heartbeat. Electrocardiography is the most commonly used diagnostic apparatus for cardiovascular diseases, as it is affordable and widely available. It is based on the electrical changes that occur temporally as the heart completes its usual cycle. These changes can be monitored in healthy sate and vary when compared to those in certain disease states (Abidov, 2016). There are two major tests that are carried out using the electrocardium: resting and exercise testing. Resting testing is commonly used to rule out rather than rule in CHD, as it has relatively low sensitivity (Abidov, 2016). Exercise testing, also known as stress testing, is performed under conditions that aim to exasperate stenosis. Patients will usually have an ECG under pharmacologically induced stress or on a treadmill. This procedure is important in the diagnosis of angina. Exercise tests are usually performed following an indicative resting ECG or to aid risk assessment and disease state following AMI to allow the clinician to tailor rehabilitation and management for individual patients. The clinician monitors the patient's condition not only by examining the graphical data but also by observation of the patient's physical condition.

3.4.2 Imaging Techniques

Magnetic resonance imaging is an imaging technique that has significant benefits in the diagnosis of CVD, as it is capable of providing information on plaque composition. According to Adame et al. (2006), there are many applications of MRI in CVD diagnosis. Contrast-enhanced magnetic resonance angiography (CE-MRA) uses materials such as gadolinium, which has desirable magnetic properties, to give greater contrast to vasculature, clearly denoting stenosis. Another MRI technique, vessel wall MRI (VWMRI), which can be used to detect vascular remodeling, has proven to be an excellent diagnostic tool. MRI can also give insight into the pathological state of an atheromatous plaque as well as elements of composition and thickness of the fibrous cap; thus, it can be used to more definitively identify patients at high risk of AMI or CVA. The major drawback of MRI is the cost. It is not suitable for patients with metallic stents or pacemakers, as it utilizes a strong magnet (Adame et al., 2006). Computed tomography, also commonly known as a CT scan, is routinely used in the diagnosis of cardiovascular diseases. It combines multiple x-ray images obtained from a ring of rotating x-rays with the aid of a computer to produce cross-sectional views of the body. Cardiac CT is a heart-imaging test that uses CT technology to visualize the heart anatomy, coronary circulation, and great vessels (Boyd and Lipton, 1983). The quality of the results obtained is similar to those of the MRI. A modified variation of CT, which is becoming more popular with the diagnosis of cardiovascular diseases, is ultrafast computer tomography (UFCT), also called electron beam CT. It varies from conventional CT scans because it does not have a rotating x-ray assembly; it has a ring of fixed crystal detectors and another fixed ring of tungsten x-ray targets, which act together. A major disadvantage of the CT scan compared to its rival, the MRI, is that it uses x-rays, which have long been associated with cancer.

3.4.3 Echocardiography

Echocardiography is a method involving ultrasound examination of the heart. Echocardiography (echo) uses sound waves to create a moving picture of the heart (Breathnach and Westphal, 2006). The taken picture shows the size and shape of the heart and how well the heart chambers and valves are working. Echo can also show areas of poor blood flow to the heart, areas of heart muscle that are not contracting normally, and previous injury to the heart muscle caused by poor blood flow.

3.4.4 Coronary Angiography

Coronary angiography (CAG) is a type of x-ray examination based on the injection of contrast substance in the coronary vessels (Stefanini and Windecker, 2015). Sometimes, doctors recommend coronary angiography if other tests or factors indicate the possibility of the patient having CHD. This test uses a dye and special x-rays to show the insides of coronary arteries. To get the dye into the coronary arteries, the doctor will use a procedure called cardiac catheterization. A thin, flexible tube called a catheter is put into a blood vessel in the arm, groin (upper thigh), or neck. The tube is threaded into the coronary arteries, and the dye is released into the bloodstream. Special x-rays are taken while the dye is flowing through the coronary arteries. The dye lets the doctor study the flow of blood through the heart and blood vessels. Cardiac catheterization usually is done in a hospital. It usually causes little or no pain, although the patent may feel some soreness in the blood vessel where the doctor inserts the catheter.

3.4.5 Immunoassay Technique

The analysis of cardiovascular markers can be carried out using immunology-based methods like immunoassays. These utilize the specificity of immunoreactive components, mostly antibodies, in binding to a specific analyte to qualitatively and quantitatively confirm the presence or absence of the antigen. The most common immunoassays are membrane-based immunoassays such as lateral flow devices (LFDs) and enzyme-linked immunosorbent assays (ELISAs). Many LFDs still use the same technology that was patented more than 25 years ago. However, better materials, detection through the use of portable readers, and more precision have made the technology very attractive as point-of-care devices. Multiplexing and simultaneous identification and quantification of multiple biomarkers in single line is also the future trend for LFDs (Zhu et al., 2011). ELISA is a method for detecting and quantifying a specific analyte in a complex mixture. In this method, the visualization of the target antigen is realized through a color-generating enzyme, covalently linked to a specific antibody. The color formation is proportional to the concentration and can be used for quantification by analyzing the absorbance. The ELISA method was made possible because of scientific advances in a number of related fields, including the production of antigen-specific monoclonal antibodies by Köhler and Milstein (1976), and the ability to chemically link antibodies to biological enzymes whose activities can be measured as a signal. Besides ELISA, there are several other immunoassay formats that were developed earlier, including radioimmunoassay, which uses antibodies labeled with radioisotopes. It was first used before alternatives were sought because of health risks associated with radiation. Several variations have been made to the original ELISA, with most changes being made to the method of transduction. A general term, "immunoassays," is more universally accepted. It is important to understand the structure and function of antibodies, as they play a major role in immunoassays. Antibodies are produced by B-cells as part of the immune response system and identify and neutralize foreign objects such as bacteria and viruses, known collectively as pathogens. Antibodies have been used as recognition elements in immunoassay and subsequently immunosensor development because of their high affinity, high specificity, versatility, and commercial availability (Allinson, 2011). A polyclonal antibody represents a collection of antibodies from different B cells that recognize multiple epitopes on the same antigen, while a monoclonal antibody represents an antibody from a single antibody-producing B cell.

There are five major classes of antibodies, IgG, IgD, IgE, IgA, and IgM. Of all these, IgG is the most abundant class in serum and has been used in immunoassay development for half a century. There are four main types of immunoassay protocols for the detection of proteins: direct, indirect, competitive, and sandwich immunoassays. The basic principles for the assays are similar, and they include capturing the analyte of interest, blocking the nonreacted surface, and recognizing the analyte. The direct immunoassay is the simplest form of analyte detection. It involves immobilizing the analyte of interest to the surface. This is followed by a washing step, drying, and blocking before a specific labeled antibody for the analyte is added for detection. Figure 3.4 shows a direct, indirect sandwich and competitive immunoassays.

The other format, called indirect immunoassay, has the ability to improve the sensitivity of the detection. In this method, the analyte of interest is bound to a specific antibody after immobilization onto the surface. A labeled secondary antibody against this primary antibody is then incubated for detection purposes. It is important that the secondary antibody be raised

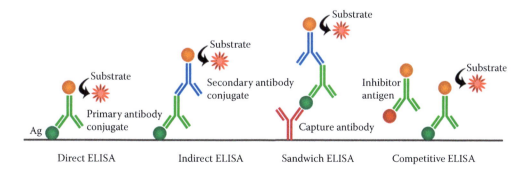

FIGURE 3.4
The illustrations of a direct, indirect, sandwich, and competitive immunoassays. (Adapted from https://www.bosterbio.com/protocol-and-troubleshooting/elisa-principle.)

in another species than the primary antibody to avoid nonspecific binding. By far the most common type of immunoassay detection is achieved by applying two specific antibodies for the analyte in a sandwich format immunoassay. Antibody-sandwich immunoassays may be the most useful of the immunosorbent assays for detecting antigens because they are frequently between two and five times more sensitive than those in which the antigen is directly bound to the solid phase (Allinson, 2011). The antigen being investigated is actually sandwiched between two antibodies, which are the coated antibody immobilized on a surface and the detection antibody, which is usually conjugated to a marker that can cause a quantifiable response proportional to its concentration. Autoanalyzers that enable automated routine biochemical tests in hospital laboratories for biomarker diagnosis are used today to perform immunoassays (Chan, 1994).

3.5 Biomarkers and Their Role in Cardiovascular Risk

The prevention and management of cardiovascular disease increasingly demands effective diagnostic testing. A method and associated diagnostic device perform a physical measurement from a patient or associated biological sample and produce a quantitative or descriptive output, known as a biomarker. A biomarker is defined as a characteristic that is objectively measured and evaluated as an indicator of normal biological processes, pathogenic processes, or pharmacologic responses to a therapeutic intervention (Biomarker Definitions Working Group, 2001). Diagnostics is strategic position at the intersection between patients and their clinically actionable data and directly impinges on the patient experience and the quality of care that individuals receive. It also furnishes valuable tools for clinical investigation and enables providers to improve upon "one-size-fits-all" treatment strategies and instead provide personalized care.

As we know, CVD is a group of different disorders that affect heart and blood vessels. CVD includes atherosclerosis or CHD, which develops when plaque builds up in the walls of the arteries, which narrows the arteries and makes it difficult for blood to flow through. It causes a heart attack or stroke. As we also understood, CVD can be caused by a range of risk factors and disorders that include genetic factors, gender, age, high blood pressure and cholesterol, diabetes, obesity and overweight, smoking, and stress. There are a number of diseases associated with CVD that have an effect on different parts of the body. Even though great progress has been made in the treatment of this disease, current medical knowledge is unable to effectively predict its risk. To predict CVD risk, use of diagnostic and prognostic biomarkers are one of the dynamic research areas recently found (Biomarker Definitions Working Group, 2001). These biomarkers can be identified in the blood and further categorized into pathogenetic and therapeutic types. In such cases, the vascular wall releases molecules into the bloodstream that can reflect the pathological processes taking place. Theoretically, the concentrations of the molecules involved in different pathological processes could be biomarkers. However, not all of these molecules are suited to this aim but should fulfill certain conditions (Biomarker Definitions Working Group, 2001). There are several important characteristics that an ideal cardiac biomarker should exhibit, including: (a) high clinical sensitivity and specificity; (b) quick release of biomarker in the blood, enabling early diagnosis; (c) capability to remain elevated for a longer time in the blood; and (d) ability to be assayed quantitatively. The most frequently studied biomarkers are summarized in relation to the different mechanisms involved in development and rupture

TABLE 3.1
A Summary of Primary Clinically Utilized Cardiac Biomarkers

Established Biomarkers	Remarks	Cutoff Levels	Initial Elevation	Time to Peak
Troponin I	Myocardial necrosis	0.01–0.1 ng/mL^{-1}	4–6 h	12–24 h
Troponin T	Myocardial necrosis	0.05–0.1 ng/mL^{-1}	4–6 h	12–24 h
Creatine kinase (CK-MB)	Myocardial necrosis Re-infarction	10 ng/mL^{-1}	4–6 h	12–24 h
C-reactive protein (CRP)	Inflammation	<10^3 to >3–15 × 10^3 ng mL^{-1} (low to high risk)	Not defined	Not defined
Brain natriuretic peptide (BNP)	Ventricular overload	Not defined	Not defined	Not defined
Myoglobin	Myocardial necrosis	70–200 ng/mL^{-1}	1–3 h	6–12 h
Fatty acid binding protein (H-FABP)	Myocardial necrosis	Patients with elevated H-FABP levels elevated stratification risk ≥ 6 ng/mL^{-1}	2–3 h	8–10 h

of atherosclerotic plaque, such as endothelial dysfunction, inflammation, oxidative stress, proteolysis, and thrombosis. Cardiac-specific biomarkers have emerged as strong and reliable risk predictors for coronary heart disease, as listed in Table 3.1. In all, C-reactive protein (CRP) has been the most frequently used single biomarker for cardiovascular risk (CVR). CVR as defined by the American Heart Association (AHA) and the Centers for Disease Control and Prevention (CDC) is regarded as low risk for a CRP concentration below 1.0 mg L^{-1}, moderate for 1.0–3.0 mg L^{-1}, and high for concentrations over 3.0 mg L^{-1} (Casas et al., 2008). CRP can rise as high as 1000-fold because of inflammation induced by infection or injury, often leading to CVR. Recent research suggests that patients with elevated basal levels of CRP are at an increased risk of diabetes and hypertension as well as CVD (Dhingra and Vasan, 2017).

Myoglobin, although not a very specific marker, is the first marker released within 4 hr of damage to myocardial muscle cells. B-type natriuretic peptide (BNP), cardiac troponin I (cTnI), and CRP are released after myoglobin, but they are specific markers for coronary events (Jaffe et al., 2010). BNP is useful for the emergency diagnosis of heart failure and for the prognosis in patients with acute coronary syndromes (ACSs) (Yang et al., 2009). CRP is an important prognostic indicator of CVR and ACS. cTnI has become a standard marker for the detection of acute myocardial infarction (Apple, 2007). During heart infarction, troponin T (TnT) is immediately released to the bloodstream. A biosensor capable of monitoring this biomarker in a short time could improve patient care by allowing a definite diagnosis of myocardial infarction in real time (Han et al., 2016). Elevated concentrations of these cardiac markers in serum are associated with recurrent CVD events and higher death rates. Simultaneous quantification of these biomarkers allows clinicians to diagnose CVD quickly and/or accurately design a patient care strategy. A fast and reliable detection of these proteins would also help medical professionals differentiate diseases among those showing similar symptoms. The clinically significant sensing ranges of myoglobin, BNP, cTnI, and CRP are extremely low (pM to nM); therefore, assay methods for these biomarkers need to be highly sensitive.

Recently, heart-type fatty acid binding protein (H-FABP), a small (15 kDa) cytoplasmic protein, was found to be a very specific biomarker that is abundant in tissues with active fatty acid metabolism, including the heart (Das, 2016). Following ischemic injury, H-FABP is quickly released into the circulation via porous endothelial membranes of cardiac myocytes. Myocardial damage can be evaluated in plasma within 1–3 hours after symptom onset. Compared to troponins, the more specific but slower-release markers of cardiac injury, H-FABP is a useful tool for rapid evaluation of ischemic injury and size. The concentration of FABP in skeletal muscle is 20 times lower than that of cardiac tissue. Myoglobin occurs in the same content for cardiac and skeletal tissue. This makes H-FABP more cardiac specific than myoglobin, and a useful biochemical marker for the early assessment or exclusion of AMI. Previous studies have suggested that H-FABP can be used as a reliable marker for hypertrophic and dilated cardiomyopathy, heart failure, early estimation of infarct size, early detector of postoperative myocardial tissue loss in patients undergoing coronary bypass surgery, stroke, and obstructive sleep apnea syndrome. In recent years, different biosensor platforms have been designed for detection of available cardiac disease biomarkers. Here, we summarize the most prominently used cardiac biomarkers and their detection by different biosensor platforms.

3.6 Detection of Cardiac Biomarkers

In the last few years, biosensors have been used extensively to detect and quantify the target molecules involved with cardiac biomarker interaction. By definition, biosensors are integrated diagnostic devices, which merge a biological or biologically-derived sensing element associated with a physicochemical transducer. Usually, the surface of a suitable transducer of a biosensor is immobilized with a biological receptor material (DNA, RNA, or antibody), which enables conversion of biochemical signal into quantifiable electronic signal through the mode of either electrochemical, optical, mass change (piezoelectric/acoustic wave), or magnetic. Biosensors possess high sensitivity, high selectivity, fast analysis, reliable pretreatment, and simple instrumentation when compared to conventional techniques such as ECG and echo. As we discussed earlier, cTnI and cTnT have become standard markers and are commonly used for the detection of acute myocardial infarction. Different methods have been developed for cardiac troponin detection and quantification, which include enzyme-linked immunosorbent assay, chemiluminescent immunoassay, fluoro-immunoassay, electrical detection, surface plasmon resonance–based detection, colorimetric protein array, and aptamer-based biosensor. This chapter further focuses on the advancement, sensitivity, and limitations of these methods for detection of cardiac troponin biomarkers.

3.6.1 Enzyme-Linked Immune Sorbent Assay

ELISA is the most common technique used for diagnostic device development. ELISA is a biochemical assay that utilizes antibodies and an enzyme-mediated color change to identify the existence of either antigen (proteins, peptides, hormones, etc.) or antibody in a given sample. Detection of very small quantities of antigens is permitted by using the fundamental concepts of immunology of an antigen binding to its specific antibody. Further, it has been applied in the detection of cardiac troponin to diagnose MI. The first report on ELISA, published by Engvall and Perlmann, demonstrated quantitative measurement of IgG in rabbit serum using alkaline phosphatase as the reporter label (1971). The antigen is allowed to bind to a specific antibody, which itself afterwards is detected by a secondary, enzyme-coupled antibody that reacts with a chromogen. The existence of the antigen is indicated by the production of a visible color change or fluorescence from a chromogenic substrate for the enzyme. This color change can be quantitatively or qualitatively measured to detect the antigen. Katus et al. developed the first generation of troponin T ELISA (TnT 1), which can be considered a more sensitive enzyme immunoassay using two troponin T-specific monoclonal antibodies (1992). It is based on the one-step sandwich assay principle, where the antigen is bound to streptavidin-coated polystyrene tubes as the solid phase by an affinity-purified polyclonal antibody from sheep and detected by peroxidase-labeled monoclonal antibody. The assay is carried out at room temperature for 90 min, and the measuring range is 0.1–15 mg/L. However, this assay is unreliable in patients with severe skeletal muscle injury because of increased false-positive results due to unspecific binding of skeletal muscle troponin T. Muller-Bardorff et al. developed the second generation of cardiac-specific troponin T ELISA (TnT 2) in which the cross-reactive antibody 1B10 has been replaced by a high-affinity cardiac-specific antibody M11.7 (1997). In terms of specificity, a considerable improvement for this ELISA is that it was found to differentiate between cardiac and skeletal muscle damage, even in patients with severe skeletal muscle injury. The detection limit is 0.012 mg/L with a coefficient of variation (CV) 5.8%. Subsequently, Hallermayer et al. reported on the Elecsyss Troponin T third-generation assay, which used recombinant human cTnT as standard material, thus allowing a reproducible and reliable standardization of troponin T assays (1999). As for cTnI, Bodor et al. made a double monoclonal sandwich enzyme immunoassay to measure cTnI in serum (1992). The assay required 4 h to perform and used an enzyme label for detection with minimum detectable dose of 1.9 mg/L. The use of monoclonal antibodies allowed greater reproducibility of reagents than the polyclonal antiserum. ELISA was quite suitable for evaluating the value of cTnI measurements, but was not rugged and precise enough for high-volume routine laboratory use. Even though ELISA has been used for the detection of cardiac troponin biomarkers, it has some hurdles like sample and reagent consumption in large-scale studies, needing to be performed in central laboratories of clinic and hospitals, and taking a very long time to perform (more than 6 h), which is incompatible with the quick decision needed to treat a heart attack patient (Zhang and Ning, 2012). Still, ELISA has some disadvantages; for example, negative controls may indicate positive results if the blocking solution is ineffective, a secondary antibody or antigen (unknown sample) can bindc to open sites in the well, and enzyme stability.

3.6.2 Chemiluminescence-Based Immunoassays

It is a fact that chemiluminescence (CL) is light emission when a molecule emits a photon in an excited state (energy is produced by chemical reaction) and relaxes to its ground state. When CL systems are integrated with immunoreactions, it becomes a method to determine the concentrations of samples according to the intensity of the luminescence and is called chemiluminescence immunoassays (CLIAs). It is possible to generate CL only by the introduction of the CL substrates, that is, luminol, isoluminol, and their derivatives, acridinium ester, peroxidase, and alkaline phosphatase (ALP), to some reagents that act as CL labels (Wang et al., 2012). In such immunoassays, the protein label in CLIA with the most commonly used labeling enzymes, that is, horseradish peroxidase (HRP) and ALP, has been used as a label in immunoassays. Schroeder et al. first described a method for monitoring competitive protein binding reactions by using CL (1976). Then, Cho et al. developed a chemiluminometric ELISA-on-chip (EOC) biosensor in combination with an image detector equipped with a cooled charge-coupled device (CCD) camera capable of detecting cTnI (Cho et al., 2009). The sensor used an immunochromatographic assay combined with an enzyme tracer that produces a light signal measureable on a simple detector. The cross-flow chromatogram is utilized in order to accomplish sequential antigen–antibody binding and signal generation. Performing ELISA on a redesigned plastic chip simplifies the fabrication process. Later on, biotin–streptavidin capture technology was implemented to enhance the sensitivity in preparing an immuno-strip that was then merged onto the chip in order to create the EOC biosensor. Marquette et al. have presented a screen-printed (SP) microarray, a platform for the achievement of multiparametric biochips (2008). It is composed of eight (0.28 mm^2) working electrodes modified with electro-addressed protein A-aryl diazonium adducts. The immobilized capture antibodies are involved in sandwich assays of cTnI together with biotinylated detection antibodies and peroxidase-labeled streptavidin in order to permit a chemiluminescent imaging of the SP platform and a sensitive detection of the assayed proteins. The performances of the system in pure buffered solutions, using a 25-min assay duration, were characterized by dynamic ranges of 0.2–20 µg/L with a limit of detection of 0.06 µg/L for cTnI. A commercialized instrument based on CL immunoassay has also been developed by Kurihara et al. (2008), where a CLIA known as Mitsubishi PATHFASTs is used to detect cTnI concentration. The system adopted a highly sensitive chemiluminescent enzyme immunoassay (CLEIA) using CDP-Star/Sapphire-II (Applied Biosystems) as chemiluminescent substrate and Magtration by which efficient bound/free (B/F) separation can be performed in a disposable pipette tip with a small volume (Obata et al., 2001). It has a quantification limit of 0.007 mg/L for cTnI, below the 99th percentile of the reference group (0.02 mg/L), and fulfilled the requirement of the guideline recommendations for imprecision (CV) at the 99th percentile of the reference group of <10% (Figure 3.5). Many researchers and clinical analysts have accepted chemiluminescent immunoassay widely, which triggers the enzymatic process because of its high sensitivity, wide linear range, and easy operation and automation.

3.6.3 Fluorescence-Based Immunoassays

It is very well known that fluorescence immunoassays (FIs) are one of the optical biosensor classifications involving the signal transduction of complexed molecules in homogeneous and heterogeneous assays. These assays are preferable due to their nondestructive, highly sensitive characteristics and the ease of modifying the biomolecules with fluorescence tags such as fluorescein isothiocyanate, rhodamine, coumarin, and cyanine. These tags are used

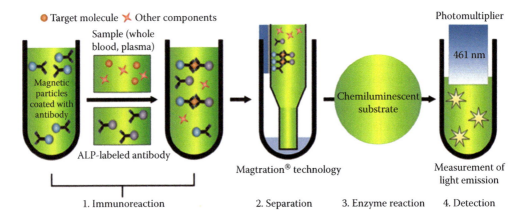

FIGURE 3.5
Mitsubishi PATHFASTs principle of operations. (Reprinted from *Biosensors and Bioelectronics*, 70, Fathil, M. F. et al. Diagnostics on acute myocardial infarction: Cardiac troponin biomarkers, 209–220, Copyright 2015, with permission from Elsevier [through RightsLink].)

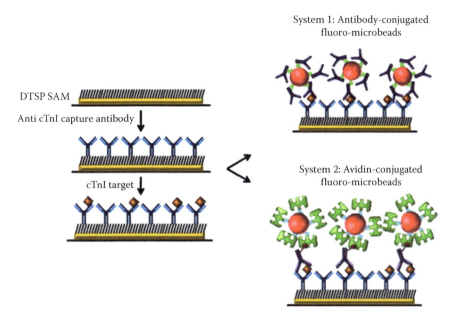

FIGURE 3.6
Schematic diagram of the sandwich immunoassay using antigen/antibody binding (system 1) and avidin/biotin affinity binding (system 2) on the FMGC. The avidin/biotin couple was used to enhance the signal. (Reprinted from *Biosensors and Bioelectronics*, 70, Fathil, M. F. et al. Diagnostics on acute myocardial infarction: Cardiac troponin biomarkers, 209–220, Copyright 2015, with permission from Elsevier [through RightsLink].)

as labels or biorecognizors, and the changes of the fluorescence signal indicate the presence of the target molecules. Significantly, the detection limit is extremely sensitive, which down to a single molecule (Fan et al., 2008). Based on FI, Song et al. (2011) developed a fluoro-microbead guiding chip (FMGC)-based sandwich immunoassay for cTnI detection. The FMGC offers the capability of using a fluorescence microscope to count the number of beads bound in the immunosensing region directly. A conjugation of antibody to fluoro-microbes is prepared as the detection component. The avidin–biotin affinity interaction is utilized to boost the antigen–antibody binding signal.

Biotin-conjugated cTnI detection antibodies were loaded onto the chip after immobilization of the capture antibody, binding of the target antigen cTnI occurred, and reacted for 30 min, as shown in Figure 3.6. In spite of being highly sensitive, detection from this assay is bulky and expensive, and requires trained personnel to perform the tests (Qureshi et al., 2012).

3.6.4 Electrical Detection-Based Assay

In earlier described immunoassays, a few limitations were observed, such as lack of portability, late detection time, and high complexity of fabrication process. To overcome the limitations of immunoassay-labeled methods, the electrical detection of biomolecular interaction proposed by Kong's group in 2012 is highly beneficial because it is suitable to become low-cost portable sensor (2012) that also can be used by nonspecialized personnel. This electrical detection is conducted by transducing the molecular binding event into a detectable electrical signal. In order to sense chemical and biological species that are very small in size, researchers have intensely studied nanostructures, such as nanowires (NWs), nanobelts, carbon nanotubes, graphene, and nanoparticles for biosensing, due to the comparable size between sensor and target. Looking toward electrical detection as the main interest, Chua and his group (2009) presented a complementary metal oxide semiconductor (CMOS)-compatible silicon nanowire (SiNW) array platform for the label-free, ultrasensitive, real-time detection of cTnT in assay buffers as well as in undiluted serum samples based on the top-down method, as shown in Figure 3.7. It is achieved through electrical measurement based on the conductance changes of the individual SiNWs.

The demonstration of human cTnT detection in assay buffer solution concentrations and the undiluted human serum environment had successfully detected down to 1 and 30 pg/L, respectively. This method has the advantage of eliminating the need for expensive, highly specialized equipment in the diagnostic screening test. Kong et al. developed detection of cTnI using the SiNW field-effect transistors (FETs) CMOS-compatible top-down method. Mab-cTnI was covalently immobilized on the SiNW surfaces (Kong et al., 2012). This device has a limit of detection down to 92 ng/L and a linear dynamic range of 0.092–46 mg/L. In a recent study, Lee et al. developed a nanowire of different material, polyaniline (PANI), for detecting four cardiac biomarkers: cTnI, Myo, CK-MB, or BNP (2012). PANI is an organic material and more easily modified with biomolecules than inorganic nanomaterial. The covalent bond between PANI and the antibody during surface functionalization of PANI allows direct measurement of physical change of conductance, capacitance, or impedance upon binding of antibodies to target proteins, along with controllable conductivity, mechanical flexibility and exceptional bioaffinity. The single PANI nanowire as shown in Figure 3.8 was fabricated by the electrochemical deposition growth method between prepatterned Au electrodes, avoiding the need for selection and alignment of the nanowire. The single PANI nanowire was functionalized by covalently attaching the monoclonal antibody on

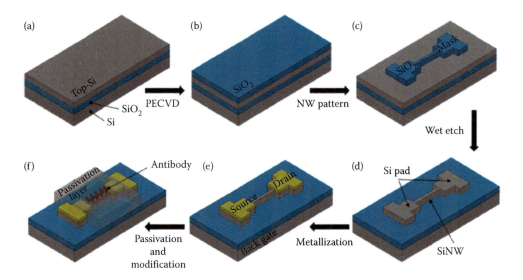

FIGURE 3.7
Key workflow steps used for the fabrication of SiNW-based FET devices. (a) Thermal oxidation to thin the device layer of the SOI wafer down to ~50 nm. (b) Thin film of SiO_2 deposited by PECVD. (c) Defining the SiO_2 pattern by lithography. (d) Crystallographic wet etching using the TMAH solution to obtain SiNWs. (e) Deposition of metal contact for source, drain, and back gate electrodes, followed by rapid thermal annealing to acquire ohmic contact. (f) Fabrication of SiO_2/SiN_x passivation layer by PECVD, lithography, and RIE processes, followed by covalent modification of cTnI antibodies. (Reprinted from *Biosensors and Bioelectronics*, 70, Fathil, M. F. et al. Diagnostics on acute myocardial infarction: Cardiac troponin biomarkers, 209–220, Copyright 2015, with permission from Elsevier [through RightsLink].)

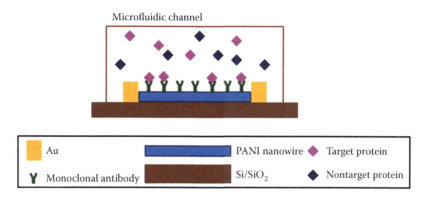

FIGURE 3.8
The cross-section view of the single PANI nanowire-based biosensor. (Adapted from Lee, I. et al. 2012. *Biosensors*, 2(4), 205–220.)

the surface. The binding between immobilized MAbs and target biomarkers changes the net surface charge of the PANI nanowire and induces carrier accumulation or depletion, which translates the changes in nanowire conductance. Detection as low as 250, 100, 150, and 50 pg/L for cTnI, Myo, CK-MB, and BNP can be achieved, respectively. Cheng's group explored the possibility of using a nanobelt field effect transistor, which showed detection of cTnI using functionalized Tin oxide (SnO_2) nanobelt field-effect transistors (2011). Biotinylation occurs with covalent 3-Aminopropyltriethoxysilane (APTES) linkage on the oxide surface of SnO_2. It is followed by biotinylation and attachment of D-biotin, and substrate passivation by incubating in streptavidin solution for 6 h. The sensing platform is utilized for detection of cTnI with a sensitivity of 2 ng/mL. The main advantage of this sensing scheme lies in its exceptional portability and detection speed. The structure of CNTs (a rolled-up tubular shell of graphite sheet with the carbon atoms covalently bound to their neighbors) and their physical and chemical properties such as electrical conductance, high mechanical stiffness, and the possibility to functionalize CNTs in order to change their intrinsic properties are well studied. The CNT electrode can reach high sensitivity with low detection limits due to rapid electron transfer, thus increasing the reaction rate of many electro-active species and then decreasing the electrode response time. In addition, CNTs increase the electro-active area by forming a nanostructured surface that promotes a greater amount of immobilized biomolecules. Due to their unusual properties, Gomes-Filho and his group developed a nanostructured immunosensor based on CNTs supported by a conductive polymer film for detection of cTnT (2013). The immunosensor achieved a low limit of detection of 0.033 ng/mL and a linear range between 1 and 10 mg/L cTnT, significant for acute myocardial infarction diagnosis. Good reproducibility and repeatability were obtained by the proposed immunosensor, supported by a coefficient of variation of 3.7% and 2.6%, respectively. In advancement, Silva et al. developed an amine functionalized CNT incorporated into the ink printing used to fabricate a screen-printed electrode (SPE) to detect cTnT (Silva et al., 2013). An enhanced stability in measurement is enabled by incorporating NH_2-CNT into the carbon ink. In addition, oriented immobilization of anti-cTnT has led to a high sensitivity and very low detection level (0.0035 mg/L), almost 10 times lower than immunosensor, as described by Gomes-Filho.

Tuteja et al. developed a highly sensitive label-free electrochemical detection platform for cTnI by using monolithic graphene sheets from a lithium ion intercalation mediated efficient exfoliation process (2014). Graphene is a monolayer of a hexagonal network of carbon atoms densely packed into a two-dimensional honeycomb crystal lattice and is characterized by fascinating properties such as an ambipolar electric field effect along with ballistic conduction of charge, elasticity, superior thermal conductivity, and excellent mechanical and electronic properties. Carboxylated graphene is used as a gate-sensing material between the source and drain gold interdigitated electrode by drop casting and incubation for 1 h at 120°C. Carboxyl groups of functionalized graphene were used for biointerface development using aqueous phase carbodiimide activation chemistry. After that, the surface of the activated carboxylated graphene is immobilized with anti-cTnI antibody for cTnI detection. A detection limit of cTnI was achieved down to 0.1 pg/mL with linear response from 1 to 1 ng/mL.

3.6.5 Surface Plasmon Resonance-Based Detection

Surface plasmon resonance is a charge–density oscillation that may exist at the interface of two media with dielectric constants of opposite signs, like a metal and a dielectric. When the surface of a thin metal film is excited by an incident beam of light with a suitable wavelength at a specific angle, an evanescent electromagnetic field is generated and is described as a charge density oscillation occurring at the interface between two media of oppositely charged dielectric constants. Dutra et al. (2007) developed SPR supported detection of only surface-confined molecular interactions occurring on the transducer surface. A basic SPR immunosensor consists of a light source, a detector, a transduction surface, a prism, a biomolecule (antigen or antibody), and a flow system. Commonly, a thin gold film with thickness from 500 to 1000 Å is deposited on a glass slide, then optically coupled to a glass prism through a refractive index matching that of oil. Plane polarized light is directed through a glass prism to the gold/solution dielectric interface over a wide range of incident angles and the intensity of the resulting reflected light is measured against the incident light angle with a detector. A minimum in the reflectivity is observed at which the light waves are coupled to the oscillation of surface plasmons at the gold/solution interface. An SPR angle is the angle at which minimum reflectivity occurs. The critical angle is very sensitive to the dielectric properties of the medium adjacent to the transducer surface, apart from its dependence on the wavelength and polarization state of the incident light.

Figure 3.9 represents a schematic view of the SPR immunoassay technique. The first SPR immunoassay was proposed by Liedberg in 1983 by letting a silver surface absorbing a water solution of antibody human γ-globulin (IgG) which specifically binds with the antihuman γ-globulin (a-IgG). It was to determine a-IgG as low as 0.2 mg/L. Similarly, the first SPR immunoassay for cardiac troponin detection was reported and developed by Wei et al. using a novel label-free sandwich immunosensing method for measuring cTnI by using three monoclonal antibodies (MAbs 9F5, 2F11, and 8C12) generated by the hybridoma technique and characterized by an SPR biosensor (2003). Two methods for detection of cTnI with the immunosensor were performed: (1) the direct detection of cTnI with a detection range of 2.5–40 μg/L and (2) the sandwich immunosensing method, in which the second antibody, 9F5, biologically amplified the sensor response. As a result, the sandwich assay showed a sensitivity of 0.25 μg/L and a detection range of 0.5–20 μg/L with within-run variation of 4.9–6.7% and between-run variation of 5.2–8.4%. This assay has greatly enhanced the sensitivity for detection compared to that previously reported in the literature. Then, Dutra and Kubota (2007) developed a quick-detection SPR immunosensor of human cTnT in real time by inventing a streptavidin-terminated self-assembled monolayer (SAM), which was used to bind biotinylated anti-cTnT monoclonal antibodies. With a linear range from 0.03 to 6.5 mg/L, the cTnT was determined. The system presented good repeatability with 3.4% variation between runs after regeneration

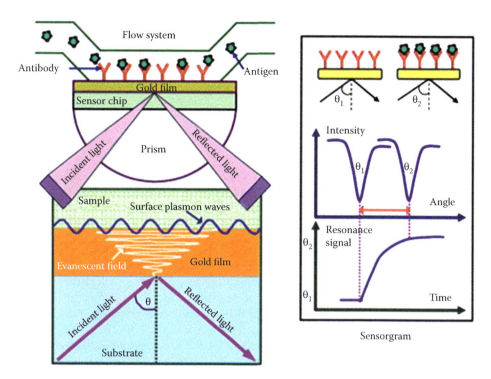

FIGURE 3.9
Basic setup for surface plasmon resonance indicating the changes in the refractive index with changes in the angles and changes in the optical waves. (Reprinted from *Sensors and Actuators B*, 121, Dhesingh, R. S. et al. Recent advancements in surface plasmon resonance immunosensors for detection of small molecules of biomedical, food and environmental interest, 158–177, Copyright 2007, with permission from Elsevier [through RightsLink].)

of the coated surface with a solution of 1% sodium dodecyl sulfate (SDS). The sensor was also able to measure cTnT without human serum dilution with good specificity and reproducibility and offered a quick response at an 800 s interval. Moreover, Dutra et al. (2007) developed an SPR sensor on a commercially available SPR AUTOLAB SPIRITs to detect the biomarker in real time. The surface of a gold substrate via a SAM of thiols by using cysteamine-coupling chemistry is where the cTnT receptor molecule was covalently immobilized. This SPR sensor has opened up new possibilities to explore using SAM to develop a regenerable immunosensor with a good reproducibility, allowing its use in clinical applications. Liu et al. employed a SAM with high antifouling ability consisting of a homogeneous mixture of oligo(ethylene glycol) (OEG)-terminated alkanethiolate and mercaptohexadecanoic acid (MHDA) on Au for immobilizing the cTnT antibody and applied in detecting cTnT by using SPR (2011). The mixed SAM has high detection capability and high accuracy and reproducibility, as well as showing strong potential to be applied in rapid clinical diagnosis for label-free detection within 2 min. Unlike many other immunoassays, such as ELISA, an SPR immunoassay is label free in that a label molecule is not required for detection of the analyte, thus reducing the number of steps, permitting real time-analysis of interaction, regeneration of the sensor surface, and low cost analysis. The SPR immunoassay can also be performed using the array technique, which has the ability to be used with high-throughput and low-cost methods. Although there are no reports related to cardiac biomarker detection using the SPR immunosensor in array technique so far, this method offers an advantage of testing in parallel with several samples in a single run. Therefore, it has opened a new opportunity for multiplexed detection of cTnI, cTnT, and H-FABP.

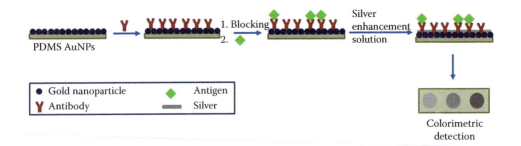

FIGURE 3.10
Schematic diagram for silver enhancement based colorimetric detection of cardiac troponin I. (Adapted from Wu, W. et al. 2010. *Sensors and Actuators B: Chemical*, 147(1), 298–303.)

3.6.6 Colorimetric Detection-Based Assay

Colorimetric sensing system is another class of optical biosensor in which chromogenic dyes are used to recognize the target. The presence of target molecules is confirmed by the intensity of the color changes. The limit of detection for this method can be narrowed down to single molecule detections; hence, it is considered a sensitive technique. Wu et al. (2010) developed a Poly(dimethylsiloxane) (PDMS)–gold nanoparticles (AuNPs) composite film-based biosensor coupled with silver enhancement colorimetric detection for cTnI. A monoclonal antibody against cTnI was immobilized on the PDMS–AuNPs composite film, followed by blocking solution and cTnI (Figure 3.10). The AuNPs act as a catalyst during the reaction of silver reduction. This catalytic ability could be inhabited when there were proteins covering the surface of AuNPs, which influenced the amount of silver metal reduction. It led to a color difference in the reaction wells. AuNPs are established as a good substrate to be functionalized with antigen, enzymes, and other biomolecules, while PDMS has very good transparency, outstanding elasticity, good thermal and oxidative stability, and ease of fabrication and is sealed with different materials. The PDMS–AuNP composite film has a shelf life of up to several month thanks to the polymer matrix of PDMS, which can protect AuNPs from aggregation. The biosensor has a detection limit down to 0.01 mg/L in less than 20 min. It has a low cost since it eliminates the need for expensive, highly specialized equipment.

3.7 Point-of-Care Assay and Technology in the Market

Point-of-care assays and technologies are enabling innovative cardiovascular diagnostics that promise to improve patient care across diverse clinical settings. The multidisciplinary working group, which includes clinicians, scientists, engineers, device manufacturers, regulatory officials, and program staff, discussed opportunities for POCT to improve cardiovascular care, realize the promise of precision medicine, and advance the clinical research enterprise, and identified barriers facing translation and integration of POCT with existing clinical systems. A roadmap in POCT development emerged to guide multidisciplinary teams of biomarker scientists, technologists, healthcare providers, and clinical trialists. For significant effects on cardiovascular care, POCT must be ensured, as they: (1) formulate needs assessments, (2) classify device design specifications, (3) develop component technologies and integrated systems, (4) execute iterative pilot testing, and (5) conduct rigorous prospective clinical testing (Wu et al., 2010).

We will be discussing POC for cTnI and cTnT only due to high specificity for and high attention to CHD compared to other biomarkers. Currently, various kinds of immune-detection methods have been used, especially for cTnI and cTnT (because of high specificity) detection, as mentioned above. Obviously, these approaches are time consuming, costly, and require labeled reagents and bulky instrumentation. A number of POC assays based on ELISA, fluorescence, chemiluminescence, and other technologies have been developed to support the diagnosis of CVD, especially AMI. It is generally required and expected that quantitative measurement of cTnI or cTnT should be provided and sensitivity of the POC systems should not deteriorate from results provided by automatic platforms in the central laboratory. cTnI and cTnT quantification of POC systems can be divided into two main types, bench-top and handheld systems (King et al., 2016). Nevertheless, the sensitivity of presently available POC assays for cTnI and cTnT is less compared to the central laboratory test, which limits the potential of POC assays for reliable diagnosis of AMI. The development of more precise and higher-sensitivity cTn POC assays has been motivated by the essential requirement of serial cTn change detection. Based on accuracy calculated to the 99th percentile and understanding, these patients are still being identified as at-risk with the present POC assays (Bingisser et al., 2012). The list of currently available quantitative POC systems for measuring cTnI and cTnT as reported by the manufacturers with their analytical characteristics is shown in Table 3.2. Based on Roche's report, the cTnT Roche Elecsys hs-cTnT POC assay has been labeled as a high-sensitive assay with a 99th percentile value at 0.013 mg/L (13 ng/L) and CV <10%. The imprecision must be between 10% and 20%, at which classified assays are clinically usable and not acceptable. Even though higher-sensitivity POC assays are available, there are significant drawbacks that need to be considered, which include user-related practical issues like training, maintenance, or accreditation high cost and reimbursement issues, particularly when laboratory services are run on a fee-for-service basis. The use of POC assays differs between ST-segment elevation myocardial infarction (STEMI) and NSTEMI patients. STEMI patients have no requirement to wait for the biomarker test result; using only symptoms and specific ECG changes is recommended for immediate referral to coronary revascularization according to the international guidelines. However, elevated cTnI or cTnT concentrations are crucial for diagnosis of NSTEMI (King et al., 2016). The characteristics of the current and the high-sensitivity cTn assays have been summarized in Table 3.3.

If we consider microfluidic and smartphone based-approaches, some challenges still exist. For developers, the challenges include: (1) increase performances of diagnostic platforms, including specificity and sensitivity; (2) decrease the cost, including time, sample, and reagents; (3) make the platform robust and simple to operate and the result easy to understand. For users, the challenges mainly include: (1) carefully follow the manufacturer's instruction; (2) carefully obtain and utilize the results; (3) if required for long-term adoption, follow and paying attention to changes. Due to the rapid development of microfluidic and nanotechnology, many portable analytical devices are developed for detection of CVDs at the point of care. Also, smartphones, as a versatile and powerful handheld tool, have been employed both as readers for microfluidic assays and analyzers for physiological indexes. POC testing refers to a laboratory assay that can be performed outside a centralized facility and shows great promise for application. Nowadays, many platforms have been developed for POCT such as optical microfluidic technologies, lab-on-a-chip systems, electrochemical biosensors, POC ultra-sonography, paper-based devices, and smartphone-based strategies

TABLE 3.2
Analytical Characteristics of the Current Cardiac Troponin I and T Assays

Company-Instrument-Assay (Generation)	Detection Limit (μg/L)	cTn at 99th Percentile (μg/L)	CV at 99th Percentile (%)	cTn at 10% CV, μg/L
Abbott AxSYM ADV (2nd)	0.02	0.04	15	0.16
Abbott ARCHITECT	<0.01	0.028	15	0.032
Abbott i-STAT	0.02	0.08	16.5	0.1
Beckman Coulter Access Accu (2nd)	0.01	0.04	14	0.06
bioMerieux Vidas Ultra (2nd)	0.01	0.01	27.7	0.11
Innotrac Aio! (2nd)	0.006	0.015	14 (at 19 ng/L)	0.036
Inverness Biosite Triage	0.05	<0.05	NA	NA
Inverness Biosite Triage (r)	0.01	0.056	17	NA
Mitsubishi Chemical PATHFAST	0.008	0.029	5	0.014
Ortho Vitros ECi ES	0.012	0.034	10	0.034
Radiometer AQT90	0.0095	0.023	17.7	0.039
Response Biomedical RAMP	0.03	<0.1	18.5	0.21
Roche E170	0.01	<0.01	18	0.03
Roche Elecsys 2010	0.01	<0.01	18	0.03
Roche Cardiac Reader	<0.05	<0.05	NA	NA
Siemens Centaur Ultra	0.006	0.04	10	0.03
Siemens Dimension RxL	0.04	0.07	20	0.14
Siemens Immulite 2500 STAT	0.1	0.2	NA	0.42
Siemens Immulite 1000 Turbo	0.15	NA	NA	0.64
Siemens Stratus CS	0.03	0.07	10	0.06
Siemens VISTA	0.015	0.045	10	0.04
Tosoh AIA II	0.06	<0.06	8.5	0.09

Source: Shu-hai, J. et al. 2014. *Progress in Biochemistry and Biophysics*, 41, 916–920.

(Fathil et al., 2015). In all of these, microfluidic-based devices have drawn intensive attention from institutes and industries due to their various useful capabilities, such as low cost, simple operation, short turnaround, and limited sample and reagent consumption. They also show the capability for instantaneous separation and detection with high resolution and sensitivity. These advantageous features make it very suitable to create portable POC medical diagnostic systems (Fathil et al., 2015). To further promote the application of microfluidic-based devices at the POC, smartphone-based data collection and analysis are being extensively explored. Furthermore, due to the development of software, hardware, and servers, smartphones have found widespread applications in monitoring physiological indexes of CVDs as well. Microfluidic- and smartphone-based devices are expected to improve healthcare at the POC. However, diagnostics in resource-limited settings is still major concern because the requirements of diagnostic techniques in resource-limited settings are quite different from those in developed countries due to constraints such as affordability, accessibility, limited staff, and laboratory infrastructure. The laboratory may also suffer from physical constraints, including limited clean water, unreliable power sources,

TABLE 3.3

Classification of the Current and Highly Sensitive cTn Assays

Current Available Assays	Detection Limit (ng/L)	Imprecision at 99th Percentile (%)	Classification According to Imprecision	% of Detectable Values in Reference Subjects
Abbott AxSYM ADV (2nd)	40	15	Clinically	<50%
Abbott ARCHITECT	28	15	Clinically	<50%
Abbott i-STAT	80	16.5	Clinically	<50%
Beckman Coulter Access Accu (2nd)	40	14	Clinically	50%–75%
bioMerieux Vidas Ultra (2nd)	10	27.7	Not acceptable	<50%
Innotrac Aio! (2nd)	15	14 (at 19 ng/L)	Clinically	<50%
Inverness Biosite Triage	<50	NA	NA	<50%
Inverness Biosite Triage (r)	56	17	Clinically	Unknown
Mitsubishi Chemical PATHFAST	29	5	Guideline	<50%
Ortho Vitros ECi ES	34	10	Guideline	<50%
Radiometer AQT90	23	17.7	Clinically	<50%
Response Biomedical RAMP	<100	18.5	Clinically	<50%
Roche E170	<10	18	Clinically	<50%
Roche Elecsys 2010	<10	18	Clinically	<50%
Roche Cardiac Reader	<50	NA	NA	<50%
Siemens Centaur Ultra	40	10	Guideline	<50%
Siemens Dimension RxL	70	20	Clinically	<50%
Siemens Immulite 2500 STAT	200	NA	NA	<50%
Siemens Immulite 1000 Turbo	NA	NA	NA	<50%
Siemens Stratus CS	70	10.0	Guideline	<50%
Siemens VISTA	45	10	Guideline	<50%
Tosoh AIA II	<60	8.5	Guideline	<50%
Beckman Coulter Access hs-cTnI	8.6	10	Guideline	>95%
Roche Elecsys hs-cTnT	13	8	Guideline	>95%
Nanosphere hs-cTnI	2.8	9.5	Guideline	75%–95%
Singulex hs-cTnI	10.1	9	Guideline	>95%

Source: Shu-hai, J. et al. 2014. *Progress in Biochemistry and Biophysics*, 41, 916–920.

and vast temperature variations. Combined, the above factors together create poor clinical sensitivity and specificity, which not only waste time and funds but also may cause serious medical accidents. Therefore, the WHO has outlined guidelines for diagnostic systems used in resource-limited settings with the ASSURED criteria (Affordable, Sensitive, Specific, User-friendly, Rapid and Robust, Equipment-free, and Deliverable to end users).

3.7.1 Point-of-Care Devices Currently Available on the Market

Presently, rapid immunochromatographic test strips or lateral flow immunoassays (LFIAs) are the most successful portable microfluidic devices. Currently, LFIAs for detecting CVDs

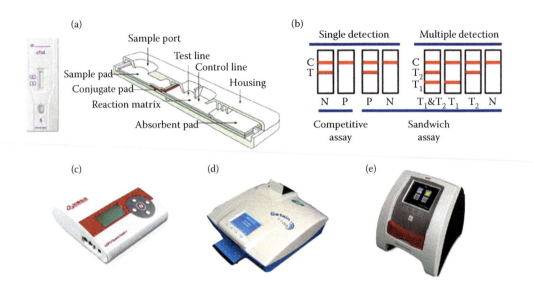

FIGURE 3.11
Typical commercially available lateral flow immunoassay-based point of care tests: (a) Image of a representative lateral flow test strip (Adapted from http://www.lepumedical.com/) and its schematic (EMD Millipore, 2013). (b) Single and multiple detection based on lateral flow immunoassays. (c) Colorimetric (Lepu Quant-Gold 1, adapted from http://www.lepumedical.com/), (d) fluorescent (Getein 1100, adapted from http://www.bio-gp.com.cn/), and (e) magnetic (MICT® Bench-Top System, adapted from http://www.magnabiosciences.com/) signals.

are used in hospitals with a device for quantification of analyses (Figure 3.11). These devices are classified into colorimetric, fluorescent, and magnetic immunoassays. The most popular devices for quantification of results are colorimetric analysis devices. They can be used for quantification of results in LFIAs with gold nanoparticles as the labels, such as Roche Cobas h 232 POC system, Getein FIA 8000 Quantitative immunoassay analyzer, and LEPU Quant-Gold-1. Though some gold nanoparticles have an ultrasensitive cut-off, most of them have limited sensitivity and do not meet the clinical level of detection (LOD), which thereby limits their application in detection of low-concentration targets like BNP. Hence, in order to improve LOD, fluorescent and magnetic particles have been introduced in LFIA as labels for replacing gold nanoparticles. Though fluorescent or magnetic assays have higher sensitivity in comparison to colorimetric assays, they are costlier and mainly used in hospitals and clinics.

In addition, microfluidic chip-based immunoassays have been found on the market. Simplex detection includes the fluorescent assay-based Triage system and electrochemical-based Abbott i-STAT system, both of which are sandwich ELISA-based, similar in principle to sandwich LFIA. Primarily, enzymes like horseradish peroxidase and alkaline phosphatase are employed instead of gold nanoparticles and corresponding substrates are added for colorimetric, fluorescent, chemiluminiscent, or electrochemical signals for readout (Hu et al., 2016). For multiplex detection, Abaxis Piccolo Xpress (for absorbance analysis) and Gyrolab xPLORE (for fluorescence analysis) are two companies that have developed centrifugation-based microfluidic compact disc (CD) and corresponding analyzers. Table 3.4 gives a summary of the commercially available products used in monitoring CVD.

TABLE 3.4
A Comprehensive Comparison of Performances of Selected Commercially Available Techniques for Monitoring of CVD

Representative Product	Specimen	Sample Pretreatment	Multiplex Detection	Featured Assay	Cost
LepuLeccurate™	Whole blood	No	Yes	cTnI: 0.5–50 ng/mL; CRP/hs-CRP: 3–50 µg/mL	~$40/strip ~$3,000/analyzer
Getein Fast Diagnostics	Whole blood	No	Yes	cTnI: 0.1–50 ng/mL; hs-CRP: 0.5–200 µg/mL	~$30–40/strip; ~$15,000/analyzer
MagnaBioSciences MICT® Test	Heparin plasma	Yes	No	cTnI: 0.2–30 ng/mL	NA
Alere Triage® system test	Whole blood in EDTA	Yes	Yes	BNP: 5–5000 pg/mL	~$50/chip; ~$7,000/analyzer
Abbott i-STAT test cartridge	Sodium or lithium heparin whole blood	Yes	No	BNP: 15–5000 pg/mL	~$50/chip; ~$8,000/analyzer
Abaxis Centrifugal based reagent CD	Whole blood	No	Yes	Blood chemistry tests	~$50/chip; ~$16,000/analyzer
Gyrolab Centrifugal based Bioaffy CD	Whole blood	No	Yes	Open platform	NA

3.7.2 Point of Care under Development in the Laboratory

Though various LFIAs for CVD tests are already available in the market, they do not fully meet clinical demands. LFIAs have two main drawbacks, poor LOD and limited detection range, which limits their clinical utility. These problems are also the barrier to applications of CVD LFIAs at home like pregnancy test kits. Several efforts have been made to generate several prototypes that have been translated into finished products. For example, cTnI for diagnosis of AMI is at ng/L, which is less than the lower limit of conventional LFIAs. Xu et al. (2009) developed LFIA using super-paramagnetic nanobeads (SPMNBs) as labels for detection of cTnI using an MICT system as magnetic assay reader, which has a LOD of 10 ng/L with a wide range of detection. Further, Ryu et al. (2011) increased the analytical sensitivity in magnetic LFIA using magnetic beads with orientation-controlled antibodies as the labels, bringing the LOD for cTnI to 10 ng/mL. Dual gold nanoparticle conjugates have also been used as labels to magnify extremely low signals (Ryu et al., 2011). Hu et al. used labels of oligonucleotides linked AuNP aggregates and are currently developing prototypes for detection of BNP, a cardiac marker (Hu et al., 2013). High-sensitivity CRP (hs-CRP), as a well-known marker for acute inflammation, has been employed for prediction of CVD risk. Kim et al. prepared a nitrocellulose (NC)-coated metal clad leaky waveguide (MCLW) sensor for one-step and label-free detection of CRP, with a detection range of 0.1–10 µg/L in diluted human serum samples (Kim et al., 2014). A novel vertical flow immunoassay (VFIA) was developed by Oh et al. for rapid and one-step detection of hs-CRP (Oh et al., 2013). Further, this work was extended and a chemiluminescence-based enzyme immunosorbent assay (EIA) system was used for CRP measurement, with working principles similar to those of Abbott i-STAT.

3.7.3 Microfluidic-Based Immunoassay and Chip-Based Devices

In comparison to LFIAs with passive control of fluid flow, microfluidic devices control flow of fluid either by passive or active means, and show better capability and flexibility in immunoassays. Based on the materials used, these assays can be classified as microfluidic chip-based (mainly active fluid flow) and paper-based (mainly passive fluid flow) immunoassays. Microfluidic chip-based capillary systems have been developed for detection of cardiac markers using fluorescent signals. Microfluidic chips using optical (e.g., fluorescence), chemiluminescent, and electrochemical readouts have been reported for detection of cardiac markers (Oh et al., 2013, Tsaloglou et al., 2014). A further development to this is a microfluidic channel with an array of gold strips for detecting and quantifying troponin T in an aqueous solution. Troponin T primary (capture) antibody is immobilized on a gold strip, and an aqueous solution containing Troponin T antigen is injected into the microchannel to allow antibody-antigen reaction to take place in less time. Later, fluorescein isothiocyanate (FITC)-tagged secondary antibody against troponin T is dispensed into the channel. The quantification of troponin T antigen is achieved by coupling the FITC system with a confocal fluorescent reader. A digital microfluidic device was used for the detection of cTnI in serum with a LOD of ~2 ng/mL in both simple (saline) as well as complex (human serum) matrices based upon a fluorogenic two-site noncompetitive heterogeneous immunoassay. The analytical performance of the system is comparable to the benchtop gold standard and other electrowetting-on-dielectric (EWOD) platforms in terms of both LOD and result time (Tsaloglou et al., 2014). Abad et al. (2012) have developed the V-chip, which opens up the possibility of highly simplified point-of-care and personalized diagnostics. Li et al. (2015) used the V-chip for qualitative detection of BNP with LOD below 5 pM. Floriano et al. (2009) have developed a saliva-based nano-biochip immunoassay to detect a panel of C-reactive protein, myoglobin, and myeloperoxidase in acute myocardial infarction patients. Originally. it was reported for the measurement of salivary biomarkers associated with AMI and explored the possible chances of using them in novel nano-biochip devices for screening patients with chest pain for AMI. Miniaturized immunoassays based on capillary forces have been used for detecting cardiac markers (Floriano et al., 2009).

3.7.4 Paper-Based Devices

Recently, paper has been functionalized as a substrate to construct microfluidic devices for use in rapid diagnostic tests. Paper-based microfluidic devices have applications in POC diagnostics with optical and electro-sensing mechanisms (Giri et al., 2016). The anticipated success of a paper-based microfluidic chip has already been confirmed with the recent development of a device by the nonprofit organization Diagnostics for All, in which a POC device measured markers of liver function (Yetisen et al., 2013).

Recently, exogenous "synthetic biomarkers" were engineered to supplement endogenous biomarkers and enable more flexible remote monitoring of thrombosis. Basically, an intravascular nanoparticle-conjugated peptide, that, when cleaved by activated thrombin, liberates a peptide fragment undergoes renal clearance that can be detectable in the urine centrally or by POC platforms like novel paper-based microfluidic assays (Vella et al., 2012).

3.7.5 Aptamer

Currently, researchers have shifted their focus to nucleic acids that bind with high-affinity ligands to the target identified through establishing an *in-vitro* screening process called "systematic evolution of ligand by exponential enrichment" (SELEX), as shown in Figure 3.10 (Wu and Kwon, 2016). These nucleic acids are termed aptamers, which means "fit" and "region" in Latin (*aptus*) and Greek (*meros*), respectively (Warren et al., 2014). Researchers have gained interest in aptamers, especially in the field of diagnosis and disease management applications. Aptamers are able to bind a wider variety of targets than antibodies, and are easier to produce and store, although antibodies have been widely used in biosensor developments. This is mainly because aptamers offer exclusive detection because of their smaller size compared to antibodies, stability, and ability of structural switching (MacKay et al., 2014). Aptamers were first reported to be used as molecular recognition elements in sensors by Davis et al. (1996), when fluorescent-tagged aptamers were utilized for optical detection of human neutrophil elastase. After several years, only one study using DNA aptamer as molecular recognition for cTnI detection has been reported. Shu-hai et al. (2014) established a new aptamer biosensor for detection of cTnI. Glassy carbon electrode (GCE) is modified with hydroxyl groups by immersing the electrode into a hydrophilic solution at 72°C for 20 min. An amination solution of APTES: H_2O, as an amino group, is introduced by attaching it to the surface of GCE. Aptamer is used in biosensor as a recognition molecule immobilized on the surface of the GCE. An electrochemical signal is generated when the cTnI in solution is bonded to the immobilized aptamer. The current produced from the binding process is dependent on the concentration of cTnI in the solution.

3.8 Challenges and Way Ahead

Across the world, cardiovascular disease is recognized as the prime cause of death due to heart disease and stroke. It is strongly believed that the key challenges *in terms of care* being faced in CVD in across the world are lower availability, accessibility, and affordability of effective and efficient treatment. This is also extended by a lack of a focused policy toward noncommunicable diseases including CVDs. Only 13% of the rural population has access to a primary healthcare facility, and less than 10% to a hospital. Affordability of quality care is a key concern for most of the population, both for preventive checkups and treatment. Big hospitals in Tier I cities are typically driven by use of advanced medical technology, thus raising the cost of treatment. The issue of affordability is further magnified by the low penetration of health insurance in the world. "Best buys" or very cost-effective interventions are highly desired to be implemented even in low-resource settings, which have even been identified by the WHO for prevention and control of cardiovascular diseases.

In terms of technology, some challenges exist in the development of POC with microfluidic and smartphone-based devices meeting with ASSURED guidelines, which requires input from various stakeholders, including the institute, the industry, the hospital, and the community. For developers, the challenges are to increase performance, including specificity and sensitivity, to decrease their cost with respect to time, sample, and reagents to make robust and simple platforms that are easy to understand.

In order to address the challenges of low accessibility, affordability, and awareness and to meet the healthcare needs of the population, Indian stakeholders or world scientists need to collaborate and innovate. Currently, rapid biosensor technology using biomarkers is playing a critical role in the diagnostic revolution of cardiovascular disease with understanding of the issues involved in its initiation, symptoms, and early detection. This reduces the high risk of sudden death associated with it. Factually, effective diagnosis is still lacking. Even though traditional methods of diagnosis such as ECG and echo remain the recommended tests to determine whether patients in emergency have suffered from MI, they still lack sensitivity. Thus, the requirement for early advanced diagnostic technology is becoming increasingly important to allow initiation of lifestyle changes or appropriate medical intervention. Providentially, cardiac biomarkers are important indicators for determination of MI for more accurate and sensitive results and help doctors make decisions and give appropriate treatment to the patient. The cTnI, cTnT, and H-FABP are the gold standard by cardiology consensus for MI detection compared to MYO and CK-MB, which has also led to the development of new devices and technologies to detect these markers with high sensitivity and specificity. A fast, affordable, accurate ability to handle small sample measurement for point-of-care testing of cardiac biomarkers is hence greatly desired. The commercialized POC assay offers portability, but the limit of detection is still behind the capability of laboratory instrumentation. Early, rapid, and sensitive POC testing of the disease state becomes a vital goal for clinical diagnoses because having a quantitative measurement as soon as the symptoms of MI arise would greatly help doctors in making the best decision based on the patient's present condition. To assist the doctor in making a decision in an accurate and rapid manner, simultaneous analysis of multiple markers with a single test using small volumes of blood sample is also greatly required, along with disease stage quantification using minimized diagnostic expenses. Multiple biomarker detections can be achieved by embedding the POC device with microfluidics that guide the blood sample containing cardiac biomarkers to flow through several sensing areas of the POC.

In order to address the challenges of new technology, it is believed that smartphones will change how healthcare is delivered and further facilitate personalized medicine by cultivating next-generation imaging, diagnostic, and measurement tools in the near future; an integrated smartphone-based system for simultaneous biochemical assays, physiological examination, and imaging may be developed for monitoring CVD at the POC level. Apparently, regulatory approval processes are a major concern in the transitional processes of such medical devices. These could be solved to rely on a standardized and regulated supply of certain mobile phones and also to make smartphone-supported adaptors for sensing.

Significantly, in terms of care and technology of CVD, generation of a large data set of patients complaining of chest pain is highly needed. Such data will be accessible to physicians and medical authorities through cloud services for accurate diagnostics and management of AMI. Cloud computing, integrated with microfluidics and smartphone-based devices, will enable personalized medicine. All the healthcare data can be uploaded to cloud storage, different platforms (e.g., personal computer), and parties (e.g., personal doctor), and then can be downloaded for further assessment.

Apart from POC development, population-wide interventions can be implemented to reduce CVDs. These include comprehensive tobacco control policies; taxation to reduce the intake of foods high in fat, sugar, and salt; encouraging people to increase physical activity by building walking and cycle paths; plans to reduce harmful use of alcohol; and providing healthy school meals to children.

Acknowledgments

This work is financially supported by SERB, India under Young Scientist Start-up Grant (Project Code: YSS/2015/000258).

MB acknowledges the Department of Science and Technology, New Delhi for the women scientist project grant (SR/WOS-A/LS-93/2013).

References

Abad, L., Javier del Campo, F., Muñoz, F., Fernández, L., Calavia, D., Colom, G., Salvador, J. et al. 2012. Design and fabrication of a COP-based microfluidic chip: Chronoamperometric detection of troponin T. *Electrophoresis*, 33(21), 3187–3194.

Abidov, A. 2016. Manual of echocardiography (1st edition). *Echocardiography*, 33(6), 950–950.

Adame, I., de Koning, P., Lelieveldt, B., Wasserman, B., Reiber, J. and van der Geest, R. 2006. An integrated automated analysis method for quantifying vessel stenosis and plaque burden from carotid MRI images: Combined postprocessing of MRA and vessel wall MR. *Stroke*, 37(8), 2162–2164.

Ahmad, N. 2005. Is coronary heart disease rising in India? A systematic review based on ECG defined coronary heart disease. *Heart*, 91(6), 719–725.

Allinson, J. 2011. Automated immunoassay equipment platforms for analytical support of pharmaceutical and biopharmaceutical development. *Bioanalysis*, 3(24), 2803–2816.

Apple, F. 2007. Cardiac troponin monitoring for detection of myocardial infarction: Newer generation assays are here to stay. *Clinica Chimica Acta*, 380(1–2), 1–3.

Bingisser, R., Cairns, C., Christ, M., Hausfater, P., Lindahl, B., Mair, J., Panteghini, M., Price, C. and Venge, P. 2012. Cardiac troponin: A critical review of the case for point-of-care testing in the ED. *The American Journal of Emergency Medicine*, 30(8), 1639–1649.

Biomarker Definitions Working Group. 2001. Biomarkers and surrogate endpoints: Preferred definitions and conceptual framework, 2001. *Clinical Pharmacology & Therapeutics*, 69(3), 89–95.

Bodor, G.S., Porter, S., Landt, Y., Ladenson, J.H. 1992. Development of monoclonal antibodies for an assay of cardiac troponin-I and preliminary results in suspected cases of myocardial infarction. *Clinical Chemistry* 38, 2203–2214.

Boyd D.P. and Lipton M.J. 1983. Cardiac computed tomography. *Proceedings of the IEEE*, 71, 298–307.

Braunwald, E. 1997. *Heart Disease* (1st edition). Philadelphia: Saunders.

Breathnach, C. and Westphal, W. 2006. Early detectors of the heart's electrical activity. *Pacing and Clinical Electrophysiology*, 29(4), 422–424.

Cai, J. 2005. In vivo quantitative measurement of intact fibrous cap and lipid-rich necrotic core size in atherosclerotic carotid plaque: Comparison of high-resolution, contrast-enhanced magnetic resonance imaging and histology. *Circulation*, 112(22), 3437–3444.

Casas, J., Shah, T., Hingorani, A., Danesh, J. and Pepys, M. 2008. C-reactive protein and coronary heart disease: A critical review. *Journal of Internal Medicine*, 264(4), 295–314.

Chan, D. 1994. Immunoassay automation: A practical guide. *Biometrics*, 50(3), 897.

Cheng, Y., Chen, K., Meyer, N., Yuan, J., Hirst, L., Chase, P. and Xiong, P. 2011. Functionalized SnO_2 nanobelt field-effect transistor sensors for label-free detection of cardiac troponin. *Biosensors and Bioelectronics*, 26(11), 4538–4544.

Cho, I.-H., Paek, E.-H., Kim, Y.-K., Kim, J.-H. and Paek, S.-H. 2009. Chemiluminometric enzyme-linked immunosorbent assays (ELISA)-on-a-chip biosensor based on cross-flow chromatography. *Anal. Chim. Acta* 632, 247–255.

Chua, J., Chee, R., Agarwal, A., Wong, S. and Zhang, G. 2009. Label-free electrical detection of cardiac biomarker with complementary metal-oxide semiconductor-compatible silicon nanowire sensor arrays. *Analytical Chemistry*, 81(15), 6266–6271.

Dahlöf, B. 2010. Cardiovascular disease risk factors: Epidemiology and risk assessment. *The American Journal of Cardiology*, 105(1), 3A–9A.

Das, U. 2016. Heart-type fatty acid-binding protein (H-FABP) and coronary heart disease. *Indian Heart Journal*, 68(1), 16–18.

Davis, K., Abrams, B., Lin, Y. and Jayasena, S. D. 1996. Use of a high affinity DNA ligand in flow cytometry. *Nucleic Acids Research*, 24(4), 702–706.

Dhesingh, R. S., Gobi, K.V. and Miura, N. 2007. Recent advancements in surface plasmon resonance immunosensors for detection of small molecules of biomedical, food and environmental interest. *Sensors and Actuators B*, 121, 158–177.

Dhingra, R. and Vasan, R. 2017. Biomarkers in cardiovascular disease: Statistical assessment and section on key novel heart failure biomarkers. *Trends in Cardiovascular Medicine*, 27(2), 123–133.

Dutra, R. and Kubota, L. 2007. An SPR immunosensor for human cardiac troponin T using specific binding avidin to biotin at carboxymethyldextran-modified gold chip. *Clinica Chimica Acta*, 376(1–2), 114–120.

Dutra, R., Mendes, R., Lins da Silva, V. and Kubota, L. 2007. Surface plasmon resonance immunosensor for human cardiac troponin T based on self-assembled monolayer. *Journal of Pharmaceutical and Biomedical Analysis*, 43(5), 1744–1750.

Engvall, E. and Perlmann, P. 1971. Enzyme-linked immunosorbent assay (ELISA) quantitative assay of immunoglobulin G. *Immunochemistry*, 8(9), 871–874.

Fan, X., White, I., Shopova, S., Zhu, H., Suter, J. and Sun, Y. 2008. Sensitive optical biosensors for unlabeled targets: A review. *Analytica Chimica Acta*, 620(1–2), 8–26.

Fathil, M.F., Md Arshad, M.K., Gopinath, S.C., Hashim U., Adzhri R., Ayub R.M., Ruslinda A.R. et al. 2015. Diagnostics on acute myocardial infarction: Cardiac troponin biomarkers. *Biosensors and Bioelectronics*, 70, 209–220.

Floriano, P., Christodoulides, N., Miller, C., Ebersole, J., Spertus, J., Rose, B., Kinane, D. et al. 2009. Use of saliva-based nano-biochip tests for acute myocardial infarction at the point of care: A feasibility study. *Clinical Chemistry*, 55(8), 1530–1538.

Gaziano, T.A. 2005. Cardiovascular disease in the developing world and its cost-effective management. *Circulation*, 112, 3547–3553.

Giri, B., Pandey, B., Neupane, B. and Ligler, F. 2016. Signal amplification strategies for microfluidic immunoassays. *TrAC Trends in Analytical Chemistry*, 79, 326–334.

Gomes-Filho, S., Dias, A., Silva, M., Silva, B. and Dutra, R. 2013. A carbon nanotube-based electrochemical immunosensor for cardiac troponin T. *Microchemical Journal*, 109, 10–15.

Hallermayer, K., Klenner, D. and Vogel, R. 1999. Use of recombinant human cardiac troponin T for standardization of third generation troponin T methods. *Scandinavian Journal of Clinical and Laboratory Investigation*, 59, 128–131.

Han, X., Li, S., Peng, Z., Othman, A. and Leblanc, R. 2016. Recent development of cardiac troponin I detection. *ACS Sensors*, 1(2), 106–114.

Hu, J., Cui, X., Gong, Y., Xu, X., Gao, B., Wen, T., Lu, T. and Xu, F. 2016. Portable microfluidic and smartphone-based devices for monitoring of cardiovascular diseases at the point of care. *Biotechnology Advances*, 34(3), 305–320.

Hu, J., Wang, L., Li, F., Han, Y., Lin, M., Lu, T. and Xu, F. 2013. Oligonucleotide-linked gold nanoparticle aggregates for enhanced sensitivity in lateral flow assays. *Lab on a Chip*, 13(22), 4352.

Jaffe, A. and Ordonez-Llanos, J. 2010. High sensitivity troponin in chest pain and acute coronary syndromes. A step forward?. *Revista Española de Cardiología (English Edition)*, 63(7), 763–769.

Katus, H.A., Looser, S., Hallermayer, K., Essig, U., Geub, U. 1992. Development and *in vitro* characterization of a new immunoassay of cardiac troponin T. *Clin. Chem.* 38, 386–393.

Kim, B.B., Im, W.J., Byun, J.Y., Kim, H.M., Kim, M.-G., Shin, Y.-B. 2014. Label-free CRP detection using optical biosensor with one-step immobilization of antibody on nitrocellulose membrane. *Sensors and Actuators B: Chemical*, 190, 243–8.

King, K., Grazette, L., Paltoo, D., McDevitt, J., Sia, S., Barrett, P., Apple, F. et al. 2016. Point-of-care technologies for precision cardiovascular care and clinical research. *JACC: Basic to Translational Science*, 1(1–2), 73–86.

Kong, T., Su, R., Zhang, B., Zhang, Q. and Cheng, G. 2012. CMOS-compatible, label-free silicon-nanowire biosensors to detect cardiac troponin I for acute myocardial infarction diagnosis. *Biosensors and Bioelectronics*, 34(1), 267–272.

Köhler, G. and Milstein, C. 1976. Derivation of specific antibody-producing tissue culture and tumor lines by cell fusion. *European Journal of Immunology*, 6(7), 511–519.

Kurihara, T., Yanagida, A., Yokoi, H., Koyata, A., Matsuya, T., Ogawa, J., Okamura, Y. and Miyamoto, D. 2008. Evaluation of cardiac assays on a benchtop chemiluminescent enzyme immunoassay analyzer, PATHFAST. *Analytical Biochemistry*, 375(1), 144–146.

Labarthe, D. and Dunbar, S. 2012. Global cardiovascular health promotion and disease prevention: 2011 and beyond. *Circulation*, 125(21), 2667–2676.

Lee, I., Luo, X., Huang, J., Cui, X. and Yun, M. 2012. Detection of cardiac biomarkers using single polyaniline nanowire-based conductometric biosensors. *Biosensors*, 2(4), 205–220.

Levine, G. 2016. *Illustrated Guide to Cardiovascular Disease* (1st edition), Jaypee Brothers Medical Publishers Private Limited, USA.

Li, Y., Xuan, J., Song, Y., Wang, P. and Qin, L. 2015. A microfluidic platform with digital readout and ultra-low detection limit for quantitative point-of-care diagnostics. *Lab on a Chip*, 15(16), 3300–3306.

Lilly, L. and Braunwald, E. 2012. *Braunwald's Heart Disease* (1st edition). Philadelphia, PA: Saunders/Elsevier.

Liu, J., Chen, C., Ikoma, T., Yoshioka, T., Cross, J., Chang, S., Tsai, J. and Tanaka, J. 2011. Surface plasmon resonance biosensor with high anti-fouling ability for the detection of cardiac marker troponin T. *Analytica Chimica Acta*, 703(1), 80–86.

MacKay, S., Wishart, D., Xing, J. and Jie Chen. 2014. Developing trends in aptamer-based biosensor devices and their applications. *IEEE Transactions on Biomedical Circuits and Systems*, 8(1), 4–14.

Marquette, C., Bouteille, F., Corgier, B., Degiuli, A. and Blum, L. 2008. Disposable screen-printed chemiluminescent biochips for the simultaneous determination of four point-of-care relevant proteins. *Analytical and Bioanalytical Chemistry*, 393(4), 1191–1198.

Muller-Bardorff, M., Hallermayer, K., Schro, A., Ebert, C., Borgya, A., Gerhardt, W., Remppis, A., Katus, H.A. and Elisa, T. 1997. Improved troponin T ELISA specific for cardiac troponin T isoform: Assay development and analytical and clinical validation, *Clinical Chemistry* 43, 458–466.

Obata, K., Segawa, O., Yakabe, M., Ishida, Y., Kuroita, T., Ikeda, K., Kawakami, B. et al. 2001. Development of a novel method for operating magnetic particles, Magtration Technology, and its use for automating nucleic acid purification. *Journal of Bioscience and Bioengineering*, 91(5), 500–503.

Oh, Y.K., Joung, H.-A., Kim, S., Kim, M.-G. 2013. Vertical flow immunoassay (VFA) biosensor for a rapid one-step immunoassay. *Lab on a Chip*, 13, 768–72.

Oliveira, G., Avezum, A. and Roever, L. 2015. Cardiovascular disease burden: Evolving knowledge of risk factors in myocardial infarction and stroke through population-based research and perspectives in global prevention. *Frontiers in Cardiovascular Medicine*, 2, 1–3.

Qureshi, A., Gurbuz, Y. and Niazi, J. 2012. Biosensors for cardiac biomarkers detection: A review. *Sensors and Actuators B: Chemical*, 171–172, 62–76.

Reddy, K. and Yusuf, S. 1998. Emerging epidemic of cardiovascular disease in developing countries. *Circulation*, 97(6), 596–601.

Reddy, K.S. 1993. Cardiovascular disease in India. *World Health Stat Q*, 46:101–107.

Ryu, Y., Jin, Z., Kang, M. and Kim, H. 2011. Increase in the detection sensitivity of a lateral flow assay for a cardiac marker by oriented immobilization of antibody. *BioChip Journal*, 5(3), 193–198.

Schroeder, H.R., Vogelhut, P.O., Carrico, R.J., Boguslaski, R.C. and Buckler, R.T. 1976. Competitive protein binding assay for biotin monitored by chemiluminescence. *Analytical Chemistry* 48, 1933–1937.

Shu-hai, J., Ting, F.A.N., Li-juan, L.I.U., Yi, C., Xiao-qing, Z. 2014. The detection of cTnI by the aptamer biosensor. *Progress in Biochemistry and Biophysics*, 41, 916–920.

Silva, B., Cavalcanti, I., Silva, M. and Dutra, R. 2013. A carbon nanotube screen-printed electrode for label-free detection of the human cardiac troponin T. *Talanta*, 117, 431–437.

Smith, S., Ralston, J. and Taubert, K. 2012. *Urbanization and Cardiovascular Disease: Raising Heart-Healthy Children in Today's Cities*. Geneva: The World Heart Federation.

Song, S., Han, Y., Kim, K., Yang, S. and Yoon, H. 2011. A fluoro-microbead guiding chip for simple and quantifiable immunoassay of cardiac troponin I (cTnI). *Biosensors and Bioelectronics*, 26(9), 3818–3824.

Srinath Reddy, K. 1999. Emerging epidemic of cardiovascular disease in the developing countries. *Atherosclerosis*, 144, 143.

Stefanini, G. and Windecker, S. 2015. Can coronary computed tomography angiography replace invasive angiography? *Circulation*, 131(4), 418–426.

Stewart, S. 2011. Tackling heart disease at the global level: Implications of the United Nations' Statement on the prevention and control of noncommunicable disease. *Circulation: Cardiovascular Quality and Outcomes*, 4(6), 667–669.

Tsaloglou, M., Jacobs, A. and Morgan, H. 2014. A fluorogenic heterogeneous immunoassay for cardiac muscle troponin cTnI on a digital microfluidic device. *Analytical and Bioanalytical Chemistry*, 406(24), 5967–5976.

Tuteja, S., Priyanka, Bhalla, V., Deep, A., Paul A. and Suri, C. 2014. Graphene-gated biochip for the detection of cardiac marker troponin I. *Analytica Chimica Acta*, 809, 148–154.

Vella, S., Beattie, P., Cademartiri, R., Laromaine, A., Martinez, A., Phillips, S., Mirica, K. and Whitesides, G. 2012. Measuring markers of liver function using a micropatterned paper device designed for blood from a fingerstick. *Analytical Chemistry*, 84(6), 2883–2891.

Vollset, S. 2006. Smoking and deaths between 40 and 70 years of age in women and men. *Annals of Internal Medicine*, 144(6), 381.

Wang, C., Wu, J., Zong, C., Xu, J. and Ju, H. 2012. Chemiluminescent immunoassay and its applications. *Chinese Journal of Analytical Chemistry*, 40(1), 3–10.

Warren, A., Kwong, G., Wood, D., Lin, K. and Bhatia, S. 2014. Point-of-care diagnostics for noncommunicable diseases using synthetic urinary biomarkers and paper microfluidics. *Proceedings of the National Academy of Sciences*, 111(10), 3671–3676.

Wei, J., Mu, Y., Song, D., Fang, X., Liu, X., Bu, L., Zhang, H., Zhang, G., Ding, J., Wang, W., Jin, Q. and Luo, G. 2003. A novel sandwich immunosensing method for measuring cardiac troponin I in sera. *Analytical Biochemistry*, 321(2), 209–216.

World Health Organization. *WHO Report on the Global Tobacco Epidemic, 2008: The MPOWER Package*. Geneva: World Health Organization.

Wu, W., Bian, Z., Wang, W., Wang, W. and Zhu, J. 2010. PDMS gold nanoparticle composite film-based silver enhanced colorimetric detection of cardiac troponin I. *Sensors and Actuators B: Chemical*, 147(1), 298–303.

Wu, Y. and Kwon, Y. 2016. Aptamers: The "evolution" of SELEX. *Methods*, 106, 21–28.

Xu, Q., Xu, H., Gu, H., Li, J., Wang, Y. and Wei, M. 2009. Development of lateral flow immunoassay system based on superparamagnetic nanobeads as labels for rapid quantitative detection of cardiac troponin I. *Materials Science and Engineering: C*, 29(3), 702–707.

Yang, Y., Lin, H., Wang, J., Shiesh, S. and Lee, G. 2009. An integrated microfluidic system for C-reactive protein measurement. *Biosensors and Bioelectronics*, 24(10), 3091–3096.

Yetisen, A., Akram, M. and Lowe, C. 2013. Paper-based microfluidic point-of-care diagnostic devices. *Lab on a Chip*, 13(12), 2210.

Zhang, G. and Ning, Y. 2012. Silicon nanowire biosensor and its applications in disease diagnostics: A review. *Analytica Chimica Acta*, 749, 1–15.

Zhu, J., Zou, N., Zhu, D., Wang, J., Jin, Q., Zhao, J. and Mao, H. 2011. Simultaneous detection of high-sensitivity cardiac troponin I and myoglobin by modified sandwich lateral flow immunoassay: Proof of principle. *Clinical Chemistry*, 57(12), 1732–1738.

Suggested Further Reading

Douglas, L.M., Douglas, P.Z., Peter, L. and Robert, O.B. 2014. *Braunwald's Heart Disease: A Textbook of Cardiovascular Medicine*, Saunders, USA.

Global Health Observatory (GHO) data at http://www.who.int/gho/ncd/mortality_morbidity/cvd/en/

Pampel, F.C. and Praeger, S.P. 2004. *Progress against Heart Disease*. Praeger, Westport, CT.

Shephard, M. *A Practical Guide to Global Point-of-Care Testing*. CSIRO Publishing, Clayton, Vic., 2016.

4

Emerging Trends in Nanotechnology for Diagnosis and Therapy of Lung Cancer

Nanda Rohra, Manish Gore, Sathish Dyawanapelly, Mahesh Tambe, Ankit Gautam, Meghna Suvarna, Ratnesh Jain, and Prajakta Dandekar

CONTENTS

4.1 Introduction 106
    4.1.1 Lung Cancer 106
    4.1.2 Types of Lung Cancer 107
    4.1.3 Causes of Lung Cancer 107
    4.1.4 Stages of Lung Cancer 107
    4.1.5 Diagnosis of Lung Cancer 108
    4.1.6 Conventional Treatment Modalities 108
        4.1.6.1 Surgery 110
        4.1.6.2 Radiation Therapy 110
        4.1.6.3 Chemotherapy 112
        4.1.6.4 Immunotherapy 112
        4.1.6.5 Small Molecule Inhibitors 115
4.2 Need for Nanotechnology in Diagnosis and Therapy of Lung Cancer 116
    4.2.1 Nanotechnology 116
    4.2.2 Advantages of Nanocarriers 118
    4.2.3 Targeting Approaches for Nanocarriers 118
        4.2.3.1 Passive Targeting for Lung Cancer 118
        4.2.3.2 Active Targeting for Lung Cancer 120
4.3 Nanoplatforms for Lung Cancer 122
    4.3.1 Polymer-Based Nanoparticles for Lung Cancer 122
        4.3.1.1 Poly (Lactic-Co-Glycolic Acid) Nanoparticles for Lung Administration 123
        4.3.1.2 Stealth Polymer: Poly Ethylene Glycol–Coated Nanoparticles for Lung Cancer 125
        4.3.1.3 Chitosan Nanoparticles for Lung Cancer 126
        4.3.1.4 Dendrimers—Architectural Nanoparticles for Lung Cancer 127
        4.3.1.5 Polymeric Micelles 128
        4.3.1.6 Polymeric Nanofibers 128
    4.3.2 Metal-Based Nanoparticles and Miscellaneous Nanoparticles for Lung Cancer 133
        4.3.2.1 Gold Nanoparticles 133
        4.3.2.2 Silver Nanoparticles 134
        4.3.2.3 Magnetic Nanoparticles 135

		4.3.2.4	Carbon-Based Nanoparticles	136
		4.3.2.5	Quantum Dots	138
		4.3.2.6	Nanoshells	139
		4.3.2.7	Mesoporous Silica Nanoparticles	140
		4.3.2.8	Lanthanide Nanoparticles	140
	4.3.3	Bio-Based Nanoparticles		141
		4.3.3.1	Apoferritin	145
		4.3.3.2	Viral Nanoparticles	145
		4.3.3.3	Protein-Based Nanoparticles	146
		4.3.3.4	Liposomes	147
		4.3.3.5	Solid Lipid Nanoparticles	148
4.4	Clinical Studies of Nanosystems for Lung Cancer			149
	4.4.1	Abraxane®		149
	4.4.2	Lipoplatin™		154
	4.4.3	Genexol®-PM		155
4.5	Conclusion and Future Perspectives			157
Acknowledgments				158
References				158

4.1 Introduction

Cancer is one of the major causes of deaths worldwide, with increasing incidences occurring every year (Wang et al. 2008). It is expected to surpass cardiac disorders and become the leading cause of mortality around the globe in the coming years (Siegel et al. 2015). In 2016, an estimated 595,690 people are expected to die from cancer in the United States, while 1,685,210 new cases of cancer are expected to be diagnosed (American Chemical Society 2016b). Moreover, the statistics for the year 2016 for India predict an estimated 14,50,000 new patients to be diagnosed with cancer, while 736,000 of these are anticipated to succumb to this disease (ICMR 2014). Among different types of cancers, lung cancer is the most common cause of cancer-related mortalities and has been affecting patients belonging to both genders for decades (Zhao et al. 2013). The graveness of this disease is alarming and has subsequently been highlighted.

4.1.1 Lung Cancer

Among several cancers, maximum mortalities can be expected due to lung cancer. The US statistics for 2016 regarding patients with lung cancer have indicated an alarming number of 224,390 patients affected by the disease, while 158,080 of these are likely to be faced with disease-related mortality by the end of this year (American Chemical Society 2016b). In India specifically, an approximate 114,000 new cases are projected to be afflicted with lung cancer (ICMR 2014).

The number of deaths due to lung cancer has already surpassed those due to breast and colorectal cancers combined (Badrzadeh et al. 2014). The diagnosis of lung cancer is usually conducted at later stages of the disease, leading to a poor survival rate among the patients. Moreover, metastatic tumors are difficult to operate on due to the limited efficiency of conventional treatment modalities. As a result, the majority of such patients succumb to the disease within a span of 2 years due to its dissemination to lung tissues and other parts of the body (Taratula et al. 2011).

4.1.2 Types of Lung Cancer

Two main types of lung cancer have been described in the literature, depending upon the size, shape, and histological information. As different types of lung cancer are treated differently, it is important to understand the prevalence and nature of each type. The main features of different types of lung cancer have been highlighted in Table 4.1.

Although Table 4.1. provides a realistic layout regarding the different forms of lung cancer, the occurrence of disease is primarily observed due to genetic, environmental, and lifestyle-related factors. Detailed elaboration on the causative factors is given in the following sections of this manuscript.

4.1.3 Causes of Lung Cancer

Lung cancer is the result of a combination of genetic alterations or mutations like deletions, point mutations, translocations, and amplifications, arising due to various environmental factors (Mehta et al. 2015). Tobacco or cigarette smoking is the most dominant cause and is responsible for around 90% of cases of lung cancer (Khan and Mukhtar 2015). Even after cessation of smoking, chances of occurrence of the disease are still very high and such people are included in the high risk category. Other causative factors of lung cancer include exposure to radon gas, asbestos, arsenic, chromium, nickel, passive smoking, and air pollution (Badrzadeh et al. 2014). The progression of lung cancer occurs in various stages, which are described as follows.

4.1.4 Stages of Lung Cancer

Most of the therapies and/or diagnostic strategies developed for lung cancer are stage specific, which thus underlines the significance of their detailed study prior to application

TABLE 4.1

Major Features of Different Types of Lung Cancer

Sr. No	Types of Lung Cancer	Major Features
1	Non-small cell lung cancer	• Most commonly occurring, affects 85%–90% of patients with lung cancer • Tumors differ in size and shape and have unusual histology
	A. Squamous cell (epidermoid) carcinoma	• Affects 25%–30% of patients with lung cancer • Begins in cells lining lung airways
	B. Adenocarcinoma	• Affects nearly 40% of patients, found in outer parts of the lung • Begins in mucus secreting cells • Tends to spread at slower rate • Occurs due to exposure to carcinogens and radiation
	C. Large cell (undifferentiated) carcinoma	• Afflicts approximately 10%–15% of patients • Tends to grow and spread quickly • Difficult to treat
2	Small cell lung cancer	• Affects nearly 10%–15% of patients with lung cancer • Tends to spread quickly throughout body • Very rare in nonsmokers

of a particular treatment and/or diagnostic approach. Lung cancer is generally classified into different stages on the basis of anatomical changes of tumor that occur inside the lung and the extent of its spread to other body tissues and lymph nodes.

Stage I: Tumor is generally localized and characterized by small tumor volume (less than 3 cm) limited to the lung tissue and has not spread to the lymph nodes. Surgery is used as main treatment approach at this stage.

Stages II and III: During these stages, local advancement and progression of the tumors occur, thereby increasing the tumor volume (larger than 3 cm). It may involve spread to nearby tissues (Stage II) as well as lymph nodes and invasion to other portions of the lungs (Stage III). Stage III is considered a locally advanced stage of the disease. Thus, surgical removal of the tumor is difficult to achieve at this stage. Stage II is usually treated by adjuvant chemotherapy (viz. postsurgical chemotherapy). A combination of chemotherapy and high-dose radiotherapy constitutes to be the main curative strategy for stage III of lung cancer.

Stage IV: At this stage, the tumor spreads beyond the affected lung to other lung regions, and metastasis occurs in distant parts of the body. Systemic radiotherapy and chemotherapy are used as the mainstay treatment strategies for advanced stages of lung cancer, and provide relief through palliation of symptoms (Wynants et al. 2007, Howington et al. 2013, Ramnath et al. 2013, Mehta et al. 2015).

Stage-specific therapy of lung cancer can be initiated following early diagnosis of the disease stage, which may be accomplished via various techniques.

4.1.5 Diagnosis of Lung Cancer

Early diagnosis of lung cancer can go a long way in improving the survival rates among the patients by allowing earlier initiation of treatments, thus reducing the fatalities associated with the disease. Also, sophisticated diagnostic techniques can help in identification of patients predisposed to the disease, thereby reducing the number of probable cases. Methods like X-ray, CT scan, and so on may fail in some cases to diagnose tumors, as the detection capacity of these techniques is solely dependent on the tumor size. Low-dose computerized tomography (LDCT) is a screening technique generally recommended for high-risk individuals, such as smokers and former smokers, as it involves a very low dose of ionizing radiation (Siegel et al. 2015). The outcomes of study conducted by National Lung Screening Trial (NLST) in 2011 and Lung Cancer module of the Cancer Risk Management Model (CRMM-LC) in 2014 showed that LDCT screening in high-risk populations had reduced lung cancer mortality by 20% and 23%, respectively. Although LDCT-mediated risk assessment was found to be effective, further studies were conducted in order to investigate better tools for screening high-risk individuals to overcome the variability in epidemiological studies (Flanagan et al. 2015).

Alternative screening tests based on graphical and histopathological assays are routinely used. These are briefly described in Table 4.2.

Once the degree of malignancy is revealed with the screening techniques, as described in Table 4.2, treatment modalities are ascertained and followed for eradication of the tumor. Approaches for the treatment of lung cancer are discussed in the following sections.

4.1.6 Conventional Treatment Modalities

Depending upon the tumor stage, various therapeutic strategies have been employed for the treatment of lung cancer. Conventional treatment methods like surgery, radiation

TABLE 4.2
Summary of Screening Techniques for Lung Cancer Diagnosis

Technique	Description	Advantages	Disadvantages	References
Fluorescence bronchoscopy	Invasive technique used to identify preinvasive lesions and detect newly growing tumors in patients who have already undergone surgery for NSCLC	More sensitive than white light Bronchoscopy	Low specificity	Colella et al. (2014), Lam et al. (2000)
Magnetic resonance imaging (MRI)	It can be used in combination with super paramagnetic iron oxide nanoparticles (SPIONs)	High spatial resolution provides both anatomical and functional information	Low sensitivity and requires expensive infrastructure	Koyama et al. (2013), Wu et al. (2011)
Positron emission tomography (PET)	Noninvasive technique, detects stages I, II, and III of NSCLC	High sensitivity and provides biochemical information	Limited anatomical information and requires expensive infrastructure	Kakhki (2007)
Endoscopic ultrasound scan	Invasive technique used in mediastinal lymph node staging as well as for detection of advanced stages of lung cancer	High quality ultrasound images and least invasive modality	Inability to assess nodes in the anterior mediastinum	Colella et al. (2014), Fickling and Wallace (2002)
Mediastinoscopy	Invasive technique with a sensitivity of approximately 78%, used for mediastinal tissue staging	Considered the gold standard in mediastinal staging	It does not cover all mediastinal lymph node stations	Annema et al. (2010)
Low-dose helical CT	Noninvasive technique, detects early stage of lung cancer	Low radiation exposure	Threat associated with exposure to radiation, expensive equipment	Marshall et al. (2013)
Biopsy	Involves excision of small part of the affected tissue for examination under the microscope to determine the type of lung cancer being investigated	High diagnostic accuracy	Chances of pulmonary infection and severe bleeding/hemorrhage	Tsukada et al. (2000)
Sputum cytology	Detects early stages of NSCLC	Simple, noninvasive, and patient-compliant technique	Time-consuming process, chances of false positive results	Muniyappa et al. (2014), Palmisano et al. (2000)

therapy, and chemotherapy (mono- and combined), as well as more advanced treatment modules, including immunotherapy and small molecule inhibitors, are used either alone or in combination for effective treatment of lung cancer.

4.1.6.1 Surgery

Surgery is the most reliable, consistent, and successful of the treatment approaches for treating the early stages of lung cancer. However, in advanced stages of the disease, the method becomes ineffective due to tumor metastasis and involvement of other organs apart from the lung. This increases the risk associated with operating on patients. Tumors can be surgically removed by the following techniques, depending on their stage (Kim et al. 2006).

Tumors in the early stages I and II of non–small cell lung cancer (NSCLC) can be successfully removed surgically by anatomic segmentectomy, wedge resection, lobectomy, or pneumonectomy (Lang-Lazdunski 2013). These techniques are described below.

Segmentectomy: This involves removal of the cancer-afflicted portion of the lung.

Wedge resection: This is used for the removal of a tumor that is surrounded by a margin of normal lung.

Lobectomy: This effective surgical technique involves removal of an entire lobe of the lung and is routinely employed even when the tumor size is very small.

Pneumonectomy: This involves removal of a part of the lung, while in cases where the tumor is situated close to the center of the chest, the entire removal of one of the lungs becomes necessary.

The surgical procedures described above possess major disadvantages such as higher rates of recurrence; morbidity and mortality have been observed in patients suffering from advanced stages of NSCLC. Apart from the techniques mentioned above, other procedures exist for successful surgical elimination of lung tumors, which are stated subsequently (Baltayiannis et al. 2013).

Radiofrequency ablation: In this technique, electric current–mediated ablation of the tumor is carried out using a needle. The time of recovery depends upon the size of the lung tissue to be removed and health of the patient before surgery (Fernando et al. 2005).

Video-assisted thoracoscopic surgery (VATS): In this technique, a small incision or cut is made in the chest through which a thoracoscope, a tube-like structure, is inserted. The thoracoscope is equipped with light and a tiny camera that is connected to a visual screen and allows the surgeon to observe the tissue during surgery. Using the thoracoscope, a lung lobe can be removed without making a large incision in the chest (Chang and Sugarbaker 2003).

Adjuvant therapy: Adjuvant therapy is a treatment strategy given after surgery to lower the risk of lung cancer reoccurrence (Wozniak and Gadgeel 2009).

Surgical complications are observed in patients with medical contraindications like emphysema, heart diseases, other lung disorders, and so on. Hence, therapy is found to be ineffective in such individuals. In addition, metastasis of tumor cells is another contributing factor for the futility of the traditional treatment modalities. Such impediments in the clinical applications of therapy can be optimally circumvented by using a combination of diverse approaches like radiotherapy and/or chemotherapy.

4.1.6.2 Radiation Therapy

Radiation therapy involves the use of high-powered energy beams, such as X-rays or ionizing radiation, to kill cancer cells. It is used as main curative option (sometimes along with chemotherapy) for conditions in which the surgical method fails to eliminate the tumor due to its location and size as well as to provide relief from symptoms such as pain, bleeding, trouble swallowing, and so on observed during advanced stages of lung cancer. During the course of therapy, accurate and large amounts of small and highly intense beams are used to deliver potent doses of radiation (Mehta et al. 2015).

Radiation therapy can be categorized as external beam and internal beam (brachytherapy) therapies. External beam radiation therapy (EBRT) is used to treat the affected tissues or the organs wherein cancer has metastasized. Most often, radiation doses are given to lung cancer patients 5 days a week over a period of 5–7 weeks, but this can vary based on the type of EBRT and the condition for which it has been prescribed (Aygun et al. 1992). The method is not specific and affects the healthy cells surrounding the cancer cells (refer to Table 4.3). In contrast to EBRT, internal beam radiation therapy, also known as brachytherapy, involves direct exposure of a radioactive source to the tumor mass by its placement within the tissue or in its proximity (Polo et al. 2011). This technique includes high dose rate (HDR) and low dose rate (LDR) brachytherapies. HDR involves radioactive emission from the applicator after placement of the radiation source in the desired location at the rate greater than 12 Gy/h. On the other hand, LDR involves a low dose emission of approximately 2 Gy/h at the tumor site during its surgical removal (Parashar et al. 2013). A radioactive implant is directly placed in the upper airway, seeded in the tumor interstitial, or implanted within the tissue through a percutaneous route. Brachytherapy is primarily used as a curative and palliative strategy in the treatment of lung cancer. For example, implantation of radioactive seeds by CT-guided brachytherapy via a percutaneous route has been reported for the difficult-to-treat NSCLC (Polo et al. 2011).

TABLE 4.3

Drawbacks Associated with the Conventional Therapies for Lung Cancer

Conventional Therapies	Limitations	Reference
Surgery	• Probability of loss of an entire organ and cancer recurrence • Lung cancer may persist, metastasize, or recur even after surgery • Postsurgical complications including dyspnoea, pain, weakness, bleeding, and pulmonary infection • Contraindicated for patients suffering from serious heart or lung disease	Bandyopadhyay et al. (2015)
Radiation	• Radiation exposure may lead to problems like pulmonary fibrosis, radiation induced pneumonitis, coughing, fatigue, skin blistering and peeling, hair loss, etc. • Chances of healthy cells being affected along with cancerous cells, often rendering the exposed part nonfunctional • Side effects include fatigue, leukopenia, and thrombocytopenia • If the digestive system gets exposed to radiations, patients may suffer from nausea, vomiting, diarrhea, and skin irritation • Expression of reactive oxygen species (ROS) scavengers leads to the development of radiation-resistant tumors	
Chemotherapy	• Drugs used for chemotherapy commonly impart bioavailability issues, stability issues, poor water solubility, systemic toxicity, etc. • Tumors are not entirely treated and cases of recurrence are very common • The drugs used may kill normal, rapidly dividing cells (similar to tumor cells) • Side effects with chemotherapy include fatigue, gastrointestinal distress, leukopenia, anemia, weight loss, hair loss, nausea, vomiting, diarrhea, and mouth sores	

However, the method is not specific and affects the healthy cells surrounding the cancer cells (Long et al. 2014). If radiation therapy irritates or inflames the lung tissue, patients may develop cough, fever, or dyspnea for months and sometimes years after the termination of radiation therapy. Radiation therapy may also cause permanent scarring of the lung tissue in the region adjoining the area where the original tumor was located. For advanced stages of NSCLC, chemotherapy is still considered better than radiation therapy (Gupta et al. 2014).

4.1.6.3 Chemotherapy

Chemotherapy uses drugs to kill cancer cells. Although chemotherapy is commonly employed to treat lung cancer, it requires administration of high doses of drugs due to factors like low bioavailability, body clearance, and so on. Chemotherapeutic drugs are generally administered through intravenous and oral routes. Both mono- and combined chemotherapy have been employed for treating lung cancer. Drugs like cisplatin and carboplatin are platinum-based drugs that are considered the standard regimen for lung cancer monotherapy (de Castria et al. 2013). Due to their dose, limiting side effects such as nephrotoxicity, cardiotoxicity, anemia, intestinal injury, peripheral neuropathy, and so on, monotherapy is no longer considered a viable option. Also in the case of monotherapy, the cancer acquires an immediate drug resistance that can be overcome by using the drugs in suitable combinations. Combination therapy involves two or more drugs to overcome the effects of monotherapy, thereby reducing the dosage required for each and increasing the therapeutic effectiveness of an individual drug. Drugs commonly used in this therapy include a combination of platinum drugs with paclitaxel, gemcitabine, etoposide, vinblastine, and so on (Babu et al. 2013). For example, a combination of cisplatin or carboplatin with at least one of the other drugs like paclitaxel (Taxol), etoposide, and doxorubicin has commonly been employed for treating lung cancer (de Castria et al. 2013).

Formulation challenges associated with chemotherapeutic agents include poor water solubility, the need for targeted delivery, the need for controlled and sustained drug release, biodistribution, and systemic toxicity. Many of the chemotherapeutic agents used are hydrophobic in nature, which makes them poorly water soluble and limits their administration in higher doses. To overcome this problem, lipid-based and other improved formulations are required so that they can easily dissolve hydrophobic drugs and subsequently enhance their bioavailability. Moreover, limitations like multidrug resistance and severe adverse effects (refer to Table 4.3) on normal healthy cells also constitute other causes of concern that need to be addressed (Long et al. 2014). Table 4.3 provides a brief overview of the limitations encountered with the conventional therapies for lung cancer.

Thus, the failure of chemotherapy to distinguish between normal (healthy) and cancerous cells is evident, which strongly highlights the need to develop advanced therapies to achieve targeted delivery of chemotherapeutic agents and offer better efficacy and reduced toxicity. Efforts are being made to exploit immunological moieties and small molecule agents for specifically attacking cancer cells by attaching to or inhibiting the cell surface receptors of cancerous cells. The subsequent section explains the significance of immunotherapy and small molecule inhibitors in lung cancer management.

4.1.6.4 Immunotherapy

The treatment of cancer using immunotherapy involves improvement of a person's immune system, thereby rendering it suitable in identifying cancer cells and reinforcing its response to combat neoplasia (American Chemical Society 2016a). This is done by stimulating and directing the patient's immune system in order to attack cancer cells, as well as strengthening the immune system by adopting a diet including proteins and requisite supplements. Immunotherapy is also referred to as biologic therapy or biotherapy (American Chemical Society 2016a).

Different immunotherapeutic modalities that have been employed for the treatment of lung cancer include (American Chemical Society 2016c)

- Monoclonal antibodies (mAbs)
- Immune checkpoint inhibitors
- Therapeutic vaccines
- Adoptive cell therapy

4.1.6.4.1 Monoclonal Antibodies

Antibodies are proteins produced by the immune system against antigens, which function in triggering an immune response. Antibodies bind to specific antigens and signal the immune system to destroy the cells expressing those antigens (American Chemical Society 2016a). Monoclonal antibodies are antibodies produced by identical immune cells that are all clones of a unique parent cell, which binds monospecifically to antigens expressed on cancer cell surfaces (American Chemical Society 2016a).

Different types of mAbs used in immunotherapy are described in following subsections.

4.1.6.4.1.1 Naked Monoclonal Antibodies Naked mAbs work individually and are the most common type of mAb used in cancer treatment. These bind to cancer-specific antigens and work either by acting as a marker for the body's immune system to destroy them or by blocking antigens on cancer cells (or other nearby cells) that help cancer cells grow or spread (American Chemical Society 2016a). Cetuximab (Erbitux) is an example of a naked mAb that targets the epidermal growth factor receptor (EGFR) and is being tested in a phase II trial for patients with stage IIIB non-small cell lung cancer (American Chemical Society 2016c). Cetuximab in combination with carboplatin and gemcitabine resulted in a higher response rate (RR) of 27.7% and longer progression-free survival (PFS) of 5.09 months as opposed to the RR (18.2%) and PFS (4.21 months) observed with chemotherapy alone in a phase II trial conducted on patients with advanced NSCLC (Sgambato et al. 2014).

4.1.6.4.1.2 Conjugated Monoclonal Antibodies mAbs bound to a chemotherapeutic drug (chemolabeled) or a radioactive particle (radiolabeled) are called conjugated monoclonal antibodies (American Chemical Society 2016a). These mAbs attach themselves to cancer-specific antigens and deliver the drug molecules or radioactive particles to the target areas, thereby reducing the damage to healthy tissues. Conjugated mAbs are also known as tagged, labeled, or loaded antibodies. Treatment with radiolabeled mAbs is also called radioimmunotherapy (RIT) (American Chemical Society 2016a). *Bavituximab* is a chemolabeled mAb given in combination with docetaxel, which targets an immune-suppressing molecule in tumors and is being evaluated in a phase III trial for patients with late-stage nonsquamous NSCLC (American Chemical Society 2016c). Bavituximab in combination with carboplatin-paclitaxel in phase II trials demonstrated an objective response rate (ORR) of 41%, median PFS of 6 months, and median overall survival (OS) of 12 months in patients with stage IIIB/IV NSCLC in an open label study (Digumarti et al. 2014).

Patritumab, in combination with erlotinib (Tarceva), is a human epidermal growth factor receptor-3 (HER3)-targeted chemolabeled mAb and was tested as a phase I trial in 24 patients with stage IIIB/IV NSCLC. The combination showed a median PFS of 44.0 days for the EGFR wild-type group and 107.0 days for the EGFR-activating mutation group. The medication was found to be well tolerated and its efficacy was found to be encouraging enough to be escalated for further studies (Nishio et al. 2015). Currently, it is being tested in a phase III trial for patients with locally advanced or metastatic NSCLC (American Chemical Society 2016c). Such mAbs are also referred to as antibody-drug conjugates (ADCs). However, conjugated mAbs such as Bevacizumab (Avastin®), a US Food and Drug Administration (FDA)-approved mAb for lung cancer treatment, pose some serious side effects such as kidney damage, high blood pressure, bleeding, poor wound healing, blood clots, and so on (American Chemical Society 2016a).

4.1.6.4.2 Immune Checkpoint Inhibitors

The immune system has a unique ability to differentiate between itself and foreign molecules. In order to distinguish between the self and nonself, it uses checkpoints (molecules on certain immune cells) that need to be activated or deactivated in order to elicit an immune response (American Chemical Society 2016a). Cancer cells use these checkpoints that render the immune system incapable of differentiating between healthy cells and cancer cells. The cancer cells, hence, evade the immune system responses. Immune checkpoint inhibitors are drugs that specifically target these checkpoints and thereby aid the immune system in identification and subsequent killing of cancer cells. PD-1 is a checkpoint protein on T cells. PD-1 binds to PD-L1, a protein found in normal cells, which acts as an off switch and helps in keeping the T cells from attacking normal cells in the body. Some cancer cells have large amounts of PD-L1, which helps them avoid immune responses (Aldarouish and Wang 2016, American Chemical Society 2016a). mAbs target and inhibit PD-1 or PD-L1 proteins, which enables identification of cancer cells and elicits an immune response against them. Examples of PD-1 inhibitors include Pembrolizumab (Keytruda) and Nivolumab (Opdivo), which are used in the treatment of NSCLC. Atezolizumab (Tecentriq) and Durvalumab are examples of PD-L1 inhibitors that are also used to treat NSCLC (Aldarouish and Wang 2016, American Chemical Society 2016c). The phase III study of *Nivolumab* carried out on patients with advanced squamous NSCLC to evaluate its efficacy and safety showed considerable improvement in RR (20% with nivolumab and 9% with docetaxel), OSR (42% with nivolumab and 24% with docetaxel), and PFS (3.5 months with nivolumab and 2.8 months with docetaxel) of the patients as compared to docetaxel (Brahmer et al. 2015).

The drawbacks associated with the use of drugs lies in their ability to attack normal, healthy cells of the body, which could evoke serious problems in the lungs, intestine, liver, kidneys, and hormone-making glands. Other common side effects include fatigue, cough, nausea and loss of appetite, skin rash, and itching (American Chemical Society 2016a).

4.1.6.4.3 Therapeutic Vaccines

A vaccine is a biomolecule that develops active acquired immunity to a particular disease (American Chemical Society 2016a). Therapeutic vaccines used for cancer treatment direct the immune system to mount an immune response against shared or tumor-specific antigens (American Chemical Society 2016c). Anticancer vaccines comprise cancer cells, parts of cells, or pure antigens. These are given in combination with other substances or cells, called adjuvants, that help in enhancing the immune response (American Chemical Society 2016a). Commonly targeted antigens include MAGE-3 (found in 42% of lung cancers), NY-ESO-1 (found in 30% of lung cancers), p53, survivin, and MUC1 (Aldarouish and Wang 2016, American Chemical Society 2016c). *Tergenpumatucel-L* (HyperAcute®) is a therapeutic vaccine that is currently being tested in phase II/III trial for patients with stage III or IV NSCLC. It consists of genetically modified human lung cancer cells comprising a mouse gene. *DRibbles* (DPV-001) is a therapeutic vaccine formulated using nine cancer antigens and toll-like receptor (TLR) adjuvants that is being tested in a phase II trial for patients with stage III NSCLC (American Chemical Society 2016c). In a phase I study, the autologous DRibbles vaccine was found to be safe when combined with docetacel and GM-CSF (Page et al. 2016).

4.1.6.4.4 Adoptive Cell Therapy

Adoptive cell therapy works by improving the immune system's response toward cancer cells. The therapy employs the use of T cells from the patient, which are subjected to genetic or chemical modification with an aim of enhancing their activity. Such modified T cells are reintroduced into the patient's body and are expected to improve the immune response. Engineered T cells are being tested in phase I/II trials to target VEGFR2 in patients with lung cancer. Tumor-infiltrating lymphocytes (TILs) are being tested in phase II trials for patients with NSCLC, followed by chemotherapy (American Chemical Society 2016c).

4.1.6.5 Small Molecule Inhibitors

Receptor tyrosine kinases (RTKs) and nonreceptor tyrosine kinases (non-RTKs) are essential to facilitate signaling pathways used for cell cycle regulation, migration, cell proliferation, angiogenesis, and differentiation (Imai and Takaoka 2006). Many such RTKs and non-RTKs are released during formation of cancerous tissue. Small molecule inhibitors are organic molecules of low molecular weight (<900 Da), with a size on the order of 10^{-9} m. These therapeutic actives work by targeting the kinases, thereby creating direct impact on the growth of tumor cells (Imai and Takaoka 2006). Small molecule inhibitors do not trigger any immune response. Moreover, owing to their small size, they are capable of permeating through the plasma membrane and interacting with the cytoplasmic domain of cell-surface receptors and intracellular signaling molecules. *Gefitinib* (Iressa) and *Erlotinib* (Tarceva) are small molecule agents that selectively inhibit EGFR expressed in NSCLC (Imai and Takaoka 2006). Crizotinib (Xalkori), another small molecule inhibitor, also selectively inhibits HGFR expressed in NSCLC (Lavanya et al. 2014).

Thus, the combinatorial method of the aforementioned therapies to treat different types of lung cancer is quite evident. The pressing need to develop alternative therapeutic strategies to overcome certain limitations posed by conventional therapy and reduce the number of mortalities associated with lung cancer is manifested. Thus, attempts are being made to design appropriate carrier molecules to direct drugs to the target sites, thereby improving their cellular uptake and therapeutic response. Among the carrier molecules investigated, nanocarriers, uniquely developed on the concepts of nanotechnology, have shown great promise in lung cancer therapy. These nanosized particulate carriers can easily facilitate targeted delivery and release of the therapeutic and/or imaging agents in a sustained and controlled manner through receptor-mediated mechanisms. This will further promote a higher systemic availability of the administered drugs and reduce their dosing frequency. The significant contributions of nanotechnology-based advances in the field of lung cancer treatment and/or diagnosis are elaborately described in the following sections.

4.2 Need for Nanotechnology in Diagnosis and Therapy of Lung Cancer

4.2.1 Nanotechnology

Nanotechnology involves the engineering of matter at the atomic and molecular level so as to provide materials and devices possessing remarkably new and varied properties and whose smallest functional organization in at least one dimension is on the nanometer scale (ranging from a few to several hundred nanometers). It is a rapidly expanding area of research with huge potential in many sectors, including the healthcare area (Jena et al. 2013, Patil et al. 2014).

This burgeoning field has contributed to significant scientific and technological advances in the field of medicine and physiology, creating more sophisticated and efficacious opportunities for disease treatment. Nanotechnology-based innovations for therapy and diagnosis are focused on design of materials and devices that can interact with the body at the subcellular (i.e., molecular) level with a high degree of specificity (Sahoo et al. 2007). In addition, it also aims at improvement of the existing strategies by achieving more effective and targeted delivery of therapeutic moieties as well as accurate diagnosis in several disease conditions like cancer, cryptococcal meningitis, and nosocomial infections. It has enabled provision of targeted and tissue-specific drug delivery carriers to achieve maximum drug efficacy with minimal side effects (Sahoo et al. 2007). Nanomedicine, a subset of nanotechnology, utilizes its concepts to develop advanced modes of therapy and diagnosis. These have gained huge popularity due to their ability to overcome biological barriers, selectively target affected sites, and effectively deliver payload to these sites. Such improved nanosized carriers are being increasingly investigated and employed in the field of neoplasm diagnosis and treatment (Babu et al. 2013).

Nanotechnology has been widely applied in the field of therapeutics, diagnostics, and theranostics for cancer therapy, as elucidated in Figure 4.1. In therapeutics, nanomedicine-based systems have been explored for selective delivery of drugs, proteins, peptides, oligonucleotides, and genes to cancer cells. They have also enabled rapid diagnosis of several types of cancers at early stages, unlike the conventional methods, which permit early arrest of carcinogenesis without affecting patient compliance. Theranostic approaches, which combine the advantages of diagnostic and therapeutic techniques, are also being studied for efficient and precise detection and elimination of cancerous tissue.

Several commercial formulations currently available in the market have proved the potential of nanoformulations in treating or diagnosing different types of cancers, especially lung cancer. Liposomal doxorubicin, a nanoformulation available by the brand name Caelyx®/Doxil®, has been used for treatment of breast cancer and has been found to have higher efficacy than doxorubicin alone, with reduced cardiotoxicity, myelosuppression, vomiting, and alopecia (O'Brien et al. 2004). Commonly used nanomedicines for lung cancer therapy include Lipoplatin™, Genexol®, PK1, Xyotax®, and so on, which are described in the following sections of this manuscript. Abraxane® is employed as the first-line treatment for patients with NSCLC. It is a nanotechnology-based formulation comprising albumin-bound paclitaxel that offers reduced toxicity, neuropathy, and a lower rate of neutropenia, as compared to paclitaxel alone. A combination of Abraxane with Carboplatin/Paraplatin, a platinum-based anticancer drug, is also being studied in clinical phase trial II for treatment of patients with SCLC as well as NSCLC (Stinchcombe et al. 2007).

Such nanoformulations possess the potential to detect and/or eliminate lung cancer by accessing the depths of cancerous tissue by the virtue of their small size. Such nanosized

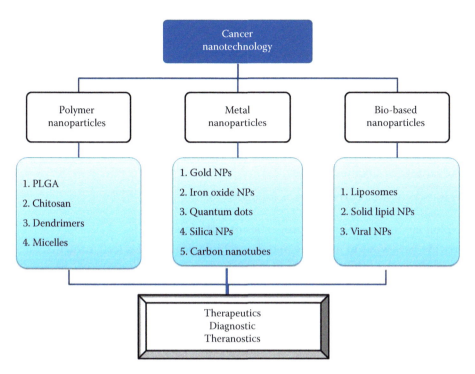

FIGURE 4.1
Schematic representation of various platforms used in lung cancer nanotechnology.

drug vehicles can also be efficiently and specifically targeted to the cancerous mass without affecting healthy tissue. Several polymer-based, metal-based, and bio-based nanocarriers are being studied for lung cancer therapy/diagnosis and have been subsequently described in detail.

As mentioned in the previous section, conventional therapies used for lung cancer treatment pose a risk of damage to the normal tissues or result in incomplete eradication of cancer. These lacunae in the existing therapy of lung cancer may be attributed to the lack of early diagnostic strategies. In almost half of the affected patients, the disease is diagnosed in advanced stages (e.g., stage IV, characterized by extensive metastasis of cancer cells to other tissues in the body) leading to a low survival rate among patients (Sukumar et al. 2013). Furthermore, inaccessibility to the deeper portions of the lung enhances the complications associated with conventional therapeutic approaches (Edwards et al. 1997, Sukumar et al. 2013). Although current treatment strategies are strongly dependent on the type and stage of malignancy at the time of diagnosis, they often involve a combination of surgery, chemotherapy, and/or radiation therapy (Babu et al. 2013). These therapies are only effective in the initial stages of treatment of SCLC, whereas NSCLCs are less sensitive to such treatment modalities. As a result, surgery (only in stages I, II, and some of IIIA) and gene therapy remain as the possible treatment alternatives (Mountain 1997). Dose-dependent side effects, patient intolerability to drug combinations, and the hydrophobic nature of major anticancer therapeutics require efficient drug delivery to target the tumor sites (Babu et al. 2013). The majority of these shortcomings involved in the therapy of lung cancer can be overcome through the use of therapeutic/diagnostic nanocarriers, which, by virtue of their small size, can effectively transcend the bronchial epithelial barrier and accumulate in the deeper lung regions (Sukumar et al. 2013).

Nanocarriers can overcome the above-mentioned limitations and offer methods to selectively destroy tumors or cancer cells, thereby causing minimal damage to healthy tissues and organs. Also, they help in the detection and elimination of cancer cells before they form tumors (Couvreur and Vauthier 2006). The scope of nanocarriers to achieve targeted therapy and effectively diagnose lung cancer can be estimated through a comprehensive understanding of the advantages offered by these carriers over conventional approaches discussed earlier.

4.2.2 Advantages of Nanocarriers

The advantages of nanocarriers are as follows:

- Exploit atomic or molecular properties of materials
- Provide new ways to measure and detect biological entities (cell, proteins, nucleic acids) through both *in vitro* and *in vivo* assays
- Allow targeting of molecular probes (DNA, RNA, peptides) and pharmaceuticals to tissues by traversing through cell membranes
- Provide ease of disease diagnosis and treatment
- Allow faster treatment, with reduced side effects, as compared to conventional therapies (for example, cisplatin, used for treatment of lung cancer, causes nephrotoxicity and has been replaced by Lipoplatin™ [liposomal cisplatin], which shows reduced nephrotoxicity)
- Offer increased bioavailability, dose proportionality (smaller drug doses), decreased toxicity, stable dosage forms of drugs, and reduction in fasted variability (Haque et al. 2010)
- Exhibit increased active agent surface area, resulting in their faster dissolution in an aqueous environment (Haque et al. 2010)
- Protect drugs from undergoing the first-pass effect before reaching the target, enhance the absorption of drugs into tumors and into cancerous cells, and provide better control over the duration and distribution of drugs to the tissue, making it easier for oncologists to assess their effectiveness

These advantages make nanocarriers a potential tool for diagnosis and treatment of lung cancer. Nanocarriers may be targeted to tumors by different approaches that are discussed in the following sections.

4.2.3 Targeting Approaches for Nanocarriers

Nanocarriers are able to selectively target lung cancer cells via two primary mechanisms, namely passive and active targeting, as shown in Figure 4.2. The figure elucidates the various pathways by which nanocarriers can gain access to the affected areas of a lung cancer patient.

4.2.3.1 Passive Targeting for Lung Cancer

Nanocarriers rely on passive targeting of lung tumors through a process known as the "enhanced permeability and retention effect" (EPR effect) (Butle and Baheti 2011).

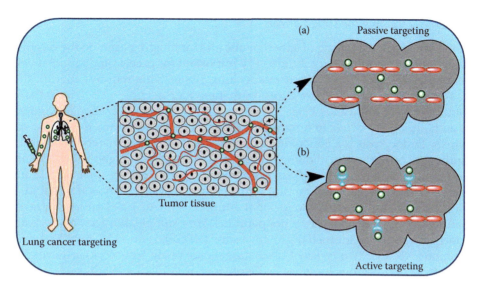

FIGURE 4.2
Schematic illustration of targeting mechanism in lung cancer cells by nanotechnology. Passive tumor targeting of NPs by enhanced permeability and retention effect (a) and active tumor targeting of NPs via receptor-mediated endocytosis (b).

It involves selective delivery of drugs or genes to the cancer-affected areas owing to the unique pathophysiological characteristics of tumor vessels. Anatomically, tumor microvasculature tends to be leaky and is characterized by abnormal branching and enlarged interendothelial gaps, breakdown of tight junctions between endothelial cells, and a disrupted basement membrane. These large interendothelial openings facilitate the extravazation of nanocarriers loaded with therapeutic and/or diagnostic moieties from the surrounding vessels into the tumor, thereby increasing accumulation and concentration of drug/imaging agent in the targeted pulmonary tissue and decreasing their circulation and toxicity in healthy tissues (Sanna et al. 2014). Some of the marketed nanocarriers that have been used for passive targeting of lung cancer have been listed in Table 4.4.

However, nanocarrier-mediated delivery of drug/imaging agent via passive targeting is not an effective option for treating lung cancer and exhibits side effects like toxicity to the surrounding healthy tissues.

TABLE 4.4

Examples of Marketed Nanocarriers Used for Passive Targeting of Lung Cancer

Nanocarrier Type	Drug	Name	Clinical Phase/Status	Intended Application	References
Polymeric micelles	Paclitaxel	Genexol®-PM	Approved	Advanced NSCLC	Danhier et al. (2009)
Polymer–drug conjugates	Paclitaxel	Opaxio™ (Xyotax® (CT-2103)	III	Recurrent and advanced NSCLC	Danhier et al. (2009), Sanna et al. (2014)
	Doxorubicin	PK1 (FCE28068)	II	NSCLC	Danhier et al. (2009)
	Camptothecin	CRLX101	II	Advanced NSCLC	Sanna et al. (2014)
Liposome	Cisplatin	Lipoplatin™	III	NSCLC	Fantini et al. (2011)

4.2.3.2 Active Targeting for Lung Cancer

Ligand-mediated active binding of a therapeutic and/or imaging agent to the receptors on cell surface and their cellular uptake are very important for precise and targeted delivery of therapeutics and/or diagnostic agents to lung tissue. This requires modification or process simulation in order to allow access of therapeutic/imaging agent within the tumor interstitium (Forssen and Willis 1998). Active targeting and internalization of nanocarriers loaded with anticancer drugs to lung tumor cells can be achieved via receptor-mediated endocytosis. Endocytosis is a vesicular-mediated process wherein materials or substances are engulfed by cells, thereby allowing their entry into the cells. Ligands that specifically bind to tumor cell surface receptors are coupled to the surface of long circulating nanocarriers to achieve active targeting of drugs to the affected areas (Patil et al. 2014). Targeting ligands can be either monoclonal antibodies (mAbs) and antibody fragments or nonantibody ligands (peptidic or nonpeptidic). For example, Bevacizumab (Avastin®), which targets vascular endothelial growth factor (VEGF), is an FDA-approved drug for nonsquamous NSCLC. Two types of tissue targets are being focused to achieve effective active targeting of cancerous lung tissue (Danhier et al. 2010).

4.2.3.2.1 Lung Cancer Cells

Targeting of internalization-prone cell-surface receptors is achieved with an aim to enhance uptake of nanocarriers by cancer cells. Some of the internalization-prone receptors on lung cancer cells include folate receptor (FR), epidermal growth factor receptor, and human epidermal growth factor receptor-3 (Danhier et al. 2010). Patritumab is a monoclonal antibody that targets human epidermal growth factor receptor-3, and its combination with erlotinib (Tarceva) is being tested in phase III clinical trial for patients suffering from locally advanced or metastatic NSCLC (Mendell et al. 2015, Yonesaka et al. 2015).

4.2.3.2.2 Tumoral Endothelium

Hypoxia has been identified to play a major role in cancer cell survival. The hypoxic microenvironment has been observed to harbor poorly differentiated tumor cells and undifferentiated stromal cells. It provides cellular interactions and environmental signals essential for tumor progression by maintenance of cancer stem cells at the tumor site. Hypoxia, thereby, presents a major therapeutic challenge due to which solid tumors have developed resistance to conventional cancer therapies (Kim et al. 2009). In order to combat this issue, active targeting of tumor endothelium has been identified as a prime strategy for killing cancerous cells. Targeting of tumor endothelial cells ultimately leads to the death of tumor cells by preventing recruitment of new blood vessels at the tumor site. Inhibition of angiogenic blood vessels results in a lack of vascular supply of oxygen and nutrients, essential for the growth of tumor cells. This results in arrest of tumor growth, as well as aiding in regulating its size and metastatic spread to other tissues. Some of these targeting receptors in lung cancer cells include vascular cell adhesion molecule-1 (VCAM-1) and matrix metalloproteinases (MMPs) (Danhier et al. 2010). van Rijt et al. have synthesized mesoporous silica nanoparticles (MSNs) that are tightly capped with avidin molecules via matrix metalloproteinase 9 (MMP9) sequence-specific linkers to allow site-specific drug delivery in high MMP9-expressing tumor areas. They observed successful MMP9-triggered release of cisplatin from MSNs in human lung tumors and in mice using the novel technology of *ex vivo* 3D lung tissue cultures, which induced apoptotic cell death only in lung tumor regions of mutant mice without causing toxicity in tumor-free areas or in healthy mice (van Rijt et al. 2015).

TABLE 4.5

Examples of Nanocarriers Used for Active Targeting of Lung Cancer

Formulation	Targeted Ligand	Bioactive Compound	Indication	Status	References
Anisamide coated lipid-protamine-hyaluronic acid NPs	CD47 (cluster of differentiation 47)	siRNA	B16F10 Lung tumor	Animal studies (B16F10 lung-tumor bearing mice)	Landesman-Milo et al. (2015)
Liposomes	EGFR (epidermal growth factor receptor)	Gemcitabine	NSCLC	Animal studies (A549 xenograft nude mice)	Master and Sen Gupta (2012)
Heparin-DDP (cis-diaminedichloroplatinum(II)) NPs	EGFR	Cisplatin	NSCLC	Animal studies (nude mice bearing H292 cell tumors)	Peng et al. (2011)

Active targeting can be combined with passive targeting to further reduce the nonspecific interaction of active moieties with healthy tissue. Drug delivery through active targeting has shown promising results in the treatment of lung cancer. Some of the nanocarriers used for active targeting of lung cancer are listed in Table 4.5.

Nanoformulations used for lung cancer treatment and/or diagnosis are delivered through several routes. However, the reported findings indicate that pulmonary disorders can be treated proficiently by maintaining high and prolonged drug concentrations in lungs (Shoyele and Slowey 2006).

Different routes of administration have been explored to deliver therapeutic and/or imaging agents to lung cancer cells. Of all the possible delivery routes, systemic delivery has been investigated for its suitability to deliver payload to the cancerous lung tissue. Jiang et al. studied the effect of a systemic route for lung cancer treatment by synthesizing polymer-based nanoparticles that were surface-modified with chitosan for oral administration of paclitaxel in lung cancer. Researchers observed that the thiolated chitosan modified poly lactic acid-polycaprolactone-d-α-tocopheryl polyethylene glycol 1000 succinate (PLA-PCL-TPGS) nanoparticles enhanced the cellular uptake and cytotoxicity of paclitaxel in the A549 lung cancer cell line (Jiang et al. 2013). However, it has been observed that even on administration of high doses of chemotherapeutic drugs, only a limited fraction efficiently targets the sites situated on the surface of lung tumor cells, making systemic drug delivery less successful in lung cancer treatment. Most chemotherapeutic drugs also affect noncancerous cells by inhibiting their growth, which can sometimes even prove lethal (Tseng et al. 2008). Therefore, other approaches such as the pulmonary route of drug delivery have been developed and studied to overcome the drawbacks of systemic drug delivery.

The pulmonary route involves delivery of drugs to the respiratory tract, either for the treatment and/or prophylaxis of diseases affecting airways, or for systemic absorption of drugs for the treatment or prophylaxis of other diseases (Shoyele and Slowey 2006). In order to investigate this route, Garbuzenko et al. have studied various lipid- and non-lipid-based nanocarriers such as micelles, liposomes, mesoporous silica nanoparticles, poly (propyleneimine) (PPI), dendrimer siRNA complexes, quantum dots, and poly (ethylene glycol) polymers for delivering doxorubicin to lungs via inhalation. In this case, it was observed that lipid-based nanocarriers (liposomes and micelles) showed higher accumulation and longer retention time in the lung tissue as compared to nonlipid-based carriers on inhalation-mediated delivery of active agents. Therefore, lipid-based nanocarriers were found to be most suitable for intravenous delivery and treatment of lung cancer (Garbuzenko et al. 2014). In another study reported by Tseng et al., conjugates of epidermal growth factor (EGF) and gelatin nanoparticles were administered by aerosol. Specific accumulation was observed after 24 h in orthotopic lung adenocarcinoma in severe combined immunodeficiency mice, as compared to unmodified nanoparticles (Tseng et al. 2007, 2008).

Apart from systemic and pulmonary routes of nanoformulation administration, chemotherapeutic drugs have been also administered intravenously to treat lung cancer. This has been conducted through the use of a syringe for a finite time or using a drip/pump to supply the drug for a longer duration.

Chen Y. et al. have developed an LPH (liposome-polycation-hyaluronic acid) nanoparticulate formulation, modified with a tumor-targeting single chain antibody fragment (scFv; GC4), to study the intravenous delivery of small-interfering RNA (siRNA) and microRNA (miRNA) to lung metastasis of murine B16F10 melanoma cells. They observed that the formulation comprising combined siRNAs in the GC4-targeted nanoparticles caused significant reduction of the tumor load in the lung, while miRNA was observed to induce apoptosis and downregulate the surviving expression in the metastatic tumor (Chen et al. 2010).

The examples cited above elucidate the role played by different administration routes whilst targeting therapeutic and/or imaging agents to the lung cancer cells. The role of nanocarriers for achieving targeting via various routes is discussed in subsequent sections.

4.3 Nanoplatforms for Lung Cancer

Research over the years has led to the development of a plethora of nanocarriers for therapy and diagnosis of lung cancer (Bharali and Mousa 2010, Mousa and Bharali 2011). The discussion of these nanocarriers, as detailed consequently, has been classified into three broad groups pertaining to polymer-based nanoparticles, metal-based nanoparticles, and bio-based nanoparticles. A schematic depiction of these nanocarriers is represented in Figure 4.3.

4.3.1 Polymer-Based Nanoparticles for Lung Cancer

Polymeric colloidal particles provide a platform for delivery of therapeutic molecules, imaging agents, and appropriate targeting moieties to obtain effective nanotheranostic systems. The flexibility in modifying physiochemical properties of polymer enables these carriers to be tuned according to the requirements of therapeutic or imaging agents. Especially in the case of biopolymer-based nanoparticles, apart from being biodegradable and biocompatible, these polymeric systems are capable of providing a sustained-release pattern of the encapsulated molecules. In addition to chemotherapeutic drugs, polymeric nanocarriers have been fabricated to deliver nucleic acids and proteins in order to effectuate their therapeutic potential at the target tumor cells (Díaz and Vivas-Mejia 2013, Sukumar et al. 2013). The most common polymers used in designing nanocarriers for lung cancer include poly (lactic-co-glycolic acid) (PLGA), poly (ε-caprolactone), polylactic acid,

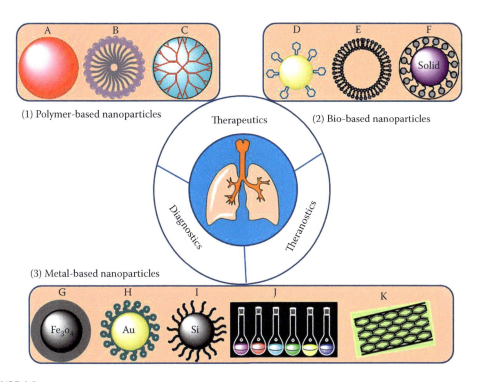

FIGURE 4.3
Schematic illustration of nanoplatforms used for lung cancer therapy and diagnosis. (1) Polymer-based nanoparticles: PLGA NPs (A), micelle (B), dendrimer (C). (2) Bio-based nanoparticles: Viral NPs (D), liposome (E), solid lipid NPs (F). (3) Metal-based nanoparticles: Iron-oxide NPs (G), gold NPs (H), silica NPs (I), quantum dots (J), carbon nanotubes (K).

alginic acid, gelatin, and chitosan (Babu et al. 2013, Sukumar et al. 2013, Bandyopadhyay et al. 2015, Bayisa et al. 2015).

These nanoparticles have gained significant attention owing to their characteristics such as targeted deposition of active moiety, bioadhesion, ability to maintain sustained release profiles, and reduction in dosing frequency, thus improving overall patient compliance (Sung et al. 2007). Moreover, encapsulation of therapeutic molecules and imaging agents using these polymeric nanocarriers has resulted in alteration of their biodistribution, leading to increased circulation half-life of these agents and thereby avoiding their degradation during transit to the delivery site. However, the toxicity of the degradation products of these polymeric nanocarriers needs to be monitored closely in case of pulmonary application, as remnant polymers may interact with the biosurfactants present in the alveoli, leading to a cascade of detrimental events. These events could eventually exhibit severe episodes of dyspnea in the patients (Sukumar et al. 2013).

The role of polymeric nanocarriers like dendrimers, micelles, nanofibers, and others formulated using PLGA, PEG, chitosan, and so on in the treatment and/or diagnosis of lung cancer is elaborately reviewed subsequently.

4.3.1.1 Poly (Lactic-Co-Glycolic Acid) Nanoparticles for Lung Administration

PLGA is an FDA-approved biodegradable polymer that has been most widely and successfully used for the formulation of nanoscale drug delivery systems. Apart from drugs, PLGA has been used for the delivery of proteins, nucleic acids, and peptides (Jain 2000, Ruhe et al. 2003, Bouissou et al. 2006).

Although paclitaxel has been demonstrated to exhibit significant activity in cases of lung cancer, the adverse effects associated with Taxol®, the commercial formulation of paclitaxel containing the surfactant Cremophor®, limit its applications. In order to enhance the therapeutic efficacy of this drug and to overcome its adverse effects, a polymeric carrier system was studied by Fonseca and colleagues. The researchers demonstrated that incorporation efficiency of paclitaxel in PLGA nanoparticles was dependent on the method of preparation, the organic phase employed, and the ratio of organic phase to aqueous phase. The researchers used the interfacial deposition method, wherein acetone was used as the organic solvent, in the ratio of 1:1 v/v to an aqueous phase consisting of Poloxamer 188 solution, to formulate nanoparticles suitable for intravenous administration. The release of paclitaxel from the developed spherical NPs (<200 nm) exhibited a biphasic profile that was characterized by an initial fast release during the first 24 h, followed by a slower and sustained release. Paclitaxel incorporated within the PLGA NPs enhanced the antitumoral efficacy as compared to Taxol® in the NCI-H69 SCLC cell line upon prolonged incubation with the cells (72 h) (Fonseca et al. 2002). Such prolonged incubation of drug-loaded NPs has also been reported in other studies involving *in vitro* cytotoxicity evaluation of paclitaxel against tumor cell lines. Also, a similar mechanism of action of paclitaxel was stated in all these reports (Liebmann et al. 1993, Raymond et al. 1997). These findings strongly suggest the capability of PLGA NPs in delivering drugs against lung cancer.

In 2005, Sengupta et al. fabricated bi-phospholipid-coated PLGA nanoparticles to simultaneously administer doxorubicin (cytotoxic drug) and comberstatin (antiangiogenic drug). The authors aimed to study the synergistic effect of these drugs in effectively eliminating NSCLC. In these nanoparticles, doxorubicin was conjugated to PLGA, while comberstatin was mixed with PEGylated phospholipid and encapsulated in the outer lipid bilayer (Sukumar et al. 2013). *In vivo* testing of this formulation was conducted in male C57/BL6 mice (implanted with Lewis lung carcinoma cells) through intravenous administration. The nanoformulation at the tumor site sequentially released the antiangiogenesis and chemotherapeutic agents present in the outer and inner envelopes, respectively. The antiangiogenesis agent resulted in cessation of tumor vasculature, while the chemotherapeutic moiety exhibited cytotoxic effect on cancer cells. This focal release within the tumor resulted in an improved therapeutic index, with reduced toxicity of the formulation, as compared to the single-drug formulations or a mix of both formulations. Through this study, the significance of a PLGA-based integrated approach in lung cancer therapy was very well demonstrated (Sengupta et al. 2005, Sukumar et al. 2013).

Delivery of nucleic acids into lung cancer cells was attempted by employing stable polyionic complexes of nucleic acids with positively charged carriers. In 2012, Benfer et al. studied siRNA delivery using PLGA NPs as the delivery vehicle. The scientists coupled PLGA to diamine derivatives of PVA (polyvinyl alcohol) to form DEAPA (3-(diethyl amino) propyl amine)-PVA-g-PLGA and used this polymer conjugate to deliver anti-GFP (green fluorescent protein) siRNA into H1299-EGFP cells (lung cancer cells expressing GFP). The cells exhibited energy-dependent and clathrin-mediated cellular uptake of siRNA-loaded nanoparticles for the initial 2 hours. The extent of cellular uptake of these particles was further enhanced by addition of lung surfactant to the carrier molecules. The *in vitro* knockdown of GFP was significantly higher for the siRNA complexes prepared with the lung surfactant, as compared to those prepared in water. The results suggested the suitability of PLGA based nanovehicles for cellular delivery of siRNA (Benfer and Kissel 2012).

Despite numerous successful attempts of utilizing PLGA nanocarriers for lung cancer treatment, the rapid clearance of PLGA from the circulatory system limits its application as a drug carrier. Scientists have attempted to overcome this drawback by coating PLGA with PEG to impart stealth properties. The details of PEG-containing nanocarriers directed against lung cancer are described in the consequent sections of this review.

4.3.1.2 Stealth Polymer: Poly Ethylene Glycol–Coated Nanoparticles for Lung Cancer

PEG, a biocompatible hydrophilic polymer, is incorporated in polymeric nanocarriers to prolong their *in vivo* residence time by lowering their predisposition to biotransformation. Incorporation of PEG also lowers immunogenicity of the resulting carriers (Sukumar et al. 2013).

Surface modification of nanocarriers with PEG has emerged as a useful strategy, as PEG minimizes accumulation and action of NPs at nonspecific sites (van Vlerken et al. 2007). In a study by Akbarzadeh et al., PEG-modified PLGA was synthesized using different molecular weights of PEG (PEG_{2000}, PEG_{3000}, and PEG_{4000}) and used as triblock polymer (PLGA-PEG-PLGA) to modify doxorubicin-loaded magnetic iron oxide (Fe_3O_4) nanoparticles (IONPs). This formulation effectively released doxorubicin in A549 cell line during *in vitro* testing, with an initial burst release for 12 h followed by sustained release. The maximum burst drug release (30.1%) was observed with PLGA-PEG_{4000}-modified IONPs owing to its higher encapsulation efficiency (78%), achieved due to its increased water uptake ability as compared to other lower molecular–weight PEGs. *In vitro* cytotoxicity evaluation revealed time-dependent cytotoxicity on A549 cells, with the IC_{50} value being the highest (0.18 mg/mL) for Fe_3O_4-PLGA-PEG_{4000} nanoparticles as against pure doxorubicin and low molecular–weight PEG formulations. Also, doxorubicin-loaded IONPs modified with PLGA-PEG triblock copolymers did not show cytotoxicity on healthy cells, thereby demonstrating biocompatibility of these formulations for the treatment of lung cancer (Akbarzadeh et al. 2012).

In a recent study by Guthi et al., a multifunctional PEG-b-PDLLA (poly D, L-lactide) micellar carrier grafted with lung cancer targeting peptide (LCTP) was loaded with superparamagnetic iron oxide nanoparticles (SPIONs) and doxorubicin. The formulation exhibited specific targeting toward $\alpha v\beta 6$-dependent H2009 lung cancer cells. The integrated multifunctional micelle (MFM) theranostic design enabled targeted delivery of therapeutic agents to lung cancer cells via an image-guided approach (Guthi et al. 2009).

In another study, Tan and colleagues employed diblock copolymers of PEG and polyethylene (PE) to encapsulate quercetin, a hydrophobic anticancer agent. The uniqueness of this system was its sensitivity toward overexpressed lactose dehydrogenase enzyme, which is the characteristic feature of A549 cells. Incorporation efficiency of quercetin was around 89% in the nanomicelles of PEG diblock copolymers. The nanomicelles were tested in the A549 cell line and administered per-orally in murine xenograft model, wherein encapsulated quercetin exhibited enhanced anticancer activity. Another significant aspect of this micellar nanocarrier was its unique stability at both highly acidic pHs (1.2) and at a pH of 7.4. As a result, this PEGylated nanocarrier helped in maintaining quercetin stability during transit through gastric fluids (pH 1.2) and intestinal fluids (pH 7.4) on per-oral administration (Tan et al. 2012, Sukumar et al. 2013). This facilitated carrier mediated delivery of quercetin specifically to lung cancer cells in stable form. Thus, PEG-based systems serve to provide stable and effective nanoformulations for lung cancer therapy.

4.3.1.3 Chitosan Nanoparticles for Lung Cancer

Chitosan is a nontoxic cationic polysaccharide synthesized by partial deacetylation of chitin, exhibiting biodegradability and biocompatibility characteristics. The presence of numerous ionizable amino groups play a key role in delivering therapeutic agents encapsulated within chitosan carriers, as positive charges on the polymer have a selective adsorption and neutralizing effect on anionic tumor cell surfaces, as well as targeting action on the lung tissue–specific receptors like EGFR, folic acid receptor, and so on (Aruna and Rajalakshmi 2013). Further, this polymeric vehicle exhibits mucoadhesiveness due to strong electrostatic interaction between the cationic amino groups and the negatively charged sites on cell surfaces (Bandyopadhyay et al. 2015). The amino and carboxyl groups on the chitosan molecule interact with mucosal glycoprotein to form a hydrogen bond, leading to an adhesive effect. Further, the positive charges on mucoprotein and chitosan have been shown to prolong the retention time of drugs and result in sustained drug release and improved drug bioavailability in animal models (Aruna and Rajalakshmi 2013). In addition, chitosan has been reported to promote the permeation of encapsulated macromolecules through well-organized lung epithelial cells (Bandyopadhyay et al. 2015).

Mehrotra and colleagues designed a chitosan-based controlled release system to deliver the potent antineoplastic agent lomustine. Lomustine-loaded chitosan nanoparticles (LChNPs) were synthesized by the homogenization assisted ionic-gelation method. The resulting nanoparticles exhibited a mean particle diameter of 98 ± 0.9 nm, with spherical morphology. LChNPs demonstrated excellent control over drug release during *in vitro* dissolution studies and enhanced *in vitro* cytotoxicity of the drug against L132 lung cancer cells, in comparison to native drug and PEG-coated LChNPs. The reduced cytotoxicity of PEG-coated LChNPs was demonstrated and attributed to the larger diameter and reduced surface charge of PEG-coated nanocarriers (Mehrotra et al. 2011).

The cationic nature of chitosan makes it a suitable vehicle for delivering nucleic acids to lung cancer cells. One such study by Jere and colleagues was carried out to efficiently deliver Akt1siRNA (siAkt) to lung cancer cells using polyethylenimine grafted chitosan (Ch-g-PEI). Overexpression of Akt1 protein is associated with survival of cancer cells, their proliferation, and further metastases (Datta et al. 1999, Brognard et al. 2001). The authors fabricated copolymer comprising chitosan and low molecular weight PEI to form stable siRNA complexes possessing spherical structures. The synthesized Ch-g-PEI-siRNA complexes were evaluated for silencing of enhanced green fluorescence protein (EGFP)—a marker siRNA—and Akt1 (oncoprotein, a therapeutic siRNA) in A549 cells. These siGFPs were successfully delivered by Ch-g-PEI complex, and silencing of EGFP was found to be nearly 2.5-fold higher than that of PEI, validating the enhanced effect of chitosan in siRNA delivery. Further, siAkt and shAkt delivery through Ch-g-PEI complex efficiently knocked down the Akt1 mRNA, resulting in its decreased protein expression, which was confirmed by Western blot assay. Also, a significant reduction (4-fold) in cell proliferation was observed with Ch-g-PEI-shAkt as compared to that without chitosan. These results suggest the importance of chitosan in effective delivery of siRNA to lung cancer cells, thereby serving as a suitable carrier system for siRNA in lung cancer treatment (Jere et al. 2009).

Another study conducted by Liu et al. demonstrated the efficacy of paclitaxel-loaded chitosan nanoparticles in a Lewis lung carcinoma mouse model. Here, following a surgical intervention, the nanocarriers delayed the local recurrence of subcutaneous lesions and modestly improved the overall survivability of the tumor-bearing animals (Liu et al. 2011).

4.3.1.4 Dendrimers—Architectural Nanoparticles for Lung Cancer

Dendrimers are branched polymeric structures with a large number of functional groups that radiate from a central core, providing the opportunity to link multiple bioactive molecules (Klajnert and Bryszewska 2000). The extensive scope for using dendrimers in effective therapy and diagnosis of lung cancer is due to their versatility to incorporate a vast range of surface functionalization groups for targeting different therapeutic and diagnostic molecules. Dendrimers of polymers such as poly (amidoamine) (PAMAM) (Liu et al. 2011), poly (glycerol succinic acid) (PGLSA) (Morgan et al. 2006), and poly (propylene imine) (Taratula et al. 2009) have been extensively used to encapsulate antineoplastic drugs and siRNA for successful delivery to NSCLC carcinoma cells (Sukumar et al. 2013).

In another study, scientists conjugated a novel NSCLC-targeting peptide (sequence RCPLSHSLICY) and fluorescein isothiocyanate (FITC) to acetylated PAMAM dendrimer to form a PAMAM-Ac-FITC-LCTP conjugate. *In vitro* studies of this conjugate in NCI-H460 NSCLC cells and 293 T cells (human embryonic kidney; HEK293 cells transformed with T antigen) exhibited both time- and dose-dependent cellular uptake. *In vivo* tissue distribution of this conjugate in athymic mice (BALB/c-nu/nu) bearing lung cancer xenografts exhibited its effective targeting to the tumors, thus demonstrating their potential as a targeted lung cancer theranostics (Liu et al. 2011).Dendrimer-based nanocarriers have been evaluated for their potential to alleviate inflammation of the lungs. In one of the studies, the dynamics of cellular entry of PAMAM dendrimers and hyperbranched polyol polymers, with and without ibuprofen, was explored. It was shown that the entry of PAMAM dendrimers with amine (NH_2) and hydroxyl (OH) end functionalities was more rapid (in approximately 1 h) as compared to the hyperbranched polyol with OH functionality (approximately 2 h) in A549 cells. Both PAMAM dendrimers and hydroxyl polyol polymers complexed with ibuprofen showed rapid entry when compared to pure ibuprofen. The complexation and encapsulation of ibuprofen inside these systems was observed to considerably lower the inflammatory response and increase the cellular uptake, thereby suggesting the efficiency of these dendrimers as drug carriers (Kannan et al. 2004, Sukumar et al. 2013).

A biocompatible polyester dendrimer formulated using PGLSA has been investigated as a carrier for camptothecin, 10-hydroxycamptothecin (10HCPT), and 7-butyl-10-amino camptothecin (BACPT) in four different lung cancer cells, namely human colorectal adenocarcinoma (HT-29), breast adenocarcinoma (MCF-7), non-small cell lung carcinoma (NCI-H460), and glioblastoma cells (SF-268). The dendrimer-mediated encapsulation method enhanced the potency of both BACPT and 10HCPT drugs in all four cancer lines. A maximum of 7-fold increase in efficacy with 10HCPT was seen in the MCF-7 cell line, while a nearly 10-fold increase in efficacy with BACPT was observed in SF-268. Cellular uptake and efflux measurements of 10HCPT in the MCF-7 cell line showed a 16-fold enhancement in cellular uptake and increase in drug retention within the cell using a dendrimer. The therapeutic potential of these dendrimers, in terms of improved cellular uptake and retention, was validated against all four cell lines, thus exhibiting the potential of dendrimer-encapsulated camptothecin in NSCLC (Morgan et al. 2006).

Dendrimers have also been used for delivering nucleic acids. Scientists have utilized surface-engineered targeted dendrimers (~100 nm), caged with a dithiol containing cross-linker molecule, followed by a PEG layer, and conjugated with a synthetic analogue of luteinizing hormone-releasing hormone (LHRH) peptide at the rear end of the polymer for targeting siRNA against the BCL2 (B-cell lymphoma 2) gene. The specific uptake of siRNA by tumor cells was evaluated by using different degrees of dendrimer modifications, and the complex with the higher modification degree exhibited nearly 1.5 times stronger siRNA uptake by the cell. *In vitro* cytotoxicity studies on A549 cells showed a nearly 20% decrease in cell viability by these engineered dendrimers. Also, *in vivo* distribution of LHRH-targeted and nontargeted dendrimer nanoformulations and encapsulated siRNA evaluated in nude mice exhibited specific siRNA delivery to the desired site by the targeted formulations of dendrimers as compared to nontargeted controls. This system demonstrated stable incorporation of siRNA through surface modification and the targeting approach of siRNA-loaded PPI dendrimers. Thus, engineered dendrimers offer promising nanostructures in systemic siRNA delivery for lung cancer therapy (Taratula et al. 2009).

4.3.1.5 Polymeric Micelles

Polymeric micelles comprising amphiphilic block copolymers, such as poly (ethylene oxide)-poly (β-benzyl-L-aspartate) and poly (N-isopropylacrylamide)-polystyrene, contain distinct hydrophobic and hydrophilic segments. The distinctive chemical nature of the two blocks brings about thermodynamic phase separation in aqueous solution and formation of nanoscopic supramolecular core/shell structures. This unique structural design enables the hydrophobic micelle core to serve as a nanoscopic depot for therapeutic or imaging agents and the shell as a biospecific surface for targeting applications. Micelles of less than 100 nm in size consisting of a hydrophobic core and hydrophilic shell are commonly used as drug carriers. Their hydrophobic core and hydrophilic shell makes them interesting carriers for poorly water-soluble anticancer drugs, including paclitaxel and docetaxel (Guthi et al. 2009, Díaz and Vivas-Mejia 2013).

The development of magnetic resonance imaging (MRI)-visible polymeric micelles that target lung cancer cells has been reported by Guthi et al. The researchers developed a prototype multifunctional micelle system wherein super paramagnetic iron oxide nanoparticles and doxorubicin were encapsulated in the micelle core and the micelle surface was functionalized with a lung cancer-targeting peptide. This peptide binds to the restrictively expressed integrin $\alpha_v\beta_6$, upregulated in many human NSCLCs as compared to normal lung tissue. The resulting MFMs showed ultrasensitivity for MR detection. LCTP-loaded MFMs showed significantly increased cell targeting and micellar uptake in $\alpha_v\beta_6$-expressing H2009 lung cancer cells as compared to $\alpha_v\beta_6$-negative H460 cells. This was verified by ^{3}H radioactivity measurement, confocal imaging, and MRI. Doxorubicin-encapsulated LCP-MFM showed significantly lower IC$_{50}$ values for $\alpha_v\beta_6$-dependent cell cytotoxicity when compared with doxorubicin as a control. The integrated $\alpha_v\beta_6$-targeting, MRI ultra sensitivity, and drug delivery functions in the MFM design open avenues for exciting opportunities for image-guided, targeted therapy of lung cancer (Guthi et al. 2009).

4.3.1.6 Polymeric Nanofibers

Polymeric nanofibers are fibers with diameters from 1 nm to 1 μm, closely matching the size scale of extracellular matrix (ECM) fibers. Nanofibers are known to load good quantities of anticancer agents and release them in a sustained manner. This is known to result in improved drug bioavailability, reduced frequency of administration, maintenance of effective drug concentration in blood for a prolonged time, reduced fluctuation in peak-trough concentrations of drug and drug-related side effects, and possibly improved specific distribution of drugs (Dixit et al. 2013).

Polymeric nanofibers may be fabricated using inorganic (i.e., titanium, silicon, or aluminum oxides) or organic (polyvinyl alcohol, gelatin, poly (N-isopropylacrylamide, polycaprolactone, or polyurethane) materials (Díaz and Vivas-Mejia 2013). Of the three available techniques for synthesizing nanofibers, namely electrospinning, phase separation, and self-assembly; electrospinning is the most commonly employed, as its simple experimental setup allows control over the thickness, porosity, and composition of the nanofiber meshes (Vasita and Katti 2006). Nanofibers possess numerous characteristics crucial for drug delivery, including large surface area-to-volume ratio (nearly 10^3 times of that of a microfiber), flexibility for diverse surface modifications, and superior mechanical performance (e.g., stiffness and tensile strength) compared to other polymeric carriers. These properties can be easily modulated by varying the voltage, capillary-collector distance, and polymer flow rate, which are the key optimization parameters involved in the electrospinning process. The surface tension and viscoelasticity of nanofibers in solution can also be modified to yield suitable fibers for drug delivery applications (Díaz and Vivas-Mejia 2013).

The use of nanofibers in drug delivery for lung cancer has been reported by Sridhar et al. For fabrication of biodegradable polycaprolactone nanofibers, the authors have made use of natural extracts like curcumin, neem, and aloe vera and their combinations owing to their anticancer activities. Different combinations of nanofibers such as PCL, 1% aloe vera-loaded PCL, 1% neem loaded-PCL, 1% cisplatin loaded PCL, 5% curcumin-loaded PCL, and combinations of neem with curcumin and aloe vera with curcumin in their respective concentrations. The evaluation of these nanofibers *in vitro* on human breast cancer (MCF7) and lung cancer (A459) cell lines showed maximum cytotoxicity when curcumin combinations with neem and aloe vera were employed. The combination of curcumin with aloe vera and that of curcumin with neem showed 23% and 18% cell viability in A549 cells, and 38% and 35% in MCF-7 cells, respectively. The amine functional groups of aloe vera were observed to inhibit cancer cell viability by forming a cyclic pyridine complex with curcumin. Inhibition of cancer cell growth after 24 h of incubation was 15% higher with these natural extract combination-loaded PCL nanofibers as compared to 1% commercial cisplatin-loaded PCL nanofibers. Through this study, the potential of these nanofibers in fabricating biocompatible drug-eluting medical devices against lung and breast cancers was demonstrated. However, studies demonstrating *in vivo* cell viability and therapeutic efficacy would be required to ensure their applicability in fabricating these devices for lung cancer treatment (Sridhar et al. 2014).

Although sizeable research has been conducted to investigate the application of nanofibers for disease treatment, limited literature is available to support the role of polymeric nanofibers as anticancer drug carriers.

The most commonly employed polymer-based nanocarriers designed for delivery to lung cancer are summarized in Table 4.6.

In a nutshell, the adaptability of polymer matrices to be tailored for a desired function (targeted drug/gene delivery) endows polymeric nanoparticles with an array of advantages for application in targeted therapy for lung cancer. Organic NPs like those fabricated using PLGA, PEG, and chitosan assist in sustained drug release; maintain stability of the drug delivery system; and exhibit biocompatibility, nontoxicity, improved circulation time, and targeting potential when surface-decorated with active targeting ligands. Imparting functionalization on dendrimers due to their complex architectural nature facilitates improved cellular uptake and drug retention for effective therapy. Moreover, polymeric micelles with a hydrophobic core and hydrophilic shell are appealing due to their capacity to load both soluble and insoluble and therapeutic and diagnostic agents. The superior mechanical strength of nanofibers as compared to other polymers can be used for the development of biomedical devices like drug-eluting implants. All these features of polymeric nanoparticles have exhibited an astounding effect in treatment of lung cancer.

TABLE 4.6
Polymer-Based Nanocarriers for Delivering Therapeutic Molecules and Imaging Agents in Lung Cancer

Nanocarriers	Polymers	Targeted Ligand	Active Ingredient	In vitro cell line	In vivo Model	Results	References
PLGA polymer NPs	PLGA NPs	–	PTX	NCI-H69 cells	–	• PTX-PLGA-NPs showed significant *in vitro* cytotoxicity compared to Taxol®	Fonseca et al. (2002)
	Chitosan-modified PLGA NPs	–	PTX	A549 cells	Lung-CDF1 metastasized Mice	• Cytotoxicity of PTX-PLGA NPs increased with increase in incubation time (24–72 h) • Chitosan-modified PTX-PLGA NPs showed increased cytotoxicity irrespective of incubation time • Distribution index of PTX specific to lung was 5.4 for Taxol, 6.6 for PTX-PLGA NPs, and 99.9 for chitosan modified PTX-PLGA NPs • Cellular uptake of Coumarin 6-loaded chitosan-PLGA NPs followed clathrin-mediated endocytosis	Yang et al. (2009)
	PEI-modified PLGA NPs	–	PTX and Stat3 siRNA	A549 cells and paclitaxel-resistant A549/T12 cells with α-tubulin mutation	–	• PLGA-PEI-TAX-Stat3 siRNA suppressed Stat3 expression and induced higher cellular apoptosis in A549 and A549/T12 cells when compared with PLGA-PEI-TAX NPs	Su et al. (2012)
	M-PLGA-b-TPGS NPs	–	DOX	A549 and H1975 cells	Xenograft BALB/c nude mice tumor model	• DOX-M-PLGA-b-TPGS NPs showed higher *in vitro* and *in vivo* cytotoxicity as compared to free DOX and PLGA-b-TPGS NPs after 48 and 72 h	Zhang et al. (2015)

(*Continued*)

TABLE 4.6 (Continued)
Polymer-Based Nanocarriers for Delivering Therapeutic Molecules and Imaging Agents in Lung Cancer

Nano-carriers	Polymers	Targeted Ligand	Active Ingredient	In vitro cell line	In vivo Model	Results	References
	Transferrin-conjugated lipid-coated PLGA NPs (TF-LP)	Transferrin receptor	DOX	A549 cells	A549 tumor-bearing mice	• TF-LP uptake was ~2.8 and 4.1 times higher compared with LP and PLGA-NP, respectively • Tumor spheroid volumes were decreased by almost 86%, 71%, and 42% for the PLGA-NP, LP, and TF-LP mice groups, respectively	Guo et al. (2015)
	(PGA-co-PCL)-b-TPGS2k	—	PTX	A-549 cells	A-549 cell-bearing nude mice	• After 48 h, cytotoxicity of A-549 cell increased by 33.7% and 160.9% for PTX-loaded NPs and (PGA-co-PCL)-b-TPGS2k NPs, respectively	Zhao et al. (2013)
Micelles	Multifunctional micelles (MFMs)	$\alpha_v\beta_6$-LCP	DOX	H2009 lung cancer cells	—	• Lung cancer-targeting peptide (LCP) encapsulated within MFM (IC_{50} = 28.3 ± 6.4 nM) displayed increased cytotoxicity over SP-encapsulated MFM (IC_{50} = 73.6 ± 6.3 nM), respectively • MRI images of H2009 cells showed increased cell targeting and uptake (3-fold) with LCP-encapsulated MFM over $\alpha_v\beta_6$-LCP-negative H460 cells incubated with SP-encapsulated MFM control	Guthi et al. (2009)
	PEG-PE	—	Quercetin	A549 lung xenograft model	Murine xenograft model	• Solubilization of quercetin was increased by 110-fold using nanomicelles • After 10 week post-tumor inoculation, per-oral quercetin nanomicellar formulation showed improved antitumor activity in lung xenograft model compared to control and per-oral quercetin suspension	Tan et al. (2012)

(Continued)

TABLE 4.6 (Continued)
Polymer-Based Nanocarriers for Delivering Therapeutic Molecules and Imaging Agents in Lung Cancer

Nanocarriers	Polymers	Targeted Ligand	Active Ingredient	In vitro cell line	In vivo Model	Results	References
PLA	PEG-modified PLA NPs	—	PTX and DTX	A549 cell line	A549 lung tumor-bearing xenograft mice model	• IV injection of PNP-PTX (10 mg/kg) with radiotherapy (5 Gy) showed about 2.2 times greater efficacy than that of PNP-PTX alone or IR alone • PNP-DTX (26.1%) showed increased antitumor efficacy compared to free DTX (40.6%)	Jung et al. (2012)
Dendrimer	Poly(glycerol-succinic acid) dendrimer	—	10-OH camptothecin (10HCPT) and 7-butyl-10-aminocamptothecin (BACPT)	NCI-H460 cell line	—	• IC_{50} value of encapsulated 10HCPT and BACPT was reduced from 32.4 to 16.7 nM/L and 1.2–0.6 nM/L compared to free 10HCPT and BACPT, respectively, in NCI-H460 cells	Morgan et al. (2006)
	Luteinizing hormone-releasing (LHRH) peptide-conjugated PEG-PPI dendrimer	LHRH peptide and BCL2 mRNA		A549 lung carcinoma cells	—	• LHRH-targeted PEG-modified PPI G5 dendrimer-siRNA complexes showed a small decrease (~20%) in cellular viability compared to free PPI G5 dendrimer	Taratula et al. (2009)
Nanofiber membrane	PCL	—	Curcumin and cisplatin	A549 lung cancer cell line	—	• After 24 h treatment, aloe vera (1%) and curcumin (5%)-loaded PCL nanofibers displayed 15% more cytotoxicity as compared to cisplatin (1%)-loaded PCL nanofibers	Sridhar et al. (2014)
Chitosan NPs	—	—	LMT	Human lung 132 cells	—	• LMT-loaded chitosan NPs exhibited increased cytotoxicity compared to native LMT	Mehrotra et al. (2011)
Expansile polymer NPs	—	—	PTX	—	Lewis lung carcinoma cells injected C57BL/6j mice	• Delayed local tumor recurrence and prolonged survival observed with PTX-expansile NPs, when compared with PTX and other groups	Liu et al. (2011)

4.3.2 Metal-Based Nanoparticles and Miscellaneous Nanoparticles for Lung Cancer

This section describes inorganic metal nanoparticles based on gold, silver, iron, lanthanide, and so on, as well as providing insights into carbon- and silica-based nanoparticles that have been explored in lung cancer therapy. The section also reviews the potential of nanoshell-based carriers for lung cancer diagnosis and/or treatment. Metal nanoparticles of gold (Au), silver (Ag), and iron (Fe) have been widely explored in the lung cancer therapy (Conde et al. 2012, Sadhukha et al. 2013). Although these systems are effective, cytotoxicity on lung cells has been observed, which is dependent on various factors including size, concentration, and time of exposure. The size and concentration of these nanoparticles can be fine-tuned during synthesis. Optimization of these parameters has been reported to improve their therapeutic and diagnostic efficacy in lung cancer (Sukumar et al. 2013).

4.3.2.1 Gold Nanoparticles

Among all inorganic nanoparticles, gold nanoparticles (Au NPs) have been extensively studied for lung cancer therapy and diagnosis. Colloidal gold has some unique properties that makes it ideal for targeted therapy. These noble metal NPs exhibit high surface area-to-volume ratio, broad optical properties, biological inertness, resistance to corrosion, low toxicity (degraded units of the particle are found to be nontoxic and either get eliminated by renal clearance or undergo biotransformation), and good antimicrobial efficacy. Gold ions have been hypothesized to interact with the sulfhydryl (S–H) groups of the biological proteins of microorganisms and inactivate them. Binding of Au NPs to DNA molecules leads to their condensation; thus, they lose their ability to replicate. This has been speculated to be the underlying mechanism by which Au NPs inhibit bacterial replications and also cancer cell growth without exerting a cytotoxic effect on normal cells (Bandyopadhyay et al. 2015).

Au NPs, either alone or in conjugation with other molecules, have been widely used in biomedical applications, bio-imaging and photothermal therapy. Cheng and colleagues developed three different Au-based nanomaterials, namely silica@Au nanoshells, Au/Ag hollow nanospheres, and Au nanorods, for photothermal destruction of malignant cells (A549 cells, HeLa cervix cancer cells, and TCC bladder cancer cells). The number of nanoparticles to efficiently destroy the malignant cells was used to assess the photodestruction efficiency of these nanoformulations. With the same particle concentration (8.36×10^8 particles/100 μL) for all three materials irradiated by NIR laser, silica@Au nanoshells showed maximum photodestruction efficiency (8.25%) against Au nanorods and nanospheres. These results suggested the effective absorbance of light and its further conversion into heat by these nanoshells for destroying malignant cells, thereby requiring a minimum number of nanoparticles for photodestruction of malignant cells. Through this study, the comparative efficiency of different Au nanoformulations for the photothermal destruction of A549 cells using a continuous-wave near-infrared laser was demonstrated. This system may serve in effective photothermal destruction of cancerous cells in the treatment for lung cancer (Cheng et al. 2009).

The effect of exhaled breath on Au NPs has been exploited for diagnosis of lung cancer. A number of sensors based on gold nanoparticles have been developed for rapid distinction between the breath of lung cancer patients and that of healthy individuals in an atmosphere of high humidity without the pretreatment of exhaled breath. Gas chromatography-mass spectrometry (GC-MS) studies have shown that various volatile organic compounds (VOCs) in breath of healthy individuals normally appear at 1–20 ppb, while they are raised to an atypical concentration (10–100 ppb) in the breath of lung cancer patients. Some VOCs are recognized as potential biomarkers for lung cancer (Peng et al. 2009). Such investigations have paved a way for noninvasive diagnosis of lung cancer.

Au NPs have also been used in cancer cell imaging and photothermal therapy in the near-infrared (NIR) region. Gold nanorods with suitable aspect ratios show strong absorbance and scattering in the NIR region (650–900 nm). Gold nanorods have been explored as contrast agents for imaging and photothermal therapy by conjugating them to anti-EGFR mAbs and incubating the conjugate with a nonmalignant epithelial cell line (HaCat) and two malignant oral epithelial cell lines (HOC 313 clone 8 and HSC 3). The anti-EGFR antibody-conjugated nanorods were reported to bind specifically to the malignant cells with much greater affinity, owing to EGFR overexpression on their surface. The emission of strongly scattered red light from gold nanorods provides a clear distinction between malignant and nonmalignant cells. It has been reported that malignant cells require about half the laser energy required for photothermal destruction than the nonmalignant cells after exposure to continuous red laser at 800 nm. This reduction in the requirement of laser energy was due to the overexpression of EGFR on the surface of malignant cells and the resultant increased binding of the anti-EGFR antibody-conjugated gold nanorods that absorbed light and converted it into heat at the cell surface (Huang et al. 2006). These studies proved the potential of Au NPs for therapy and diagnosis of lung cancer.

4.3.2.2 Silver Nanoparticles

Silver nanoparticles have also been extensively investigated for their antiproliferative and cytotoxic effects, as well as accurate diagnostic potential in lung cancer (Conde et al. 2012, Stoehr et al. 2011). Studies on exposure of silver nanoparticles (Ag NPs) to different cell lines such as BRL4A (rat liver cells) (Hussain et al. 2005), PC-12 (neuroendocrine cells) (Hussain et al. 2005, 2006), germline stem cells (Braydich-Stolle et al. 2005), and rat alveolar macrophages (Carlson et al. 2008) have indicated their ability to produce membrane leakage, decrease viability, and inhibit mitochondrial activity in these cells. These studies demonstrated that the cytotoxicity of Ag NPs in various cell lines is effectuated by apoptotic and necrotic mechanisms. These mechanisms are stimulated due to alteration in membrane structure, upregulation of apoptotic signaling molecules and generation of reactive oxygen species (ROS), thereby resulting in cell death (Foldbjerg et al. 2011, Sukumar et al. 2013).

In another study, the effect of Ag NPs on the A549 cell line was investigated by Foldbjerg and colleagues, which provided insight into use of these NPs as drug carriers in lung cancer treatment. When the effect of polyvinyl pyrrolidone (PVP)-coated Ag NPs were evaluated in the A549 cell line, dose-dependent cellular toxicity and elevated cellular ROS levels were observed. However, the use of the N-acetyl-L-cysteine (NAC) antioxidant reduced the cellular ROS level, highlighting the need of a radical scavenger to protect the cells from toxic effect of Ag NPs. Also, increased DNA damage induced by Ag NP-generated ROS was observed in the form of bulky DNA adducts. Moreover, the levels of these bulky DNA adducts were found to correlate with cellular ROS levels. These results led to speculation that Ag NPs act as mediators of ROS-induced genotoxicity (Foldbjerg et al. 2011).

However, another study conducted by Stoeher and colleagues demonstrated that the cytotoxic effect of Ag NPs was dependent on their shape, size, surface chemistry, and so on. This investigation was carried out in order to study the toxic and immunotoxic effects of silver nanowires and spherical silver nanoparticles on A549 cells. It was observed that exposure of PVP-coated silver nanowires and nanospheres resulted in varied effects on the cells. Although spherical NPs did not exert any toxic impact, silver wires exhibited a strong cytotoxic effect. However, the absence of an immunotoxin effect of silver wires was speculated to occur due to the unique needle-shaped structure of these wires (Stoehr et al. 2011). Despite the favorable anticancer action of Ag NPs, the *in vitro* exposure of A549 cells to silver nanoparticles resulted in ROS-induced genotoxicity, which raised concern of an unfavorable risk-to-benefit ratio regarding their medical application (Foldbjerg et al. 2011).

In another study, the biosynthetic route was explored to organically modify Ag NPs by capping them with stem latex from medicinal plant, *Euphorbia nivulia*. This route enabled researchers to synthesize NPs free from toxic chemicals. These NPs were capped with nontoxic stem latex to maintain their stability and enhance their cellular uptake. The peptide and terpenoid contents of the latex helped in modification and synthesis of stable latex-capped silver nanoparticles (LAg NPs). These NPs were found to be cytotoxic against A549 cells in a dose-dependent manner. The weak chemical interaction between latex components and LAg NPs assisted in enhancing the cellular permeability of the capped NPs. The researchers have proposed that this system may serve as a biocompatible carrier for drug delivery to lung cancer cells (Valodkar et al. 2011).

Overall literature reports indicate that the anticancer activity of Ag NPs in the case of lung cancer is dependent on factors like their surface chemistry (to ensure their biocompatibility) or the time and/or dose of exposure of Ag NPs to the cells and/or *in vivo* models.

4.3.2.3 Magnetic Nanoparticles

The inherent properties of magnetic nanoparticles (MNPs) allow flexibility to control their activity on the tumor tissue by applying an external magnetic field (Bandyopadhyay et al. 2015). They have been widely explored in theranostic applications (Babu et al. 2013, Sukumar et al. 2013). Superparamagnetic iron oxide nanoparticles, when subjected to alternating currents, generate sublethal heat that causes heat-induced ablation of desired tumor tissue. This noninvasive process is termed magnetic hyperthermia. The use of magnetic nanoparticles for hyperthermic destruction of lung tumors was studied by Sadhukha et al. (Sadhukha et al. 2013). The researchers developed inhalable SPIONs that were targeted to EGFR, an overexpressed receptor in nearly 70% of NSCLC patients. The *in vitro* studies of EGFR-targeted SPIONs in A549 cells exhibited 4.5-fold higher cell uptake of targeted SPIONs as compared to the nontargeted ones. For *in vivo* studies, inhalable EGFR-targeted SPIONs were administered in an orthotopic lung tumor murine model (Fox Chase SCID® Beige). Administration of inhalable SPIONs showed enhanced tumor retention and significant inhibition of lung tumor growth in mice treated with EGFR-targeted SPIONs as compared to the ones treated with nontargeted SPIONs. Thus, this magnetic nanoparticulate system holds the ability to be a promising anticancer treatment modality for NSCLC (Sadhukha et al. 2013).

Superparamagnetic cores in core-shell nanoformulation exhibit higher biocompatibility and stability as compared to other metal oxides. These properties of superparamagnetic cores were explored for their antineoplastic potential with mesoporous core-shell structure in the study conducted by Chen et al. Multifunctional mesoporous nanocomposites loaded with doxorubicin were synthesized by encapsulating magnetic iron oxide nanoparticles within a silica shell. The porous, nontoxic silica shell allowed amine modification of the nanocomposite followed by subsequent grafting with fluorescent polymethacrylic acid (PMA) and folic acid (FA). The PMA and FA functionalities on silica shell assisted in optical imaging and selective targeting of folic acid receptor on the tumor cells for drug release. *In-vitro* studies of these nanocomposites on human hepatoma 7402 cell lines and H446 lung cancer cell lines showed sustained release of doxorubicin over 100 h and that the carrier exhibited enhanced cellular internalization for drug release. The higher fluorescence intensity of FA-positive cells as compared to the FA-negative ones exhibited effective targeting of tumor cells and cellular uptake of the drug with these nanocomposites (Chen et al. 2010). These results of multifunctional magnetic nanocomposites revealed their potential as a biocompatible and water-soluble theranostic for lung cancer.

Another research group developed magnetic nanoparticles capable of detecting micrometastasis in lung cancer. IONPs were conjugated with the epithelial tumor cell marker pan-cytokeratin (pan-ck) for efficient isolation of circulating tumor cells (CTCs) from patients diagnosed with NSCLC. Cadmium telluride (CdTe) quantum dots (QDs) were coupled to the NSCLC micrometastasis marker lung-specific X protein (LUNX) and prominent pulmonary surfactant protein-A (SP-A) antibody for identification. The very low concentration of CTCs in blood (1 per 10^{6-7} leukocytes) requires enrichment of cells with labeled MNPs (MNP-pan-ck). After enrichment of four epithelial cell lines (lung cancer A549 and SPC-A-1, liver cancer HepG2, and colon cancer HCT-8), A549 and SPC-A-1 were successfully identified by QDs with double-labeled antibodies. Further, a trial on 32 patients (16 NSCLC-affected individuals and 16 healthy volunteers) resulted in successful identification of 21 cases, thus proving this technique to be promising for simultaneous isolation and identification of CTCs from NSCLC patients (Wang et al. 2012). This is the first study that reported the detection of micrometastasis in peripheral blood of lung cancer patients and signified the ability of MNPs in isolation of CTCs from the blood of NSCLC patients.

Magnetic nanoparticles have also been used to overcome resistance for cisplatin (DDP). Li et al. explored the potential of cisplatin-loaded spherical IONPs in reversing cisplatin resistance in lung cancer cells and in investigating mechanisms of multidrug resistance. The cytotoxicity studies in A549 cells showed time-dependent and dose-dependent inhibition with DDP and DDP-IONP, the inhibitory rate being higher (90% after 48 h, 50 µmol/L) with IONP than with only DDP (60% after 48 h, 50 µmol/L). A group of cisplatin-resistant BALB/c nude mice implanted with A549 lung tumor xenografts grown to 200 mm^3 tumor volume were treated with IONP, DDP, and DDP-IONPs. Tumor inhibition in terms of its volume was found to be increased in the IONP and untreated group of mice (554 ± 38 mm^3 and 417 ± 31 mm^3, respectively) controls, while DDP- and DDP-IONP-treated mice showed significant tumor inhibition (tumor volumes: 265 ± 23 mm^3 and 149 ± 16 mm^3, respectively). These results suggested effective suppression of the tumor growth in presence of IONPs. Molecular studies demonstrated a significant reduction in localization of proteins associated with lung resistance and enhanced cytotoxicity of cisplatin in the treated tumors. This showed the clinical potential of iron oxide NPs loaded with cisplatin to reverse drug resistance in lung cancer cells (Li et al. 2013).

4.3.2.4 Carbon-Based Nanoparticles

Carbon-based nanoparticles (CNPs) can be categorized as carbon nanotubes (CNTs), carbon nanofibers, nanodiamond (ND), and carbon nanohorn (CNH) (Bianco et al. 2008). The role of CNTs and ND in the treatment for lung cancer is discussed as follows.

4.3.2.4.1 Carbon Nanotubes

Carbon nanotubes are either made up of a single layer of graphene sheet (single-walled carbon nanotube [SWNT]: with diameter of 0.4–2 nm) or concentric multiple layers of SWNTs (multiwalled carbon nanotube [MWNT]: with inner diameter of 1–3 nm and outer diameter of 2–100 nm), commonly held together through sp^2 bonds. The layers are rolled into a seamless cylinder that can be open ended or capped at the extremities with a buck ball structure. These cylinders have a high aspect ratio, diameters as small as 1 nm, and a length of several micrometers (Bianco et al. 2005, Heister et al. 2009, Surendiran et al. 2009, Meng et al. 2012, Bandyopadhyay et al. 2015).

Pristine carbon nanotubes are insoluble in commonly used organic or aqueous solvents. Therefore, for better functionality and hydrophilicity, the nanotube surface needs to be modified for utility in biological applications (Kostarelos et al. 2009). The cylindrical structure of CNTs provides a large surface area that can be modified as per the end application and the large empty internal space can be exploited for encapsulation, which allows binding to a multitude of therapeutic and biologically active macromolecules. Such CNTs are potential candidates for efficient drug/gene delivery (Sobhani et al. 2011).

A CNT-based drug delivery system was formulated by Sohani and colleagues for effective treatment in lung cancer. The authors synthesized MWNTs, which were oxidized and covalently functionalized with hyperbranched poly (citric acid) (PCA) due to their high capacity for conjugation to drug molecules. This modified MWNT was conjugated with paclitaxel molecules forming an MWNT-g-PCA-PTX conjugate. The conjugate being acid sensitive successfully released paclitaxel in the tumor microenvironment (pH 6.8 and 5) through enzymatic hydrolysis, as demonstrated in an *in-vitro* study. Further, *in-vitro* cytotoxicity of the conjugate was investigated in SKOV3 ovarian cancer cells and A549 cells. The conjugate showed a higher cytotoxic effect than the drug over a shorter incubation time, which was significant at 48 h (Sobhani et al. 2011). Thus, the effective cytotoxicity in a lung cancer cell line using functionalized CNT reveals its ability for effective delivery of drug in lung cancer.

In another study, delivery of an apoptotic siRNA with MWNT was investigated for lung cancer studies. The polo-like Kinase (PLK1) gene is overexpressed in variety of tumors, including lung (Golsteyn et al. 1994, Holtrich et al. 1994). The PLK1 gene was targeted by delivering the apoptotic siRNA (siPLK1) using an ammonia-functionalized MWNT (MWNT-NH$_3^+$) vector. The gene-silencing efficacy of synthesized siPLK1-MWNT-NH$_3^+$ complex was checked in *in-vitro* human lung carcinoma Calu6 cell lines, and *in vivo* in female Swiss nude mice bearing Calu6 xenograft tumors by intratumoral delivery. *In-vitro* studies showed ineffective gene silencing by siPKL1: MWNT complex, as compared to that shown by siPKL1 and cationic liposomal (DOTAP: cholesterol) complex. However, *in-vivo* studies of the siPLK1-MWNT-NH$_3^+$ complex significantly improved the animal survival as compared to liposomal complex. PKL1 knockdown-mediated tumor suppression using biologically active siRNA was confirmed through intratumoral studies. Tumor inhibition was observed due to longer retention time of CNT complexes than that of liposomal complex. These results demonstrated the tremendous potential of MWNTs as vectors for gene therapy in lung cancer (Guo et al. 2015).

4.3.2.4.2 Nanodiamonds

Nanodiamonds are nontoxic and biocompatible carbon-based nanomaterials that have been investigated for biomedical applications such as cancer cell imaging (Liu et al. 2007, 2009). NDs can be conjugated with various chemicals, biomolecules, and antineoplastic drugs via covalent or noncovalent linkages.

The covalent linkage between ND and paclitaxel has been evaluated for use in lung cancer therapy. The surface-modified carboxylated nanodiamond was conjugated with paclitaxel to form ND-paclitaxel powder. Treatment of A549 cells with ND-drug conjugate reduced the cell viability significantly after 48 h, as compared to only drug. The effect of conjugate administered in xenograft of CB17/Icr-Prkdc^{scid}/Crl (severe combined immune-deficient-SCID) mice resulted in prominent inhibition of tumor growth in lung tissue.

The mechanism responsible for this inhibition was proposed to be through the induction of mitotic arrest, apoptosis, and antitumorigenesis (Sukumar et al. 2013). The covalent conjugation of ND was speculated to be the reason for effective drug release and tumor growth inhibition in the therapy of lung cancer.

In another study by Cui et al., utilization of NDs as sustained chemotherapeutic drug delivery carriers was investigated. The researchers developed cisplatin-coated sodium alginate-functionalized NDs (fNDs) and investigated them in different tumor cell lines: HepG2 liver carcinoma cells, BEAS-2 bronchial epithelial, HeLa cervical cancer cells, A549 lung cancer cells, and RAW264.7 macrophages. Drug release from the cisplatin-coated fNDs complex was observed to occur in a sustained manner along with significantly improved drug accumulation within tumor cells. The tumor inhibition effect of the complex was subsequently evaluated and showed significant accumulation of cisplatin in the cells, leading to continuous killing of tumor cells, the cell viability being nearly 60% after 72 h. Further, the antitumor activity of the complex was found to be very similar to that of only the drug. However, the adverse side effects caused by cisplatin were obviated with this complex. The sustained drug release in tumor cells and significant decrease in cytotoxicity to normal cells attained with fNDs thereby exhibited their remarkable capability to be used in lung cancer treatment (Cui et al. 2016).

Despite the potential of carbon nanoparticles in drug/gene delivery through their various functionalized forms (CNTs, ND) (Bandyopadhyay et al. 2015), their clinical applications are rather limited on account of the oxidative damage caused by them on the cellular membranes. Although strong oxidative damage by CNTs and ND has been reported for murine myoblastoma cells (Cancino et al. 2013) and brain of juvenile largemouth bass (Oberdörster 2004), efforts still need to be carried out to explore the same for lung cancer.

4.3.2.5 Quantum Dots

Quantum dots are novel semiconductor-based nanosized crystals/particles with unique optical properties, which make them suitable for *in-vitro* and *in-vivo* fluorophore-based biological investigations like imaging, labeling, and sensing (Medintz et al. 2005). The structure of QDs is a core-shell type, composed of an inorganic elemental core (e.g., cadmium selenide; CdSe) with a surrounding metal shell (e.g., zinc sulfide; ZnS). They have an intrinsic fluorescence emission spectra wavelength between 400 nm and 2000nm, depending on their size and composition (Cuenca et al. 2006).

The unique chemical or physical properties of QDs, such as size-tunable light emission, enhanced signal brightness, resistance to photobleaching, and excitation of multiple fluorescence (multiplexing phenomenon) colors simultaneously (Medintz et al. 2005), have enabled QD labeling for the detection of various biological analytes like DNA, protein, cells, and small organic molecules. Also, the recent reports on QD surface chemistry for biocompatibility and bioconjugation have shown the potential of QDs in the detection of the above-mentioned analytes (Li et al. 2011).

Another striking feature of QDs is their ability to be associated with multiple cancer antigens (CAs), which enables development of a multicomponent diagnostic system for rapid and sensitive diagnosis of cancer-related abnormalities at an early stage (Bandyopadhyay et al. 2015). This feature of QDs has been effectively utilized in the work carried out by Li et al. Here, a multiplexing method was employed for simultaneous detection of two lung cancer biomarkers, namely carcinoembryonic antigen (CEA) and neuron-specific enolase (NSE) using dual-color QDs. The three-step method followed for detection included: (i) mixing of biotinylated CEA and NSE capture antibodies, CEA and NSE antigen, and FITC-modified

NSE and rabbit anti-CEA detection antibodies, and addition of streptavidin beads after the reaction; (ii) addition and reaction of dual-color QDs, that is, antirabbit and anti-FITC QDs with respective antibodies; and (iii) dissociation of bead-QD conjugates and fluorescence detection of CEA and NSE using free QDs. The assay results indicated sensitive analysis of both biomarkers, the precision being within 0.53% for each and comparable to those of the commercial single-analyte-based ELISA detection technique. Further, evaluation of CEA and NSE from 25 human serum samples using this method showed good correlation with the conventional assay for these biomarkers, suggesting applicability of this method for routine assay of clinical samples (Li et al. 2011). QDs, along with magnetic IONPs, have also been employed for the detection of micrometastasis in lung cancer (Wang et al. 2012). This study has been elaborately discussed in the subsection on magnetic nanoparticles.

Despite these successful reports of QD-based detection, the presence of heavy metals and cytotoxicity-related issues limit the potential use of QDs in humans. Uncoated QDs are water insoluble and unstable when exposed to ultraviolet (UV) radiation, causing release of toxic cadmium in the cellular environment (Derfus et al. 2004). However, modification/coating of quantum dots (i.e., PEGylation and micelle encapsulation) may limit the release of toxic metals in response to UV radiation, making them dispersible in the cellular environment (Cuenca et al. 2006).

4.3.2.6 Nanoshells

Nanoshells comprise a dielectric core, generally silica, surrounded by a thin metal shell, typically gold. They are approximately 10–300 nm in size. The optical properties of nanoshells are different from quantum dots. Nanoshells depend on the plasmon-mediated conversion of electrical energy into light to exert their effect. Nanoshells can be optically tuned similarly to QDs and also have emission/absorption properties that range from the UV to the infrared region. Nanoshells are appealing as they offer imaging and therapeutic opportunities in the healthcare sector without being associated with heavy metal toxicity. Through several studies, it has been demonstrated that nanoshells may be potentially attractive vehicles for *in-vivo* and *in-vitro* imaging (Alper 2005, Cuenca et al. 2006).

For the early diagnosis of lung cancer, organic fluorophores used in the fluorescence *in situ* hybridization (FISH) assay have exhibited certain shortcomings in terms of weak emission signal, rapid photobleaching, strong photoblinking, and lifetimes similar to that of cellular fluorescence (2–10 ns), hampering their application in effective diagnosis. Based on these facts, Zhang and colleagues investigated the use of fluorescent metal nanoshells as fluorophores to evaluate the expression of miRNA-486 in lung cancer cells. This miRNAs have emerged as tissue-specific biomarkers for lung cancer diagnosis and identification (Takamizawa et al. 2004). Moreover, the miRNA-486 molecule, a biomarker in the case of lung cancer (Peng et al. 2013), was targeted in lung cancer positive cell lines using the FISH method. The scientists synthesized fluorescent silica spheres to encapsulate Tris (bipyridine) ruthenium (Ru (bpy)$_3^{2+}$) complex as the fluorophore agent and coated the silica spheres with a silver metal shell using the layer-by-layer approach. The silver metal shell was used to prevent undesired photochemical reactions, thereby maintaining photostability. Oligonucleotides targeting miRNA-486 molecules were covalently bound to the nanoshells for targeting these vehicles to the cancerous cells. The cell imaging studies were carried out in three human lung cancer cell lines, viz. H460 and H1499, positive for miRNA-486 expression and A549 that exhibited a negative miRNA-486 expression. The fluorescent cell images exhibited a stark distinction between the fluorescence generated by the conjugated metal nanoshells and the cellular autofluorescence. This was speculated to be due to the stronger emission intensity and longer lifetime of the metal nanoshells. The strong emission spots enabled accurate counting and analysis of the cells containing nanoshells hybridized to the miRNA-positive cells (Zhang et al. 2010). The effective isolation and detection of the fluorescent spots with metal nanoshells reflected their prospective in early diagnosis of lung cancer.

4.3.2.7 Mesoporous Silica Nanoparticles

MSNs include the mesoporous form of silica, which has a high surface area and porous interior (Bandyopadhyay et al. 2015). They offer unique properties such as tunable pore sizes and volumes for high drug-loading capacity and efficient encapsulation of a wide variety of drug molecules (Li et al. 2012). Other advantages of these carriers include their biocompatibility, ease of preparation, lower cost, and possibility of selective functionalization of MSNs at specific sites within the nanoparticles. These attributes enable their application as solid supports for immobilizing biological entities such as enzymes, proteins, and DNA, thereby making them suitable for sensing, cellular imaging, drug delivery, and gene transfection in therapy and/or diagnosis (Cauda et al. 2009, Feifel and Lisdat 2011). Attempts have also been made to target these nanovehicles to the biomarkers overexpressed on the surface of cancer cells.

Matrix metalloproteinases 2 and 9 (MMP2 and MMP9) are highly expressed in the advanced stages of lung cancer, of which MMP9 significantly increases the metastases of malignant cells in the tumor microenvironment. Overexpression of MMP9 is speculated to be linked with lower life expectancy in lung cancer cases (Martins et al. 2009). Uses of protease-sensitive linkers have been exploited for controlled drug release from NPs as they respond to altered enzyme levels in the biological microenvironment (Turk et al. 2001). Overexpression of MMP9 has been exploited by researchers for effective delivery of MSNs. Controlled drug delivery using MSNs has been investigated by van Rijt et al. The authors synthesized cisplatin-loaded MSNs tightly capped by avidin molecules through MMP9 sequence-specific linkers (cleavable MSNs-cMSNs) to facilitate their site-selective drug delivery in tumor areas with elevated levels of MMP9. MSNs without MMP9-specific linkers (noncleavable MSNs-ncMSNs) were synthesized and used as control. Avidin complexation of MSN was explored for pore sealing to achieve controlled drug release. The drug release from these MMP9-responsive MSNs was evaluated in human lung cancer cell lines (A549 and H1299) and *ex vivo* lung tissue models of mice (*Kras* mutant) and humans, keeping MSNs as control. These evaluations confirmed the MMP9-dose-responsive release of cisplatin from the cMSNs and subsequent induction of dose-dependent cell death. The therapeutic efficacy of cMSNs was proved by induction of apoptosis specifically in the tumor areas, revealed by 10- to 25-fold higher cell death than in the nontumor area. Combinatorial drug delivery studies with cMSNs using cisplatin and proteasome inhibitor bortezomib exhibited synergistic effects. The enhanced therapeutic efficacy of these combinatorial MMP9-responsive MSNs was evident in 25-fold higher tumor cell apoptosis than bortezomib alone. Further, the authors also confirmed MMP9-controlled drug release applicability in human lung tumors (van Rijt et al. 2015). The studies manifested the promising capability of functionalized MSNs to treat human lung cancer.

4.3.2.8 Lanthanide Nanoparticles

Lanthanides, also called rare earth metals, comprise the metallic elements with atomic numbers 57 (lanthanum) to 71 (lutetium). Lanthanide nanoparticles (LNPs) have been primarily utilized for tagging receptors on the cell surface to facilitate diagnosis and/or delivery of biological macromolecules like nucleic acids and proteins (Wang et al. 2010). Their inherent photoluminescent properties facilitate upconversion of low energy light (NIR) to high-energy light (UV–Vis), due to which they have been named upconverting NPs (UCNPs). UCNPs generate a higher-energy output photon from two or more low-energy photons, which is one of the key reasons for upconversion. The photoexcitable property of UCNPs in the NIR (biological window) region limits any background cellular absorption and autofluorescence (Bandyopadhyay et al. 2015).

The extremely narrow emission spectra of UCNPs facilitate their easy separation, enhance the selectivity of imaging protocol, and exhibit emission irrespective of their size. Single-phase lanthanide oxide nanoparticles are usually considered for cellular imaging on account of their narrow fluorescent bandwidth, unlike organic dyes; very high photostability; and nontoxicity in biological systems. As against QDs, LNPs can be excited in the NIR region, due to which they can safely penetrate into the deeper areas of cancerous tissues and enable high-resolution imaging.

Excitation of Ln^{3+} ions occurs through real electronic states of defined as well as relatively long lifetimes. Thus, weak and cheap laser diodes can be used to excite LNPs. Upconverting or downconverting fluorescent Ln (III) ions are often doped with paramagnetic MRI agents to utilize traditional MRI contrast agents in MRI-FI (fluorescence imaging) (Xu et al. 2012). Additional functionalization of paramagnetic LNPs with target-specific anticancer drug molecules facilitates improved cancer treatment with simultaneous diagnosis/imaging (Bandyopadhyay et al. 2015).

The commonly used metal-based and miscellaneous nanoparticles in lung cancer therapy and diagnosis are listed in Table 4.7.

Thus, metal-based nanoparticles have been studied to play a major role in diagnosis and treatment of lung cancer. The antimicrobial properties of noble metals (gold and silver) are manifested in their nanoparticulate form, enhancing cytotoxicity in cancerous cells. Iron oxide nanoparticles have revealed their theranostic applications in causing a hyperthermic effect on cancerous cells, diagnosis of the metastases condition, and overcoming drug resistance. The functionalization possibility offered by various forms of carbon nanoparticles such as CNTs and NDs has been utilized for effective conjugation of drug molecules to target these vehicles to cancer-affected tissues. The inherent optical properties of QDs and nanoshells facilitate detection of various analytes, which has assisted in diagnosis of lung cancer. High drug loading and potential of mesoporous silica nanoparticles to be used as solid supports has been effectively employed to induce apoptosis/cytotoxicity in cancer cells and to enable signal sensing for diagnosis. However, the toxicity and biocompatibility issues of metal-based nanoparticles (Ag and QDs) require further research to gain insights into their formulation chemistry so that they may be routinely used for successful treatment of lung cancer.

4.3.3 Bio-Based Nanoparticles

The toxicity and biocompatibility issues of polymeric and metal-based nanoparticles can be overcome through the use of bionanotechnology-based therapeutic nanocarriers, incorporating pre-existing biological systems/components with therapeutic nanoparticles. The inclusion of such biological systems/components in nanoparticles has been reported to enhance their specificity, permeability, stability, and biocompatibility (Sukumar et al. 2013). Such systems have been successfully devised and targeted specifically to lung cancer cells.

TABLE 4.7
Metal-Based and Miscellaneous Nanoparticles Used in Lung Cancer

Nanocarrier	Type of Metal	Target Receptor	Drug	In Vitro Cell Line	In Vivo Model	Results	References
Metal-based NPs	Gold NPs	—	Oxaliplatin	A549 human lung epithelial cancer cell line	—	• Oxaliplatin-tethered gold NPs ($IC_{50} = 0.135$ μM) have shown 6-fold higher cytotoxicity than oxaliplatin ($IC_{50} = 0.775$ μM) and also a higher cellular uptake (21.34 ± 0.89 ng Pt/10^6 cells) compared to free oxaliplatin (10.19 ± 0.95 ng Pt/10^6 cells)	Brown et al. (2010)
		—	Methotrexate	LL2 Lewis lung carcinoma cells	Mice (C57BL/6) with LL2	• MTX–AuNP ($IC_{50} = 0.046$) showed 17-fold improved cytotoxicity compared to free MTX ($IC_{50} = 0.816$ μg/mL) in LL2 cells • In *in vivo* studies in mice, no significant difference observed in the volume of ascites between MTX–AuNP-treated and MTX-treated or AuNP-treated mice	Chen et al. (2007)
	Silver NPs	—	—	A549 cell line	—	• Decreased cytotoxicity of Ag NPs observed with the antioxidant N-acetyl-cysteine • Ag NPs found to act as a mediator of ROS-induced genotoxicity	Foldbjerg et al. (2011)
		—	Artemisia leaf extract as a bioreductant	L132 human lung and A549 human lung carcinoma cells	—	• After 24 h, Ag NPs showed a dose-dependent decrease in cell viability compared to control group (media without Ag NPs) • No significant difference observed between Ag NP-treated L132 cells and control group	Gurunathan et al. (2015)

(*Continued*)

TABLE 4.7 (Continued)
Metal-Based and Miscellaneous Nanoparticles Used in Lung Cancer

Nanocarrier	Type of Metal	Target Receptor	Drug	In Vitro Cell Line	In Vivo Model	Results	References
	—	—	Euphorbia nivuliaplant stem latex-capped Ag NPs	A549 cells	—	• Dose-dependent cytotoxicity observed • Proven as biocompatible carriers for tumor apoptosis in A549 lung carcinoma cells	Valodkar et al. (2011)
	Ag wires and micro/nanoparticles	—	—	A549 cells	—	• After 24–48 h, Ag wires (length: 1.5–25 μm, diameter 100–160 nm) significantly reduced (60%–80%) cell viability and increased lactate dehydrogenase (LDH) release from lung carcinoma cells • Silver wires induced a strong cytotoxicity, loss in cell viability, and early calcium influx compared to spherical Ag nanoparticles (30 nm) and microparticles (<45 μm)	Stoehr et al. (2011)
Magnetic NPs (MNPs)	—	EGFR	—	A549 and A549-luc (luciferase transfected) cells	A549-luc cells injected fox chase SCID® beige female mice model	• Aerosol delivery (inhalation) of EGFR-targeted SPIONs showed enhances tumor retention and resulted in a significant inhibition of *in vivo* tumor growth • Targeted SPIONs showed 4.5-fold higher cellular uptake in A549 cells. After instillation, iron content found to be 4.5- to 5-fold higher in lungs compared to inhalation (400 vs 80 μg for EGFR-targeted SPIONs)	Sadhukha et al. (2013)

(*Continued*)

TABLE 4.7 (Continued)
Metal-Based and Miscellaneous Nanoparticles Used in Lung Cancer

Nanocarrier	Type of Metal	Target Receptor	Drug	In Vitro Cell Line	In Vivo Model	Results	References
	Fluorescent polymethacrylic acid embedded magnetic SiO_2/IONPs	Folic acid receptor	Doxorubicin (DOX)	H446 lung cancer cell, HeLa cells (with folic acid receptor, FR (+)) and H446 lung cancer cells (without folic acid receptor FR (−))	—	• DOX loading; 105 ± 8 µg • IONPs showed 84% and 90% of DOX release after 18 h and 100 h, respectively • Both targeted and nontargeted composites showed 85% of cell viability • DOX loaded SiO_2IONPs showed increased mean fluorescent intensity in FR (+) cells compared FR (−) cells	Chen et al. (2010)
	IONPs	—	Cisplatin	—	A549 lung tumor xenografted BALB/c nude mouse model	• Cisplatin-IONPs showed cytotoxicity effect in both A549 cells and DDP-resistant A549 cells in a time-dependent and dose-dependent manner • Accumulation of intracellular cisplatin in the cisplatin-IONPs treated group (15.28 ± 1.16 µg/L) 3-fold greater than DDP treated group (5.53 ± 0.09 µg/L) • Cisplatin resistant A549 xenografted BALB/c nude mice model showed increased tumor growth inhibition by decreasing localization of lung resistance-related protein	Li et al. (2013)
Carbon-based NPs	Multiwalled carbon nanotubes with polycitric acid	—	Paclitaxel	A549 cells	—	• MWNT-g-PCA-PTX showed 20% higher cytotoxic effect than free PTX	Sobhani et al. (2011)
Nano-diamond (ND)	—	—	Paclitaxel	A549 cells	CB17/Icr-Prkdc^{scid}/CrlSCIDmice model	• ND-PTX conjugation showed significant antitumor activity in both A549 and xenograft SCID mice • Localization of ND-PTX in the microtubules and cytoplasm of A549 cells observed by confocal microscopy	Sukumar et al. (2013)

The following section describes different bio-based nanocarrier systems, namely apoferritin, viral nanoparticles, protein-based, liposomes, and solid lipid nanoparticles that have been explored for treatment and/or diagnosis of lung cancer.

4.3.3.1 Apoferritin

Ferritin (Ft) is an iron-containing protein complex comprising 24 self-assembled polypeptide subunits having internal and external diameters of 7.6 and 12 nm, respectively. These protein-based cage-like networks exhibit three distinct interfaces, the interior, exterior, and the interface between the subunits, for functionalization (Li et al. 2012). Removal of the iron core from the inner cavity leaves the hollow protein cage called the apoferritin (apoFt) nanocage, which undergoes assembling and disassembling with the change in pH of the environment of the molecule. The apoFt nanocage can be used to incorporate inorganic metals inside its cavity for scavenging ROS generated through several mechanisms in the cellular environment (Zhang et al. 2009). This property has been exploited as a template for synthesizing an array of nanomaterials for theranostic applications in cancer therapy. The apoFt nanoparticles enter into the target tumor cells by receptor-mediated endocytosis (Liu et al. 2011), clathrin-mediated endocytosis, and macropinocytosis processes (Liu et al. 2012).

Li et al. developed a ferritin-based multifunctional nanostructure for fluorescence and MR imaging of lung cancer cells simultaneously (Li et al. 2012). The authors engineered human H-chain ferritin (rHF) with green fluorescent protein (GFP) for stable fluorescence in the cells. Further, RGD (arginylglycilaspartic acid) peptide was fused on the exterior surface of the ferritin cage for targeting $\alpha_v\beta_3$ integrin receptors on tumor cells (human glioblastoma U87MG cells and A549 cells). Multifunctional nanostructures based on ferritin (rHF/Fe$_3$O$_4$, GFP-rFH/Fe$_3$O$_4$ and RGD-GFP-ferritin [RGF]/Fe$_3$O$_4$) were developed by synthesizing iron oxide (Fe$_3$O$_4$) nanoparticles in the engineered ferritin cages. The fluorescence imaging of these cages targeted to $\alpha_v\beta_3$ integrin-positive U87MG and A549 cells showed high-intensity fluorescence with RGF, as compared to GFP-rHF control cells. Also, magnetic resonance imaging with RGF significantly enhanced the signal to facilitate accurate diagnosis, as compared to GFP-rHF/Fe$_3$O$_4$ or no contrast agent (Li et al. 2012). Therefore, effective targeting and fluorescence imaging of lung cancer cells using engineered nanocages was proven to be a useful vehicle among the various multifunctional, nanostructured, protein-based tools for fluorescent imaging.

The antioxidant enzymes present inside the human body, such as superoxide dismutase (SOD), catalase, and peroxidase, fail to protect the cells under sudden oxidative damage/stress conditions (Zhang et al. 2009, Fan et al. 2011). Thus, many ongoing studies have focused on the development of artificial antioxidants capable of reducing oxidative stress during lung cancer therapy. Apoferritin-encapsulated Pt nanoparticles (Liu et al. 2011) have been explored as artificial antioxidants on account of their catalase-, peroxidase-, and SOD-mimicking activity. Apoferritin-CeO$_2$ nanotruffle (Liu et al. 2012) has been utilized as artificial redox enzyme due to its ability to mimic SOD activity. This property can be exploited to combat ROS-mediated lung cancer by scavenging hydrogen peroxide, superoxide, and other small molecules triggered in sudden oxidative damage. Thus, these systems possess potential for successful application in lung cancer therapy (Zhang et al. 2009, Fan et al. 2011, Sukumar et al. 2013).

4.3.3.2 Viral Nanoparticles

Virus-based nanoparticles (Viral NPs) have been obtained from different sources such as plant viruses, animal viruses, and bacteriophages (Steinmetz 2010). Viruses of sizes ranging from 20 to 400 nm are infectious pathogens with their genome (nucleic acid) encapsulated within a protein coat. Viral NPs (VNPs) are composed of virus capsids (shells) that lack a genome, thereby making them noninfectious and incapable of replication. This confers upon them the advantages of nonimmunogenicity and biodegradability (Cormode et al. 2010). Capsid, the protective coating of the virus, protects viral NPs by imparting them with stability against extremes of temperatures, pH, and chemical exposure and making them suitable as nanoparticulate carriers. Structural symmetry, polyvalence, and monodispersity are the striking features of these dynamic and self-assembling VNPs (Cormode et al. 2010, Steinmetz 2010). VNPs have been explored in many biomedical applications ranging from biosensing and bioimaging to drug/gene delivery and also in vaccine development because of their biocompatible nature, availability in a wide range of shapes and sizes, and ease of surface modification/functionalization (Cormode et al. 2010, Steinmetz 2010, Sukumar et al. 2013). Inherent and acquired drug resistance for the majority of the currently used small molecule-based antineoplastic agents observed in lung cancer patients has created a need to integrate conventional therapies with immunotherapeutic approaches. These approaches have played a significant role in minimizing the chances of developing drug resistance.

Beljanski and Hiscott have investigated the use of conventional chemotherapeutic drugs, along with genetically modified oncolytic viruses (OVs), for lung cancer therapy. In this combination, the inability of chemotherapeutic agents to kill cancer stem cells was found to be well complemented by OV-mediated gene therapy (Beljanski and Hiscott 2012). TG4010, an OV-based anticancer vaccine, is a targeted immunotherapy based on a poxvirus (modified vaccinia virus Ankara)-encoding MUC1 tumor-associated antigen and interleukin 2 (IL-2) as an immunoadjuvant to reverse the suppression of T-cell response. This genetically modified vaccine was evaluated for its efficacy in combination with first-line chemotherapy in advanced NSCLC in a phase IIB trial. MUC1-expressing advanced NSCLC patients (n = 148; Stage IIIB or IV) were allocated for this trial. Half of these patients received combination therapy (TG4010 and cisplatin), while the remaining half received chemotherapy alone. Results indicated that 43.2% of the patients administered with combination therapy were free from disease progression at the end of 6 months, against only 35.1% of the patients administered with only chemotherapy. The results indicated the synergistic potential of TG4010 along with chemotherapeutic drugs (Quoix et al. 2011).

Apart from the therapeutic application of VNPs, researchers have also investigated their potential as fluorophores in imaging of lung cancer cells. A study headed by Robertson demonstrated the use of engineered T4 viral nanoparticles as fluorescent imaging probes for cell imaging and flow cytometry analysis in A549 cell lines. In this study, amine groups located in the head region of the T4 bacteriophage were conjugated with fluorescent dyes (Cy3 and Alexa Fluor 546). These dye-tagged T4 VNPs were found to be retained in A549 cells for 72 h and enabled their tracking, as uptake was visible by fluorescence microscopy and quantifiable by flow cytometry. Incorporation of the T4 bacteriophage provided a larger surface area for the accumulation of dye molecules (about 19,000) per viral nanoparticle. Also, the use of Cy3 dye with T4 VNPs enhanced fluorescence up to 90%, as compared to the untagged dye (Robertson et al. 2011). These studies revealed the functionalization capability of T4 VNPs owing to their reactive amine head surface and suggested their potential as a molecular probe for lung cancer imaging.

4.3.3.3 Protein-Based Nanoparticles

Protein nanoparticles fabricated using naturally occurring proteins like albumin, gelatin, gliadin, and legumin, either alone or in combination with biodegradable polymers, have been used for drug delivery (Jahanshahi and Babaei 2008, MaHam et al. 2009). The suitability of these nanoparticles for drug/gene delivery to human airway epithelium was evaluated by Brzoska and colleagues using porcine gelatin and human serum albumin (HSA)-based nanoparticles. *In-vitro* studies conducted on primary airway epithelium cells and the human bronchial epithelial cell line (16HBE14o-) showed no evidence of cytotoxicity or inflammation due to interleukin 8 (IL-8) release. The investigation demonstrated the suitability of protein-based nanoparticles as drug and gene delivery carriers due to their biocompatibility, high cellular uptake, and lack of inflammation in human bronchial epithelial cells (Brzoska et al. 2004). These studies exhibited the potential of protein-based nanoparticles as gene or drug carriers for lung epithelial cells.

In another study, human serum albumin, a biocompatible and biodegradable protein, was explored for rapid delivery of anticancer drugs and subsequent therapeutic effect to drug-resistant lung cancer. Choi and group fabricated inhalable HSA nanoparticles that were conjugated with doxorubicin and octyl aldehyde and with adsorbed apoptotic tumor necrosis factor (TNF)-related apoptosis-inducing ligand (TRAIL) protein. The octyl aldehyde modification of HSA was done to improve hydrophobicity, while the ability of the TRAIL protein to specifically bind to death receptors on cancer cells without killing normal cells was employed in this formulation. Doxorubicin-conjugated HSA NPs coated with TRAIL were synthesized. Significant apoptosis and cytotoxicity were observed in TRAIL-resistant H226 cells (a human lung squamous cell carcinoma cell line) on account of the synergistic effect of TRAIL and doxorubicin of this nanoformulation compared to only drug or TRAIL-based HSA nanoformulation. Further, the antitumor efficacy of the formulation was evaluated by administering it by insufflation in BALB/c nu/nu mice bearing H226 cell-induced metastatic tumors. A remarkable decrease in lesion size and number, along with reduced tumor weights (337.5 ± 7.5 mg), was observed in mice treated with the TRAIL/DOX HSA NP, as compared to those treated with TRAIL HSA NP (678.2 ± 51.5 mg) and DOX HSA NP (598.9 ± 24.8 mg). These results were attributed to the synergistic effect of TRAIL and doxorubicin coated on HSA NPs (Choi et al. 2015). The functionalization ability of albumin-based nanoparticles presents an effective inhalable drug delivery approach for the treatment of resistant lung cancer.

4.3.3.4 Liposomes

Liposomes are self-assembling, concentric, closed colloidal structures composed of lipid bilayers enclosing an aqueous core (Wang et al. 2008). The hydrophobicity of lipid bilayer allows loading of various drug molecules, proteins, nucleic acid, or plasmids for drug/gene delivery, along with encapsulation of fluorophores (dyes, enzymes) for early detection of cancer (Bandyopadhyay et al. 2015).

Liposomes have been used as drug and gene delivery carriers for the treatment of lung cancer. Cisplatin, a drug used for the treatment of NSCLC for the last two decades, is known to develop nephrotoxicity in 20% of patients receiving high doses for treatment (Yao et al. 2007). In 2004, Boulikas developed a liposomal formulation of cisplatin called Lipoplatin™ to reduce the systemic toxicity of cisplatin. The scientists formulated Lipoplatin (SPI-077) by encapsulating cisplatin in sterically-stabilized liposomes and tested it against NSCLC. In Phase I clinical trials, SPI-077 exhibited a different pharmacokinetic behavior than cisplatin, with a half-life of 134 h and renal elimination reaching only 4% of the total dose after 72 h, thereby prolonging circulation time of drug, reducing its elimination from the system for effective treatment. SPI-077, administered at doses of 260 mg/m^2 against NSCLC, demonstrated a mild toxicity profile as compared to cisplatin alone (Boulikas 2004).

Furthermore, in a multiple rat tumor model, Lipoplatin™ injection significantly reduced nephrotoxicity (to negligible levels) as compared to standard therapy (Babu et al. 2013).

Pulmonary delivery is generally not preferred for this system because of low bioavailability of drugs. In an attempt to address this problem, Jain and colleagues conducted a layer-by-layer assembly of polyelectrolytes over liposomes for administration of paclitaxel. Paclitaxel coupled with stearoyl amine formed the core of the nanoparticles, which were further coated with subsequent layers of anionic poly (acrylic acid) (PAA), followed by coating with cationic poly (allyl amine hydrochloride) (PAH), to form layersomes. *In-vitro* studies carried out on A549 cells revealed a sustained release of paclitaxel over 24 h. The higher uptake of liposomal complexes, as compared to free drugs, was revealed through the high intensity of green fluorescence of coumarin-6–encapsulated systems in A549 cells. *In-vivo* pharmacokinetics showed a nearly 4.07-fold increase in the overall oral bioavailability of paclitaxel as compared to that of free drugs (Jain et al. 2012). The results confirmed the efficacy of the liposomal system for lung cancer treatment.

In another study, Mainelis et al. developed an integrated drug carrier system (DCS) for inhalation-mediated delivery of anticancer agents directly to the lung, together with suppression of cancer cell resistance. This system was proposed to enhance treatment efficiency and limit the adverse side effects of chemotherapeutics on healthy organs. This vehicle contained doxorubicin as the cell death inducer, along with either p-ethoxy-modified antisense oligonucleotide or siRNA against multidrug resistance-associated protein 1 (MRP1) mRNA as the suppressor of cancer cell resistance. This liposomal system was evaluated in nude nu/nu mice by using a small nose-only exposure chamber, wherein mice were positioned in such a way that their noses were situated at the "inhalation point" of the chamber. Lower amounts of the drug were observed to reach to the lungs via the inhalation-based therapy as compared to an intravenous injection of the drug. However, the retention time was significantly higher (a considerable amount of drug remained after 3 days) in lungs with inhalation-based delivery than intravenous injection (complete elimination of drug from lungs in 2 days). Further, tumor volume reduction of more than 90% was observed in the case of inhalation delivery as compared to intravenous injection. A significant improvement in retention time of the drug and reduction of tumor volume by the liposomal system offers a better method for the treatment of lung cancer (Mainelis et al. 2013).

4.3.3.5 Solid Lipid Nanoparticles

Solid lipid nanoparticles (SLNs) are natural or synthetic lipid-based systems composed of a solid hydrophobic core surrounded by a monolayer of phospholipid coating, usually in the nanometer (submicron) size range (50–1000 nm). The solid hydrophobic lipid core is generally made from lipid molecules of triglycerides (e.g., Compritol 888 ATO and Dynasan 112), beeswax, carnauba wax, cetyl alcohol, and cholesterol. Though SLNs are structurally quite similar to liposomes and nanoemulsions, they are more stable in biological systems due to the presence of a rigid hydrophobic core (Sukumar et al. 2013, Bandyopadhyay et al. 2015).

Mehnert and Muller et al. published a comprehensive review on the synthesis and characterization of SLNs, emphasizing their scope in drug delivery (Müller et al. 2000, Mehnert and Mäder 2001). By the virtue of the inherent hydrophobic environment of their solid core, they act as suitable carriers for delivery of various anticancer drugs including doxorubicin, idarubicin, paclitaxel, camptothecin, and etoposide. The cationic and amphiphilic nature of SLNs ensures effective *in-vitro* gene transfer to lung cancer cells.

This gene transfer is the result of stable complex formation between the hydrophilic amino group of SLNs and negatively charged DNA plasmids (Bandyopadhyay et al. 2015).

Mutations in the p53 tumor suppressor gene are the most common tumorigenic processes in human cancers, including lung cancer (Johnson et al. 1995). Wild-type p53 causes arrest of the cell growth at the G1 stage of the cell cycle and leads to cell apoptosis (Sidransky and Hollstein 1996). In p53 mutant cells, reintroduction or overexpression of the wild-type p53 gene has been known to induce apoptosis and growth arrest in various lung cancer cell lines (Takahashi et al. 1992). Choi et al. investigated p53 gene delivery by using stable cationic SLN formulations. The authors developed a cationic SLN formulation by mixing four components in varying ratios, namely (i) tricaprin (TC) as the core material of SLNs, (ii) 3β-[N-(N,Ndimethylaminoethane)-carbamoyl] cholesterol (DC-chol) as the cationic lipid, (iii) dioleoylphosphatidyl ethanolamine (DOPE) as the helper lipid, and (iv) Tween 80 as surfactant. The plasmid DNA-encoding p53-EGFP (pp53-EGFP) signaling probe, the latter being a fusion of p53 and EGFP, was used for the study. Further, plasmid DNA-encapsulated SLNs were studied for transfection in H1299 cells (a human NSCLC cell). Of the various SLNs fabricated, the formulation containing a combination of TC:DC-Chol:DOPE:Tween 80 in the weight ratio of 0.3:0.3:0.3:1 demonstrated the highest transfection efficiency. The transfection efficiency was found to be even higher than the commercially available Lipofectin®. This transfection resulted in enhanced levels of wild-type p53 mRNA and protein expression in H1299 cells, thereby leading to re-establishment of p53 gene function in lung cancer and restoration of apoptotic pathway (Choi et al. 2008). These results signified the potential of SLNs for successful lung cancer therapy by virtue of their apoptotic induction based on gene re-establishment and tumor growth inhibition.

Some of these bio-based systems targeted to lung cancer cells are listed in Table 4.8.

Overall, the nontoxicity and biocompatibility of bio-based nanoparticles enhance their specificity and permeability in the targeted cancerous cells, making them suitable candidates for drug delivery. Despite the efficacy of these polymer-based, metal-based, and bio-based nanocarriers, only a few systems have made it to clinical trials, which are described in subsequent sections of the manuscript.

4.4 Clinical Studies of Nanosystems for Lung Cancer

Three of the nanoparticle-based systems, namely Abraxane®, Lipoplatin™, and Genexol® PM, have made it to clinical trials and/or the commercialization stage for the treatment of lung cancer. Figure 4.4 illustrates the various stages involved in the clinical translation of nanomedicines, from their formulation to their FDA approval.

4.4.1 Abraxane®

Abraxane® (ABI-007), developed by American Bio Science, Inc., USA, is a Cremophor-free, 130-nanometer, albumin-bound particle form of paclitaxel. This formulation has been manufactured using patented nab® (nab-nanoparticle albumin-bound) technology. It is indicated as the first-line treatment of locally advanced or metastatic NSCLC, in combination with carboplatin, in patients who cannot undergo curative surgery or radiation therapy.

The widespread employment of paclitaxel is limited because of its poor solubility and the toxicity associated with Cremophor® EL, a lipid-based solvent (polyoxyethylated castor oil)

TABLE 4.8

Bio-Based Nanoparticles for Lung Cancer

Nanocarrier	Target	Active Ingredient	In Vitro Cell Line	In Vivo Model	Results	References
Viral nanoparticles	Tailless T4 bacteriophage head viral NPs	Cy3 and Alexa Fluor 546 Fluorescent dyes	A549 cells	—	• Cy3 Dye-T4 NPs conjugates led to fluorescent enhancements of up to 90% compared to free Cy3T4 nanoparticles conjugated • Cellular uptake of dye-T4 NPs clearly visualized for at least 72 h confirmed by fluorescent microscopy	Robertson et al. (2011)
		PTX and cisplatin	NSCLC cell lines: NCI-H522, NCI-H1703, NCI-H157, NCI-H838, and NCI-H2347	—	• Lung cancer cells (NCI-H522 and NCI-H1703) with nonfunctional p53 at least 10 times more sensitive to ONYX-015 cytolysis (100%) than lung cancer cells (NCI-H2347 and NCI-H838) with wild-type p53 • Combination of ONYX-015 with paclitaxel and cisplatin displayed a synergistic effect on cancer cells and primary cultures	You et al. (2000)
		Adriamycin and cisplatin	Human lung cancer cell lines (A549 and NCI-H460)	NCI-H460 xenografted female BALb/c nude mice	• Combinational therapy with ZD55-IL24 and ADM/DDP showed enhanced antitumor effect in lung cancer both in *in vitro* and *in vivo* models	Zhong et al. (2009)
Gelatin nanoparticles (GPs)	Biotinylated EGF conjugated GPs	EGFR	A549 and HFL1 cell lines	A549 cells injected CB-17/lcrCrl-SCID-bg mice	• Fluorescence spots and cellular internalization of GP-Av-bEGF NPs displayed higher efficiency on EGFR-expressed A549 than that on non-EGFR normal lung (HFL1) cells • *In vivo* aerosol administration of NeutrAvidin^{FITC}-biotinylated EGF GPs showed higher fluorescence intensity in A549 mouse model than that of the control group	Tseng et al. (2007)

(*Continued*)

TABLE 4.8 (Continued)
Bio-Based Nanoparticles for Lung Cancer

Nanocarrier	Target	Active Ingredient	In Vitro Cell Line	In Vivo Model	Results	References
Solid liposome nanoparticle (SLN)	—	—	A549 cells	Murine precision-cut lung slices (ex vivo)	• IC_{50} of SLN20 (20% lipid matrix): 4080 μg/mL • IC_{50} of SLN50 (50% lipid matrix): 1520 μg/mL • In ex vivo model, SLN50 showed higher cytotoxicity in murine precision-cut lung slices compared to SLN20	Nassimi et al. (2009)
	Folate receptor	Paclitaxel	M109, a murine lung cancer and FR(+) and in FR(−) Chinese hamster ovary cells	M109 engrafted BALB/c mice tumor model	• F-PTX lipid NPs showed significant tumor growth inhibition and animal survival compared to nontargeted LN	Stevens et al. (2004)
Albumin nanoparticles	^{99m}Tc chelated HP-ANP	—	A549 cell line	—	• ^{99m}Tc-HP-ANP demonstrated good scintigraphic imaging properties in the rabbit with an extended biological half-life compared to ^{99m}Tc-HP	Yang et al. (2010)
	TRAIL protein (TRAIL/Dox HSA-NP) and doxorubicin	DOX	H226 lung cancer cells	H226 cell bearing BALB/c nu/nu metastatic tumors mice	• Inhalable apoptotic TRAIL protein-adsorbed DOX-conjugated HAS NPs displayed enhanced antitumor activity in tumor mouse model compared to TRAIL or Dox HSA-NP alone • TRAIL/Dox HSA-NP treated mice model displayed lower average lung weight (337.5 ± 7.5 mg) compared to DOX alone (678.2 ± 51.5 mg) and TRAIL HSA-NP alone (598.9 ± 24.8 mg) compared to nontreated implanted mice (746.1 ± 65.2)	Choi et al. (2015)

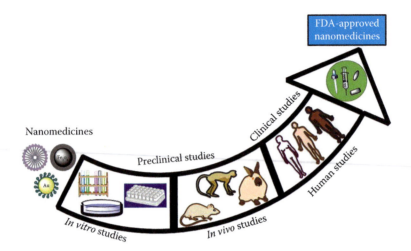

FIGURE 4.4
Schematic diagram of clinical translation of nanomedicines.

used as a commercial vehicle to deliver paclitaxel. Moreover, use of this solvent has been known to result in leaching of plasticizers from standard intravenous tubing, leading to histamine release; severe anaphylaxis; hyperlipidemia; abnormal lipoprotein patterns; aggregation of erythrocytes; and prolonged, sometimes irreversible sensory neuropathy (Lorenz et al. 1977, Gelderblom et al. 2001). Furthermore, administration of Taxol® (paclitaxel) requires a long infusion period (typically 3–24 h), inline filters, and premedication with steroids and antihistamines to minimize the risk of hypersensitivity reactions (Bristol-Myers Squibb Company 2012). In addition, Cremophor® EL also forms micelles that entrap paclitaxel in the plasma compartment, resulting in nonlinear pharmacokinetics for Taxol® (Sparreboom et al. 1999).

On the other hand, Abraxane® delivers paclitaxel as a suspension of albumin particles in saline, which evades the need for Cremophor® EL, facilitates shorter duration of infusion, and mitigates solvent-related toxicities. Thus, use of albumin as a vehicle circumvents the requirement of premedication with steroids and antihistamines. Furthermore, albumin has the potential to increase drug delivery to tumors by facilitating albumin gp60 receptor-mediated transcytosis across endothelial cells. Gp60 is a 60-kDa endothelial cell surface-localized glycoprotein that binds to native albumin with a high affinity in the nanomolar range. The binding of albumin activates gp60 clustering and association with caveolar-scaffolding protein caveolin-1, resulting in the formation of vesicles called caveolae. These vesicles thereby carry either gp60-bound and fluid-phase albumin or albumin-bound drugs in a process known as transcytosis from the apical to the basal membrane, leading to the release of vesicle contents into the subendothelial space (John et al. 2003).

A phase I/II clinical study for Abraxane® was conducted to define the maximum tolerated dose (MTD) and dose-limiting toxicities (DLTs), as well as to evaluate the safety and antitumor activity of ABI-007 in patients with advanced stage IV NSCLC. In a phase I study, the Abraxane® formulation was safely administered as a short intravenous infusion without steroid or antihistamine premedication. Here, the MTD was 300 mg/m^2, which was given over 30 min once every 3 weeks. Alopecia occurred in all patients, whereas neurotoxicity and mucositis were observed to be dose-limiting effects. Grade IV granulocytopenia and an episode of superficial keratopathy were seen at the dose of 375 mg/m^2. The area under the

TABLE 4.9

Clinical Trials of Abraxane® Prior to Commercialization for Lung Cancer Treatment

Sr. No	Phase	Age	Condition	Organizations/ Sponsors	Trial Sites	Year	Clinical Trial Identifier
1	I	18 and above	Advanced NSCLC	Celgene Corporation	USA	2003–2007	NCT00073723
2	II	18 and above	Advanced NSCLC	Celgene Corporation	USA	2003–2007	NCT00073723
3	III	18 and above	Advanced NSCLC	Celgene Corporation	USA, Canada, Japan, Australia, Russia, and Ukraine	2007–2009	NCT00540514

curve (AUC) for Abraxane® was found to be lower as compared to the same dose of paclitaxel solubilized with Cremophor® EL. This potentially reflects improved tissue distribution of albumin-based paclitaxel. In a phase II clinical trial, a dose of 260 mg/m^2 of Abraxane® was administered as intravenous infusion over 30 min without premedication. In this case, the formulation demonstrated significant tumor responses and was found to be well-tolerated in NSCLC patients (Ibrahim et al. 2002, Green et al. 2006). A phase III clinical trial of Abraxane® was conducted by using it in combination with carboplatin, wherein a dose of 100 mg/m^2 of Abraxane® was administered intravenously over a period of half an hour. Abraxane® was finally approved by the FDA for treatment of advanced NSCLC in 2012 due to improved therapeutic profile and lower side effects. Table 4.9 describes the clinical trials conducted before the approval of Abraxane® for the treatment of lung cancer.

In the United States, Abraxane® was first approved in January 2005 for the treatment of breast cancer, after failure of combination chemotherapy or remission within 6 months of adjuvant chemotherapy for the metastatic disease. It has been globally approved for the treatment of metastatic breast cancer (MBC) in more than 40 countries. In October 2012, Abraxane® was approved by the FDA as the first-line therapy for locally advanced or metastatic NSCLC, in combination with carboplatin, in patients who cannot undergo curative surgery or radiation therapy. An overview of approval proceedings followed for Abraxane® by FDA is given in Table 4.10.

Inherent properties of albumin like biocompatibility and bioavailability, exploited in Abraxane® formulation, assisted in improved delivery of paclitaxel, which otherwise faces problems of poor solubility and toxicity. Thus, the choice of carrier is important in a drug delivery system so as to increase the potency and decrease side effects of the active ingredient.

TABLE 4.10

Events Associated with FDA Approval of Abraxane®

Sr.No	Date	Highlights
1	March 8, 2004	New drug application filed for FDA approval of Abraxane® for the treatment of metastatic breast cancer
2	January 7, 2005	FDA approved Abraxane® for treatment of metastatic breast cancer
3	October 12, 2012	FDA approved Abraxane® for the first-line treatment of advanced NSCLC
4	September 6, 2013	FDA approved Abraxane® for late-stage pancreatic cancer

4.4.2 Lipoplatin™

Cisplatin is one of the important drugs in chemotherapy, with applications in more than 50% of human cancers, including NSCLC. Cisplatin can be combined with radiation therapy and a variety of other anticancer drugs such as gemcitabine, taxanes, or vinca alkaloids. However, the wider use of cisplatin is deterred due to severe damage to kidneys, peripheral nerves, bone marrow, the gastrointestinal tract, hair follicles, and other tissues.

Lipoplatin™, or Liposomal cisplatin, composed of lipids and cisplatin, has an average diameter of 110 nm. It is composed of a lipid shell made of dipalmitoyl-phosphatidyl glycerol (DPPG), soy, phosphatidyl-choline, cholesterol, methoxy-polyethylene, and glycol-distearoyl phosphatidyl ethanolamine lipid conjugate (mMPEG 2000-DSPE). The central core consists of 8.9% of Cisplatin and 91.1% total lipids. Lipoplatin™ is synthesized by forming reverse micelles of cisplatin and DPPG under special conditions of pH, solvent (ethanol), ionic strength, and so on. Cisplatin-DPPG reverse micelles are subsequently converted into liposomes by interaction with neutral lipids.

Lipoplatin™ has been reported to substantially reduce the renal toxicity, peripheral neuropathy, ototoxicity, and myelo-toxicity, as well as nausea/vomiting and asthenia associated with cisplatin in phase I, II, and III clinical studies, with enhanced efficacy as compared to cisplatin. During clinical development in lung cancer patients, 10- to 200-fold higher accumulation of Lipoplatin™ was observed in solid tumors as compared to the adjacent normal tissue (Boulikas 2009).

Targeting Lipoplatin™ to tumor vasculature in animals suggested its antiangiogenic potential. Further, Lipoplatin™ was proposed to serve dual actions, viz. chemotherapeutic and antiangiogenic effects. In combination with paclitaxel, it has demonstrated superior effects as compared to cisplatin in the chemotherapy of NSCLC adenocarcinomas. This unique formulation is not detected by macrophages and immune cells and therefore possesses longer circulation time in body fluids and tissues, with preferential accumulation in primary tumor sites and metastases.

Lipoplatin™ nanoparticles within the tumor cell mass can fuse with the cell membrane owing to the presence of the fusogenic lipid DPPG in their lipid bilayer. An alternative mechanism involves endocytosis of Lipoplatin™ within the tumor. These processes occurring at the cell membrane level are further promoted by the lipid-based shell of the nanoparticles. Lipoplatin™ delivers its toxic payload (cisplatin) within the cellular cytoplasm to induce tumor cell apoptosis (Stathopoulos and Boulikas 2012).

Preclinical studies with Lipoplatin™ have demonstrated effective anticancer activity and lower nephrotoxicity as compared to cisplatin. Studies conducted in different *in-vivo* preclinical models involving rat, mouse, dog, and so on have suggested that a dose of up to 150 mg/m^2 of Lipoplatin™ can be safely administered with minimum side effects. The lower toxicity of Lipoplatin™ may be due to alterations in its pharmacokinetics, preferential localization in tumors containing compromised vasculature, and enhanced cellular uptake. In a phase I clinical evaluation, Lipoplatin™ was administered at doses ranging between 25 and 125 mg/m^2 to cancer patients, wherein the drug demonstrated mild hematological and gastrointestinal toxicity. It did not exhibit any nephrotoxicity, neurotoxicity, or hair loss in these patients. Also, in a phase II clinical evaluation, Lipoplatin™ was administered along with Gemcitabine at doses of 120 and 1000 mg/m^2, respectively, in NSCLC patients. It demonstrated improved anticancer activity over cisplatin and significant reduction in nephrotoxicity. A phase III clinical investigation compared the efficacy of a combination of Lipoplatin™ and gemcitabine with that of cisplatin and gemcitabine. This study demonstrated a distinct advantage of the former combination in terms of therapeutic profile and safety.

TABLE 4.11
Clinical Studies Conducted for Lipoplatin™

Phase of Study	Condition	No. of Patients	Dose (mg/m²)	Results	References
Phase 1	Stage IV (pancreatic carcinomas, renal cell carcinomas, gastric cancer, and squamous cell carcinoma of the head and neck)	27	25–125	Lipoplatin™ showed mild hematological and gastrointestinal toxicity, did not exhibit any nephrotoxicity, neurotoxicity, or hair loss	Stathopoulos et al. (2010)
Phase 2 (Lipoplatin™ with gemcitabine as first-line in NSCLC)	NSCLC	88	Lipoplatin™ 120 and gemcitabine 1000	Significant reduction in nephrotoxicity	Mylonakis et al. (2010)
Phase 3 (Lipoplatin™ with gemcitabine) vs (cisplatin with gemcitabine)	NSCLC	200	Lipoplatin™ 120	Significant reduction in nephrotoxicity, nausea, vomiting, neurotoxicity, and asthenia was observed in patients administered with a combination of Lipoplatin™ and gemcitabine, as compared to those administered with cisplatin and gemcitabine	Boulikas et al. (2007)

Preclinical and clinical studies conducted with Lipoplatin™ are described in Table 4.11. Overall, these investigations demonstrated reduction in side effects such as renal toxicity, peripheral neuropathy, ototoxicity, myelotoxicity, and so on with Lipoplatin™ as compared to cisplatin. Lipoplatin™ is currently under phase III clinical trials and has not yet been approved for treatment of NSCLC.

From the literature, it can be inferred that the lipid content of Lipoplatin enhances its ability to fuse with cell membrane and improves its ability to deliver its toxic cargo to the tumor mass, in turn inducing apoptosis. This formulation is able to evade immune surveillance and extravagate through the compromised endothelium of the vasculature in tumors that is created during the process of neoangiogenesis. Combined with its antiangiogenic properties and reduced nephrotoxicity, Lipoplatin™ has been proven to be superior to cisplatin alone.

4.4.3 Genexol®-PM

Genexol®-PM is a novel Cremophor® EL-free polymeric micelle formulation of paclitaxel (Taxol) consisting of two block copolymers: PEG, which is useful as a nonimmunogenic carrier, and the core-forming PDLLA, which allows the solubilization of the hydrophobic drug (Kim et al. 2004). Preclinical *in-vivo* studies in murine B16 melanoma-induced female mice revealed that the biodistribution of Genexol®-PM was 2–3 times higher than paclitaxel in various tissues including liver, spleen, kidneys, lungs, heart, and tumor. In addition, Genexol®-PM showed a 3-fold increase in MTD and significantly increased antitumor efficacy compared to free paclitaxel (Kim et al. 2001).

A phase I study of Genexol®-PM, administered intravenously for 3 h every 3 weeks, established an MTD of 390 mg/m^2, which is higher than the MTD for paclitaxel in a 3-week regimen (175 mg/m^2) (Kim et al. 2004). Dose-limiting toxic effects such as neuropathy, myalgia, and neutropenia were observed. No hypersensitivity reactions were observed without premedication.

Phase II clinical studies of Genexol®-PM have demonstrated the safety and efficacy of this formulation, with high response rates in patients with advanced pancreatic cancer and metastatic breast cancer (Kim et al. 2006). Hypersensitivity reactions were observed in 19.5% of patients out of the total population affected with metastatic breast cancer (Kim et al. 2007). Moreover, Genexol®-PM and cisplatin combination chemotherapy has shown significant antitumor activity and facilitated administration of higher doses of paclitaxel in comparison with Cremophor® EL-based formulation in patients with advanced NSCLC. Genexol®-PM is currently approved and marketed in several countries, including South Korea, for metastatic breast cancer, NSCLC, and ovarian cancer. A combination of Genexol®-PM and cisplatin showed significant antitumor activity, with relatively low incidence and severity of toxicity, despite the high dose of paclitaxel, in patients with advanced NSCLC.

Being partly hydrophilic and partly hydrophobic, Genexol®-PM facilitates solubilization and administration of paclitaxel, which is otherwise limited by its low solubility and bioavailability. Also, a combination of nanoparticulate systems is being employed for the better management of treatment-associated side effects.

Various nanocarrier-based formulations have been approved, as has been discussed above, while some are in final stages of approval for their regular employment in lung cancer. Table 4.12 represents a review of the status of the approved drugs and those in the development stages for lung cancer therapy. Paclitaxel bound with albumin nanoparticles is currently used for treatment of NSLC and is also being studied for its possible use in

TABLE 4.12

Nanocarrier-Based Drugs for Lung Cancer: Approved and Those under Various Stages of Development

Carrier	Active Ingredient (Brand Name)	Status	References
Albumin NPs	Paclitaxel (Abraxane)	Approved by US FDA in October 2012	Babu et al. (2013)
	Combination of paclitaxel with carboplatin and Bevacizumab	Phase II trial in NSCLC	Reynolds et al. (2009)
	Combination of paclitaxel with carboplatin	Phase III trial in NSCLC	Socinski et al. (2012)
Liposomes	Paclitaxel (Lipusu)	Approved by state FDA of China	Babu et al. (2013)
	Cisplatin (Lipoplatin™)	Clinical phase III	Stathopoulos et al. (2010)
	Cisplatin (SPI-77)	Clinical phase III	Kim et al. (2001)
	Paclitaxel (LEP-ETU)	Clinical phase I/II	Zhang et al. (2005)
	Lurtotecan (OSI-211)	Clinical phase II	Gelmon et al. (2004)
	Liposomal/FUS1 (INGN-401)	Clinical phase I	McNeil (2009)
Polymeric micelle	Paclitaxel (Genexol®-PM)	Approved in South Korea	Wang et al. (2013)

combination with carboplatin, bevacizumab, and radiation therapy so as to increase the efficacy of therapy in NSCLC patients. Combinations of albumin-bound nanoparticles with carboplatin and bevacizumab are being studied in phase II clinical studies, while their combination with carboplatin is being investigated in phase III trials. Genexol®-PM has been approved in South Korea in cases of NSCLC for its notable performance in reducing the severity of high paclitaxel doses in micellar combinations.

4.5 Conclusion and Future Perspectives

Lung cancer is considered one of the most lethal malignancies of all cancer types due to its unique biological portrait and lack of effective screening techniques for the same. Although significant research has been conducted to explore strategies for the development of various diagnostic and therapeutic modalities for this malignancy, it still constitutes a major cause of cancer-related mortalities. The major lacunae in the current treatment of lung cancer include lack of availability of tools for conducting early diagnosis of the disease and ineffective strategies for drug targeting and delivery. Thus, improvement in these areas would significantly help in revolutionizing lung cancer management.

Nanotechnology has the potential to offer great avenues for early diagnosis of lung cancer, when the disease is still curable. As evident, nanodimensional materials hold promising potential for formulating effective and safe theranostic systems for lung malignancies. Materials consisting of polymers, inorganic (i.e., metals), and organic carriers and several bionano strategies have been investigated for developing safe and effective nanotheranostic regimes for lung cancer. Polymers continue to provide significant advantages as carriers for pulmonary drug delivery on account of their versatile fabrication characteristics, modification capabilities, and drug-loading abilities. Metal nanoparticles also find wide application in treatment of small cell lung cancer, specifically in theranostic approaches, as they can simultaneously act as imaging agents as well as drug carriers. Toxicity associated with these metal-based systems continues to be a major impediment in their clinical application. Considering these toxicity issues, biologically derived nanoparticulate approaches have gained significant attention in the recent past. As obvious from examples to this date, it may be stated that polymers still hold a better scope as a carrier for therapeutic agents. Moreover, recently explored routes of administration, such as inhalation of aerosolized chemotherapeutic drugs, have gained a huge impetus in lung cancer therapy. Nanocarrier-mediated delivery of chemotherapeutic moieties via the inhalation route would thereby further improve effectiveness of the therapy.

In the current scenario, lack of knowledge of the underlying mechanism of the initiation and prognosis of lung cancer prevents research from obtaining a realistic outlook on therapy at the clinical level. The future theranostic systems targeted for lung cancer therapy would thereby aim at resolving these unanswered mechanisms to improve the efficacy of therapy. Besides these concerns, additional aspects like nanotoxicological issues remain to be tackled in order to develop efficacious strategies for lung cancer diagnosis and/or treatment. Therefore, it is essential that fundamental research be carried out to address these limitations if fruitful and efficient application of these technologies is to be achieved in the near future. The future nanomedicine for lung cancer therapy will rely on lucid design of nanotechnology-based materials and tools on the basis of a detailed understanding of the pathophysiological processes involved in the carcinogenesis of the pulmonary tissue. Moreover, extensive research has to be conducted for the development of better nanomedicines for lung cancer that may survive the challenges of clinical trials and the transition from laboratories to markets.

Acknowledgments

The authors would like to appreciate Uday Koli, Tejal Pant, Aanshu Deokuliar, and Akhil Krishnan for their valuable suggestions and timely inputs in writing this manuscript. Prajakta Dandekar is thankful to Ramanujan fellowship, Department of Science and Technology (DST), Government of India (SR/S2/RJN-139/2011), Ratnesh Jain is grateful to Ramalingaswami Fellowship, Department of Biotechnology (DBT), Government of India (BT/RLF/Re-entry/51/2011) for financial assistance. Sathish Dyawanapelly, Mahesh Tambe, and Ankit Gautam would like to thank DBT, while Nanda Rohra would like to acknowledge University Grants Commission (UGC), Government of India, for their fellowships. Manish Gore is thankful to DST for awarding him the INSPIRE fellowship. Meghna Suvarna is thankful to All India Council for Technical Education (AICTE), Government of India, for the fellowship.

References

Akbarzadeh, A., H. Mikaeili, N. Zarghami, R. Mohammad, A. Barkhordari, and S. Davaran. 2012. Preparation and *in vitro* evaluation of doxorubicin-loaded Fe_3O_4 magnetic nanoparticles modified with biocompatible copolymers. *International Journal of Nanomedicine*. 7:511.

Aldarouish, M. and C. Wang. 2016. Trends and advances in tumor immunology and lung cancer immunotherapy. *Journal of Experimental & Clinical Cancer Research*. 35(1):157.

Alper, J. 2005. Shining a light on cancer research. *NCI Alliance for Nanotechnology in Cancer USA*, 1–3.

Annema, J. T., J. P. van Meerbeeck, R. C. Rintoul, C. Dooms, E. Deschepper, O. M. Dekkers, P. De Leyn, J. Braun, N. R. Carroll, and M. Praet. 2010. Mediastinoscopy vs endosonography for mediastinal nodal staging of lung cancer: A randomized trial. *JAMA*. 304(20):2245–2252.

Aruna, U., R. Rajalakshmi, Y. Indira Muzib, V. Vinesh, M. Sushma, K. R. Vandana, and N. Vijay Kumar. 2013. Role of chitosan nanoparticles in cancer therapy. *International Journal of Innovative Pharmaceutical Research*. 4(3):318–324.

Aygun, C., S. Weiner, A. Scariato, D. Spearman, and L. Stark. 1992. Treatment of non-small cell lung cancer with external beam radiotherapy and high dose rate brachytherapy. *International Journal of Radiation Oncology* Biology* Physics*. 23(1):127–132.

Babu, A., A. K. Templeton, A. Munshi, and R. Ramesh. 2013. Nanoparticle-based drug delivery for therapy of lung cancer: Progress and challenges. *Journal of Nanomaterials*. 2013:14.

Badrzadeh, F., M. Rahmati-Yamchi, K. Badrzadeh, A. Valizadeh, N. Zarghami, S. M. Farkhani, and A. Akbarzadeh. 2014. Drug delivery and nanodetection in lung cancer. *Artificial Cells, Nanomedicine, and Biotechnology*. 44(2):618–634.

Baltayiannis, N., M. Chandrinos, D. Anagnostopoulos, P. Zarogoulidis, K. Tsakiridis, A. Mpakas, N. Machairiotis, N. Katsikogiannis, I. Kougioumtzi, and N. Courcoutsakis. 2013. Lung cancer surgery: An up to date. *Journal of Thoracic Disease*. 5(Suppl 4):S425.

Bandyopadhyay, A., T. Das, and S. Yeasmin. 2015. *Nanoparticles in Lung Cancer Therapy-Recent Trends*: Springer, India.

Bayisa, T. K., M. H. Bule, and J. L. Lenjisa. 2015. The potential of nano technology based drugs in lung cancer management. *The Pharma Innovation*. 4(4):77–81.

Beljanski, V. and J. Hiscott. 2012. The use of oncolytic viruses to overcome lung cancer drug resistance. *Current Opinion in Virology*. 2(5):629–635.

Benfer, M. and T. Kissel. 2012. Cellular uptake mechanism and knockdown activity of siRNA-loaded biodegradable DEAPA-PVA-g-PLGA nanoparticles. *European Journal of Pharmaceutics and Biopharmaceutics*. 80(2):247–256.

Bharali, D. J. and S. A. Mousa. 2010. Emerging nanomedicines for early cancer detection and improved treatment: Current perspective and future promise. *Pharmacology & Therapeutics*. 128(2):324–335.

Bianco, A., K. Kostarelos, and M. Prato. 2005. Applications of carbon nanotubes in drug delivery. *Current Opinion in Chemical Biology*. 9(6):674–679.

Bianco, A., K. Kostarelos, and M. Prato. 2008. Opportunities and challenges of carbon-based nanomaterials for cancer therapy. *Expert opinion on drug delivery*. 5(3):331–342.

Bouissou, C., J. J. Rouse, R. Price, and C. F. Van der Walle. 2006. The influence of surfactant on PLGA microsphere glass transition and water sorption: Remodeling the surface morphology to attenuate the burst release. *Pharmaceutical Research*. 23(6):1295–1305.

Boulikas, T. 2004. Low toxicity and anticancer activity of a novel liposomal cisplatin (Lipoplatin) in mouse xenografts. *Oncology Reports*. 12(1):3–12.

Boulikas, T. 2009. Clinical overview on Lipoplatin™: A successful liposomal formulation of cisplatin. *Expert Opinion on Investigational Drugs*. 18(8):1197–1218.

Boulikas, T., N. Mylonakis, G. Sarikos, J. Angel, A. Athanasiou, G. Politis, A. Rapti, A. Rassidakis, M. Karabatzaki, and N. Anyfantis. 2007. Lipoplatin plus gemcitabine versus cisplatin plus gemcitabine in NSCLC: Preliminary results of a phase III trial. *Journal of Clinical Oncology*. 25(18 suppl.):18028–18028.

Brahmer, J., Kr. L. Reckamp, P. Baas, L. Crinò, W. E. E. Eberhardt, E. Poddubskaya, S. Antonia, A. Pluzanski, E. E. Vokes, and E. Holgado. 2015. Nivolumab versus docetaxel in advanced squamous-cell non–small-cell lung cancer. *New England Journal of Medicine*. 373(2):123–135.

Braydich-Stolle, L., S. Hussain, J. J. Schlager, and M.-C. Hofmann. 2005. *In vitro* cytotoxicity of nanoparticles in mammalian germline stem cells. *Toxicological Sciences*. 88(2):412–419.

Bristol-Myers Squibb Company. 2012. TAXOL (paclitaxel) injection prescribing information. Princeton, USA. Retrieved from https://www.accessdata.fda.gov/drugsatfda_docs/label/2011/020262s049lbl.pdf, accessed on November 3, 2016.

Brognard, J., A. S. Clark, Y. Ni, and P. A. Dennis. 2001. Akt/protein kinase B is constitutively active in non-small cell lung cancer cells and promotes cellular survival and resistance to chemotherapy and radiation. *Cancer Research*. 61(10):3986–3997.

Brown, S. D., P. Nativo, J.-A. Smith, D. Stirling, P. R. Edwards, B. Venugopal, D. J. Flint, J. A. Plumb, D. Graham, and N. J. Wheate. 2010. Gold nanoparticles for the improved anticancer drug delivery of the active component of oxaliplatin. *Journal of the American Chemical Society*. 132(13):4678–4684.

Brzoska, M., K. Langer, C. Coester, S. Loitsch, T. O. F. Wagner, and C. V. Mallinckrodt. 2004. Incorporation of biodegradable nanoparticles into human airway epithelium cells—In vitro study of the suitability as a vehicle for drug or gene delivery in pulmonary diseases. *Biochemical and Biophysical Research Communications*. 318(2):562–570.

Butle, S. R. and P. R. Baheti. 2011. Role of gold nanoparticles in the detection and treatment of cancer. *Int J Pharm Sci Rev Res*. 10:54–59.

Cancino, J., I. M. M. Paino, K. C. Micocci, H. S. Selistre-de-Araujo, and V. Zucolotto. 2013. *In vitro* nanotoxicity of single-walled carbon nanotube–dendrimer nanocomplexes against murine myoblast cells. *Toxicology Letters*. 219(1):18–25.

Carlson, C., S. M. Hussain, A. M. Schrand, L. K. Braydich-Stolle, K. L. Hess, R. L. Jones, and J. J. Schlager. 2008. Unique cellular interaction of silver nanoparticles: Size-dependent generation of reactive oxygen species. *The Journal of Physical Chemistry B*. 112(43):13608–13619.

Cauda, V., A. Schlossbauer, J. Kecht, A. Zürner, and T. Bein. 2009. Multiple core– shell functionalized colloidal mesoporous silica nanoparticles. *Journal of the American Chemical Society*. 131(32):11361–11370.

Chang, M. Y. and D. J. Sugarbaker. 2003. Surgery for early stage non-small cell lung cancer. *Seminars in Surgical Oncology.* 21(2):74–84.

Chen, D., M. Jiang, N. Li, H. Gu, Q. Xu, J. Ge, X. Xia, and J. Lu. 2010. Modification of magnetic silica/iron oxide nanocomposites with fluorescent polymethacrylic acid for cancer targeting and drug delivery. *Journal of Materials Chemistry.* 20(31):6422–6429.

Chen, Y.-H., C.-Y. Tsai, P.-Y. Huang, M.-Y. Chang, P.-C. Cheng, C.-H. Chou, D.-H. Chen, C.-R. Wang, A.-L. Shiau, and C.-L. Wu. 2007. Methotrexate conjugated to gold nanoparticles inhibits tumor growth in a syngeneic lung tumor model. *Molecular Pharmaceutics.* 4(5):713–722.

Chen, Y., X. Zhu, X. Zhang, B. Liu, and L. Huang. 2010. Nanoparticles modified with tumor-targeting scFv deliver siRNA and miRNA for cancer therapy. *Molecular Therapy.* 18(9):1650–1656.

Cheng, F.-Y., C.-T. Chen, and C.-S. Yeh. 2009. Comparative efficiencies of photothermal destruction of malignant cells using antibody-coated silica@ Au nanoshells, hollow Au/Ag nanospheres and Au nanorods. *Nanotechnology.* 20(42):425104.

Choi, S. H., H. J. Byeon, J. S. Choi, L. Thao, I. Kim, E. S. Lee, B. S. Shin, K. C. Lee, and Y. S. Youn. 2015. Inhalable self-assembled albumin nanoparticles for treating drug-resistant lung cancer. *Journal of Controlled Release.* 197:199–207.

Choi, S. H., S.-E. Jin, M.-K. Lee, S. J. Lim, J.-S. Park, B.-G. Kim, W. S. Ahn, and C.-K. Kim. 2008. Novel cationic solid lipid nanoparticles enhanced p53 gene transfer to lung cancer cells. *European Journal of Pharmaceutics and Biopharmaceutics.* 68(3):545–554.

Colella, S., P. Vilmann, L. Konge, and P. F. Clementsen. 2014. Endoscopic ultrasound in the diagnosis and saging of lung cancer. *Endoscopic Ultrasound.* 3(4):205.

Conde, J., G. Doria, and P. Baptista. 2012. Noble metal nanoparticles applications in cancer. *Journal of Drug Delivery.* 2012:1–12.

Cormode, D. P., P. A. Jarzyna, W. J. M. Mulder, and Z. A. Fayad. 2010. Modified natural nanoparticles as contrast agents for medical imaging. *Advanced Drug Delivery Reviews.* 62(3):329–338.

Couvreur, P. and C. Vauthier. 2006. Nanotechnology: Intelligent design to treat complex disease. *Pharmaceutical Research.* 23(7):1417–1450.

Cuenca, A. G., H. Jiang, S. N. Hochwald, M. Delano, W. G. Cance, and S. R. Grobmyer. 2006. Emerging implications of nanotechnology on cancer diagnostics and therapeutics. *Cancer.* 107(3):459–466.

Cui, Z., Y. Zhang, J. Zhang, H. Kong, X. Tang, L. Pan, K. Xia, A. Aldalbahi, A. Li, and R. Tai. 2016. Sodium alginate-functionalized nanodiamonds as sustained chemotherapeutic drug-release vectors. *Carbon.* 97:78–86.

Danhier, F., O. Feron, and V. Préat. 2010. To exploit the tumor microenvironment: Passive and active tumor targeting of nanocarriers for anti-cancer drug delivery. *Journal of Controlled Release.* 148(2):135–146.

Danhier, F., N. Lecouturier, B. Vroman, C. Jérôme, J. Marchand-Brynaert, O. Feron, and V. Préat. 2009. Paclitaxel-loaded PEGylated PLGA-based nanoparticles: *In vitro* and *in vivo* evaluation. *Journal of Controlled Release.* 133(1):11–17.

Datta, S. R., A. Brunet, and M. E. Greenberg. 1999. Cellular survival: A play in three Akts. *Genes & Development.* 13(22):2905–2927.

de Castria, T. B., E. M. da Silva, A. F. Gois, and R. Riera. 2013. Cisplatin versus carboplatin in combination with third-generation drugs for advanced non-small cell lung cancer. *Cochrane Database Syst Rev.* 8:1–59.

Derfus, A. M., W. C. W. Chan, and S. N. Bhatia. 2004. Probing the cytotoxicity of semiconductor quantum dots. *Nano Letters.* 4(1):11–18.

Digumarti, R., P. P. Bapsy, A. V. Suresh, G. S. Bhattacharyya, L. Dasappa, J. S. Shan, and D. E. Gerber. 2014. Bavituximab plus paclitaxel and carboplatin for the treatment of advanced non-small-cell lung cancer. *Lung Cancer.* 86(2):231–236.

Dixit, N., S. D. Maurya, and B. P. Sagar. 2013. Sustained release drug delivery system. *Indian Journal of Research in Pharmacy and Biotechnology.* 1(3):305.

Díaz, M. R. and P. E. Vivas-Mejia. 2013. Nanoparticles as drug delivery systems in cancer medicine: Emphasis on RNAi-containing nanoliposomes. *Pharmaceuticals.* 6(11):1361–1380.

Edwards, D. A., J. Hanes, G. Caponetti, J. Hrkach, A. Ben-Jebria, M. L. Eskew, J. Mintzes, D. Deaver, N. Lotan, and R. Langer. 1997. Large porous particles for pulmonary drug delivery. *Science*. 276(5320):1868–1872.

Fan, J., J.-J. Yin, B. Ning, X. Wu, Y. Hu, M. Ferrari, G. J. Anderson, J. Wei, Y. Zhao, and G. Nie. 2011. Direct evidence for catalase and peroxidase activities of ferritin–platinum nanoparticles. *Biomaterials*. 32(6):1611–1618.

Fantini, M., L. Gianni, C. Santelmo, F. Drudi, C. Castellani, A. Affatato, M. Nicolini, and A. Ravaioli. 2011. Lipoplatin treatment in lung and breast cancer. *Chemotherapy Research and Practice*. 2011:1–7.

Feifel, S. C. and F. Lisdat. 2011. Silica nanoparticles for the layer-by-layer assembly of fully electroactive cytochrome c multilayers. *J. Nanobiotechnol*. 9:59.

Fernando, H. C., A. De Hoyos, R. J. Landreneau, S. Gilbert, W. E. Gooding, P. O. Buenaventura, N. A. Christie, C. Belani, and J. D. Luketich. 2005. Radiofrequency ablation for the treatment of non-small cell lung cancer in marginal surgical candidates. *The Journal of Thoracic and Cardiovascular Surgery*. 129(3):639–644.

Fickling, W. and M. B. Wallace. 2002. EUS in lung cancer. *Gastrointestinal Endoscopy*. 56(4):S18–S21.

Flanagan, W. M., W. K. Evans, N. R. Fitzgerald, J. R. Goffin, A. B. Miller, and M. C. Wolfson. 2015. Performance of the cancer risk management model lung cancer screening module. *Health Reports*. 26(5):11–18.

Foldbjerg, R., D. A. Dang, and H. Autrup. 2011. Cytotoxicity and genotoxicity of silver nanoparticles in the human lung cancer cell line, A549. *Archives of Toxicology*. 85(7):743–750.

Fonseca, C., S. Simoes, and R. Gaspar. 2002. Paclitaxel-loaded PLGA nanoparticles: Preparation, physicochemical characterization and *in vitro* anti-tumoral activity. *Journal of Controlled Release*. 83(2):273–286.

Forssen, E. and M. Willis. 1998. Ligand-targeted liposomes. *Advanced Drug Delivery Reviews*. 29(3):249–271.

Garbuzenko, O. B., G. Mainelis, O. Taratula, and T. Minko. 2014. Inhalation treatment of lung cancer: The influence of composition, size and shape of nanocarriers on their lung accumulation and retention. *Cancer Biology & Medicine*. 11(1):44.

Gelderblom, H., J. Verweij, K. Nooter, and A. Sparreboom. 2001. Cremophor EL: The drawbacks and advantages of vehicle selection for drug formulation. *European Journal of Cancer*. 37(13):1590–1598.

Gelmon, K., H. Hirte, B. Fisher, W. Walsh, M. Ptaszynski, M. Hamilton, N. Onetto, and E. Eisenhauer. 2004. A phase 1 study of OSI-211 given as an intravenous infusion days 1, 2, and 3 every three weeks in patients with solid cancers. *Investigational New Drugs*. 22(3):263–275.

Golsteyn, R. M., S. J. Schultz, J. Bartek, A. Ziemiecki, T. Ried, and E. A. Nigg. 1994. Cell cycle analysis and chromosomal localization of human Plk1, a putative homologue of the mitotic kinases *Drosophila polo* and *Saccharomyces cerevisiae* Cdc5. *Journal of Cell Science*. 107(6):1509–1517.

Green, M. R., G. M. Manikhas, S. Orlov, B. Afanasyev, A. M. Makhson, P. Bhar, and M. J. Hawkins. 2006. Abraxane®, a novel Cremophor®-free, albumin-bound particle form of paclitaxel for the treatment of advanced non-small-cell lung cancer. *Annals of Oncology*. 17(8):1263–1268.

Guo, C., W. T. Al-Jamal, F. M. Toma, A. Bianco, M. Prato, K. T. Al-Jamal, and K. Kostarelos. 2015. Design of cationic multiwalled carbon nanotubes as efficient siRNA vectors for lung cancer xenograft eradication. *Bioconjugate Chemistry*. 26(7):1370–1379.

Gupta, N., H. Hatoum, and G. K. Dy. 2014. First line treatment of advanced non-small-cell lung cancer–Specific focus on albumin bound paclitaxel. *International Journal of Nanomedicine*. 9:209.

Gurunathan, S., J.-K. Jeong, J. W. Han, X.-F. Zhang, J. H. Park, and J.-H. Kim. 2015. Multidimensional effects of biologically synthesized silver nanoparticles in *Helicobacter pylori*, *Helicobacter felis*, and human lung (L132) and lung carcinoma A549 cells. *Nanoscale Research Letters*. 10(1):1–17.

Guthi, J. S., S.-G. Yang, G. Huang, S. Li, C. Khemtong, C. W. Kessinger, M. Peyton, J. D. Minna, K. C. Brown, and J. Gao. 2009. MRI-visible micellar nanomedicine for targeted drug delivery to lung cancer cells. *Molecular Pharmaceutics*. 7(1):32–40.

Haque, N., R. R. Khalel, N. Parvez, S. Yadav, N. Hwisa, M. S. Al-Sharif, B. Z. Awen, and K. Molvi. 2010. Nanotechnology in cancer therapy: A review. *J Chem Pharm Res*. 2:161–8.

Heister, E., V. Neves, C. Tîlmaciu, K. Lipert, V. S. Beltrán, H. M. Coley, S. Ravi P. Silva, and J. McFadden. 2009. Triple functionalisation of single-walled carbon nanotubes with doxorubicin, a monoclonal antibody, and a fluorescent marker for targeted cancer therapy. *Carbon.* 47(9):2152–2160.

Holtrich, U., G. Wolf, A. Bräuninger, T. Karn, B. Böhme, H. Rübsamen-Waigmann, and K. Strebhardt. 1994. Induction and down-regulation of PLK, a human serine/threonine kinase expressed in proliferating cells and tumors. *Proceedings of the National Academy of Sciences.* 91(5):1736–1740.

Howington, J. A., M. G. Blum, A. C. Chang, A. A. Balekian, and S. C. Murthy. 2013. Treatment of stage I and II non-small cell lung cancer: Diagnosis and management of lung cancer: American College of Chest Physicians evidence-based clinical practice guidelines. *CHEST Journal.* 143(5_suppl):e278S–e313S.

Huang, X., I. H. El-Sayed, W. Qian, and M. A. El-Sayed. 2006. Cancer cell imaging and photothermal therapy in the near-infrared region by using gold nanorods. *Journal of the American Chemical Society.* 128(6):2115–2120.

Hussain, S. M., A. K. Javorina, A. M. Schrand, H. M. Duhart, S. F. Ali, and J. J. Schlager. 2006. The interaction of manganese nanoparticles with PC-12 cells induces dopamine depletion. *Toxicological Sciences.* 92(2):456–463.

Hussain, S. M., K. L. Hess, J. M. Gearhart, K. T. Geiss, and J. J. Schlager. 2005. In vitro toxicity of nanoparticles in BRL 3A rat liver cells. *Toxicology in vitro.* 19(7):975–983.

Ibrahim, N. K., N. Desai, S. Legha, P. Soon-Shiong, R. L. Theriault, E. Rivera, B. Esmaeli, S. E. Ring, A. Bedikian, and G. N. Hortobagyi. 2002. Phase I and pharmacokinetic study of ABI-007, a Cremophor-free, protein-stabilized, nanoparticle formulation of paclitaxel. *Clinical Cancer Research.* 8 (5):1038–1044.

ICMR. 2014. Press Release: Report to the nation on the status of cancer in India. *Cancer Registry Reports, Indian Council of Medical Research.*

Imai, K. and A. Takaoka. 2006. Comparing antibody and small-molecule therapies for cancer. *Nature Reviews Cancer.* 6(9):714–727.

Jahanshahi, M. and Z. Babaei. 2008. Protein nanoparticle: A unique system as drug delivery vehicles. *African Journal of Biotechnology.* 7(25):4926–4934.

Jain, R. A. 2000. The manufacturing techniques of various drug loaded biodegradable poly (lactide-co-glycolide)(PLGA) devices. *Biomaterials.* 21(23):2475–2490.

Jain, S., D. Kumar, N. K. Swarnakar, and K. Thanki. 2012. Polyelectrolyte stabilized multilayered liposomes for oral delivery of paclitaxel. *Biomaterials.* 33(28):6758–6768.

Jena, M., S. Mishra, S. Jena, and S. S. Mishra. 2013. Nanotechnology-future prospect in recent medicine: A review. *International Journal of Basic & Clinical Pharmacology.* 2(4):353–359.

Jere, D., H.-L. Jiang, Y.-K. Kim, R. Arote, Y.-J. Choi, C.-H. Yun, M.-H. Cho, and C.-S. Cho. 2009. Chitosan-graft-polyethylenimine for Akt1 siRNA delivery to lung cancer cells. *International Journal of Pharmaceutics.* 378(1):194–200.

Jiang, L., X. Li, L. Liu, and Q. Zhang. 2013. Thiolated chitosan-modified PLA-PCL-TPGS nanoparticles for oral chemotherapy of lung cancer. *Nanoscale Research Letters.* 8(1):66–77.

John, T. A., S. M. Vogel, C. Tiruppathi, A. B. Malik, and R. D. Minshall. 2003. Quantitative analysis of albumin uptake and transport in the rat microvessel endothelial monolayer. *American Journal of Physiology-Lung Cellular and Molecular Physiology.* 284(1):L187–L196.

Johnson, B. E., R. I. Linnoila, J. P. Williams, D. J. Venzon, P. Okunieff, G. B. Anderson, and G. E. Richardson. 1995. Risk of second aerodigestive cancers increases in patients who survive free of small-cell lung cancer for more than 2 years. *Journal of Clinical Oncology.* 13(1):101–111.

Jung, J., S.-J. Park, H. K. Chung, H.-W. Kang, S.-W. Lee, M. H. Seo, H. J. Park, S. Y. Song, S.-Y. Jeong, and E. K. Choi. 2012. Polymeric nanoparticles containing taxanes enhance chemoradiotherapeutic efficacy in non-small cell lung cancer. *International Journal of Radiation Oncology* Biology* Physics.* 84(1):e77–e83.

Kakhki, V. R. D. 2007. Positron emission tomography in the management of lung cancer. *Annals of Thoracic Medicine.* 2(2):69.

Kannan, S., P. Kolhe, V. Raykova, M. Glibatec, R. M. Kannan, M. Lieh-Lai, and D. Bassett. 2004. Dynamics of cellular entry and drug delivery by dendritic polymers into human lung epithelial carcinoma cells. *Journal of Biomaterials Science, Polymer Edition.* 15(3):311–330.

Khan, N. and H. Mukhtar. 2015. Dietary agents for prevention and treatment of lung cancer. *Cancer Letters*. 359(2):155–164.

Kim, C., S. Bae, N. Lee, K. Lee, S. Park, D. Kim, J. Won, D. Hong, and H. Park. 2006. Phase II study of Genexol (paclitaxel) and carboplatin as first-line treatment of advanced or metastatic non-small-cell lung cancer (NSCLC). *Journal of Clinical Oncology*. 24(18 suppl.):17049–17049.

Kim, D.-W., S.-Y. Kim, H.-K. Kim, S.-W. Kim, S. W. Shin, J. S. Kim, K. Park, M. Y. Lee, and D. S. Heo. 2007. Multicenter phase II trial of Genexol-PM, a novel Cremophor-free, polymeric micelle formulation of paclitaxel, with cisplatin in patients with advanced non-small-cell lung cancer. *Annals of Oncology*. 18(12):2009–2014.

Kim, E. S., C. Lu, F. R. Khuri, M. Tonda, B. S. Glisson, D. Liu, M. Jung, W. K. Hong, and R. S. Herbst. 2001. A phase II study of STEALTH cisplatin (SPI-77) in patients with advanced non-small cell lung cancer. *Lung Cancer*. 34(3):427–432.

Kim, H. S., Y. S. Choi, K. Kim, Y. M. Shim, and J. Kim. 2006. Surgical resection of recurrent lung cancer in patients following curative resection. *Journal of Korean Medical Science*. 21(2):224–228.

Kim, S. C., D. W. Kim, Y. H. Shim, J. S. Bang, H. S. Oh, S. W. Kim, and M. H. Seo. 2001. In vivo evaluation of polymeric micellar paclitaxel formulation: Toxicity and efficacy. *Journal of Controlled Release*. 72(1):191–202.

Kim, T.-Y., D.-W. Kim, J.-Y. Chung, S. G. Shin, S.-C. Kim, D. S. Heo, N. K. Kim, and Y.-J. Bang. 2004. Phase I and pharmacokinetic study of Genexol-PM, a cremophor-free, polymeric micelle-formulated paclitaxel, in patients with advanced malignancies. *Clinical Cancer Research*. 10(11):3708–3716.

Kim, Y., Q. Lin, P. M. Glazer, and Z. Yun. 2009. Hypoxic tumor microenvironment and cancer cell differentiation. *Current Molecular Medicine*. 9(4):425–434.

Klajnert, B. and M. Bryszewska. 2000. Dendrimers: Properties and applications. *Acta Biochimica Polonica*. 48(1):199–208.

Kostarelos, K., A. Bianco, and M. Prato. 2009. Promises, facts and challenges for carbon nanotubes in imaging and therapeutics. *Nature Nanotechnology*. 4(10):627–633.

Koyama, H., Y. Ohno, S. Seki, M. Nishio, T. Yoshikawa, S. Matsumoto, and K. Sugimura. 2013. Magnetic resonance imaging for lung cancer. *Journal of Thoracic Imaging*. 28(3):138–150.

Lam, S., C. MacAulay, J. C. leRiche, and B. Palcic. 2000. Detection and localization of early lung cancer by fluorescence bronchoscopy. *Cancer*. 89(S11):2468–2473.

Landesman-Milo, D., S. Ramishetti, and D. Peer. 2015. Nanomedicine as an emerging platform for metastatic lung cancer therapy. *Cancer and Metastasis Reviews*. 34(2):291–301.

Lang-Lazdunski, L. 2013. Surgery for nonsmall cell lung cancer. *Eur Respir Rev*. 22(129):382–404. doi: 10.1183/09059180.00003913.

Lavanya V., Mohamed Adil A. A., N. Ahmed, A. K. Rishi, S. Jamal. 2014. Small molecule inhibitors as emerging cancer therapeutics. *Integrative Cancer Science and Therapeutics*. 1(3):39–46.

Li, H., Z. Cao, Y. Zhang, C. Lau, and J. Lu. 2011. Simultaneous detection of two lung cancer biomarkers using dual-color fluorescence quantum dots. *Analyst*. 136(7):1399–1405.

Li, K., B. Chen, L. Xu, J. Feng, G. Xia, J. Cheng, J. Wang, F. Gao, and X. Wang. 2013. Reversal of multidrug resistance by cisplatin-loaded magnetic Fe_3O_4 nanoparticles in A549/DDP lung cancer cells *in vitro* and *in vivo*. *International Journal of Nanomedicine*. 8:1867.

Li, K., Z.-P. Zhang, M. Luo, X. Yu, Y. Han, H.-P. Wei, Z.-Q. Cui, and X.-E. Zhang. 2012. Multifunctional ferritin cage nanostructures for fluorescence and MR imaging of tumor cells. *Nanoscale*. 4(1):188–193.

Li, Z., J. C. Barnes, A. Bosoy, J. Fraser Stoddart, and J. I. Zink. 2012. Mesoporous silica nanoparticles in biomedical applications. *Chemical Society Reviews*. 41(7):2590–2605.

Liebmann, J. E., J. A. Cook, C. Lipschultz, D. Teague, J. Fisher, and J. B. Mitchell. 1993. Cytotoxic studies of paclitaxel (Taxol) in human tumour cell lines. *British Journal of Cancer*. 68(6):1104.

Liu, J., L. Chu, Y. Wang, Y. Duan, L. Feng, C. Yang, L. Wang, and D. Kong. 2011. Novel peptide-dendrimer conjugates as drug carriers for targeting nonsmall cell lung cancer. *Int J Nanomedicine*. 6:59–69.

Liu, K.-K., C.-L. Cheng, C.-C. Chang, and J.-I. Chao. 2007. Biocompatible and detectable carboxylated nanodiamond on human cell. *Nanotechnology*. 18(32):325102.

Liu, K.-K., C.-C. Wang, C.-L. Cheng, and J.-I. Chao. 2009. Endocytic carboxylated nanodiamond for the labeling and tracking of cell division and differentiation in cancer and stem cells. *Biomaterials.* 30(26):4249–4259.

Liu, R., O. V. Khullar, A. P. Griset, J. E. Wade, K. Ann V. Zubris, M. W. Grinstaff, and Y. L. Colson. 2011. Paclitaxel-loaded expansile nanoparticles delay local recurrence in a heterotopic murine non-small cell lung cancer model. *The Annals of Thoracic Surgery.* 91(4):1077–1084.

Liu, X., W. Wei, C. Wang, H. Yue, D. Ma, C. Zhu, G. Ma, and Y. Du. 2011. Apoferritin-camouflaged Pt nanoparticles: Surface effects on cellular uptake and cytotoxicity. *Journal of Materials Chemistry.* 21(20):7105–7110.

Liu, X., W. Wei, Q. Yuan, X. Zhang, N. Li, Y. Du, G. Ma, C. Yan, and D. Ma. 2012. Apoferritin–CeO_2 nano-truffle that has excellent artificial redox enzyme activity. *Chemical Communications.* 48(26):3155–3157.

Long, J.-T., T.-y. Cheang, S.-Y. Zhuo, R.-F. Zeng, Q.-S. Dai, H.-P. Li, and S. Fang. 2014. Anticancer drug-loaded multifunctional nanoparticles to enhance the chemotherapeutic efficacy in lung cancer metastasis. *Journal of Nanobiotechnology.* 12(1):37.

Lorenz, W., H.-J. Reimann, A. Schmal, P. Dormann, B. Schwarz, E. Neugebauer, and A. Doenicke. 1977. Histamine release in dogs by Cremophor El® and its derivatives: Oxethylated oleic acid is the most effective constituent. *Agents and Actions.* 7(1):63–67.

MaHam, A., Z. Tang, H. Wu, J. Wang, and Y. Lin. 2009. Protein-based nanomedicine platforms for drug delivery. *Small.* 5(15):1706–1721.

Mainelis, G., S. Seshadri, O. B. Garbuzenko, T. Han, Z. Wang, and T. Minko. 2013. Characterization and application of a nose-only exposure chamber for inhalation delivery of liposomal drugs and nucleic acids to mice. *Journal of Aerosol Medicine and Pulmonary Drug Delivery.* 26(6):345–354.

Marshall, H. M., R. V. Bowman, I. A. Yang, K. M. Fong, and C. D. Berg. 2013. Screening for lung cancer with low-dose computed tomography: A review of current status. *Journal of Thoracic Disease.* 5(Suppl 5):S524.

Martins, S. J., T. Y. Takagaki, A. G. P. Silva, C. P. Gallo, F. B. A. Silva, and V. L. Capelozzi. 2009. Prognostic relevance of TTF-1 and MMP-9 expression in advanced lung adenocarcinoma. *Lung Cancer.* 64(1):105–109.

Master, A. M. and A. S. Gupta. 2012. EGF receptor-targeted nanocarriers for enhanced cancer treatment. *Nanomedicine.* 7(12):1895–1906.

McNeil, S. E. 2009. Nanoparticle therapeutics: A personal perspective. *Wiley Interdisciplinary Reviews: Nanomedicine and Nanobiotechnology.* 1(3):264–271.

Medintz, I. L., H. Tetsuo Uyeda, E. R. Goldman, and H. Mattoussi. 2005. Quantum dot bioconjugates for imaging, labelling and sensing. *Nature Materials.* 4(6):435–446.

Mehnert, W. and K. Mäder. 2001. Solid lipid nanoparticles: Production, characterization and applications. *Advanced Drug Delivery Reviews.* 47(2):165–196.

Mehrotra, A., R. C. Nagarwal, and J. K. Pandit. 2011. Lomustine loaded chitosan nanoparticles: Characterization and *in-vitro* cytotoxicity on human lung cancer cell line L132. *Chemical and Pharmaceutical Bulletin.* 59(3):315–320.

Mehta, A., S. Dobersch, A. J. Romero-Olmedo, and G. Barreto. 2015. Epigenetics in lung cancer diagnosis and therapy. *Cancer and Metastasis Reviews.* 34(2):229–241.

Mendell, J., D. J. Freeman, W. Feng, T. Hettmann, M. Schneider, S. Blum, J. Ruhe, J. Bange, K. Nakamaru, and S. Chen. 2015. Clinical translation and validation of a predictive biomarker for patritumab, an anti-human epidermal growth factor receptor 3 (HER3) monoclonal antibody, in patients with advanced non-small cell lung cancer. *EBioMedicine.* 2(3):264–271.

Meng, L., X. Zhang, Q. Lu, Z. Fei, and P. J. Dyson. 2012. Single walled carbon nanotubes as drug delivery vehicles: Targeting doxorubicin to tumors. *Biomaterials.* 33(6):1689–1698.

Morgan, M. T., Y. Nakanishi, D. J. Kroll, A. P. Griset, M. A. Carnahan, M. Wathier, N. H. Oberlies, G. Manikumar, M. C. Wani, and M. W. Grinstaff. 2006. Dendrimer-encapsulated camptothecins: Increased solubility, cellular uptake, and cellular retention affords enhanced anticancer activity *in vitro*. *Cancer Research.* 66(24):11913–11921.

Mountain, C. F. 1997. Revisions in the international system for staging lung cancer. *Chest.* 111(6):1710–1717.

Mousa, S. A. and D. J. Bharali. 2011. Nanotechnology-based detection and targeted therapy in cancer: Nano-bio paradigms and applications. *Cancers.* 3(3):2888–2903.

Müller, R. H., A. Dingler, T. Schneppe, and S. Gohla. 2000. Large scale production of solid lipid nanoparticles (SLN™) and nanosuspensions (DissoCubes™). *Handbook of Pharmaceutical Controlled Release Technology*, edited by D.L. Wise, 359–376. Marcel Dekker, New York.

Muniyappa, M., V. S. Rao, and K. B. Vidya. 2014. To evaluate the role of sputum in the diagnosis of lung cancer in south Indian population. *International Journal of Research in Medical Sciences.* 2(2):545–550.

Mylonakis, N., A. Athanasiou, N. Ziras, J Angel, A. Rapti, S. Lampaki, N. Politis, C. Karanikas, and C. Kosmas. 2010. Phase II study of liposomal cisplatin (Lipoplatin™) plus gemcitabine versus cisplatin plus gemcitabine as first line treatment in inoperable (stage IIIB/IV) non-small cell lung cancer. *Lung Cancer.* 68(2):240–247.

Nassimi, M., C. Schleh, H.-D. Lauenstein, R. Hussein, K. Lübbers, G. Pohlmann, S. Switalla, K. Sewald, M. Müller, and N. Krug. 2009. Low cytotoxicity of solid lipid nanoparticles in *in vitro* and *ex vivo* lung models. *Inhalation Toxicology.* 21(sup1):104–109.

Nishio, M., A. Horiike, H. Murakami, N. Yamamoto, H. Kaneda, K. Nakagawa, H. Horinouchi, M. Nagashima, M. Sekiguchi, and T. Tamura. 2015. Phase I study of the HER3-targeted antibody patritumab (U3-1287) combined with erlotinib in Japanese patients with non-small cell lung cancer. *Lung Cancer.* 88(3):275–281.

O'Brien, M. E., N. Wigler, M. Inbar, R. Rosso, E. Grischke, A. Santoro, R. Catane, D. G. Kieback, P. Tomczak, and S. P. Ackland. 2004. Reduced cardiotoxicity and comparable efficacy in a phase III trial of pegylated liposomal doxorubicin HCl (CAELYX™/Doxil®) versus conventional doxorubicin for first-line treatment of metastatic breast cancer. *Annals of Oncology.* 15(3):440–449.

Oberdörster, E. 2004. Manufactured nanomaterials (fullerenes, C60) induce oxidative stress in the brain of juvenile largemouth bass. *Environmental Health Perspectives.* 112(10):1058.

Page, D. B., T. W. Hulett, T. L. Hilton, H.-M. Hu, W. J. Urba, and B. A. Fox. 2016. Glimpse into the future: Harnessing autophagy to promote anti-tumor immunity with the DRibbles vaccine. *Journal for Immunotherapy of Cancer.* 4(1):1.

Palmisano, W. A., K. K. Divine, G. Saccomanno, F. D. Gilliland, S. B. Baylin, J. G. Herman, and S. A. Belinsky. 2000. Predicting lung cancer by detecting aberrant promoter methylation in sputum. *Cancer Research.* 60(21):5954–5958.

Parashar, B., S. Arora, and A. G. Wernicke. 2013. Radiation therapy for early stage lung cancer. *Seminars in Interventional Radiology.* 30(2):185–190.

Patil, G., S. S. Patil, S. Sonawane, A. Khalore, and P. M. Chouragade. 2014. Comparative study of nanotechnology based cancer treatment: A systematic approach. *International Journal of Scientific & Engineering Research.* 5(12):7.

Peng, G., U. Tisch, O. Adams, M. Hakim, N. Shehada, Y. Y. Broza, S. Billan, R. Abdah-Bortnyak, A. Kuten, and H. Haick. 2009. Diagnosing lung cancer in exhaled breath using gold nanoparticles. *Nature Nanotechnology.* 4(10):669–673.

Peng, X.-H., Y. Wang, D. Huang, Y. Wang, H. J. Shin, Z. Chen, M. B. Spewak, H. Mao, X. Wang, and Y. Wang. 2011. Targeted delivery of cisplatin to lung cancer using ScFvEGFR-heparin-cisplatin nanoparticles. *Acs Nano.* 5(12):9480–9493.

Peng, Y., Y. Dai, C. Hitchcock, X. Yang, E. S. Kassis, L. Liu, Z. Luo, H.-L. Sun, R. Cui, and H. Wei. 2013. Insulin growth factor signaling is regulated by microRNA-486, an underexpressed microRNA in lung cancer. *Proceedings of the National Academy of Sciences.* 110(37):15043–15048.

Polo, A., M. Castro, A. Montero, and P. Navío. 2011. Brachytherapy for lung cancer. In *Advances in Radiation Oncology in Lung Cancer*, edited by Branislav Jeremic, 477–488. Springer, Berlin Heidelberg.

Quoix, E., R. Ramlau, V. Westeel, Z. Papai, A. Madroszyk, A. Riviere, P. Koralewski, J.-L. Breton, E. Stoelben, and D Braun. 2011. Therapeutic vaccination with TG4010 and first-line chemotherapy in advanced non-small-cell lung cancer: A controlled phase 2B trial. *The Lancet Oncology.* 12(12):1125–1133.

Ramnath, N., T. J. Dilling, L. J. Harris, A. W. Kim, G. C. Michaud, A. A. Balekian, R. Diekemper, F. C. Detterbeck, and D. A. Arenberg. 2013. Treatment of stage III non-small cell lung cancer: Diagnosis and management of lung cancer: American College of Chest Physicians evidence-based clinical practice guidelines. *CHEST Journal.* 143(5_suppl):e314S–e340S.

Raymond, E., A. Hanauske, S. Faivre, E. Izbicka, G. Clark, E. K. Rowinsky, and D. D. Von Hoff. 1997. Effects of prolonged versus short-term exposure paclitaxel (Taxol®) on human tumor colonyforming units. *Anti-Cancer Drugs.* 8(4):379–385.

Reynolds, C., D. Barrera, R. Jotte, A. I. Spira, C. Weissman, K. A. Boehm, S. Pritchard, and L. Asmar. 2009. Phase II trial of nanoparticle albumin-bound paclitaxel, carboplatin, and bevacizumab in first-line patients with advanced nonsquamous non-small cell lung cancer. *Journal of Thoracic Oncology.* 4(12):1537–1543.

Robertson, K. L., C. M. Soto, M. J. Archer, O. Odoemene, and J. L. Liu. 2011. Engineered T4 viral nanoparticles for cellular imaging and flow cytometry. *Bioconjugate Chemistry.* 22(4):595–604.

Ruhe, P. Q., E. L. Hedberg, N. Torio Padron, P. H. M. Spauwen, J. A. Jansen, and A. G. Mikos. 2003. rhBMP-2 release from injectable poly (DL-lactic-co-glycolic acid)/calcium-phosphate cement composites. *The Journal of Bone & Joint Surgery.* 85(suppl 3):75–81.

Sadhukha, T., T. S. Wiedmann, and J. Panyam. 2013. Inhalable magnetic nanoparticles for targeted hyperthermia in lung cancer therapy. *Biomaterials.* 34(21):5163–5171.

Sahoo, S. K., S. Parveen, and J. J. Panda. 2007. The present and future of nanotechnology in human health care. *Nanomedicine: Nanotechnology, Biology and Medicine.* 3(1):20–31.

Sanna, V., N. Pala, and M. Sechi. 2014. Targeted therapy using nanotechnology: Focus on cancer. *International Journal of Nanomedicine.* 9:467.

Sengupta, S., D. Eavarone, I. Capila, G. Zhao, N. Watson, T. Kiziltepe, and R. Sasisekharan. 2005. Temporal targeting of tumour cells and neovasculature with a nanoscale delivery system. *Nature.* 436(7050):568–572.

Sgambato, A., F. Casaluce, P. Maione, A. Rossi, F. Ciardiello, and C. Gridelli. 2014. Cetuximab in advanced non-small cell lung cancer (NSCLC): The showdown? *Journal of Thoracic Disease.* 6(6):578–580.

Shoyele, S. A. and A. Slowey. 2006. Prospects of formulating proteins/peptides as aerosols for pulmonary drug delivery. *International Journal of Pharmaceutics.* 314(1):1–8.

Sidransky, D. and M. Hollstein, 1996. Clinical implications of the p53 gene. *Annual Review of Medicine.* 47(1):285–301.

Siegel, R. L., K. D. Miller, and A. Jemal. 2015. Cancer statistics, 2015. *CA: a Cancer Journal for Clinicians.* 65(1):5–29.

Sobhani, Z., R. Dinarvand, F. Atyabi, M. Ghahremani, and M. Adeli. 2011. Increased paclitaxel cytotoxicity against cancer cell lines using a novel functionalized carbon nanotube. *Int J Nanomedicine.* 6:705–719.

Socinski, M. A., C. J. Langer, I. Okamoto, J. K. Hon, V. Hirsh, S. R. Dakhil, R. D. Page, J. Orsini, H. Zhang, and M. F. Renschler. 2012. Safety and efficacy of weekly nab®-paclitaxel in combination with carboplatin as first-line therapy in elderly patients with advanced non-small-cell lung cancer. *Annals of Oncology.* 24(2):314–321.

Sparreboom, A., L. van Zuylen, E. Brouwer, W. J. Loos, P. de Bruijn, H. Gelderblom, M. Pillay, K. Nooter, G. Stoter, and J. Verweij. 1999. Cremophor EL-mediated alteration of paclitaxel distribution in human blood clinical pharmacokinetic implications. *Cancer Research.* 59(7):1454–1457.

Sridhar, R., S. Ravanan, J. R. Venugopal, S. Sundarrajan, D. Pliszka, S. Sivasubramanian, P. Gunasekaran, M. Prabhakaran, K. Madhaiyan, and A. Sahayaraj. 2014. Curcumin-and natural extract-loaded nanofibres for potential treatment of lung and breast cancer: *In vitro* efficacy evaluation. *Journal of Biomaterials Science, Polymer Edition.* 25(10):985–998.

Stathopoulos, G. P., D. Antoniou, J. Dimitroulis, P. Michalopoulou, A. Bastas, K. Marosis, J. Stathopoulos, A. Provata, P. Yiamboudakis, and D. Veldekis. 2010. Liposomal cisplatin combined with paclitaxel versus cisplatin and paclitaxel in non-small-cell lung cancer: A randomized phase III multicenter trial. *Annals of Oncology.* 21(11):2227–2232.

Stathopoulos, G. P., and T. Boulikas. 2012. Lipoplatin formulation review article. *Journal of Drug Delivery*. 2012:1–10.

Steinmetz, N. F. 2010. Viral nanoparticles as platforms for next-generation therapeutics and imaging devices. *Nanomedicine: Nanotechnology, Biology and Medicine*. 6(5):634–641.

Stevens, P. J., M. Sekido, and R. J. Lee. 2004. A folate receptor–targeted lipid nanoparticle formulation for a lipophilic paclitaxel prodrug. *Pharmaceutical Research*. 21(12):2153–2157.

Stinchcombe, T. E., M. A. Socinski, C. M. Walko, B. H. O'Neil, F. A. Collichio, A. Ivanova, H. Mu, M. J. Hawkins, R. M. Goldberg, and C. Lindley. 2007. Phase I and pharmacokinetic trial of carboplatin and albumin-bound paclitaxel, ABI-007 (Abraxane®) on three treatment schedules in patients with solid tumors. *Cancer Chemotherapy and Pharmacology*. 60(5):759–766.

Stoehr, L. C., E. Gonzalez, A. Stampfl, E. Casals, A. Duschl, V. Puntes, and G. J. Oostingh. 2011. Shape matters: Effects of silver nanospheres and wires on human alveolar epithelial cells. *Part Fibre Toxicol*. 8(36):3–15.

Su, W.-P., F.-Y. Cheng, D.-B. Shieh, C.-S. Yeh, and W.-C. Su. 2012. PLGA nanoparticles codeliver paclitaxel and Stat3 siRNA to overcome cellular resistance in lung cancer cells. *International Journal of Nanomedicine*. 7:4269.

Sukumar, U. K., B. Bhushan, P. Dubey, I. Matai, A. Sachdev, and G. Packirisamy. 2013. Emerging applications of nanoparticles for lung cancer diagnosis and therapy. *International Nano Letters*. 3(1):1–17.

Sung, J. C., B. L. Pulliam, and D. A. Edwards. 2007. Nanoparticles for drug delivery to the lungs. *Trends in Biotechnology*. 25(12):563–570.

Surendiran, A., S. Sandhiya, S. C. Pradhan, and C. Adithan. 2009. Novel applications of nanotechnology in medicine. *The Indian Journal of Medical Research*. 130(6):689–701.

Takahashi, T., D. Carbone, T. Takahashi, M. M. Nau, T. Hida, I. Linnoila, R. Ueda, and J. D. Minna. 1992. Wild-type but not mutant p53 suppresses the growth of human lung cancer cells bearing multiple genetic lesions. *Cancer Research*. 52(8):2340–2343.

Takamizawa, J., H. Konishi, K. Yanagisawa, S. Tomida, H. Osada, H. Endoh, T. Harano, Y. Yatabe, M. Nagino, and Y. Nimura. 2004. Reduced expression of the let-7 microRNAs in human lung cancers in association with shortened postoperative survival. *Cancer Research*. 64(11):3753–3756.

Tan, B.-J., Y. Liu, K.-L. Chang, B. K. W. Lim, and G. N. C. Chiu. 2012. Perorally active nanomicellar formulation of quercetin in the treatment of lung cancer. *International Journal of Nanomedicine*. 7:651.

Taratula, O., O. B. Garbuzenko, A. M. Chen, and T. Minko. 2011. Innovative strategy for treatment of lung cancer: Targeted nanotechnology-based inhalation co-delivery of anticancer drugs and siRNA. *Journal of Drug Targeting*. 19(10):900–914.

Taratula, O., O. B. Garbuzenko, P. Kirkpatrick, I. Pandya, R. Savla, V. P. Pozharov, H. He, and T. Minko. 2009. Surface-engineered targeted PPI dendrimer for efficient intracellular and intratumoral siRNA delivery. *Journal of Controlled Release*. 140(3):284–293.

Tseng, C.-L., T.-W. Wang, G.-C. Dong, S. Y.-H. Wu, T.-H. Young, M.-J. Shieh, P.-J. Lou, and F.-H. Lin. 2007. Development of gelatin nanoparticles with biotinylated EGF conjugation for lung cancer targeting. *Biomaterials*. 28(27):3996–4005.

Tseng, C.-L., S. Y.-H. Wu, W.-H. Wang, C.-L. Peng, F.-H. Lin, C.-C. Lin, T.-H. Young, and M.-J. Shieh. 2008. Targeting efficiency and biodistribution of biotinylated-EGF-conjugated gelatin nanoparticles administered via aerosol delivery in nude mice with lung cancer. *Biomaterials*. 29(20):3014–3022.

Tsukada, H., T. Satou, A. Iwashima, and T. Souma. 2000. Diagnostic accuracy of CT-guided automated needle biopsy of lung nodules. *American Journal of Roentgenology*. 175(1):239–243.

Turk, B. E., L. L. Huang, E. T. Piro, and L. C. Cantley. 2001. Determination of protease cleavage site motifs using mixture-based oriented peptide libraries. *Nature Biotechnology*. 19(7):661–667.

Valodkar, M., R. N. Jadeja, M. C. Thounaojam, R. V. Devkar, and S. Thakore. 2011. *In vitro* toxicity study of plant latex capped silver nanoparticles in human lung carcinoma cells. *Materials Science and Engineering: C*. 31(8):1723–1728.

van Rijt, S. H., D. A. Bölükbas, C. Argyo, S. Datz, M. Lindner, O. Eickelberg, M. Königshoff, T. Bein, and S. Meiners. 2015. Protease-mediated release of chemotherapeutics from mesoporous silica nanoparticles to ex vivo human and mouse lung tumors. *ACS Nano.* 9(3):2377–2389.

van Vlerken, L. E., T. K. Vyas, and M. M. Amiji. 2007. Poly (ethylene glycol)-modified nanocarriers for tumor-targeted and intracellular delivery. *Pharmaceutical Research.* 24(8):1405–1414.

Vasita, R. and D. S. Katti. 2006. Nanofibers and their applications in tissue engineering. *International Journal of Nanomedicine.* 1(1):15.

Wang, A., N. D. Cummings, M. Sethi, E. Wang, R. Sukumar, D. T. Moore, and M. E. Werner. 2013. Preclinical evaluation of genexol-PM, a nanoparticle formulation of paclitaxel, as a novel radiosensitizer for the treatment of non-small cell lung cancer. *Journal of Clinical Oncology.* 31(15 suppl.):e135456–e135456.

Wang, F., D. Banerjee, Y. Liu, X. Chen, and X. Liu. 2010. Upconversion nanoparticles in biological labeling, imaging, and therapy. *Analyst.* 135(8):1839–1854.

Wang, X., L. Yang, Z. G. Chen, and D. M. Shin. 2008. Application of nanotechnology in cancer therapy and imaging. *CA: A Cancer Journal for Clinicians.* 58(2):97–110.

Wang, Y., Y. Zhang, Z. Du, M. Wu, and G. Zhang. 2012. Detection of micrometastases in lung cancer with magnetic nanoparticles and quantum dots. *International Journal of Nanomedicine.* 7:2315.

Wozniak, A. J. and S. M. Gadgeel. 2009. Adjuvant therapy for resected non-small cell lung cancer. *Therapeutic Advances in Medical Oncology.* 1(2):109–118.

Wu, N.-Y., H.-C. Cheng, J. S. Ko, Y.-C. Cheng, P.-W. Lin, W.-C. Lin, C.-Y. Chang, and D.-M. Liou. 2011. Magnetic resonance imaging for lung cancer detection: Experience in a population of more than 10,000 healthy individuals. *BMC Cancer.* 11(1):242.

American Chemical Society. 2016a. What is Cancer Immunotherapy?. Atlanta: American Chemical Society. Retrieved from https://www.cancer.org/treatment/treatments-and-side-effects/treatment-types/immunotherapy/what-is-immunotherapy.html, accessed on November 3, 2016.

American Chemical Society. 2016b. Cancer Facts and Figures 2016. Atlanta: American Chemical Society. Retrieved from https://www.cancer.org/research/cancer-facts-statistics/all-cancer-facts-figures/cancer-facts-figures-2016.html, accessed on November 3, 2016.

American Chemical Society. 2016c. Monoclonal antibodies to treat cancer. Atlanta: American Chemical Society. Retrieved from https://www.cancer.org/treatment/treatments-and-side-effects/treatment-types/immunotherapy/monoclonal-antibodies.html, accessed on November 3, 2016.

Wynants, J., S. Stroobants, C. Dooms, and J. Vansteenkiste. 2007. Staging of lung cancer. *Radiologic Clinics of North America.* 45(4):609–625.

Xu, W., K. Kattel, J. Y. Park, Y. Chang, T. J. Kim, and G. H. Lee. 2012. Paramagnetic nanoparticle T_1 and T_2 MRI contrast agents. *Physical Chemistry Chemical Physics.* 14(37):12687–12700.

Yang, R., S.-G. Yang, W.-S. Shim, F. Cui, G. Cheng, I.-W. Kim, D.-D. Kim, S.-J. Chung, and C.-K. Shim. 2009. Lung-specific delivery of paclitaxel by chitosan-modified PLGA nanoparticles via transient formation of microaggregates. *Journal of Pharmaceutical Sciences.* 98(3):970–984.

Yang, S.-G., J.-E. Chang, B. Shin, S. Park, K. Na, and C.-K. Shim. 2010. 99mTc-hematoporphyrin linked albumin nanoparticles for lung cancer targeted photodynamic therapy and imaging. *Journal of Materials Chemistry.* 20(41):9042–9046.

Yao, X., K. Panichpisal, N. Kurtzman, and K. Nugent. 2007. Cisplatin nephrotoxicity: A review. *The American Journal of the Medical Sciences.* 334(2):115–124.

Yonesaka, K., K. Hirotani, H. Kawakami, M. Takeda, H. Kaneda, K. Sakai, I. Okamoto, K. Nishio, P. A. Jänne, and K. Nakagawa. 2015. Anti-HER3 monoclonal antibody patritumab sensitizes refractory non-small cell lung cancer to the epidermal growth factor receptor inhibitor erlotinib. *Oncogene.* 35(7):878–886.

You, L., C.-T. Yang, and D. M. Jablons. 2000. ONYX-015 works synergistically with chemotherapy in lung cancer cell lines and primary cultures freshly made from lung cancer patients. *Cancer Research.* 60(4):1009–1013.

Zhang, J. A., G. Anyarambhatla, L. Ma, S. Ugwu, T. Xuan, T. Sardone, and I. Ahmad. 2005. Development and characterization of a novel Cremophor® EL free liposome-based paclitaxel (LEP-ETU) formulation. *European Journal of Pharmaceutics and Biopharmaceutics.* 59(1):177–187.

Zhang, J., Y. Fu, Y. Mei, F. Jiang, and J. R. Lakowicz. 2010. Fluorescent metal nanoshell probe to detect single miRNA in lung cancer cell. *Analytical Chemistry*. 82(11):4464–4471.

Zhang, J., W. Tao, Y. Chen, D. Chang, T. Wang, X. Zhang, L. Mei, X. Zeng, and L. Huang. 2015. Doxorubicin-loaded star-shaped copolymer PLGA-vitamin E TPGS nanoparticles for lung cancer therapy. *Journal of Materials Science: Materials in Medicine*. 26(4):1–12.

Zhang, L., L. Laug, W. Munchgesang, E. Pippel, U. Gösele, M. Brandsch, and M. Knez. 2009. Reducing stress on cells with apoferritin-encapsulated platinum nanoparticles. *Nano Letters*. 10(1):219–223.

Zhao, T., H. Chen, Y. Dong, J. Zhang, H. Huang, J. Zhu, and W. Zhang. 2013. Paclitaxel-loaded poly (glycolide-co-ε-caprolactone)-bD-α-tocopheryl polyethylene glycol 2000 succinate nanoparticles for lung cancer therapy. *International Journal of Nanomedicine*. 8:1947.

Zhao, T., H. Chen, L. Yang, H. Jin, Z. Li, L. Han, F. Lu, and Z. Xu. 2013. DDAB-modified TPGS-b-(PCL-ran-PGA) Nanoparticles as oral anticancer drug carrier for lung cancer chemotherapy. *Nano*. 8(02):1350014.

Zhong, S., D. Yu, Y. Wang, S. Qiu, S. Wu, and X. Y. Liu. 2009. An armed oncolytic adenovirus ZD55-IL-24 combined with ADM or DDP demonstrated enhanced antitumor effect in lung cancer. *Acta Oncologica*. 49(1):91–99.

5

Surface Modification of Nanomaterials for Biomedical Applications: Strategies and Recent Advances

Ragini Singh and Sanjay Singh

CONTENTS

5.1 Introduction	172
5.2 Ligand Exchange Reactions	173
5.2.1 Gold NPs	174
5.2.2 Iron Oxide NPs	175
5.2.3 Quantum Dots	176
5.2.4 Carbon Nanotubes	177
5.3 Silanization of NP Surface	178
5.3.1 Gold NPs	178
5.3.2 Iron Oxide NPs	179
5.3.3 Carbon Nanotubes	180
5.3.4 Quantum Dots	180
5.4 Use of Click Chemistry for Surface Modification	181
5.4.1 Gold NPs	181
5.4.2 Iron Oxide NPs	183
5.4.3 Carbon Nanotubes	184
5.4.4 Quantum Dots	185
5.5 Amphiphilic Polymer Coating	186
5.5.1 Gold NPs	186
5.5.2 Iron Oxide NPs	187
5.5.3 Carbon Nanotubes	188
5.5.4 Quantum Dots	189
5.6 Use of Polyethylene Glycol	190
5.6.1 Gold NPs	190
5.6.2 Iron Oxide NPs	191
5.6.3 Carbon Nanotubes	192
5.6.4 Quantum Dots	193
5.7 Surface Modification with Biomolecules	193
5.7.1 Gold NPs	194
5.7.2 Iron Oxide NPs	195
5.7.3 Carbon Nanotubes	196
5.7.4 Quantum Dots	197
5.8 Use of Block Copolymers	197
5.8.1 Gold NPs	199

    5.8.2   Iron Oxide NPs 199
    5.8.3   Carbon Nanotubes 200
    5.8.4   Quantum Dots 200
5.9 Lipid Coating over NP Surfaces 201
    5.9.1   Gold NPs 202
    5.9.2   Iron Oxide NPs 202
    5.9.3   Carbon Nanotubes 203
    5.9.4   Quantum Dots 204
5.10 Conclusion and Future Perspectives 204
Acknowledgments 205
Conflict of Interest 205
References 205

5.1 Introduction

Recent research developments in the area of nanoscience and nanotechnology have led to nanoparticle (NP) applications in the area of biomedicines. These applications include use of NPs as drug/gene carriers, cell labels and trackers (Jamieson et al. 2007, Boisselier and Astruc 2009), hyperthermia-based therapy, and imaging contrast agents (Pankhurst et al. 2003, Pankhurst et al. 2009). The shape- and size-dependent unique physicochemical properties of NPs are fundamental to these applications. Certain nanomaterials such as gold, iron oxide, cerium oxide, and carbon-based nanomaterials have recently been investigated to possess intrinsic biological enzyme-like characteristics, which are being exploited for potential applications in sensitive disease diagnostic techniques based on colorimetric assays (Gao et al. 2007, Ambrosi et al. 2010, Karakoti et al. 2010, Cui et al. 2011, Singh et al. 2011). The success of medical treatment depends on the precise control over the delivery of drugs at the targeted organs. Additionally, to understand the efficacy of treatment of complex diseases, it is imperative to comprehend the complex spatiotemporal interplay of biomolecules in various cellular processes and signaling pathways (Patra et al. 2007). Key to successful implementation of NPs in biomedicines is precise control over the synthesis and surface modification of the particles (Wang and Hong 2011). In myriad reports from nanomaterial safety assessment, it is recommended that NPs should possess certain specific criteria for their use in biomedicine, which are: (i) possess minimum toxicity, (ii) prevent nonspecific interactions, (iii) be stable in different physiological conditions, and (iv) avoid premature release of drug and nucleotides (in gene therapy). Keeping the above parameters in mind, modification of NP surfaces becomes an essential parameter for their interaction with specific cell surface targets (Thanha and Green 2010). In this chapter, we have discussed the methods commonly being used for NP surface modification intended for biological applications. The common methods used to achieve controlled NP surface modification are ligand exchange, surface silanization, coating with polymers, and functionalization through different biomolecules (Figure 5.1) (Wang and Hong 2011). The arrangement of atoms on the surface of NPs and their interactions with ligands are essential criteria to ensure the surface modification of NPs. It is well documented that NPs can be synthesized in both aqueous and organic media with the same ease using suitable capping molecules. Therefore, these capped NPs can be further simply modified with either organic or inorganic molecules. For example, hydroxyl groups are commonly used in NP coating, where it reacts with different silane groups through the –O-Si bond; however, with the carboxyl group, its oxygen atoms are involved in the reaction (Kyoungja and Hong 2005).

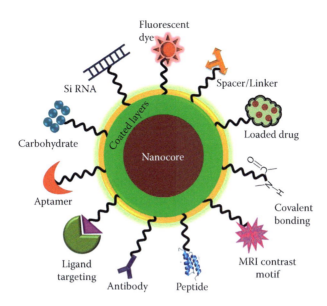

FIGURE 5.1
Potential modifications of NPs with ligands to impart targeting for various biomedical applications.

Surface modification of NPs generally involves physical interactions like electrostatic (De Campos et al. 2003) and van-der Waals (Ding and Jiang 2007) interaction; however, chemical reactions are also needed where strong binding is required. Metal NPs like silver (AgNPs) and gold (AuNPs) show a strong affinity toward thiol groups; thus, sulfur groups are a key factor for their functionalization (Aslan et al. 2004). Silanization and polymerization are other methods through which modification of NPs can be achieved (Wang et al. 2003). NPs synthesized in nonhydrolytic solutions require attachment of functional groups like the hydroxyl and mercapto groups on their surfaces for further conjugation reactions (Duguet et al. 2006). The amine and oxysilane functional groups are commonly used for functionalization of NP surfaces in aqueous suspensions (Shen et al. 2008). In order to comprehensively cover the possible surface modification methods, we have divided the strategies into the following broad categories.

5.2 Ligand Exchange Reactions

Through ligand exchange reactions, the properties and functionality of NPs can be tuned, which improves aqueous solubility and cell/tissue targeting efficiency. Generally, NPs are capped with ligands, which provide them stability and help in size control. During the ligand exchange process, these ligands or some nonfunctional ligands are replaced with mono- or bifunctional ligands (Sperling and Parak 2010, Karakoti et al. 2015). Often, it has been found that inorganic NP surfaces bind more strongly to the incoming ligand molecule (Sperling and Parak 2010). Many polymers such as poly (vinyl pyrollidone) (PVP), α-acetylene-poly (tert-butyl acrylate) (ptBA), poly (acrylic acid) (PAA), hexadecyl amine (HDA), oleic acid (OA) and tetradecyl phosphoric acid (TDPA) have been used to coat the surfaces of different NPs (Graf et al. 2006, Lin et al. 2005). Ligands on NP surfaces are in

FIGURE 5.2
Schematic diagram showing lipid exchange process in NPs.

dynamic equilibrium, leading to continuous exchange with the free ligands present in the solvent through simple mass action (Mei et al. 2008, Olsson 2009). It is suggested that during the ligand exchange process, the replacing ligand concentration should be equal to that of existing ligands with greater affinity for NP surfaces than the existing ones (Kim and Bawendi 2003, Zhang and Clapp 2011). Additionally, a higher concentration of the replacing ligand is required if its affinity is lower than the existing one, to have increased attachment probability (Zhang and Clapp 2011). In order to find biomedical applications of nanomaterials, ligand exchange methods have been tried on many NP systems. Among them, NPs of gold, iron oxide, quantum dots, and carbon-based materials have been investigated in detail. Therefore, in this and other sections, we will mainly focus on these nanomaterials. A schematic diagram of the ligand exchange process is shown in Figure 5.2.

5.2.1 Gold NPs

Synthesis of AuNPs coated with citrate or tetra octylammonium bromide (TOAB) present as an assembled monolayer of alkane thiol and their derivatives on the surface, which is due to increased binding strength during ligand exchange process. The gold-sulfur (Au-S) bond has a bond strength of 210 kJ mol^{-1} and serves as a common method for functionalization of AuNPs (Vericat et al. 2005). Alternatively, strongly attached ligands can also be replaced by ligands present in higher concentrations in the solution during the exchange process. This is in accordance with Le Chatellier's principle (Equation 5.1), where n and m are the number of molecules of X and Y, respectively (Thanha and Green 2010).

$$K = [NP\!-\!X_n][Y]^m / [NP\!-\!X_m][X]^n \tag{5.1}$$

AuNPs can find application in various fields like electronics (Lawrence et al. 2014), sensors (Nishi et al. 2015), catalysis (Zhang et al. 2015), and in biomedical fields for therapy and diagnostic purposes (Nair et al. 2015) due to their tunable electronic and optical properties (Sauerbeck et al. 2014). AuNPs are extensively synthesized by the well-known Turkevich method (Turkevich et al. 1951), which uses $HAuCl_4$ as a precursor and citrate as a reducing and capping agent. The size of AuNPs can be tuned from 10–100 nm, and they are stabilized by the citrate (Park and Shumaker-Parry 2015). These NPs need to be chemically modified as per the need of desired applications. Citrate molecules are loosely attached to the AuNP surface (6.7 kJ/mol) (Tarazona-Vasquez and Balbuena 2004) and therefore can be easily exchanged with the suitable molecules such as amines, carboxylates, thiols, or phosphine moieties. Particles can be made biocompatible through ligand exchange methods involving surface modifications and can be conjugated with molecules for biomedical interest (Dreaden et al. 2012). Thiols are commonly used as a modification agent because they can easily get chemisorbed at room temperature and remain adsorbed for ∼35 days under relevant physiological conditions (Dreaden et al. 2012). It has been reported (Dinkel et al.

2015) that 3-mercapto-1-propanesulfonate (MPS) can also be used for ligand exchange on citrate layer-coated AuNP surfaces by second-harmonic light scattering (SHS). SHS provides high surface sensitivity and also the time resolution to observe *in situ* and real-time exchange reaction processes on NP surfaces. Another strategy for surface modification of AuNPs involves the use of thiols, which undergo ligand exchange by an associative (SN_2-type) reaction mechanism. Since disulfides are less reactive, the ligand exchange reaction of thiolated AuNPs with disulfides was found to follow an SN_1-type reaction mechanism. It has been suggested that after dissociation of thiol ligand from AuNPs surface, the vacant site may break the –S-S- (disulfide) bond of the incoming ligand in thiols (Isaacs et al. 2005). The intrinsic properties of AuNPs are also altered using suitable ligand exchange molecules. In this context, Crawford et al. have shown that ligand exchange between phosphine-coated AuNPs and sulfur-containing molecules leads to the initiation of photoluminescence in AuNPs (Crawford et al. 2015).

5.2.2 Iron Oxide NPs

Iron oxide nanoparticles (IONPs) possess the superparamagnetic property (Amstad et al. 2009, da Costa et al. 2014); therefore, they are used in various biological applications, that is, magnetic resonance imaging (MRI), biomedicine (Li et al. 2013), catalysis (Hudson et al. 2014), data storage (Uenuma et al. 2013), and magnetic fluids (Shultz et al. 2007). However, synthesis of biocompatible IONPs remains a challenging task due to the use of highly hydrophobic ligands such as oleic acid in the synthesis (Tong et al. 2014). For any biological application, these IONPs need to be dispersed in an aqueous suspension. In this context, it has been shown that the oleic acid ligand can be exchanged with highly polar compounds such as Tiron (4,5-dihydroxy-1,3-benzenedisulfonic acid disodium salt monohydrate). During first step of this process, dopamine is required, which is compatible with both polar and nonpolar solvents. In the second step of ligand exchange, IONPs are stabilized by Tiron and dopamine residues, which makes them water soluble. These results reveal that solvent incompatibility is the major hindrance in the preparation of water-soluble and biomedical applications of superparamagnetic IONPs (Korpany et al. 2016). The ligand exchange process can also alter the polarity of the hydrophobic layer and make it hydrophilic (Wang, Neoh et al. 2009; Yang et al. 2013). It has also been observed that the addition of an excess amount of ligand in the NP suspension leads to replacement of the original ligand from the particle surface and results in aggregation. The advantage with the ligand exchange process is that the size and shape of IONPs generally does not change during surface modification and the phase transfer process. Nitrosonium tetrafluoroborate ($NOBF_4$) is reported to replace the organic ligands from NP surfaces and stabilize them in polar and hydrophilic solvents for longer periods of time (up to several years) without any aggregations (Figure 5.3) (Wu et al. 2015). Mechanistically, weak binding of BF_4^- anions on the particle surfaces allows secondary surface modification, resulting in the phase transfer of NPs from hydrophobic to hydrophilic solvents (Dong et al. 2010). It has also been studied that the magnetic behavior and relaxivity of IONPs depend on the functionality of the attached group and the method followed for surface modification. Smolensky et al. (2011) reported that IONPs coated with polyethylene glycol (PEG) through a catecholate-type anchoring moiety retain the relaxivity and saturation magnetization of magnetite NPs suspended in a hydrophobic suspension medium. Similarly, other functional moieties, such as phosphonate, carboxylate, and dopamine, are found to decrease the magnetization and relaxivity of NPs, thus the produced contrast. When explored for the protocol of functionalization, the NPs were found to be influenced by the behavior of the material. IONPs refunctionalized by a direct

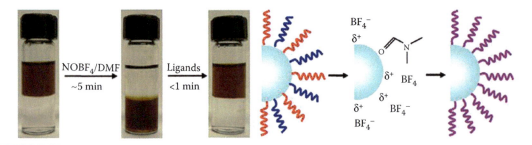

FIGURE 5.3
Ligand-exchange approach; maintains size and shape of nanocolloids and enables them for surface functionalization and phase transfer. (Reprinted with permission from Dong, A. et al. 2010. A generalized ligand-exchange strategy enabling sequential surface functionalization of colloidal nanocrystals. *J Am Chem Soc* 133(4):998–1006. Copyright 2011 American Chemical Society.)

biphasic protocol resulted in enhanced relaxivity compared to those refunctionalized by a two-step procedure in which the first step involved stripping of the functional moieties from the NPs. Such observations suggest that a systematic study of both the binding moiety and the functionalization protocol on the relaxivity and magnetization of aqueous dispersed IONPs must be completed before their application in MRIs as contrast agents or other applications of similar sorts.

5.2.3 Quantum Dots

Thiols like dithiothreitol, mercaptoacetic acid (MAA), and carbodithiolates are commonly used in the ligand exchange process to replace the original ligands from the quantum dot (QD) surface. The thiol group present in the incoming ligands binds to QDs, whereas the carboxyl group stabilizes QDs in aqueous suspension through electrostatic repulsion due to its negative charge. QDs possess a ZnS layer on their surface for most biological applications. Experimental analysis and computational modeling show that thiol groups present in the incoming ligands bind primarily with zinc metal present on the QD surface through Zn-S bonds and also with a weaker sulfur-sulfur bond (Grapperhaus et al. 1998, Chiou et al. 2000, Pejchal and Ludwig 2005). The affinity of sulfur for metal plays a major role in the exchange process, where its affinity can be improved by the disulfide linking group. The ligand exchange process offers several advantages, and one of them is that the diameter of QDs remains unchanged. However, in silanization and ligand attachment methods, the size of QDs is altered significantly, which leads to changes in properties. Due to this advantage, many exchange ligands are synthesized to provide stability and functionality to QDs for biomedical applications. Apart from several advantages, the ligand exchange process also offer certain disadvantages, such as weaker interaction between QD surface and thiol group. In this context, thiol molecules sometimes form disulfide bond and get detached from the QD surface, which results in aggregation and surface oxidation in aqueous environments (Karakoti et al. 2015). These changes lead to a decrease in quantum yields exhibited by QDs. The use of dihydrolipoic acid (DHLA) has been reported to enhance the stability and yield of QDs for longer storage with minimum loss of properties. These particles are shown to be stable at alkaline pH; however, they undergo agglomeration at even slight acidic conditions. It has been suggested that in acidic conditions, the carboxylate anions get protonated, which leads to a decrease in the electrostatic stability, thus producing aggregation (Zhang and Clapp 2011). Alternatively,

PEG-terminated DHLA can be used, which provides stability to QDs at acidic pH. Further, carboxyl-, biotin-, hydroxyl-, and amine-terminated QDs can also be synthesized by grafting thioctic acid to these PEG-terminated DHLA-capped QDs (Susumu et al. 2009). Peptides, cross-linked dendrons, and phosphine are also used for imparting stability; however, these methods compromise the unique advantage of the ligand exchange process by increasing the size of the QDs and making the whole process more complicated (Xu and Chen 2012). Wang et al. (2004) have reported a novel method for surface passivation of QDs with polydimethylaminoethyl methacrylate (PDMAEMA) through a ligand exchange process. In this method, PDMAEMA was used as a multidentate ligand to modify the surface of QDs (CdSe/ZnS) in toluene with trioctylphosphine oxide. PDMAEMA was exchanged with trioctylphosphine oxide on the QD surface. This simple process is free of aggregation, preserved photoluminescence properties well, and had solubility in methanol. The ligand exchange process is also reported to synthesize QDs with the desired charge. In this context, Yeh et al. (2011) have reported cationic CdSe/ZnS QDs synthesis using quaternary ammonium derivatives. The resulting QDs are soluble in aqueous as well as biological media, which is an essential prerequisite for biomedical applications.

5.2.4 Carbon Nanotubes

Carbon nanotubes (CNTs) were discovered in 1991, and since then have been extensively studied to show applications in many fields, such as preparation of nanodevices and composite materials for biomedicines (Islam et al. 2003, Paredes and Burghard 2004, Kim, Han et al. 2008). CNTs possess strong van der Waals interaction among tubes and therefore exhibit poor solubility in aqueous solvents (Islam et al. 2003). However, various biocompatible polymeric materials are developed that can efficiently modify CNTs and impart solubility in aqueous suspensions (Zhang et al. 2004). Methods like plasma treatment, reaction with strong acids, irradiation with ion beam, and oxidation reaction are among some of the common methods used for grafting chemical functional groups on CNT surfaces. However, due to the intrinsic stability of CNTs in harsh conditions, chemical functionalization methods are not successful. Additionally, due to harsh physiological conditions, the hexagonal array of carbon atoms on the CNT surface can also be damaged, which poses further limitations (Cui et al. 2009, Lee et al. 2009). Therefore, a chemical modification method is needed that can avoid the distortion of translational symmetry of the carbon arrays on the CNTs. In a report, authors have grafted polyvinyl ferrocene-co-styrene [poly (Vf-co-St)] on the surface of CNTs by the ligand exchange process to improve its dispersibility (Shi et al. 2014). Poly(Vf-co-St) is a copolymer of styrene and vinyl ferrocene monomer prepared by copolymerization reaction between two π-conjugated structures formed by the ferrocene group of the vinyl ferrocene group exchanged with the hexagonal ring on the CNT surface (Shi et al. 2014). This leads to covalent grafting of molecular chains on the CNT surface without altering its original structure. Due to this modification, the dispersibility of CNTs increases with their hexagonal arrays remaining intact. CNTs can also be covalently functionalized with the tetra-manganese complexes by the ligand exchange process (Frielinghaus et al. 2015). In this case, ligand exchange occurs with the carboxylic group present on the oxidized tubes, and the number of carboxylic groups present on the surface is directly proportional to the functionalization on the CNT surface. Functionalization of CNTs with a complex of $[Mn_4L_2(OAc)_4]$ occurs at room temperature by the reaction mentioned below:

$$CNT\text{-}COOH + [Mn_4L_2(OAc)_4]CNT\text{-}COO \rightarrow Mn_4L_2(OAc)_3 + AcOH$$

The grafting density of Mn over the CNT surface is governed by the oxidation time and different methods of oxidation. When oxidation time is small (<3 min.), tube coverage with [Mn$_4$] is low and the distance between the two complexes was found to be several tens of nanometers.

5.3 Silanization of NP Surface

Silanization of NPs is a process of creating an amorphous silica layer over NP surfaces, as shown in Figure 5.4. In this method, surface self-assembly is covered with organo-functional alkoxysilane molecules. Materials containing hydroxyl groups, such as metal-oxides and glass, are silanized as their hydroxyl groups are displaced by the alkoxy groups from the silanes, which results in the formation of the –Si-O-Si- covalent bond. A coating of silica enhances the solubility of NPs in aqueous media due to its electrochemical properties. The advantage of silica is that it is very stable at neutral pH and, unlike other oxides, it possesses a strange behavior and does not coalesce over the isoelectric point (IEP). Further, strong steric repulsion caused due to hydrogen bonding of the water layer with silica surface reduces agglomeration.

5.3.1 Gold NPs

Bahadur et al. (2011) demonstrated a method for rapid synthesis of silica-coated AuNPs (Au@SiO$_2$) with a tunable property to control silica shell thickness over the AuNP surface. Using microwave irradiation, these particles can be synthesized within 5 min by simple mixing of an Au solution, tetraethoxysilane (TEOS), and ammonia. The thickness of the silica shell can be tuned from the range of 5–105 nm by varying the concentration of TEOS only. Au@SiO$_2$ can also be further functionalized with amino and carboxyl groups easily via silane chemistry. In another attempt, Mine et al. (2003) showed silica coating over AuNPs through the seeded polymerization technique. Here, uniform coating was obtained by addition of tetraethylorthosilicate and water prior to the addition of NH$_4$OH. It was found that by varying the concentration of TEOS in the range of 0.0005–0.02 M, the thickness of the silica shell can be tuned from 30 to 90 nm. The synthesized Au@SiO$_2$ particles showed a difference in the surface plasmon resonance (SPR) pattern, which is

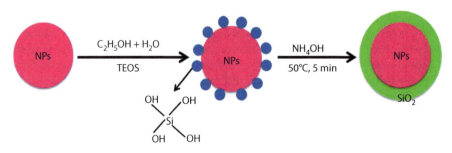

FIGURE 5.4
Schematic diagram showing surface silanization procedure for NP coating with SiO$_2$.

dependent on the silica thickness. The change in SPR absorbance wavelength correlates well with the scattering theory and theoretical predictions of conventional absorption. Several applications have also been shown using Au@SiO$_2$ NPs. Recently, Cerruti et al. (2006) showed the application of silica-coated AuNPs in thermographic DNA detection at low concentration of 10 pM, whereas bare AuNPs could detect only 100 pM of DNA. It is suggested that due to the greater IR emissivity of silica compared to Au, silica-coated AuNPs have 2–5 times more T-jumps (difference in temperature before and after laser irradiation), which is dependent on silica shell thickness when compared to bare AuNPs. Further, the good biocompatibility of Au and SiO$_2$ also offers a suitable NP type to be used for biomedical applications. In this context, Hayashi et al. (2013) have shown that Au@SiO$_2$ NPs can be used for CT (computed tomography) and fluorescence-based dual imaging modalities. They also observed anatomical information, which included the location and size of lymph nodes (LNs) and lymph vessels (LVs), for determining a surgery plan, and provided intraoperative imaging of LNs and LVs to make the operation simpler. Upon toxicity evaluation, Au@SiO$_2$ NPs did not show any sign of hepatotoxicity or nephrotoxicity.

5.3.2 Iron Oxide NPs

Recent research has led to the widespread use of IONPs for therapeutic benefits; however, some raise concerns about their possible toxicity to human health and the environment (Malvindi et al. 2014). It has also been suggested that redox active materials undergo physicochemical changes leading to free radical generation in mammalian system. Additionally, redox-active nanomaterials are known to have "hot spots" that undergo one electron transition, thus underlying the toxicity. Therefore, surface passivation of IONPs using silanization is an attractive strategy to overcome these issues. Several attempts have been made in this direction. Kim, Kim et al. (2008) have synthesized a monodisperse Fe$_3$O$_4$ nanocrystal core with a mesoporous silica shell and showed many applications such as MRI, fluorescence imaging, and drug delivery. Santra et al. (2001) synthesized uniformly distributed silica-coated IONPs of extremely small size (<5 nm) using the water-in-oil microemulsion method. The microemulsion was prepared by using three different nonionic surfactants (Triton X-100, Igepal, and Brij-97). Using the base-catalyzed hydrolysis reaction and polymerization of TEOS in microemulsion, 1-nm-thin uniform silica coating was obtained over bare IONPs. Silica-coated IONPs can easily be dispersed into organic and aqueous solutions, as compared to polystyrene-based magnetic particles. Additionally, Hakami et al. (2012) prepared thiol-functionalized silica-coated magnetite NPs (TF-SCMNPs) and used it for adsorption of Hg(II) dissolved in water. Further, it was shown that thiourea dissolved in 3M HCl solution can easily desorb Hg(II) from the NP surface without affecting the activity of particles. Thus, TF-SCMNPs can be reused in Hg removal and are also equally useful for other repetitive adsorption tests. The adsorption capacity was more than that of commonly used adsorbents, and ~90% of Hg(II) can be removed within 5 min. Interference by other ions was also investigated, and in tap water, no interference in the adsorption of Hg(II) on IONPs was observed. It is also reported that silica-coated IONPs are efficiently internalized by mammalian cells up to a significantly higher degree than dextran-coated IONPs at all sizes. Cellular internalization of silica-coated IONPs was reported to be through an active actin cytoskeleton-mediated process (Kunzmann et al. 2014). These results conclude that silica-coated IONPs are promising candidates for medical imaging, cell tracking, isolation, and other biological applications.

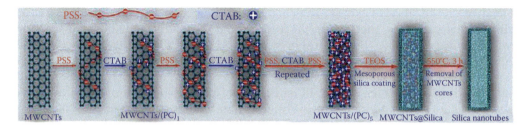

FIGURE 5.5
Procedure of synthesizing mesoporous silica shell-coated MWCNTs and mesoporous silica nanotubes. (Reprinted with permission from Deng, X. et al. 2013. Mesoporous silica coating on carbon nanotubes: Layer-by-layer method. *Langmuir* 29(23):6815–22. Copyright 2013 American Chemical Society.)

5.3.3 Carbon Nanotubes

CNTs are known to be insoluble in aqueous suspensions, which hiders their application in biomedicines. However, silica coating over CNTs is one of the best modalities to render water solubility to CNTs. Deng et al. (2013) prepared mesoporous silica-coated, CTAB-wrapped multiwalled carbon nanotubes (MWCNTs) via the layer-by-layer self-assembly method, as shown in Figure 5.5. Whitsitt and Barron (2003) prepared silica-coated single walled carbon nanotubes (SWNTs) by using silica/H_2SiF_6 in a surfactant-stabilized solution of SWNTs. The thickness of silica coating over the particles can be controlled by the reaction time, whereas coating of individual SWNTs versus small ropes can be controlled by surfactant choice. The exposed ends of tubes can show spontaneous interconnection between the isolated SWNTs, whereas the exposed center parts of tubes have found application in the construction of sensors and devices. Stynoski et al. (2013) functionalized MWNTs with the silica via sol-gel method; in this case, silica binds to CNTs through covalent bonding, makes them more dispersive in water, and provides interfacial bond sites for matrix substances. Point-of-zero-charge (PZC) measurement further augments that the steric effect of silica is responsible for better stability of coated CNTs. Liu, Tang et al. (2006) demonstrated the preparation of silica-coated CNTs via the seeded sol-gel method. Silica particles strung along particles and complexes form stable identities due to the interaction of silica particles with CNT wires. Two types of binding force may occur between silica and CNTs. CNTs possesses a porous nature, which leads to the adsorption of silica on their walls through the sol-gel process and ultimately silica NP anchors on CNTs. Secondly, the walls of CNTs may have several defects (e.g., hydroxyl groups existing); thus, silica precursors show covalent binding with CNTs. Changes in the weight ratio of CNTs to siloxane govern the morphology of the CNT-silica product. Ma et al. (2006) presented a method for chemical functionalization of CNTs via a process consisting of both UV/O_3 exposure and silanization. The silane molecule possesses an epoxy group at its end and covalently binds to the CNT surface, which in turn modifies the compatibility of CNTs with epoxy. CNTs were exposed to UV light along with ozone to oxidize them and to generate active moieties. Further, oxidized CNTs were reduced in the presence of lithium aluminum hydride followed by 3-glycidoxypropyltrimethoxy silane treatment for silanization.

5.3.4 Quantum Dots

The silanization of QDs requires attachment of the initial layer of silane molecules to form a primer layer over the QD surface. Correa-Duarte et al. (Correa-Duarte et al. 2001) demonstrated the exchange of citrate coating over QDs with silica coating by using (3-sulfanylprosulfanylpropyl) trimethoxysilane to form the first layer of silica, which is followed by solvent exchange in ethanol. Further, by adding TEOS in the solution, the thickness of the silica shell can be increased by the stober process. Ma et al. (2014) synthesized highly stable and small-sized (<30 nm) silica-coated QDs (QD@SiO$_2$) by using the stober method. In this method, before silica coating, hydrophobic ligands on QDs were replaced with adenosine 5'-monophosphate (AMP) to make oil-soluble QDs dispersible in an alcohol-water suspension. MPS were used as the silane nucleation primer, and the thickness of the silica shell over the surface of the QDs was controlled by condensation/hydrolysis of tetraethyl orthosilicate. As prepared, QD@SiO$_2$ exhibited same luminescent efficiency (50–65%) as exhibited by uncoated oil-soluble QD. It exhibited better colloidal and photostability, even in long-term exposure to harsh conditions (wide pH range of 3–13, thermal exposure of 100°C, and saturated NaCl solution). Nann et al. (2005) reported the phase transfer of QDs in hydrophobic-hydrophilic solvent via exchange of trioctylphosphine oxide (TOPO) and HDA ligands at the QD surface with MPS in tetrahydrofuran (THF). Stable colloids are formed by MPS-coated QDs in methanol and ethanol. The presence of less MPS can cause coagulation of QDs, whereas an excess amount of MPS increased the nucleation rate of silica particles. The size of silica particles depends on the concentration of QDs: silica particle size increases with a decrease in the concentration of QDs. The amount of TEOS added and the concentration of QDs can be used to calculate the size of silica, when per silica one QD is available. The diameter (D) of particles can be calculated with the formula given below:

$$D = 2r = 2 \cdot \sqrt[3]{3}/\sqrt[3]{4\pi}\sqrt[3]{V_{tot}}$$

where V_{tot} = total volume of one particle. $V_{tot} = V_{SiO_2} + V_{QD}$. The silica per particle can be calculated by $V_{SiO_2} = [TEOS]/(N_{QD} \cdot D)$, where [TEOS] = amount of TEOS added (in mol), N_{QD} = number of QDs, and D = the molar density of silica.

5.4 Use of Click Chemistry for Surface Modification

Click chemistry is a process that possesses certain unique criteria such as reactions giving a high yield and safe byproducts that can be easily removed by nonchromatographic techniques, and it is stereo specific. The conditions of reaction are simple and therefore can be performed in a benign or easily removable solvent. The product should be stable in physiological conditions and secluded by nonchromatographic techniques like distillation or crystallization (Kolb et al. 2001). Click chemistry requires high thermodynamic energy (>20 kcalmol^{-1}). Common examples of click chemistry are (i) unsaturated species such as 1,3-dipolar cycloaddition of azides with alkynes, catalyzed by copper (Li et al. 2013); (ii) electrophilic species such as aziridines, epoxides, episulfonium ions, and aziridinium ions that undergo nucleophilic substitution reactions of ring opening; and (iii) urea, thiourea, oximes, and several other nonaldol type substance carbonyl groups (Kolb et al. 2001). Schemes 5.1 and 5.2 show the cycloaddition reaction process and the mechanism involved in it.

5.4.1 Gold NPs

Due to myriad applications, surface functionalization of AuNPs is of great concern (Mulvaney 1996, Daniel and Astruc 2004). Although many methods of functionalization of AuNPs are reported, many are still incompatible with processes like the ligand exchange reaction (Templeton et al. 2000), high temperature functionalization, and direct synthesis

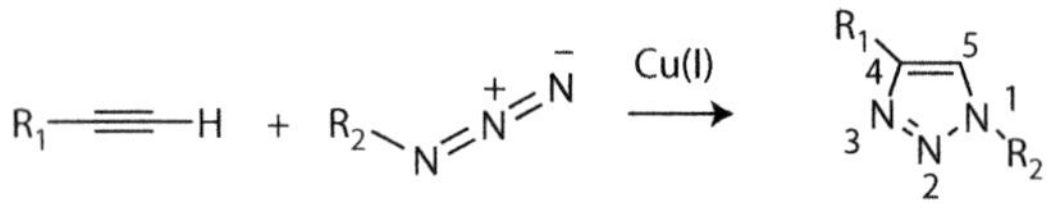

SCHEME 5.1
Reaction schematic showing Cu(I)-catalyzed azide-alkyne cycloaddition reaction (CuAAC). In such reactions, either the alkyne or azide is attached to a solid-phase support.

(Brust et al. 1994). Therefore, better methods of AuNP functionalization need to be worked on for its better use in biomedical applications. Due to the above-mentioned advantages, click chemistry stands out as one of the best methods of surface functionalization of AuNPs with required ligands (Boisselier et al. 2008). Thode and Williams (2008) have reported that 52% alkylthiolate ligands can be substituted by reaction with bromo-undecane thiols. Triazole rings can be produced by click reaction with numerous terminal alkynes, which can be activated by carbonyl linkage and produce 1,2,3-triazole with a yield of 1–22%. AuNPs can be assembled with alkyne-modified DNA templates using click chemistry, which can form DNA NPs conjugate in a chain-like manner (Fischler et al. 2008) in which AuNPs are glutathione azide-functionalized, whereas the DNA duplex is alkyne modified and forms traizole linkage in the copper(I)-catalyzed click reaction. Boisselier et al. (2008) reported the specific conditions to increase the yield of functionalized AuNPs: the reaction should contain water and THF in equal ratio with a large quantity of Cu$^+$ taken from CuSO$_4$ and sodium ascorbate to prevent Cu$^+$ aggregation in the medium (Figure 5.6). The whole process takes place at 20°C in an inert atmosphere to obtain the reaction between various alkynes (hydrophilic [PEG] and hydrophobic [organometallic and organic]) and AuNPs. Li et al. (2014) synthesized bifunctional 1,2,3-triazole derivatives containing a PEG chain and other functional species (i.e., fluorescence dye, carboxylic acid, a polymer, redox-robust metal complex, alcohol, allyl, dendron, or a b-cyclodextrin unit) by click chemistry, which can

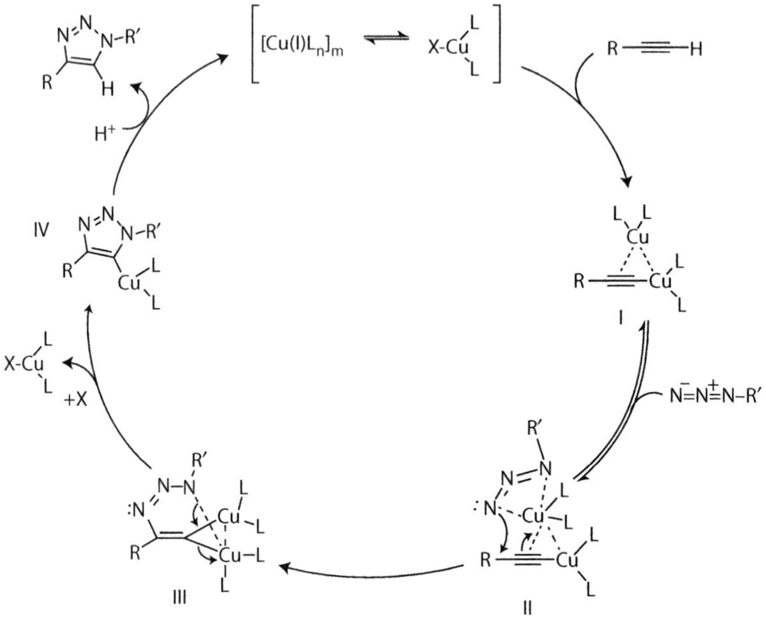

SCHEME 5.2
Schematic reaction mechanism showing Cu(I)-catalyzed azide-alkyne cycloaddition.

Surface Modification of Nanomaterials for Biomedical Applications

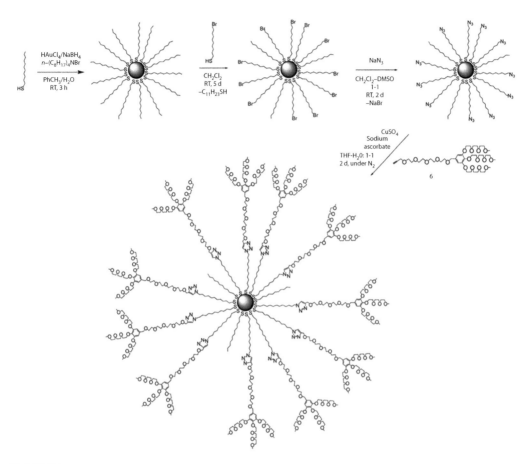

FIGURE 5.6
Overall synthetic scheme for the efficient functionalization of AuNPs under ambient conditions using the "click" reaction with the dendritic alkyne. (Boisselier, E. et al. 2008. How to very efficiently functionalize gold NPs by "click" chemistry. *Chemical Communications* (44):5788–90. Reproduced by permission of The Royal Society of Chemistry.)

be further conjugated with AuNPs as a catalysis and biosensor. A triazole ring can attach with AuNPs via lone pair of electrons in sp2-hybridizednitrogen atoms. Finetti et al. (2016) developed a system in which azido-modified antibodies are covalently linked to AuNPs coated with a thin silicon layer. In the derivatization step, first, silica was coated by a linear polymer copoly (N,N-dimethylacrylamide [DMA]–3-trimethylsilyl-prop-2-ynyl methacrylate [Protected PMA]–γ-methacryloxypropyltrimethoxysilane [MAPS]) by mixing copolymer with particle suspension. An external polymer-containing alkyne functional group formed a covalent bond with PEG-modified IgGs on the AuNP surface through the CuAAC reaction.

5.4.2 Iron Oxide NPs

Superparamagnetic iron oxide nanoparticles (SPIONS) can be easily metabolized and degraded during the metabolism of biological systems. Iron can also be incorporated into a serum-Fe system to form hemoglobin; therefore, Fe-based materials are biocompatible compared to other inorganic NPs. Like other nanomaterials, SPIONS too possess a large surface area, which has been utilized by several investigators for carrying specific genes and drugs (Liu et al. 2010, Huang et al. 2011). Methods are reported to conjugate SPIONs with peptides (Wunderbaldinger et al. 2002), monoclonal antibodies (Baio et al. 2010) for intended use in biomedical applications such as MRI (Lin et al. 2009), targeted drug delivery (Yang et al. 2011), and treatment of tumors through hyperthermia (Hayashi et al. 2009, 2010). Due to the reaction simplicity and high yield, click chemistry is the preferred surface modification method compared to many other conventional methods (De et al. 2008). The biggest advantage of this method is that the reaction can take place both in aqueous and organic media and exhibit more selectivity, with slight or no side reactions (De et al. 2008, Hayashi et al. 2009). Since the magnetic properties of SPIONs are mainly dependent on their spatial arrangement (Poddar et al. 2002), a surface conjugation method that does not alter the spatial arrangement is the most suitable one. Click chemistry can be used to control the interparticle distance in IONPs (SPIONs) with 2D assembly. Alkyne-terminated self-assembled monolayers (SAMs) were used to immobilize azide-terminated IONPs through a reaction. Interparticle distance can be controlled on the basis of reaction time and reaction kinetics, which govern assembly of NPs and their density on SAMs (Toulemon et al. 2011). Due to the compatibility of CuAAC with DNA (Kumar et al. 2007, Gramlich et al. 2008), a click reaction can be used to form a thick oligonucleotide (alkyne-modified) monolayer over SPIONs to prepare polyvalent conjugates that are stable and have highly dense oligonucleotides on their surfaces. With a thick DNA coating, these particles can easily penetrate a HeLa cell membrane without any transfection agent (Cutler et al. 2010). High heat formation and bio-orthogonality during click chemistry minimize columbic repulsion between the oligonucleotides; therefore, conjugation can be promoted even in salty milieus. Huang et al. (2011) applied click chemistry to conjugate folic acid to SPIONs for targeted drug delivery in cancerous cells in which folic acid receptors are overexpressed (i.e., breast, kidney, lung, and ovarian cancer). Poly (glycidyl methacrylate-co-poly(ethylene glycol) methyl ether methacrylate) was grafted on SPIONs via atom transfer radical polymerization (ATRP) to form SPIONs-P(GMA-co-PEGMA). The amount of folic acid grafted on particle surfaces and its water solubility depend on the GMA [Poly(glycidyl methacrylate)]/PEGMA [poly(ethylene glycol) methyl ether methacrylate] ratio, respectively. Folic acid was conjugated to GMA in the polymer via click chemistry, whereas PEGMA provides stability to the particles and increases their circulation time *in vivo*. The epoxy group in GMA reacts with NaN_3 to generate the azide group on the SPIONs-P(GMA-co-PEGMA) surface via ring-opening reaction to form SPIONs-P(GMA-co-PEGMA)-N_3-FA functionalized NPs. SPIONs-P(GMA-co-PEGMA)-FA was synthesized via a click reaction of SPIONs-P(GMA-co-PEGMA)-N_3 with propargyl folate.

5.4.3 Carbon Nanotubes

Due to their good conducting and electrocatalytic behavior, CNTs have been used to construct excellent immune sensors (Agui et al. 2008). Sanchez-Tirado et al. (Sanchez-Tirado et al. 2016) have recently reported the preparation of immune sensors for detection of transforming growth factor β1 (TGF-β1) cytokine via click chemistry. It is a proficient method to immobilize immune reagents without compromising their biological activity and configurations. CNTs were functionalized with the azide group and assembled with alkyne-functionalized IgG to prepare a scaffold for preparation of the immune sensor. It was comparable to conventional enzyme-linked immunosorbent assay (ELISA) and used for sensing cytokines present in clinical samples with high sensitivity and a low detection limit of 1.3 pg mL^{-1}. Zabihi et al. (2015) demonstrated that interaction of CNTs with epoxy matrix can be improved by the functionalization of multiwalled carbon nanotubes by a thiol-ene click chemistry-based method for efficient interaction between the epoxy matrix and CNTs. This polymer nanocomposite was prepared by using 2–aminoethanethiol hydrochloride radicals as engineered filler. CNTs were aminated by free radicals produced by thermal decomposition of dicumyl peroxide; further, these radicals acquire hydrogen radicals from the thiol group of 2–aminoethanethiol hydrochloride to form thiol radicals. These thiol radicals are grafted on the CNT surface by breaking its $C = C$ bond. SWCNTs can be functionalized with protein via click chemistry and used in immunoassays as a sensing platform. They also show better water dispersion as well retaining their bioactivity (Qi et al. 2012). Cu(I)-catalyzed formation of 1,2,3-triazole was done efficiently by pairing azide-functionalized SWNTs with alkyne-modified protein (Chao et al. 2009). Carboxylated SWCNTs are attached with the azide group via carbodiimide chemistry, and subsequently alkyne groups were introduced to protein. Immunoglobulin-functionalized SWCNTs fixed on a glass plate can be used for detection of anti-immunoglobulin (anti-IgG) with horseradish peroxidase (HRP) as an enzyme tag. CNTs were reported to be functionalized with zinc-phthalocyanine (ZnPc) via click chemistry (Campidelli et al. 2008). First, CNTs were functionalized with 4-(2-trimethylsilyl) ethynylaniline after deprotection of the triple bond, which was further attached to a phthalocyanine having an azide functional group. The Cu(I) catalyst was produced *in situ* from $CuSO_4$ and sodium ascorbate. Voggu et al. (2007) showed the functionalization of CNTs with AuNPs through click chemistry. Alkyne-functionalized AuNPs are allowed to react with CNTs containing the azide functional group. Oxidized CNTs reacted with 4-azidobutylamine to form nanotube derivatives. Further, they reacted with 5-hexynethiol-capped gold nanocrystals in the presence of sodium ascorbate and $CuSO_4$. Su et al. (2013) prepared functionalized CNTs covalently grafted with thermoresponsive poly(N-isopropylacrylamide) (PNIPAM) via click chemistry. Reversible addition fragmentation chain-transfer (RAFT) was used for synthesis of azide-terminated poly(N-isopropylacrylamide) (N3-PNIPAM); further, N3-PNIPAM moiety was attached with CNTs via click chemistry. MWNT-PNIPAM shows excellent solubility in water, and, due to low grafting density, no temperature-dependent behavior was observed.

5.4.4 Quantum Dots

QDs are excellent materials used for imaging of cells and tissues under *in vitro* and *in vivo* experimental conditions. Bernardin et al. (2010) demonstrated the functionalization of QDs via copper-free click chemistry to bypass the inhibition of the luminescence properties of QDs by copper present as a catalyst in a click reaction. They developed highly luminescent conjugates via reaction between cyclooctyne functionalized QD and azido-biomolecules. Kotagiri et al. (2014) also developed antiepidermal growth factor receptor (EGFR) antibody-conjugated QDs via copper-free click reaction. In comparison to earlier methods, the F_c and F_{ab} regions of this conjugate retain its binding ability. In contrast, reports are also available for functionalization of QDs via copper-mediated click chemistry. Cheng et al. (2014) prepared silicon quantum dots (SiQDs) by reduction of halogenated silane precursor. Further, dialkyne molecules were used to functionalize SiQDs via ultraviolet (UV)-based hydrosilylation and then were further modified with azides by CuAAC click reaction. Functionalized QDs are monodisperse with a quantum yield of 5.2% and show excellent resistance to photo bleaching. Photo-stable QDs serve biomedical applications both *in vitro* and *in vivo*. Cheng et al. (2012) reported the preparation of colloidal silicon QDs, which can be functionalized with thiol-ene click chemistry to attach to some specific functional groups like $-NH_2$, $-COOH$, $-SO^{3-}$, alkane, and alkene. Allyl trichlorosilane acts both as surfactant and reactant, self-assembled to halogenated silane precursor (SiX_4 $X = Cl$, Br) in toluene. SiX_4 core and allyl trichlorosilanes contained in prepared reverse micelles and treated with

LiAlH$_4$ form alkene group-contained SiQDs. These particles can be functionalized with thiol-containing molecules via thiol-ene click reaction.

5.5 Amphiphilic Polymer Coating

Various methods like ligand exchange, surface chemicals, and covalent modification are reported for the functionalization and stability of NPs, but these techniques have several drawbacks: (i) one head group containing ligand binding on the NP surface can desorb easily and weaken the stabilization of particles, particularly in suspension media free of surplus unbound ligands, and (ii) even though thiol-containing ligands bind comparatively strongly to the particles, ligands should be carefully selected for core materials (Sperling and Parak 2010). Van der Waals forces and hydrophobic interaction are primarily responsible for adsorption of amphiphilic layer over the NPs, and it does not depend on the inorganic core material. Amphiphilic polymers and ligands have several contact points; thus, polymer molecules cannot be easily detached from the particles. The surface properties of coated particles do not depend on their core substance and have same chemical and physical properties. PAA-based polymers are commonly used as a functionalization agent for nanomaterials. It is an extensively charged, liner polyelectrolyte that could be modified by amide bond formation between aliphatic amines and its carboxylic group. Another class of amphiphilic polymers is Poly (maleic anhydride) copolymers, which can be synthesized by maleic anhydride and olefin copolymerization as an alternating copolymer. In comparison to PAA, PMA hydrophobic side chains are aligned in alternating layers in spite of a random arrangement, having a carboxylic acid functional group with higher density (Wang et al. 1988). Table 5.1 shows different polymers in general use to coat nanomaterials.

5.5.1 Gold NPs

Jans et al. (2010) demonstrated the synthesis of PAA-coated AuNPs via replacing citrate with PAA at the particle surface and their properties under various conditions. The optimal concentration for efficient coating of AuNPs with PAA without aggregation was found to be 0.025 mM PAA for 90 nM AuNPs. PAA are very sensitive to the presence of Na$^+$ ions in solutions. PAA-coated AuNPs aggregate in NaCl solution due to dehydration and shrinkage of the PAA layer over particles. PAA-coated AuNPs are shown to be temperature sensitive because of the PAA property of shrinkage on heating and expansion on cooling. Contrary to citrate-stabilized particles, PAA-coated particles are shown to be stable even at acidic pHs. Deng et al. have reported the molecular interaction and binding kinetics of PAA-coated gold NPs (PAA-GNPs) with human fibrinogen in a size-dependent manner (7–22 nm) (Deng et al. 2012). Large particles have a higher affinity and low dissociation rate for bound fibrinogen. Each fibrinogen molecule can bind to two 7 nm particles, but at a small size of 10 nm, only one particle is able to bind with fibrinogen. Particles greater than 12 nm in size bond with multiple fibrinogens in a positively cooperative manner, but particles shows aggregation when present in excess, as they can bond with more than one protein molecule. Protein binding on NP surfaces and within the protein corona is influenced by the size of particles. PAA-coated AuNPs are also reported to selectively interact with human plasma fibrinogen and induce unfolding (Deng et al. 2011). Gold

TABLE 5.1
Various Polymers Used in NPs Coating

Polymer	Ligand Addition Method	Core NP	Chemical Functional Group
PEG-phosphine (Thanha and Green 2010)	Ligand addition	Au	Multidentate-PO groups binds with Au, while hydrophobic tail makes soluble in organic solvents.
PVP (Graf et al. 2006)	Ligand exchange via hydrophobic interaction	CdSe/ZnS QDs, Au, Fe_2O_3	PVP provides stability to couple with amino-functionalized colloids.
Poly[2-(methacryloyloxy) ethyl phosphorylcholine]-*block*- (glycerol monomethacrylate) (Yuan et al. 2006)	Ferric and ferrous salt co-precipitate in block polymer	Fe_3O_4	PO_4^- and NMe_4+ act as cell membrane phospholipid head groups, while glycerol residue attached to NPs via 1,2-diol group.
PAA (Lin et al. 2005)	Co-precipitation in presence of polymer	γ-Fe_2O_3 and Fe_3O_4	—COOH provides pH solubility, adjustability, and stability in water at pH > 5.
Poly(amidoamine) (Lin et al. 2008)	Ligand addition in methanol	Au	Amine, amide groups. Dendrimer offers a steric framework.
PAA-Octylamine (Kairdolf et al. 2013)	Carbodiimide coupling in the presence of QDs	CdSe, CdS, ZnS QDs	—OH group on surface reduces nonspecific cellular binding
Poly(2-methoxyethyl methacrylate) (Gelbrich et al. 2006)	Surface activation and atom transfer radical	Fe_3O_4	—COC— and —COOC—
Poly(methacrylic acid)-dodecanethiol (Hussain et al. 2005)	Reduction of $CoCl_2$ and chloroauric acid in the presence of polymer	Co, Au	—COOH group provides solubility, stability, and morphology control.
Poly(glycolide)-poly(lactide)-poly(ethylene glycol) (Moore et al. 2014)	Ring opening polymerization of glycoloide	MWNTs	Isocyanate-terminated PEG reacts with terminal OH groups. PEG improved the stability of coated CNTs.

nanorod surfaces can be modified by PAA and utilized as a template for metallic NP (Pd, Au and Pt) formation, creating binary bimetallic configurations. This binary metallic NP system provides an enhanced surface area of each metal NP, which is stabilized on the nanorod surface as well as in close proximity (Hotchkiss et al. 2010). Brubaker et al. (2016) demonstrated the application of PAA-coated AuNPs as a pH sensor based on the electromagnetic properties of AuNPs and the pH response variation property of PAA. pH change in the surrounding environment results in swelling and deswelling of PAA polymer present over the particle surface, and the interaction of individual AuNPs leads to a shift in the optical properties. At low pH (pH 4), PAA shows a stretched conformation that leads to increased lossy mode resonance (LMR) peak intensity and higher LMR/local surface plasmon resonance (LSPR) peak ratio in comparison with higher pHs (water and pH 10), at which PAA possesses a coiled structure.

5.5.2 Iron Oxide NPs

PAA-coated magnetite nanoparticles (PAA-MNPs) synthesized via the thermal decomposition route are used in combination with the anti-TB drug rifampicin against *Mycobacterium smegmatis* as an efflux inhibitor, shown in Figure 5.7. PAA-MNPs attached to the bacterial

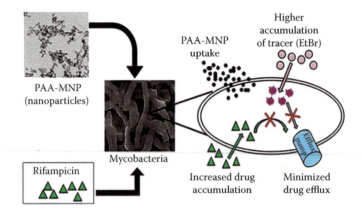

FIGURE 5.7
PAA-MNP enhanced antimicrobial effect of drug rifampicin against mycobacteria. (Reprinted with permission from Padwal, P. et al. 2014. Polyacrylic acid-coated iron oxide NPs for targeting drug resistance in mycobacteria. *Langmuir* 30(50):15266–76. Copyright 2014 American Chemical Society.)

cell surface can be efficiently internalized by cells. This combination showed ~3-fold high accumulation of the drug inside cells and thus exhibited a synergistic effect with ~4-fold higher inhibition of bacterial growth when compared to rifampicin alone (Padwal et al. 2014). This strategy is of great importance because mycobacterium shows resistance to the majority of drugs due to its active efflux pump, which leads to low concentration of drug retention inside cells (Loddenkemper et al. 2002). Thus, PAA-MNPs show a novel strategy to bypass the efflux of drugs and improve their efficacy (Padwal et al. 2014). PAA-coated IONPs are also reported to be excellent adsorbents for methylene blue and show a high saturation point of 830 mg/g. PAAs of molecular weight ~2100 g/mol were used to coat IONPs of 6–12 nm size. An external magnetic field may be used to separate pollutant/NP complexes in which iron oxide is present as a magnetic core (Fresnais et al. 2013). Another report by Zhou et al. (2013) concerned the efficient removal of methylene blue (MB) dye through PAA-coated IONP nanocomposites synthesized by the polyol-media solovothermal method. PAA-coated IONPs remove MB dye more readily in comparison to carbon granules and bare IONPs at alkaline pH (pH 9). Iversen et al. (2013) showed that PAA-coated IONPs can be prepared by polycondensation of metallic salts in aqueous media. Upon animal exposure, it was found that IONPs accumulate in endothelial and Kupffer cells of liver within 1 hr of exposure. Although the accumulation of IONPs does not alter the kidney function of healthy mice, it decreases the blood pressure temporarily. Recently, Couto et al. (2015) have shown the effect of bare and PAA-coated IONPs on the immune system to unravel the effect of IONPs on MAP kinase and the NF-κB signaling pathway. Cytokines, interferon gamma (IFN-γ), interleukin 1 beta (IL-1β), interleukin10 (IL-10), interleukin 6 (IL-6), interleukin 8 (IL-8), and tumor necrosis factor alpha (TNF-α) were found to be induced after exposure to both bare and PAA-coated IONPs. It was also reported that the activation of p38 mitogen-activated protein kinases (p38 MAPK), c-Jun N-terminal kinases (JNK), and transforming growth factor beta (TGF-β)-activated kinase (TAK1) takes place.

5.5.3 Carbon Nanotubes

MWNTs form a stable composite with PAA that is soluble and stable in aqueous media. Since PAA contains the reactive carboxyl moiety, PAA-MWNT composites can be used to attach biomacromolecules such as DNA, RNA, proteins, and enzymes and thus can be very efficiently used for the development of bioelectronic devices and biosensing applications. Additionally, water-suspended PAA-MWNT composite can also be used in gene and drug delivery systems as well as in other chemical and biological fields (Liu et al. 2006). Zhang et al. (2015) reported the preparation of PAA-coated CNTs by the *in situ* precipitation polymerization method. PAA grafting on the CNT surface can be efficiently controlled by adjusting the ratio of CNTs to monomer, and the maximum value of grafting is reported up to 30wt%. Chen et al. (2006) prepared water-soluble MWCNTs by *in situ* radiation-induced polymerization in a two-step γ-radiation process. During irradiation of MWNTs in ethanol, many reactive species ($\cdot CH_2CH_2OH$ and $\cdot CH(CH_3)OH$) are produced by the radiolysis of ethanol; these radicals interact with the MWNT surface via a $C=C$ bond, which results in the formation of hydroxyethylated MWNTs (HOEt-MWNTs). Further, MWNTs get irradiated in the presence of acrylic acid monomer and PAA covalently attached to the modified MWNT surface. As prepared, MWCNTs are highly soluble in water and in other polar solvents, further opening aspects for manipulation through other functional polymer groups. Huang et al. (2010) prepared PAA-coated MWCNTs by sonication and used them for modification of a screen-printed carbon electrode (SPCE) surface for determination of norepinephrine (NE), uric acid (UA), and ascorbic acid (AA) by cyclic voltammetry in their mixed solution. PAA shows affinity adsorption for UA and NE via hydrogen bonding and ion exchange mechanisms, respectively, but inhibits the adsorption of AA due to electrostatic repulsion, resulting in decreased oxidation potential of AA and enhanced oxidation peak of NE and UA at PAA-MWCNT. This property of PAA serves for better determination when compared to SPCE alone with resolved oxidation potential due to repulsive effects on AA. A pH-dependent study depicted that the oxidation of AA, NE, and UA at the PAA-MWCNT/SPCE involved two-electron transfer, and their interaction with PAA depends on their electrochemical behavior in different pH conditions. The lowest detection of NE, UA, and AA was found to be 0.131, 0.458, and 49.8 μM, respectively.

5.5.4 Quantum Dots

CdTe/CdSe QDs can be synthesized in the presence of PAA, with 40% modification with dodecyl amine. This strategy could lead to the synthesis of amphiphilic QDs that can be solubilized in water as well as in different organic solvents; a double layer of polymer is expected to be formed around the particle (Kairdolf et al. 2008). Celebi et al. (2007) prepared cadmium sulfide (CdS) QDs coated with PAA of molecular weight 2100 and 5000 g/mol. The highest photoluminescence (PL) intensity and quantum yield (QY) of ~17% was obtained at an optimum COOH/Cd ratio of 1.5–2 for molecular weight and pH (5.5 and 7.5). The luminescence intensity was found to be stable for up to 8 months. Acar et al. (2009) showed the effect of reaction of temperature and coating composition on properties (stability, particle size distribution, and luminescence intensity) of CdS QDs coated with either PAA, MAA, or a binary mixture of both. In comparison with CdS-MAA QDs, CdS PAA QDs were found to be less luminescent; however, they maintain their size and QY in high temperature conditions and exhibit excellent stability over a period of several months. A mixture of PAA/MAA coating provide QDs with high stability, QY, and tuning of size and color of emission properties. QDs coated with a PAA/MAA mixture (60:40 ratio) showed the highest quantum yield (50% of rhodamine B). Luccardini et al. (2006) reported the synthesis of alkyl-modified PAA-coated QDs and studied their colloidal properties in various buffers. They found that at neutral pH, these particles do not show binding with lipid molecules present in cell membrane, regardless of their lipid composition. Further, at a pH of ~6.0, these particles show abrupt aggregation on the lipid membrane, suggesting the tendency of the polymer to trigger critical binding. Additionally, Wang et al. (2013) evaluated the comparative effect of four coatings on QDs, carboxyl derivatized polymer (COOHP), polyethylene glycol functionalized polymer (PEGP), PAA-octylamine (PAA-OA), and linoleic acid (LA) and studied the effect of coating on retention and transport of QDs in porous media under biologically relevant conditions. It was also observed that the mobility of QDs was found to be dependent on the type of polymer coating and zeta potential of suspended QDs. The relative mobility of different coated QDs were found in the following sequence: LA > PAA-OA > COOHP > PEGP. QDs coated with carboxylic (−COOH) group containing polymers possessed higher mobility in comparison with nonionic polymer coating (PEG) under the same experimental conditions (ionic strength of 3 mM at pH 7). The mobility of PAA-OA coated particles can be increased by increasing pH and decreasing ionic strength, and correlated with their zeta potential. LA-coated QDs with the highest negative zeta potential (−54 mV) exhibited greater mobility.

5.6 Use of Polyethylene Glycol

PEG is a biocompatible polymer, and its structural simplicity and chemical stability offer several applications in biomedicines. PEG-coated (PEGylated) NPs are highly stable in biologically relevant conditions and their structure and properties are unaffected by high salt concentration (Daou et al. 2009). PEGylated NPs are less prone to nonspecific binding to proteins, which makes them successfully escape from the reticuloendothelial system (RES), thus providing more circulation time in the blood. At present, numerous carboxylate, thiol, or lipoamide terminal groups containing PEG molecules are available as a monofunctional, bifunctional, or heterogeneous bifunctional terminal groups (Bentzen et al. 2005). Using these molecules, several PEGylated NPs have been developed, and some of them are summarized below and in Table 5.2.

5.6.1 Gold NPs

A uniform PEG coating over AuNPs (PEG-AuNPs) can be prepared rapidly by microwave heating with an average diameter of 14.3 nm by tuning the microwave power in the initial phase of synthesis (Seol et al. 2013). PEG-coated AuNPs shows stable dispersions in phosphate buffer saline (PBS), water, PBS with bovine serum albumin (PBS-BSA), and PBS with dichloromethane (PBS-DCM). In comparison with citrate-capped AuNPs, which show immediate aggregation in salt, PEG coating provides stability in high salt concentrations ranging from 0.15–1 M. A linear relationship is reported between stability and capping density. A high PEG coating results in lower stability in PBS and PBS/BSA, in comparison with DCM and H_2O due to the NaCl effect. Comparable to a PEG surface density of ~1.13 chains/nm^2, a maximum 14wt% of PEG loading level was achieved, and the PEG chain acquired a random coil configuration. A lower loading level enhances the stability of AuNPs and allows their cofunctionalization with various specific targeted drugs as a drug carrier system (Manson et al. 2011). Yang et al. (2016) have shown the biological accumulation and safety of chronic exposure of AuNPs to mouse heart. They reported that small-sized PEG-AuNPs can be used for biomedical applications in cardiac disease. Bioaccumulation of AuNPs in the heart depends on its size, and 10 nm particle

TABLE 5.2
PEGylation of Various NPs

NPs	R1 (PEG Unit for Annealing to NPs Surface)	R2 (PEG Unit to Interface with Solvent)	Applications
AuNPs (Wuelfing et al. 1998)	Thiol	MeO (methoxy group)	Imaging
Silica QDs (Koole et al. 2008)	DSPE (lipid; 1,2-distearoyl-*sn*-glycero-3-phosphoethanolamine)	MeO	Imaging
Gold nanorod (von Maltzahn et al. 2009)	Thiol	MeO	Phototherapy
NaYF$_4$ (Boyer et al. 2010)	Phosphate	MeO	Imaging
Poly(hexadecyl cyanoacrylate) (Peracchia et al. 1999)	Cyanoacetate	MeO	Drug delivery
Fe$_3$O$_4$ (Xie et al. 2007)	Dopamine	COOH	Magnetic resonance contrast
Magnetic silica (Yoon et al. 2006)	Si(OCH3)$_3$	MeO	Imaging
AuNPs (Mei et al. 2008)	DHLA	MeO, NH$_2$, COOH	Imaging
Poly(lactic-co-glycolic acid)/PEG (Farokhzad et al. 2006)	NH$_2$	COOH	Drug delivery

accumulation induces reversible cardiac hypertrophy in mice in 2 weeks. Zhang et al. (2012) synthesized covalently modified PEG-AuNPs and demonstrated *in vivo* and *in vitro* radio-sensitization of PEG-AuNPs in a size-dependent manner (4.8, 12.1, 27.3, and 46.6 nm). In an *in-vitro* study, all particles showed reduction in cancer cell viability after gamma radiation, but particles of 12.1 and 27.3 nm size possessed dispersive distribution and a higher sensitization effect than 4.8- and 46.6-nm particles. *In-vivo* studies also showed that 12.1- and 27.3-nm particles showed more sensitization than 4.8- and 46.6-nm particles, which led to complete removal of tumors in mice. Rahme et al. (2013) demonstrated the synthesis of AuNPs of different sizes ranging from 15 to 170 nm by using hydroxyl amine as a reducing agent and thiolated PEG (mPEG-SH) (MWt. 2100–51,000 g mol^{-1}) as a stabilizing agent. It was found that due to the conformational entropy of the polymer and increased steric hindrance with increased PEG length, the grafting density of mPEG-SH reduced to 0.31 PEG nm^{-2} from 3.93 nm^{-2} when the MWt. of PEG increased from 2100–51,400 g/mol. The grafting density of polymer also depends on the diameter of the coating particles; when the mean diameter of the Au core increased from 15 to 115 nm, the grafting density decreased from 1.57 to 0.8 nm^{-2}, respectively, due to the fact that particles with small diameters provide a higher surface area for polymer loading. These particles are shown to be nontoxic toward CT-26 cells.

5.6.2 Iron Oxide NPs

PEG-coated IONPs (PEG-IONPs) are found to be stable at a broad pH range of 3–10, and their ionic strength reaches up to 0.3 M NaCl. They are also resistant to protein-mediated alteration in structure and function under physiological conditions. PEG-IONPs shows efficient MRI contrast characteristics in *in-vitro* studies and also possess the required T2 negative contrast effect. Thus, PEG-IONPs can be used in various biomedical fields like magnetic fluid hyperthermia and MRI (Yue-Jian et al. 2010). The most optimal length of PEG to minimize the protein adsorption over poly(lactic acid) NPs was reported as 5000 Da (Gref et al. 2000). IONPs coated with 20,000 Da PEG exhibit the lowest uptake in cells in comparison to low molecular weight PEG (333, 750, 2000, 5000, and 10,000 Da) (Larsen et al. 2012). Jimeno et al. have reported that by the coprecipitation method, magnetite-based ferrofluid was stabilized by using unmodified PEG. There is a dipole-cation binding interaction between the positively charged magnetite and the ether group. Ferrofluid maintained the same magnetic and colloidal property after a duration of 2 years (García-Jimeno and Estelrich 2012). Hoa et al. (2009) used a 1:2 molar ratio of $Fe^{2+}:Fe^{3+}$ for preparation of magnetic NPs by the coprecipitation method. Initially, particles were coated with sodium oleate as a primary layer followed by coating with PEG (MWt. 6000) as second layer. Particles depicted thermal stability due to the interaction between particles and their protective layer. Khoee and Kavand 2014 demonstrated the coating of IONPs with methoxy-PEG (mPEG) as a hydrophilic shell. First, mPEG was acrylated and the Michael addition reaction took place between acrylated PEG and 3-aminopropy; triethoxysilane served as a coupling agent to form triethoxysilyl-terminated poly(ethylene glycol)[$(EtO)_3$-Si-mPEG]. Further, the formed complex was grafted to the IONP surface by a condensation reaction between [$(EtO)_3$-Si-mPEG and the hydroxyl group on the surface of the particle (sol-gel reaction). Mukhopadhyay et al. (2012) reported the interaction of PEG-coated and bare IONPs with cytochrome C. Bare IONPs show an interaction with cytochrome C protein, which leads to a reduction of protein, whereas PEG-IONPs do not show such an effect with protein.

5.6.3 Carbon Nanotubes

PEG coating on carbon nanotubes (PEG-CNTs) can be done by both covalent and noncovalent functionalization strategies (Ravelli et al. 2013). Covalent attachment of PEG on CNTs can be achieved by following reported methods: (i) using the cycloaddition reaction {[2 + 1], [3 + 2] and [4 + 2]}, in which different-sized rings are directly attached with CNTs; that is, the CNT surface can be directly fused with three-membered rings [2 + 1] in the presence of extremely reactive nitrenes (Nie et al. 2010) or acyl nitrenes (Jiang et al. 2011) produced due to the heating of azides and azido carbonates at 130–160°C. (ii) Using amidation or esterification of the carboxylic acid functional group at the nanotube surface with the amino or -OH group of PEG, respectively; that is, first, the –COOH group is inserted on the surface of nanotubes in the presence of strong oxidizing agent HNO_3 (with H_2SO_4). Further, CNTs are allowed to react with PEG-OH or PEG-NH_2 to obtain PEGylated CNTs. The last reaction takes place by classic Fischer esterification (by refluxing an acidic, for example, HCl, alcoholic solution of CNT–COOH and the PEG derivatives) (Kim et al. 2010) or by use of a suitable condensing agent (e.g., a carbodiimide, R–NLCLN–R, such as DCC, EDAC, and EDC) (Heister et al. 2012). (iii) By addition of PEG chain via polymerization reaction to prefunctionalized CNTs by anionic (by $Ph_3C^-K^+$) polymerization of ethylene oxide (Sakellariou et al. 2008). (iv) Using the addition of PEG-containing radicals, which can be achieved by using a photocatalyst that can absorb light and activate the PEG chain by hemolytic C-H cleavage to manage alpha-oxy radicals. It is further captured by nanotubes, and PEGylated derivatives are obtained (Georgakilas et al. 2002). In comparison to chemical functionalization, the noncovalent functionalization strategy has advantages: an accurate graphitic structure of CNTs can be preserved, and it can be operated in mild conditions. Noncovalent functionalization depends on supramolecular complexion employing several adsorption forces like van der Waals, electrostatic force, hydrogen bonds, and π-stacking interaction (Meng et al. 2009). PEG-SWNTs show discrete behavior in pharmacokinetic and biodistribution studies and have higher blood circulation time, which leads to a higher uptake of PEG-SWNTs in tumor cells (Yang et al. 2008). In comparison with oxidized

SWCNTs, PEG-SWNTs are highly aqueous-dispersible and become more hydrophilic. PEG does not interfere with the drug delivery property of SWNTs, increases the drug delivery potential, and facilitates its slow excretion *in vivo*. Bhirde et al. synthesized more hydrophilic SWCNTs and demonstrated that PEG-SWNTs can be used as a therapeutic agent for targeted drug delivery and can successfully inhibit tumor progression in mice (Bhirde et al. 2010). PEG wrapping over SWCNTs is not 100%, and Jung et al. reported through thermogravimetric analysis that PEG is covalently wrapped around the SWCNT surface with approximately 18% of the weight portion (Jung et al. 2004).

5.6.4 Quantum Dots

To utilize the fluorescent property of QDs and overcome its toxic effect due to the release of cadmium ions, Poulose et al. (2012) prepared PEG-functionalized cadmium chalcogenides based on three differently fluorescent QDs (CdS, CdSe, and CdTe) for cancer cell detection by the folate-targeting strategy. These particles are internalized into cells via folate-mediated endocytosis and thus can be efficiently used in an *in vivo* model for tumor detection. Lv et al. (2013) reported preparation of QDs encapsulated in PEGylated chitosan derivatives (mPEG-poly(ethylene glycol) monomethyl ether; OPEGC, N-octyl-N-mPEG-chitosan), formed via a Schiff base reduction reaction between chitosan and mPEG-aldehyde. Chitosan serves as a base of grafted copolymers and mPEG aldehyde acts as a hydrophilic chain. As prepared, QDs are uniformly dispersed in water and exhibit a low toxic effect and high quantum yield. Ulusoy et al. (2016) synthesized CdTe/CdS/ZnS QDs functionalized with 3-mercaptopropionic acid (MPA) in the presence of surface coordinating agent thiol-terminated mPEG molecules. First, CdTe/CdS QDs with a small core and thick shell were synthesized with emissions at the 650–660 nm wavelength. Further, the outer ZnS shell was synthesized from this crude solution in the presence of MPA (as a sulfur source and coordinating agent) and mPEGthiol (as a co-coordinating agent). The ratio of Zn/mPEG thiol in the reaction system was maintained at 1:5 and 1:10. The presence of mPEG in the solution leads to a reduced deposition rate of the ZnS shell and maintains its efficiency and monodispersity. The colloidal stability of QDs increased with respect to the grafting density of PEG on their surface, and QDs having a higher mPEG density at the surface (1:10) possessed a distinguished cellular binding feature with 2D and 3D tumor-like spheroid structures. Dif et al. (2009) developed QDs coated with tricysteine, PEG, and aspartic acid ligand, which can bind with the poly-histidine tag protein *in vitro* and in living cells.

5.7 Surface Modification with Biomolecules

For development of efficient nanoprobes, conjugation chemistry between NPs and biomolecules needs to be considered essentially. Biomolecules should be bound to NP surfaces very firmly via covalent or coordinate bonds. NP functionalization with biomolecules requires some additional precautions in terms of preserving the functionality and three-dimensional molecular structure as well as the orientations of biomolecules in aqueous and buffered media (Oh et al. 2015). Various biomolecules used for functionalization of nanomaterials are shown in Table 5.3. A schematic diagram for surface modification by biomolecules is shown in Figure 5.8.

TABLE 5.3
Different Biomolecule-Coated NPs

Nanomaterials	Interface/Linking Agent	Biomolecules for Functionalization	Applications
Starch AuNPs (Veerapandian and Yun 2011)	Thiotic acid	Annexin V, Bombesin	Apoptosis and cellular interaction.
AuNPs-NH$_2$ (Norman et al. 2008; Veerapandian and Yun 2011)	EDC	Anti-PA3	Antibacterial (*P. aeruginosa*)
MWNT-COOH (Pangule et al. 2010)	1-Ethyl-3-(3-dimethylaminopropyl) carbodiimide/N-hydroxysuccinimide	Lysostaphin and ALE1	Bactericide (*S. aureus* and *S. epidermidis*)
AgNPs (An et al. 2009)	Polyethylene oxide	Chitosan fiber	Bacteriostatic agent (*E. coli*)
CdTe QDs (Lu et al. 2014)	Amine	PEG/folate	Cancer cell imaging
CdSe/ZnS (Hushiarian et al. 2014)	MPA	DNA	Biosensor
MWNTs-COOH (Gómez et al. 2005)	Electrostatic interaction	β-glucosidase	Bio catalysis
MWNTs/chitosan/ionic liquid-modified GCE (Zhang and Zheng 2008)	–	Cytochrome C	H$_2$O$_2$ detector
Fe$_3$O$_4$ (He et al. 2006)	–	Human serum albumin	MRI
IONPs (Huang et al. 2013)	–	Casein	MRI

5.7.1 Gold NPs

Thiols can form covalent bonds to various metal NPs such as AuNPs, AgNPs, and copper nanoparticles (CuNPs). Covalent bonds can be used to attach cell-targeting biomolecules to AuNPs via the sulfur-gold bond between cysteine-terminated molecules and AuNPs and can be employed in various biomedical applications (Tkachenko et al. 2005). The binding of the thiols group to AuNPs is a simple strategy and can be done in an aqueous solution at room temperature (Templeton et al. 2000). An approximate molar ratio of 2500 thiol molecules per NP should be maintained for 20-nm particles. For monolayer attachment of thiols, this ratio can be varied according to the core size of particles. The availability of adsorption site has a direct correlation with the total surface area and also with the square of the particle diameter (Tkachenko et al. 2005). AuNPs can be conjugated with

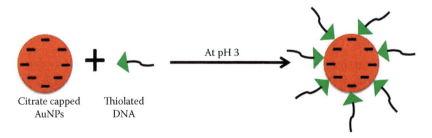

FIGURE 5.8
Schematic diagram showing the process of surface functionalization of NPs with DNA using DNA thiolation method.

Surface Modification of Nanomaterials for Biomedical Applications

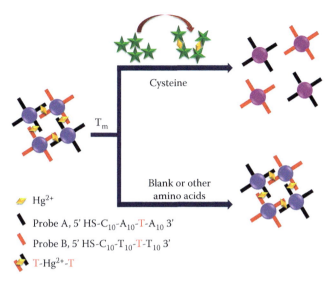

FIGURE 5.9
Schematic diagram for detection of cysteine by using oligonucleotide-coated AuNPs in presence of Hg^{2+}.

antibodies by means of physical and chemical interactions, which depend on the following different phenomena: (i) electrostatic attraction between positively charged antibodies and negatively charged AuNPs, and (ii) hydrophobic attraction between their surfaces. Chemical interaction can be done by (i) chemisorption by thiol derivatives, (ii) bifunctional linkers, (iii) adapter molecules such as biotin and streptavidin (Jazayeri et al. 2016). Genetic biosensors have also been developed by interaction between modified colloidal AuNPs and DNA. Oligonucleotides consist of a thiol group at one end that can bind to the AuNP surface via covalent binding. AuNPs of ~13 nm are known for changing from red (stable) to blue (aggregated) color when assembled with a thiolated DNA probe. This characteristic red-to-blue color change can be used for quantification of wide range of biomolecules (Wang, Li et al. 2009). Lee et al. developed a rapid and specific technique for colorimetric detection of cysteine (100 nM) by using AuNP-oligonucleotide conjugates. This assay depends on the thiophilicity of Hg^{2+}, the selectivity of the thymidine-thymidine (T-T) mismatch for Hg^{2+}, its unique optical properties, and the sharp melting properties of DNA-Au NPs in a competition assay format (Figure 5.9) (Lee et al. 2008). Sandstorm et al. demonstrated that despite their negative charge, single- and double-stranded DNA can strongly bind to AuNPs also via non-thiol-mediated (nonspecific) binding. The mechanism behind this is suggested to be ion-induced dipole dispersive interactions, where a dipole is induced on highly polarizable AuNPs by a negatively charged phosphate group. The immobilization type decides the geometry of the bound strand; thiol-attached oligonucleotides are more extended in comparison to nonspecifically bound molecules (Sandström et al. 2003). Sometimes, peptides can be initially bound with BSA via bifunctional cross-linker MBS and then attached to AuNPs through electrostatic interactions. This process makes the peptides more soluble and leads to sol stability (Lanford et al. 1986, Tkachenko et al. 2005).

5.7.2 Iron Oxide NPs

Different biological molecules like antibodies (Nam et al. 2004) and proteins (He et al. 2006) tend to bind with the surface of IONPs to make them target specific and biocompatible.

These molecules can attach to the IONP surface directly or indirectly by chemical interaction via some specific functional terminal group (Wu et al. 2008). Zhang et al. demonstrated the synthesis of human serum albumin (HSA)-coated IONPs of diameter ~200 nm by the microemulsion method as a radioisotope carrier labeled ^{188}Re. These particles show excellent stability in bovine serum albumin (BSA) up to 72 hrs and can be employed as *in vivo* magnetically targeted therapy (Chunfu et al. 2004). Magnetic particles of average diameter 8.8 nm with a carboxylic group attached on the surface were prepared by Lee et al. (2006) in which 1-ethyl-3-(3-dimrthylaminopropyl) carbodiimide hydrochloride (EDC) was used as a linker agent to immobilize the protein streptavidin on the particle surface. This complex can be used for separation of biotin-functionalized molecules from solution by utilizing the strong affinity between biotin and streptavidin. Recently, Schwaminger et al. (Schwaminger et al. 2015) studied the interaction of seven different amino acids (L-alanine, L-glutamic acid, L-cysteine, glycine, L-serine, Lysine, and L-histidine) with magnetite NPs in a solution of pH 6. These amino acids bind to NPs through different mechanisms depending on their side chains. Histidine and serine were assumed to bind to magnetite NPs through their imidazole and hydroxyl groups, respectively, whereas glutamic acid can utilize the alpha or side-chain carboxylic group for binding purposes. The order of adsorption capacity of amino acids was found as Cys > Glu > Ser ~ His ~ Gly > Lys > Ala. The dependence of the binding behavior of amino acids on their side chains can be utilized to predict the binding of other larger molecules like protein (Xia et al. 2011). Adsorption of amino acids also depends on their polarity (Gao et al. 2008). The carboxylic group is considered a key factor for binding of amino acids to particles through unipolar interaction occurring on the surface of IONPs. Instead of the thiol group, there is bond formation of S-S bonds to IONPs. In addition to the carboxyl and amino groups, adsorption of biomolecules is governed by the redox behavior of Fe^{3+} and Fe^{2+} and the reactivity of IONPs. Hydrogen bond formation and ionic interactions are majorly influenced by the surface charge of IONPs (Schwaminger et al. 2015). Lee and Lee. (Lee and Lee 2008) reported the synthesis of superparamagnetic NPs immobilized with Ni^{2+}, which provide affinity toward histidine-tagged proteins. These proteins can be effectively isolated from the solution by applying the appropriate magnetic field.

5.7.3 Carbon Nanotubes

CNTs are modified to find potential application in the biomedical field, such as biosensing and drug delivery (Nagaraju et al. 2015). Oxidation is considered an initial approach to functionalize CNTs, which was done by using alkali and acid through photo oxidation, plasma, or gas phase treatment (Datsyuk et al. 2008). Exfoliation (during the oxidation process) due to the attachment of oxygen-carrying groups (hydroxyl, carboxyl) on the CNT surface makes CNTs soluble in aqueous and organic solvents (Balasubramanian and Burghard 2005). After oxidation, other functional groups can also be attached to the CNT surface by different processes like amidation, esterification, and so on. Treatment with acid creates a hole on the surface of CNTs, which leads to attachment of different functional groups on its surface. This process make nanotubes hydrophilic and can be easily solubilized in different solvents and buffers with noncytotoxicity to mammalian cells (Carrero-Sanchez et al. 2006). Oxidized CNTs can absorb various proteins and bind nonspecifically to their sidewalls, and act as transporters to deliver protein inside mammalian cells with unaltered biological activity (Kam and Dai 2005). CNTs can also be modified with covalent functionalization via the diimide-activated amidation process through direct coupling of carboxylic acid to proteins by applying a coupling agent, that is, N-ethyl-N-(3-dimethyl-aminopropyl) carbodiimide hydrochloride (EDAC) or N, N'-dicyclohexyl carbodiimide (DCC). Second, CNTs can also be functionalized by grafting (covalently attaching) polyetherketones on their surface in polyphosphoric acid media (Nagaraju et al. 2015). Proteins can covalently attach to CNTs via 1-ethyl-3-(3-dimethylaminopropyl) carbodiimide as a cross-linker, but interaction between protein and CNTs can also be due to absorption. Small molecules like peptides are known to be conjugated to CNTs (Gao and Kyratzis 2008). Sometimes noncovalent methods are preferred for functionalization because they do not affect the tube network (Zhang et al. 2007). Macroscopic materials can be assembled to nanotubes through π–π interactions (Zhu et al. 2004). CNTs exposed to the vapor of materials (to be functionalized) also cause noncovalent functionalization. This was further stabilized by chemical stabilizer and by exposure to other material to protect functionalized materials (Zhang et al. 2007). CNTs show interaction with the human serum proteins, with different adsorption capacities. Binding ability was mainly influenced by the π–π stacking interactions between aromatic amino acids and CNTs. Protein-coated CNTs show altered cellular interactions and also decrease in their cytotoxic behavior depending on their adsorption capacity. This would lead to the design of the safest NPs for biomedical applications (Ge et al. 2011).

5.7.4 Quantum Dots

QDs can be solubilized in water by attaching different ligands on the surface for further attachment of different biomolecules (Figure 5.10) (Zrazhevskiy et al. 2010) by providing a terminal functional group (usually –NH_2, –OH, –SH, or –COOH). Carboxyl-terminated functional groups are generally known to form amide bonds with free amines present in proteins, peptides, and antibodies (Pereira and Lai 2008). QDs are also found to attach to thiol/amine groups of peptides and proteins. Sulfo-SMCC (succinimide 4-(N-maleimidomethyl) cyclohexane-1-carboxylate) serves as both the spacer and linker molecule, which leads to increases in the overall size of the QDs. Derfus et al. (2007) demonstrated the delivery of siRNA and tumor-homing peptides (F3) to tumor cells via conjugation with PEG-coated QDs as a scaffold. Schumacher et al. (2009) showed the conjugation of synthetic peptides and R-phycoerythrin with QDs for the detection of *Bacillus anthracis* spores. Tiwari et al. (2009) synthesized QDs conjugated with three different types of anti-HER2 antibody through different coupling agents [EDC/sulfo-NHS(1-ethyl-3-(3-dimethyl-aminopropyl) carbodiimide hydrochloride/3-sulfo-N-hydroxysuccinimide sodium salt), iminothiolane/sulfo-SMCC, and sulfo-SMCC]. Coated QDs were conjugated with anti-HER2 antibody for imaging of breast cancer cells overexpressing HER2.

5.8 Use of Block Copolymers

Block copolymers are amphiphilic and used for functionalization of NPs that contain a hydrophobic and hydrophilic part. In contrast to dispersive solvents, these polymers have their hydrophilic or hydrophobic sides inside and form a micellar structure (Figure 5.11) (Sperling and Parak 2010). This structure has found a potential application for synthesis of nanoparticles (Möller et al. 1996), for its coating (Berret et al. 2006), and for phase transfer from one solvent to another. Some block copolymers are laterally cross-linked (Cheng et al. 2008), and their thickness over the NPs can be controlled by choosing a polymer of proper block length (Kang and Taton 2005).

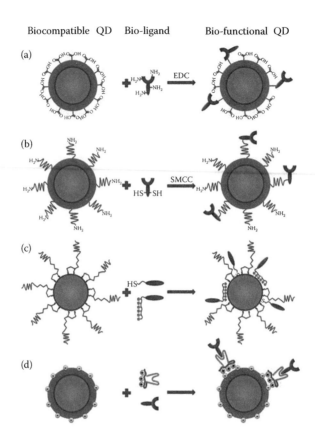

FIGURE 5.10
Routes for QD biofunctionalization. Decoration of QD surface with bioligands can be achieved via (a,b) covalent conjugation, (c) noncovalent coordination of thiol groups or poly histidine tags with the QD surface metal atoms, or (d) electrostatic deposition of charged molecules on the QD organic shell. (Zrazhevskiy, P. et al. 2010. Designing multifunctional quantum dots for bioimaging, detection, and drug delivery. *Chemical Society Reviews* 39(11):4326–54. Reproduced by permission of The Royal Society of Chemistry.)

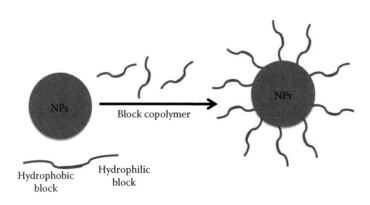

FIGURE 5.11
Schematic diagram showing synthesis of block copolymer-modified NPs.

5.8.1 Gold NPs

Chen et al. (Chen et al. 2014) demonstrated the formation of AuNPs coated with thiol reactive block copolymer poly(ethylene oxide)-block-poly(pyridyldisulfide ethylmethacrylate) (PEO-b-PPDSM). This copolymer forms a neutral surface by creating a coat over dispersed AuNPs, making their surface ideal for biomedical applications through functionalization. This formulation is very stable in different physiological environments and can be used for carrying hydrophobic drugs like doxorubicin. Due to the presence of multiple pyridyldisulfide groups on the PPDSM block, these uniformly dispersed nanocomplexes can show direct functionalization by disulfide-thiol exchange chemistry. Housni et al. (Housni and Zhao 2010) prepared diblock copolymer-coated AuNPs that can show either upper critical solution temperature (UCST) or lower critical solution temperature (LCST) depending on the surrounding temperature without aggregation in particles. It can be achieved by coating AuNPs with a diblock copolymer poly(ethylene oxide) (PEO) and poly-(N,N-dimethylaminoethyl methacrylate) (PEO-b-PDMAEMA), and after reaction with 1,3-propane sulfone quaternized PEO-b-PDMAEMA, chains were formed on the AuNP surface. PEO-b-PDMAEMA-coated AuNPs were heated and showed LCST-type solubility, whereas a quaternized PDMAEMA block demonstrated a UCST-type solubility change. In both conditions, a water-soluble PEO block protected the AuNPs from aggregation even when polymer chains on the AuNPs became dehydrated at temperature > LCST or temperature < UCST. Thus, polymer-coated AuNPs can go through hydration/dehydration transition upon an increase/decrease in the temperature of the solution.

5.8.2 Iron Oxide NPs

Poly(hydroyethyl acrylate-b-N-isopropylacrylamide) block copolymer-coated iron oxide NPs [Fe_3O_4@P(HEA-b-NIPAm)] are prepared by consecutive surface-initiated atom transfer radical polymerization (SI-ATRP) from NP surfaces. As prepared, Fe_3O_4 NPs are temperature sensitive and thus are used for controlled drug delivery. The loading capacity of NPs for drugs was modified by an inner PHEA block and the temperature was controlled by the PNIPA block, which acted as a controller for release of the encapsulated drug (Mu et al. 2011). Colloidal ultrasmall superparamagnetic iron oxide nanoparticles (USNPs) proved to be an excellent, promising MRI contrast agent for vascular MRI diagnosis when coated with mussel-inspired multidentate block copolymer. Multidentate block copolymer with catechol groups (Cat-MDBC) is prepared by controlled radical polymerization and postmodification methods. Cat-MDBC possesses an excellent affinity for the USNP surface and makes the surface hydrophilic. These particles are stable over a wide pH range, and no significant adsorption of proteins is observed on their surface. They possess excellent magnetic properties and *in vitro* MRI (Li et al. 2015). Hemmati et al. (2016) demonstrated the synthesis of nanocarriers for drug delivery by using a diblock or triblock copolymer of PEG and poly(ε-caprolactone) (PCL)-coated IONPs. Here, the surface of particles was first modified with PAA and propargyl alcohol (MNP–PAA–C≡CH), and, further, a click reaction was employed between azide-terminated block copolymers and particles to synthesize a magnetic nanocarrier, as shown in Figure 5.12. Li et al. (2016) reported the synthesis of stabilized extremely small iron oxide nanoparticles (ESNPs) with a mussel-inspired multidentate block copolymer strategy, with pendant anchoring catechol groups (Cat-MDBC). Cat-MDBC consists of an anchoring block containing 29 pendant catechol groups and a hydrophilic block with 27 pendant oligo(ethylene oxide)moieties. The biphasic ligand exchange process was used to fabricate ESNPs with stabilized Cat-MDBC in ratio of 1/5 wt/wt in aqueous solution (aq. Cat-MDBC/ESNP colloids) by sonication.

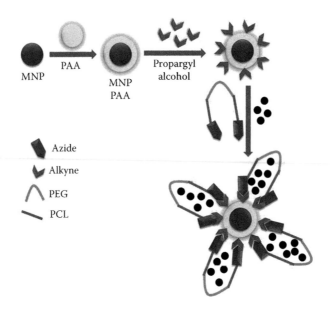

FIGURE 5.12
Schematic diagram for synthesis of MNP–PAA@PCL–PEG–PCL loaded with drug.

5.8.3 Carbon Nanotubes

Hydrophobic and hydrophilic polymer-coated CNTs are known to decrease in toxicity and have high therapeutic efficacy. Biocompatible block copolymer poly(lactide)-poly(ethylene glycol) (PLA-PEG)-coated CNTs exhibit less toxicity both *in vitro* and *in vivo* and prevent aggregation of particles. The PEG layer improves CNT solubility and pharmacokinetic properties, whereas the PLA layer encapsulates the particles and leads to controlled hydrophobic drug release. Polymer-coated CNTs showed less aggregation in tissues and thus were tolerated in tested animal models at a maximum dose of approximately 50 mg kg^{-1}, in comparison with pristine CNTs, which can only be tolerated at 25 mg kg^{-1} (Moore et al. 2013). Han et al. (2016) demonstrated triblock functionalized carbon nanotubes by using thermosensitive poly(ethylene oxide) (PEO) and poly(propylene oxide) (PPO) as PEO-PPO-PEO in aqueous solution. These ampiphilic blocks exhibited intermolecular interactions between their molecules, were adsorbd on CNT surfaces, and formed a continuous layer. These CNTs were characterized by small-angle neutron scattering experiments and model fit analyses, which provide an efficient method for synthesis and characterization of carbon nanotubes with structure-tunable encapsulation.

5.8.4 Quantum Dots

Jia et al. (2012) demonstrated the incorporation of QDs into micelles to make a unique microreactor. Glucose oxidase (GOX) and horseradish peroxidase enzymes were labeled with fluorescent dye, where FRET occurs between QDs and labeled enzymes present in the system. This study provides a basic platform for multienzyme colocalization and an efficient approach for characterization of multienzyme immobilization and colocalization. In this context, Ku et al. (2013) developed microspheres containing a multicolored hybrid of block copolymer and QDs through controlled location of different-colored QDs in micelles to manage FRET efficiency among different-colored QDs. Further, Park et al.

(2015) developed block copolymer-integrated graphene quantum dots (bcp-GQDs) as colorimetric multifunctional sensors for detection of temperature, pH, and different metal ions. Blue-emitting, thermally responsive poly(7-(4-(acryloyloxy)butoxy)coumarin)-b-poly(N-isopropylacrylamide) (P7AC-b-PNIPAAm) polymers were grafted on 10-nm green color-emitting QDs. A relative amount of P7AC-b-PNIPAAm blue emission and green emission from the GQD core governs its colorimetric multisensing behavior, which strongly depends on different stimuli. Temperature response was produced by adjusting blue and green emitter distance due to FRET, whereas the change in pH and different metal ions were detected by luminescence of green-emitting QDs. Chae et al. (2014) demonstrated the dual functionalization of block copolymers by incorporating two simultaneous functionalities into the nanostructure by two block copolymer blendings possessing different functional moieties but similar chemical structure. They synthesized block copolymers with reactive blocks, with activated esters to attach thiol and fluorophore functional groups; further, QDs were incorporated in this lamellar nanostructure. Poly(pentafluorophenylmethacrylate) (PPFPMA) with an activated ester was used as a reactive block (which can react with various amines to form amide bonds) (Eberhardt et al. 2005), whereas Poly(methylmethacrylate) (PMMA) was used as second block, synthesized by RAFT to be optically clear.

5.9 Lipid Coating over NP Surfaces

Liposomes have found a potential application in the field of biomedicine as passive as well as targeted drug delivery systems, with an increase in the efficacy and concomitant decrease in the adverse effect of delivered drugs (Muthu and Feng 2010). Liposomes mimic the simplest artificial biological cells, as they are composed of natural lipids such as cholesterol and phospholipids. Water-soluble and water-insoluble drugs can be encapsulated with the same ease in either the hydrophilic or hydrophobic core of liposomes, respectively (Figure 5.13). Thermosensitive liposomes are also reported, which act via external thermo-stimulated content release; however, photosensitive liposomes function to release drugs after achieving lower critical solution temperature (Kang and Ko 2015).

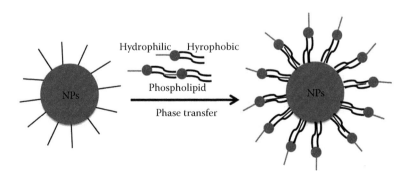

FIGURE 5.13
Schematic diagram showing the orientation of phospholipids coating on NP surface during phase transfer process.

5.9.1 Gold NPs

Kang et al. (Kang and Ko 2015) demonstrated the incorporation of drug docetaxel (DTX) into phospholipid-coated AuNPs to increase the efficacy of delivered drugs. Both formulations, DTX-loaded and lipid-coated nanohybrids, are synthesized by the thin-film formation, hydration, and sonication method. Docetaxel-loaded anionic lipid-coated gold nanorods (AL-AuNR-DTX) and cationic lipid-coated gold nanoparticle (CL-AuNP-DTX) show an increase in internalization and increase in their toxicity when compared to uncoated AuNRs, AuNPs, and DTX. Between the two formulations, CL-AuNP-DTX ensures better encapsulation of drugs and persistent release of drugs in physiological conditions. As formulated, the nanocarrier systems can be used as efficient cancer chemotherapeutic agents. Wang et al. (Wang and Petersen 2015) reported that AuNPs can be coated with lipids during synthesis; thus, their size and shape can be tuned by reaction conditions. Fluorescent probe-grafted lipids can be used to track AuNPs in a cellular system. The lipid coating serves as a platform for further surface modification for targeted drug delivery systems. Kong et al. (2012) reported the synthesis of cationic lipid-coated AuNPs (L-AuNPs) by the emulsification/solvent evaporation method as an efficient therapeutic siRNA delivery vehicle. AuNPs and three lipid components were constituted of {3β-[N-(N′,N′-dimethylaminoethane)-carbamoyl]-cholesterol (DC-Chol, 54% w/w), cholesterol (19% w/w), and L-α-dioleoyl phosphatidylethanolamine (DOPE, 27% w/w)}. DC-chol (chloroform analogue) possesses a tertiary amine group and provides a cationic charge to the L-AuNP surface, which leads to its interaction with anionic siRNA molecules to form a polyelectrolyte complex. Bhattacharya et al. (Bhattacharya and Srivastava 2003) synthesized AuNPs with cationic single-chain, double-chain, and cholesterol-based amphiphilic units. Various other cationic lipids attach to the AuNP core through their thiol groups. Single-chain doped NPs cause phase formation with a high melting point, whereas NPs with double-chain doping interact and thus decrease the melting point of 1, 2 dipalmitoylphosphatidylcholine (DPPC) vesicles. Cationic cholesteryl NPs interact with the DPPC membrane and cause DPPC melting transition. Cai et al. (Cai 2015) reported the use of phosphatidylcholine-coated AuNPs (PC-AuNPs) for measurement of lipoxygenase activity. PC-AuNPs can detect the change in the refractive index due to lipoxygenase and are also able to differentiate between lipoxygenase inhibitors having different inhibition mechanisms.

5.9.2 Iron Oxide NPs

Lipid-coated IONPs are prepared by surface immobilization with an equimolar mixture of DPPC and L-α-dipalmitoylphosphatidyl glycerol (DPPG). Coating through avidin-biotin chemistry offers high affinity interaction and provides colloidal stability to particles in different buffer ions. As a result of negative surface charge in citrate buffer (pH 7.4), particles are in an extensively dispersed state. In contrast to uncoated particles, the thermal behavior of lipid-coated particles does not depend on suspension but is influenced by the particle concentration in an alternating magnetic field. Thus, a lipid coating provides colloidal stability without altering the hyperthermia properties (Allam et al. 2013). Jiang et al. (2013) developed a system composed of IONPs and cationic lipid-like materials, also termed lipoids, for efficient delivery of nucleic acid, as shown in Figure 5.14. IONPs and lipoids along with oleic acid are dispersed in chloroform, followed by an N-methyl-2-pyrrolidone (NMP) addition, to encourage adhesion among lipid and NP surfaces. To make these particles compatible in aqueous solution, chloroform is evaporated, and the sample was

Surface Modification of Nanomaterials for Biomedical Applications

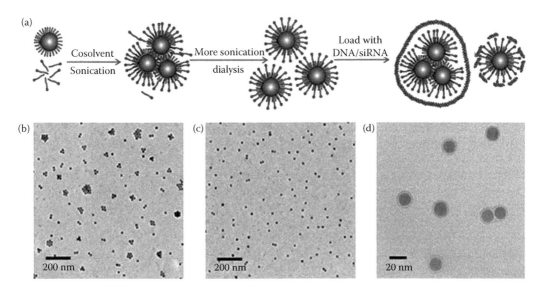

FIGURE 5.14
(a) Schematic plot of the procedure of coating iron oxide NPs; transmission electron microscopy (TEM) images of (b) particle clusters; (c) individual particles; (d) coating on the NP surface. (Reprinted with permission from Jiang, S. et al. 2013. *Nano Lett* 13(3):1059–64. Copyright 2013 American Chemical Society.)

dialyzed to remove extra lipid from the solution. Finally, the positively charged NP surfaces are attached to negatively charged DNA and siRNA through electrostatic interaction. These nucleic acids does not present on the particles; rather, they are encapsulated inside the lipid coating to ensure protection from enzymatic degradation. The entrapment of DNA depends on the size of NPs; for small particles of ~40 nm, delivery efficiency decreases up to 34%, whereas binding of siRNA is not affected by the size of particles and remains 90% in all cases.

5.9.3 Carbon Nanotubes

Wu et al. (2006) proposed a "half-cylinder" model for binding lipids with a large dipole moment and conical geometry to SWNTs. Coating with lysophospholipids or a single-chain lipid leads to enhanced solubility and hence stability of SWNTs, whereas double-chained phospholipids did not show any effect on the solubility of SWNTs. Additionally, binding of lysophospholipids to SWNTs depends on the polarity of the lipids. SWNTs are wrapped by lysophospholipids when they are in the lipid phase and not in the vacuum phase, illustrating that the environment serves an important role in their binding. Phospholipid-coated MWNTs are used as electrochemical labels in immunoassay techniques. Self-assembled monolayers of two phospholipids such as phosphocholine (PC) and phosphoethanolamine (PE) are used for coating of MWNTs, which can be achieved by van der Waals and hydrophobic interactions among phospholipid alkyl side chains and MWNT sidewalls. Further, this complex is conjugated with monoclonal antibodies via amine terminal. Taken together, this strategy finds a potential application in the field of clinical diagnostics and proteomic research (Nie et al. 2009). Roberts et al. (Roberts et al. 2007) reported that lipid-coated CNTs can easily be ingested by daphnia as normal food and utilize lysophophatidylcholine coating as a food source.

5.9.4 Quantum Dots

Phospholipid encapsulation is beneficial for QDs in several aspects: (i) the optical properties of QDs are not altered by the type of coating; (ii) the surface of QDs is not altered due to the binding of lipids; and (iii) phospholipid coating prevents nonspecific adsorption at the QD surface, which proves to be one of the major problem in *in-vivo* drug delivery (Karakoti et al. 2015). Aime et al. (2013) demonstrated the functionalization of QDs by lipid oligonucleotide conjugates (LON-QDs) targeting microRNA. LON is attached at the QD surface through hydrophobic moieties with an amphiphilic layer of QDs. Thermodynamics and kinetics of the duplex formed at surface of QDs remains the same with respect to the duplex formed in solution or at the liposome surface. Alloyed $CdTe_{1-x}Se_x/CdS$-type QDs are encapsulated in PEG-grafted phospholipid micelles. These coated QDs were attached with cyclic arginine–glycine–aspartic acid (cRGD) peptide for targeted delivery to $\alpha_v\beta_3$ integrin, overexpressed in the angiogenic tumor vasculature cells (Hu et al. 2010). Erogbogbo et al. (2010) showed magneto-fluorescent nanoprobe formation via co-encapsulating silicon QDs and IONPs in phospholipid-polyethylene glycol (DSPE-PEG) micelles. These particles have stable luminescence in the *in-vivo* condition of prostate cancer. Murcia et al. (2008) demonstrated the synthesis of QDs conjugated with lipids and their use as a cell surface tracker. First, CdSe/ZnS QDs were coated with two different hydrophilic compounds, that is, 2-(2-aminoethoxy) ethanol (AEE) and phospholipid/lipopolymer [60:40 molar mixture of 1,2-dipalmitoyl-sn-glycero-3-phosphocholine and 1,2-dipalmitoyl-sn-glycero-3-phosphoethanolamine-N-[methoxy(polyethyleneglycol-2000] (LIPO). Further, these AEE and LIPO-capped QDs are conjugated with sulfhydryl lipids through maleimide reactive groups present on their surface. This system can be used for a cell surface lipid-tracking experiment with the help of high-speed single molecule fluorescence microscopy. Galloway et al. (2012) synthesized and evaluated the physicochemical properties of QDs coated with PEGylated double acyl chain lipids, single and double acyl chain lipids, and single alkyl chain surfactant molecules with charged head groups. QDs attached to a single acyl chain lipid with charged surfactant are monodispersed and have high stability, whereas the addition of a double acyl chain zwitterion lipid (more than 20 mol%) to a single acyl chain lipid encapsulation layer results in decreased stability because of high curvature. PEGylated double acyl chain lipids provide negative charge and moderate stability, whereas a mixture of double acyl chain PEGylated lipids, acyl chain lipids, and charged surfactants leads to better stability in water and other biologically relevant buffers.

5.10 Conclusion and Future Perspectives

The recent advent of nanoscience and technology has shown tremendous promise in biomedical applications such as drug/gene delivery, therapeutics, theranostics, and imaging. In order to explore the full potential use of nanotechnology, nanomaterials must be tailored to a specific target. Further, surface modification of nanomaterials is a key to the success of any aimed biomedical application. Several methods for ligation of ligands on the surface of nanomaterials have been developed to keep in mind that the unique intrinsic properties of NPs must not be altered. Such developments have opened several avenues of collaboration between experts of diverse research areas such as material science, biotechnology, chemistry, toxicology, genetics, and so on. The collective efforts from these experts have led to the development of nanomaterials with tunable solubility, stability, and reduced toxicity. Multifunctional nanomaterials have also recently been discovered, which involves the conjugation of several biomolecules on a single NP surface. However, the purification of the final nanocomposite becomes difficult, as the large surface area of NPs can adsorb impurities from the reaction constituents, which may be difficult to remove using a single purification process. Additionally, multiple purification processes may not be a viable option for nanomaterial synthesis at the industrial scale. Therefore, nanomaterial synthesis and the surface modification process that do not involve lengthy procedures and are inexpensive for commercial-scale production with minimum batch-to-batch variation in physicochemical properties are imperative. Toxicity is another factor to be given considerable attention during synthesis and surface modification of nanomaterials. Weakley attached or simply adsorbed ligands may easily be desorbed from the surface and cause a toxic response in mammalian cells. For example, QDs are generally made up of heavy metals such as Cd or Se ions, which may leach out due to the dissolution of QDs after dispersion in biologically relevant media or in contact with mammalian cells. Such events pose one of the major hindrances to the successful application of QDs in biomedical applications. There are mixed reports about toxicity concerns of nanomaterials. Such discrepancies are probably due to the lack of a well-defined protocol of steps of purification and testing toxicity. It is clear that the unique properties of nanomaterials could be merged with the recent discoveries of molecular biology and explored for several cutting-edge applications such as gene profiling and optical bio-bar coding for high throughput gene expression analysis and novel drug discovery strategies. While endless possibilities can be explored using nanomaterials, the prime need is the design of surface capping/targeting molecules/biomolecules and a concomitant ligation strategy to develop new multifunctional nanomaterials with minimum toxicity.

Acknowledgments

R. Singh would like to thank the Department of Science and Technology, New Delhi, for providing INSPIRE Senior Research Fellowship (SRF). The financial assistance for the Centre for Nanotechnology Research and Applications (CENTRA) by the Gujarat Institute for Chemical Technology (GICT) is acknowledged. The funding from the Department of Science and Technology—Science and Engineering Research Board (SERB) (Grant No.: ILS/SERB/2015-16/01) to Dr. Sanjay Singh under the scheme of a Start-Up Research Grant (Young Scientists) in Life Sciences is also gratefully acknowledged. This chapter carries a DBLS communication number DBLS-081.

Conflict of Interest

The authors declare no competing financial interest.

References

Acar, H. Y., S. Celebi, N. I. Serttunali, and I. Lieberwirth. 2009. Development of highly stable and luminescent aqueous CdS quantum dots with the poly(acrylic acid)/mercaptoacetic acid binary coating system. *J Nanosci Nanotechnol* 9(5):2820–9.

Agui, L., P. Yanez-Sedeno, and J. M. Pingarron. 2008. Role of carbon nanotubes in electroanalytical chemistry: A review. *Anal Chim Acta* 622(1–2):11–47.

Aime, A., N. Beztsinna, A. Patwa, A. Pokolenko, I. Bestel, and P. Barthelemy. 2013. Quantum dot lipid oligonucleotide bioconjugates: Toward a new anti-microRNA nanoplatform. *Bioconjug Chem* 24(8):1345–55.

Allam, A. A., M. E. Sadat, S. J. Potter et al. 2013. Stability and magnetically induced heating behavior of lipid-coated Fe_3O_4 nanoparticles. *Nanoscale Res Lett* 8(1):426.

Ambrosi, A., F. Airo, and A. Merkoci. 2010. Enhanced gold nanoparticle based ELISA for a breast cancer biomarker. *Anal Chem* 82(3):1151–6.

Amstad, E., T. Gillich, I. Bilecka, M. Textor, and E. Reimhult. 2009. Ultrastable iron oxide nanoparticle colloidal suspensions using dispersants with catechol-derived anchor groups. *Nano Lett* 9(12):4042–8.

An, J., H. Zhang, J. Zhang, Y. Zhao, and X. Yuan. 2009. Preparation and antibacterial activity of electrospun chitosan/poly(ethylene oxide) membranes containing silver nanoparticles. *Colloid and Polymer Science* 287(12):1425–34.

Aslan, K., C. C. Luhrs, and V. H. Pérez-Luna. 2004. Controlled and reversible aggregation of biotinylated gold nanoparticles with streptavidin. *The Journal of Physical Chemistry B* 108(40):15631–39.

Bahadur, N. M., S. Watanabe, T. Furusawa et al. 2011. Rapid one-step synthesis, characterization and functionalization of silica coated gold nanoparticles. *Colloids and Surfaces A: Physicochemical and Engineering Aspects* 392(1):137–44.

Baio, G., M. Fabbi, S. Salvi et al. 2010. Two-step *in vivo* tumor targeting by biotin-conjugated antibodies and superparamagnetic nanoparticles assessed by magnetic resonance imaging at 1.5 T. *Mol Imaging Biol* 12(3):305–15.

Balasubramanian, K., and M. Burghard. 2005. Chemically functionalized carbon nanotubes. *Small* 1(2):180–92.

Bentzen, E. L., I. D. Tomlinson, J. Mason et al. 2005. Surface modification to reduce nonspecific binding of quantum dots in live cell assays. *Bioconjug Chem* 16(6):1488–94.

Bernardin, Aude, A. Cazet, L. Guyon et al. 2010. Copper-free click chemistry for highly luminescent quantum dot conjugates: Application to *in vivo* metabolic imaging. *Bioconjugate Chemistry* 21(4):583–8.

Berret, J. F., A. Sehgal, M. Morvan, O. Sandre, A. Vacher, and M. Airiau. 2006. Stable oxide nanoparticle clusters obtained by complexation. *J Colloid Interface Sci* 303(1):315–8.

Bhattacharya, S., and A. Srivastava. 2003. Synthesis and characterization of novel cationic lipid and cholesterol-coated gold nanoparticles and their interactions with dipalmitoylphosphatidylcholine membranes. *Langmuir* 19(10):4439–47.

Bhirde, A. A., S. Patel, A. A. Sousa et al. 2010. Distribution and clearance of PEG-single-walled carbon nanotube cancer drug delivery vehicles in mice. *Nanomedicine (Lond)* 5(10):1535–46.

Boisselier, E., and D. Astruc. 2009. Gold nanoparticles in nanomedicine: Preparations, imaging, diagnostics, therapies and toxicity. *Chem Soc Rev* 38(6):1759–82.

Boisselier, E., L. Salmon, J. Ruiz, and D. Astruc. 2008. How to very efficiently functionalize gold nanoparticles by "click" chemistry. *Chemical Communications* (44):5788–90.

Boyer, J. C., M. P. Manseau, J. I. Murray, and F. C. van Veggel. 2010. Surface modification of upconverting NaYF4 nanoparticles with PEG-phosphate ligands for NIR (800 nm) biolabeling within the biological window. *Langmuir* 26(2):1157–64.

Brubaker, C. D., T. M. Frecker, I. Njoroge, G. K. Jennings, and D. E. Adams. 2016. Poly(acrylic acid) coated gold nanoparticles for pH sensing applications. *8th European Workshop on Structural Health Monitoring (EWSHM 2016)*.

Brust, M., M. Walker, D. Bethell, D. J. Schiffrin, and R. Whyman. 1994. Synthesis of thiol-derivatised gold nanoparticles in a two-phase liquid-liquid system. *Journal of the Chemical Society, Chemical Communications* (7):801–2.

Cai, Y. 2015. Lipid-coated gold nanoparticle as a biosensor for lipid-protein interactions. *Electronic Theses and Dissertations* 105. University of Denver, Denver, U.S.

Campidelli, S., B. Ballesteros, A. Filoramo et al. 2008. Facile decoration of functionalized single-wall carbon nanotubes with phthalocyanines via "click chemistry." *Journal of the American Chemical Society* 130(34):11503–9.

Carrero-Sanchez, J. C., A. L. Elias, R. Mancilla et al. 2006. Biocompatibility and toxicological studies of carbon nanotubes doped with nitrogen. *Nano Lett* 6(8):1609–16.

Celebi, S., A. Koray Erdamar, A. Sennaroglu, A. Kurt, and H. Y. Acar. 2007. Synthesis and characterization of Poly(acrylic acid) stabilized cadmium sulfide quantum dots. *The Journal of Physical Chemistry B* 111(44):12668–75.

Cerruti, M. G., M. Sauthier, D. Leonard et al. 2006. Gold and silica-coated gold nanoparticles as thermographic labels for DNA detection. *Anal Chem* 78(10):3282–8.

Chae, S., J.-H. Kim, P. Theato, R. Zentel, and B.-H. Sohn. 2014. Dual functionalization of nanostructures of block copolymers with quantum dots and organic fluorophores. *Macromolecular Chemistry and Physics* 215(7):654–61.

Chao, J., W. Y. Huang, J. Wang, S. J. Xiao, Y. C. Tang, and J. N. Liu. 2009. Click-chemistry-conjugated oligo-angiomax in the two-dimensional DNA lattice and its interaction with thrombin. *Biomacromolecules* 10(4):877–83.

Chen, H., H. Zou, H. J. Paholak et al. 2014. Thiol-reactive amphiphilic block copolymer for coating gold nanoparticles with neutral and functionable surfaces. *Polym Chem* 5(8):2768–73.

Chen, S., G. Wu, Y. Liu, and D. Long. 2006. Preparation of Poly(acrylic acid) grafted multiwalled carbon nanotubes by a two-step irradiation technique. *Macromolecules* 39(1):330–4.

Cheng, X., R. Gondosiswanto, S. Ciampi, P. J. Reece, and J. Justin Gooding. 2012. One-pot synthesis of colloidal silicon quantum dots and surface functionalization via thiol-ene click chemistry. *Chemical Communications* 48(97):11874–6.

Cheng, X., S. B. Lowe, S. Ciampi et al. 2014. Versatile "click chemistry" approach to functionalizing silicon quantum dots: Applications toward fluorescent cellular imaging. *Langmuir* 30(18):5209–16.

Cheng, Z., S. Liu, H. Gao et al. 2008. A facile approach for transferring hydrophobic magnetic nanoparticles into water-soluble particles. *Macromolecular Chemistry and Physics* 209(11):1145–51.

Chiou, S. J., J. Innocent, C. G. Riordan, K. C. Lam, L. Liable-Sands, and A. L. Rheingold. 2000. Synthetic models for the zinc sites in the methionine synthases. *Inorg Chem* 39(19):4347–53.

Chunfu, Z., C. Jinquan, Y. Duanzhi, W. Yongxian, F. Yanlin, and T. Jiaju. 2004. Preparation and radiolabeling of human serum albumin (HSA)-coated magnetite nanoparticles for magnetically targeted therapy. *Appl Radiat Isot* 61(6):1255–9.

Correa-Duarte, M. A., Y. Kobayashi, R. A. Caruso, and L. M. Liz-Marzan. 2001. Photodegradation of SiO_2-coated CdS nanoparticles within silica gels. *J Nanosci Nanotechnol* 1(1):95–9.

Couto, D., M. Freitas, G. Porto et al. 2015. Polyacrylic acid-coated and non-coated iron oxide nanoparticles induce cytokine activation in human blood cells through TAK1, p38 MAPK and JNK pro-inflammatory pathways. *Arch Toxicol* 89(10):1759–69.

Crawford, S. E., C. M. Andolina, A. M. Smith et al. 2015. Ligand-mediated "turn on," high quantum yield near-infrared emission in small gold nanoparticles. *J Am Chem Soc* 137(45):14423–9.

Cui, H.-F., Jian-Shan Y., Wei-De Z., and Fwu-Shan S. 2009. Modification of carbon nanotubes with redox hydrogel: Improvement of amperometric sensing sensitivity for redox enzymes. *Biosensors & Bioelectronics* 24(6):1723–9.

Cui, R., Z. Han, and J. J. Zhu. 2011. Helical carbon nanotubes: Intrinsic peroxidase catalytic activity and its application for biocatalysis and biosensing. *Chemistry* 17(34):9377–84.

Cutler, J. I., D. Zheng, X. Xu, D. A. Giljohann, and C. A. Mirkin. 2010. Polyvalent oligonucleotide iron oxide nanoparticle "click" conjugates. *Nano Lett* 10(4):1477–80.

da Costa, G. M., C. Blanco-Andujar, E. De Grave, and Q. A. Pankhurst. 2014. Magnetic nanoparticles for *in vivo* use: A critical assessment of their composition. *J Phys Chem B* 118(40):11738–46.

Daniel, M. C., and D. Astruc. 2004. Gold nanoparticles: Assembly, supramolecular chemistry, quantum-size-related properties, and applications toward biology, catalysis, and nanotechnology. *Chem Rev* 104(1):293–346.

Daou, T. J., L. Li, P. Reiss, V. Josserand, and I. Texier. 2009. Effect of poly(ethylene glycol) length on the *in vivo* behavior of coated quantum dots. *Langmuir* 25(5):3040–4.

Datsyuk, V., M. Kalyva, K. Papagelis et al. 2008. Chemical oxidation of multiwalled carbon nanotubes. *Carbon* 46(6):833–40.

De Campos, A. M., A. Sanchez, R. Gref, P. Calvo, and M. J. Alonso. 2003. The effect of a PEG versus a chitosan coating on the interaction of drug colloidal carriers with the ocular mucosa. *Eur J Pharm Sci* 20(1):73–81.

De, P., S. R. Gondi, and B. S. Sumerlin. 2008. Folate-conjugated thermoresponsive block copolymers: Highly efficient conjugation and solution self-assembly. *Biomacromolecules* 9(3):1064–70.

Deng, X., P. Qin, M. Luo et al. 2013. Mesoporous silica coating on carbon nanotubes: Layer-by-layer method. *Langmuir* 29(23):6815–22.

Deng, Z. J., M. Liang, M. Monteiro, I. Toth, and R. F. Minchin. 2011. Nanoparticle-induced unfolding of fibrinogen promotes Mac-1 receptor activation and inflammation. *Nat Nanotechnol* 6(1):39–44.

Deng, Z. J., M. Liang, I. Toth, M. J. Monteiro, and R. F. Minchin. 2012. Molecular interaction of poly(acrylic acid) gold nanoparticles with human fibrinogen. *ACS Nano* 6(10):8962–9.

Derfus, A. M., A. A. Chen, D.-H. Min, E. Ruoslahti, and S. N. Bhatia. 2007. Targeted quantum dot conjugates for siRNA delivery. *Bioconjugate Chemistry* 18(5):1391–6.

Dif, A., F. Boulmedais, M. Pinot et al. 2009. Small and stable peptidic PEGylated quantum dots to target polyhistidine-tagged proteins with controlled stoichiometry. *J Am Chem Soc* 131(41):14738–46.

Ding J.-F., and Jiang J.-S. 2007. Surface modification of Fe_3O_4 nanoparticles prepared in high temperature organic solution by sodium oleate. *Journal of Inorganic Materials* 22(5):859–63.

Dinkel, R., B. Braunschweig, and W. Peukert. 2015. Fast and slow ligand exchange at the surface of colloidal gold nanoparticles. *The Journal of Physical Chemistry C* 120:1673–82.

Dong, A., X. Ye, J. Chen et al. 2010. A generalized ligand-exchange strategy enabling sequential surface functionalization of colloidal nanocrystals. *J Am Chem Soc* 133(4):998–1006.

Dreaden, E. C., A. M. Alkilany, X. Huang, C. J. Murphy, and M. A. El-Sayed. 2012. The golden age: Gold nanoparticles for biomedicine. *Chem Soc Rev* 41(7):2740–79.

Duguet, E., S. Vasseur, S. Mornet, and J. M. Devoisselle. 2006. Magnetic nanoparticles and their applications in medicine. *Nanomedicine (Lond)* 1(2):157–68.

Eberhardt, M., R. Mruk, R. Zentel, and P. Théato. 2005. Synthesis of pentafluorophenyl(meth)acrylate polymers: New precursor polymers for the synthesis of multifunctional materials. *European Polymer Journal* 41(7):1569–75.

Erogbogbo, F., K.-T. Yong, R. Hu et al. 2010. Biocompatible magnetofluorescent probes: Luminescent silicon quantum dots coupled with superparamagnetic iron(III) oxide. *ACS Nano* 4(9):5131–8.

Farokhzad, O. C., J. Cheng, B. A. Teply et al. 2006. Targeted nanoparticle-aptamer bioconjugates for cancer chemotherapy *in vivo*. *Proc Natl Acad Sci USA* 103(16):6315–20.

Finetti, C., L. Sola, M. Pezzullo et al. 2016. Click chemistry immobilization of antibodies on polymer coated gold nanoparticles. *Langmuir* 32(29):7435–41.

Fischler, M., A. Sologubenko, J. Mayer et al. 2008. Chain-like assembly of gold nanoparticles on artificial DNA templates via "click chemistry". *Chemical Communications* (2):169–71.

Fresnais, J., M. Yan, J. Courtois, T. Bostelmann, A. Bee, and J. F. Berret. 2013. Poly(acrylic acid)-coated iron oxide nanoparticles: Quantitative evaluation of the coating properties and applications for the removal of a pollutant dye. *J Colloid Interface Sci* 395:24–30.

Frielinghaus, R., C. Besson, L. Houben, A. K. Saelhoff, C. M. Schneider, and C. Meyer. 2015. Controlled covalent binding of antiferromagnetic tetramanganese complexes to carbon nanotubes. *RSC Advances* 5(102):84119–24.

Galloway, J. F., A. Winter, K. H. Lee et al. 2012. Quantitative characterization of the lipid encapsulation of quantum dots for biomedical applications. *Nanomedicine* 8(7):1190–9.

Gao, L., J. Zhuang, L. Nie et al. 2007. Intrinsic peroxidase-like activity of ferromagnetic nanoparticles. *Nat Nanotechnol* 2(9):577–83.

Gao, Q., W. Xu, Y. Xu et al. 2008. Amino acid adsorption on mesoporous materials: Influence of types of amino acids, modification of mesoporous materials, and solution conditions. *J Phys Chem B* 112(7):2261–7.

Gao, Y., and I. Kyratzis. 2008. Covalent immobilization of proteins on carbon nanotubes using the cross-linker 1-ethyl-3-(3-dimethylaminopropyl)carbodiimide—A critical assessment. *Bioconjug Chem* 19(10):1945–50.

García-Jimeno, S., and J. Estelrich. 2012. Ferrofluid based on polyethylene glycol-coated iron oxide nanoparticles: Characterization and properties. *Colloids and Surfaces A: Physicochemical and Engineering Aspects* 420:74–81.

Ge, C., J. Du, L. Zhao et al. 2011. Binding of blood proteins to carbon nanotubes reduces cytotoxicity. *Proc Natl Acad Sci USA* 108(41):16968–73.

Gelbrich, T., M. Feyen, and A. M. Schmidt. 2006. Magnetic thermoresponsive core-shell nanoparticles. *Macromolecules* 39:3469–72.

Georgakilas, V., N. Tagmatarchis, D. Pantarotto, A. Bianco, J. P. Briand, and M. Prato. 2002. Amino acid functionalisation of water soluble carbon nanotubes. *Chem Commun (Camb)* (24):3050–1.

Graf, C., S. Dembski, A. Hofmann, and E. Ruhl. 2006. A general method for the controlled embedding of nanoparticles in silica colloids. *Langmuir* 22(13):5604–10.

Gramlich, P. M., C. T. Wirges, A. Manetto, and T. Carell. 2008. Postsynthetic DNA modification through the copper-catalyzed azide-alkyne cycloaddition reaction. *Angew Chem Int Ed Engl* 47(44):8350–8.

Grapperhaus, C. A., T. Tuntulani, J. H. Reibenspies, and M. Y. Darensbourg. 1998. Methylation of tethered thiolates in [(bme-daco)Zn](2) and [(bme-daco)Cd](2) as a model of zinc sulfur-methylation proteins. *Inorg Chem* 37(16):4052–8.

Gref, R., M. Luck, P. Quellec et al. 2000. "Stealth" corona-core nanoparticles surface modified by polyethylene glycol (PEG): Influences of the corona (PEG chain length and surface density) and of the core composition on phagocytic uptake and plasma protein adsorption. *Colloids Surf B Biointerfaces* 18(3–4):301–13.

Gómez, J. M., M. D. Romero, and T. M. Fernández. 2005. Immobilization of Î²-Glucosidase on carbon nanotubes. *Catalysis Letters* 101(3):275–8.

Hakami, O., Y. Zhang, and C. J. Banks. 2012. Thiol-functionalised mesoporous silica-coated magnetite nanoparticles for high efficiency removal and recovery of Hg from water. *Water Res* 46(12):3913–22.

Han, Y., S. K. Ahn, Z. Zhang, G. S. Smith, and C. Do. 2016. Functionalization of single-walled carbon nanotubes with thermo-reversible block copolymers and characterization by small-angle neutron scattering. *J Vis Exp* (112):e53969.

Hayashi, K., M. Moriya, W. Sakamoto, and T. Yogo. 2009. Chemoselective synthesis of folic acid—Functionalized magnetite nanoparticles via click chemistry for magnetic hyperthermia. *Chemistry of Materials* 21(7):1318–25.

Hayashi, K., M. Nakamura, and K. Ishimura. 2013. Near-infrared fluorescent silica-coated gold nanoparticle clusters for x-ray computed tomography/optical dual modal imaging of the lymphatic system. *Adv Healthc Mater* 2(5):756–63.

Hayashi, K., K. Ono, H. Suzuki et al. 2010. High-frequency, magnetic-field-responsive drug release from magnetic nanoparticle/organic hybrid based on hyperthermic effect. *ACS Appl Mater Interfaces* 2(7):1903–11.

He, H.-W., H.-J. Liu, K.-C. Zhou, W. Wang, and P.-F. Rong. 2006. Characteristics of magnetic Fe_3O_4 nanoparticles encapsulated with human serum albumin. *Journal of Central South University of Technology* 13(1):6–11.

Heister, E., V. Neves, C. Lamprecht, S. R. P. Silva, H. M. Coley, and J. McFadden. 2012. Drug loading, dispersion stability, and therapeutic efficacy in targeted drug delivery with carbon nanotubes. *Carbon* 50(2):622–32.

Hemmati, K., R. Alizadeh, and M. C. P. A. T. Ghaemy. 2016. Synthesis and characterization of controlled drug release carriers based on functionalized amphiphilic block copolymers and super-paramagnetic iron oxide nanoparticles. *Polymers for Advanced Technologies* 27(4):504–14.

Hoa, L. T. M., T. T. Dung, T. M. Danh, N. H. Duc, and C. D. Mau. 2009. Preparation and characterization of magnetic nanoparticles coated with polyethylene glycol. *Journal of Physics: Conference Series* 187(1):012048.

Hotchkiss, J. W., B. G. R. Mohr, and S. G. Boyes. 2010. Gold nanorods surface modified with poly(acrylic acid) as a template for the synthesis of metallic nanoparticles. *Journal of Nanoparticle Research* 12(3):915–30.

Housni, A., and Y. Zhao. 2010. Gold nanoparticles functionalized with block copolymers displaying either LCST or UCST thermosensitivity in aqueous solution. *Langmuir* 26(15):12933–9.

Hu, R., K. T. Yong, I. Roy et al. 2010. Functionalized near-infrared quantum dots for *in vivo* tumor vasculature imaging. *Nanotechnology* 21(14):145105.

Huang, C., K. G. Neoh, and E. T. Kang. 2011. Combined ATRP and "click" chemistry for designing stable tumor-targeting superparamagnetic iron oxide nanoparticles. *Langmuir* 28(1):563–71.

Huang, J., L. Wang, R. Lin et al. 2013. Casein-coated iron oxide nanoparticles for high MRI contrast enhancement and efficient cell targeting. *ACS Appl Mater Interfaces* 5(11):4632–9.

Huang, S. H., H. H. Liao, and D. H. Chen. 2010. Simultaneous determination of norepinephrine, uric acid, and ascorbic acid at a screen printed carbon electrode modified with polyacrylic acid-coated multi-wall carbon nanotubes. *Biosens Bioelectron* 25(10):2351–5.

Hudson, R., Y. Feng, R. S. Varma, and A. Moores. 2014. Bare magnetic nanoparticles: Sustainable synthesis and applications in catalytic organic transformations. *Green Chemistry* 16(10):4493–505.

Hushiarian, R., N. A. Yusof, A. H. Abdullah, S. A. Ahmad, and S. W. Dutse. 2014. A novel DNA nanosensor based on CdSe/ZnS quantum dots and synthesized Fe_3O_4 magnetic nanoparticles. *Molecules* 19(4):4355–68.

Hussain, I., S. Graham, Z. Wang et al. 2005. Size-controlled synthesis of near-monodisperse gold nanoparticles in the 1–4 nm range using polymeric stabilizers. *J Am Chem Soc* 127(47):16398–9.

Isaacs, S. R., E. C. Cutler, J. S. Park, T. R. Lee, and Y. S. Shon. 2005. Synthesis of tetraoctylammonium-protected gold nanoparticles with improved stability. *Langmuir* 21(13):5689–92.

Islam, M. F., E. Rojas, D. M. Bergey, A. T. Johnson, and A. G. Yodh. 2003. High weight fraction surfactant solubilization of single-wall carbon nanotubes in water. *Nano Letters* 3(2):269–73.

Iversen, N. K., S. Frische, K. Thomsen et al. 2013. Superparamagnetic iron oxide polyacrylic acid coated gamma-Fe_2O_3 nanoparticles do not affect kidney function but cause acute effect on the cardiovascular function in healthy mice. *Toxicol Appl Pharmacol* 266(2):276–88.

Jamieson, T., R. Bakhshi, D. Petrova, R. Pocock, M. Imani, and A. M. Seifalian. 2007. Biological applications of quantum dots. *Biomaterials* 28(31):4717–32.

Jans, H., K. Jans, L. Lagae, G. Borghs, G. Maes, and Q. Huo. 2010. Poly(acrylic acid)-stabilized colloidal gold nanoparticles: Synthesis and properties. *Nanotechnology* 21(45):455702.

Jazayeri, M. H., H. Amani, A. A. Pourfatollah, H. Pazoki-Toroudi, and B. Sedighimoghaddam. 2016. Various methods of gold nanoparticles (GNPs) conjugation to antibodies. *Sensing and Bio-Sensing Research* 9:17–22.

Jia, F., Y. Zhang, B. Narasimhan, and S. K. Mallapragada. 2012. Block copolymer-quantum dot micelles for multienzyme colocalization. *Langmuir* 28(50):17389–95.

Jiang, S., A. A. Eltoukhy, K. T. Love, R. Langer, and D. G. Anderson. 2013. Lipidoid-coated iron oxide nanoparticles for efficient DNA and siRNA delivery. *Nano Lett* 13(3):1059–64.

Jiang, Y., C. Jin, F. Yang et al. 2011. A new approach to produce amino-carbon nanotubes as plasmid transfection vector by [2 + 1] cycloaddition of nitrenes. *Journal of Nanoparticle Research* 13(1):33–38.

Jung, D.-H., Y. K. Ko, and H.-T. Jung. 2004. Aggregation behavior of chemically attached poly(ethylene glycol) to single-walled carbon nanotubes (SWNTs) ropes. *Materials Science and Engineering: C* 24(1–2):117–21.

Kairdolf, B. A., A. M. Smith, and S. Nie. 2008. One-pot synthesis, encapsulation, and solubilization of size-tuned quantum dots with amphiphilic multidentate ligands. *Journal of the American Chemical Society* 130(39):12866–7.

Kairdolf, B. A., A. M. Smith, T. H. Stokes, M. D. Wang, A. N. Young, and S. Nie. 2013. Semiconductor quantum dots for bioimaging and biodiagnostic applications. *Annu Rev Anal Chem (Palo Alto Calif)* 6:143–62.

Kam, N. W., and H. Dai. 2005. Carbon nanotubes as intracellular protein transporters: Generality and biological functionality. *J Am Chem Soc* 127(16):6021–6.

Kang, J. H., and Y. T. Ko. 2015. Lipid-coated gold nanocomposites for enhanced cancer therapy. *Int J Nanomedicine* 10(Spec Iss):33–45.

Kang, Y., and T. Andrew Taton. 2005. Controlling shell thickness in core–Shell gold nanoparticles via surface-templated adsorption of block copolymer surfactants. *Macromolecules* 38(14):6115–21.

Karakoti, A., S. Singh, J. M. Dowding, S. Seal, and W. T. Self. 2010. Redox-active radical scavenging nanomaterials. *Chem Soc Rev* 39(11):4422–32.

Karakoti, A. S., R. Shukla, R. Shanker, and S. Singh. 2015. Surface functionalization of quantum dots for biological applications. *Adv Colloid Interface Sci* 215:28–45.

Khoee, S., and A. Kavand. 2014. A new procedure for preparation of polyethylene glycol-grafted magnetic iron oxide nanoparticles. *Journal of Nanostructure in Chemistry* 4(3):111.

Kim, J., H. S. Kim, N. Lee et al. 2008. Multifunctional uniform nanoparticles composed of a magnetite nanocrystal core and a mesoporous silica shell for magnetic resonance and fluorescence imaging and for drug delivery. *Angew Chem Int Ed Engl* 47(44):8438–41.

Kim, J. Y., S. I. Han, and S. Hong. 2008. Effect of modified carbon nanotube on the properties of aromatic polyester nanocomposites. *Polymer* 49(15):3335–45.

Kim, M. J., J. Lee, D. Jung, and S. E. Shim. 2010. Electrospun poly(vinyl alcohol) nanofibers incorporating PEGylated multi-wall carbon nanotube. *Synthetic Metals* 160(13–14):1410–4.

Kim, S., and M. G. Bawendi. 2003. Oligomeric ligands for luminescent and stable nanocrystal quantum dots. *J Am Chem Soc* 125(48):14652–3.

Kolb, H. C., M. G. Finn, and K. B. Sharpless. 2001. Click chemistry: Diverse chemical function from a few good reactions. *Angew Chem Int Ed Engl* 40(11):2004–21.

Kong, W. H., K. H. Bae, S. D. Jo, J. S. Kim, and T. G. Park. 2012. Cationic lipid-coated gold nanoparticles as efficient and non-cytotoxic intracellular siRNA delivery vehicles. *Pharm Res* 29(2):362–74.

Koole, R., M. M. van Schooneveld, J. Hilhorst et al. 2008. Paramagnetic lipid-coated silica nanoparticles with a fluorescent quantum dot core: A new contrast agent platform for multimodality imaging. *Bioconjug Chem* 19(12):2471–9.

Korpany, K. V., C. Mottillo, J. Bachelder et al. 2016. One-step ligand exchange and switching from hydrophobic to water-stable hydrophilic superparamagnetic iron oxide nanoparticles by mechanochemical milling. *Chem Commun (Camb)* 52(14):3054–7.

Kotagiri, N., Z. Li, X. Xu, S. Mondal, A. Nehorai, and S. Achilefu. 2014. Antibody quantum dot conjugates developed via copper-free click chemistry for rapid analysis of biological samples using a microfluidic microsphere array system. *Bioconjugate Chemistry* 25(7):1272–81.

Ku, K. H., M. P. Kim, K. Paek et al. 2013. Multicolor emission of hybrid block copolymer–quantum dot microspheres by controlled spatial isolation of quantum dots. *Small* 9(16):2667–72.

Kumar, R., A. El-Sagheer, J. Tumpane, P. Lincoln, L. M. Wilhelmsson, and T. Brown. 2007. Template-directed oligonucleotide strand ligation, covalent intramolecular DNA circularization and catenation using click chemistry. *J Am Chem Soc* 129(21):6859–64.

Kunzmann, A., B. Andersson, C. Vogt et al. 2014. Efficient internalization of silica-coated iron oxide nanoparticles of different sizes by primary human macrophages and dendritic cells. *Toxicol Appl Pharmacol* 253(2):81–93.

Kyoungja, W., and J. Hong. 2005. Surface modification of hydrophobic iron oxide nanoparticles for clinical applications. *IEEE Transactions on Magnetics* 41(10):4137–9.

Lanford, R. E., P. Kanda, and R. C. Kennedy. 1986. Induction of nuclear transport with a synthetic peptide homologous to the SV40 T antigen transport signal. *Cell* 46(4):575–82.

Larsen, E. K., T. Nielsen, T. Wittenborn et al. 2012. Accumulation of magnetic iron oxide nanoparticles coated with variably sized polyethylene glycol in murine tumors. *Nanoscale* 4(7):2352–61.

Lawrence, J., J. T. Pham, D. Y. Lee, Y. Liu, A. J. Crosby, and T. Emrick. 2014. Highly conductive ribbons prepared by stick-slip assembly of organosoluble gold nanoparticles. *ACS Nano* 8(2):1173–9.

Lee, C. W., K. T. Huang, P. K. Wei, and Y. D. Yao. 2006. Conjugation of $γ$-Fe_2O_3 nanoparticles with single strand oligonucleotides. *Journal of Magnetism and Magnetic Materials* 304(1):e412–4.

Lee, J. S., P. A. Ulmann, M. S. Han, and C. A. Mirkin. 2008. A DNA-gold nanoparticle-based colorimetric competition assay for the detection of cysteine. *Nano Lett* 8(2):529–33.

Lee, K. S., and I. S. Lee. 2008. Decoration of superparamagnetic iron oxide nanoparticles with Ni2+: Agent to bind and separate histidine-tagged proteins. *Chem Commun (Camb)* (6):709–11.

Lee, S., T. Oda, P.-K. Shin, and B.-J. Lee. 2009. Chemical modification of carbon nanotube for improvement of field emission property. *Microelectronic Engineering* 86(10):2110–3.

Li, L., W. Jiang, K. Luo et al. 2013. Superparamagnetic iron oxide nanoparticles as MRI contrast agents for non-invasive stem cell labeling and tracking. *Theranostics* 3(8):595–615.

Li, N., P. Zhao, N. Liu et al. 2014. "Click" chemistry mildly stabilizes bifunctional gold nanoparticles for sensing and catalysis. *Chemistry* 20(27):8363–9.

Li, P., P. Chevallier, P. Ramrup et al. 2015. Mussel-inspired multidentate block copolymer to stabilize ultrasmall superparamagnetic Fe_3O_4 for magnetic resonance imaging contrast enhancement and excellent colloidal stability. *Chemistry of Materials* 27(20):7100–9.

Li, P., W. Xiao, P. Chevallier et al. 2016. Extremely small iron oxide nanoparticles stabilized with catechol-functionalized multidentate block copolymer for enhanced MRI. *ChemistrySelect* 1(13):4087–91.

Li, Q., T. Dong, X. Liu, and X. Lei. 2013. A bioorthogonal ligation enabled by click cycloaddition of o-quinolinone quinone methide and vinyl thioether. *J Am Chem Soc* 135(13):4996–9.

Lin, C. I., C. F. Lee, and W. Y. Chiu. 2005. Preparation and properties of poly(acrylic acid) oligomer stabilized superparamagnetic ferrofluid. *J Colloid Interface Sci* 291(2):411–20.

Lin, J. J., J. S. Chen, S. J. Huang et al. 2009. Folic acid-Pluronic F127 magnetic nanoparticle clusters for combined targeting, diagnosis, and therapy applications. *Biomaterials* 30(28):5114–24.

Lin, S. Y., N. T. Chen, S. P. Sum, L. W. Lo, and C. S. Yang. 2008. Ligand exchanged photoluminescent gold quantum dots functionalized with leading peptides for nuclear targeting and intracellular imaging. *Chem Commun (Camb)* (39):4762–4.

Liu A., I. Honma, M. Ichihara, and Z. Haoshen. 2006. Poly(acrylic acid)-wrapped multi-walled carbon nanotubes composite solubilization in water: Definitive spectroscopic properties. *Nanotechnology* 17(12):2845.

Liu, G., M. Swierczewska, S. Lee, and X. Chen. 2010. Functional nanoparticles for molecular imaging guided gene delivery. *Nano Today* 5(6):524–39.

Liu, Y., J. Tang, X. Chen et al. 2006. Carbon nanotube seeded sol-gel synthesis of silica nanoparticle assemblies. *Carbon* 44(1):165–7.

Loddenkemper, R., D. Sagebiel, and A. Brendel. 2002. Strategies against multidrug-resistant tuberculosis. *Eur Respir J Suppl* 36:66s–77s.

Lu, M., W. Zhang, Y. Gai et al. 2014. Folate-PEG functionalized silica CdTe quantum dots as fluorescent probes for cancer cell imaging. *New Journal of Chemistry* 38(9):4519–26.

Luccardini, C., C. Tribet, F. Vial, V. Marchi-Artzner, and M. Dahan. 2006. Size, charge, and interactions with giant lipid vesicles of quantum dots coated with an amphiphilic macromolecule. *Langmuir* 22(5):2304–10.

Lv, Y., K. Li, and Y. Li. 2013. Surface modification of quantum dots and magnetic nanoparticles with PEG-conjugated chitosan derivatives for biological applications. *Chemical Papers* 67(11):1404–13.

Ma, P. C., J.-K. Kim, and B. Z. Tang. 2006. Functionalization of carbon nanotubes using a silane coupling agent. *Carbon* 44(15):3232–8.

Ma, Y., Y. Li, S. Ma, and X. Zhong. 2014. Highly bright water-soluble silica coated quantum dots with excellent stability. *Journal of Materials Chemistry B* 2(31):5043–51.

Malvindi, M. A., V. De Matteis, A. Galeone et al. 2014. Toxicity assessment of silica coated iron oxide nanoparticles and biocompatibility improvement by surface engineering. *PLoS One* 9(1):e85835.

Manson, J., D. Kumar, B. J. Meenan, and D. Dixon. 2011. Polyethylene glycol functionalized gold nanoparticles: The influence of capping density on stability in various media. *Gold Bulletin* 44(2):99–105.

Mei, B. C., K. Susumu, I. L. Medintz, J. B. Delehanty, T. J. Mountziaris, and H. Mattoussi. 2008. Modular poly(ethylene glycol) ligands for biocompatible semiconductor and gold nanocrystals with extended pH and ionic stability. *Journal of Materials Chemistry* 18(41):4949–58.

Meng, L., C. Fu, and Q. Lu. 2009. Advanced technology for functionalization of carbon nanotubes. *Progress in Natural Science* 19(7):801–10.

Mine, E., A. Yamada, Y. Kobayashi, M. Konno, and L. M. Liz-Marzan. 2003. Direct coating of gold nanoparticles with silica by a seeded polymerization technique. *J Colloid Interface Sci* 264(2):385–90.

Moore, T. L., S. W. Grimes, R. L. Lewis, and F. Alexis. 2014. Multilayered polymer-coated carbon nanotubes to deliver dasatinib. *Molecular Pharmaceutics* 11(1):276–82.

Moore, T. L., J. E. Pitzer, P. Podila et al. 2013. Multifunctional polymer-coated carbon nanotubes for safe drug delivery. *Particle & Particle Systems Characterization* 30(4):365–73.

Mu, B., T. Wang, Z. Wu, H. Shi, D. Xue, and P. Liu. 2011. Fabrication of functional block copolymer grafted superparamagnetic nanoparticles for targeted and controlled drug delivery. *Colloids and Surfaces A: Physicochemical and Engineering Aspects* 375(1–3):163–8.

Mukhopadhyay, A., N. Joshi, K. Chattopadhyay, and G. De. 2012. A facile synthesis of PEG-coated magnetite (Fe_3O_4) nanoparticles and their prevention of the reduction of cytochrome C. *ACS Applied Materials & Interfaces* 4(1):142–9.

Mulvaney, P. 1996. Surface plasmon spectroscopy of nanosized metal particles. *Langmuir* 12(3):788–800.

Murcia, M. J., D. E. Minner, G. M. Mustata, K. Ritchie, and C. A. Naumann. 2008. Design of quantum dot-conjugated lipids for long-term, high-speed tracking experiments on cell surfaces. *J Am Chem Soc* 130(45):15054–62.

Muthu, M. S., and S. S. Feng. 2010. Nanopharmacology of liposomes developed for cancer therapy. *Nanomedicine (Lond)* 5(7):1017–9.

Möller, M., J. P. Spatz, and A. Roescher. 1996. Gold nanoparticles in micellar poly(styrene)-b-poly(ethylene oxide) films—Size and interparticle distance control in monoparticulate films. *Advanced Materials* 8(4):337–40.

Nagaraju, K., R. Reddy, and N. Reddy. 2015. A review on protein functionalized carbon nanotubes. *J Appl Biomater Funct Mater* 13(4):e301–12.

Nair, L. V., S. S. Nazeer, R. S. Jayasree, and A. Ajayaghosh. 2015. Fluorescence imaging assisted photodynamic therapy using photosensitizer-linked gold quantum clusters. *ACS Nano* 9(6):5825–32.

Nam, J. M., S. I. Stoeva, and C. A. Mirkin. 2004. Bio-bar-code-based DNA detection with PCR-like sensitivity. *J Am Chem Soc* 126(19):5932–3.

Nann, T., J. Riegler, P. Nick, and P. Mulvaney. 2005. Quantum dots with silica shells. *Proc SPIE* 5705: 77–84.

Nie, H., W. Guo, Y. Yuan et al. 2010. PEGylation of double-walled carbon nanotubes for increasing their solubility in water. *Nano Research* 3(2):103–9.

Nie, H., S. Liu, R. Yu, and J. Jiang. 2009. Phospholipid-coated carbon nanotubes as sensitive electrochemical labels with controlled-assembly-mediated signal transduction for magnetic separation immunoassay. *Angew Chem Int Ed Engl* 48(52):9862–6.

Nishi, H., S. Hiroya, and T. Tatsuma. 2015. Potential-scanning localized surface plasmon resonance sensor. *ACS Nano* 9(6):6214–21.

Norman, R. S., J. W. Stone, A. Gole, C. J. Murphy, and T. L. Sabo-Attwood. 2008. Targeted photothermal lysis of the pathogenic bacteria, *Pseudomonas aeruginosa*, with gold nanorods. *Nano Lett* 8(1):302–6.

Oh, J. H., H. Park do, J. H. Joo, and J. S. Lee. 2015. Recent advances in chemical functionalization of nanoparticles with biomolecules for analytical applications. *Anal Bioanal Chem* 407(29):8627–45.

Olsson, C. 2009. Synthesis of bifunctional poly(ethylene oxides) and their use as ligands to nanoparticles.

Padwal, P., R. Bandyopadhyaya, and S. Mehra. 2014. Polyacrylic acid-coated iron oxide nanoparticles for targeting drug resistance in mycobacteria. *Langmuir* 30(50):15266–76.

Pangule, R. C., S. J. Brooks, C. Z. Dinu et al. 2010. Antistaphylococcal nanocomposite films based on enzyme-nanotube conjugates. *ACS Nano* 4(7):3993–4000.

Pankhurst, Q. A., J. Connolly, S. K. Jones, and J. Dobson. 2003. Applications of magnetic nanoparticles in biomedicine. *Journal of Physics D: Applied Physics* 36(13):R167.

Pankhurst, Q. A., N. T. K. Thanh, S. K. Jones, and J. Dobson. 2009. Progress in applications of magnetic nanoparticles in biomedicine. *Journal of Physics D: Applied Physics* 42(22):224001.

Paredes, J. I., and M. Burghard. 2004. Dispersions of individual single-walled carbon nanotubes of high length. *Langmuir* 20(12):5149–52.

Park, C. H., H. Yang, J. Lee et al. 2015. Multicolor emitting block copolymer-integrated graphene quantum dots for colorimetric, simultaneous sensing of temperature, pH, and metal ions. *Chemistry of Materials* 27(15):5288–94.

Park, J. W., and J. S. Shumaker-Parry. 2015. Strong resistance of citrate anions on metal nanoparticles to desorption under thiol functionalization. *ACS Nano* 9(2):1665–82.

Patra, C. R., R. Bhattacharya, S. Patra, S. Basu, P. Mukherjee, and D. Mukhopadhyay. 2007. Lanthanide phosphate nanorods as inorganic fluorescent labels in cell biology research. *Clin Chem* 53(11):2029–31.

Pejchal, R., and M. L. Ludwig. 2005. Cobalamin-independent methionine synthase (MetE): A face-to-face double barrel that evolved by gene duplication. *PLoS Biol* 3(2):e31.

Peracchia, M. T., E. Fattal, D. Desmaele et al. 1999. Stealth PEGylated polycyanoacrylate nanoparticles for intravenous administration and splenic targeting. *J Control Release* 60(1):121–8.

Pereira, M., and E. P. C. Lai. 2008. Capillary electrophoresis for the characterization of quantum dots after non-selective or selective bioconjugation with antibodies for immunoassay. *Journal of Nanobiotechnology* 6(1):1–15.

Poddar, P., T. Telem-Shafir, T. Fried, and G. Markovich. 2002. Dipolar interactions in two- and three-dimensional magnetic nanoparticle arrays. *Physical Review B* 66(6):060403.

Poulose, A. C., S. Veeranarayanan, M. S. Mohamed et al. 2012. PEG coated biocompatible cadmium chalcogenide quantum dots for targeted imaging of cancer cells. *J Fluoresc* 22(3):931–44.

Qi, H., C. Ling, R. Huang et al. 2012. Functionalization of single-walled carbon nanotubes with protein by click chemistry as sensing platform for sensitized electrochemical immunoassay. *Electrochimica Acta* 63:76–82.

Rahme, K., L. Chen, R. G. Hobbs, M. A. Morris, C. O'Driscoll, and J. D. Holmes. 2013. PEGylated gold nanoparticles: Polymer quantification as a function of PEG lengths and nanoparticle dimensions. *RSC Advances* 3(17):6085–94.

Ravelli, D., D. Merli, E. Quartarone, A. Profumo, P. Mustarelli, and M. Fagnoni. 2013. PEGylated carbon nanotubes: Preparation, properties and applications. *RSC Advances* 3(33):13569–82.

Roberts, A. P., A. S. Mount, B. Seda et al. 2007. In vivo biomodification of lipid-coated carbon nanotubes by Daphnia magna. *Environ Sci Technol* 41(8):3025–9.

Sakellariou, G., H. Ji, J. W. Mays, and D. Baskaran. 2008. Enhanced polymer grafting from multiwalled carbon nanotubes through living anionic surface-initiated polymerization. *Chemistry of Materials* 20(19):6217–30.

Sanchez-Tirado, E., A. Gonzalez-Cortes, P. Yanez-Sedeno, and J. M. Pingarron. 2016. Carbon nanotubes functionalized by click chemistry as scaffolds for the preparation of electrochemical immunosensors. Application to the determination of TGF-beta 1 cytokine. *Analyst* 141(20):5730–7.

Sandström, P., M. Boncheva, and B. Åkerman. 2003. Nonspecific and thiol-specific binding of DNA to gold nanoparticles. *Langmuir* 19(18):7537–43.

Santra, S., R. Tapec, N. Theodoropoulou, J. Dobson, A. Hebard, and W. Tan. 2001. Synthesis and characterization of silica-coated iron oxide nanoparticles in microemulsion: The effect of nonionic surfactants. *Langmuir* 17(10):2900–6.

Sauerbeck, C., M. Haderlein, B. Schurer, B. Braunschweig, W. Peukert, and R. N. Klupp Taylor. 2014. Shedding light on the growth of gold nanoshells. *ACS Nano* 8(3):3088–96.

Schumacher, W. C., A. J. Phipps, and P. K. Dutta. 2009. Detection of *Bacillus anthracis* spores: Comparison of quantum dot and organic dye labeling agents. *Advanced Powder Technology* 20(5):438–46.

Schwaminger, S. P., P. F. García, Georg K. Merck et al. 2015. Nature of interactions of amino acids with bare magnetite nanoparticles. *The Journal of Physical Chemistry C* 119(40):23032–41.

Seol, S. K., D. Kim, S. Jung, W. S. Chang, and J. T. Kim. 2013. One-step synthesis of PEG-coated gold nanoparticles by rapid microwave heating. *Journal of Nanomaterials* 2013:6.

Shen, R., P. H. Camargo, Y. Xia, and H. Yang. 2008. Silane-based poly(ethylene glycol) as a primer for surface modification of nonhydrolytically synthesized nanoparticles using the Stober method. *Langmuir* 24(19):11189–95.

Shi, R., H. Wang, P. Tang, and Y. Bin. 2014. Synthesis of vinylferrocene and the ligand-exchange reaction between its copolymer and carbon nanotubes. *Frontiers of Chemical Science and Engineering* 8(2):171–8.

Shultz, M. D., J. U. Reveles, S. N. Khanna, and E. E. Carpenter. 2007. Reactive nature of dopamine as a surface functionalization agent in iron oxide nanoparticles. *J Am Chem Soc* 129(9):2482–7.

Singh, S., T. Dosani, A. S. Karakoti, A. Kumar, S. Seal, and W. T. Self. 2011. A phosphate-dependent shift in redox state of cerium oxide nanoparticles and its effects on catalytic properties. *Biomaterials* 32(28):6745–53.

Smolensky, E. D., H. Y. Park, T. S. Berquo, and V. C. Pierre. 2011. Surface functionalization of magnetic iron oxide nanoparticles for MRI applications—Effect of anchoring group and ligand exchange protocol. *Contrast Media Mol Imaging* 6(4):189–99.

Sperling, R. A., and W. J. Parak. 2010. Surface modification, functionalization and bioconjugation of colloidal inorganic nanoparticles. *Philos Trans A Math Phys Eng Sci* 368(1915):1333–83.

Stynoski, P., P. Mondal, E. Wotring, and C. Marsh. 2013. Characterization of silica-functionalized carbon nanotubes dispersed in water. *Journal of Nanoparticle Research* 15(1):1–10.

Su, X., Y. Shuai, Z. Guo, and Y. Feng. 2013. Functionalization of multi-walled carbon nanotubes with thermo-responsive azide-terminated poly(N-isopropylacrylamide) via click reactions. *Molecules* 18(4):4599–612.

Susumu, K., B. C. Mei, and H. Mattoussi. 2009. Multifunctional ligands based on dihydrolipoic acid and polyethylene glycol to promote biocompatibility of quantum dots. *Nat Protoc* 4(3):424–36.

Tarazona-Vasquez, F., and P. B. Balbuena. 2004. Complexation of the lowest generation poly(amidoamine)-NH_2 dendrimers with metal ions, metal atoms, and Cu(II) hydrates: An *ab initio* study. *J. Phys. Chem. B* 108:15992–16001.

Templeton, A. C., W. P. Wuelfing, and R. W. Murray. 2000. Monolayer-protected cluster molecules. *Acc Chem Res* 33(1):27–36.

Thanha, N. T. K., and L. A.W. Green. 2010. Functionalisation of nanoparticles for biomedical applications. *Nano Today* 5:213–30.

Thode, C. J., and M. E. Williams. 2008. Kinetics of 1,3-dipolar cycloaddition on the surfaces of Au nanoparticles. *J Colloid Interface Sci* 320(1):346–52.

Tiwari, D. K., S. Tanaka, Y. Inouye, K. Yoshizawa, T. M. Watanabe, and T. Jin. 2009. Synthesis and characterization of anti-HER2 antibody conjugated CdSe/CdZnS quantum dots for fluorescence imaging of breast cancer cells. *Sensors (Basel)* 9(11):9332–64.

Tkachenko, A., H. Xie, S. Franzen, and D. L. Feldheim. 2005. Assembly and characterization of biomolecule-gold nanoparticle conjugates and their use in intracellular imaging. *Methods Mol Biol* 303:85–99.

Tong, C. L., E. Eroglu, and C. L. Raston. 2014. *In situ* synthesis of phosphate binding mesocellular siliceous foams impregnated with iron oxide nanoparticles. *RSC Advances* 4(87):46718–22.

Toulemon, D., B. P. Pichon, X. Cattoen, M. W. Man, and S. Begin-Colin. 2011. 2D assembly of non-interacting magnetic iron oxide nanoparticles via "click" chemistry. *Chem Commun (Camb)* 47(43):11954–6.

Turkevich, J., Stevenson, P. C., and Hillier, J. 1951. A study of the nucleation and growth processes in the synthesis of colloidal gold. *Discuss. Faraday Soc* 11:55–75.

Uenuma, M., T. Ban, N. Okamoto et al. 2013. Memristive nanoparticles formed using a biotemplate. *RSC Advances* 3(39):18044–8.

Ulusoy, M., R. Jonczyk, J. G. Walter et al. 2016. Aqueous synthesis of PEGylated quantum dots with increased colloidal stability and reduced cytotoxicity. *Bioconjug Chem* 27(2):414–26.

Veerapandian, M., and K. Yun. 2011. Functionalization of biomolecules on nanoparticles: Specialized for antibacterial applications. *Appl Microbiol Biotechnol* 90(5):1655–67.

Vericat, C., M. E. Vela, and R. C. Salvarezza. 2005. Self-assembled monolayers of alkanethiols on Au(111): Surface structures, defects and dynamics. *Phys Chem Chem Phys* 7(18):3258–68.

Voggu, R., P. Suguna, S. Chandrasekaran, and C. N. R. Rao. 2007. Assembling covalently linked nanocrystals and nanotubes through click chemistry. *Chemical Physics Letters* 443(1–3):118–21.

von Maltzahn, G., J. H. Park, A. Agrawal et al. 2009. Computationally guided photothermal tumor therapy using long-circulating gold nanorod antennas. *Cancer Res* 69(9):3892–900.

Wang, K. T., I. Iliopoulos, and R. Audebert. 1988. Viscometric behaviour of hydrophobically modified poly(sodium acrylate). *Polymer Bulletin* 20(6):577–82.

Wang, L., J. Li, S. Song, D. Li, and F. Chunhai. 2009. Biomolecular sensing via coupling DNA-based recognition with gold nanoparticles. *Journal of Physics D: Applied Physics* 42(20):203001.

Wang, L., K. G. Neoh, E. T. Kang, B. Shuter, and S.-C. Wang. 2009. Superparamagnetic hyperbranched polyglycerol-grafted Fe_3O_4 nanoparticles as a novel magnetic resonance imaging contrast agent: An *in vitro* assessment. *Advanced Functional Materials* 19(16):2615–22.

Wang, L. S., and R. Y. Hong. 2011. Synthesis, surface modification and characterization of nanoparticles. In *Advances in Nano-composites—Synthesis, Characterization and Industrial Applications*, Boreddy Reddy (ed.), INTECH, China, pp. 289–323.

Wang, M., and N. O. Petersen. 2015. Characterization of phospholipid-encapsulated gold nanoparticles: A versatile platform to study drug delivery and cellular uptake mechanisms. *Canadian Journal of Chemistry* 93(2):265–71.

Wang, X. S., T. E. Dykstra, M. R. Salvador, I. Manners, G. D. Scholes, and M. A. Winnik. 2004. Surface passivation of luminescent colloidal quantum dots with poly(dimethylaminoethyl methacrylate) through a ligand exchange process. *J Am Chem Soc* 126(25):7784–5.

Wang, Y., X. Teng, J.-S. Wang, and H. Yang. 2003. Solvent-free atom transfer radical polymerization in the synthesis of Fe_2O_3@polystyrene core-shell nanoparticles. *Nano Letters* 3(6):789–93.

Wang, Y., H. Zhu, M. D. Becker et al. 2013. Effect of surface coating composition on quantum dot mobility in porous media. *Journal of Nanoparticle Research* 15(8):1805.

Whitsitt, E. A., and A. R. Barron. 2003. Silica coated single walled carbon nanotubes. *Nano Letters* 3(6):775–8.

Wu, W., Q. He, and C. Jiang. 2008. Magnetic iron oxide nanoparticles: Synthesis and surface functionalization strategies. *Nanoscale Res Lett* 3(11):397–415.

Wu, W., Z. Wu, T. Yu, C. Jiang, and K. Woo-Sik. 2015. Recent progress on magnetic iron oxide nanoparticles: Synthesis, surface functional strategies and biomedical applications. *Science and Technology of Advanced Materials* 16(2):023501.

Wu, Y., J. S. Hudson, Q. Lu et al. 2006. Coating single-walled carbon nanotubes with phospholipids. *J Phys Chem B* 110(6):2475–8.

Wuelfing, W. P., S. M. Gross, D. T. Miles, and R. W. Murray. 1998. Nanometer gold clusters protected by surface-bound monolayers of thiolated poly(ethylene glycol) polymer electrolyte. *Journal of the American Chemical Society* 120(48):12696–7.

Wunderbaldinger, P., L. Josephson, and R. Weissleder. 2002. Tat peptide directs enhanced clearance and hepatic permeability of magnetic nanoparticles. *Bioconjug Chem* 13(2):264–8.

Xia, X. R., N. A. Monteiro-Riviere, S. Mathur et al. 2011. Mapping the surface adsorption forces of nanomaterials in biological systems. *ACS Nano* 5(11):9074–81.

Xie, J., C. Xu, N. Kohler, Y. Hou, and S Sun. 2007. Controlled PEGylation of monodisperse Fe_3O_4 nanoparticles for reduced non-specific uptake by macrophage cells. *Adv. Mater.* 19:3163–6.

Xu, L., and C. Chen. 2012. Physiological behavior of quantum dots. *Wiley Interdiscip Rev Nanomed Nanobiotechnol* 4(6):620–37.

Yang, C., A. Tian, and Z. Li. 2016. Reversible cardiac hypertrophy induced by PEG-coated gold nanoparticles in mice. *Sci Rep* 6:20203.

Yang, D., J. Ma, M. Gao et al. 2013. Suppression of composite nanoparticle aggregation through steric stabilization and ligand exchange for colorimetric protein detection. *RSC Advances* 3(25):9681–6.

Yang, S. T., K. A. Fernando, J. H. Liu et al. 2008. Covalently PEGylated carbon nanotubes with stealth character *in vivo*. *Small* 4(7):940–4.

Yang, X., H. Hong, J. J. Grailer et al. 2011. cRGD-functionalized, DOX-conjugated, and (6)(4)Cu-labeled superparamagnetic iron oxide nanoparticles for targeted anticancer drug delivery and PET/MR imaging. *Biomaterials* 32(17):4151–60.

Yeh, Y. C., D. Patra, B. Yan et al. 2011. Synthesis of cationic quantum dots via a two-step ligand exchange process. *Chem Commun (Camb)* 47(11):3069–71.

Yoon, T. J., K. N. Yu, E. Kim et al. 2006. Specific targeting, cell sorting, and bioimaging with smart magnetic silica core-shell nanomaterials. *Small* 2(2):209–15.

Yuan, J. J., S. P. Armes, Y. Takabayashi et al. 2006. Synthesis of biocompatible poly[2-(methacryloyloxy) ethyl phosphorylcholine]-coated magnetite nanoparticles. *Langmuir* 22(26):10989–93.

Yue-Jian, C., T. Juan, X. Fei et al. 2010. Synthesis, self-assembly, and characterization of PEG-coated iron oxide nanoparticles as potential MRI contrast agent. *Drug Dev Ind Pharm* 36(10):1235–44.

Zabihi, O., M. Ahmadi, M. Akhlaghi bagherjeri, and M. Naebe. 2015. One-pot synthesis of aminated multi-walled carbon nanotube using thiol-ene click chemistry for improvement of epoxy nanocomposites properties. *RSC Advances* 5(119):98692–9.

Zhang, P., Z. A. Qiao, X. Jiang, G. M. Veith, and S. Dai. 2015. Nanoporous ionic organic networks: Stabilizing and supporting gold nanoparticles for catalysis. *Nano Lett* 15(2):823–8.

Zhang, W. D., L. Shen, I. Y. Phang, and T. Liu. 2004. Carbon nanotubes reinforced nylon-6 composite prepared by simple melt-compounding. *Macromolecules* 37(2):256–9.

Zhang, X. D., D. Wu, X. Shen et al. 2012. Size-dependent radiosensitization of PEG-coated gold nanoparticles for cancer radiation therapy. *Biomaterials* 33(27):6408–19.

Zhang, Y. 2015. Carbon nanotubes/polyacrylic acid coating materials prepared by *in situ* polymerization technique. *Polymer Bulletin* 72(10):2519–26.

Zhang, Y., and A. Clapp. 2011. Overview of stabilizing ligands for biocompatible quantum dot nanocrystals. *Sensors (Basel)* 11(12):11036–55.

Zhang, Y., and J. Zheng. 2008. Direct electrochemistry and electrocatalysis of cytochrome c based on chitosan room-temperature ionic liquid-carbon nanotubes composite. *Electrochimica Acta* 54(2):749–54.

Zhang, Y. B., M. Kanungo, A. J. Ho et al. 2007. Functionalized carbon nanotubes for detecting viral proteins. *Nano Lett* 7(10):3086–91.

Zhou, C., W. Zhang, M. Xia et al. 2013. Synthesis of poly(acrylic acid) coated-Fe_3O_4 superparamagnetic nano-composites and their fast removal of dye from aqueous solution. *J Nanosci Nanotechnol* 13(7):4627–33.

Zhu, J., M. Yudasaka, M. Zhang, and S. Iijima. 2004. Dispersing carbon nanotubes in water: A noncovalent and nonorganic way. *The Journal of Physical Chemistry B* 108(31):11317–20.

Zrazhevskiy, P., M. Sena, and X. Gao. 2010. Designing multifunctional quantum dots for bioimaging, detection, and drug delivery. *Chemical Society Reviews* 39(11):4326–54.

6
Nanoparticle Contrast Agents for Medical Imaging

Rabee Cheheltani, Johoon Kim, Pratap C. Naha, and David P. Cormode

CONTENTS

6.1 Introduction 220
    6.1.1 Quantum Dots 221
    6.1.2 Nanocrystals 222
    6.1.3 Micelles 223
    6.1.4 Liposomes 223
    6.1.5 Polymeric Nanoparticles 224
    6.1.6 Natural Nanoparticles 224
    6.1.7 Dendrimers 224
    6.1.8 Emulsions 224
    6.1.9 Imaging Modalities and Contrast Agents 224
6.2 Computed Tomography 225
    6.2.1 Iodine-Containing Nanoparticles 226
    6.2.2 Gold Nanoparticles 227
    6.2.3 Upconverting Nanoparticles 227
    6.2.4 Other Heavy Metal Nanoparticles 227
6.3 Magnetic Resonance Imaging 228
    6.3.1 T_1 Contrast Agents 229
    6.3.2 T_2 Contrast Agents 229
    6.3.3 Chemical Exchange Saturation Transfer Agents 230
    6.3.4 Fluorine Contrast Agents 231
6.4 Fluorescence Imaging 232
    6.4.1 Quantum Dots 232
    6.4.2 Gold Nanoparticles 233
    6.4.3 Carbon Nanostructures 234
    6.4.4 Fluorescently Labeled Liposomes 235
    6.4.5 Cerenkov Luminescence Imaging 235
6.5 Nuclear Imaging 235
    6.5.1 Liposomes 236
    6.5.2 Polymeric Nanoparticles 236
    6.5.3 Radiolabeled Inorganic Nanoparticles 236
6.6 Photoacoustic Imaging 237
    6.6.1 Metal Nanoparticles 238
    6.6.2 Fluorophores 238
6.7 Surface-Enhanced Raman Spectroscopy 239
6.8 Discussion 241
Acknowledgments 242
References 242

6.1 Introduction

Contrast agents, or contrast media, are materials that are introduced into the body to enhance the detectability of target tissue, disease sites, or tumors in various medical imaging applications. They are most commonly administered intravenously as blood pool agents for imaging blood vessels or orally to image the gastrointestinal (GI) tract. Developing nano-sized contrast agents is one of the major applications of nanotechnology in biomedical research since nanoparticles can have several advantages over conventional contrast agents (Cormode et al. 2009). They can contain higher payloads of contrast material to create higher intensity image contrast. They can also be designed to have various beneficial properties such as size and surface modifications for a longer circulation time in the body for blood pool contrast agents; targeting moieties for accumulation in specific disease sites; passive transport and accumulation into tumors via the enhanced permeability and retention (EPR) effect; and size and shape modifications for specific optical, magnetic, nuclear, or x-ray properties. Furthermore, nanoparticle platforms allow easy development of multimodal and multifunctional properties.

A general depiction of a nanoparticle contrast agent is shown in Figure 6.1 (Cormode et al. 2009). The contrast material can be incorporated in the core, in the shell, or on the surface of the nanoparticle. The nanoparticle may be coated for stability in biological media. The surface of the nanoparticle may be further modified to include targeting ligands for disease site accumulation. Multiple types of contrast-generating material can be included in the same nanoparticle platform, such as an iron oxide core for magnetic resonance imaging (MRI) contrast and fluorophores for fluorescence imaging.

Contrast agent nanoparticles should be biocompatible with acceptable pharmacokinetic properties in the body; therefore, nanoparticles should be extensively studied for their biocompatibility with cells *in vitro* and in animals before further translation and clinical studies on humans. One major aspect of biocompatibility is excretion from the body (Peer et al. 2007). While it is advantageous for contrast agents to stay in the body for the duration of the imaging session, they should ultimately either clear from the body or degrade into biologically compatible components. Another major challenge for nanoparticles is their detection and phagocytosis by the reticuloendothelial system (RES). The RES can rapidly clear nanoparticles from the blood, preventing accumulation at target sites. There are several strategies for synthesizing stealth particles that can evade the RES. One of the

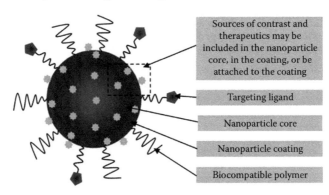

FIGURE 6.1
Generalized schematic of a nanoparticle contrast agent. (Adapted with permission from Cormode D. P. et al. *Arteriosclerosis, Thrombosis, and Vascular Biology.* 2009;29(7):992–1000.)

most common methods is to attach polyethylene glycol (PEG) chains on the surface of nanoparticles, a process known as PEGylation (Jokerst et al. 2011).

The size of nanoparticles can affect their behavior in the body. Nanoparticles smaller than about 6 nm are swiftly excreted via the kidneys and eliminated via the urine (Choi et al. 2007). Nanoparticles that are 6 nm or larger in all dimensions will not experience rapid renal clearance and can circulate for a long time given the appropriate coating (van Schooneveld et al. 2008). Nanoparticles of 400 nm or more start to become comparable in size to capillaries in the lungs and liver, so nanoparticles in that size range can have rapid clearance from the circulation due to accumulation in these capillaries. Therefore, depending on the application, nanoparticle contrast agents are usually in the 1–400 nm size range. Furthermore, some tissues such as tumors are permeable to nanoparticles, while tissues such as muscle or brain are not, since tumors have gaps in their endothelium in the hundreds of nanometers range, while in brain or muscle, the endothelial gaps are so small as to essentially prevent entry to nanoparticles. Smaller nanoparticles can penetrate further into tumors than larger nanoparticles (Sykes et al. 2014), which can be a consideration in contrast agent design. On the other hand, if the goal is to image the vasculature or vascular targets, larger nanoparticles, whose extravasation is poor, may have better performance. For a specific nanoparticle platform or imaging technique, size may have an effect; for example, the emission wavelength of quantum dots depends on the nanoparticle size (Medintz et al. 2005).

Nanoparticle contrast agents are used in conjunction with several medical imaging modalities, each of which has advantages and disadvantages for *in vivo* imaging. Table 6.1 summarizes the spatial resolution, sensitivity, and penetration depth of commonly used imaging techniques.

A brief overview of many of the different kinds of nanoparticles relevant to biomedical imaging is provided below. These nanoparticles can be divided into two major groups: contrast-generating components (quantum dots, nanocrystals) and contrast agent carriers (micelles, liposomes, polymeric nanoparticles, natural nanoparticles, and so forth). A schematic of each nanoparticle type is shown in Figure 6.2. In the sections to follow, we will describe the major medical imaging modalities and the most important forms of nanoparticle contrast agents used with these modalities.

6.1.1 Quantum Dots

Quantum dots (QDs) are semiconductor nanocrystals 1–10 nm in diameter that have fluorescent properties with tunable emission wavelengths (Figure 6.2a). They have very wide excitation bands, are nonphotobleaching, and have narrow and efficient emission (Medintz et al. 2005). They can be conjugated with targeting ligands to allow imaging

TABLE 6.1

Comparison of Commonly Used Medical Imaging Modalities

Imaging Modalities	Spatial Resolution	Sensitivity (mol)	Penetration Depth
Computed Tomography (CT)	50 um	10^{-6}	No limit
Magnetic Resonance Imaging (MRI)	50 um	10^{-9} to 10^{-6}	No limit
Nuclear Imaging	1–2 mm	10^{-14}	No limit
Fluorescence Imaging	1–3 mm	10^{-12}	<1 to <5 cm
Photoacoustic Imaging	50 um	10^{-12}	<5 cm

Source: Adapted with permission from Hahn M.A. et al. *Analytical and Bioanalytical Chemistry*. 2011;399(1):3–27; Chinen A.B. et al. *Chemical Reviews*. 2015;115(19):10530–10574.

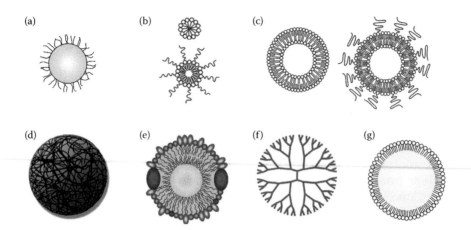

FIGURE 6.2
Schematics of different nanoparticle types developed as contrast agents for medical imaging: (a) nanocrystals or quantum dots, (b) micelles, (c) liposomes, (d) polymeric nanoparticles, (e) multifunctional lipoproteins, (f) dendrimers and (g) emulsions. (Adapted with permission from Mulder W. J. M. et al. *NMR in Biomedicine.* 2006;19(1):142–164; Griset A. P. et al. *Journal of the American Chemical Society.* 2009;131(7):2469–2471; Cormode D. P. et al. *Nano Letters.* 2008;8(11):3715–3723; Myers V. S. et al. *Chemical Science.* 2011;2(9):1632.)

of specific targets at the molecular level. Apart from their applications in fluorescent imaging, quantum dots have been utilized to create multimodal imaging agents, especially by combining them with magnetic materials. Recent advances in improving the imaging capabilities of quantum dots include precise atom placement to reduce size and shape variation (Fölsch et al. 2014), doping of different ions to avoid self-quenching and to lower the toxicity (Wu and Yan 2013), and synthesis of different shape and size QDs to optimize quantum mechanical properties (Ghosh Chaudhuri and Paria 2012).

6.1.2 Nanocrystals

Several metal elements and their oxides have been explored for the synthesis of nanostructures, which can be in the form of nanospheres, nanorods, plates, stars, cubes, and other polyhedral shapes (Tao et al. 2008). Nanoparticles synthesized from gold (AuNP) (Mieszawska et al. 2013, Li and Chen 2015), silver (Homan et al. 2012), iron oxide (IONP) (Xi et al. 2014), copper (Ku et al. 2012, Pan et al. 2012), and bismuth (Brown et al. 2014, Naha et al. 2014) are some examples of nanocrystals studied for biomedical imaging applications. These nanocrystals are typically synthesized from metal salts either directly or by substituting their coating for improved biocompatibility (Cormode et al. 2013). They can be the major component of the nanoparticle contrast agent or a relatively minor component embedded within a larger carrier structure. For example, iron oxides have been delivered using liposomes, polymeric nanoparticles, and emulsions. Among different nanocrystals, AuNPs have been used extensively in both imaging and nonimaging biomedical applications because of their controllable size, shape, and surface composition, as well as their applications in several different imaging modalities (Dreaden et al. 2012, Myroshnychenko et al. 2012, Saha et al. 2012, Mieszawska et al. 2013). Furthermore, AuNPs have several therapeutic applications in drug delivery, gene delivery, photothermal therapy, and radiotherapy, which makes them attractive candidates as theranostic agents (Mieszawska et al. 2013). Figure 6.3 provides an overview of different structures and coatings of gold nanoparticles.

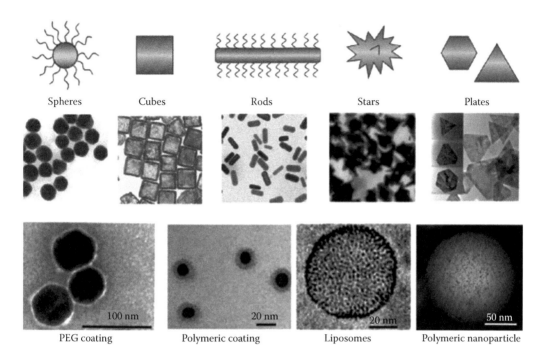

FIGURE 6.3
Schematic of gold nanoparticles as contrast agents for non-invasive imaging. (Adapted with permission from Mieszawska A. J. et al. *Molecular Pharmaceutics*. 2013;10(3):831–847.)

6.1.3 Micelles

Micelles are nanoparticles that are formed from amphiphiles, with a hydrophobic core composed of the hydrophobic tails of the amphiphiles and a hydrophilic surface (Figure 6.2b). They are between 5 and 200 nm in size and are formed by lipids with wider hydrophilic headgroups than their hydrophobic tails. These lipids self-assemble into a micelle when their concentration in an aqueous solution exceeds their critical micelle concentration (CMC) (Cormode et al. 2014). The hydrophobic core can carry various water-insoluble compounds that are either contrast agents or a combination of contrast agents and therapeutic compounds for simultaneous imaging and drug delivery (Lee et al. 2013). In addition, the amphiphiles that make up the micelle may be diagnostically active themselves (Vucic et al. 2009).

6.1.4 Liposomes

Liposomes are hollow amphiphilic spheres with varying numbers of lipid bilayers that encapsulate an aqueous core (Figure 6.2c). They were one of the first nanomaterials to be developed, with their discovery by Bangham in 1965 (Bangham et al. 1965). They are synthesized via several methods, including mechanical dispersion, solvent dispersion, and emulsion preparation. Typically, the lipid is dried into a thin film then dispersed in a solution, allowing the lipid to self-assemble in various layers and configurations. Liposomes are easy to modify and can encapsulate drugs and contrast agents, as they can contain both hydrophilic and hydrophobic payloads by loading them either into the hydrophilic core or inside the lipid bilayer, respectively (Mulder et al. 2009).

6.1.5 Polymeric Nanoparticles

Several natural and synthetic polymers have been studied to form nanoparticle platforms to carry contrast agents (Figure 6.2d). Polymers can either form a shell to contain the contrast agent payload (Ghosh Chaudhuri and Paria 2012) or be the core of the nanostructure, carrying the contrast agents on their surface (Chhour et al. 2014). Dextran, poly(lactic-co-glycolic acid) (PLGA), polycaprolactone (PCL) and alginate are examples of Food and Drug Administration (FDA)-approved polymers that have been further studied for forming contrast agent nanoparticles (Tassa et al. 2011, Naha et al. 2014, Wang et al. 2014).

6.1.6 Natural Nanoparticles

Natural nanoparticles can be modified to perform as delivery vehicles for contrast agents (Cormode et al. 2010a). Lipoproteins, whose primary function is carrying cholesterol and other lipids in the body, are nanoparticles with the size range of 7–1000 nm (Figure 6.2e). Contrast-generating components can be included either in the core or in the coating of the lipoprotein (Cormode et al. 2008b). Virus shells or capsids have also been studied for carrying contrast-generating material either in their core, in the interfaces between their subunits, or on their surface (Li and Wang 2014). Ferritin is another natural nanoparticle. Its function is to store and manage iron in the body. As a consequence, it has potential as a magnetic resonance imaging contrast agent. Magnetic properties of ferritin can either be harnessed from its endogenous iron oxide core or by replacing the iron oxide core with alternative payloads (Zhen et al. 2014).

6.1.7 Dendrimers

Dendrimers are three-dimensional tree-like structures formed by polymeric repeatedly branching units linked to the core molecule (Figure 6.2f). The number of the branching units, or generations, determines the complexity of the dendrimers (Lee et al. 2013). Contrast-generating agents can be conjugated on the surface of dendrimers, which are currently commercially available (Louie 2010, Röglin et al. 2011).

6.1.8 Emulsions

Emulsions are formed by suspension of hydrophobic oils in aqueous solution, stabilized by amphiphilic coating (Figure 6.2g). They have been studied as nongaseous ultrasound contrast agents (Lanza et al. 1996). They can be labeled to act as contrast agents via the use of oils that generate contrast, dispersion of contrast-generating media in the oil, or inclusion of contrast-generating lipids in the amphiphile coating (Cormode et al. 2009, Lee et al. 2013).

6.1.9 Imaging Modalities and Contrast Agents

Nanoparticle contrast agents of many different forms are used together with several imaging modalities. In this chapter, contrast agents for computed tomography (CT), magnetic resonance imaging, fluorescence imaging, nuclear imaging, and photoacoustic imaging will be discussed in detail, as they are the most widely used imaging modalities for which nanoparticles are relevant. Nanoparticle contrast agents in CT have been based on many heavy elements, such as iodine, gold, bismuth, bromine, and tantalum, to name a few. Most of the studies in MRI contrast agents are done with different types of gadolinium chelate-labeled nanoparticles or iron oxide-based nanoparticles. Fluorescence imaging uses a wide

TABLE 6.2

Summary of Different Imaging Modalities and Relevant Nanoparticle Technologies Discussed in This Chapter

Imaging Modality	Nanoparticle Contrast Agents Discussed
Computed tomography	Iodine-containing nanoparticles, gold nanoparticles, upconverting nanoparticles, other heavy metal nanoparticles
Magnetic resonance imaging	Gadolinium chelates, ultrasmall iron oxides, superparamagnetic iron oxides, chemical exchange saturation transfer (CEST) agents, fluorine contrast agents
Fluorescence imaging	Quantum dots, gold nanoparticles, carbon nanostructures, fluorescently labeled liposomes, Cerenkov luminescence imaging
Nuclear imaging	Liposomes, polymeric nanoparticles, radiolabeled inorganic nanoparticles
Photoacoustic imaging	Metal nanoparticles, fluorophores
Surface-enhanced Raman spectroscopy	Gold nanostructures

variety of quantum dots, such as Ag_2S, CdTe, or CdSe, as well as gold and carbon-based nanostructures and fluorophore tagged polymer-based nanoparticles. Nuclear imaging utilizes different radioisotopes embedded in structures of nanoparticles. ^{99m}Tc, ^{111}In, ^{125}I for SPECT and ^{18F}-FDG, ^{64}Cu, ^{67}Ga, and ^{89}Zr for PET imaging are few of the frequently used radioisotopes. For photoacoustic imaging, gold and silver nanoparticles, and dye-embedded nanoparticles (e.g., Prussian blue), are most commonly used.

Although these widely used nanoparticle contrast agents have many advantages over molecule-based contrast agents, such as longer blood retention and better signal generation, considerable scope remains for their improvement. For instance, complete clearance is a challenge for some nanoparticles such as those formed from gold (Cheheltani et al. 2016). Increasing quantum yield is a constant goal in fluorescence imaging, but quantum dots that have high quantum yields often have poor biocompatibility. Gadolinium chelates, one of the most commonly studied MRI nanoparticle contrast agents, suffer from biocompatibility and stability issues.

The field of nanoparticle contrast agents for medical imaging is continuously expanding, and discoveries of new materials and synthetic methods are frequently made. We present to the reader an overview of the field for each imaging modality. Table 6.2 summarizes major nanoparticle technologies discussed in this chapter for their application in medical imaging.

6.2 Computed Tomography

Computed tomography is one of the most common and well-established medical imaging modalities. The costs of CT imaging are reasonable, and it has the advantage of deep tissue penetration. CT scanners are composed of two main components: the x-ray source and the detectors, which are positioned opposite each other and rotate 360° around the patient. In the x-ray source, electrons are fired at a target (typically tungsten) inside a high-voltage vacuum tube, which results in x-ray emission. This x-ray beam is directed into the patient and, after interacting with the atoms in the organs of the patient's body, is attenuated through absorption and scattering. This attenuation of the transmitted radiation is measured via the detector array. As the source and detector rotate, a 360° dataset is built up, which is processed into a tomographic image set that makes up the CT scan (Figure 6.4).

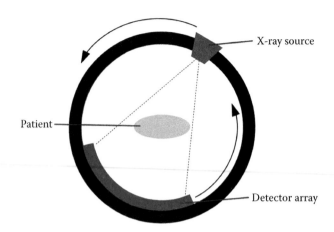

FIGURE 6.4
Schematic of a CT scanner. (Adapted with permission from Cormode D. P. et al. *Contrast Media and Molecular Imaging.* 2014;9(1):37–52.)

Different materials attenuate x-rays differently. This attenuation is measured in Hounsfield units (HU), where the signal in water is defined as 0 HU and that in air as −1000 HU. Consequently, in the patient's body, the lungs have the weakest attenuation, as they are mostly composed of air, and the bones the strongest attenuation (∼400–1000 HU), with the soft tissues attenuating x-rays in between and close to the attenuation of water (0–200 HU). Because soft tissues have similar attenuation properties, making it difficult to distinguish between different tissues, contrast materials with high atomic numbers and thus high x-ray attenuation are utilized. For example, water-soluble iodinated small molecules are injected in the bloodstream for cardiovascular angiography, and barium sulfate suspensions are used for imaging gastrointestinal tract. While iodinated contrast agents have strong CT attenuations and are well established for use in the clinic, they are excreted rapidly through the kidneys, and sometimes several doses have to be administered in order to image the target organ over time. Furthermore, these agents are contraindicated for use in patients with renal insufficiency, as the use of these agents in such patients leads to kidney damage, an event known as contrast-induced nephropathy (Jun et al. 2005, Solomon and Dumouchel 2006, Tepel et al. 2006, Stacul et al. 2011). Lastly, iodinated compounds are nonspecific blood pool agents, and they are limited in their application when imaging a specific organ or tumor is required. To overcome these limitations, nanoparticle CT contrast agents have been developed. The sections below provide an overview of the different types of these nanoparticles.

6.2.1 Iodine-Containing Nanoparticles

To take advantage of well-established CT contrast properties of iodinated contrast agents, several nanoparticle platforms, with longer circulation time and specificity to certain organs, have been developed to contain iodinated molecules. Micelles and nanoemulsions can contain hydrophobic iodinated contrast agent materials by stabilizing them in their hydrophobic core, which can also contain hydrophobic drugs for simultaneous CT imaging and drug delivery (Hallouard et al. 2010). Liposomes can carry water-soluble iodinated contrast agents in their core, and a liposome-encapsulated formulation of iodixanol was shown to create contrast in liver tissue and tumors (Leander et al. 2001). Dendrimers, polymers, and metal-organic frameworks are other platforms that have been used to carry iodinated agents for CT imaging (Lee et al. 2013).

6.2.2 Gold Nanoparticles

Gold is an inert element with a high atomic number that makes it a superior contrast material for CT. Gold nanoparticle synthesis is very well established in the literature; gold nanoparticles are highly biocompatible (Thakor et al. 2011, Mieszawska et al. 2013, Naha et al. 2015), and they have been found to be highly potent x-ray contrast agents (Cai et al. 2007, Cormode et al. 2008a, 2010b, Arifin et al. 2011, Galper et al. 2012, Naha et al. 2015). They produce stronger contrast than iodinated agents (Jackson et al. 2010, Galper et al. 2012), can circulate longer as blood pool agents (Cai et al. 2007, Kim et al. 2007), can specifically be detected with techniques such as dual or spectral CT (Cormode et al. 2010b, Galper et al. 2012, Schirra et al. 2012), and have been found to be highly biocompatible. Furthermore, clinical trials for AuNPs as therapeutic agents have reported no adverse effects (Libutti et al. 2010), which makes them promising agents for translation into clinics. Gold nanoparticles represent by far the largest category of nanoparticle contrast agent for CT, with new reports being published frequently. For example, gold nanoparticles have recently been reported to allow cell tracking for CT (Meir et al. 2015).

6.2.3 Upconverting Nanoparticles

Upconverting nanoparticles (UCNPs) take advantage of lanthanide ions to emit light at a lower wavelength than excitation wavelength. One of the advantages of these agents is the high efficiency of the upconversion process that leads to high imaging sensitivity (Lee et al. 2013). Due to the high atomic number and high x-ray attenuation of lanthanides, such UCNPs have been developed with stronger CT contrast properties than iodinated agents. Er^{3+}/Yb^{3+} or Tm^{3+}/Yb^{3+} co-doped $NaYF_4$ particles, as well as Er^{3+}-doped Yb_2O_3, are examples of such CT contrast agents. By including Gd^{3+} in the formulation, particles can also have MRI applications (Lee et al. 2013).

6.2.4 Other Heavy Metal Nanoparticles

Other heavy metals have been proposed as contrast agents, such as tantalum, bismuth, platinum, ytterbium, and so forth (Cormode et al. 2014). Bonitatibus et al. developed a tantalum oxide nanoparticle platform that has a hydrophilic siloxane molecular coating with low tissue retention and no toxicity for the kidneys (Bonitatibus et al. 2012). Bismuth nanoparticles are another class of nanoparticles with high x-ray attenuation. Bismuth is an inexpensive and biocompatible heavy metal. Naha et al. developed dextran-coated bismuth iron oxide nanoparticles (BIONs) as multifunctional nanoparticles for CT and MRI imaging (Naha et al. 2014). Dextran-coated iron oxides are clinically approved for MRI imaging; therefore, basing contrast agents on this platform may lead to a translatable agent. BIONs were synthesized through co-precipitation of ferrous chloride, ferric chloride, and bismuth nitrate in the presence of dextran (Figure 6.5a). BIONs did not show toxicity *in vitro* after incubation with human hepatocellular liver carcinoma and human foreskin fibroblast cells, and CT and MRI imaging of these particles after injection into the bloodstream confirmed their strong contrast in both modalities (Figure 6.5b,d,e). The increase in CT contrast in the bladder over time (Figure 6.5c) and in *in vitro* experiments confirmed the biodegradation of the BION composite nanohybrid.

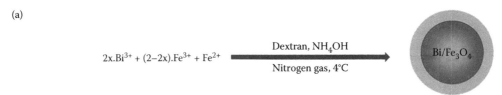

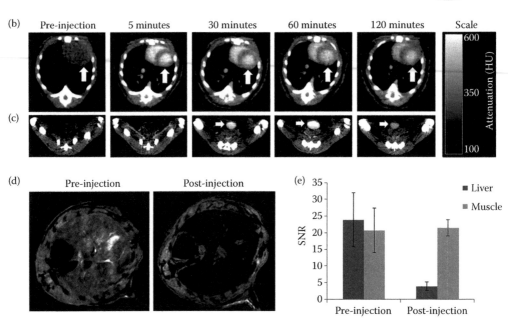

FIGURE 6.5
(a) Schematic of the synthesis of dextran coated BION as multifunctional contrast agents for CT and MRI. (b) CT images of a mouse injected with dextran coated BION 5 minutes to 120 minutes post injection, showing strong contrast in the ventricles of the heart. (c) Development of contrast in the bladder indicated the excretion of particles through kidneys. (d) T_2 MRI images of the liver of a mouse pre- and 2 hours post injection of BION particles. (e) Quantification of the MRI intensity of the images in (d). (Adapted with permission from Naha P. C. et al. *Journal of Materials Chemistry B*. 2014;2(46):8239–8248.)

6.3 Magnetic Resonance Imaging

MRI is a commonly used imaging technique that is capable of giving high-resolution images of the anatomy, and provides particularly good soft tissue contrast. However, it is a relatively slow imaging technique; therefore, imaging regions where there is motion can be challenging, and it is expensive. It relies on the use of very powerful magnetic fields, such as 1.5–3 T for routine clinical use, and higher fields, such as 9.4 T, for small animal imaging. In comparison, the earth's magnetic field is 25–65 μT, roughly 100,000 times lower. The enormous magnetic fields used in MRI produce a small energy differential in the spin states of protons. This gap results in a small difference in the number of protons in the upper state versus the lower state. Radio waves are used to excite more protons into the upper state. Over a short period of time, the excited protons relax into their ground state, emitting radio waves. These radio waves are recorded and MR images are derived from this data. Small variations within the magnetic field of the scanner, known as gradients, provide spatial information, since the frequency of the emitted radio waves depends on the strength of the magnetic field. The relaxation rate of protons is highly dependent on their environment, so water in different tissues will relax at different rates and thereby produce different intensities in images, which is the source of soft tissue contrast in MRI. The rate of relaxation can be described by two parameters known as T_1 and T_2. The two most established classes of MRI contrast agents are known as T_1 and T_2 contrast agents and affect these parameters. Nanoparticle contrast agents benefit MRI since their high payloads can overcome MRI's relatively poor sensitivity. While the costs of manufacture of MRI contrast agents vary depending on the agent, the material costs for dextran-coated iron oxide agents are cents per gram; therefore, these agents are low cost (Gao et al. 2016). Nanoparticle contrast agents for MRI range widely in size, from 5 to 400 nm (Mulder et al. 2007a). The size used for a given application depends on a complex interplay of factors such as contrast generation mechanism, rotational time, location and type of target, and physical chemistry of the delivery platform.

6.3.1 T_1 Contrast Agents

T_1 contrast agents are paramagnetic and interact with water to locally increase the relaxation rate, which results in a signal increase. Most T_1 agents are based on gadolinium chelates (Caravan 2006), although some are based on manganese (MacDonald et al. 2014) or ultrasmall iron oxides (Zeng et al. 2012). Gadolinium is chelated to prevent its release into biological fluids, since free Gd^{3+} ions produce toxicity. Gadolinium produces MRI contrast via direct contact with water, so the chelates leave 1–2 coordination sites free for water molecules (Aime et al. 2004). Several gadolinium chelates are FDA-approved for use in patients. To form MRI contrast agents, gadolinium chelates have been appended to almost every type of nanoparticle structure known, such as liposomes (Mulder et al. 2007b), micelles (Mulder et al. 2006), lipoproteins (Briley-Saebo et al. 2009), viruses (Allen et al. 2005), and emulsions (Winter et al. 2003), to name just a few examples. These structures have been modified to be targeted (Mulder et al. 2007b), deliver drugs/genes (Writer et al. 2012), and produce additional contrast for other types of imaging modalities such as fluorescence or computed tomography (Cormode et al. 2008a). For example, liposomes that include gadolinium chelate-modified lipids have been used to measure levels of the $\alpha_v\beta_3$-integrin in a mouse model of cancer. A fluorescent label allowed the targeting to be confirmed via microscopy (Mulder et al. 2007b). The main effect of size on gadolinium chelate-carrying nanoparticles is an increase in payload; therefore, nanoparticles of 250 nm have been used to image sparse targets (Morawski et al. 2004). Despite these achievements, progression in this area has slowed due to concerns over the potentially toxic effects of long-term retention of gadolinium in the body (Sieber et al. 2008).

6.3.2 T_2 Contrast Agents

The second major class of MRI contrast agents are those that affect T_2 (or T_2^*). T_2 contrast agents produce what is known as negative contrast; that is, they reduce the intensity in images. The majority of these contrast agents are based on superparamagnetic iron oxides (Cormode et al. 2013), although other types of agent have been reported, most notably doped iron oxides (Lee et al. 2007). Several iron oxide nanoparticle formulations have been approved for human use (Corot et al. 2006). For example, Feridex I.V.® (ferumoxides injectable solution) is an FDA-approved superparamagnetic iron oxide nanoparticle (Mahajan et al. 2013). These and other formulations have been used as labels for cell tracking in numerous studies (Bulte and Kraitchman 2004, Shapiro 2015). Iron oxides are often coated with a polysaccharide known as dextran, although many other coatings have been reported (Cormode et al. 2013). This coating can be aminated to provide moieties to allow attachment of targeting ligands and fluorophores. Such nanoparticles have been used to image a variety of biological disease markers, such as macrophages, VCAM-1, VEGFr, and apoptosis (Schellenberger et al. 2002, Kelly et al. 2005, Reichardt et al. 2005, Weissleder et al. 2014). Iron oxides have also been used in a variety of multifunctional platforms, such as drug or gene delivery agents (Cormode et al. 2011, Gianella et al. 2011). In the case of iron oxide-based nanoparticles, contrast generation increases with increases in core size up to 50 nm (Lee et al. 2007, Huang et al. 2012).

6.3.3 Chemical Exchange Saturation Transfer Agents

The third type of MRI contrast agents is based around chemical exchange saturation transfer (CEST). Balaban and Ward first proposed CEST contrast agents in 2000 (Ward et al. 2000). A CEST contrast agent has numerous protons whose resonance is several ppm offset from that of water. Another criteria for a CEST agent is that the protons must be rapidly exchangeable with bulk water (Castelli et al. 2013). An example of material on which to base CEST agents is polylysine, whose amine protons fit both criteria (McMahon et al. 2006). Contrast generation is produced via CEST by applying a radiopulse at the frequency of the contrast agent protons. This suppresses signal from the contrast agent protons, which is rapidly transferred to the bulk water protons, suppressing the signal from water. CEST agents are therefore negative contrast agents. They have the advantage that preinjection imaging is not needed, as postinjection imaging can be done with and without the saturation pulse—the differences between the two images indicate the presence of the contrast agent. In addition, multiplex imaging can be done, where several different CEST agents can be detected in the same field of view if the agents have different offset frequencies (Liu et al. 2012).

Nanoparticle CEST agents have mostly been formed from liposomes. The liposome carries a cargo that can generate contrast in CEST, such as organic molecules that have protons suitable for use in CEST, such as polylysine, arginine, or glycogen (Liu et al. 2012). Alternatively, the payload can be chelates of metals such as thulium or dysprosium (Terreno et al. 2007). These complexes are known as shift reagents, since they shift the resonance frequency of water; therefore, the water within the liposome is the source of the CEST protons. Alternative platforms have been based on dendrimers (Pikkemaat et al. 2007), perfluorocarbon emulsions (Winter et al. 2006), viruses (Vasalatiy et al. 2008), micelles (Evbuomwan et al. 2012a), and silica (Evbuomwan et al. 2012b). Those agents typically had lanthanide chelates attached. McMahon et al. (2006) published a particularly notable use of a liposomal CEST agent (Chan et al. 2013). An approach to treat diabetes is to implant pancreatic islet cells obtained from donors (Shapiro et al. 2003). These cells will produce the needed insulin, but are vulnerable to attack from the patient's immune system. To prevent this attack, hydrogel encapsulation systems have been developed, which protect the cells, but allow the passage of nutrients to the cells and insulin into the rest of the body (Lim and Sun 1980). Liposomes containing arginine were encapsulated in hydrogels together with cells and were injected into mice. Since the magnitude of CEST effects is reduced when pH drops due to changes in the rate of water exchange and a pH drop accompanies cell death, this system can be used to monitor cell viability *in vivo* (Figure 6.6a). As can be seen in Figure 6.6b, the CEST contrast is reduced over time when immunosuppression is not given to mice, compared to when immunosuppression was given; that is, the CEST signal can be used as a marker for cell viability and thus can be used to monitor the treatment

Nanoparticle Contrast Agents for Medical Imaging

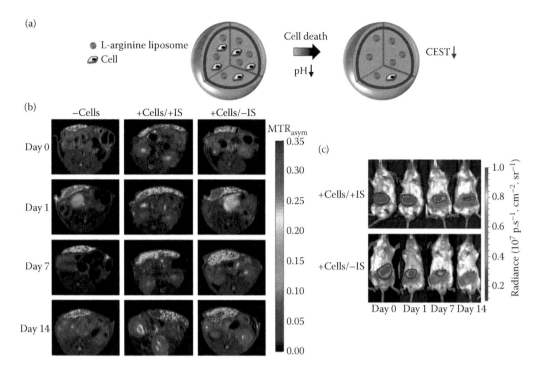

FIGURE 6.6
(a) Schematic depiction of CEST liposomes encapsulated together with cells, to allow detection of cell death via pH changes. (b) MRI images of mice transplanted with CEST liposome-loaded capsules. (c) Bioluminescence images of mice injected with CEST liposome-loaded capsules, indicating the level of cell viability. (Adapted with permission from Chan K. W. et al. *Nature Materials*. 2013;12(3):268–275.)

in vivo. This result was corroborated by bioluminescence measurements, since the cells used expressed luciferase (Figure 6.6c).

6.3.4 Fluorine Contrast Agents

The last type of MRI contrast agent is based on fluorine. In addition to protons, MRI can be used to image certain other kinds of nuclei. Fluorine MRI has the advantage that there is no endogenous fluorine; therefore, any signal arises solely from the administered agent. Despite this advantage, fluorine MRI suffers from low sensitivity. Nevertheless, some interesting results have been published in this area. The majority of fluorine MRI nanoparticle contrast agents have been based on perfluorocarbon emulsions (Ruiz-Cabello et al. 2011). Perfluorocarbons are highly biocompatible—doses of grams per kilogram can be tolerated, and they have been explored as blood substitutes (Mattrey 1989). Ahrens published one of the first examples of fluorine MRI nanoparticle contrast agents (2005). Perfluoro-15-crown-5 ether was formed into an emulsion with lecithin (a mixture of phospholipids derived from foodstuffs). This emulsion was used to label dendritic cells, and it was found that the location of the cells could be imaged *in vivo*. A number of other studies have been performed where perfluorocarbon (PFC) emulsions are used for cell tracking (Flogel et al. 2008, Barnett et al. 2010, Hitchens et al. 2011, Gaudet et al. 2015), and the results of a clinical trial of this approach have recently been published (Ahrens et al. 2014). In the latter study, detection and quantification of PFC-labeled cells in human subjects were successfully performed using images acquired from a clinical 3 T MRI scanner. Perfluorocarbon emulsions have also been used for targeted fluorine MRI imaging (Waters et al. 2008). Some other platforms, such as dendrimers, have been reported as fluorine MRI contrast agents (Criscione et al. 2009).

6.4 Fluorescence Imaging

Fluorescence imaging is a minimally invasive imaging modality with high temporal and spatial resolution (Fernández-Suárez and Ting 2008, Yuan et al. 2013b). It has wide applications in disease diagnosis and visualization for biodistribution, targeting pathways, and pharmacokinetics studies. There are four essential components needed for fluorescence imaging: excitation light source, emission light filter, fluorescent agent, and detector. The light from the excitation light source is directed to the sample, which gets excited at a specific spectrum of wavelength and emits light energy in another specific spectrum of wavelength. Subsequently, the emitted light travels through one or a set of emission filters before reaching the detector. For optimal image acquisition, many factors need to be considered since each fluorescent agent has its own unique photophysical properties, such as absorption spectrum, emission spectrum, extinction coefficient, and quantum yield. Thorough characterization of extinction coefficient and quantum yield is required to have optimal signal intensity. The wavelengths of the excitation light source and the emission light filter also need to be selected carefully to have good intensity and high signal-to-noise ratio.

The near infrared (NIR) window includes wavelengths in the range of 650–1350 nm, and fluorescence probes and systems for imaging biological tissues are commonly designed to function within these frequencies. This is because tissue has low scattering and absorption of light in the NIR window, resulting in a higher efficiency and depth penetration (Smith et al. 2009). The NIR window is used for fluorescent intraoperative imaging applications, for example, where the target anatomical location is labeled with fluorescent probes to guide the surgeon in the procedure, such as the resection of a tumor (Cui et al. 2015). The NIR window is divided into the first NIR (650–950 nm) and second NIR (950–1350 nm) windows. The first NIR window has lower absorption of light by water and lipids, and the second NIR window has higher penetration depth and lower background noise due to tissue autofluorescence (Smith et al. 2009).

Some of the most widely used types of fluorescence agents include fluorescent proteins and organic or synthetic fluorophores (Fernández-Suárez and Ting 2008). Fluorescent proteins are commonly used to carry out genetic studies, whereas fluorophores are widely used for immunofluorescence (van de Linde et al. 2012). Fluorescent agents often have limitations, such as rapid degradation and photobleaching. These limitations restrict the time window of studies and lead to poor signal brightness, which narrows the usefulness of fluorescence imaging for *in vivo* applications. Some nanoparticle contrast agents, such as quantum dots, address these issues. Below, a brief description of the most important applications of nanoparticle contrast agents in fluorescent imaging is provided.

6.4.1 Quantum Dots

Quantum dots have diverse applications and potential in fluorescence imaging due to their narrow and symmetric emission, broad and strong absorption, stability, and low light scattering (Wu and Yan 2013). The surface of QDs can be covalently linked to biorecognition molecules, such as peptides or antibodies, and/or different polymers to form multifunctional probes for cancer targeting and *in vivo* imaging (Dubertret et al. 2002, Larson et al. 2003, Gao et al. 2004). The potential of QDs in cancer diagnostics appears to be even more promising due to their advantages in sensitivity and multicolor fluorescence imaging capability (Gao et al. 2004). Another application of QDs is in fluorescence imaging of tissue vasculature. Imaging skin and adipose tissues is extremely difficult with conventional organic dyes, but QD-containing vasculature can be imaged with higher spatial resolution and depth penetration (Larson et al. 2003). However, most syntheses of QDs result in hydrophobic nanoparticles that are not soluble in biological fluids (Dubertret et al. 2002, Larson et al. 2003). To address this issue, many researchers have coated QDs with amphiphilic copolymers and micelles (Dubertret et al. 2002, Larson et al. 2003, Gao et al. 2004). These coatings not only make QDs water soluble, but also improve their biocompatibility.

QDs also hold great promise as a potential fluorescent agent for intraoperative tumor imaging (Li et al. 2012). QDs that emit in the near-infrared are drawing particular interest due to their improved sensitivity, better tissue penetration depth, and minimal photodamage (Vahrmeijer et al. 2013). Since it is hypothesized that intraoperative imaging will improve the outcomes of cancer surgery, and that palpation and visual inspection are the two currently dominant methods to decide which tissue needs to be resected, NIR QDs are expected to draw continuous attention. Recently, Li et al. synthesized tumor-specific NIR CdTe QDs with active tumor targeting peptides, called cyclic Arg-Gly-Asp (cRGD). With maximum fluorescence emission peak at 728 nm, that is, in the NIR region, and high quantum yield, these QDs allowed distinguished visualization of tumor and allowed successful tumor resection using a Fluobeam-700 NIR imaging system (Li et al. 2012).

6.4.2 Gold Nanoparticles

Gold nanoparticles are attracting growing attention as fluorescent contrast agents. Gold nanoparticles can be either inherently fluorescent or can be made fluorescent by conjugation of fluorophores to the surface. Inherent fluorescence of gold nanoparticles is highly dependent on the size and shape. For example, most spherical gold nanoparticles are not fluorescent; however, small gold nanospheres (~2 nm) are fluorescent (Huang et al. 2007). Gold nanorods are another form of gold nanoparticles that exhibit fluorescent properties. Of note, their emission maximum has been found to be dependent on their aspect ratio. For example, Mohamed et al. reported that increasing the aspect ratio of gold nanorods from 2 to 5.4 led to a shift in the emission maximum from 540 to 740 nm (2000). This has led to many applications in imaging cancer, such as targeted imaging of cancer cells (Durr et al. 2007). Other structures have been found to have fluorescence properties, such as nanoshells and nanocubes (Jaesook et al. 2008, Wu et al. 2010). Gold nanostars have also proven to be effective for fluorescence imaging in both targeted and blood pool imaging applications (Yuan et al. 2012).

Easy and controllable surface modification of AuNPs with thiol-group functionalized ligands allows other fluorescent agents to be easily conjugated to AuNPs, broadening their application in fluorescence imaging.

For example, AuNPs can act as activatable contrast agents, since they have an exceptional ability to quench the near-infrared-fluorescence (NIRF) of chromophores in close proximity (Yuan et al. 2013a). If the linker between NIRF probes and AuNPs is designed to be sensitive to biological molecules (i.e., proteases), AuNP can act as a biosensor with high sensitivity and specificity. This is because fluorescence will increase in the presence of the protease, as the fluorophores are cleaved from the surface of the gold nanoparticle and their fluorescence is no longer quenched (Lee et al. 2008, Yuan et al. 2013a).

6.4.3 Carbon Nanostructures

Carbon nanostructures have gained considerable attention in the field of fluorescence imaging. Carbon nanodots have favorable properties for fluorescence imaging, such as low toxicity, excellent biocompatibility, excellent water solubility, versatile surface modification, and photostability (Chen et al. 2013, Ko et al. 2013, Xu et al. 2013, Thorek et al. 2014). Carbon nanodots generally consist of carbon, hydrogen, and oxygen, but there is ongoing research on incorporating different atoms, such as nitrogen, to optimize their optical properties (Xu et al. 2013). The emission spectra of carbon nanodots are broad, ranging between the UV/visible and NIR regions. Their emission spectra are also dependent on the excitation wavelength, providing great flexibility to excite the nanodots with varying wavelengths of light. Carbon nanostructures can be used in fluorescence imaging in the second near-infrared window fluorescent light (950–1400 nm) in the form of carbon nanotubes. Yi et al. designed single-walled carbon nanotubes (SWNTs) that are stably assembled on genetically engineered M13 phages, along with prostate-specific membrane antigen (PSMA) antibody for second-window NIR fluorescence imaging (Figure 6.7a) (Yi et al. 2012). In phantom experiments, M13-SWNTs proved to be an efficient fluorescent probe for deep tissue imaging even at low dosage. *In vivo* imaging was performed with PSMA targeted and untargeted M13-SWNTs in a mouse model of prostate cancer (Figure 6.7b,c). As can be seen, the targeted formulation resulted in markedly greater signal intensity in the tumor than the untargeted version (Figure 6.7c). Their results demonstrate that fluorescent SWNTs can

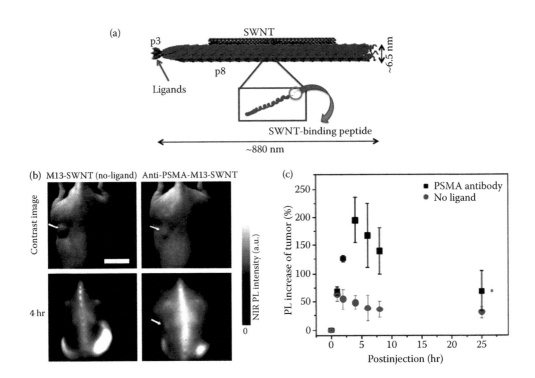

FIGURE 6.7
(a) Schematic depiction of M13-SWNT nanoprobe. (b) *In vivo* second NIR fluorescence images of control (M13-SWNT) and anti-PSMA-M13-SWNT probes in mice with tumor. (c) Photoluminescence level of anti-PSMA-M13-SWNT and control (M13-SWNT) showing highest tumor uptake at 4 hr post-injection. (Adapted with permission from Yi H. et al. *Nano Letters*. 2012;12(3):1176–1183.)

yield sufficient quantum yield to be utilized in second NIR window fluorescence imaging of tumors *in vivo*.

6.4.4 Fluorescently Labeled Liposomes

Liposomes are typically produced by forming a dry lipid film, which is then dispersed in an aqueous solution, allowing the lipids to self-assemble into various layers and configurations. Additional processing, such as sonication, homogenization, or extrusion, is frequently used to reduce polydispersity, control the number of layers, and form the final formulation. During the dispersion, fluorescent probes can be encapsulated at the core without any additional processes to label the liposomes. Compared to small molecule fluorophores, liposomes provide longer circulation times and less clearance by the kidneys for more efficient accumulation in the target tissues (Louie 2010). The surface of the liposomes can also be modified to control the biodistribution and to directly label the lipids with fluorescent dyes. In recent studies, multiple types of contrast agents have been incorporated into liposomes to allow detection and diagnosis with several imaging modalities. Any combination of CT, MRI, and fluorescent imaging agents can be either encapsulated at the core and/or conjugated to the surface of the lipids. Fluorescently labeled biological molecules (i.e., siRNA) can also be encapsulated in liposomes to allow verification of molecule delivery to the target (Mikhaylova et al. 2008).

6.4.5 Cerenkov Luminescence Imaging

Cerenkov luminescence imaging (CLI) has recently emerged as a promising technique that uses light emitted by the decay of radioactive nuclei to excite fluorophores *in situ*. CLI has great potential especially for *in vivo* tumor imaging as a cheaper and simpler modality, with higher throughput and resolution than its nuclear counterparts (Ruggiero et al. 2010, Xu et al. 2011, Wang et al. 2013). Cerenkov radiation in the visible and NIR ranges can also be used to excite fluorescent nanoparticles, such as quantum dots, *in situ*. This eliminates the need for an external excitation source, which can limit tissue penetration depth due to absorbance and scattering in tissue. Consequently, autofluorescence and light scattering are minimized to increase the signal-to-noise ratio (Wang et al. 2013). Most examples of CLI use α- or β-emitter radionuclides, such as ^{18}F, ^{131}I, and ^{225}Ac included in the structure of the nanoparticle, and the detected emission is either caused by the radiation directly or from the excitation of the nearby fluorophores excited by the radiation (Wang et al. 2013).

6.5 Nuclear Imaging

In nuclear imaging, radiotracers are used as contrast agents and the gamma radiation emitted from the tracer is detected and used to form medical images. In single-photon emission computed tomography (SPECT), these gamma radiations are detected directly after emission from the radiotracer. In positron emission tomography (PET), radiotracers emit a positron while decaying, which annihilates with an electron and releases two gamma rays at 180° angles. The gamma rays are detected by cameras collecting the energy of gamma rays followed by electronic processing to create images. Higher positron energy results in higher resolution. In order to shorten the body's exposure to radiation from the radionuclide, it is desirable that they have short half lives. ^{11}C, ^{15}O, ^{13}N, and ^{18}F are some of the most commonly used radionuclides. PET using ^{18}FDG (deoxyglucose labeled with ^{18}F) as the radiotracer is one of the most powerful imaging modalities available in the clinic for detecting tumors; however, it is still limited in detecting highly differentiated neuroendocrine tumors, brain tumors, and prostatic tumors (Srivatsan and Chen 2014). Nanoparticle platforms have been introduced to increase the dose efficiency of radiotracers and targeting strategies for more specificity in tumor detection. Alternatively, radiolabeling and imaging can be used as a straightforward method to study the biodistribution and targeting of nanoparticles designed for drug delivery. These nanoparticles can either be directly radiolabeled or encapsulate and carry a radiolabeled cargo (Liu and Welch 2012). Chelators such as 1,4,7-triazacyclononane-1,4,7-triacetic acid (NOTA), 1,4,7,10-tetraazacyclododecane-1,4,7,10-tetraacetic acid (DOTA), p-isothiocyanatobenzyl-desferrioxamine (Df-Bz- NCS), and diethylene triamine pentaacetic acid (DTPA) are commonly used to form stable complexes with the radioisotopes and the nanoparticle platform; however, chelator-free techniques for radiolabeling of nanoparticles have also been developed (Goel et al. 2014).

6.5.1 Liposomes

Radiotracers can be used to label liposomes by inclusion inside the hydrophilic core, in the lipid bilayer, or on the surface of the liposomes (Seo et al. 2015). Long-circulating liposomes labeled with [^{18}F]fluorodipalmitin ([^{18}F]FDP) into the phospholipid bilayer were developed for *in vivo* cell trafficking imaging applications (Marik et al. 2007). Liposomes have been labeled with ^{67}Ga as long-circulating blood pool contrast agents (Woodle 1993). ^{124}I has also been used to label liposomes that target tumors (Medina et al. 2011). ^{64}Cu is another radionuclide that has been used for liposome labeling (Seo et al. 2010). Perez-Medina et al. developed long-circulating bimodal ^{89}Zr-labeled liposomes (DiIC@^{89}Zr-SCL) for PET/optical imaging by surface chelation of ^{89}Zr (Figure 6.8a) and incorporation of the NIR fluorophore DiIC (Figure 6.8b) (2014). Using NIR fluorescence and PET imaging, they demonstrated the accumulation of these particles in the breast tumor of mice (Figure 6.8c,d). PET imaging also shows a high accumulation in liver and spleen (Figure 6.8d), which is not apparent in the fluorescence image (Figure 6.8c) due to autofluorescence of tissue and limited depth penetration of light.

6.5.2 Polymeric Nanoparticles

Polymers are another category of nanoparticle platforms that can carry radiolabels as cargo to lengthen the circulation time or enhance the targeting capabilities with the advantage of versatility in their synthetic chemistry (Liu and Welch 2012). Poly-(4-vinyl)phenol nanoparticles were labeled with ^{125}I or ^{124}I radioiodines and were coated with antibodies for targeting endothelial cells as vascular imaging probes (Simone et al. 2012). Dextran nanoparticles radiolabeled with ^{89}Zr were used to image macrophages in a mice model of atherosclerosis (cholesterol buildup in arteries) and create a potential platform for tracking anti-inflammatory therapies for this disease (Majmudar et al. 2013).

6.5.3 Radiolabeled Inorganic Nanoparticles

Radiolabeling has also been applied to study *in vivo* pharmacokinetics of inorganic fluorescent and magnetic contrast agent nanoparticles and theranostic agents. Quantum dots were radiolabeled with ^{64}Cu using DOTA chelator on their surface to image their accumulation

Nanoparticle Contrast Agents for Medical Imaging

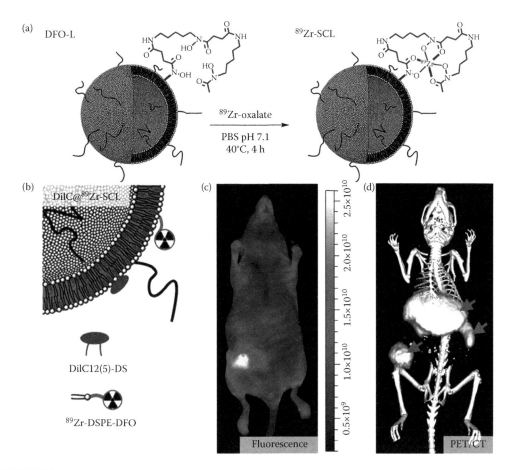

FIGURE 6.8
(a) Surface chelation approach of labeling liposomes with ^{89}Zr labels. (b) Schematic depiction of the final liposome product, labeled with both fluorophore and radionuclide. (c) Epifluorescence NIR image and (d) PET-CT fusion image acquired 24 hours post administration of DiIC@89Zr-SCL showing high accumulation in liver (top arrow), spleen (right arrow) and tumor (left arrow). (Adapted with permission from Pérez-Medina C. et al. *Journal of Nuclear Medicine*. 2014;55(10):1706–1712.)

using a small animal PET scanner (Michalet et al. 2005). Iron-oxide nanoparticles have been radiolabeled with ^{64}Cu, ^{68}Ga, and ^{124}I to study their biodistribution *in vivo* and their targeting efficacy (Choi et al. 2008, Glaus et al. 2010, Stelter et al. 2010, Yang et al. 2011). Gold nanoshells used for photothermal cancer treatment were labeled with ^{64}Cu for PET imaging of their accumulation in tumor tissue (Xie et al. 2010). Since gold nanoshells developed by Chen et al. (2014) are in clinical trials for photothermal ablation of prostate cancer, the importance of nanoparticle radiolabeling and imaging is underscored (Nanospectra Biosciences Inc.).

6.6 Photoacoustic Imaging

Photoacoustic (PA) imaging is a hybrid biomedical imaging modality that takes advantage of high resolution and depth penetration of ultrasound imaging with contrast capabilities of optical imaging. In this imaging modality, pulsed laser irradiation creates thermoelastic expansion in the sample, inducing acoustic waves that can be detected by ultrasound. A major advantage of PA imaging is the lack of ionizing radiation. Although certain endogenous biological materials, such as hemoglobin, have high optical absorbance and can be PA imaging targets for imaging blood vessels, a wide range of nanoparticle contrast agents have been developed to enhance the imaging capabilities of PA imaging. Similar to fluorescence imaging, PA imaging performs best when near-infrared light is used. Therefore, agents that have an absorbance in this region are preferred.

6.6.1 Metal Nanoparticles

In some metal nanoparticles, the electromagnetic field induces oscillations in the free charges on the surface of the metal particle. These oscillations are known as surface plasmon resonance (SPR) (Dixon et al. 2015, Li and Chen 2015). SPR results in strong absorbance of light, which is highly dependent on the size, morphology, and environment of the nanoparticle, which can render these nanoparticles highly useful as PA contrast agents. Among metal nanoparticles, AuNPs are one of the most studied exogenous contrast agents for PA imaging. Several different types of gold nanostructures have been used as PA contrast agents, such as gold nanorods, gold nanocages, gold nanospheres, and gold nanoshells, with a premium placed on formulations that have NIR absorbance (Lu et al. 2010). The depth of penetration for PA imaging using AuNP can be as deep as 5–6 cm (Li and Chen 2015). One advantage of this class of nanoparticles is the ability to tune their absorption spectra by changes in their structure during synthesis. Increasing the aspect ratio in gold nanorods (Nikoobakht and El-Sayed 2003) and reducing shell thickness for gold nanocages (Oldenburg et al. 1999) shift their absorption spectra deeper into the NIR region and thus makes them more potent PA imaging contrast agents. One challenge of using Au nanostructures for PA imaging is their photostability. Intense heating of these structures during the imaging can induce melting and fragmentation, which changes the shape and therefore the contrast enhancement properties can be lost (Ungureanu et al. 2011). Melting and fragmentation depend on the shape of the gold nanoparticles. Spherical nanoparticles fragment at high laser fluences of $\sim$1 J cm^{-2}, whereas melting is observed at laser fluence rates as low as 0.01 J cm^{-2} for gold nanorods (Link et al. 2000). Since deep tissue imaging is typically done at $\leq$10 mJ cm^{-2} level, care needs to be taken to use laser levels low enough to avoid melting/fragmentation (Jathoul et al. 2015).

6.6.2 Fluorophores

NIR fluorophores have been used for combined NIR and PA imaging agents, although this represents a compromise for PA contrast since some of the input light is converted to emission light, as opposed to heat, and therefore PA contrast. Encapsulating squarine dyes in micelles prevents their aggregation in biological media and creates a stable platform with low toxicity while maintaining the photophysical properties of squarine dye (Sreejith et al. 2015). Prussian blue, which the FDA has approved for treatment of exposure to radioactive materials, has been formed into nanoparticles and studied as a contrast agent for PA imaging of mouse brain (Liang et al. 2013). Frozen naphthalocyanine micelles were developed as nanoparticles for photoacoustic imaging that are stable even in the harsh conditions of the GI tract and are not absorbed systemically. These particles were orally administered in mice as GI contrast agents for functional intestinal imaging of dynamic gut processes (Zhang et al. 2014).

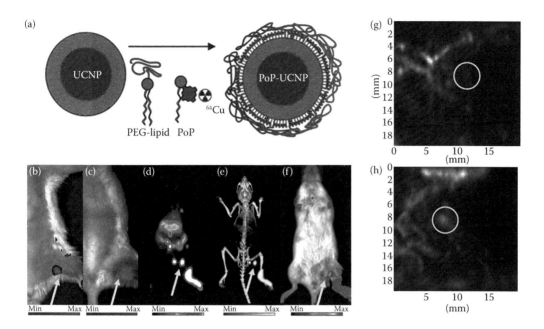

FIGURE 6.9
(a) Schematic synthesis of PoP-coated UNCP. (b–h) Lymph node imaging of nanoparticles in mice using fluorescence (b), upconversion (c), PET (d), PET/CT (e), Cerenkov luminescence (f) and PA (g/h). The first lymph node is indicated by arrows in (b–f). (g) The PA imaging site before injection. (h) PA image acquired 1 hour post injection. Circles indicate the first lymph node. (Adapted with permission from Rieffel J. et al. *Advanced Materials*. 2015;27(10):1785–1790.)

Porphyrin-phospholipid (PoP)-coated UCNPs were developed as contrast agents for *in vivo* lymphatic imaging of mice using six modalities. Their structure is depicted in Figure 6.9a (Rieffel et al. 2015). UCNPs with $NaYbF_4$:Tm core and $NaYF_4$ shells allowed for upconversion luminescence in NIR for traditional fluorescence, upconversion, and PA imaging, while the electron-dense cores provided contrast for CT. Particles were synthesized by coating oleic acid-capped UCNPs with PoP and a PEG-lipid for dispersion in aqueous media and were postlabeled with ^{64}Cu, which allowed PET and Cerenkov luminescence imaging (Figure 6.9a). Particles were injected in the rear foot pad of mice, and 1 hour post injection (enough time for the lymphatic drainage of the particles), images were acquired using the different imaging modalities (Figure 6.9b–h). As can be seen, strong signals can been seen in the foot, but the first draining lymph node can be detected in each case. PA signal for these particles was detectable for up to 30 mm tissue depth *in vitro*.

6.7 Surface-Enhanced Raman Spectroscopy

Raman spectroscopy is the technique of characterizing molecules by recording their Raman scattering after interaction with visible, near infrared, or near ultraviolet light. Raman spectra, which can be considered a fingerprint of a molecule, are very weak compared to the other scattering modes of a molecule. Rough metal structures or nanostructures can enhance the intensity of Raman spectra by several orders of magnitude, a phenomenon known as surface-enhanced Raman scattering (SERS). For example, gold nanostructures have been utilized as contrast agents for SERS imaging of macrophages (Fujita et al. 2009), lymphocytes (Eliasson et al. 2005), and tumors (Qian et al. 2007). SERS effects are strongest at vertices of nanoparticles or at points where several nanoparticles come together in close proximity. Therefore, particular interest has focused on structures such as nanostars or nanoparticles that contain several cores (Liu et al. 2015).

Kircher et al. synthesized triple-modality nanoparticles for MRI, photoacoustic imaging, and Raman imaging (MPR) of brain tumors (Kircher et al. 2012). They coated 60 nm gold nanoparticles with a thin layer of the Raman active molecule trans-1,2-bis(4-pyridyl)-ethylene, protected the structure with a 30-nm-thick silica layer, and further functionalized the silica surface by conjugating maleimide-DOTA-Gd^{3+} to thiol (SH) groups that had been added to the surface of the silica (Figure 6.10a,b). The absorbance of the gold nanoparticle cores provides the photoacoustic contrast, the molecules absorbed at the gold surface provide the SERS contrast, and the gadolinium chelates provide MRI contrast. Using an orthotopic glioblastoma brain tumor model in mice, they demonstrated that MPR particles accumulate in the tumor through the EPR effect, which allowed for visualization of the tumor in all three modalities (Figure 6.10c). Furthermore, the authors used SERS imaging

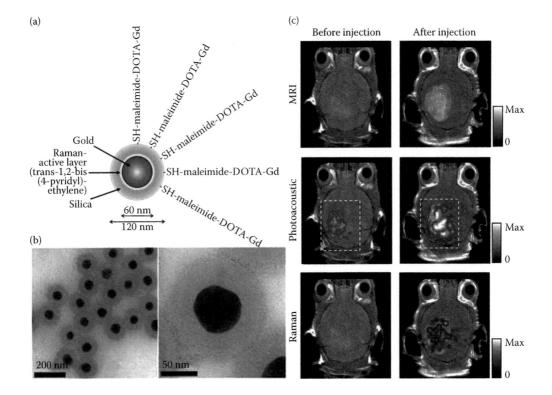

FIGURE 6.10
(a) Schematic depiction of MPR nanoparticles. (b) Electron microscopy of MPR nanoparticles in which the gold core (smaller, darker circles) and the silica coating (larger, light gray circles) are visible. (c) Accumulation of MPR particles in a glioblastoma mouse tumor model detected through three modalities. (Adapted with permission from Kircher M.F. et al. *Nature Medicine*. 2012;18(5):829–834.)

as an aid to tumor resection in these mice, which resulted in improved tumor margins compared with surgery without image guidance.

6.8 Discussion

The currently available medical imaging techniques each have advantages and disadvantages. MRI and CT provide high spatial resolution images without limitation in penetration depth. Nuclear imaging has much improved sensitivity over MRI and CT, but suffers from much lower resolution; photoacoustic imaging has high resolution and high sensitivity, but is limited by penetration depth. It is important to determine which imaging modality is best suited to image the target, and nanoparticle contrast agents are then designed, taking into consideration both the imaging modality and the target. Many nanoparticle platforms have been developed upon which improved contrast agents have been based, in terms of their increased sensitivity and accuracy of detection, longer circulation, and specificity to diseases sites. Nanoparticle platforms provide flexibility over size, loading, and surface modification to achieve desired pharmacokinetics, organ uptake, and delivery efficacy. Certain types of nanoparticles provide unique forms of contrast that are not available from small molecule agents, such as the T_2 MRI contrast provided by iron oxides or the fluorescence properties of quantum dots. Nanoparticle platforms also make possible a variety of strategies for multimodal imaging applying more than one imaging modality, or multifunctional applications for imaging and therapy.

The main challenge in designing nanoparticle contrast agent platforms is to confirm their ability to safely carry the contrast-generating cargo and stay stable in biological media. Furthermore, depending on the application, clearance of nanoparticles from the body in a time that is adequate for imaging purposes and at the same time does not expose the body to potential toxicity risks is an essential element of translatability to the clinic. There was a burst of activity from the mid-1990s to the mid-2000s when several iron oxide nanoparticle MRI contrast agents were FDA approved (Corot et al. 2006). Translational activity has been diminished since then, although the poor economic conditions in the intervening period have likely played a role. However, it is important that the drive to translate novel formulations to the clinic restarts. Academics should build connections with industry, patent their technologies, and form startup companies around their intellectual property. This will help access the investment that is needed to perform the toxicity studies needed to obtain Investigational New Drug (IND) status from the FDA. IND-enabling studies entail extensive toxicological testing in two species (one nonrodent) with a range of doses, repeat doses in the same animal, LD50 determination, and so forth. Once IND status is obtained, clinical trials may begin, with the phase I trial focusing on safety in humans, while later phase trials (II/III) focus on efficacy.

Overall, tremendous advances in this field have been made over the last two decades, with the development of targeted agents, long-circulating agents, and labels for cell tracking. Contrast media for novel techniques such as CEST, photoacoustics, SERS, and intraoperative imaging have been developed. Several nanoparticle contrast agents have been FDA approved and more are in clinical trials. Over the next years, we expect to see more agents being FDA approved and continued innovation that allow contrast to be produced in existing and novel imaging techniques.

Acknowledgments

We acknowledge funding from NIH grants R01HL131557, R00 EB012165, and R03 CA171661, as well as startup funds from the University of Pennsylvania.

References

Ahrens E. T., Flores R., Xu H., Morel P. A. In vivo imaging platform for tracking immunotherapeutic cells. *Nature Biotechnology.* 2005;23(8):983–987.

Ahrens E. T., Helfer B. M., O'Hanlon C. F., Schirda C. Clinical cell therapy imaging using a perfluorocarbon tracer and fluorine-19 MRI. *Magnetic Resonance in Medicine.* 2014;72(6):1696–1701.

Aime S., Calabi L., Cavallotti C., Gianolio E., Giovenzana G. B., Losi P., Maiocchi A., Palmisano G., Sisti M. [Gd-AAZTA]: A new structural entry for an improved generation of MRI contrast agents. *Inorganic Chemistry.* 2004;43(24):7588–7590.

Allen M., Bulte J. W. M., Liepold L., Basu G., Zywicke H. A., Frank J. A., Young M., Douglas T. Paramagnetic viral nanoparticles as potential high-relaxivity magnetic resonance contrast agents. *Magnetic Resonance in Medicine.* 2005;54(4):807–812.

Arifin D. R., Long C. M., Gilad A. A., Alric C., Roux S., Tillement O., Link T. W., Arepally A., Bulte J. W. M. Trimodal gadolinium-gold microcapsules containing pancreatic islet cells restore normoglycemia in diabetic mice and can be tracked by using US, CT, and positive-contrast MR imaging. *Radiology.* 2011;260(3):790–798.

Bangham A. D., Standish M. M., Watkins J. C. Diffusion of univalent ions across the lamellae of swollen phospholipids. *Journal of Molecular Biology.* 1965;13(1):238–IN27.

Barnett B. P., Ruiz-Cabello J., Hota P., Liddel R., Walczak P., Howland V., Chacko V. P., Kraitchman D. L., Arepally A., Bulte J. W. M. Fluorocapsules for improved function, immunoprotection, and trimodal visualization of cellular therapeutics with magnetic resonance, ultrasound and X-ray imaging. *Radiology.* 2010;258(1):182–191.

Bonitatibus P. J., Torres A. S., Kandapallil B., Lee B. D., Goddard G. D., Colborn R. E., Marino M. E. Preclinical assessment of a zwitterionic tantalum oxide nanoparticle X-ray contrast agent. *ACS nano.* 2012;6(8):6650–8.

Briley-Saebo K. C., Geninatti-Crich S., Cormode D. P., Barazza A., Mulder W. J. M., Chen W., Giovenzana G. B., Fisher E.A., Aime S., Fayad Z.A. High-relaxivity gadolinium-modified high-density lipoproteins as magnetic resonance imaging contrast agents. *The Journal of Physical Chemistry. B.* 2009;113(18):6283–6289.

Brown A. L., Naha P. C., Benavides-Montes V., Litt H. I., Goforth A. M., Cormode D. P. Synthesis, x-ray opacity, and biological compatibility of ultra-high payload elemental bismuth nanoparticle x-ray contrast agents. *Chemistry of Materials.* 2014;26(7):2266–2274.

Bulte J. W. M., Kraitchman D. L. Iron oxide MR contrast agents for molecular and cellular imaging. *NMR in Biomedicine.* 2004;17:484–499.

Cai Q.-Y., Kim S. H., Choi K. S., Kim S. Y., Byun S. J., Kim K. W., Park S. H., Juhng S. K., Yoon K.-H. Colloidal gold nanoparticles as a blood-pool contrast agent for x-ray computed tomography in mice. *Investigative Radiology.* 2007;42(12):797–806.

Caravan P. Strategies for increasing the sensitivity of gadolinium based MRI contrast agents. *Chemical Society Reviews.* 2006;35(6):512.

Castelli D. D., Terreno E., Longo D., Aime S. Nanoparticle-based chemical exchange saturation transfer (CEST) agents. *NMR in Biomedicine.* 2013;26(7):839–849.

Chan K. W., Liu G., Song X., Kim H., Yu T., Arifin D. R., Gilad A. A. et al. MRI-detectable pH nanosensors incorporated into hydrogels for *in vivo* sensing of transplanted-cell viability. *Nature Materials*. 2013;12(3):268–275.

Cheheltani R., Ezzibdeh R. M., Chhour P., Pulaparthi K., Kim J., Jurcova M., Hsu J. C. et al. Tunable, biodegradable gold nanoparticles as contrast agents for computed tomography and photoacoustic imaging. *Biomaterials*. 2016;102:87–97.

Chen B., Li F., Li S., Weng W., Guo H., Guo T., Zhang X. et al. Large scale synthesis of photoluminescent carbon nanodots and their application for bioimaging. *Nanoscale*. 2013;5(5):1967.

Chen W., Ayala-Orozco C., Biswal N. C., Perez-Torres C., Bartels M., Bardhan R., Stinnet G. et al. Targeting pancreatic cancer with magneto-fluorescent theranostic gold nanoshells. *Nanomedicine*. 2014;9(8):1209–1222.

Chhour P., Gallo N., Cheheltani R., Williams D., Al-Zaki A., Paik T., Nichol J. L. et al. Nanodisco balls: Control over surface versus core loading of diagnostically active nanocrystals into polymer nanoparticles. *ACS Nano*. 2014;8(9):9143–9153.

Chinen A. B., Guan C. M., Ferrer J. R., Barnaby S. N., Merkel T. J., Mirkin C. A. Nanoparticle probes for the detection of cancer biomarkers, cells, and tissues by fluorescence. *Chemical Reviews*. 2015;115(19):10530–10574.

Choi H. S., Liu W., Misra P., Tanaka E., Zimmer J. P., Itty Ipe B., Bawendi M. G., Frangioni J. V. Renal clearance of quantum dots. *Nature Biotechnology*. 2007;25(10):1165–1170.

Choi J., Park J. C., Nah H., Woo S., Oh J., Kim K. M., Cheon G. J., Chang Y., Yoo J., Cheon J. A hybrid nanoparticle probe for dual-modality positron emission tomography and magnetic resonance imaging. *Angewandte Chemie International Edition*. 2008;47(33):6259–6262.

Cormode D. P., Jarzyna P. A., Mulder W. J. M., Fayad Z. A. Modified natural nanoparticles as contrast agents for medical imaging. *Advanced Drug Delivery Reviews*. 2010a;62(3):329–338.

Cormode D. P., Naha P. C., Fayad Z. A. Nanoparticle contrast agents for computed tomography: A focus on micelles. *Contrast Media and Molecular Imaging*. 2014;9(1):37–52.

Cormode D. P., Roessl E., Thran A., Skajaa T., Gordon R. E., Schlomka J.-P., Fuster V. et al. Atherosclerotic plaque composition: Analysis with multicolor CT and targeted gold nanoparticles. *Radiology*. 2010b;256(3):774–782.

Cormode D. P., Sanchez-Gaytan B. L., Mieszawska A. J., Fayad Z. A., Mulder W. J. M. Inorganic nanocrystals as contrast agents in MRI: Synthesis, coating and introduction of multifunctionality. *NMR in Biomedicine*. 2013;26(7):766–780.

Cormode D. P., Skajaa G. O., Delshad A., Parker N., Jarzyna P. A., Calcagno C., Galper M. W. et al. A versatile and tunable coating strategy allows control of nanocrystal delivery to cell types in the liver. *Bioconjugate Chemistry*. 2011;22(3):353–361.

Cormode D. P., Skajaa T., Fayad Z. A., Mulder W. J. M. Nanotechnology in medical imaging: Probe design and applications. *Arteriosclerosis, Thrombosis, and Vascular Biology*. 2009;29(7):992–1000.

Cormode D. P., Skajaa T., van Schooneveld M. M., Koole R., Jarzyna P., Lobatto M. E., Calcagno C. et al. Nanocrystal core high-density lipoproteins: A multimodality contrast agent platform. *Nano Letters*. 2008a;8(11):3715–3723.

Cormode D. P., Skajaa T., van Schooneveld M. M., Koole R., Jarzyna P., Lobatto M. E., Calcagno C. et al. Nanocrystal core high-density lipoproteins: A multimodal molecular imaging contrast agent platform. *Nano Letters*. 2008b;8(11):3715–3723.

Corot C., Robert P., Idee J., Port M. Recent advances in iron oxide nanocrystal technology for medical imaging. *Advanced Drug Delivery Reviews*. 2006;58(14):1471–1504.

Criscione J. M., Le B. L., Stern E., Brennan M., Rahner C., Papademetris X., Fahmy T. M. Self-assembly of pH-responsive fluorinated dendrimer-based particulates for drug delivery and noninvasive imaging. *Biomaterials*. 2009;30(23–24):3946–3955.

Cui L., Lin Q., Jin C. S., Jiang W., Huang H., Ding L., Muhanna N. et al. A PEGylation-free biomimetic porphyrin nanoplatform for personalized cancer theranostics. *ACS Nano*. 2015;9(4):4484–4495.

Dixon A. J., Hu S., Klibanov A. L., Hossack J. A. Oscillatory dynamics and *in vivo* photoacoustic imaging performance of plasmonic nanoparticle-coated microbubbles. *Small*. 2015;11(25):3066–3077.

Dreaden E. C., Alkilany A. M., Huang X., Murphy C. J., El-Sayed M. A. The golden age: Gold nanoparticles for biomedicine. *Chemical Society Reviews.* 2012;41(7):2740–2779.

Dubertret B., Skourides P., Norris D. J., Noireaux V., Brivanlou A. H., Libchaber A. In vivo imaging of quantum dots encapsulated in phospholipid micelles. *Science (New York, N.Y.).* 2002;298(5599):1759–1762.

Durr N. J., Larson T., Smith D. K., Korgel B. A., Sokolov K., Ben-Yakar A. Two-photon luminescence imaging of cancer cells using molecularly targeted gold nanorods. *Nano Letters.* 2007;7(4):941–945.

Eliasson C., Lorén A., Engelbrektsson J., Josefson M., Abrahamsson J., Abrahamsson K. Surface-enhanced Raman scattering imaging of single living lymphocytes with multivariate evaluation. *Spectrochimica Acta Part A: Molecular and Biomolecular Spectroscopy.* 2005;61(4):755–760.

Evbuomwan O. M., Kiefer G., Sherry A. D. Amphiphilic EuDOTA-tetraamide complexes form micelles with enhanced CEST sensitivity. *European Journal of Inorganic Chemistry.* 2012a;(12):2126–2134.

Evbuomwan O. M., Merritt M. E., Kiefer G. E., Dean Sherry A. Nanoparticle-based PARACEST agents: The quenching effect of silica nanoparticles on the CEST signal from surface-conjugated chelates. *Contrast Media & Molecular Imaging.* 2012b;7(1):19–25.

Fernández-Suárez M., Ting A. Y. Fluorescent probes for super-resolution imaging in living cells. *Nature Reviews. Molecular Cell Biology.* 2008;9(12):929–943.

Flogel U., Ding Z., Hardung H., Jander S., Reichmann G., Jacoby C., Schubert R., Schrader J. In vivo monitoring of inflammation after cardiac and cerebral ischemia by fluorine magnetic resonance imaging. *Circulation.* 2008;118(2):140–148.

Fujita K., Ishitobi S., Hamada K., Smith N. I., Taguchi A., Inouye Y., Kawata S. Time-resolved observation of surface-enhanced Raman scattering from gold nanoparticles during transport through a living cell. *Journal of Biomedical Optics.* 2009;14(2):24038.

Fölsch S., Martínez-Blanco J., Yang J., Kanisawa K., Erwin S. C. Quantum dots with single-atom precision. *Nature Nanotechnology.* 2014;9(7):505–508.

Galper M. W., Saung M. T., Fuster V., Roessl E., Thran A., Proksa R., Fayad Z. A., Cormode D. P. Effect of computed tomography scanning parameters on gold nanoparticle and iodine contrast. *Investigative Radiology.* 2012;47(8):475–481.

Gao L., Liu Y., Kim D., Li Y., Hwang G., Naha P. C., Cormode D. P., Koo H. Nanocatalysts promote *Streptococcus mutans* biofilm matrix degradation and enhance bacterial killing to suppress dental caries in vivo. *Biomaterials.* 2016;101:272–284.

Gao X., Cui Y., Levenson R. M., Chung L. W. K., Nie S. In vivo cancer targeting and imaging with semiconductor quantum dots. *Nature Biotechnology.* 2004;22(8):969–976.

Gaudet J. M., Ribot E. J., Chen Y., Gilbert K. M., Foster P. J. Tracking the fate of stem cell implants with fluorine-19 MRI. *PloS One.* 2015;10(3):E.0118544.

Ghosh Chaudhuri R., Paria S. Core/shell nanoparticles: Classes, properties, synthesis mechanisms, characterization, and applications. *Chemical Reviews.* 2012;112(4):2373–2433.

Gianella A., Jarzyna P. A., Mani V., Ramachandran S., Calcagno C., Tang J., Kann B. et al. Multifunctional nanoemulsion platform for imaging guided therapy evaluated in experimental cancer. *ACS Nano.* 2011;5(6):4422–4433.

Glaus C., Rossin R., Welch M. J., Bao G. In vivo evaluation of 64 cu-labeled magnetic nanoparticles as a dual-modality PET/MR imaging agent. *Bioconjugate Chemistry.* 2010;21(4):715–722.

Goel S., Chen F., Ehlerding E. B., Cai W. Intrinsically radiolabeled nanoparticles: An emerging paradigm. *Small.* 2014;15:3825–3830.

Griset A. P., Walpole J., Liu R., Gaffey A., Colson Y. L., Grinstaff M. W. Expansile nanoparticles: Synthesis, characterization, and in vivo efficacy of an acid-responsive polymeric drug delivery system. *Journal of the American Chemical Society.* 2009;131(7):2469–2471.

Hahn M. A., Singh A. K., Sharma P., Brown S. C., Moudgil B. M. Nanoparticles as contrast agents for in-vivo bioimaging: Current status and future perspectives. *Analytical and Bioanalytical Chemistry.* 2011;399(1):3–27.

Hallouard F., Anton N., Choquet P., Constantinesco A., Vandamme T. Iodinated blood pool contrast media for preclinical x-ray imaging applications: A review. *Biomaterials.* 2010;31(24):6249–6268.

Hitchens T. K., Ye Q., Eytan D. F., Janjic J. M., Ahrens E. T., Ho C. 19F MRI detection of acute allograft rejection with *in vivo* perfluorocarbon labeling of immune cells. *Magnetic Resonance in Medicine*. 2011;65(4):1144–1153.

Homan K. A., Souza M., Truby R., Luke G. P., Green C., Vreeland E., Emelianov S. Silver nanoplate contrast agents for *in vivo* molecular photoacoustic imaging. *ACS Nano*. 2012;6(1):641–650.

Huang C.-C., Yang Z., Lee K.-H., Chang H.-T. Synthesis of highly fluorescent gold nanoparticles for sensing mercury(II). *Angewandte Chemie International Edition*. 2007;46(36):6824–6828.

Huang J., Zhong X., Wang L., Yang L., Mao H. Improving the magnetic resonance imaging contrast and detection methods with engineered magnetic nanoparticles. *Theranostics*. 2012;2(1):86–102.

Jackson P. A., Rahman W. N. W. A., Wong C. J., Ackerly T., Geso M. Potential dependent superiority of gold nanoparticles in comparison to iodinated contrast agents. *European Journal of Radiology*. 2010;75(1):104–109.

Jaesook P., Estrada A., Sharp K., Sang K., Schwartz J. A., Smith D. K., Coleman C. et al. Two-photon-induced photoluminescence imaging of tumors using near-infrared excited gold nanoshells. In: *2008 Conference on Quantum Electronics and Laser Science Conference on Lasers and Electro-Optics, CLEO/QELS,* San Jose, CA, USA, 2008.

Jathoul A. P., Laufer J., Ogunlade O., Treeby B., Cox B., Zhang E., Johnson P. et al. Deep *in vivo* photoacoustic imaging of mammalian tissues using a tyrosinase-based genetic reporter. *Nature Photonics*. 2015;9:8.

Jokerst J. V., Lobovkina T., Zare R. N., Gambhir S. S. Nanoparticle PEGylation for imaging and therapy. *Nanomedicine (London, England)*. 2011;6(4):715–728.

Jun Y.-W., Huh Y.-M., Choi J.-S., Lee J.-H., Song H.-T., Kim S., Yoon S. et al. Nanoscale size effect of magnetic nanocrystals and their utilization for cancer diagnosis via magnetic resonance imaging. *Journal of the American Chemical Society*. 2005;127(16):5732–5733.

Kelly K. A., Allport J. R., Tsourkas A., Shinde-Patil V. R., Josephson L., Weissleder R. Detection of vascular adhesion molecule-1 expression using a novel multimodal nanoparticle. *Circulation Research*. 2005;96:327–336.

Kim D., Park S., Lee J. H., Jeong Y. Y., Jon S. Antibiofouling polymer-coated gold nanoparticles as a contrast agent for *in vivo* x-ray computed tomography imaging. *Journal of the American Chemical Society*. 2007;129(24):7661–7665.

Kircher M. F., de la Zerda A., Jokerst J. V., Zavaleta C. L., Kempen P. J., Mittra E., Pitter K. et al. A brain tumor molecular imaging strategy using a new triple-modality MRI-photoacoustic-Raman nanoparticle. *Nature Medicine*. 2012;18(5):829–834.

Ko H. Y., Chang Y. W., Paramasivam G., Jeong M. S., Cho S., Kim S. *In vivo* imaging of tumour bearing near-infrared fluorescence-emitting carbon nanodots derived from tire soot. *Chemical Communications*. 2013;49(87):10290.

Ku G., Zhou M., Song S., Huang Q., Hazle J., Li C. Copper sulfide nanoparticles as a new class of photoacoustic contrast agent for deep tissue imaging at 1064 nm. *ACS Nano*. 2012;6(8):7489–7496.

Lanza G. M., Wallace K. D., Scott M. J., Cacheris W. P., Abendschein D. R., Christy D. H., Sharkey A. M., Miller J. G., Gaffney P. J., Wickline S. A. A novel site-targeted ultrasonic contrast agent with broad biomedical application. *Circulation*. 1996;94(12):3334–3340.

Larson D. R., Zipfel W. R., Williams R. M., Clark S. W., Bruchez M. P., Wise F. W., Webb W. W. Water-soluble quantum dots for multiphoton fluorescence imaging *in vivo*. *Science (New York, N.Y.)*. 2003;300(5624):1434–1436.

Leander P., Höglund P., Børseth A., Kloster Y., Berg A. A new liposomal liver-specific contrast agent for CT: First human phase-I clinical trial assessing efficacy and safety. *European Radiology*. 2001;11(4):698–704.

Lee J. H., Huh Y. M., Jun Y., Seo J., Jang J., Song H. T., Kim S. et al. Artificially engineered magnetic nanoparticles for ultra-sensitive molecular imaging. *Nature Medicine*. 2007;13(1):95–99.

Lee N., Choi S. H., Hyeon T. Nano-sized CT contrast agents. *Advanced Materials*. 2013;25(19):2641–2660.

Lee S., Cha E.-J., Park K., Lee S.-Y., Hong J.-K., Sun I.-C., Kim S. Y. et al. A near-infrared-fluorescence-quenched gold-nanoparticle imaging probe for *in vivo* drug screening and protease activity determination. *Angewandte Chemie*. 2008;120(15):2846–2849.

Li F., Wang Q. Fabrication of nanoarchitectures templated by virus-based nanoparticles: Strategies and applications. *Small*. 2014;10(2):230–245.

Li W., Chen X. Gold nanoparticles for photoacoustic imaging. *Nanomedicine*. 2015;10(2):299–320.

Li Y., Li Z., Wang X., Liu F., Cheng Y., Zhang B., Shi D. In vivo cancer targeting and imaging-guided surgery with near infrared-emitting quantum dot bioconjugates. *Theranostics*. 2012;2(8):769–776.

Liang X., Deng Z., Jing L., Li X., Dai Z., Li C., Huang M. Prussian blue nanoparticles operate as a contrast agent for enhanced photoacoustic imaging. *Chemical Communications (Cambridge, England)*. 2013;49(94):11029–11031.

Libutti S. K., Paciotti G. F., Byrnes A. A., Alexander H. R., Gannon W. E., Walker M., Seidel G. D., Yuldasheva N., Tamarkin L. Phase I and pharmacokinetic studies of CYT-6091, a novel PEGylated colloidal gold-rhTNF nanomedicine. *Clinical Cancer Research : An Official Journal of the American Association for Cancer Research*. 2010;16(24):6139–6149.

Lim F., Sun A. M. Microencapsulated islets as bioartificial endocrine pancreas. *Science*. 1980;210(4472):908–910.

Link S., Burda C., Nikoobakht B., El-Sayed M. A. Laser-induced shape changes of colloidal gold nanorods using femtosecond and nanosecond laser pulses. *The Journal of Physical Chemistry B*. 2000;104(26):6152–6163.

Liu G. S., Moake M., Har-el Y. E., Long C. M., Chan K. W. Y., Cardona A., Jamil M. et al. In vivo multicolor molecular MR imaging using diamagnetic chemical exchange saturation transfer liposomes. *Magnetic Resonance in Medicine*. 2012;67(4):1106–1113.

Liu Y., Ashton J. R., Moding E. J., Yuan H., Register J. K., Fales A. M., Choi J. et al. A plasmonic gold nanostar theranostic probe for *in vivo* tumor imaging and photothermal therapy. *Theranostics*. 2015;5(9):946–960.

Liu Y., Welch M. J. Nanoparticles labeled with positron emitting nuclides: Advantages, methods, and applications. *Bioconjugate Chemistry*. 2012;23(4):671–682.

Louie A. Multimodality imaging probes: Design and challenges. *Chemical Reviews*. 2010;110(5):3146–3195.

Lu W., Huang Q., Ku G., Wen X., Zhou M., Guzatov D., Brecht P. et al. Photoacoustic imaging of living mouse brain vasculature using hollow gold nanospheres. *Biomaterials*. 2010;31(9):2617–2626.

MacDonald T. D., Liu T. W., Zheng G. An MRI-sensitive, non-photobleachable porphysome photothermal agent. *Angewandte Chemie*. 2014;126(27):7076–7079.

Mahajan K. D., Fan Q., Dorcéna J., Ruan G., Winter J. O. Magnetic quantum dots in biotechnology—Synthesis and applications. *Biotechnology Journal*. 2013;8(12):1424–1434.

Majmudar M. D., Yoo J., Keliher E. J., Truelove J. J., Iwamoto Y., Sena B., Dutta P. et al. Polymeric nanoparticle PET/MR imaging allows macrophage detection in atherosclerotic plaques. *Circulation Research*. 2013;112(5):755–761.

Marik J., Tartis M. S., Zhang H., Fung J. Y., Kheirolomoom A., Sutcliffe J. L., Ferrara K. W. Long-circulating liposomes radiolabeled with [18F]fluorodipalmitin ([18F]FDP). *Nuclear Medicine and Biology*. 2007;34(2):165–171.

Mattrey R. Perfluorooctylbromide: A new contrast agent for CT, sonography, and MR imaging. *American Journal of Roentgenology*. 1989;152(2):247–252.

McMahon M. T., Gilad A. A., Zhou J., Sun P. Z., Bulte J. W. M., van Zijl P. C. M. Quantifying exchange rates in chemical exchange saturation transfer agents using the saturation time and saturation power dependencies of the magnetization transfer effect on the magnetic resonance imaging signal (QUEST and QUESP): Ph calibration for poly. *Magnetic Resonance in Medicine*. 2006;55(4):836–847.

Medina O. P., Pillarsetty N., Glekas A., Punzalan B., Longo V., Gönen M., Zanzonico P., Smith-Jones P., Larson S. M. Optimizing tumor targeting of the lipophilic EGFR-binding radiotracer SKI 243 using a liposomal nanoparticle delivery system. *Journal of Controlled Release*. 2011;149(3):292–298.

Medintz I. L., Uyeda H. T., Goldman E. R., Mattoussi H. Quantum dot bioconjugates for imaging, labelling and sensing. *Nature Materials*. 2005;4(6):435–446.

Meir R., Shamalov K., Betzer O., Motiei M., Horovitz-Fried M., Yehuda R., Popovtzer A., Popovtzer R., Cohen C. J. Nanomedicine for cancer immunotherapy: Tracking cancer-specific T-cells in vivo with gold nanoparticles and CT imaging. *ACS Nano*. 2015;9(6):6363–6372.

Michalet X., Pinaud F. F., Bentolila L. A., Tsay J. M., Doose S., Li J. J., Sundaresan G., Wu A. M., Gambhir S. S., Weiss S. Quantum dots for live cells, *in vivo* imaging, and diagnostics. *Science.* 2005;307(5709):538–544.

Mieszawska A. J., Mulder W. J. M., Fayad Z. A., Cormode D. P. Multifunctional gold nanoparticles for diagnosis and therapy of disease. *Molecular Pharmaceutics.* 2013;10(3):831–847.

Mikhaylova M., Stasinopoulos I., Kato Y., Artemov D., Bhujwalla Z. M. Imaging of cationic multifunctional liposome-mediated delivery of COX-2 siRNA. *Cancer Gene Therapy.* 2008;16(3):217–226.

Mohamed M. B., Volkov V., Link S., El-Sayed M. A. The "lightning" gold nanorods: Fluorescence enhancement of over a million compared to the gold metal. *Chemical Physics Letters.* 2000;317(6):517–523.

Morawski A. M., Winter P. M., Crowder K. C., Caruthers S. D., Fuhrhop R. W., Scott M. J., Robertson J. D., Abendschein D.R., Lanza G.M., Wickline S.A. Targeted nanoparticles for quantitative imaging of sparse molecular epitopes with MRI. *Magnetic Resonance in Medicine.* 2004;51(3):480–486.

Mulder W. J., Griffioen A. W., Strijkers G. J., Cormode D. P., Nicolay K., Fayad Z. A. Magnetic and fluorescent nanoparticles for multimodality imaging. *Nanomedicine.* 2007a;2(3):307–324.

Mulder W. J. M., van der Schaft D. W. J., Hautvast P. A. I., Strijkers G. J., Koning G. A., Storm G., Mayo K. H., Griffioen A. W., Nicolay K. Early *in vivo* assessment of angiostatic therapy efficacy by molecular MRI. *FASEB Journal: Official Publication of the Federation of American Societies for Experimental Biology.* 2007b;21(2):378–383.

Mulder W. J. M., Strijkers G. J., van Tilborg G. A. F., Cormode D. P., Fayad Z. A., Nicolay K. Nanoparticulate assemblies of amphiphiles and diagnostically active materials for multimodality imaging. *Accounts of Chemical Research.* 2009;42(7):904–914.

Mulder W. J. M., Strijkers G. J., van Tilborg G. A. F., Griffioen A. W., Nicolay K. Lipid-based nanoparticles for contrast-enhanced MRI and molecular imaging. *NMR in Biomedicine.* 2006;19(1):142–164.

Myers V. S., Weir M. G., Carino E. V., Yancey D. F., Pande S., Crooks R. M. Dendrimer-encapsulated nanoparticles: New synthetic and characterization methods and catalytic applications. *Chemical Science.* 2011;2(9):1632.

Myroshnychenko V., Nelayah J., Adamo G., Geuquet N., Rodríguez-Fernández J., Pastoriza-Santos I., MacDonald K. F. et al. Plasmon spectroscopy and imaging of individual gold nanodecahedra: A combined optical microscopy, cathodoluminescence, and electron energy-loss spectroscopy study. *Nano Letters.* 2012;12(8):4172–4180.

Naha P. C., Al Zaki A., Hecht E., Chorny M., Chhour P., Blankemeyer E., Yates D. M. et al. Dextran coated bismuth–iron oxide nanohybrid contrast agents for computed tomography and magnetic resonance imaging. *Journal of Materials Chemistry B.* 2014;2(46):8239–8248.

Naha P. C., Chhour P., Cormode D. P. Systematic *in vitro* toxicological screening of gold nanoparticles designed for nanomedicine applications. *Toxicology in Vitro.* 2015;29(7):1445–1453.

Nanospectra Biosciences Inc. MRI/US fusion imaging and biopsy in combination with nanoparticle directed focal therapy for ablation of prostate tissue. https://clinicaltrials.gov/show/NCT02680535 [Internet].

Nikoobakht B., El-Sayed M. A. Preparation and growth mechanism of gold nanorods (NRs) using seed-mediated growth method. *Chemistry of Materials.* 2003;15(10):1957–1962.

Oldenburg S. J., Jackson J. B., Westcott S. L., Halas N. J. Infrared extinction properties of gold nanoshells. *Applied Physics Letters.* 1999;75(19):2897.

Pan D., Cai X., Yalaz C., Senpan A., Omanakuttan K., Wickline S. A., Wang L.V., Lanza G. M. Photoacoustic sentinel lymph node imaging with self-assembled copper neodecanoate nanoparticles. *ACS Nano.* 2012;6(2):1260–1267.

Peer D., Karp J. M., Hong S., Farokhzad O. C., Margalit R., Langer R. Nanocarriers as an emerging platform for cancer therapy. *Nature Nanotechnology.* 2007;2(12):751–760.

Pikkemaat J. A., Wegh R. T., Lamerichs R., van de Molengraaf R. A., Langereis S., Burdinski D., Raymond A. Y. F. et al. Dendritic PARACEST contrast agents for magnetic resonance imaging. *Contrast Media Mol. Imaging.* 2007;2(5):229–239.

Pérez-Medina C., Abdel-Atti D., Zhang Y., Longo V. A., Irwin C. P., Binderup T., Ruiz-Cabello J. et al. A modular labeling strategy for in vivo PET and near-infrared fluorescence imaging of nanoparticle tumor targeting. *Journal of Nuclear Medicine*. 2014;55(10):1706–1712.

Qian X., Peng X.-H., Ansari D. O., Yin-Goen Q., Chen G. Z., Shin D. M., Yang L., Young A. N., Wang M. D., Nie S. In vivo tumor targeting and spectroscopic detection with surface-enhanced Raman nanoparticle tags. *Nature Biotechnology*. 2007;26(1):83–90.

Reichardt W., Hu-Lowe D., Torres D., Weissleder R., Bogdanov A. Imaging of VEGF receptor kinase inhibitor-induced antiangiogenic effects in drug-resistant human adenocarcinoma model. *Neoplasia*. 2005;7(9):847–853.

Rieffel J., Chen F., Kim J., Chen G., Shao W., Shao S., Chitgupi U. et al. Hexamodal imaging with porphyrin-phospholipid-coated upconversion nanoparticles. *Advanced Materials*. 2015;27(10):1785–1790.

Ruggiero A., Holland J. P., Lewis J. S., Grimm J. Cerenkov luminescence imaging of medical isotopes. *Journal of Nuclear Medicine*. 2010;51(7):1123–1130.

Ruiz-Cabello J., Barnett B. P., Bottomley P. A., Bulte J. W. Fluorine (19F) MRS and MRI in biomedicine. *NMR in Biomedicine*. 2011;24(2):114–129.

Röglin L., Lempens E. H. M., Meijer E. W. A synthetic "Tour de Force": Well-defined multivalent and multimodal dendritic structures for biomedical applications. *Angewandte Chemie International Edition*. 2011;50(1):102–112.

Saha K., Agasti S. S., Kim C., Li X., Rotello V. M. Gold nanoparticles in chemical and biological sensing. *Chemical Reviews*. 2012;112(5):2739–2779.

Schellenberger E. A., Hogemann D., Josephson L., Weissleder R. Annexin V-CLIO: A nanoparticle for detecting apoptosis by MRI. *Academic Radiology*. 2002;9:S.310–S311.

Schirra C. O., Senpan A., Roessl E., Thran A., Stacy A. J., Wu L., Proska R., Pan D. Second generation gold nanobeacons for robust K-edge imaging with multi-energy CT. *Journal of Materials Chemistry*. 2012;22(43):23071–23077.

Seo J. W., Mahakian L. M., Kheirolomoom A., Zhang H., Meares C.F., Ferdani R., Anderson C. J., Ferrara K. W. Liposomal Cu-64 labeling method using bifunctional chelators: Poly(ethylene glycol) spacer and chelator effects. *Bioconjugate Chemistry*. 2010;21(7):1206–1215.

Seo J. W., Mahakian L. M., Tam S., Qin S., Ingham E. S., Meares C. F., Ferrara K. W. The pharmacokinetics of Zr-89 labeled liposomes over extended periods in a murine tumor model. *Nuclear Medicine and Biology*. 2015;42(2):155–163.

Shapiro A. M. J., Ricordi C., Hering B. Edmonton's islet success has indeed been replicated elsewhere. *The Lancet*. 2003;362(9391):1242.

Shapiro E. M. Biodegradable, polymer encapsulated, metal oxide particles for MRI-based cell tracking. *Magnetic Resonance in Medicine*. 2015;73(1):376–389.

Sieber M.A., Pietsch H., Walter J., Haider W., Frenzel T., Weinmann H.-J. A preclinical study to investigate the development of nephrogenic systemic fibrosis: A possible role for gadolinium-based contrast media. *Investigative Radiology*. 2008;43(1):65–75.

Simone E. A., Zern B. J., Chacko A. M., Mikitsh J. L., Blankemeyer E. R., Muro S., Stan R. V., Muzykantov V. R. Endothelial targeting of polymeric nanoparticles stably labeled with the PET imaging radioisotope iodine-124. *Biomaterials*. 2012;33(21):5406–5413.

Smith A. M., Mancini M. C., Nie S. Bioimaging: Second window for in vivo imaging. *Nature Nanotechnology*. 2009;4(11):710–711.

Solomon R., Dumouchel W. Contrast media and nephropathy: Findings from systematic analysis and food and drug administration reports of adverse effects. *Investigative Radiology*. 2006;41(8):651–660.

Sreejith S., Joseph J., Lin M., Menon N. V., Borah P., Ng H. J., Loong Y. X., Kang Y., Yu S. W.-K., Zhao Y. Near-infrared squaraine dye encapsulated micelles for *in vivo* fluorescence and photoacoustic bimodal imaging. *ACS Nano*. 2015;9(6):5695–5704.

Srivatsan A., Chen X. Recent advances in nanoparticle-based nuclear imaging of cancers. In: Pomper M., Fisher P., eds. *Advances in Cancer Research*. Vol. 124. 1st ed. Elsevier Inc., London, UK, 2014, pp. 83–129.

Stacul F., van der Molen A. J., Reimer P., Webb J. A. W., Thomsen H. S., Morcos S. K., Almén T. et al. Contrast induced nephropathy: Updated ESUR contrast media safety committee guidelines. *European radiology.* 2011;21(12):2527–2541.

Stelter L., Pinkernelle J. G., Michel R., Schwartländer R., Raschzok N., Morgul M. H., Koch M. et al. Modification of aminosilanized superparamagnetic nanoparticles: Feasibility of multimodal detection using 3T MRI, small animal PET, and fluorescence imaging. *Molecular Imaging and Biology.* 2010;12(1):25–34.

Sykes E. A., Chen J., Zheng G., Chan W. C. W. Investigating the impact of nanoparticle size on active and passive tumor targeting efficiency. *ACS Nano.* 2014;8(6):5696–5706.

Tao A. R., Habas S., Yang P. Shape control of colloidal metal nanocrystals. *Small.* 2008;4(3):310–325.

Tassa C., Shaw S. Y., Weissleder R. Dextran-coated iron oxide nanoparticles: A versatile platform for targeted molecular imaging, molecular diagnostics, and therapy. *Accounts of Chemical Research.* 2011;44(10):842–852.

Tepel M., Aspelin P., Lameire N. Contrast-induced nephropathy: A clinical and evidence-based approach. *Circulation.* 2006;113(14):1799–806.

Terreno E., Cabella C., Carrera C., Castelli D. D., Mazzon R., Rollet S., Stancanello J., Visigalli M., Aime S. From spherical to osmotically shrunken paramagnetic liposomes: An improved generation of LIPOCEST MRI agents with highly shifted water protons. *Angewandte Chemie.* 2007;46(6):966–968.

Thakor A. S., Jokerst J., Zavaleta C., Massoud T. F., Gambhir S. S. Gold nanoparticles: A revival in precious metal administration to patients. *Nano Letters.* 2011;11(10):4029–4036.

Thorek D. L. J., Das S., Grimm J. Molecular imaging using nanoparticle quenchers of Cerenkov luminescence. *Small (Weinheim an der Bergstrasse, Germany).* 2014;10(18):3729–3734.

Ungureanu C., Kroes R., Petersen W., Groothuis T. A. M., Ungureanu F., Janssen H., Van Leeuwen F. W. B., Kooyman R. P. H., Manohar S., Van Leeuwen T. G. Light interactions with gold nanorods and cells: Implications for photothermal nanotherapeutics. *Nano Letters.* 2011;11(5):1887–1894.

van de Linde S., Heilemann M., Sauer M. Live-cell super-resolution imaging with synthetic fluorophores. *Annual Review of Physical Chemistry.* 2012;63(1):519–540.

van Schooneveld M. M., Vucic E., Koole R., Zhou Y., Stocks J., Cormode D. P., Tang C. Y. et al. Improved biocompatibility and pharmacokinetics of silica nanoparticles by means of a lipid coating: A multimodality investigation. *Nano Letters.* 2008;8(8):2517–2525.

Vahrmeijer A. L., Hutteman M., van der Vorst J. R., van de Velde C. J. H., Frangioni J. V. Image-guided cancer surgery using near-infrared fluorescence. *Nature Reviews Clinical Oncology.* 2013;10(9):507–518.

Vasalatiy O., Gerard R. D., Zhao P., Sun X. K., Sherry A. D. Labeling of adenovirus particles with PARACEST agents. *Bioconjugate Chemistry.* 2008;19(3):598–606.

Vucic E., Sanders H. M. H. F., Arena F., Terreno E., Aime S., Nicolay K., Leupold E. et al. Well-defined, multifunctional nanostructures of a paramagnetic lipid and a lipopeptide for macrophage imaging. *Journal of the American Chemical Society.* 2009;131(2):406–407.

Wang Y., Liu Y., Luehmann H., Xia X., Wan D., Cutler C., Xia Y. Radioluminescent gold nanocages with controlled radioactivity for real-time *in vivo* imaging. *Nano Letters.* 2013;13(2):581–585.

Wang Y. J., Strohm E. M., Sun Y., Niu C., Zheng Y., Wang Z., Kolios M. C. PLGA/PFC particles loaded with gold nanoparticles as dual contrast agents for photoacoustic and ultrasound imaging. *Proceedings of SPIE.* 2014;8943:89433M.

Ward K. M., Aletras A. H., Balaban R. S. A new class of contrast agents for MRI based on proton chemical exchange dependent saturation transfer (CEST). *Journal of Magnetic Resonance.* 2000;143(1):79–87.

Waters E. A., Chen J. J., Allen J. S., Zhang H. Y., Lanza G. M., Wickline S. A. Detection and quantification of angiogenesis in experimental valve disease with integrin-targeted nanoparticles and 19-fluorine MRI/MRS. *Journal of Cardiovascular Magnetic Resonance.* 2008;10:43.

Weissleder R., Nahrendorf M., Pittet M. J. Imaging macrophages with nanoparticles. *Nature Materials.* 2014;13(2):125–138.

Winter P. M., Cai K., Chen J., Adair C. R., Kiefer G. E., Athey P. S., Gaffney P. J. et al. Targeted PARACEST nanoparticle contrast agent for the detection of fibrin. *Magnetic Resonance in Medicine*. 2006;56(6):1384–1388.

Winter P. M., Caruthers S. D., Kassner A., Harris T. D., Chinen L. K., Allen J. S., Lacy E. K. et al. Molecular Imaging of Angiogenesis in Nascent Vx-2 Rabbit Tumors Using a Novel {alpha}{nu}{beta}3-targeted Nanoparticle and 1.5 Tesla Magnetic Resonance Imaging. *Cancer Research*. 2003;63(18):5838–5843.

Woodle M. C. 67Gallium-labeled liposomes with prolonged circulation: Preparation and potential as nuclear imaging agents. *Nuclear Medicine and Biology*. 1993;20(2):149–155.

Writer M. J., Kyrtatos P. G., Bienemann A. S., Pugh J. A., Lowe A. S., Villegas-Llerena C., Kenny G. D. et al. Lipid peptide nanocomplexes for gene delivery and magnetic resonance imaging in the brain. *Journal of Controlled Release*. 2012;162(2):340–348.

Wu P., Yan X.-P. Doped quantum dots for chemo/biosensing and bioimaging. *Chemical Society Reviews*. 2013;42(12):5489–521.

Wu X., Ming T., Wang X., Wang P., Wang J., Chen J. High-photoluminescence-yield gold nanocubes: For cell imaging and photothermal therapy. *ACS Nano*. 2010;4(1):113–120.

Xi L., Satpathy M., Zhao Q., Qian W., Yang L., Jiang H. HER-2/neu targeted delivery of a nanoprobe enables dual photoacoustic and fluorescence tomography of ovarian cancer. *Nanomedicine: Nanotechnology, Biology, and Medicine*. 2014;10(3):669–677.

Xie H., Wang Z. J., Bao A., Goins B., Phillips W. T. In vivo PET imaging and biodistribution of radiolabeled gold nanoshells in rats with tumor xenografts. *International Journal of Pharmaceutics*. 2010;395(1–2):324–330.

Xu Y., Liu H., Cheng Z. Harnessing the power of radionuclides for optical imaging: Cerenkov luminescence imaging. *Journal of Nuclear Medicine*. 2011;52(12):2009–2018.

Xu Y., Wu M., Liu Y., Feng X.-Z., Yin X.-B., He X.-W., Zhang Y.-K. Nitrogen-doped carbon dots: A facile and general preparation method, photoluminescence investigation, and imaging applications. *Chemistry—A European Journal*. 2013;19(7):2276–2283.

Yang X., Hong H., Grailer J. J., Rowland I. J., Javadi A., Hurley S. A., Xiao Y. et al. cRGD-functionalized, DOX-conjugated, and 64Cu-labeled superparamagnetic iron oxide nanoparticles for targeted anticancer drug delivery and PET/MR imaging. *Biomaterials*. 2011;32(17):4151–4160.

Yi H., Ghosh D., Ham M.-H., Qi J., Barone P. W., Strano M. S., Belcher A. M. M13 phage-functionalized single-walled carbon nanotubes as nanoprobes for second near-infrared window fluorescence imaging of targeted tumors. *Nano Letters*. 2012;12(3):1176–1183.

Yuan H., Khoury C. G., Hwang H., Wilson C. M., Grant G. A., Vo-Dinh T. Gold nanostars: Surfactant-free synthesis, 3D modelling, and two-photon photoluminescence imaging. *Nanotechnology*. 2012;23(7):75102.

Yuan L., Lin W., Zheng K., He L., Huang W. Far-red to near infrared analyte-responsive fluorescent probes based on organic fluorophore platforms for fluorescence imaging. *Chemical Society Reviews*. 2013a;42(2):622–661.

Yuan L., Lin W., Zheng K., Zhu S. FRET-based small-molecule fluorescent probes: Rational design and bioimaging applications. *Accounts of Chemical Research*. 2013b;46(7):1462–1473.

Zeng L., Ren W., Zheng J., Cui P., Wu A. Ultrasmall water-soluble metal-iron oxide nanoparticles as T1-weighted contrast agents for magnetic resonance imaging. *Physical Chemistry Chemical Physics: PCCP*. 2012;14(8):2631–2636.

Zhang Y., Jeon M., Rich L. J., Hong H., Geng J., Zhang Y., Shi S. et al. Non-invasive multimodal functional imaging of the intestine with frozen micellar naphthalocyanines. *Nature Nanotechnology*. 2014;9(8):631–638.

Zhen Z., Tang W., Todd T., Xie J. Ferritins as nanoplatforms for imaging and drug delivery. *Expert Opinion on Drug Delivery*. 2014;11(12):1913–1922.

7
Recapitulating Tumor Extracellular Matrix: Design Criteria for Developing Three-Dimensional Tumor Models

Dhaval Kedaria and Rajesh Vasita

CONTENTS

7.1 Introduction ... 251
7.2 Tumor Microenvironment ... 252
    7.2.1 Components of Tumor Extracellular Matrix ... 253
    7.2.2 Role of Extracellular Matrix in Regulation of Cancer Cell Behavior ... 256
7.3 Mimicking Properties of Tumor Extracellular Matrix ... 257
    7.3.1 Geometrical Properties ... 257
        7.3.1.1 Topography ... 258
        7.3.1.2 Dimensionality ... 258
        7.3.1.3 Porosity ... 260
    7.3.2 Mechanical Properties ... 261
        7.3.2.1 Stiffness ... 261
    7.3.3 Biochemical Properties ... 263
        7.3.3.1 Surface Chemistry ... 264
        7.3.3.2 Chemical Composition ... 265
7.4 Mimicking Approaches by Three-Dimensional Platforms for Drug Screening ... 265
    7.4.1 Hydrogel ... 266
    7.4.2 Nanofibers ... 268
    7.4.3 Microfluidics Models ... 269
7.5 Conclusion and Future Perspectives ... 270
Acknowledgments ... 271
References ... 271

7.1 Introduction

Cancer is a disease characterized by out-of-control growth of abnormal cells, forming a mass of tissue called a malignant tumor, adept at invading and propagating to other parts of the body. This is achieved by the creation of supporting stroma, which not only contain cellular components but also other biomolecules, including growth factors, cytokines, and extracellular matrix (ECM) constituents, forming an intricate tumor microenvironment.

ECM components interact with variety of surface receptors specifically present on each cell type, which influences cell fate and the response to cytokines and growth factors (Lin and Bissell 1993). Indeed, in a context-dependent manner, the physicochemical and biomechanical properties of the ECM can regulate cell viability, adhesion, invasion, differentiation, and gene expression (Ingber and Folkman 1989, Juliano and Haskill 1993). These lead to control physiological processes such as tissue morphogenesis and angiogenesis along with pathological conditions including transformation and metastasis (Yurchenco and Schittny 1990, Damsky and Werb 1992, Mosher et al. 1992, Pupa et al. 2002). Since this evidence necessitates ECM components as key players, it is indispensable to consider them during targeting cancer by chemotherapeutic drugs.

Chemotherapy is one of the common cancer treatment strategies involving the use of natural and/or synthetic therapeutic molecules. The molecules can be used clinically as chemotherapeutic agents after the preclinical screening and testing stages of development. In the last decade, there has been an escalation in the number of potential chemotherapeutic agents being screened for anticancer drug development; however, only about 10% of them show productivity throughout clinical development (Arrondeau et al. 2010, Hait 2010). The most commonly used methods for screening include the cell culture of human-derived tumor cell lines in two dimensional petri dishes compromised with inappropriate microenvironmental cues that produce altered cell–cell and cell–matrix interactions (Birgersdotter et al. 2005, Smalley et al. 2006, Levental et al. 2009). These may raise the issues of reduced clinical efficacy and undesirable toxicity of drugs in the clinical phase of development, two of the main reasons for drug failures (Hopkins 2008). This technical uncertainty could be reduced before the beginning of expensive preclinical and clinical phase to reduce the cost of the drug development process (Paul et al. 2010). It is therefore essential to improve *in vitro* cell culture assays for more predictive and reliable preclinical drug screening. Three-dimensional cultures perhaps will be the answer to use as a model to explain how cell matrix interaction affects the morphology, proliferation, gene expressions, and drug responses (Smalley et al. 2006).

In recent years, 3D cultures have gained attention, which has been established on prefabricated scaffolds (Shakibaei et al. 2015). Natural and synthetic biomaterials have been used to fabricate 3D porous scaffolds. Interconnected pores in the scaffold provide a surface area for cellular attachment to a greater extent, and cells seeded into such scaffold can migrate, grow, and anchor to form 3D cellular structures (Moscato et al. 2015). Moreover, a poriferous network allows oxygen, nutrients, and drugs to reach the multiplying cells and enables removal of waste molecules, providing suitable cell culture conditions (Schmidt et al. 2008). Therefore, mimicking native tumor ECM properties is of the utmost importance and can be achieved by controlling the properties of biomaterials and the scaffold fabrication method. This progress report summarizes the contemporary understanding of biochemical and biomechanical properties of native tumor ECMs that influence cellular behavior, an inspiration for developing 3D tumor models. This understanding encourages innovative approaches to mimicking the properties of ECMs owing to the biomaterials and will be discussed in depth. Success in recreating 3D tumors on scaffolds has been explored for drug screening, and its concurrent view has been deliberated in detail.

7.2 Tumor Microenvironment

The tumor microenvironment resides not only in malignant cells but, what's more, has multiple nonmalignant cell types such as fibroblasts, endothelial cells, and leukocytes, as well as a multifaceted noncellular part comprising the extracellular matrix. The

ECM is a complex structure with a diverse range of proteins and polysaccharides that compartmentalize cells and regulate nutrient and bioactive molecule trafficking. The ECM not only provides a platform for cellular attachment and growth, but also serves as a local depot for growth factors.

ECM components are produced uniquely by cells homing in it. Transformed cells can activate the surrounding ECM, initiate local signaling within the microenvironment, and exponentially increase the ECM turnover and remodeling by nonmalignant cells. For example, a tumor is characterized by stiffer tissue than normal as a result of excess ECM production and collagen cross-linking (Paszek et al. 2005, Butcher et al. 2009) produced by the resident fibroblast (Levental et al. 2009). Every single component of the ECM has unique features, and altered expression of these components provides a facilitating ground for cancer progression (see Schematic 7.1).

7.2.1 Components of Tumor Extracellular Matrix

Two main classes of macromolecules make up the matrix, glycosaminoglycans (GAGs), bound to protein in the form of proteoglycans, and protein such as collagen, elastin, fibronectin, and laminin as fibril proteins and enzymes such as matrix metalloproteinase (MMPs) (see Table 7.1) (Bosman and Stamenkovic 2003, Schaefer and Schaefer 2010). GAGs are linear, negatively charged polysaccharides made up of repeating disaccharide units comprising acetylated amino sugars and uronic acid. GAG chains are classified into chondroitin sulfate (CS), dermatan sulfate (DS), keratan sulfate (KS), and heparin sulfate (HS), and nonsulfated GAGs such as hyaluronan (HA). The polyanionic nature of GAG pulls water in, causing GAGs to swell. This swelling can open up pathways for invasion and migration of cells, which is attributed to cancer metastasis, as suggested for HA (Toole 2002).

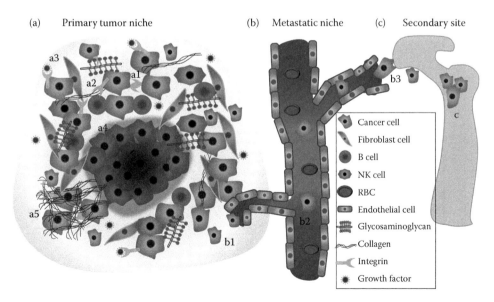

SCHEMATIC 7.1
Tumor microenvironment. (a) Primary tumor niche, (a1) cell–cell interactions through integrin, (a2) cell–matrix component interaction by integrin, (a3) GF attachment and signaling, (a4) hypoxia, (a5) matrix cross-linking and stiffness. (b) Metastatic niche, (b1) detachment from primary site, (b2) migration through blood vessel, (b3) extravasation in to secondary site. (c) Secondary tumor formation at distant site (bone).

TABLE 7.1
Components of ECM: The Structure and Functions of Various ECM Molecules and Their Expression/Role in Tumor Microenvironment

ECM Component	Structure	Properties/Function	Expression in Tumor	References
Chondroitin sulfate and Dermatan sulfate	Repeating disaccharides of N-acetylgalactosamine and d-glucuronic acid with sulfation pattern for chondroitin sulfate and repeating disaccharides of N-acetylgalactosamine and l-iduronic acid with sulfation pattern for dermetan sulfate	Large polyanionic chains provide hydrated gel-space filling structure, contribute to the physical strength of cartilaginous tissue	Enhances the assembly of a pericellular matrix and promotes cancer cell motility and invasion; CD44 mediates cell adhesion	Asimakopoulou et al. (2008), Nikitovic et al. (2008), Wegrowski and Maquart (2006), Malavaki et al. (2008), Raman et al. (2005).
Heparin sulfate	Repeating disaccharide moities containing glucosamine and hexuronic acid (glucuronic acid and iduronic acid)	Modulator of cell–cell and cell–matrix interactions, bioavailability of FGF-2	Degradation promotes growth factor release and also enhances invasion and metastasis	Guimond et al. (1993), Ashikari-Hada et al. (2004), Sanderson (2001), Sasisekharan et al. (2002), Raman et al. (2005).
Keratan sulfate	3 Galactose β1-4 N-acetylglucosamine β1-repeating disaccharide unit, C6 sulfated	Maintenance of tissue hydration in cornea, antiadhesive property during cell migration	CD44-mediated regulation of cancer progression	Takahashi et al. (1996), Afratis et al. (2012), Gesslbauer et al. (2007), Raman et al. (2005) Yip et al. (2006).
Hyaluronan	Nonsulfated, disaccharides repeating unit of d-glucuronic and N-acetylglucosamine	Principle receptor for CD44, RHAMM	Creates a viscoelastic environment, promoting cancer cell proliferation and migration	Toole (2002), Afratis et al. (2012).
Collagen	Repeating unit of Gly-X-Y, tripeptide chain	Fibrous structural protein	Increased density/ cross-linking-produced stiffness could enhance FAK-dependent tumor growth	Provenzano et al. (2008), Shoulders and Raines (2009), Gelse (2003), Bretscher et al. (2001), Koster et al. (2008), Lucero and Kagan (2006).
Elastin	Made up of several cross-linked tropoelastin fibers	Fibrous structural protein gives flexibility to tissue, mostly found in skin, lung, and cartilage	Increased cross-linking with collagen enhances tumor progression	Butcher et al. (2009), Jarvelainen et al. (2009), Payne et al. (2007).

(Continued)

TABLE 7.1 (*Continued*)

Components of ECM: The Structure and Functions of Various ECM Molecules and Their Expression/Role in Tumor Microenvironment

ECM Component	Structure	Properties/Function	Expression in Tumor	References
Fibronectin	Glycoprotein having protein dimer, each monomer made up of antiparallel β sheets	Mediates cellular interactions with the collagen, heparin, and fibrin	Increased expression, specifically in NSCLC, enhanced tumorigenicity through integrin signaling	Han et al. (2006), Tanzer (2006), Xu et al. (2006), White et al. (2008).
Laminin	Glycoprotein, heterotrimer assembled from α, β, and γ subunits	From network structure of BM with adhesive, migration-promoting, and signaling functions	Decreased expression or modification of laminin causes BM structure during invasion	Patarroyo et al. (2002), Xu et al. (2006), Mochizuki et al. (2003).
Matrix metalloproteinase	Zink-dependent endopeptidase enzymes, present as zymogen form	Proteinases that participate in ECM remodeling	Mediates ECM degradation for tumor invasion and metastasis	Kessenbrock et al. (2010), Bosman and Stamenkovic (2003), Visse and Nagase (2003), Kajita et al. (2001), Zaman et al. (2006).

The sulfation pattern is the determining factor for the binding of growth factors to GAG. For example, the binding of FGF2 to HS chains follows through N-sulfated glucosamine and 2-O-sulfated iduronic acid units or 6-O-sulfated glucosamine (Guimond et al. 1993, Gallagher 2001), while highly sulfated HS can trigger FGF2-stimulated cell proliferation during cancer progression (Nikitovic et al. 2008).

In the case of proteins, fibrous proteins including collagen, elastin, and laminin make a network structure that provides mechanical properties to the matrix, whereas adhesive proteins such as fibronectin are largely responsible for cell binding to matrix. ECM protein expression regulates intracellular signaling and its altered expression significantly affects the cell fate (see Table 7.1).

7.2.2 Role of Extracellular Matrix in Regulation of Cancer Cell Behavior

For successive transition from a neoplasm containing transformed cells to invasive carcinoma, the alteration of homeostatic balance is essential. Successive events that potentially create "reactive" stroma include undue production of growth factors and exchanging stromal cells with tumor cells. For instance, production of transforming growth factor-β as well as epidermal growth factor in turn induces the selection and expansion of neoplastic cells (Liotta and Kohn 2001). Consecutive clonal expansion of transformed cells alters the microenvironment composition (Table 7.1). It has been suggested that the "seeds" of cancer cells with their altered "soil" form a dynamic 3D "organ-like tissue" that repetitively changes as malignancy progresses (Fidler 2003). These include tumor cells driving elevated expression of lysyl oxidase (an enzyme necessary for collagen I cross-linking) and differentiation of fibroblasts into matrix-depositing myofibroblasts, promoting enhanced deposition, linearization, and cross-linking of collagen type I and fibronectin-rich matrices (Otranto et al. 2012). The resultant occurrence of integrin gathering due to collagen and phosphorylation of integrin-regulated effectors, such as focal adhesion kinase, leads to the promotion of invasive behavior (Paszek et al. 2005, Levental et al. 2009). In remodeled ECM, an extensive number of proliferated cells and ECM cross-linking together may fall out in increased tissue density, hypoxia, and interstitial acidosis (Paszek and Weaver 2004, Cardone et al. 2005, Brahimi-Horn et al. 2007). Proteoglycans are likewise abnormally expressed in a wide variety of malignant tumors. Proteoglycans appear to be associated with cancer through direct involvement in cellular functions or by activity of other effective molecules, such as growth factors and cytokines. Elevated levels of chondroitin sulfate proteoglycan-versican have been observed in various tumor types, such as stroma of prostate cancer and breast cancer; malignant melanomas; testicular tumors; and pancreatic, laryngeal, and gastric cancer, where highly negatively charged CS side chains enhance the assembly of a pericellular matrix and promote prostate cancer cell motility and invasion *in vitro* (Ricciardelli et al. 1998, Labropoulou et al. 2006, Wegrowski and Maquart 2006).

As a tumor progresses, MMPs are secreted and activated, which alters the microenvironment with sequentially released and activated ECM embedded growth factors (Bosman and Stamenkovic 2003, Kessenbrock et al. 2010). The release of growth factors, such as vascular endothelial growth factor (VEGF), improves vascular permeability and stimulates new vessel growth, which creates interstitial tissue pressure. This induces angiogenesis along with disrupted basement membrane support invasion and, eventually, brings about metastasis. This in turn leads to a positive feedback loop to enhance tumor growth and survival (Paszek et al. 2005, Butcher et al. 2009, Erler and Weaver 2009).

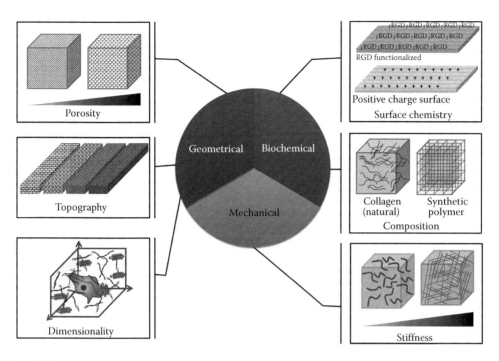

SCHEMATIC 7.2
Schematic illustration of mimicking properties of tumor ECM. Three principle properties, namely geometrical, mechanical, and biochemical properties, could be represented individually or in combination in engineered microenvironments.

7.3 Mimicking Properties of Tumor Extracellular Matrix

The ECM has unique physicochemical features, which specifically regulate tumor cell activity. Inspired by the native tumor microenvironment, the geometrical, mechanical, and biochemical properties of the ECM have been attempted to be reconstituted by developing scaffold-based *in vitro* models to recapitulate *in vivo* tumors (see Schematic 7.2) (Holen et al. 2015).

7.3.1 Geometrical Properties

Solid tumors are similar to organs, although functionally abnormal, and are made up of different cells embedded in a specialized ECM arranged in such a way that forms a three-dimensional tissue-like structure (Aumailley and Gayraud 1998, Egeblad et al. 2010). Altered ECM components in solid tumors maneuver architectural properties including the density and porosity of matrix. For example, tumor initiation may reflect an increase in collagen density and cross-linking, which produce a compact matrix environment (McKegney et al. 2001, Provenzano et al. 2008). An increase in ECM collagen density not only decreases porosity but will increase focal adhesion for cells. As a consequence, increased density of integrin in the surrounding microenvironment initiates intracellular signaling pathways among cells for tumor progression (Levental et al. 2009). Therefore, understanding the effect of ECM molecules, their density, and packing will contribute significantly to creating a 3D atmosphere for cancer cells.

7.3.1.1 Topography

The distribution of features on the surface of or within a scaffold is described by its topography for micro- to nanoscale features. Surface topography, such as roughness and grooves of scaffolds, regulates adhesion and the spreading characteristics of cells. A study done by Zang et al. has shown that increasing surface roughness improves the cellular attachment and growth. Poly lactide co-glycolic acid (PLGA) films with smooth or nanopatterned surface have been used for this study and their results demonstrated that films with nanopatterned surfaces have a significantly greater number of viable cells of lung adenocarcinoma (Zhang and Webster 2013). Furthermore, the scale of roughness, which influences the cell behavior, was also determined by Zhang et al. Similar groups have fabricated PLGA films with different roughnesses by polystyrene bead-based casting. Surface roughness was measured as root-mean-squared (RMS) roughness. A549 lung adenocarcinoma cells have been seeded with fixed density on PLGA films. Interestingly, in this study, cell adhesion was not correlated to increasing surface roughness. It was observed that after 4 hours postseeding, the maximum adhesion was 44.83% of the seeding density with a 5.03 nm RMS value compared to 24.54% for nearly smooth surfaces with a 0.62 nm RMS value. Interestingly, cell proliferation was pursued with 2.23, 5.03, 5.42, and 36.89 nm RMS values, and the PLGA films had 239%, 270%, 220%, and 223%, respectively, cell density compared to on 0.62 nm RMS PLGA surfaces after 3 days (Zhang et al. 2010). These studies demonstrate that cell adhesion and proliferation rate were not linear with increasing roughness. This could be attributed to a precise topographic pattern generated by increasing surface roughness that regulates cancer cell attachment and subsequent growth.

Various ECM molecules combined to form micro- to nanoscale architectures, including grooves, which establish cell compartmentalization and contact guidance migrations. The grooves on the surface may confine cell movement or fine-tune its actin filaments to adjust to topography (Walboomers et al. 1999). Studies have been done in which a nanofibrous scaffold has been used with random or aligned fiber orientation; cancer cells change their morphology according to the fibers. It has been observed that when breast cancer cells were cultured on aligned polycaprolactone fibers, the cytoskeleton was aligned and nuclei were elongated, forming spindle-like shapes on the fiber axes. Similar cells were grown on a fiber with random orientation, and its cellular cytoskeleton was elongated to criss-cross fibers, making a star-like shape. Moreover, epithelial mesenchymal transition (EMT) was also upregulated in cancer cells on aligned fibers, indicating that topographical clues of the tumor microenvironment may play an important role in cancer progression (Saha et al. 2012).

Biophysical clues may recapitulate the native tumor microenvironment and be sensed by tumor cells; for example, increased collagen cross-linking could linearize collagen fibers, which is sensed by tumor cells during cancer progression. This mechanism might be imitated in a nanofibrous scaffold that promotes cancer progression through EMT activation, along with the rough surface that would be required for improved cellular adhesion. These studies clearly indicate the significance of the topographical features of the matrix on tumor cell morphology, invasive properties, and gene expression.

7.3.1.2 Dimensionality

The dimensionality is the major factor that is altered in scaffold-based culture of cells. A polymeric scaffold offers three-dimensionality to the cells that greatly influences cellular behavior in the microenvironment (Bai et al. 2015, Zhang et al. 2015). While considering 2D to 3D, complete spatial distribution has been altered, including cell receptors, binding

Recapitulating Tumor Extracellular Matrix 259

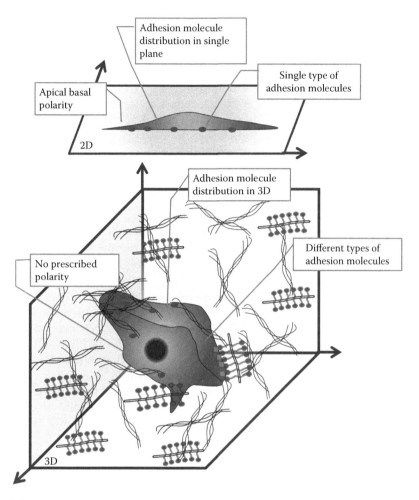

SCHEMATIC 7.3
Difference between morphology of cell in 2D vs. 3D. In 2D (top), apical-basal polarity of cell, adhesion molecule distribution in single plane, single type of adhesion molecules. In 3D (bottom), no prescribed polarity in 3D environment, adhesion molecule distribution in three dimensions, different types of adhesion molecules due to composition of scaffold.

sites, and growth factor distribution (see Schematic 7.3) (Baker and Chen 2012, Kim et al. 2015, Moscato et al. 2015).

A variety of studies have reported a change in tumor cell behavior when grown on 3D scaffolds. Serebriiskii and co-workers have studied the effects of dimensionality on various cancer cells. They have used an NIH 3T3-derived fibroblast matrix to prepare a 2D substrate or 3D matrix. Though with similar matrix protein composition, four selected cancer cells, including NCI-H460, HCT116, PA-1, and COLO 205, gave specific proliferation responses in the 3D matrix as NCI-H460 and HCT116 growth ceased, PA-1 showed moderately inhibited growth, while COLO 205 was unaffected, with equal growth compared to the 2D substrate (2008). As soon as tumor cells were seeded on the 3D scaffold, their morphology adjusted in accordance with their integrins that sense dimensionality (Schematic 2) (Larsen et al. 2006, Moscato et al. 2015). Moreover, there were altered collagen spatial distribution outcomes with regional variations in 3D-matrix exogenous tension that altogether revised cell–cell and cell–matrix interactions (Ingber 2006). Studies have shown that epithelial tumor cells, when cultured in three dimensions, lose polarity and form acini-like structures of disorganized proliferative masses or aggregates with perversely overexpressing oncogenic signals, analogous to those observed in *in vivo* tumor progression (Kenny et al. 2007). Given appropriate spatial and biochemical cues from the microenvironment, cancer cells acquire apoptotic resistance stimuli. As demonstrated by Weaver et al., when grown in 2D conditions, the malignant mammary epithelial cells (MECs) are sensitive to death signals. Conversely, when cells are grown on reconstituted basement membrane (rBM) 3D gel, they adopt a polarized architecture. Laminin-1, the main component of the rBM interaction via $\alpha 6\beta 4$ integrin, induces polarity to cancer cells. As downstream signaling, this in turn regulates NFκB activation to cause apoptosis resistance (Weaver et al. 2002).

Tumor cell invasion and migration are fundamental processes for the metastatic behavior of cancer. In 2D conditions, migration is reliant on adhesion. When migrating cells are retained in a 3D matrix, they become embedded in the scaffold and associate with the contiguous matrix (Kurniawan et al. 2015). Cell migration is exclusively an integrin-mediated traction in 2D support, while in the case of a 3D microenvironment, tumor cells have been remunerated by this route (Friedl and Wolf 2003). Cell migration studies in 3D collagen matrices have revealed that embedded tumor cells have displayed stress fibers radiating from the nucleus toward the plasma membrane that form pseudopodial protrusions (Bloom et al. 2008). Moreover, tumor cells initiated MMP-mediated local pericellular proteolysis, followed by activation of cytoskeleton architectural regulator ROCK and myosin II, which further induced large-scale tension and successive mechanical breakdown of the matrix. This matrix flaw releases the tension at the posterior end of the cell and helps the pseudopodial protrusions grow and, in turn, allow the cell to move forward.

These studies indicate that increased dimensionality from 2D to 3D could directly alter the architecture of the surrounding microenvironment of tumor cells, which is reflected by altered cell matrix adhesion-based cellular shape adaptation. This will in turn regulate tumor cell intracellular signaling, proliferation, and apoptosis resistance. Beyond that, cell-induced spatial asymmetry of the matrix and specialized architecture of the cell have been adopted during migration in 3D that could not be achieved on flattened stiff substrata even coated with ECM components.

7.3.1.3 Porosity

Porosity is the amount of the open pore volume within the scaffolds. Nutrient transport and waste removal processes require a preferably highly porous scaffold (Ramanujan et al. 2002). Porosity can also provide an adequate surface for cell colonization, ECM production, and subsequent formation of tumor spheroids. For instance, when lung cancer cells are grown on porous scaffold Variotis™ with 100-μm pores, the single layer of proliferative cells on the scaffold encourage them to form multilayer cell clumps of up to 500 μm in size (Zhang et al. 2013).

Pore size is major factor that affects cellular behavior within the porous matrix. Pore size is described as the diameter of circular pores or the longest length for noncircular pores. Suitable pore size provides structural advantages to cells that allow them to adhere to pores and link with neighboring cells. There is an optimum size range for cell growth above which cells fail to bridge between neighboring cells; this depends upon the cell type and polymeric materials used (Zeltinger et al. 2001). Cancer cell migration is accomplished through pores of the scaffold and is influenced by the pore size of matrix (Carey et al. 2015). For example, Balzer et al. have studied cancer cell migration in polydimethylsiloxane (PDMS) microchannels with specific channel width. It has been demonstrated that in 50-μm channels, which exceeds the cell nucleus, proteolysis of ECM is not required and tumor cells adapt "standard" integrin-dependent movement. Interestingly, confining cells to much a smaller pore size by switching to integrin-independent migration relies instead on microtubule dynamics, which has been demonstrated in 3-μm channels, where β1 integrin blockage has no effect on the migration of cancer cells (Balzer et al. 2012). Membrane type1 MMPs would serve as dominant molecules to remodel the ECM by proteolysis and increase pore size for facilitating migration of tumor cells (Rowe and Weiss 2009).

The porosity of the scaffold holds substantial requirements that do not merely provide suitable support for cellular attachment, but also facilitate cell migration, which is required for tumor formation and growth. However, an increase in pore size may decrease the strength. Therefore, constructing a highly porous polymeric scaffold with an optimum porosity profile is an advisable parameter in order to grow tumors *in vitro*.

7.3.2 Mechanical Properties

Tumors are commonly examined as rigid tissue through physical palpation. Indeed, diagnosis of tumors by monitoring tissue stiffness has been exploited for cancer screening (Khaled et al. 2004). An increase in the matrix stiffening is linked to fibrosis, characterized by the formation of excessive fibrous connective tissue and chiefly by an increased deposition of collagen I (Paszek and Weaver 2004). Aberrant collagen synthesis by myofibroblast in tumor stroma, accompanied by lysyl oxidase synthesis, could cause linearizing and cross-linking of collagen fibers. This all together could promote an increase in matrix rigidity. Typically, in the case of lungs, the stiffness of lung tissue increases from 10 kPa in normal condition to 25~35 kPa in lung fibrosis (Ebihara et al. 2000). The fibrotic ECM disturbs cellular polarity and provokes cell proliferation, creating a context for tumor development and progression. Consequently, a cultured microenvironment provides a platform to understand cell–ECM interaction or, more precisely, cell–ECM mechanobiology.

7.3.2.1 Stiffness

Stiffness, also called rigidity, of the matrix is defined as the amount to which it resists disruption against an applied force. An increase in collagen density and lysyl oxidase mediated cross-linking could be attributed to tissue fibrosis, which results in an elevated level of stiffness. Tumor cells experience this stiffness via mechanotransduction through integrin signaling (Levental et al. 2009). A study has been done in which it was observed that breast and prostate cancer cell lines formed smoother spherical tumoroids compared to those grown on soft matrigel (Bray et al. 2015). The work of Tilghman et al. showed that when cancer cell lines were grown in a collagen-coated matrix with increasing elastic moduli, ranging from 150 to 9600 Pa, nine cell lines revealed a matrix rigidity reliance for growth, growing considerably better on stiff matrices than on soft matrices (Tilghman et al. 2010). Not only that, they showed that cellular adenosine triphosphate (ATP) and protein synthesis including glycolytic enzymes were slower in the cells grown in soft gels. The decrease in metabolic activity led to staying in a dormant stage, and the reverse was observed in stiffer gels, resulting in a more aggressive phenotype (Tilghman et al. 2012). Matrix stiffness could also influence cancer stem cell maintenance; however, the optimum stiffness could depend upon the type of cells (Jabbari et al. 2015). The increase in matrix rigidity disturbs tissue architecture and enhances integrin adhesions that induce focal adhesions and Rho GTPase activity. Rho GTPases are guanosine triphosphate (GTP)-based molecular switches that stimulate actin polymerization as a downstream target. The resultant elevated cytoskeletal tension and focal adhesion formation could eventually stimulate growth factor (GF)-dependent extracellular signal-regulated kinase (ERK) activation that enhances tumor cell growth and progression (see Figure 7.1) (Paszek et al. 2005).

A stiffer matrix can not only induce tumorigenic behavior but also enhance tumor invasiveness. Migration studies of glioblastoma stem cells on 3D collagen gel with different collagen formulations have revealed that by increasing the stiffness gel in different collage types, the migration distance and velocity were increased to their optimum level at 225–397 Pa, similar to native brain tissue stiffness of 100–600 Pa (Herrera-Perez et al. 2015). Another experiment revealed that mammary tumor cells grown in cross-linked, stiffened rBM matrix activate oncogenes such as ErbB2 to promote invasive behavior detected by the loss of β-catenin and disorganized β4 integrin staining with cells observed to invade into the gels (Levental et al. 2009). Moreover, high proliferation of cells along with a rigid matrix may increase compressive stress, which could further increase the migratory potential of tumor cells. For instance, work done by Tse et al. (2012) showed that increased applied compressive stress (ACS) could enhance the migratory potential of tumor cells. For that, they constructed a system in which a membrane contains cells on which specific compression has been applied through a weighted piston. Interestingly, keeping ACS at 5.8 mmHg, similar to the native breast tumor microenvironment, enhances the motility

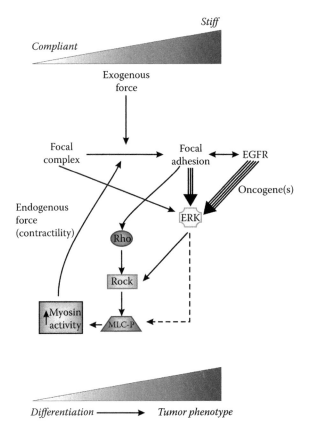

FIGURE 7.1
Model of tensional homeostasis and force-dependent tumor progression. The exogenous force can trigger ERK and ROCK via focal adhesion kinases, resulting in malignant phenotype. (Reprinted from *Cancer Cell*, 8(3), Paszek, M. J. et al. Tensional homeostasis and the malignant phenotype, 241–54, Copyright 2005, with permission from Elsevier.)

Recapitulating Tumor Extracellular Matrix

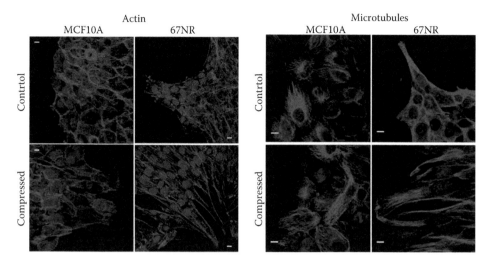

FIGURE 7.2
Cytoskeletal staining of cells at the edge of the cell-denuded area. Phalloidin and tubulin staining for actin filaments and microtubules, respectively. At increased applied compressive stress, highly aggressive 67NR cells show extended actin filaments (left panel) and microtubule reorganization (right panel); however, noninvasive MCF10A cells do not. (Scale bar = 10 μm). (Reproduced with permission from Tse, J. M., et al. Mechanical compression drives cancer cells toward invasive phenotype. *Proc Natl Acad Sci USA* 109(3):911–6, Copyright 2012, National Academy of Sciences, USA.)

of highly aggressive 4T1, 67NR, and MDA-MB-231 cells in comparison to noninvasive, well-differentiated MCF7 and normal mammary epithelial MCF10A cells. In support of that, aligned actin stress fibers and microtubule rearrangement were observed in highly aggressive cells 67NR compared to MCF10A cells at that ACS (see Figure 7.2) (Tse et al. 2012).

Increased stromal stiffness is a characteristic feature of tumors. All of the above studies revealed that matrix rigidity perturbs tissue architecture and increases cytoskeletal tension-dependent growth and supports tumor progression by ameliorating the migratory potential of tumor cells (Kurniawan et al. 2015). Therefore, considering rigidity while designing scaffolds along with physical attributes could be the superior choice to recapitulate the tumor microenvironment.

7.3.3 Biochemical Properties

The biochemical composition and orientation of the basement membrane matrix provide clues to cells for normal functions. Changes in biochemical composition can bring about modifications in intracellular signaling, which in turn induces alterations in the binding activities or spatial distribution of cell-surface receptors that consequently regulate cell proliferation and tumorigenesis (Radisky et al. 2001, Benton et al. 2015). For example, an elevated level of collagen molecules increases availability of integrin binding sites, which further increases by lysyl oxidase. This will lead to exponentially activating integrin-dependent signaling of growth and invasion (Levental et al. 2009). MMPs are activated, which in turn remodel the ECM by proteolysis of various proteins, resulting in exposure of adhesion sites and releasing of growth-stimulating hormones, and stimulate the autocrine loop of cell proliferative signaling. As a whole, aberrant proliferation of cancer cells produces enlarged tumors that cross the oxygen diffusion limit. This could successively create a heterogeneous hypoxic microenvironment that activates hypoxia inducible factor (HIF-1α), which induces expression of various tumor progression molecules, including

VEGF (Brahimi-Horn et al. 2007). Additionally, HIF-1α and abnormal metabolic activity could lead to adaptation of the glycolytic phenotype with generation of lactate, which creates an acidic microenvironment (Gatenby and Gillies 2004, De Milito and Fais 2005). Therefore, understanding the biochemical properties of the scaffold and their effect on cancer cells will be helpful in constructing the tumor microenvironment.

7.3.3.1 Surface Chemistry

The surface is the outermost layer of scaffold, which primarily engages with cells. Surface charge and hydrophobicity may alter as the surface chemistry of the scaffold changes, leading to modification of the cellular attachment to the scaffold. A study done by Zhang and Webster observed that when PLGA film was modified with alginate or chitosan, the wettability of PLGA changed and proliferation of epithelial lung carcinoma (A549) cells was increased by 35% and 30%, respectively, compared to that of pure PLGA films. Similar results were obtained with the use of patterned PLGA films and alginate- or chitosan-modified PLGA films (Zhang and Webster 2013). Another work by Girard et al. showed that a change in surface properties could hinder tumor formation capacity of cancer cells. They used a block co-polymer mPEG/LA and PLGA (3P) nanofibrous scaffold for tumor formation, which supports tumoroids formation. The addition of chitosan to this scaffold changed the surface charge and hydrophilicity of the scaffold, and it was observed that Lewis lung carcinoma (LLC1) cancer cells grew on the 3P/chitosan scaffold, but failed to form tumoroids (see Figure 7.3) (Girard et al. 2013). This could be because of the presence of cationic chitosan, which favors negatively charged cellular membranes, resulting in increased adhesion with the scaffold and hindering cell–cell interaction to form tumoroids.

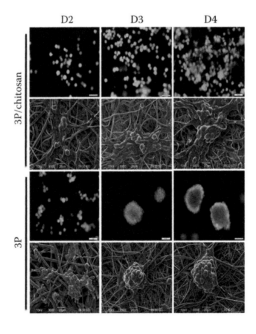

FIGURE 7.3
Effects of surface chemistry on tumoroid formation. LLC1 cells cultured on 3P scaffold alone or with chitosan for 4 days. Top panel: calcein AM/EthD-1 staining showing live/dead cells on scaffolds having tumoroids formation on 3P. Scale bar = 50 mm. Bottom panel: SEM of LLC1 cells grown on scaffolds. (Courtesy of Girard, Y. K. et al. 2013. *PLoS ONE* 8(10):e75345.)

These studies indicate that surface charge and wettability are key features required for cell adhesion to the scaffold. This adhesive property not only supports the cell, but could also facilitate the cellular signaling pathways for proliferation and tumor formation.

7.3.3.2 Chemical Composition

Cells can attach to their matrix substrates by integrin molecules, which recognize the RGD (Arginine-Glycin-Aspartate) peptide sequence present in the matrix proteins, such as fibronectin and laminin (Ruoslahti and Pierschbacher 1987). Providing the RGD sequence in the scaffold not only improves cell adhesion but also survival and proliferation of cells (Mi and Xing 2015). Fischbach et al. used alginate gel for culturing oral squamous cell carcinoma cells (OSCC-3). It was observed that though alginate is typically nonadhesive to cells, attaching the RGD sequence allowed OSCC-3 cells to adhere to the substrate and proliferate. Not only that, OSCC-3 cells showed increased cell survival, proliferation, and secretion of several proangiogenic factors, including VEGF, interleukin 8 (IL-8), and basic fibroblast growth factor (bFGF) in a significantly increased manner (Fischbach et al. 2009). Similarly, hyaluronan has been reported for its interaction with cell surface proteins for cancer stem cells. The incorporation of hyaluronan in the scaffold could lead to a change in the cellular morphology and increase stem cell marker expression (Martinez-Ramos and Lebourg 2015).

Invasive properties could also be adopted according to the biochemical composition of the matrix (Jeon et al. 2015). In one study, Glioma U373-MG cells invading fibrous 3D collagen-based matrices exhibited nonmesenchymal motility, which is devoid of lamellipodia at the moving edge. When these cells were cultured on hyaluronic acid-based gels with RGD peptides, they pursued a series of projection, retraction, and branching and subsequently rapid forward cell movement (Ananthanarayanan et al. 2011). Similarly, in another study, the addition of hyaluronan in the collagen gel induced a morphological change of cells. Additionally, detachment of cells from the matrix and migration were also reduced by presence of hyaluronan, with increasing concentrations reducing them further (Herrera-Perez et al. 2015). Fibronectin is an ECM protein that binds to heparin sulfate proteoglycans on the surface of tumor cells. It has been demonstrated that incorporation of fibronectin-derived heparin binding peptide (FHBP) into polyethylene glycol diacrylate (PEGDA) gel has enhanced tumorsphere formation of 4T1 breast cancer cells *in vitro* compared to PEGDA gel alone. Interestingly, flow cytometric analysis reveal that the subpopulation of CD^{44+}/CD^{24-} cancer stem cells (CSCs) was increased from 12% to about 21% in FHBP-incorporated gels, which shows that 4T1 cancer cell tumoroid formation was associated with the CSC subpopulation in the gel (Yang et al. 2013).

These studies indicate that biochemical composition of the scaffold comprising various cell-binding peptides has a profound effect on tumor cell adhesion. Beyond that, cancer cell behavior, including cell morphology, growth, tumor formation, cancer cell stemness, and migration, is also altered depending upon the biochemical properties of the scaffold.

7.4 Mimicking Approaches by Three-Dimensional Platforms for Drug Screening

Each property of tumor ECMs has been studied in depth as an individual or combined on various platforms, which have been discussed above. These properties have a profound effect on tumor formation, growth, and metastasis. As a downstream effect, when

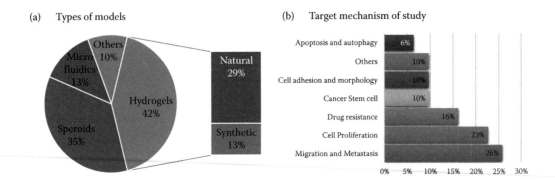

FIGURE 7.4
Statistics for articles regarding 3D tumor models published in 2015. (a) Occurrence of different types of 3D tumor models used for study. (b) Percentage of studied mechanisms in these studies.

chemotherapeutic drug screening will be done, heterogeneous populations of tumors have been formed, which effectively changes the drug distribution profile. The improved cell and cell–ECM interaction could alter drug diffusion kinetics and the bioavailability of the drug to each tumor cell. Moreover, enhanced bio-interaction would initiate biosignaling pathways for growth and drug resistance. All together, these would form a distinct tumor with its "niche" with more relevant tumor formation in *in-vivo* conditions. Considering mimicking approaches, certain devices were explored for drug screening assays where a significant change in drug response had been observed. To further represent this field, an extensive literature survey has been performed on 3D tumor models for drug screening assays in 2015. Among them, the types of models and studied targets have been evaluated. As shown in Figure 7.4a, hydrogel is the most extensively used model for drug screening assays, especially collagen-based hydrogel, because of the major constituents of the basement membrane. Spheroids have also been used in several cases for drug screening; however, they have been excluded from discussion because they lack any design properties for 3D tumors. The major target mechanism that has been evaluated is migration studies, which could be done exclusively on 3D models (Figure 7.4b). The other mechanisms, including proliferation and drug resistance, directly reflect drug effects on tumor cells in such an environment. These studies clearly indicate the importance of 3D tumor models in drug screening assays. Hydrogel, nanofibers, and microfluidic-based 3D platforms that have been used recently for drug screening assays are discussed in this section in detail.

7.4.1 Hydrogel

Hydrogels are cross-linked networks formed by physical, ionic, or covalent interactions between hydrophilic polymers (Elisseeff 2008). Water absorbed by polymers successively penetrates between the polymer chains, causing swelling and formation of a hydrogel (Peppas et al. 2006). Hydrogels can be fabricated by natural or synthetic polymers with the advantage of biocompatibility or controlled physiochemical nature, respectively (Hamdi et al. 2015, Martinez-Ramos and Lebourg 2015). These polymer-engineered structures can be used as a support matrix for *in vitro* cell culture to generate 3D tumor spheroids (Bray et al. 2015, Veiga et al. 2015). The key benefit of hydrogels is their tunable physicochemical properties that can suitably mimic the biochemical and mechanical properties of the ECM. Cells can be deeply embedded in a porous hydrogel that closely recapitulates native 3D tumor morphology and facile oxygen diffusion (Peppas et al. 2006). Such hydrogel models have been developed and employed to evaluate drug toxicity profiling for tumors (Gomes et al. 2015).

Recently, a hyaluronan hydrogel-based 3D tumor model has been developed and evaluated for *in vitro* drug-loaded nanoparticles by Xian Xu et al. (2014). HA is a natural polysaccharide enriched in a tumor microenvironment. HA has been chemically modified, carrying acrylate groups (HA-AC) or thiols (HA-SH), which are cross-linked to form a network with pore size of 70–100 nm, similarly found in native tumor tissue (Pluen et al. 2001). LNCaP prostate cancer cells were grown in these hydrogels and organized into 3D tumoroids. Interestingly, cells grown in 3D exhibited 3.2 ± 1.2 and 1.4 ± 0.1 times more expression of multidrug resistance (MDR) proteins MRP1 and LRP as compared to cells cultured on 2D. MRP1 is major drug efflux pump that extrudes drugs from the cell, while LRP mediates compartmentalization of drugs to the cytoplasm from the nucleus, equally decreasing drug accumulation at the active target. These observations represent the native tumor microenvironment, indicating it as a more relevant model for drug response study. Interestingly, cellular uptake studies showed that Doxorubicin (Dox) and Dox-nanoparticles (Dox-NPs) were localized predominantly in cytoplasm compared to nuclear localization in 2D, which could be in line for overexpression of MDR proteins. Accordingly, drug dosage studies revealed a similar trend, with IC50 values for Dox and Dox-NP increasing from 5.0 ± 0.1 mM and 11.0 ± 0.8 mM in 2D to 15.0 ± 0.4 and 12.3 ± 0.8 mM in 3D cultures, respectively.

In a different approach, integrated semisynthetic hydrogel has been utilized for 3D tumor formation and drug evaluation (Ki et al. 2014). The authors used poly (ethylene glycol)-tetra-norbornene attached to terminal cysteine containing MMP-sensitive peptide substrate. Collagen 1 was physically entrapped during photo-cross-linking of scaffold. Pancreatic ductal adenocarcinoma (PDAC) cells cultured on this scaffold showed not only active proliferation in 3D hydrogel—that is, 60% of the cells were Edu positive—they were also resistant to chemotherapeutic drugs. Gemcitabine (Gem), commonly used for PDAC treatment and given with 1-μM dosage, showed significant cell death in 2D cultures, whereas a similar dosage in 3D conditions demonstrated <20% reduction in metabolic activity and slight cell death at the periphery of cell clusters. Moreover, a 6-fold increase in caspase activity for apoptosis was observed in 2D, while in the case of 3D, it was less than a 3-fold increase. Cancer stem cells have shown superior resistance to chemotherapy; therefore, to examine this possibility, pancreatic cancer stem cell markers have been evaluated. Strikingly, CD24 mRNA expression was elevated 6-fold in cells cultured in 3D, which was also confirmed with high protein expression. Flow cytometry analysis also supports this observation, where only 0.05% of the CD24+ cell population was observed in the case of 2D, which rose to 9.71% in 3D culture. Similarly, other stem cells markers were also elevated in cells grown in 3D. These results strongly support drug-resistance phenomena in 3D culture conditions, which might be due to altered cell–cell and cell–ECM interactions in the 3D environment. Hydrogels have also been fabricated with a combination of synthetic and natural polymers and evaluated for drug toxicity studies. Star PEG has been cross-linked with maleimide-functionalized heparin to form a hydrogel with tunable stiffness (Bray et al. 2015). The angiogenesis microenvironment was created by co-culturing breast or prostate carcinoma cells with human umbilical vein endothelial cells (HUVEC) and mesenchymal stem cells. After 3 days of post-treatment with either epirubicin or paclitaxel drugs, the metabolic activity of cells grown in 3D was similar to the untreated control, while a significant difference was observed with 2D. These observations clearly indicate higher resistance to drugs due to 3D distribution, altered diffusion kinetics, and bioavailability of drugs in 3D models.

These studies clearly demonstrate the utility of hydrogel scaffolds for 3D tumor development for drug screening assays. Several natural and synthetic materials are explored for hydrogel fabrication for tissue engineering application. Hydrogel is perhaps the most widely used as scaffolds among all types of 3D tumor models; however, few of them are yet explored for drug screening assays. The major challenge with hydrogel is to develop techniques that allow spatiotemporal control of scaffold properties, such as the production of patterned hydrogels.

7.4.2 Nanofibers

A nanofibrous scaffold is a thin sheet made up of randomly aligned nanometer-sized polymeric fibers. A nanofibrous substrate can represent the fibrous ECM environment and provide topographical features to cells for adhesion and growth. These nanofibers can be fabricated by several techniques, including electrospinning, phase separation, and self-assembly (Vasita and Katti 2006). These fibers can be fabricated with varying chemistry, diameter, length, porosity, and mechanical properties. These properties can provide different surface topography, including pores, ridges, fibers, and charge to cells, which can regulate cellular adhesion (Glass-Brudzinski et al. 2002), morphology (Karuri et al. 2004), proliferation (Dalby et al. 2002), and gene expression (Glass-Brudzinski et al. 2002). Several nanofibrous scaffolds have been used recently for 3D tumor development and drug screening (Santoro et al. 2015).

Synthetic polymers have been extensively used for fabrication of nanofibrous substrate because of their ease of use and scalable properties. Recently, nanofibrous substrate made up of three synthetic polymers, viz. PLGA random copolymer, poly(lactide)-polyethylene glycol block copolymer (referred to as 3P), have been utilized to develop a 3D tumor model for drug screening (Girard et al. 2013). LLC1 lung cancer cells have been grown on this scaffold and, interestingly, have formed tumoroids on this scaffold, while cells grown on monolayer PLGA scaffolds or PLGA-PEG scaffolds did not form tumoroids (Figure 7.3). The cells on 3P nanofibrous substrate showed tumorogenesis on the third day by showing loss of E-cadherin (cell–cell adhesion) and expression of vimentin protein (mesenchymal marker), which indicated EMT activation and the invasive potential of the cells. Furthermore, the authors have evaluated the utility of this model for drug screening assays by growing tumor biopsies on 3P scaffolds and determined drug effects on it. Single-cell suspensions of tumor biopsy and LLC1 cells have been cultured on a 3P scaffold in the presence or absence of LY294002 and U0126, inhibitors of EMT. LY294002 and U0126 inhibitors effectively impeded the growth of cells and tumoroid formation, but, interestingly, the IC_{50} values were 0.1 mM and 6.72 nM for monolayer culture, respectively. These dosages were not significant for 3D cultures, which showed IC_{50} values were 1.092 mM and 652 nM, respectively, for LLC1 cells grown on 3P scaffold. Furthermore, the IC_{50}s of both inhibitors were found to be >1 mM in demonstrating a superior resistance of biopsy-induced tumoroids to drugs compared to LLC1 tumoroids. These findings suggest that the 3D microenvironment not only alters drug diffusion kinetics but the cellular signaling for drug resistance as well, which would be according to the cellular heterogeneity of tumor cells.

These nanofiber-based models have several key features, for instance, providing topographical features to cancer cells for 3D tumoroid development and reproducibility with ease of imaging of tumoroids could establish it as an elegant approach for drug screening assays. However, the very marginal pore size of the nanofibrous scaffold could not allow cells to infiltrate the 3D environment of the scaffold. Incorporating nanofibers into hydrogel could minimize the limitations and provide a reasonable structure to cancer cells for 3D tumor development.

7.4.3 Microfluidics Models

Microfluidic technology includes silicon/elastomer-based devices with microchannels with proportions of 1–1000 μm that operate with a very low (10^{-9} to 10^{-18} L) amount of fluids (Zhang and Nagrath 2013). In such systems, fluid flow is strictly laminar rather than turbulent; consequently, the concentrations of molecules can be well controlled both in space and time. Microfluidic technologies offer features including small quantities of sample and media requirements, short processing times, high resolution, and sensitivity with time-lapse cell imaging Whitesides 2006). Recently, microfluidics has also been used to simulate biological signals in the cellular environment by precisely controlling spatial and temporal gradients of soluble biological factors and cell–cell contacts (El-Ali et al. 2006, Pisano et al. 2015). These attractive features make microfluidics a powerful tool to study tumor progression, invasion, and angiogenesis, as well as high throughput drug screening assays (Zhang and Nagrath 2013, Jeon et al. 2015).

Recently, Yang and co-workers fabricated a PDMS-based microfluidic device and evaluated photodynamic therapy (PDT) (Yang et al. 2015). The eight-chamber microfluidic device was made using a soft lithography technique with a PDMS-integrated glass slide (75 mm × 55 mm × 1 mm). As shown in the figure, the chambers were fabricated as elongated hexagonal prism shapes (12 × 6 × 0.2 mm L × W × H) with a camber capacity of 10 μL (see Figure 7.5). Human breast cancer cells (MCF-7) and primary adipose derived stromal cells (ASCs) were co-cultured on this device to develop an *in vitro* 3D tumor, followed by PDT, which activates the photosensitive molecule 5-aminolevulinic acid (5-ALA) given to the cells. A particular-wavelength light exposure would result in activation of a photosensitive (PS) molecule and generate reactive oxygen species (ROS). PDT efficacy has been assessed by injecting therapeutic agents directly or with gold nanoparticles (Au NPs) and studying their distribution profiles in 3D tumors and monolayers. The results showed that after PDT

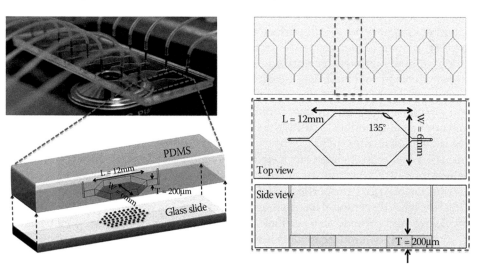

FIGURE 7.5
Design of microfluidic device. Left top: transparent microfluidic device on the microscope stage inside a temperature control incubator. Left bottom: design of PDMS-based microfluidic device for 3D breast cancer tissue formation. Right: schematic of eight chambers on a 3″ × 1″ glass slide with defined lengths of individual chambers. (Yang, Y. et al. 2015. Evaluation of photodynamic therapy efficiency using an *in vitro* three-dimensional microfluidic breast cancer tissue model. *Lab Chip* 15(3):735–44. Reproduced by permission of The Royal Society of Chemistry.)

exposure of 1 min. with 5-ALA, about 50% and 70% of cell destruction were observed in 2D with direct and Au NPs, respectively. Interestingly, the cell destruction was reduced to 17% and 50% in 3D microfluidics. Histological analysis indicated that the outer layer of cells in 3D microfluidic cancer tissue were destroyed after PDT, while the inner mass remained healthy, presenting retained morphology and distinct nuclei. These results indicated that PDT efficacy is dependent on restricted light penetration through present 3D microfluidic tissues, which represents a more realistic representation of *in-vivo* conditions. Furthermore, the microenvironmental gradient could acquire cellular heterogeneity in 3D microfluidic tumors, which results in a PDT-resistant effect reliant on the spatial distribution of cells. The continuous flow of nutrients with PS established a physiologically relevant profile of blood circulation of drugs that generated a drug concentration gradient in 3D microfluidic tumors and resulted in reduced PDT efficacy.

The microfluidic *in vitro* 3D cancer model has the capability to mimic the heterogeneous and physiologic environment of cancer. Therapeutic agents delivered via perfusion closely emulate *in-vivo* intravenous or local injections and provide suitable dimensionality of cancer tissue, while allowing convenient endpoint characterization through live cell imaging. One of the key challenges with microfluidics is that the microfluidic environment needs special design consideration. The complexities involving patient samples are far ahead of present models. However, these models can be used for high-throughput screening that reduces the burden of the initial screening stages of therapeutic agents.

7.5 Conclusion and Future Perspectives

It is clearly evident that the ECM plays a significant role in tumor cell behavior. Specific ECM components append to cell surface receptors and stimulate peculiar responses for cell polarity, growth, drug resistance, or metastatic potential. In recent years, the tumor microenvironment has been recognized as a major contributor to tumor progression and has become the focus of extensive research as a potential target for chemoprevention. It is now well recognized that to effectively investigate the pathobiology of human cancers, it is necessary to recreate the appropriate tumor microenvironment; 3D tumor models perhaps could be appropriate models to fulfill the need. The native ECM can be deconstructed and serve as a clue for reconstructing artificial scaffolds for generation of 3D tumor models. The discussed models have been fabricated based on fundamental knowledge of the tumor microenvironment. These culture models have delivered new understandings regarding the contribution of matrix properties and their outcome on drug delivery, diffusion, toxicity, and drug response. Each model has distinct ECM-mimicking properties that regulate tumor cell physiology and provide a closer look toward *in vivo* tumor drug-response behaviors. Along with that, these culture models are quite economical and will possibly reduce the usage of *in vivo* animal models as well as the cost of clinical phase trials.

Nevertheless, several additional considerations are essential to expand the utility of such models. In particular, cancer is a multiscale process involving a number of types of stromal cells and signaling pathways orchestrated at a time, and reorienting such an environment is essential for predictive drug screening. The biological advancement could include (i) co-culturing of stromal cells with tumor cells to provide a complex cell signaling environment, and (ii) functionalization of devices with signaling molecules such as growth factors or functional peptides to improve the biomimicking of the tumor microenvironment.

Furthermore, the biofabrication of two or more integrated devices could be achieved with focusing on (i) mimicking angiogenesis by fabricating microcapillaries or hollow nanofibers inside hydrogel scaffolds, and (ii) creating hypoxic conditions by fabricating the dense network of polymers or limiting oxygen diffusion inside the scaffolds. In the end, the future motivation should be engineering models for high-throughput screenings that are easily accessible for clinicians and researchers in the cancer biology field. Looking forward, engineering microenvironment-based 3D cancer models could enhance both basic as well as translation cancer research and chemoprevention.

Acknowledgments

The author Dhaval Kedaria would like to thank the Department of Biotechnology (Govt. of India) for research fellowship. Dr. Rajesh Vasita would like to acknowledge the Department of Science and Technology (Govt. of India) and Gujarat State Biotechnology Mission (Govt. of Gujarat) for the financial support.

References

Afratis, N., C. Gialeli, D. Nikitovic, T. Tsegenidis, E. Karousou, A. D. Theocharis, M. S. Pavao, G. N. Tzanakakis, and N. K. Karamanos. 2012. Glycosaminoglycans: Key players in cancer cell biology and treatment. *FEBS J* 279(7):1177–97. doi: 10.1111/j.1742-4658.2012.08529.x.

Ananthanarayanan, B., Y. Kim, and S. Kumar. 2011. Elucidating the mechanobiology of malignant brain tumors using a brain matrix-mimetic hyaluronic acid hydrogel platform. *Biomaterials* 32(31):7913–23. doi: 10.1016/j.biomaterials.2011.07.005.

Arrondeau, J., H. K. Gan, A. R. Razak, X. Paoletti, and C. Le Tourneau. 2010. Development of anticancer drugs. *Discov Med* 10(53):355–62.

Ashikari-Hada, S., H. Habuchi, Y. Kariya, N. Itoh, A. H. Reddi, and K. Kimata. 2004. Characterization of growth factor-binding structures in heparin/heparan sulfate using an octasaccharide library. *J Biol Chem* 279(13):12346–54. doi: 10.1074/jbc.M313523200.

Asimakopoulou, A. P., A. D. Theocharis, G. N. Tzanakakis, and N. K. Karamanos. 2008. The biological role of chondroitin sulfate in cancer and chondroitin-based anticancer agents. *In Vivo* 22(3):385–9.

Aumailley, M. and B. Gayraud. 1998. Structure and biological activity of the extracellular matrix. *J Mol Med* 76(3–4):253–65. doi: 10.1007/s001090050215.

Bai, C., M. Yang, Z. Fan, S. Li, T. Gao, and Z. Fang. 2015. Associations of chemo- and radio-resistant phenotypes with the gap junction, adhesion and extracellular matrix in a three-dimensional culture model of soft sarcoma. *J Exp Clin Cancer Res* 34:58. doi: 10.1186/s13046-015-0175-0.

Baker, B. M. and C. S. Chen. 2012. Deconstructing the third dimension: How 3D culture microenvironments alter cellular cues. *J Cell Sci* 125(Pt 13):3015–24. doi: 10.1242/jcs.079509.

Balzer, E. M., Z. Tong, C. D. Paul, W. C. Hung, K. M. Stroka, A. E. Boggs, S. S. Martin, and K. Konstantopoulos. 2012. Physical confinement alters tumor cell adhesion and migration phenotypes. *FASEB J* 26(10):4045–56. doi: 10.1096/fj.12-211441.

Benton, G., G. DeGray, H. K. Kleinman, J. George, and I. Arnaoutova. 2015. *In vitro* microtumors provide a physiologically predictive tool for breast cancer therapeutic screening. *PLoS One* 10(4):e0123312. doi: 10.1371/journal.pone.0123312.

Birgersdotter, A., R. Sandberg, and I. Ernberg. 2005. Gene expression perturbation *in vitro*—A growing case for three-dimensional (3D) culture systems. *Semin Cancer Biol* 15(5):405–12. doi: 10.1016/j.semcancer.2005.06.009.

Bloom, R. J., J. P. George, A. Celedon, S. X. Sun, and D. Wirtz. 2008. Mapping local matrix remodeling induced by a migrating tumor cell using three-dimensional multiple-particle tracking. *Biophys J* 95(8):4077–88. doi: 10.1529/biophysj.108.132738.

Bosman, F. T. and I. Stamenkovic. 2003. Functional structure and composition of the extracellular matrix. *J Pathol* 200(4):423–8. doi: 10.1002/path.1437.

Brahimi-Horn, M. C., J. Chiche, and J. Pouyssegur. 2007. Hypoxia and cancer. *J Mol Med (Berl)* 85(12):1301–7. doi: 10.1007/s00109-007-0281-3.

Bray, L. J., M. Binner, A. Holzheu, J. Friedrichs, U. Freudenberg, D. W. Hutmacher, and C. Werner. 2015. Multi-parametric hydrogels support 3D *in vitro* bioengineered microenvironment models of tumour angiogenesis. *Biomaterials* 53:609–20. doi: 10.1016/j.biomaterials.2015.02.124.

Bretscher, L. E., C. L. Jenkins, K. M. Taylor, M. L. DeRider, and R. T. Raines. 2001. Conformational stability of collagen relies on a stereoelectronic effect. *J Am Chem Soc* 123(4):777–8.

Butcher, D. T., T. Alliston, and V. M. Weaver. 2009. A tense situation: Forcing tumour progression. *Nat Rev Cancer* 9(2):108–22.

Cardone, R. A., V. Casavola, and S. J. Reshkin. 2005. The role of disturbed pH dynamics and the Na+/H+ exchanger in metastasis. *Nat Rev Cancer* 5(10):786–95. doi: 10.1038/nrc1713.

Carey, S. P., A. Rahman, C. M. Kraning-Rush, B. Romero, S. Somasegar, O. M. Torre, R. M. Williams, and C. A. Reinhart-King. 2015. Comparative mechanisms of cancer cell migration through 3D matrix and physiological microtracks. *Am J Physiol Cell Physiol* 308(6):C436–47. doi: 10.1152/ajpcell.00225.2014.

Dalby, M. J., S. J. Yarwood, M. O. Riehle, H. J. Johnstone, S. Affrossman, and A. S. Curtis. 2002. Increasing fibroblast response to materials using nanotopography: Morphological and genetic measurements of cell response to 13-nm-high polymer demixed islands. *Exp Cell Res* 276(1):1–9. doi: 10.1006/excr.2002.5498.

Damsky, C. H. and Z. Werb. 1992. Signal transduction by integrin receptors for extracellular matrix: Cooperative processing of extracellular information. *Curr Opin Cell Biol* 4(5):772–81.

De Milito, A. and S. Fais. 2005. Tumor acidity, chemoresistance and proton pump inhibitors. *Future Oncol* 1(6):779–86. doi: 10.2217/14796694.1.6.779.

Ebihara, T., N. Venkatesan, R. Tanaka, and M. S. Ludwig. 2000. Changes in extracellular matrix and tissue viscoelasticity in bleomycin-induced lung fibrosis. Temporal aspects. *Am J Respir Crit Care Med* 162(4 Pt 1):1569–76. doi: 10.1164/ajrccm.162.4.9912011.

Egeblad, M., E. S. Nakasone, and Z. Werb. 2010. Tumors as organs: Complex tissues that interface with the entire organism. *Dev Cell* 18(6):884–901. doi: 10.1016/j.devcel.2010.05.012.

El-Ali, J., P. K. Sorger, and K. F. Jensen. 2006. Cells on chips. *Nature* 442(7101):403–11. doi: 10.1038/nature05063.

Elisseeff, J. 2008. Hydrogels: Structure starts to gel. *Nat Mater* 7(4):271–3.

Erler, J. T. and V. M. Weaver. 2009. Three-dimensional context regulation of metastasis. *Clin Exp Metastasis* 26(1):35–49. doi: 10.1007/s10585-008-9209-8.

Fidler, I. J. 2003. The pathogenesis of cancer metastasis: The "seed and soil" hypothesis revisited. *Nat Rev Cancer* 3(6):453–8.

Fischbach, C., H. J. Kong, S. X. Hsiong, M. B. Evangelista, W. Yuen, and D. J. Mooney. 2009. Cancer cell angiogenic capability is regulated by 3D culture and integrin engagement. *Proc Natl Acad Sci USA* 106(2):399–404. doi: 10.1073/pnas.0808932106.

Friedl, P. and K. Wolf. 2003. Tumour-cell invasion and migration: Diversity and escape mechanisms. *Nat Rev Cancer* 3(5):362–74. doi: 10.1038/nrc1075.

Gallagher, J. T. 2001. Heparan sulfate: Growth control with a restricted sequence menu. *J Clin Investig* 108(3):357–61. doi: 10.1172/jci200113713.

Gatenby, R. A. and R. J. Gillies. 2004. Why do cancers have high aerobic glycolysis? *Nat Rev Cancer* 4(11):891–9. doi: 10.1038/nrc1478.

Gelse, K. 2003. Collagens—Structure, function, and biosynthesis. *Adv Drug Delivery Rev* 55(12):1531–46. doi: 10.1016/j.addr.2003.08.002.

Gesslbauer, B., A. Rek, F. Falsone, E. Rajkovic, and A. J. Kungl. 2007. Proteoglycanomics: Tools to unravel the biological function of glycosaminoglycans. *Proteomics* 7(16):2870–80. doi: 10.1002/pmic.200700176.

Girard, Y. K., C. Wang, S. Ravi, M. C. Howell, J. Mallela, M. Alibrahim, R. Green, G. Hellermann, S. S. Mohapatra, and S. Mohapatra. 2013. A 3D fibrous scaffold inducing tumoroids: A platform for anticancer drug development. *PLoS ONE* 8(10):e75345. doi: 10.1371/journal.pone.0075345.

Glass-Brudzinski, J., D. Perizzolo, and D. M. Brunette. 2002. Effects of substratum surface topography on the organization of cells and collagen fibers in collagen gel cultures. *J Biomed Mater Res* 61(4):608–18. doi: 10.1002/jbm.10243.

Gomes, L. R., A. T. Vessoni, and C. F. Menck. 2015. Three-dimensional microenvironment confers enhanced sensitivity to doxorubicin by reducing p53-dependent induction of autophagy. *Oncogene* 34(42):5329–40. doi: 10.1038/onc.2014.461.

Guimond, S., M. Maccarana, B. B. Olwin, U. Lindahl, and A. C. Rapraeger. 1993. Activating and inhibitory heparin sequences for FGF-2 (basic FGF). Distinct requirements for FGF-1, FGF-2, and FGF-4. *J Biol Chem* 268(32):23906–14.

Hait, W. N. 2010. Anticancer drug development: The grand challenges. *Nat Rev Drug Discov* 9(4):253–4. doi: 10.1038/nrd3144.

Hamdi, D. H., S. Barbieri, F. Chevalier, J. E. Groetz, F. Legendre, M. Demoor, P. Galera, J. L. Lefaix, and Y. Saintigny. 2015. *In vitro* engineering of human 3D chondrosarcoma: A preclinical model relevant for investigations of radiation quality impact. *BMC Cancer* 15:579. doi: 10.1186/s12885-015-1590-5.

Han, S. W., F. R. Khuri, and J. Roman. 2006. Fibronectin stimulates non–small cell lung carcinoma cell growth through activation of Akt/mammalian target of Rapamycin/S6 kinase and inactivation of LKB1/AMP-activated protein kinase signal pathways. *Cancer Res* 66(1):315–23. doi: 10.1158/0008-5472.can-05-2367.

Herrera-Perez, M., S. Voytik-Harbin, and J. L. Rickus. 2015. Extracellular matrix properties regulate the migratory response of glioblastoma stem cells in 3D culture. *Tissue Eng Part A* 21(19–20):2572–8. doi: 10.1089/ten.TEA.2014.0504.

Holen, I., F. Nutter, J. M. Wilkinson, C. A. Evans, P. Avgoustou, and P. D. Ottewell. 2015. Human breast cancer bone metastasis *in vitro* and *in vivo*: A novel 3D model system for studies of tumour cell-bone cell interactions. *Clin Exp Metastasis* 32(7):689–702. doi: 10.1007/s10585-015-9737-y.

Hopkins, A. L. 2008. Network pharmacology: The next paradigm in drug discovery. *Nat Chem Biol* 4(11):682–90.

Ingber, D. E. and J. Folkman. 1989. Mechanochemical switching between growth and differentiation during fibroblast growth factor-stimulated angiogenesis *in vitro*: Role of extracellular matrix. *J Cell Biol* 109(1):317–30.

Ingber, D. E. 2006. Cellular mechanotransduction: Putting all the pieces together again. *FASEB J* 20(7):811–27. doi: 10.1096/fj.05-5424rev.

Jabbari, E., S. K. Sarvestani, L. Daneshian, and S. Moeinzadeh. 2015. Optimum 3D matrix stiffness for maintenance of cancer stem cells is dependent on tissue origin of cancer cells. *PLoS One* 10(7):e0132377. doi: 10.1371/journal.pone.0132377.

Jarvelainen, H., A. Sainio, M. Koulu, T. N. Wight, and R. Penttinen. 2009. Extracellular matrix molecules: Potential targets in pharmacotherapy. *Pharmacol Rev* 61(2):198–223. doi: 10.1124/pr.109.001289.

Jeon, J. S., S. Bersini, M. Gilardi, G. Dubini, J. L. Charest, M. Moretti, and R. D. Kamm. 2015. Human 3D vascularized organotypic microfluidic assays to study breast cancer cell extravasation. *Proc Natl Acad Sci USA* 112(1):214–9. doi: 10.1073/pnas.1417115112.

Juliano, R. L. and S. Haskill. 1993. Signal transduction from the extracellular matrix. *J Cell Biol* 120(3):577–85.

Kajita, M., Y. Itoh, T. Chiba, H. Mori, A. Okada, H. Kinoh, and M. Seiki. 2001. Membrane-type 1 matrix metalloproteinase cleaves Cd44 and promotes cell migration. *J Cell Biol* 153(5):893–904. doi: 10.1083/jcb.153.5.893.

Karuri, N. W., S. Liliensiek, A. I. Teixeira, G. Abrams, S. Campbell, P. F. Nealey, and C. J. Murphy. 2004. Biological length scale topography enhances cell-substratum adhesion of human corneal epithelial cells. *J Cell Sci* 117(Pt 15):3153–64. doi: 10.1242/jcs.01146.

Kenny, P. A., G. Y. Lee, C. A. Myers, R. M. Neve, J. R. Semeiks, P. T. Spellman, K. Lorenz et al. 2007. The morphologies of breast cancer cell lines in three-dimensional assays correlate with their profiles of gene expression. *Mol Oncol* 1(1):84–96. doi: 10.1016/j.molonc.2007.02.004.

Kessenbrock, K., V. Plaks, and Z. Werb. 2010. Matrix metalloproteinases: Regulators of the tumor microenvironment. *Cell* 141(1):52–67. doi: 10.1016/j.cell.2010.03.015.

Khaled, W., S. Reichling, O. T. Bruhns, H. Boese, M. Baumann, G. Monkman, S. Egersdoerfer et al. 2004. Palpation imaging using a haptic system for virtual reality applications in medicine. *Stud Health Technol Inform* 98:147–53.

Ki, C. S., T.-Y. Lin, M. Korc, and C.-C. Lin. 2014. Thiol-ene hydrogels as desmoplasia-mimetic matrices for modeling pancreatic cancer cell growth, invasion, and drug resistance. *Biomaterials* 35(36):9668–77. doi: 10.1016/j.biomaterials.2014.08.014.

Kim, S. A., E. K. Lee, and H. J. Kuh. 2015. Co-culture of 3D tumor spheroids with fibroblasts as a model for epithelial-mesenchymal transition *in vitro*. *Exp Cell Res* 335(2):187–96. doi: 10.1016/j.yexcr.2015.05.016.

Koster, S., H. M. Evans, J. Y. Wong, and T. Pfohl. 2008. An *in situ* study of collagen self-assembly processes. *Biomacromolecules* 9(1):199–207. doi: 10.1021/bm700973t.

Kurniawan, N. A., P. K. Chaudhuri, and C. T. Lim. 2015. Concentric gel system to study the biophysical role of matrix microenvironment on 3D cell migration. *J Vis Exp* (98):52735. doi: 10.3791/52735.

Labropoulou, V. T., A. D. Theocharis, P. Ravazoula, P. Perimenis, A. Hjerpe, N. K. Karamanos, and H. P. Kalofonos. 2006. Versican but not decorin accumulation is related to metastatic potential and neovascularization in testicular germ cell tumours. *Histopathology* 49(6):582–93. doi: 10.1111/j.1365-2559.2006.02558.x.

Larsen, M., V. V. Artym, J. A. Green, and K. M. Yamada. 2006. The matrix reorganized: Extracellular matrix remodeling and integrin signaling. *Curr Opin Cell Biol* 18(5):463–71. doi: 10.1016/j.ceb.2006.08.009.

Levental, K. R., H. Yu, L. Kass, J. N. Lakins, M. Egeblad, J. T. Erler, S. F. T. Fong et al. 2009. Matrix crosslinking forces tumor progression by enhancing integrin signaling. *Cell* 139(5):891–906. doi: 10.1016/j.cell.2009.10.027.

Lin, C. Q. and M. J. Bissell. 1993. Multi-faceted regulation of cell differentiation by extracellular matrix. *FASEB J* 7(9):737–43.

Liotta, L. A. and E. C. Kohn. 2001. The microenvironment of the tumour-host interface. *Nature* 411(6835):375–9. doi: 10.1038/35077241.

Lucero, H. A. and H. M. Kagan. 2006. Lysyl oxidase: An oxidative enzyme and effector of cell function. *Cell Mol Life Sci* 63(1–20):2304–16. doi: 10.1007/s00018-006-6149-9.

Malavaki, C., S. Mizumoto, N. Karamanos, and K. Sugahara. 2008. Recent advances in the structural study of functional chondroitin sulfate and dermatan sulfate in health and disease. *Connect Tissue Res* 49(3–4):133–9. doi: 10.1080/03008200802148546.

Martinez-Ramos, C. and M. Lebourg. 2015. Three-dimensional constructs using hyaluronan cell carrier as a tool for the study of cancer stem cells. *J Biomed Mater Res B Appl Biomater* 103(6):1249–57. doi: 10.1002/jbm.b.33304.

McKegney, M., I. Taggart, and M. H. Grant. 2001. The influence of crosslinking agents and diamines on the pore size, morphology and the biological stability of collagen sponges and their effect on cell penetration through the sponge matrix. *J Mater Sci Mater Med* 12(9):833–44.

Mi, K. and Z. Xing. 2015. CD44(+)/CD24(−) breast cancer cells exhibit phenotypic reversion in three-dimensional self-assembling peptide RADA16 nanofiber scaffold. *Int J Nanomedicine* 10:3043–53. doi: 10.2147/IJN.S66723.

Mochizuki, M., Y. Kadoya, Y. Wakabayashi, K. Kato, I. Okazaki, M. Yamada, T. Sato, N. Sakairi, N. Nishi, and M. Nomizu. 2003. Laminin-1 peptide-conjugated chitosan membranes as a novel approach for cell engineering. *FASEB J* 17(8):875–7. doi: 10.1096/fj.02-0564fje.

Moscato, S., F. Ronca, D. Campani, and S. Danti. 2015. Poly(vinyl alcohol)/gelatin hydrogels cultured with HepG2 cells as a 3D model of hepatocellular carcinoma: A morphological study. *J Funct Biomater* 6(1):16–32. doi: 10.3390/jfb6010016.

Mosher, D. F., J. Sottile, C. Wu, and J. A. McDonald. 1992. Assembly of extracellular matrix. *Curr Opin Cell Biol* 4(5):810–8.

Nikitovic, D., M. Assouti, M. Sifaki, P. Katonis, K. Krasagakis, N. K. Karamanos, and G. N. Tzanakakis. 2008. Chondroitin sulfate and heparan sulfate-containing proteoglycans are both partners and targets of basic fibroblast growth factor-mediated proliferation in human metastatic melanoma cell lines. *Int J Biochem Cell Biol* 40(1):72–83. doi: 10.1016/j.biocel.2007.06.019.

Otranto, M., V. Sarrazy, F. Bonte, B. Hinz, G. Gabbiani, and A. Desmouliere. 2012. The role of the myofibroblast in tumor stroma remodeling. *Cell Adh Migr* 6(3):203–19. doi: 10.4161/cam.20377.

Paszek, M. J. and V. M. Weaver. 2004. The tension mounts: Mechanics meets morphogenesis and malignancy. *J Mammary Gland Biol Neoplasia* 9(4):325–42. doi: 10.1007/s10911-004-1404-x.

Paszek, M. J., N. Zahir, K. R. Johnson, J. N. Lakins, G. I. Rozenberg, A. Gefen, C. A. Reinhart-King et al. 2005. Tensional homeostasis and the malignant phenotype. *Cancer Cell* 8(3):241–54. doi: 10.1016/j.ccr.2005.08.010.

Patarroyo, M., K. Tryggvason, and I. Virtanen. 2002. Laminin isoforms in tumor invasion, angiogenesis and metastasis. *Semin Cancer Biol* 12(3):197–207. doi: 10.1016/S1044-579X(02)00023-8.

Paul, S. M., D. S. Mytelka, C. T. Dunwiddie, C. C. Persinger, B. H. Munos, S. R. Lindborg, and A. L. Schacht. 2010. How to improve R&D productivity: The pharmaceutical industry's grand challenge. *Nat Rev Drug Discov* 9(3):203–14. http://www.nature.com/nrd/journal/v9/n3/suppinfo/nrd3078_S1.html.

Payne, S. L., M. J. Hendrix, and D. A. Kirschmann. 2007. Paradoxical roles for lysyl oxidases in cancer—A prospect. *J Cell Biochem* 101(6):1338–54. doi: 10.1002/jcb.21371.

Peppas, N. A., J. Z. Hilt, A. Khademhosseini, and R. Langer. 2006. Hydrogels in biology and medicine: From molecular principles to bionanotechnology. *Adv Mater* 18(11):1345–60. doi: 10.1002/adma.200501612.

Pisano, M., V. Triacca, K. A. Barbee, and M. A. Swartz. 2015. An *in vitro* model of the tumor-lymphatic microenvironment with simultaneous transendothelial and luminal flows reveals mechanisms of flow enhanced invasion. *Integr Biol (Camb)* 7(5):525–33. doi: 10.1039/c5ib00085h.

Pluen, A., Y. Boucher, S. Ramanujan, T. D. McKee, T. Gohongi, E. di Tomaso, E. B. Brown et al. 2001. Role of tumor-host interactions in interstitial diffusion of macromolecules: Cranial vs. subcutaneous tumors. *Proc Natl Acad Sci USA* 98(8):4628–33. doi: 10.1073/pnas.081626898.

Provenzano, P. P., D. R. Inman, K. W. Eliceiri, J. G. Knittel, L. Yan, C. T. Rueden, J. G. White, and P. J. Keely. 2008. Collagen density promotes mammary tumor initiation and progression. *BMC Med* 6:11. doi: 10.1186/1741-7015-6-11.

Pupa, S. M., S. Menard, S. Forti, and E. Tagliabue. 2002. New insights into the role of extracellular matrix during tumor onset and progression. *J Cell Physiol* 192(3):259–67. doi: 10.1002/jcp.10142.

Radisky, D., C. Hagios, and M. J. Bissell. 2001. Tumors are unique organs defined by abnormal signaling and context. *Semin Cancer Biol* 11(2):87–95. doi: 10.1006/scbi.2000.0360.

Raman, R., V. Sasisekharan, and R. Sasisekharan. 2005. Structural insights into biological roles of protein-glycosaminoglycan interactions. *Chem Biol* 12(3):267–77. doi: 10.1016/j.chembiol.2004.11.020.

Ramanujan, S., A. Pluen, T. D. McKee, E. B. Brown, Y. Boucher, and R. K. Jain. 2002. Diffusion and convection in collagen gels: Implications for transport in the tumor interstitium. *Biophys J* 83(3):1650–60. doi: 10.1016/S0006-3495(02)73933-7.

Ricciardelli, C., K. Mayne, P. J. Sykes, W. A. Raymond, K. McCaul, V. R. Marshall, and D. J. Horsfall. 1998. Elevated levels of versican but not decorin predict disease progression in early-stage prostate cancer. *Clin Cancer Res* 4(4):963–71.

Rowe, R. G. and S. J. Weiss. 2009. Navigating ECM barriers at the invasive front: The cancer cell-stroma interface. *Annu Rev Cell Dev Biol* 25:567–95. doi: 10.1146/annurev.cellbio.24.110707.175315.

Ruoslahti, E. and M. D. Pierschbacher. 1987. New perspectives in cell adhesion: RGD and integrins. *Science* 238(4826):491–7.

Saha, S., X. Duan, L. Wu, P. K. Lo, H. Chen, and Q. Wang. 2012. Electrospun fibrous scaffolds promote breast cancer cell alignment and epithelial-mesenchymal transition. *Langmuir* 28(4):2028–34. doi: 10.1021/la203846w.

Sanderson, R. D. 2001. Heparan sulfate proteoglycans in invasion and metastasis. *Semin Cell Dev Biol* 12(2):89–98. doi: 10.1006/scdb.2000.0241.

Santoro, M., S. E. Lamhamedi-Cherradi, B. A. Menegaz, J. A. Ludwig, and A. G. Mikos. 2015. Flow perfusion effects on three-dimensional culture and drug sensitivity of Ewing sarcoma. *Proc Natl Acad Sci USA* 112(33):10304–9. doi: 10.1073/pnas.1506684112.

Sasisekharan, R., Z. Shriver, G. Venkataraman, and U. Narayanasami. 2002. Roles of heparan-sulphate glycosaminoglycans in cancer. *Nat Rev Cancer* 2(7):521–8. doi: 10.1038/nrc842.

Schaefer, L. and R. M. Schaefer. 2010. Proteoglycans: From structural compounds to signaling molecules. *Cell Tissue Res* 339(1):237–46. doi: 10.1007/s00441-009-0821-y.

Schmidt, J. J., J. Rowley, and H. J. Kong. 2008. Hydrogels used for cell-based drug delivery. *J Biomed Mater Res A* 87(4):1113–22. doi: 10.1002/jbm.a.32287.

Serebriiskii, I., R. Castello-Cros, A. Lamb, E. A. Golemis, and E. Cukierman. 2008. Fibroblast-derived 3D matrix differentially regulates the growth and drug-responsiveness of human cancer cells. *Matrix Biol* 27(6):573–85. doi: 10.1016/j.matbio.2008.02.008.

Shakibaei, M., P. Kraehe, B. Popper, P. Shayan, A. Goel, and C. Buhrmann. 2015. Curcumin potentiates antitumor activity of 5-fluorouracil in a 3D alginate tumor microenvironment of colorectal cancer. *BMC Cancer* 15:250. doi: 10.1186/s12885-015-1291-0.

Shoulders, M. D. and R. T. Raines. 2009. Collagen structure and stability. *Annu Rev Biochem* 78(1):929–58. doi: 10.1146/annurev.biochem.77.032207.120833.

Smalley, K. S., M. Lioni, and M. Herlyn. 2006. Life isn't flat: Taking cancer biology to the next dimension. *In Vitro Cell Dev Biol Anim* 42(8–9):242–7. doi: 10.1290/0604027.1.

Takahashi, K., I. Stamenkovic, M. Cutler, A. Dasgupta, and K. K. Tanabe. 1996. Keratan sulfate modification of CD44 modulates adhesion to hyaluronate. *J Biol Chem* 271(16):9490–6. doi: 10.1074/jbc.271.16.9490.

Tanzer, M. L. 2006. Current concepts of extracellular matrix. *J Orthop Sci* 11(3):326–31. doi: 10.1007/s00776-006-1012-2.

Tilghman, R. W., E. M. Blais, C. R. Cowan, N. E. Sherman, P. R. Grigera, E. D. Jeffery, J. W. Fox et al. 2012. Matrix rigidity regulates cancer cell growth by modulating cellular metabolism and protein synthesis. *PLoS One* 7(5):e37231. doi: 10.1371/journal.pone.0037231.

Tilghman, R. W., C. R. Cowan, J. D. Mih, Y. Koryakina, D. Gioeli, J. K. Slack-Davis, B. R. Blackman, D. J. Tschumperlin, and J. T. Parsons. 2010. Matrix rigidity regulates cancer cell growth and cellular phenotype. *PLoS One* 5(9):e12905. doi: 10.1371/journal.pone.0012905.

Toole, B. P. 2002. Hyaluronan promotes the malignant phenotype. *Glycobiology* 12(3):37R–42R. doi: 10.1093/glycob/12.3.37R.

Tse, J. M., G. Cheng, J. A. Tyrrell, S. A. Wilcox-Adelman, Y. Boucher, R. K. Jain, and L. L. Munn. 2012. Mechanical compression drives cancer cells toward invasive phenotype. *Proc Natl Acad Sci USA* 109(3):911–6. doi: 10.1073/pnas.1118910109.

Vasita, R. and D. S. Katti. 2006. Nanofibers and their applications in tissue engineering. *Int J Nanomedicine* 1(1):15–30.

Veiga, C., T. Long, B. Siow, M. Loizidou, G. Royle, and K. Ricketts. 2015. MO-F-CAMPUS-I-04: Magnetic resonance imaging of an *in vitro* 3D tumor model. *Med Phys* 42(6):3579. doi: 10.1118/1.4925470.

Visse, R. and H. Nagase. 2003. Matrix metalloproteinases and tissue inhibitors of metalloproteinases: Structure, function, and biochemistry. *Circ Res* 92(8):827–39. doi: 10.1161/01.RES.0000070112.80711.3D.

Walboomers, X. F., H. J. Croes, L. A. Ginsel, and J. A. Jansen. 1999. Contact guidance of rat fibroblasts on various implant materials. *J Biomed Mater Res* 47(2):204–12.

Weaver, V. M., S. Lelievre, J. N. Lakins, M. A. Chrenek, J. C. Jones, F. Giancotti, Z. Werb, and M. J. Bissell. 2002. beta4 integrin-dependent formation of polarized three-dimensional architecture confers resistance to apoptosis in normal and malignant mammary epithelium. *Cancer Cell* 2(3):205–16.

Wegrowski, Y. and F. X. Maquart. 2006. Chondroitin sulfate proteoglycans in tumor progression. *Adv Pharmacol* 53:297–321. doi: 10.1016/s1054-3589(05)53014-x.

White, E. S., F. E. Baralle, and A. F. Muro. 2008. New insights into form and function of fibronectin splice variants. *J Pathol* 216(1):1–14. doi: 10.1002/path.2388.

Whitesides, G. M. 2006. The origins and the future of microfluidics. *Nature* 442(7101):368–73.

Xu, X., C. R. Sabanayagam, D. A. Harrington, M. C. Farach-Carson, and X. Jia. 2014. A hydrogel-based tumor model for the evaluation of nanoparticle-based cancer therapeutics. *Biomaterials* 35(10):3319–30. doi: 10.1016/j.biomaterials.2013.12.080.

Xu, Y., Y. Zhao, B. Su, Y. Chen, and C. Zhou. 2006. Expression of collagen IV, fibronectin, laminin in non-small cell lung cancer and its correlation with chemosensitivities and apoptosis. *Chinese-German J Clin Oncol* 5(1):58–62. doi: 10.1007/s10330-005-0440-3.

Yang, X., S. K. Sarvestani, S. Moeinzadeh, X. He, and E. Jabbari. 2013. Effect of CD44 binding peptide conjugated to an engineered inert matrix on maintenance of breast cancer stem cells and tumorsphere formation. *PLoS One* 8(3):e59147. doi: 10.1371/journal.pone.0059147.

Yang, Y., X. Yang, J. Zou, C. Jia, Y. Hu, H. Du, and H. Wang. 2015. Evaluation of photodynamic therapy efficiency using an *in vitro* three-dimensional microfluidic breast cancer tissue model. *Lab Chip* 15(3):735–44. doi: 10.1039/c4lc01065e.

Yip, G. W., M. Smollich, and M. Gotte. 2006. Therapeutic value of glycosaminoglycans in cancer. *Mol Cancer Ther* 5(9):2139–48. doi: 10.1158/1535-7163.MCT-06-0082.

Yurchenco, P. D. and J. C. Schittny. 1990. Molecular architecture of basement membranes. *FASEB J* 4(6):1577–90.

Zaman, M. H., L. M. Trapani, A. L. Sieminski, D. Mackellar, H. Gong, R. D. Kamm, A. Wells, D. A. Lauffenburger, and P. Matsudaira. 2006. Migration of tumor cells in 3D matrices is governed by matrix stiffness along with cell-matrix adhesion and proteolysis. *Proc Natl Acad Sci USA* 103(29):10889–94. doi: 10.1073/pnas.0604460103.

Zeltinger, J., J. K. Sherwood, D. A. Graham, R. Mueller, and L. G. Griffith. 2001. Effect of pore size and void fraction on cellular adhesion, proliferation, and matrix deposition. *Tissue Eng* 7(5):557–72. doi: 10.1089/107632701753213183.

Zhang, L., Y. W. Chun, and T. J. Webster. 2010. Decreased lung carcinoma cell density on select polymer nanometer surface features for lung replacement therapies. *Int J Nanomedicine* 5:269–75.

Zhang, L. and T. J. Webster. 2013. Effects of chemically modified nanostructured PLGA on functioning of lung and breast cancer cells. *Int J Nanomedicine* 8:1907–19. doi: 10.2147/IJN.S41570.

Zhang, M., P. Boughton, B. Rose, C. S. Lee, and A. M. Hong. 2013. The use of porous scaffold as a tumor model. *Int J Biomater* 2013:396056. doi: 10.1155/2013/396056.

Zhang, M., B. Rose, C. S. Lee, and A. M. Hong. 2015. *In vitro* 3-dimensional tumor model for radiosensitivity of HPV positive OSCC cell lines. *Cancer Biol Ther* 16(8):1231–40. doi: 10.1080/15384047.2015.1056410.

Zhang, Z. and S. Nagrath. 2013. Microfluidics and cancer: Are we there yet? *Biomed Microdevices* 15(4):595–609. doi: 10.1007/s10544-012-9734-8.

8

Understanding the Interaction of Nanomaterials with Living Systems: Tissue Engineering

Shashi Singh

CONTENTS

8.1	Fabrication of Nanoscaffolds	281
8.2	Types of Scaffolds	283
8.3	Two Main Categories of Bioscaffolds	283
	8.3.1 Injectables	283
	8.3.2 Custom Scaffold	284
8.4	Interactions with Cells	284
	8.4.1 Biocompatibility	284
	8.4.2 Surface of the Nanomaterial	285
	8.4.3 Nanoscale Features	286
	8.4.4 Nanoscale Composition	288
	8.4.5 Porosity	289
	8.4.6 Mechanosensing	291
	8.4.7 Responses of Cells and Signaling Pathways	291
	8.4.8 Mitogen-Activated Protein Kinase Pathway	292
	8.4.9 Rho-Associated Protein Kinases	292
	8.4.10 Transforming Growth Factor-β Pathways	292
	8.4.11 The PI3 kinase/Akt	292
	8.4.12 Wnt/β-Catenin	292
References		293

The quest to live longer and lead a healthy, comfortable life has led to myths of elixirs and queer devices. Experimentations with atypical materials have had accidental successes, and more followed, leading to present-day biomaterials. The earliest recorded use of materials as sutures is as old as 3000 BC from Egypt. Cat gut, silk, and even ant's pincers were used as "Sutures" by Egyptians, Indians, Romans, and Greeks. Apart from sutures, implants for teeth have been in use for a long time, followed by metal implants for skeletal injuries (Ratner 2013). These materials were not called biomaterials until the 1960s, though the concept has been around since prehistoric times.

Biomaterials are a group of heterogeneous materials that interact with biological systems without any adverse effects. Usually used in medicine to replace an affected structure or augment a function, biomaterials have come a long way. There is huge variety of biomaterials, with the family ever growing adding newer material frequently. Presently in use are metals; glass; ceramics; bio-, synthetic, and composite polymers for prosthetics; sensors and monitors; electrodes or drug delivery systems; and now tissue-engineered devices.

Tissue engineering has come in as an alternative and/or supplementation to transplantation. Tissue engineering is an amalgam of material sciences/engineering and biology. Tissue engineering could be as simple as injecting cells or growth factors at the site of injury to as complex as building a three-dimensional organoid or tissue construct. Making a construct has become almost synonymous with tissue engineering that requires assembly of biomaterials, cells, and growth factors (Figure 8.1). The idea of tissue engineering is derived from the tissue structure itself. In the body, cells are surrounded by a scaffolding of extracellular matrix that supports the cells mechanically and physiologically to maintain homeostasis function and integrity of the tissue/organ. The structures formed by the extracellular matrix have a nanotexture that influences the cellular state and function. For a successful construct to be able to mimic the function of the native tissue/organ, the scaffold design should be as close as possible to the native extracellular matrix in terms of physical cues (at the nano and micro scales) and composition chemically. Imitation of the components or structures to design material or scaffolds to suit the needs of the extracellular matrix is also called biomimetism. Here, the nanotechnological advances of recent years have made inroads to uphold the potential of tissue engineering.

Biomaterials for tissue engineering scaffolds have to fulfill certain criteria. The material should be biocompatible. These materials should be able to exist without causing catastrophes like cell death, infections, extreme immune response, or the like. Biocompatibility means the ability of a material to perform with an adequate host response in a specific situation (Williams 2008).

Stability should preferably last until complete normalization. Material should not degrade so quickly as to prevent complete healing but should degrade over a period of time, leaving behind no signs, scars, and so on. It is thus expected that the material should be degradable and allow the host's own cells to replace the implanted material (Brown et al. 2010, Badylak et al. 2012) Another property expected of these materials is bioactivity: the implant should integrate with the natural surroundings; it should be able to react to signals or provide signaling for wound healing and normalization. An important consideration is that the mechanical properties have to match the tissue in repair. Until the healing is complete, the scaffold should be able to provide mechanical strength and incorporate the necessary cues for normal functioning. Apart from these, porosity both macro for cell penetration and micro for nutrient/gaseous/waste exchanges, and so on, and moldability to specific shapes are some site specific requirements (Hutmacher 2000). So, a fine balance between

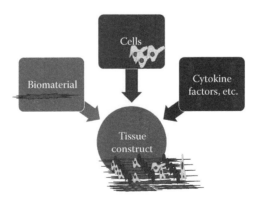

FIGURE 8.1
Schematic of tissue engineering: the component essentials in making a construct involve cells, compatible biomaterial, and some factors to promote growth and differentiation.

porosity to allow infiltration of cells and mechanical properties of the material has to be found while designing scaffolds.

Nanofabrication of materials or scaffolds provides us with the potential to incorporate many of these requirements into a construct that can direct cell activity. Recent advances in nanotechnology have facilitated the design of biocompatible materials into nanoparticles (Kim and Fisher 2007, Baranes et al. 2015, Łukasiewicz et al. 2015, Shi et al. 2016), nanoporous material, nanofibers (Mota et al. 2015, Raspa et al. 2016, Szymanski et al. 2014), nanotubes (Childs et al. 2013, Hopley et al. 2014, Zhang et al. 2009b), and so on. All these forms of bionanomaterials serve special roles in tissue engineering. The success of tissue engineering also depends on the right kind of cells used in preparing the construct. Selection of a different cell type may dictate a subtle change in the design of scaffold to achieve the desired results. After all, it is a two-way interaction between the cells and the scaffolds. Generally speaking, a successful construct requires cells that are easy to obtain in desired numbers, proliferative, multipotential to differentiate, and preferably nonimmunogenic. Stem cells have the potential to self-renew and differentiate into many kinds of cells. Stem cells are of many kinds themselves, from embryonic and adult stem cells to tissue-specific progenitor cells. Each has its own pros and cons, but stem cells appear to be promising candidates for tissue engineering. The latest to add in this category are induced pluripotent stem cells (iPSCs). iPSCs are adult somatic cells induced to become pluripotent by transgenic induction of transcription factors that drive the embryonic cell-like characteristics. iPSCs are important for tissue engineering due to their unlimited proliferation and differentiation potential, like embryonic cells, and they can also be custom made for patients without any ethical issues attached. There is a caveat that stem cells of the embryonic or iPSC kind would be difficult for use in human beings due to their immense tumorogenic potential.

8.1 Fabrication of Nanoscaffolds

Nanomaterials have become important for tissue engineering in the last few decades due to their bulk and surface properties and the techniques available to design them in a custom way. Apart from standard fabrication tools of nanomaterial preparation, advanced microfabrication and a range of nanoenabling tools support accomplishment of substrates or 3D engineered constructs for *in vitro* studies as well as for implantation (Ainslie and Desai 2008, Park et al. 2010a, Cesca et al. 2014).

Nanoparticles as scaffolds: Nanoparticles have been significant in altering the physicochemical properties of the scaffold. Precise topographical changes can be produced by decorating the surfaces of the scaffolds or incorporating them during scaffold synthesis to impart mechanical strength, electrical conductivity, or some catalytical/biochemical activity. Nanoparticles belonging to different categories have been used in scaffold genesis, like carbon nanotubes, metal and metal oxide nanoparticles, ceramic nanoparticles, and organic nanoparticles. Nanoparticles are synthesized either by top-down methods like attrition or bottom-up methods like pyrolysis, thermoreactions, sol-gel, or structured media. Once the particles are characterized for their size, purity, and other characteristics, they are utilized with synthetic or biopolymers for scaffold synthesis (Abdelrasoul et al. 2015). Using a novel method, metal-vapor synthesis, for Ag and Au nanoparticles, collagen-chitosan scaffolds have been modified for dermal wound healing with antibacterial properties

(Rubina et al. 2016) or cross-linked with collagen using bioactive groups like succinate, with end results of better bioactivity (Mandal et al. 2015).

For bone tissue engineering, ceramic nanoparticles have been synthesized *in situ* during block formation using polymer-based reduction of calcium salts to form precipitated hydroxyapatite in nanoforms. Due to their similarity to bone mineral, calcium phosphate ceramics, including hydroxyapatite (HA), have drawn interest in bone tissue constructs. This results in *en bloc* synthesis of nanoparticles (Sinha et al. 2003) and shows good ostoegenic activity when implanted *in vivo* (Reddy et al. 2013, Komakula et al. 2015). Biotemplating is another common procedure of producing nanoparticles for biological uses. These biologically based fabrication schemes can usually be designed to work at room temperature and pressure, and have been successfully implemented (Hamada et al. 2004, Sinha et al. 2003).

Nanofibrous scaffolds: Nanofibrous scaffolds can be fabricated by self-assembly, phase separation, and electrospinning. As the name suggests, self assembly is the spontaneous arrangement of unit blocks by noncovalent interactions, and the process depends on external physical and chemical factors. The skill lies in the design of building block to obtain a structure that mimics the features of a biological system. Generally, small peptides are used as building blocks with a custom design to fabricate scaffolds (Guo et al. 2007, Shah et al. 2010). These systems are not very popular due to the inability to control pore size and unclear degradation profiles (Smith et al. 2009). Phase-separation techniques for fabrication of fibrous scaffolds involve separation of phases due to temperature-induced physical incompatibility (Ramakrishna 2005). This is a promising technique for developing nanofibrous scaffolds with well-defined pore shape and size, but the orientation of fibers cannot be controlled and lacks mechanical properties. Electrospinning is a well-established process capable of producing fibers in the nanoscale range. Nanofibers of different architectures and shapes can be fabricated upon different surfaces, patterned, molded, or fabricated in solution using electrospinning (Figure 8.2). The fiber diameter is generally less than the cell surface area and helps in creating a nanotopographical surface. The fibers have been formed using polymers proteins and others successfully (Zeugolis et al. 2008, Zafar et al. 2016). The fibers can be randomly aligned or ordered in specific alignments (Vasita

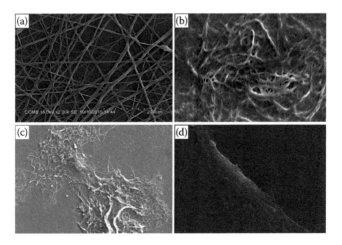

FIGURE 8.2
Electrospun matrices of bio- and synthetic polymers. (a) Fibers spun from a solution of poly vinyl alcohol. (b–d) Fibrous scaffold spun from solutions of collagen with other matrix molecules like fibronectin (b) and chondroitin sulfate (c). Cell spreading on such spun surfaces (d).

and Katti 2006). The fibers can be fabricated into complex architectures of stacked arrays tubes, multilayered film, and so on (Chen et al. 2013). Even electrospinning is beset with problems of finer control over uniformity of size.

Nanoporous material: Different fabrication methods that have been used to prepare appropriate nanoporous materials are salt leaching, gas foaming, phase separation, and freeze-drying. These techniques do not ensure thorough control of architecture, porosity, and inappropriate mechanical strength. These methods have some disadvantages, such as expensive processing devices, poor pore interconnectivity, and producing abnormal microstructure. The sol-gel process has potential that permits us to obtain structures at the nanometric scale (Hesaraki et al. 2007, Sohrabi et al. 2013). Simple porogens like sugar or salt are used and are eliminated from the scaffold by dissolution, finally leaving a porous scaffold. Gas foaming involves use of gases as porogens; solid polymer disks are exposed to high-pressure carbon dioxide to allow saturation of carbon dioxide in the polymer that leads to thermodynamic instability by rapidly releasing carbon dioxide gas from the polymer system. This forms closed surfaces and pore sizes of about 100 nm (Cooper 2000, Harris et al. 1998, Nam et al. 2000).

Surface modification is sometimes mandatory, as scaffolds may lack biological recognition due to a variety of synthetic biodegradable polymers used for scaffold making. Polymers may not be ideal for cell material interaction, so many pretreatments are designed prior to use (Mikos et al. 1994, Neff et al. 1998). Sometimes the nanomaterial can elicit unexpected responses with disrupted functionalities due to surface modifications. Pristine nanomaterial surfaces are hence engineered to make them inert or more biocompatible.

To achieve better designs in nanoporous scaffolds, rapid prototyping techniques with computer-aided design (CAD) modeling allow fabrication of complex architectures with precise architecture mimicking the natural organ/tissue.

8.2 Types of Scaffolds

The latest in fabrication of tissue building blocks is achieved via multiple approaches, like fabrication of cell-encapsulating microscale hydrogels (microgels), self-assembled cell aggregation, generation of cell sheets, and direct printing of cells. Laser stereolithography techniques or three-dimensional printing that involves layerwise printing of material to make a 3D model have better control due to better design repeatability, part consistency, and control of scaffold architecture at macro- to nano levels. Some 3D printing techniques allow cells to be laid during the scaffold printing process rather than incubating with cells after synthesizing the scaffolds. This process can be carried out only with cytocompatible material.

8.3 Two Main Categories of Bioscaffolds

8.3.1 Injectables

Sometimes nonhealing sites can be repaired by injecting material at the site of injury with or without cells; this is especially true for skeletal injuries. Injectable scaffolds are clinically relevant and reduce morbidity, risk of infections, scar formation, and also cost of treatment. Scaffold components are laid at the site of injury in form of slurry or a suspension that will set *in situ* as a solid. After solidification at the site, the construct offers a temporary scaffold to the cells for recovery of the injured site. The material can also serve as a carrier for bioactive molecules or controlled release device for drugs. Nanoparticles of ceramics and others are most used in these types of tissue constructs (Komakula et al. 2015). Many biopolymers have also emerged as injectables like alginate, agarose, gelatin, and fibrin, as they exhibit biocompatibility, support cell adhesion, and growth and have been used as scaffolds to repair cardiac and cartilage tissues (Li et al. 2015). Injectable gels of fibrin containing vascular endothelial growth factor (VEGF) have been prepared to promote angiogenesis and show prolonged presentation of growth factor within a carrier matrix (Zisch et al. 2001). As expected of any tissue construct, the injectables are supposed to be nontoxic, and any releases during the setting process, whether chemical or physical (thermal), should be nontoxic to cells/tissues. The solidification process should not be hampered in physiological conditions. After solidification, the mechanical properties of the construct should be compatible with the tissue in repair.

8.3.2 Custom Scaffold

If the repair site is not being injected, it has to be filled/ replaced with an exact replica of the tissue or excised portion; the constructs for engineering in this case are to be custom molded. These kinds of scaffolds are like 3D structures and could be printed (laser stereolithography technique) or cast molded to match the repair site (Limongi et al. 2015a,b). This microfabrication can be achieved *in vivo* also (Curtis et al. 2005). The nanomaterials used here could be in any form to be incorporated into the super-3D-like structure and further modified to enhance biocompatibility or bioactivity. The problems faced here could be of biointegration at times: how the human body reacts to these structures.

8.4 Interactions with Cells

So far, biomaterials have not developed with design but with trial and error, and to dictate a specific response from cells/tissue would be a huge challenge. Further, how these nanomaterials interact with the cell becomes important in the wake of size that completely alters the properties of materials. Making a scaffold means these materials will interact with cells, whether in petri dishes or once implanted in the body. Still, a few issues could be considered while discussing interactions with cells.

8.4.1 Biocompatibility

The choice of material becomes imperative while designing a scaffold. Material that will remain in the body for a period of time until the healing is complete or even later needs to be safe. The material itself or its degradation components should be nontoxic in every sense. The materials in question are subjected to a battery of test to look for cytotoxicity, genotoxicity, and immunotoxicity both at bulk and nanoforms. Nanoforms of material can induce unexpected cellular responses. For cytocompatibility, one also looks for cell adherence and proliferation on these surfaces or particles (Ishihara et al. 2016, Ying et al. 2016).

8.4.2 Surface of the Nanomaterial

Surface chemistry is important to the extent that different grades of titania used for bone implants can affect the response of cells to the implants. Though the surface often gets coated with proteins when implanted or being tested, the material comes in contact with the biological fluids often containing proteins, lipids, salts, carbohydrates, and so on. Nanomaterial also displays a coating of proteins on the surface once exposed to body fluids (Monopoli et al. 2011, 2012, Lynch et al. 2009). So, actually, the cells interact with the protein coatings on the surface. These kinds of protein coatings have been named a corona (Lynch et al. 2009; Walczyk et al. 2010). The corona could be the defining factor for the kind of response the biomaterial may invoke; of course, the kind of proteins that bind to the nanomaterial will be determined by the physicochemical properties of the nanomaterial. We have shown by infrared spectroscopy titania particles coated with various types of body fluids, and the culture medium shows signatures of peptide and amide bonds within 5 minutes of exposure to body fluids (Figure 8.2). Deng et al. 2011 have shown negatively charged gold nanoparticles conjugated with poly(acrylic acid) bind to fibrinogen and change its conformation. This new surface can promote interaction of nanoparticles with the integrin receptor and, consequently, release inflammatory cytokines. The binding of proteins may also depend on time and incubation temperature (Mahmoudi et al. 2011, 2014, Tenzer et al. 2013). These studies are mostly performed with pristine nanomaterials; protein adhering to scaffolds, per se, has not been explored extensively. These interactions have implications, as the molecules adhering to the scaffolds will create a specific kind of niche that may influence the ultimate fate of cells. Using fibrillar collagen scaffolds as patches, Seerpooshan et al. 2015 have shown the formation of complex protein shells on fibrous constructs when introduced into various biological environments. The corona composition varied with the tissue type, health conditions, or *in vitro* conditions in different media (Figure 8.3).

Not only proteins but many osteoinducing nanomaterials have been tested in simulated body fluids and show better performance. There is a change in form of deposition of apatite or resorbtion of certain phosphates, making interaction more viable for bioactivity (Landi et al. 2004, Sánchez-Salcedo et al. 2009, Morimune-Moriya et al. 2015). Surface-activated poly(L-lactic acid) shows formation of hydroxyapaptite in simulated body fluids and increases in total cellularity when seeded with fibroblasts, chondrocytes, and osteoblasts (Park et al. 2010b).

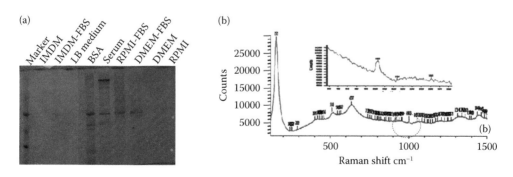

FIGURE 8.3
(a) Proteins could be recovered from TiO_2 samples coated with biological fluids confirming adsorption of proteins. (b) Raman analysis of coated eNPS showed presence of tyrosine peak in the TiO_2 sample.

In a nutshell, there is tremendous change in the surface of the scaffold once exposed to body fluids that ultimately affects the outcome of the graft. Since most of the studies are trial and error here also, we have always looked at the endpoint, i.e., the kind of function we get. Graphene substrates show promise for neural stem cell differentiation into neurons after being fabricated by chemical vapor deposition. Laminin-related cellular pathways were found to be enhanced; these modified graphene substrates allowed cell adhesion and were found to be suitable for long-term differentiation (Park et al. 2011). Working with mouse hippocampal neurons cultured on graphene, Li et al. (2011) showed enhanced neurite sprouting and outgrowth via the GAP43-related pathways. We may be in for surprises when we study interactions in detail, and that would help in tweaking to design a precise custom scaffold. Fluorinated graphene sheets have been developed as scaffolds to guide neural growth. Further follow-up using mesenchymal stem cells has been shown by Wang et al. (2012), where they added retinoic acid to fluorinated graphene and achieved neuronal differentiation. Different derivatives of graphene have been produced as substrates that show enhanced differentiation of mesenchymal stem cells (MSCs) into osteogenic, myogenic, chondrogenic, cardiomyogenic, and so on (Ku and Park 2013, Park et al. 2014, Yoon et al. 2014, Zhang et al. 2016). A significant amount of research has been carried out on interactions between cells and scaffolds, and evaluation of coronas is definitely going to add to the understanding of the response of hosts to biomaterials. The addition of appropriate cellular signals also aids in achieving the desired response, and inclusion of factors such as active proteins or even DNA has been critical to success.

8.4.3 Nanoscale Features

Cells in a 3D matrix respond to the various architectural features at different scales from nano to micro to millimeters with differing functions in apoptosis, proliferation, migration, and differentiation. Cells prefer to grow on rough surfaces (Gentile et al. 2010; Kim et al. 2010). The dominating effect of matrix geometrical force on cell-fate induction is of prime importance in tissue-specific regeneration. It is believed that the introduction of well-defined nanometer-scale surface topography would generate appropriate surface signals to manipulate cell function akin to the natural extracellular matrix. This has led to advances in surface fabrication techniques (Figure 8.4). Cells in the body are generally in the micrometer range with features, components, and associated environments at the nanometer or submicron level. As stated earlier, once the bionanomaterials are exposed to body fluids, they are coated with biomolecules. All the interactions between the cell and nanometer-scale structures such as proteins are crucial for controlling a variety of cell functions such as proliferation, migration, and the production of the extracellular matrix (ECM). How the physical, geometrical, or chemical features influence this entire process of corona formation or cell adhesion is still a matter of investigation.

Cells cultivated on scaffolds containing 10–100 μm-sized ridges and grooves promoted elongated cell growth that was oriented in the direction of the surface feature (Thakar et al. 2009). Dalby et al. (2004) showed that the interaction of fibroblast cell filopodia with surface features as small as 10 nm directly influenced cell adhesion. The natural scaffold structure of the ECM also has architectural features such as fibers, pores, and ridges of varying sizes, which are capable of providing physical signs that can directly affect cell behavior (Geiger et al. 2001, Stevens and George 2005).

In addition, the physical structure and chemistry of nanometer-scale structures directly influence growth parameters responsible for cell adhesion, apoptosis, differentiation, genetic expression, migration, morphology, orientation, and proliferation (Norman and Desai 2006).

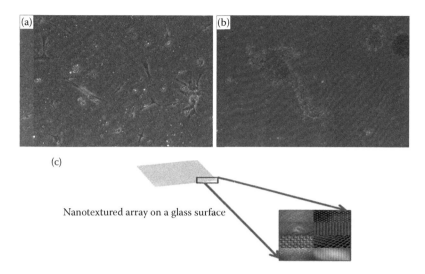

FIGURE 8.4
(a,b) Cells plated on substrates with different topology and stiffness. (b) Cells do not adhere on to extremely smooth surfaces. (c) Schematic to show arrays on a substrate to assay cell adherence, proliferation, or behavior on textured surfaces.

Collagen is the most abundant protein of the body and is seen in every organ /tissue; it is also the most common protein used for making scaffolds in the form of tropocollagen (after cleavage of a portion of the native collagen). Upon self-assembly, tropocollagen forms larger collagen fibrils around a diameter of ~50 nm that have additional features of adhesive ridges alternately with grooves (Figure 8.5). These are important features that direct cell fate (Orgel et al. 2006, Baselt et al. 1993, Meshel et al. 2005). Similarly, collagen with regular decoration of apatite crystals in nanoscale tends to induce osteogenic transformation of cells. Collagen can be adopted into many scaffold types.

Many studies have shown that surface nanofeatures directly influence cellular adhesion, proliferation, and morphology when grown on different arrays with protrusions or nanotextured surfaces. Topological features provide migration and growth guidance to cells called contact guidance, brought about by adhesion of the cell and influencing its cytoskeletal arrangement. Spider silk fibers promoted successful cell adhesion, migration, and growth of Schwann cells as a possible nerve conduit, as was shown by Allmeling et al. 2006. Cells also transform their microenvironment on these features by the ECM they produce, cytokines, and so on. The topographical features can also dictate the fate of cells, as seen for mesenchymal stem cells. Multipotent mesenchymal stem cells display lineage-specific differentiation when cultured on substrates that mimic the stiffness of native tissue environments (Figure 8.6). In the case of MSCs cultured on substrate that mimics the bone environment, the cells become osteogenic (Rowlands et al. 2008), while MSCs exposed to substrates that mimic a myogenic tissue environment become muscle cells (Kloxin et al. 2010). Local nanotopography inducing differential response is shown in a study by (Kilian et al. 2010); human MSCs grown on microcontact-printed polydimethylsiloxane (PDMS) show musculoskeletal characteristics at the edges of the matrix and adipogenic/neuro nature toward the inner region of the scaffold (Figure 8.7).

Mohtaram et al. (2015) have investigated the effect of micro- and nanoscale topography on promoting neuronal differentiation of human iPSCs and directing the resulting neuronal outgrowth. Loop mesh and biaxial-aligned microscale retinoic acid functionalized-PCL

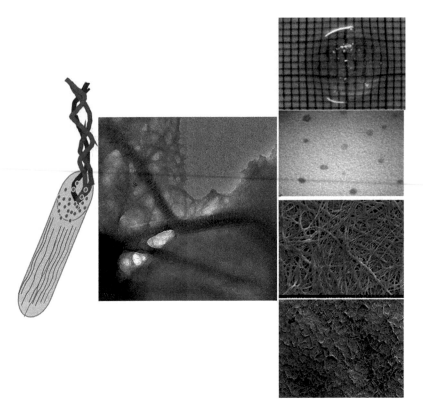

FIGURE 8.5
Collagen fibers are the most abundant proteins of the natural extracellular matrix. Collagen is made up of three helical polypeptide chains that aggregate into fibrils of about 50 nm. Collagen can be adopted as tissue constructs in many forms. Here we show collagen in form of gels, nanoparticles, fibers, and porous hydrogel.

nanofibers were seeded with neural progenitors derived from human iPSCs. From the cell culture study, it was noticed that the maximum neurite outgrowth length of these cells occurred on the biaxial aligned scaffolds compared to the loop mesh topography, giving insight into how physical and chemical cues can be used to engineer neural tissue. This study confirms the importance of nanoscale features in the designing of scaffolds for directing stem cell differentiation. Evidently, it has been proven that incorporation of various growth factors or bioactive molecules in nanocarrier systems as biological signals can promote the desired differentiation lineage within the seeded stem cells.

8.4.4 Nanoscale Composition

In continuation with nanotextures introduced in scaffolds to increase efficacy of the final construct, there are instances of fine modulation in components of the scaffold. Ligand binding and controlled release of regulatory factors can improve the degree of control over spatial organization and differentiation of stem cells within scaffolds. Cells adhere to surfaces through integrins, and while we are using synthetic material, the cells may not find suitable binding sites. Investigators have ligated small peptides like RGD (arginin-glycine-aspartic acid) or other peptides like AGD and IKVAV to the polymers to make the surfaces more bioconductive. RGD is an attachment site for integrins, and over 20 known integrins are known to recognize this sequence in their adhesion protein ligands. Incorporating RGD

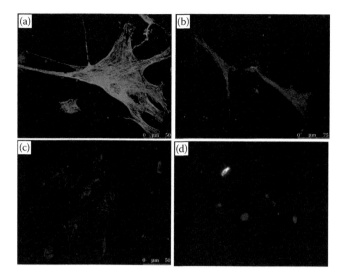

FIGURE 8.6
Beta tubulin and alpha smooth muscle staining of mesenchymal stem cells grown on substrates with different stiffness. (a) α-SMA expresses in cells grown on stiff substrate and β-tubulin expression is seen in cells grown on less stiff substrates (b). Cells grown on these substrates were also stained for osteogenic markers like osteocalcin (c) and RUNX2 (d) that was not expressed.

into a polyethylene glycol scaffold resulted in enhanced osteogenic differentiation of rat bone marrow-derived mesenchymal stem cells even in the absence of dexamethasone. Also, regulating the release of regulatory molecules from porous poly (D,L Lactide co-glycolide) scaffold induced rabbit bone marrow derived mesenchymal stem cells (BMSCs) within the scaffold to mineralize *in vitro* into bone (Iwakura et al. 2001, Hosseinkhani et al. 2006).

There are instances where specific growth factors are incorporated in scaffolds to dictate a particular cell lineage. Beta fibroblast growth factor (bFGF) encapsulated within alginate, gelatin, collagen, or polymer carriers promoted neoangiogenesis in tissue constructs (Edelman et al. 1991, Chen et al. 2006). VEGF has been incorporated into many polymer scaffolds to ease vasculogenesis. In poly lactic acid (PLA) scaffolds to induce angiogenesis, VEGF has been shown to remain active for long periods of time (Kanczler et al. 2007). One possible way of enhancing the *in vivo* efficacy is to achieve its controlled release over an extended time period by incorporating the growth factor in a polymer carrier or nanocarrier. Scaffolds can serve a dual purpose of mechanical support and differentiation as well as drug delivery in certain cases to expedite the process of recovery. Growth factor incorporation can also be enhanced by site-specific incorporation to cater to local cell populations rather than the whole scaffold.

8.4.5 Porosity

Porosity is one of the essential requirements of a tissue scaffold, and it concerns macro-, micro-, and nanoporosity. 3D scaffolds are generally made highly porous with interconnected pore networks to facilitate cell impregnation, nutrient and oxygen diffusion, and waste removal. The porosity of the scaffolds is determined by various means before setting out for construct use. Porosity also has to be accompanied by easy pass ability/penetrability that is achieved by interconnected pores. Porosity and interconnectivity in a scaffold facilitate cell migration within its porous milieu so that cell proliferation and growth can take place

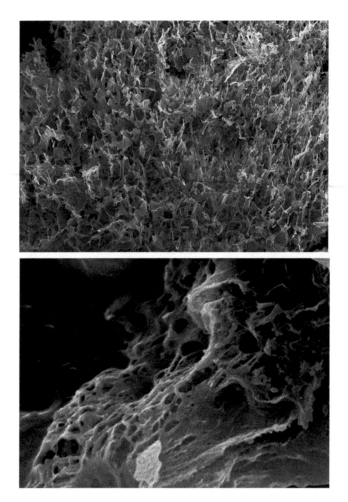

FIGURE 8.7
Porosity introduced in block gel by a foaming process. The pore sizes are variable from macro to nano and allow cell adhesion and migration of cells to different parts of the scaffold and also metabolite mobility.

and overcrowding is avoided. The incorporation of nanopores enhances the surface area, which, in principle, can provide an increased number of sites for protein adsorption, as well as ion concentrations optimized for more efficient cell growth (Woo et al. 2003).

Experiments using scaffolds with low porosity have shown changes in the proliferation rate and differentiation potential of cells. Cells cultured in a low-density matrix were able to spread across adjacent fibers easily and had a higher proliferation rate. However, due to smaller pore size, the formation of large cell aggregates was limited, leading to reduced differentiation of the cells. In a high-porosity scaffold, on the other hand, differentiation of cells was higher with higher aggregation (Ma et al. 2000). In woven matrices using synthetic polymers or silk fibers, it was shown that porosity with higher fiber diameter promoted cell proliferation. Thus porosity, pore size, and the wall (fiber) thickness affect cellular behavior (Mandal and Kundu, 2009). There is a limit to porosity that can be introduced in a scaffold lest it compromise the mechanical attributes of the construct. This is especially so for skeletal tissues. Different pore sizes did not affect mineralization, but osteoblasts have been shown to prefer small pore sizes for adhering and show migration in large pore sizes. Advantages of large pores by cell migration seem to outweigh the initial attachment preferences to small pores. One likely explanation for the decline in response to small pore size is that at later time points, a scaffold surface coated with proteins will influence the underlying topology. That's why a graded size range is expected in the pore size while designing a construct rather than a uniform pore size.

Porosity has also been shown to affect gene expression. Especially modulated are the secretions of the extracellular matrix. The amount of collagen and the proteoglycans secreted are affected by the pore size. During cartilage bioengineering, Lien et al. (2009) showed that pore size around 200–500 μm favored ECM secretion and maintenance of chondrogenic status, but a reduction in pore size to below 200 μm resulted in dedifferentiation. What is important, ultimately, is a fine tuning between porosity, cell type, scaffold material, and fabrication conditions that determine the ultimate fate of cell.

Porosity also supports angiogenesis, which is important for sustenance of the new tissue/organ we are trying to create. Scaffolds are seeded with endothelial cells or angiogenic factors to enable formation of new blood vessels, and the minimum porosity required for this process has been shown to be approximately 40 to 400 μm to enable the exchange of metabolic components and to facilitate endothelial cell entrance in different studies (Madden et al. 2010, Oliviero et al. 2012). It is really tough to determine the actual requirement unless one goes back to natural scaffolds and designs pores by the printing method.

Generally, scaffolds are first crafted and then cells are allowed to seep in or are laid over; sometimes it may not be easy for us to see cell penetration in these scaffolds due to inappropriate porosity or the material itself needing some kind of functionalization. So for the newly developed methods of printing scaffolds or tissue constructs, keep in mind the requirements of the cells and ECM are the best way to model.

8.4.6 Mechanosensing

Tissue stiffness is determined by the kind of extracellular matrix they reside in and the kind of functions they carry out. Cells respond to microenvironmental stiffness through the dynamics of the cytoskeletal network. Each cell has a maximal sensitivity for a particular value of stiffness, which depends on the cytoskeletal structure and the F-actin organization. Cells respond to the substrates by modulating the actin filament bundling (DeSantis et al. 2011). MSCs cultured on matrices of high stiffness present a large spreading size and tend to differentiate into musculoskeletal lineages, and on low-stiffness surfaces, these cells tend to form the neuronal phenotype. With soft matrices, cell types downregulate their actin myosin contractile machinery and exert much less tension, and the opposite happens in stiff matrices. These stresses are sensed by an equally force-sensitive nuclear membrane (Alam et al. 2015).

8.4.7 Responses of Cells and Signaling Pathways

The link between a cell and its surrounding is through focal adhesions and large protein assemblies as a direct mechanical link between extracellular matrix and cytoskeleton. Cells respond to local environmental cues using signaling pathways. Cell density and cell–cell interactions also play a role in fate determination. Cell membranes sense nanofeatures through the focal adhesions composed of integrins and respond by changes and rearrangement in the cortical cytoskeleton. Integrins are adhesion receptors that allow cells to sense and respond to microenvironmental signals encoded by the extracellular matrix. Many of the signaling pathways in the cells, like mitogen-activated protein kinase (MAPK), rho-associated protein kinase (RhoA/ROCK), the PI3K/Akt, Wnt/β-catenin, and the transforming growth factor (TGF)-β pathway, are incited by cell environment interactions through integrins, and are coupled to other gene expression pathways, leading to changes in cell fate.

8.4.8 Mitogen-Activated Protein Kinase Pathway

MAPK modules consist of three distinct kinases, an upstream MAP kinase kinase kinase (MAP3 K), a MAP kinase kinase (MAP2 K), and a downstream MAPK. This cascade is initiated by extracellular cues and leads to activation of a particular sequential MAPK kinase kinase and MAPK kinase of a series of vital signal transduction pathways that regulate processes such as cell proliferation, cell differentiation, and cell death all in eukaryotes (Provenzano et al. 2009, Shih et al. 2011). The MAPK thus activated translocates to the nucleus and initiates various target gene expressions through phosphorylations. MAPK has been shown to be involved in many lineage differentiations (Oeztuerk-Winder and Ventura, 2012).

8.4.9 Rho-Associated Protein Kinases

Rho kinases regulate stress fibers, contractions, formation of filopodia, and so on. Specific integrin activity and RhoA signaling mediate the generation of cytoskeleton tension, and the activity of the actin tension is important for the cell to sense mechanical stiffness of the surroundings and respond appropriately by gene regulation. It is the actin structure that is responsible for the transfer of strain to the nucleus. Inhibitors of the rho/ROCK pathway have been shown to abrogate the responses mediated by substrates. Stress fiber-focal adhesion complexes generate forces against the ECM, causing morphological change, migration, and differentiation of the cell (Mathieu and Loboa 2012).

8.4.10 Transforming Growth Factor-β Pathways

TGF-β and the related bone morphogenetic proteins (BMPs) signal through transmembrane serine/threonine kinase receptors and regulate diverse developmental and homeostatic processes. Integrin binding and subsequent cell-generated traction forces are shown to be involved in activation of TGF-β that is bound to the extracellular matrix (Annes et al. 2004, Wipff et al. 2007). Mechanical activation is also shown in fluid environments in response to the application of fluid shear stress or stirring forces (Ahamed et al. 2008, Albro et al. 2012).

8.4.11 The PI3 kinase/Akt

AKT signaling is adhesion-dependent in many cell types and activated via integrins. ECM stiffness regulates cells function via its impact on the contraction force in actomyosin fibers, the subcellular allocation of integrin, and the PI3 K pathways (Jiang et al. 2016). Upon binding to the cell surface receptors, growth factors such as insulin-like growth factor (IGF)-1 and neurotrophins (NGF) can trigger the activation of the PI3 K pathway, promoting survival and self-renewal of stem cells (Sudha et al. 2009). Activation of PI3K was also seen in cells that interact via RGD (Zhang et al. 2009a).

8.4.12 Wnt/β-Catenin

Wnt proteins are a large family of 19 secreted glycoproteins that trigger multiple signaling cascades essential for embryonic development and tissue regeneration. The Wnt/β-catenin pathway is a rather ubiquitous mechanism in controlling diverse cell functions and behaviors, including cell adhesion, migration, differentiation, and proliferation, and these cellular behaviors respond significantly to ECM stiffness. Wnt signaling is also responsive to matrix rigidity (Barbolina et al. 2013) The integrin-activated β-catenin/Wnt pathway connects with the canonical Wnt/β-catenin pathway to form a positive feedback loop, which is crucial to the promotion of the Wnt signal by stiff ECM and the regulation of mesenchymal stem cell differentiation and primary chondrocyte phenotype maintenance. (Ezzell et al. 1997, Barbolina et al. 2013). Biophysical signals have been shown to directly regulate Wnt signaling, as demonstrated in osteoblasts in a time-dependent manner (Jansen et al. 2010).

It is the signaling crosstalk between multiple pathways that finally decides the cell fate. With the passage of time, due to cell activity in the scaffolds, cell density and the mechanical properties of the scaffold may change, and the interaction of cells with the matrix will become less. Over a period of time, as the implanted cells are expected to secrete their own extracellular matrix, this new ECM will become the determinant.

Tissue engineering/regenerative medicine strategies require interaction and integration with tissue and cells through incorporation of appropriate physical and cellular signals. Therefore, inclusion of modifying factors like biologically active proteins, growth factors, and so on are also critical to success. Many simpler procedures have been more successful in tissue reconstruction, like replacement of skin in burn cases with sheets of skin grown in a lab on collagen or polymer matrices, corneal epithelium for ocular surface disorders, some natural recellularized scaffolds for trachea and bladder, and so on offering hope for more complex tissue-engineering procedures in the future.

References

Abdelrasoul, G. N., B. Farkas, I. Romano, A. Diaspro, S. Beke, Nanocomposite scaffold fabrication by incorporating gold nanoparticles into biodegradable polymer matrix: Synthesis, characterization, and photothermal effect; *Materials Science and Engineering: C*. 2015; 56:305–310.

Ahamed, J., N. Burg, K. Yoshinaga et al., *In vitro* and *in vivo* evidence for shear-induced activation of latent transforming growth factor-beta1; *Blood*. 2008; 112:3650–3660.

Ainslie, K. M., T. A. Desai, Microfabricated implants for applications in therapeutic delivery, tissue engineering, and biosensing; *Lab Chip*. 2008; 8(11):1864–1878, doi: 10.1039/b806446f

Alam, S. G., D. Lovett, D. I. Kim, K. J. Roux, R. B. Dickinson, T. P. Lele, The nucleus is an intracellular propagator of tensile forces in NIH 3T3 fibroblasts; *J. Cell Sci*. 2015; 128:1901–1911, doi: 10.1242/jcs.161703

Albro, M. B., A. D. Cigan, R. J. Nims et al., Shearing of synovial fluid activates latent TGF-beta; *Osteoarthr Cartil*. 2012; 20:1374–1382.

Allmeling, C., A. Jokuszies, K. Reimers, S. Kall, P. M. Vogt, Use of spider silk fibres as an innovative material in a biocompatible artificial nerve conduit; *Journal of Cellular and Molecular Medicine*. 2006; 10:770–777.

Annes, J. P., Y. Chen, J. S. Munger et al., Integrin alphaVbeta6- mediated activation of latent TGF-beta requires the latent TGF-beta binding protein-1; *J Cell Biol*. 2004; 165:723–734.

Badylak, S. F., D. J. Weiss, A. Caplan, P. Macchiarini, Engineered whole organs and complex tissues; *Lancet*. 2012; 379(9819):943–52.

Baranes, K., M. Shevach, O. Shefi, T. Dvir, Gold nano particle decorated scaffolds promote neuronal differentiation and maturation; *Nano Lett*. 2015; 16(5):2916–2920, doi: 10.1021/acs.nanolett.5b04033.

Barbolina M. V. et al., Matrix rigidity activates Wnt signaling through down-regulation of Dickkopf-1 protein; *J Biol Chem*. 2013; 288:141–151.

Baselt, D. R., J. P. Revel, J. D. Baldeschwieler, Subfibrillar structure of type I collagen observed by atomic force microscopy; *Biophys J*. 1993; 65(6):2644–2655.

Brown, B. N., C. A. Barnes, R. T. Kasick et al., Surface characterization of extracellular matrix scaffolds; *Biomaterials*. 2010; 31 428–437.

Cesca, F., T. Limongi, A. Accardo, A. Rocchi, M. Orlando et al. Fabrication of biocompatible free-standing nanopatterned films for primary neuronal cultures; *RSC Adv*. 2014; 4(86):45696–45702, doi: 10. 1039/c4ra08361j

Chen, L., Y. Bai, G. Liao, E. Peng, B. Wu, Y. Wang, X. Zeng, X. Xie, Electrospun poly(l-lactide)/poly(ε-caprolactone) blend nanofibrous scaffold: Characterization and biocompatibility with human adipose-derived stem cells; *PLOS* 2013; http://dx.doi.org/10.1371/journal.pone.0071265

Chen, Y. W., S. H. Chiou, T. T. Wong et al., Using gelatin scaffold with coated basic fibroblast growth factor as a transfer system for transplantation of human neural stem cells; *Transplantation Proceedings*, 2006; 38:1616–1617.

Childs, A., U. D. Hemraz, N. J. Castro, H. Fenniri, L. G. Zhang, Novel biologically-inspired rosette nanotube PLLA scaffolds for improving human mesenchymal stem cell chondrogenic differentiation; *Biomed. Mater*. 2013; 8(6), 065003, doi: 10.1088/1748-6041/8/6/065003.

Cooper, A. I., Polymer synthesis and processing using supercritical carbon dioxide; *J. Mater. Chem*. 2000; 10:207–234.

Curtis, A. S., C. D. Wilkinson, J. Crossan, C. Broadley, H. Darmani, K. K. Johal, H. Jorgensen, W. Monaghan, An *in vivo* microfabricated scaffold for tendon repair; *Eur. Cell Mater*. 2005; 9:50–57.

Dalby, M. J., M. O. Riehle, H. Johnstone, S. Affrossman, A. S. G. Curtis, Investigating the limits of filopodial sensing: A brief report using SEM to image the interaction between 10 nm high nano-topography and fibroblast filopodia; *Cell Biology International*. 2004; 28:229–236.

Deng, Z. J., M. Liang, M. Monteiro, I. Toth, R. F. Minchin, Nanoparticle-induced unfolding of fibrinogen promotes Mac-1 receptor activation and inflammation; *Nat. Nanotechnol*. 2011; 6:39.

DeSantis, G., A. B. Lennon, F. Boschetti, B. Verhegghe, P. Verdonck, P. J. Prendergast, How can cells sense the elasticity of a substrate? An analysis using a cell tensegrity model; *Eur Cells Mater*. 2011; 22:202–213.

Edelman, E. R., E. Mathiowitz, R. Langer, M. Klagsbrun, Controlled and modulated release of basic fibroblast growth factor; *Biomaterials* 1991; 12:619–6626.

Ezzell, R. M., W. H. Goldmann, N. Wang, N. Parasharama, D. E. Ingber, Vinculin promotes cell spreading by mechanically coupling integrins to the cytoskeleton; *Exp Cell Res*. 1997; 231:14–26.

Geiger, B., A. Bershadsky, R. Pankov, K. M. Yamada, Transmembrane extracellular matrix-cytoskeleton crosstalk; *Nature Reviews Molecular Cell Biology* 2001; 2(11):793–805.

Gentile, F., L. Tirinato, E. Battista, F. Causa, C. Liberale, E. M. di Fabrizio et al., Cells preferentially grow on rough substrates; *Biomaterials*. 2010; 31(28):7205–7212, doi: 10.1016/j. biomaterials.2010.06.016

Guo, J., H. Su, Y. Zeng, L. X. Liang, W. M. Wong, R. G. Ellis-Behnke, K. F. So, W. Wu. Reknitting, the injured spinal cord by self-assembling peptide nanofiber scaffold; *Nanomedicine: Nanotechnology, Biology and Medicine* 2007; 3:311–321.

Hamada, D., I. Yanagihara, K. Tsumoto, Engineering amyloidogenicity towards the development of nanofibrillar materials; *Trends Biotechnol*. 2004; 22, 93–97.

Harris, L. D., B. S. Kim, D. J. Mooney, Open pore biodegradable matrices formed with gas foaming; *J. Biomed. Mater. Res*, 1998; 42:396–402.

Hesaraki, S., F. Moztarzadeh, D. Sharifi, Formation of interconnected macropores in apatitic calcium phosphate bone cement with the use of an effervescent additive; *J. Biomed. Mater. Res. Part A* 2007; 83(1) 80–87.

Hopley, E. L., S. Salmasi, D. M. Kalaskar, A. M. Seifalian, Carbon nanotubes leading the way forward in new generation 3D tissue engineering; *Biotechnol. Adv*. 2014; 32(5), 1000–1014, doi:10. 1016/j.biotechadv.2014.05.003.

Hosseinkhani, H., M. Hosseinkhani, A. Khademhosseini, H. Kobayashi, Y. Tabata, Enhanced angiogenesis through controlled release of basic fibroblast growth factor from peptide amphiphile for tissue regeneration; *Biomaterials*. 2006; 27:5836–44.

Hutmacher, D. W., Scaffolds in tissue engineering bone and cartilage; *Biomaterials*. 2000; 21(24): 2529–43.

Ishihara, K., W. Chen, Y. Liu, Y. Tsukamoto, Y. Inoue, Cytocompatible and multifunctional polymeric nanoparticles for transportation of bioactive molecules into and within cells; *Science and Technology of Advanced Materials*. 2016; 17(1):300–312.

Iwakura, A., Y. Tabata, N. Tamura, K. Nishimura, T. Nakamura et al., Gelatin sheet incorporating basic fibroblast growth factor enhances healing of devascularized sternum in diabetic rats; *Circulation*. 2001; 104:325–9.

Jansen, J. H., M. Eijken, H. Jahr, H. Chiba, J. A. Verhaar, J. A., van Leeuwen, H. Weinans, Stretch-induced inhibition of Wnt/beta-catenin signaling in mineralizing osteoblasts; *J Orthop Res*. 2010; 28:390–396.

Jiang, L., Z. Sun, X. Chen, J. Li, Y. Xu, J. Zu, J. Hu, D. Han, C. Yang, Cells sensing mechanical cues: Stiffness influences the lifetime of cell–extracellular matrix interactions by affecting the loading rate; *ACS Nano*. 2016; 10(1):207–217, doi: 10.1021/acsnano.5b03157

Kanczler, J. M., J. Barry, P. Ginty, S. M. Howdle, K. M. Shakesheff, R. O. Oreffo, Supercritical carbon dioxide generated vascular endothelial growth factor encapsulated poly(DL-lactic acid) scaffolds induce angiogenesis *in vitro*; *Biochem Biophys Res Commun*. 2007; 352:135–141.

Kilian, K. A., B. Bugarija, B. T. Lahn, M. Mrksich, Geometric cues for directing the differentiation of mesenchymal stem cells; *Proc Natl Acad Sci U S A*. 2010; 107:4872–4877.

Kim, E. J., C. A. Boehm, A. Mata, A. J. Fleischman, G. F. Muschler, S. Roy, Post microtextures accelerate cell proliferation and osteogenesis; *Acta Biomaterialia*. 2010; 6:160–169, ISSN 1742-7061, https://doi.org/10.1016/j.actbio.2009.06.016.

Kim, K., J. P. Fisher, Nanoparticle technology in bone tissue engineering; *J Drug Target*. 2007; 15(4):241–2252.

Kloxin, M., J. A. Benton, K. S. Anseth, *In situ* elasticity modulation with dynamic substrates to direct cell phenotype; *Biomaterials*. 2010; 31:1–8.

Komakula, S. S. B., S. Raut, N. P. Verma, T. A. Raj, M. J. Kumar, A. Sinha, S. Singh, Assessment of injectable and cohesive nanohydroxyapatite composites for biological functions; *Prog Biomater*. 2015; 4:31, doi: 10.1007/s40204-014-0034-7

Ku, S. H., C. B. Park, Myoblast differentiation on graphene oxide; *Biomaterials*. 2013; 34:2017–2023.

Landi, E., A. Tampieri, G. Celotti, R. Langenati, M. Sandri, S. Sprio, Nucleation of bio-mimetic apatite in synthetic body fluids, dense and porous scaffold development; *Biomaterials*. 2004; 08.010, doi: 10.1016/j.biomaterials.2004.08.010

Li, N., X. Zhang, Q. Song, R. Su, Q. Zhang et al. The promotion of neurite sprouting and outgrowth of mouse hippocampal cells in culture by graphene substrates; *Biomaterials*. 2011; 32:9374–9382.

Li, Y., H. Meng, Y. Liu, B. P. Lee, Fibrin gel as an injectable biodegradable scaffold and cell carrier for tissue engineering; *The Scientific World Journal*. 2015; Article ID 685690, 10 pages. doi: 10.1155/2015/685690

Lien, S. M., L. Y. Ko, T. J. Huang, Effect of pore size on ECM secretion and cell growth in gelatin scaffold for articular cartilage tissue engineering; *Acta Biomater*. 2009; 5:670.

Limongi, T., E. Miele, V. Shalabaeva, R. La Rocca, R. Schipani et al., Development, characterization and cell cultural response of 3d biocompatible micro-patterned poly-e-caprolactone scaffolds designed and fabricated integrating lithography and micromolding fabrication techniques; *J. Tissue Sci. Eng*. 2015a; 6(1):5. doi: 10.4172/2157-7552.1000145

Limongi, T., R. Schipani, A. Di Vito, A. Giugni, M. Francardi et al., Photolithography and micromolding techniques for the realization of 3D polycaprolactone scaffolds for tissue engineering applications; *Microelectron. Eng*. 2015b; 141:135–139, doi: 10.1016/j.mee.2015.02.030

Łukasiewicz, S., K. Szczepanowicz, E. Błasiak, M. Dziedzicka-Wasylewska, Biocompatible polymeric nanoparticles as promising candidates for drug delivery; *Langmuir*. 2015; 31(23):6415–6425.

Lynch, I., A. Salvati, K. A. Dawson, Protein-nanoparticle interactions, what does the cell see; *Nat. Nanotechnology.* 2009; 4:546.

Ma, T., Y. Li, S. T. Yang, D. A. Kniss, Effects of pore size in 3-D fibrous matrix on human trophoblast tissue development; *Biotechnol Bioeng.* 2000; 70:606.

Madden, L. R., D. J. Mortisen, E. M. Sussman, S. K. Dupras, J. A. Fugate, J. L. Cuy, K. D. Hauch, M. A. Laflamme, C. E. Murry, B. D. Ratner, Proangiogenic scaffolds as functional templates for cardiac tissue engineering; *Proc Natl Acad Sci U S A.* 2010; 107:15211.

Mahmoudi, M., S. E. Lohse, C. J. Murphy, A. Fathizadeh, A. Montazeri, K. S. Suslick, Variation of protein corona composition of gold nanoparticles following plasmonic heating; *Nano Lett.* 2014; 14:6.

Mahmoudi, M., I. Lynch, M. R. Ejtehadi, M. P. Monopoli, F. B. Bombelli, S. Laurent, Protein-nanoparticle interactions: Opportunities and challenges; *Chem Rev.* 2011; 111:5610–5637.

Mandal, A., S. Sekar, N. Chandrasekaran, A. Mukherjee, T. P. Sastry, Synthesis characterization and evaluation of collagen scaffolds crosslinked with aminosilane functionalized silver nanoparticles *in vitro* and *in vivo* studies; *J. Mater. Chem. B*, 2015; 3:3032–3043.

Mandal, B., S. Kundu, Cell proliferation and migration in silk fibroin 3D scaffolds; *Biomaterials.* 2009; 30:2956.

Mathieu, P. S., E. G. Loboa, Cytoskeletal and focal adhesion influences on mesenchymal stem cell shape, mechanical properties, and differentiation down osteogenic, adipogenic, and chondrogenic pathways; *Tissue Eng B Rev.* 2012; 18:436–444. doi: 10.1089/ten.teb.2012.0014

Meshel, A. S., Q. Wei, R. S. Adelstein, M. P. Sheetz, Basic mechanism of three-dimensional collagen fibre transport by fibroblasts; *Nat Cell Biol.* 2005; 7(2):157–164.

Mikos, A. G., M. D. Lyman, L. E. Freed, R. Langer, Wetting of poly (L-lactic acid) and poly(DL-lactic-co-glycolic acid) foams for tissue culture; *Biomaterials.* 1994; 15:55–68.

Mohtaram, N. K., J. Ko, C. King, L. Sun, N. Muller, M. B. Jun, S. M. Willerth, Electrospun biomaterial scaffolds with varied topographies for neuronal differentiation of human-induced pluripotent stem cells; *J Biomed Mater Res Part A.* 2015; 103A:2591–2601.

Monopoli, M. P., F. B. Bombelli, K. A. Dawson, Rapid formation of plasma protein corona critically affects nanoparticle pathophysiology; *Nat. Nanotechnol.* 2011; 6:11.

Monopoli, M. P., C. Åberg, A. Salvati, K. A. Dawson, Biomolecular coronas provide the biological identity of nanosized materials; *Nat. Nanotechnol.* 2012; 7:779.

Morimune-Moriya, S., S. Kondo, A. Sugawara-Narutaki, T. Nishimura, T. Kato, C. Ohtsuki, Hydroxyapatite formation on oxidized cellulose nanofibers in a solution mimicking body fluid; *Polymer Journal.* 47, 158–163 (February 2015), doi:10.1038/pj.2014.127

Mota, C., S. Danti, D. D'Alessandro, L. Trombi, C. Ricci, D. Puppi, D. Dinucci et al., Multiscale fabrication of biomimetic scaffolds for tympanic membrane tissue engineering; *Biofabrication* 2015; 7(2):025005. doi: 10.1088/1758-5090/7/2/025005.

Nam, Y. S., J. J. Yoon, T. G. Park, A novel fabrication method of macroporous biodegradable polymer scaffolds using gas foaming salt as a porogen additive; *Journal of Biomedical Materials Research.* 2000; 53:1–7.

Neff, J. A., K. D. Caldwell, P. A. Tresco, A novel method for surface modification to promote cell attachment to hydrophobic substrates; *J. Biomed. Mater. Res.* 1998; 40:511–519.

Norman, J. J., T. A. Desai, Methods for fabrication of nanoscale topography for tissue engineering scaffolds; *Annals of Biomedical Engineering.* 2006; 34:89–101.

Oeztuerk-Winder, F., J.-J. Ventura, The many faces of p38 mitogen-activated protein kinase in progenitor/stem cell differentiation; *Biochem. J.* 2012; 445:1–10, doi:10.1042/BJ20120401

Oliviero, O., M. Ventre, P. A. Netti, Functional porous hydrogels to study angiogenesis under the effect of controlled release of vascular endothelial growth factor; *Acta Biomater.* 2012; 8:3294.

Orgel, J. P., T. C. Irving, A. Miller, T. J. Wess, Microfibrillar structure of type I collagen in situ; *Proc Natl Acad Sci U S A.* 2006; 103(24):9001–9005.

Park, C. W., Y. S. Rhee, S. H. Park, S. D. Danh, S. H. Ahn et al. *In vitro/in vivo* evaluation of NCDS-microfabricated biodegradable implant; *Arch. Pharm. Res.* 2010a; 33(3):427–432, doi:10.1007/s12272-010-0312-4

Park, J., S. Park, S. Ryu, S. H. Bhang, J. Kim et al. Graphene-regulated cardiomyogenic differentiation process of mesenchymal stem cells by enhancing the expression of extracellular matrix proteins and cell signaling molecules; *Adv. Healthc. Mater.* 2014; 3:176–181.

Park, K., H. J. Jung, J.-J. Kim, D. K. Han, Effect of surface-activated PLLA scaffold on apatite formation in simulated body fluid; *Journal of Bioactive and Compatible Polymers February 3*, 2010b; 25:27–39, doi:10.1177/0883911509353677

Park, S. Y., J. Park, S. H. Sim, M. G. Sung et al., Enhanced differentiation of human neural stem cells into neurons on grapheme; *Adv. Mater.* 2011; 23:H263–267.

Provenzano, P., D. R. Inman, K. W. Eliceiri, P. J. Keely, Matrix density-induced mechanoregulation of breast cell phenotype, signaling and gene expression through a FAK–ERK linkage; *Oncogene*. 2009; 28:4326–4343.

Ramakrishna S., An introduction to electrospinning and nanofibers; Singapore; World Scientific Pub Co Inc 2005.

Raspa, A., A. Marchini, R. Pugliese, M. Mauri et al., A biocompatibility study of new nanofibrous scaffolds for nervous system regeneration; *Nanoscale* 2016; 8:253–265.

Ratner, B. D., A history of biomaterials; in B. D. Ratner, A. S. Hoffman, F. J. Schoen, and J. E. Lemons (Eds.), *Biomaterials Science: An Introduction to Materials in Medicine*, 2013, pp. xli–liii. Amsterdam: Elsevier/Academic Press. https://doi.org/10.1016/B978-0-08-087780-8.00154-6

Reddy, S., S. Wasnik, A. Guha, J. M. Kumar, A. Sinha, S. Singh, Evaluation of nano-biphasic calcium phosphate ceramics for bone tissue engineering applications: *In vitro* and preliminary *in vivo* studies; *J Biomater Appl*. 2013; 27:565–575.

Rowlands, S., P. A. George, J. J. Cooper-White, Directing osteogenic and myogenic differentiation of MSCs: Interplay of stiffness and adhesive ligand presentation; *American Journal of Physiology*. 2008; 295:C.1037–C1044.

Rubina, M. S., E. E. Kamitov, Y. V. Zubavichus, G. S. Peters, A. V. Naumkin, S. Suzer, A. Y. Vasil'kov, Collagen-chitosan scaffold modified with Au and Ag nanoparticles: Synthesis and structure; *Applied Surface Science*. 2016; 366:365–371.

Sánchez-Salcedo, S., F. Balas, I. Izquierdo-Barba, M. Vallet-Regí, *In vitro* structural changes in porous HA/β-TCP scaffolds under simulated body fluid; *Acta Biomaterialia*. 2009; 5(7):2738–2751.

Seerpooshan V. et al., Protein corona influences cell–biomaterial interactions in nanostructured tissue engineering scaffolds; *Adv. Funct Mater*. 2015; 25(28):4379–4389.

Shah, R. N., N. A. Shah, M. M. D. R. Lim, C. Hsieh, G. Nuber, S. I. Stupp, Supramolecular design of self-assembling nanofibers for cartilage regeneration; *Proceedings of the National Academy of Sciences* 2010; 107:3293–3298.

Shi, D., X. Xu, Y. Ye, K. Song, Y. Cheng et al. Photo-cross-linked scaffold with kartogeninen capsulated nanoparticles for cartilage regeneration; *ACS Nano*. 2016; 10(1):1292–1299, doi: 10.1021/acsnano.5b06663

Shih, Y. R. V., K. F. Tseng, H. Y. Lai, C. H. Lin, C. O. K. Lee, Matrix stiffness regulation of integrin-mediated mechanotransduction during osteogenic differentiation of human mesenchymal stem cells; *J Bone Miner Res*. 2011; 26:730–738.

Sinha, A., S. Nayar, A. Agarwal, D. P. Bhattacharya, J. Ramachandrarao, Synthesis of nanosized and microporous precipitated hydroxyapatite in synthetic and biopolymers; *J Am Ceram Soc*. 2003; 86:357–362.

Smith, I., X. Liu, L. M. A. P. Smith, Nanostructured polymer scaffolds for tissue engineering and regenerative medicine; *Wiley Interdisciplinary Reviews: Nanomedicine and Nanobiotechnology*. 2009; 1:226–236.

Sohrabi, M., S. Hesaraki, A. Kazemzadeh, M. Alizadeh, Development of injectable biocomposites from hyaluronic acid and bioactive glass nano-particles obtained from different sol-gel routes; *Mater. Sci. & Eng.C*, 2013; 33:3730–3744.

Stevens, M., J. H. George, Exploring and engineering the cell surface interface; *Science*. 2005; 310:1135–1138.

Sudha, B., S. Jasty, S. Krishnan, S. Krishnakumar, Signal transduction pathway involved in the *ex vivo* expansion of limbal epithelial cells cultured on various substrates; *Indian J Med Res*. 2009; 129:382–389.

Szymanski, J. M., Q. Jallerat, A. W. Feinberg ECM protein nanofibers and nanostructures engineered using surface-initiated assembly; *J Vis Exp*. 2014; 86:51176, doi: 10.3791/51176.

Tenzer, S., D. Docter, J. Kuharev, A. Musyanovych, V. Fetz et al., Rapid formation of plasma protein corona critically affects nanoparticle pathophysiology; *Nat. Nanotechnol*. 2013; 8:772.

Thakar, R. G., Q. Cheng, S. Patel et al., Cell-shape regulation of smooth muscle cell proliferation; *Biophysical Journal*, 2009; 96:3423–3432.

Vasita, R., D. S. Katti, Nanofibers and their applications in tissue engineering; *Int J Nanomedicine*. 2006; 1:15–30.

Walczyk, D., F. B. Bombelli, M. P. Monopoli, I. Lynch, K. A. Dawson, What the cell sees in bionanoscience; *J. Am. Chem. Soc*. 2010; 132:5761.

Wang, Y., W. C. Lee, K. K. Manga, P. K. Ang et al., Fluorinated graphene for promoting neuro-induction of stem cells; *Adv. Mater*. 2012; 24:4285–4290.

Williams, D. F., On the mechanisms of biocompatibility; *Biomaterials*. 2008; 29:2941–2953.

Wipff, P J., D. B. Rifkin, J. J. Meister et al., Myofibroblast contraction activates latent TGF-beta1 from the extracellular matrix; *J Cell Biol*. 2007; 179:1311–1323.

Woo, K.M., V.J. Chen, P.X. Ma, Nano-fibrous scaffolding architecture selectively enhances protein adsorption contributing to cell attachment; *J Biomed Mater Res Part A*. 2003; 67A:531.

Ying, Y., A. Haiyong, W. Yugang et al., Cytocompatibility with osteogenic cells and enhanced *in vivo* anti-infection potential of quaternized chitosan-loaded titania nanotubes; *Bone Research*. 4:16027 doi:10.1038/boneres.2016.27

Yoon, H. H., S. H. Bhang, T. Kim, T. Yu, T. Hyeon, B. S. Kim, Dual roles of graphene oxide in chondrogenic differentiation of adult stem cells: Cell adhesion substrate and growth factor-delivery carrier; *Adv. Funct. Mater*. 2014; 24:6455–6464.

Zafar, M., S. Najeeb, Z. Khurshid, M. Vazirzadeh, S. Zohaib, B. Najeeb, F. Sefat, Potential of electrospun nanofibers for biomedical and dental applications; *Materials*. 2016; 9:73.

Zeugolis, D. I., S. T. Khew, E. S. Yew, A. K. Ekaputra, Y. W. Tong, L. Y. Yung, D. W. Hutmacher, C. Sheppard, M. Raghunath, Electro-spinning of pure collagen nanofibres—Just an expensive way to make gelatin; *Biomaterials*. 2008; 29:2293–22305.

Zhang, H., C. Y. Lin, S. J. Hollister, The interaction between bone marrow stromal cells and RGD-modified three-dimensional porous polycaprolactone scaffolds; *Biomaterials*. 2009a; 30:4063.

Zhang, L., J. Rodriguez, J. Raez, A. J. Myles, H. Fenniri, T. J. Webster, Biologically inspired rosette nanotubes and nanocrystalline hydroxyapatite hydrogel nanocomposites as improved bone substitutes; *Nanotechnology*. 2009b; 20:17.

Zhang, T., N. Li, K. Y. Li, R. F. Gao, W. Gu et al., Enhanced proliferation and osteogenic differentiation of human mesenchymal stem cells on biomineralized three-dimensional graphene foams; *Carbon*. 2016; 105:233–243.

Zisch, A. H., U. Schenk, J. C. Schense, S. E. Sakiyama-Elbert, J. A. Hubbell, Covalently conjugated VEGF–fibrin matrices for endothelialization; *J. Control Release*. 2001; 72:101–113.

9

Nanoparticles and the Aquatic Environment: Application, Impact and Fate

Violet A. Senapati and Ashutosh Kumar

CONTENTS

9.1 Introduction	299
9.2 Environmental Applications of Engineered Nanoparticles	301
9.2.1 Degradation of Organic Pollutants	302
9.2.2 Engineered Nanoparticles as Adsorbents	303
9.2.3 Treatment and Remediation	303
9.2.4 Abatement of Pollution	304
9.2.5 Nanosensors	304
9.2.6 Construction	304
9.2.7 Nanomembranes	304
9.2.8 Catalysis	305
9.3 Ecotoxicological Impacts of Engineered Nanoparticles	305
9.4 Fate and Accumulation of Engineered Nanoparticles	307
9.4.1 Water	307
9.4.2 Air	309
9.4.3 Soil	309
9.5 Safety of Nanotechnology	309
9.5.1 Risk Assessment	314
9.5.2 Risk Management	314
9.6 Conclusion	315
Acknowledgments	315
References	316

9.1 Introduction

Nanotechnology is a field of applied science focused on the design, synthesis, characterization, and application of materials and devices on the nanoscale. It is a "converging technology," which amalgamates various scientific disciplines, such as physics, chemistry, information technology, medicine, and biology, to provide new and innovative solutions. It is also referred to as "enabling technology," since it opens new avenues in various disciplines of science and technology. Nanotechnology is considered the next logical step in science. This is due to the fact that size reduction leads to increased surface area, imparting new optical, magnetic, and quantum properties to the material. These properties cannot be explained with the conventional assays used to understand biological effects.

Nanoscience is the study of phenomena and manipulation of materials at the atomic, molecular, and macromolecular scales, where the properties differ significantly from those at a larger scale. According to the US National Nanotechnology Initiative, nanotechnology is "the understanding and control of matter at dimensions between approximately 1 and 100 nanometers, where unique phenomena enable novel applications" (NNI 2004), while the US Environmental Protection Agency (USEPA) defines nanotechnology as "the creation and use of structures, devices, and systems that have novel properties and functions because of their small size" (USEPA 2007). Nanomaterials with one dimension consists of thin films and surface coatings, which are used in the electronics industry. Nanomaterials with two dimensions are composed of nanofibers and nanowires. Nanoscale materials with three dimensions are known as nanoparticles (Depledge et al. 2010). Nanoparticles have been used since ancient times, when carbon black was used in cave paintings and nanoscale gold and silver flakes were used to color transparent glass red. Currently, nanoparticles have found wide applications in medicine, electronics, cosmetics, and food packaging, among others (Patel et al. 2014).

Natural nanoparticles came into existence when life began on earth. The most common natural nanoparticles are silicate clay minerals, humic organic matter, natural ammonium sulfate particles, fine sand, dust, and black carbon. Also, the evaporation of sea water spray generates nanocrystals of sea salts (Hartland et al. 2013). However, incidental nanoparticles are formed as a result of anthropogenic activity like grinding of primary or secondary minerals or through the wear of metal and mineral surfaces, combustion, and emission from the smoke of volcanoes or fires (Buzea et al. 2007, Sharma et al. 2012a,b,c). Engineered nanoparticles are human-made particles having specific properties (Sharma et al. 2012a,b,c). These are made up of either single elements like silver or carbon or a mixture of elements/molecules. ENPs include metals; metal oxides and alloys; carbon-based materials, for example, titanium dioxide and zinc oxide; fullerenes; nanotubes; nanofibers; silicates; dendrimers; and quantum dots (Aitken et al. 2006, Colvin et al. 2003, Lin et al. 2011).

ENPs are of great interest for wide applications due to their unique physicochemical properties, that is, increased surface area per unit mass, changes in surface reactivity and charge, color, solubility, conductivity, catalytic activity, and modified electronic characteristics compared to bulk material (Ebbesen et al. 1996, Pal et al. 1997, Powell et al. 1988, Reed 1993). The heightened production and use of ENPs in consumer products has led to increased release of ENPs into aquatic, terrestrial, and atmospheric biomes, which is a growing concern for the safety and health of available life forms (Kessler 2011, Nazarenko et al. 2011, Pirela et al. 2015). With the increase in nano consumer products, consumers are more likely to be exposed to ENPs. The unavailability of analytical techniques to detect and quantify nanomaterials in soil, water, and air has led to the increase in harmful effects of ENPs (Phillips et al. 2010). It is predicted that the rate of production of metal oxide ENPs incorporated into cosmetics is 103 tons/year (Clausen et al. 2010).

Zinc oxide ENPs have a production rate of 528 tons/year (Gumus et al. 2014, Zhang and Saebfar 2010), and fullerenes have an annual production of 1500 million tons (Kumar et al. 2012). The Project on Engineering Nanotechnologies (PEN) stated that 212 nanotechnology-based products were present in March 2006 (Peralta-Videa et al. 2011) and the number is predicted to be ~3400 by 2020 (PEN 2017).

Currently, ENPs are being incorporated into commercial products at a faster rate than the development of knowledge and regulations to mitigate potential health and environmental impacts associated with their manufacturing, application, and disposal. Governing bodies and scientific communities are concerned about the inadvertent release of ENPs into the environment. A European Commission—Nanoforum joint workshop was organized to determine the potential impacts of nanotechnology on the environment (Morrison 2006). The United States Environmental Protection Agency has considered the boon of nanotechnology along with the need to study the fate, transport, and health effects of nanomaterials in the environment (Sharma et al. 2012a,b,c). Environmentalists have also documented guidance for the safe handling of nanomaterials (Dhawan et al. 2011). The EPA has led the mandate to protect health and the environment from the risks of ENP-induced toxicity (USEPA 2007). The European Commission's Scientific Committee on Consumer Products (SCCP) issued a document titled "Opinion on Safety of Nanomaterials in Cosmetic Products" and raised a concern to validate the data on toxicity with reference to nanomaterial used in cosmetics for its safety evaluation and emphasized considering the implications on animal testing (SCCP 2007). The Scientific Committee on Emerging and Newly Identified Health Risks (SCENIHR) brought to light the need for proper ENP characterization (SCENIHR 2006).

The behavior of ENPs in air is mostly governed by diffusion, agglomeration, and potential resuspension of aerosol from deposited nanomaterials. It is also reported that in traditional aerosol science, particle size, inertia, and gravitational and diffusion forces govern aerosol behavior in the environment. As the particle size decreases, diffusional forces become increasingly important and nanoscale particles are thus likely to behave in a manner more alike to a gas or vapor. The particle diffusion coefficient is inversely proportional to the particle diameter. Particles with a high diffusion coefficient such as ENPs have high mobility and mix rapidly in an aerosol. After their release into the environment, atmospheric diffusion facilitates the ENPs migrating rapidly from a higher to a lower concentration, thus resulting in rapid dispersion and potential for particles to travel a great distance from the source. Humans get exposed to ENPs at various steps of its synthesis (laboratory), manufacture (industry), use (consumer products, devices, medicines, etc.), and the environment (through disposal).

ENPs are also reported to be cytotoxic, genotoxic, and immunotoxic (Senapati et al. 2015a,b, Sharma et al. 2012a,b, Shukla et al. 2013) in *in-vivo* and *in-vitro* scenarios. The behavior and bioavailability of ENPs in freshwater/marine ecosystems depends on their interaction with the aquatic colloids, such as natural organic matters (NOMs), humic substances, and salt ions. NOMs usually get adsorbed on the surface of the ENPs by different electrostatic, hydrogen bonding, and hydrophobic interactions, which affects their dispersity and bioavailability. ENPs are weakly water soluble, increasing their persistence in the aquatic environment, which further leads to their bioaccumulation, bioconcentration, and biomagnification in aquatic systems (Adams et al. 2006, Ahn et al. 2014, Bai et al. 2010, Batley et al. 2013, Maenosono et al. 2009). The physicochemical characteristics like size, surface area, chemical composition, crystalline structure, surface charge, and aggregation of ENPs determine their interactions with the environment (Figure 9.1). These characteristics impart novel properties to ENPs which in turn are responsible for their induced toxicity (Kumar et al. 2011a,b,c).

Hence, in the present article, we have reviewed the applications of ENP in environmental scenarios and an attempt has been made to understand the different physicochemical characteristics of ENPs for determining their toxicity. In addition, the data on ecotoxicity of ENPs have also been gathered, and approaches to prevent imposed toxicity to human and health are addressed.

9.2 Environmental Applications of Engineered Nanoparticles

ENPs have tremendous potential in improving the quality of life due to their high surface-to-volume ratio and quantum effects. ENPs are directly or indirectly used in air filters, tailpipes of automobiles, chimneys of factories to remove air contaminants before the air

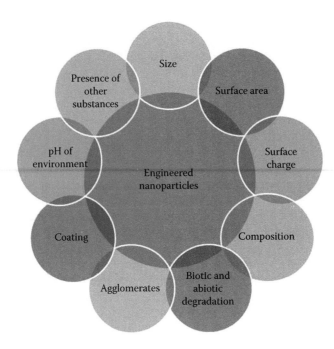

FIGURE 9.1
Parameters determining the fate of ENPs in the environment.

enters the atmosphere (Mansoori et al. 2008), and to detect the pollutants present in the environment and industrial products and processes. Understanding the fundamental properties of ENPs has led to their applications in environmental technology.

9.2.1 Degradation of Organic Pollutants

ENPs have been demonstrated to remove or degrade environmental pollutants (He et al. 2011, Sinha and Ahmaruzzaman 2015). Copper nanoparticles have been used for the degradation of methylene blue dye from aqueous solutions due to their photocatalytic activity (Sinha and Ahmaruzzaman 2015). Additionally, the doping of copper and silver on ZnO nanostructures is also expected to modify absorption, resulting in higher photocatalytic activity and special characteristics compared with single-element doping into semiconductor oxides. The property of suppression of electron/hole pairs and acceleration of surface charge transfer by Ag ENPs has been harnessed in Ag/TiO_2-nanotube-composed nanoelectrodes for photocatalytic degradation of pentachlorophenol and methylene blue dye (He et al. 2011). It was observed that as the photocatalytic reaction proceeds, there is a peak shift from around 668 nm to around 656 nm.

The photocatalytic property of ZnO ENPs also helps degrade the harmful chemicals of pesticides into harmless molecules (Ion et al. 2011). ZnO crystals have a band gap energy of 3.3 eV at wavelengths under 375 nm. Electrons are released from the valence band, which get transferred to the conduction band to form photo-excited electrons and holes. Hence, it results in formation of highly active free radicals such as superoxide and hydroxyl radicals, which decompose organic pollutants (He et al. 2013). Yu et al. (2007) determined the potential of nano-TiO_2 film in photocatalytic degradation of organochlorine pesticides and organic pollutants (Artale et al. 2001) like atrazine (Ion et al. 2011). Rhenium-doped nano TiO_2 has been effectively used in photocatalytical degradation of organophosphorus and carbamate pesticides in tomato leaves and soil (Zeng et al. 2010). Toxic gases like toluene are photocatalytically converted by TiO_2 ENPs into less harmful compounds (Rickerby and Morrison 2007, Sánchez et al. 2011). Also, FeS ENPs are used in the degradation of organic contaminants like lindane from water (Paknikar et al. 2005).

9.2.2 Engineered Nanoparticles as Adsorbents

ENPs hold greater surface area and functional groups on their surface, which enhances their adsorption capacity. Magnetic ENPs hold promise for efficient adsorbents due to their large surface area, small diffusion resistance, and superparamagnetic properties. Magnetite ENPs are more effective in adsorption of As(V) at lower pH due to a positive charge at pH values below 6.8. The adsorption mechanism is explained by the Langmuir adsorption isotherm. The kinetic model of the organic degradation rate (Wang et al. 2013) can be described as follows:

$$dc/dt = K_a K_r C / 1 + K_a C$$

where dc/dt = rate of degradation, K_a is the contribution of organic adsorption, and K_r designates the contribution of organic degradation.

Gold ENPs supported on alumina were also used as adsorbents to remove inorganic mercury from drinking water (Lisha et al. 2009). Cr(VI) has effectively been removed through adsorption onto the surface of cerium-oxide ENPs (Recillas et al. 2010).

9.2.3 Treatment and Remediation

Environmental contamination of chlorinated phenols occurs from industrial effluents through their manufacture, agricultural runoff, incineration, and spontaneous formation during chlorination of water for disinfection. The chlorinated phenol hence finds its way into the soil biosphere, further contaminating the drinking water. The photocatalytic properties of large band gap semiconductors like ZnO and TiO_2 ENPs are useful in the remediation of organic contaminants from air, water, and soil (Masciangioli and Zhang 2003). ZnO ENPs are applied on semiconductor films to detect chlorinated phenols by the quenching of visible emission (Kamat et al. 2002). Also, carbon ENPs have been shown to have a high affinity for Cu(II) and Cd(II) present in groundwater due to their large specific surface areas, hollow, and layered structures, and hence are used in water treatment plants.

Environmental nanotechnology also offers the potential to effectively treat contaminants *in situ*, thus avoiding the pumping of contaminated water from the ground, and it is highly cost effective. *In situ* water remediation is enhanced due to the small size and high surface area per unit mass, which increases the likelihood of contact with the target contaminant underground. Nanoscale zero-valent iron is most widely used, as it degrades the chlorinated solvents by beta elimination and reductive chlorination pathways due to its small size and random motion (Singh and Kumar 2016). Ponder et al. reported the immobilization of Cr(VI) and Pb(II) from an aqueous solution by reducing the Cr(VI) to Cr(III) and the Pb(II) to Pb(0) with simultaneous oxidation of Fe. The surface area of zerovalent iron is a crucial parameter for the rate of remediation of analytes (Ponder et al. 2000). Amphiphilic polyurethane ENPs are also used in soil remediation, as they enhance the desorption of polynuclear aromatic hydrocarbons (Tungittiplakorn et al. 2004). Pan et al. (2010) put forth that nanomagnetite ENPs immobilize phosphate in soil by adsorption and easily penetrate in the soil column.

9.2.4 Abatement of Pollution

TiO_2 nanotube arrays are developed for generating hydrogen by splitting water using sunlight as the source of energy (Wiesner and Bottero 2007). Generating hydrogen from water at a larger scale has great potential as a clean energy resource. TiO_2 ENPs are also employed in organic solar cells to convert light into energy (Rickerby and Morrison 2007). Due to antireflection, surface passivation properties, and their ability to trap light, silicon ENPs have been used for solar cell production (Rickerby and Morrison 2007). Nanoscale calcium peroxide has recently been used for the cleaning of oil spills (Karn et al. 2009). It acts as an oxidant for organic contaminants, such as gasoline, heating oil, methyl tertiary butyl ether (MTBE), ethylene glycol, and solvents.

Pollution is also controlled by using less toxic and renewable reagents. Graphene layers are used in hydrogen-fueled cars to enhance the binding energy of hydrogen to the graphene surface. This property imparts increased hydrogen storage and also provides a lighter-weight fuel tank (Wang et al. 2009), and therefore is also used in airplanes. It also reduces the weight of wind turbine blades, hence increasing the production of renewable energy sources (Zhang et al. 2011a,b).

9.2.5 Nanosensors

The development of new and sensitive sensors is the most emerging application of nanoparticles due to their unique physicochemical properties. Nanoparticles such as iron are used for the detection and removal of contaminants from soil and ground water (Mansoori et al. 2008). Immobilization of enzymes on the surface of nanoparticles has also been used to detect pollutants such as organochlorines, organophosphates, and carbamates efficiently by using nano-TiO_2/Ti (Khot et al. 2012) and acetylcholinesterase and choline oxidase immobilized on Au-Pt bimetallic ENPs (Sassolas et al. 2012). The fluorescence properties of quantum dots have also been used in the detection of pollutants in water. Mn-doped ZnS quantum dots have been used to detect pentachlorophenol with phosphorescence measurements carried at an excitation wavelength of 316 nm (Zhang et al. 2011a,b). The aggregation of ENPs leads to a color change or alteration of the refractive index, which has been exploited as an optical sensing method for the detection of heavy metals, toxins, and other pollutants in water and soil (Wang et al. 2013). Platinum ENP-modified glassy carbon electrodes have been used for oxidative determination of trace As(II) with a detection limit of 0.12 ppb (Zhang and Fang 2010).

9.2.6 Construction

TiO_2 ENPs possess semiconducting, photocatalytic, energy-converting, and gas-sensing properties. The photocatalytic property of TiO_2 has been explored in paints to coat surfaces. However, the hydrophobicity of TiO_2 ENPs is being exploited to coat water metal pipes to prevent corrosion (Arivalagan et al. 2011). SiO_2 ENPs are present as one of the ingredients of modern concrete, as they increase packing. They also prevent the degradation of calcium silicate hydrate and block the penetration of water, thus enhancing the building's durability (Arivalagan et al. 2011).

9.2.7 Nanomembranes

ENPs like zeolites are being embedded in filtration membranes, which makes them hydrophilic so that water passes through them more easily. Moreover, these particles have a diameter of around 200 nm, which is similar to the thickness of the membrane. Carbon nanotube-based membranes are used to desalinate seawater to provide drinking water due to exhausted fresh water sources. The salient feature of using nanomembranes is their capacity to remove divalent ions like calcium and magnesium that contribute to the hardness of water. Based on the transmembrane pressure difference, the ions get transported through the membranes.

9.2.8 Catalysis

Multiwall carbon nanotubes (MWCNTs) are finding applications in catalysis as the components of the electrodes of batteries (Wang et al. 2013). Nanocatalysts have high specific surface area and possess high catalysis activities. Also, high quantities of dyes like malachite green and methylene blue find their way into water bodies from industries during the discharge of unconsumed dyes. The contamination of dyes increases the chemical oxygen demand (COD) and biochemical oxygen demand (BOD). Heterogeneous photocatalysis is the method of choice for dye degradation due to its advantages over conventional technologies. TiO_2 nanotubes are used as photocatalysts because of their high catalytic efficiency, high chemical stability, and low cost and toxicity. They can also be recycled and reapplied in many photo-degradation cycles. The solid-base KF/CaO nanocatalyst is also used for biodiesel production with a yield of more than 96%. It is used to convert oil with higher acid values into biodiesel (Chaturvedi et al. 2012).

9.3 Ecotoxicological Impacts of Engineered Nanoparticles

Although ENPs possess beneficial characteristics, they are released into the environment along the life cycle of consumer products containing ENPs. This includes entry into the biosphere through accidental leakage or during the disposal of ENP-incorporated products and raises issues of threats to the environment (Figure 9.2). The soil and sediments in water bodies are the sink of ENPs; thus, organisms present in these environmental conditions are most likely to be affected by ENPs.

The surfaces of ENPs undergo constant modifications during their movement through the environment, and it is through their surfaces that they interact with environmental components. Thus, the toxicity of ENPs to human health and the environment depends not only on the initial physicochemical characteristics of ENPs but also the physicochemical changes/evolution of ENPs in environmental surroundings (Sánchez et al. 2011).

Plants are an important component of the ecosystem, and nanoparticles such as C^{70} fullerenes and multiwalled nanotubes have been reported to reduce rice yields (Lin et al. 2009); carbon nanotubes also diminish rice yields and make wheat more vulnerable to uptake of pollutants (Wild and Jones 2009) by piercing the cell wall of roots. The ENPs present in the environment can directly affect plants or find their way to crops and vegetables and accumulate, which in turn are consumed by humans and animals.

The toxicity of ENPs in the environment is estimated based on the comparison between the predicted environmental concentration (PECs) of the ENPs and the predicted no-effect concentrations (PNECs) derived from the literature. When PEC/PNEC <1, the substance is not of immediate concern and has to be reconsidered only when new information is

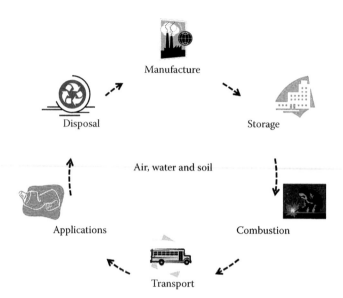

FIGURE 9.2
Sources of release of ENPs into the environment.

presented. When PEC/PNEC >1 and no additional data are available, the substance is of concern and the authorities in charge need to take measures immediately to reduce the risk. The ratio of PEC/PNEC for CNT and nano-Ag were less than one, so they were considered environmentally compatible (Mueller and Nowack 2008). For most nanoparticles, the data for PEC and PNEC are unavailable (Sánchez et al. 2011).

Living organisms can have ENP exposure through different routes, such as terrestrial and aerial organisms that breathe air; hence, ENPs enter through the respiratory tract and lodge in the lungs, causing respiratory and cardiovascular ailments. ENPs can also enter into the blood circulation from the pulmonary interstitial sites and can be translocated into the central nervous system (Genc et al. 2012) and to the secondary target organs like the liver and spleen.

In aquatic organisms, ENPs can enter through the skin, gills, or surface epithelia (Oberdörster et al. 2006). When adult medaka fish were exposed to 39.4 nm fluorescent ENPs, these particles were detected at high levels in the gills and intestine. It was also observed that the ENPs reached the liver, gallbladder, and kidney, probably through the circulatory system, further penetrating the blood–brain barrier (Kashiwada 2006). The exposure of copper ENPs in zebrafish showed that the gills were its primary target organ (Griffitt et al. 2007). In earthworms, ZnO ENPs were reported to cross the membranes at the gut/intestinal tract (Li et al. 2011). Silver ENPs were observed within the gut lumen, close to the microvilli, and in the extracellular matrix (Garcia-Alonso et al. 2011) in the polychaete.

Due to their reduced size and increased surface-to-volume ratio, ENPs interact directly with bacterial membranes (Ren et al. 2009). ENPs get internalized into the bacteria (Kumar et al. 2011a,b) and protozoa through their cell membranes, leading to biomagnifications of the ENPs. Brayner et al. (2006) showed that when ZnO ENPs were exposed to *Escherichia coli* cells, the membrane was disrupted. Membrane permeability was increased, leading to accumulation of ZnO ENPs in the bacteria.

9.4 Fate and Accumulation of Engineered Nanoparticles

It is thought prudent to understand the environmental behavior of ENPs to accurately assess the environmental risk. The direct use of ENPs in environmental remediation also accounts for their presence and accumulation in the environment. The environmental fate of ENPs is strongly influenced by their stability and tendency to form homo- or heteroaggregates. The stability of ENP colloids follows models of extended Derjaguin-Landau-Verwey-Overbee, steric, and electrosteric stabilization (Schaumann et al. 2014). ENPs, when they enter the environment, may be aggregated or embedded inside other materials, or can be dissolved and modified (Figure 9.3). It is reported that 600 μg/L of nano-TiO_2 runs off from paints on building facades (Kaegi et al. 2008), and 30% of nano-Ag is released from surface paints within a year of paint application (Kaegi et al. 2010). The characteristics of the environment (water, air, and soil) and the physicochemical properties of ENPs determine their fate (Lowry et al. 2010) (Table 9.1).

9.4.1 Water

Single-walled carbon nanotubes are reported to pass through a porous medium more rapidly and to a greater extent than colloidal n-C60; hence, the charge and type of ENP determines its fate in water. Additionally, if ENPs have a coating, they avoid aggregation and show increased mobility (Wiesner et al. 2006). Dunphy Guzman et al. (2006) reported that size is an important factor that controls ENP transport in water. ENPs have small size distribution and therefore settle slowly compared to bulk particles, but, due to an increase in surface area, they have high potential to sorb to soil (Oberdörster et al. 2005). When ENPs in aquatic systems get attached to naturally available organic matter, it results in the formation of stable colloidal suspensions (Velzeboer et al. 2014). The presence of humic acids decreases the bioavailability and bioaccumulation of TiO_2 NPs in fish (Lopez-Serrano Oliver et al. 2015).

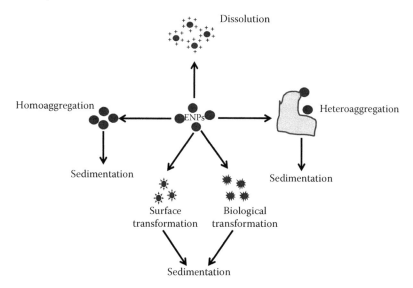

FIGURE 9.3
Schematic representation of the fate of ENPs in the aquatic environment.

TABLE 9.1

The Techniques Used for Measuring the Parameters Responsible for Fate and Transport of ENPs in the Environment

Parameter	Techniques	Significance
Size	Dynamic light scattering (DLS), Transmission electron microscopy (TEM), Scanning electron microscopy (SEM)	As the size decreases, the mobility of ENPs increases
Surface area	Brunauer-Emmet-Teller (BET)	As the size of ENPs decreases, the corresponding surface area increases, further leading to higher sorption and greater sedimentation
Surface charge	Phase analysis light scattering	Surface charge of the ENPs determines their solubility in aqueous medium
Composition	Fourier transform infrared spectroscopy (FTIR), Raman spectroscopy, Nuclear magnetic resonance (NMR), X-ray photo electron spectroscopy (XPS), Synchrotron	Different types of ENPs have different rates of mobility
Biotic and abiotic degradation and presence of other substances	Bright field microscope, TEM, Inductively coupled plasma mass spectrometry (ICPMS), Atomic absorption spectroscopy (AAS), UV-visible spectroscopy, fluorescence spectroscopy, SEM with energy dispersive X-ray analysis (EDAX)	The other chemicals/organisms present in the environment degrade and hydrolyze the ENPs
Agglomerates	TEM, SEM, epiphaniometer, DLS, UV-visible spectroscopy, dark and bright field microscopy	The agglomerated ENPs tend to get trapped in soil or sludge
Coating	FTIR, UV-visible spectroscopy, Raman spectroscopy	Uncoated ENPs are less stable and get agglomerated
pH of environment	pH meter	It affects the dissolution and aggregation of ENPs, thus affecting their mobility

The transformation of ENPs in water can be due to dissolution or physical and chemical changes. pH can also influence the sorption, agglomeration, and settling of ENPs (EPA 2007). Changes in the ionic strength strongly affect the stability of ENPs. In saline environments, ENPs have a tendency to aggregate and thus get settled, and may sorb to carbohydrate, lipid, and proteins present on the surface microlayers and be transported in aquatic environments over long distances (Nurmi et al. 2005). The presence of clay particles in the aquatic system also alters the distributions of ENPs (Gupta et al. 2016, 2017).

When substances enter the environment, they experience partitioning, which is the distribution of particles between the various phases of the system, predicted by the octanol water partition coefficient (K_{ow}). However, due to the interactions of ENPs with various components of water, ENPs are at near zero concentration in their original form; hence, K_{ow} values are not calculated. Though little predictive modeling works for TiO_2, carbon nanotubes, and silver ENPs (Boxall et al. 2007, Mueller and Nowack 2008), due to insufficient knowledge of the behavior of ENPs in waters, the concentrations of ENPs cannot be assessed. Therefore, there is a need to advance the studies in this area. It was recently observed that even solar light has the potential to induce physicochemical transformation of ENPs, further enhancing their toxicity (George et al. 2014) and therefore emphasizing the need for toxicity studies under environmentally relevant exposure conditions.

9.4.2 Air

The natural sources of NPs in the atmosphere are volcanic eruptions, physical and chemical weathering of rocks, precipitation reactions, forest fires, sea sprays, viruses, and so on. Human-made sources of NPs in the atmosphere are vehicle emissions or industrial combustion processes. As of now, it is reported that air contains 10–500 ENPs per cubic cm (Tardif 2007). ENPs released into the atmosphere may remain in the atmosphere for many years or react with other substances and find their way into the food chain. The aerosol behavior of ENPs is due to dominance of diffusional forces, along with inertial and gravitational forces. The rate of ENPs settling in air due to gravitational forces is proportional to particle weight, while the rate of diffusion is inversely proportional to the particle diameter (Aitken et al. 2004). Particles, on entering the atmosphere, undergo increased diffusion, which results in multiple collisions amounting to coagulation and agglomeration of ENPs. This is a principle mechanism in NP formation, which undergoes nucleation and growth. Nucleation is defined as formation of molecular embryos or clusters before the formation of a new phase during the transformation of vapor to liquid and to solid (Zhang et al. 2011a,b). Condensation occurs when gaseous ENPs are present in the air, and when two or more ENPs combine, aggregation or coagulation of ENPs is reported. First, nucleation-mode particles are formed in the atmosphere within a size range of 20–50 nm (Aiken 1984). The atmosphere contains sulfuric acid, nitric acid, and gases, of which sulfuric acid is a key player in nucleation. These nucleation-mode particles further grow by coagulation and vapor condensation to form Aiken mode and accumulation mode particles (Biswas and Wu 2005).

9.4.3 Soil

The mobility of ENPs in soil depends on the physicochemical characteristics of the ENPs, along with the properties of the soil (EPA 2007). The interactions of ENPs with natural clay colloids influence the fate of ENPs in soil. Cai et al. (2014) showed that the presence of clay particles in NaCl solutions increased the transport of TiO_2 ENPs. The coatings of organic substances also alter the mobility and dispersion of ENPs. The aggregation of ENPs depends on the charge and surface potential of the ENPs. As ENPs reach the soil, the negatively charged components of the soil, for example, humic and fulvic acids and clay particles, bind to the aggregated ENPs (Waalewijn-Kool et al. 2013). The high surface area of ENPs accounts for the immediate sorption and accumulation of the particles in the soil matrix. Due to their small size, ENPs fits in the spaces in the soil matrix and travel long distances. Studies have reported that zinc sorption increased with increasing pH, where the Freundlich kf values ranged from 98.9 (L/kg)1/n to 333 (L/kg)1/n for 30 nm ZnO (Waalewijn-Kool et al. 2013). Hence, the physical and chemical properties and surface coatings of ENPs determine the strength of sorption (EPA 2007).

9.5 Safety of Nanotechnology

The wealth of information gained on the toxicity of ENPs to the environment and health raises the concerns of regulatory and governing bodies about the detrimental effects of ENPs. Therefore, it is important to ensure that the potential health and environmental impacts of ENPs are well within control. A summary of the ecotoxicological effects of ENPs based on previously published data can be seen in Table 9.2.

TABLE 9.2
The ecotoxicity effects of ENPs based on previously published data

Types of ENPs	ENPs	Organism	Concentration	Effects	References
Metal Oxide ENPs	Zinc oxide	*Escherichia coli*	15.63–1000 mg/mL	Antibacterial activity was observed at 500 mg/mL	Premanathan et al. (2011)
			2–40 mg/L	The exposure of ZnO NPs via electrospray showed a higher antibacterial activity	Wu et al. (2010)
			10–5000 ppm	At 1000 ppm 48% growth reduction was obtained	Adams et al. (2006)
		Pseudomonas putida	0–500 mg/L	Inhibition of the growth of *P. putida* was observed with decreasing agglomerate size and increasing concentration	Vielkind et al. (2013)
		Vibrio fischeri	1.9 mg/L	The L(E)C50 value was 1.9 mg/L	Heinlaan et al. (2008)
		Bacillus subtilis	10–5000 ppm	90% growth reduction was observed at 10 ppm	Adams et al. (2006)
		Acartia tonsa	10–1000 mg/L	Copepod survival and reproduction was declined with increasing dietary ZnO ENPs exposed *T. weissflogii* over a 7 day period. Exposure	Jarvis et al. (2013)
		Daphnia magna	3.2 mg/L	The L(E)C50 value was 3.2 mg/L	Heinlaan et al. (2008)
		Thamnocephalus platyurus	0.18 mg/L	The L(E)C50 value was 0.18 mg/L	Heinlaan et al. (2008)
		Caenorhabditis elegans	0–2000 mg/L	Zinc oxide ENPs and $ZnCl_2$ exhibited lethality, change in behavior, decrease in reproduction and increase in transgene expression	Ma et al. (2009)
		Chlorella vulgaris	10–300 mg/L	Exhibited cytotoxicity, decrease in cell viability, and induced oxidative stress	Suman et al. (2015)
		Danio rerio	1–100 mg/L	Zinc oxide ENPs led to zebrafish embryonic mortality at 50 and 100 mg/L, retarded embryo hatching at 1–25 mg/L and reduced the body length of larvae	Bai et al. (2010)
		Thalassiosira pseudonana, Skeletonema marinoi, Dunaliella tertiolecta, Isochrysis galbana	0–1000 μg/L	Inhibition of growth rate of *T. pseudonana* at 500 μg/L, *S. marinoi* at 1000 μg/L, *D. tertiolecta* and *I. galbana* at 1 mg/L	Miller et al. (2010)

(Continued)

TABLE 9.2 (Continued)
The ecotoxicity effects of ENPs based on previously published data

Types of ENPs	ENPs	Organism	Concentration	Effects	References
	Copper oxide	*Daphnia magna* and *Thamnocephalus platyurus*	Not available	The L(E)C50 values for both crustaceans ranged from 90 to 224 mg Cu/L	Blinova et al. (2010)
		Tetrahymena thermophila	Not available	Accumulation of the Copper oxide ENPs was observed in the food vacuoles and EC50 was not determined due to interference with optical density measurements	Blinova et al. (2010)
	Titanium oxide	*Escherichia coli, Bacillus subtilis*	10–5000 mg/L	Exhibited increase in toxicity showing 72% growth reduction after 5000 mg/L while in *B. subtilis* 75% growth reduction was observed at 1000 mg/L	Adams et al. (2006)
		Danio rerio	0.1 and 1 mg/L for ionic titanium and 2 and 10 mg/L for TiO$_2$ ENPs	Ionic and TiO$_2$ ENPs showed no bioaccumulation in zebrafish eleutheroembryos	Lopez-Serrano Oliver et al. (2015)
		Rainbow trout	1 mg/L	Exhibited increase in TiO$_2$ ENPs concentrations in gills with significant decrease in swimming speed	Boyle et al. (2013)
		Thalassiosira pseudonana, Skeletonema marinoi, Dunaliella tertiolecta, Isochrysis galbana	0–1000 μg/L	No effect on growth	Miller et al. (2010)
	Silicon oxide	*Bacillus subtilis*	10–5000 mg/L	5000 mg/L showed 99% growth reduction of *B. subtilis*	Adams et al. (2006)
	Nickel oxide	*Chlorella vulgaris*	0–50 mg/L	It showed an EC50 value of 32.28 mg/L NiO after 72 h and caused plasmolysis, cytomembrane breakage and thylakoids disorder in the cells	Gong et al. (2011)
	Alumina	*Daphnia magna*	10–1000 mg/L	It exhibited dose dependent immobilization and mortality	Zhu et al. (2009)
Metal ENPs	Silver	*Escherichia coli*	10–100 μg/cm^3	It accumulated in the bacterial membrane and damaged the cell wall	Bilberg et al. (2012a)

(Continued)

TABLE 9.2 (Continued)

The ecotoxicity effects of ENPs based on previously published data

Types of ENPs	ENPs	Organism	Concentration	Effects	References
		Caenorhabditis elegans	0.05–0.5 mg/L	It showed oxidative stress mediated toxicity, decrease in reproductivity and increased expression of the sod-3 and daf-12 gene	Roh et al. (2009)
			0.025–0.075 μg/mL	Both bare silver ENPs and 8 nm polyvinylpyrrolidone coated silver ENPs induced oxidative DNA damage	Ahn et al. (2014)
		Oncorhynchus mykiss	0.032–100 mg/L	96 hr LC50 values were 0.25, 0.71, and 2.16 mg/L for the early life stages, with reduction of chloride and potassium in plasma and increase of cortisol and cholinesterase in silver treated fish	Johari et al. (2013)
		Danio rerio	1.25–10 mg/L	LC50 values after 48 h is 2.9 mg/L in adult and 2.7 mg/L in eggs of the fish and induced yolk deformation and body shortening of the embryos	Kovriznych et al. (2013)
		Danio rerio	0–160 μg/mL	The LC50 value is 84 μg/L and showed respiratory toxicity with increase in operculum movement and surface respiration	Bilberg et al. (2012b)
	Copper	Danio rerio	2.5–40 mg/L	LC50 values after 48 h is 4.2 mg/L in adult and 24 mg/L in eggs of the fish and exhibited body shortening and malformation of embryos	Kovriznych et al. (2013)
		Danio rerio	0.01–10 mg/L	It shows acute toxicity with a 48 h LC50 of 1.5 mg/L	Griffitt et al. (2007)
	Gold	Oryza sativa, Lolium perenne L, Raphanus Sativus, Cucurbita mixta	31.25 nM	Lolium perenne L, Raphanus sativus accumulated higher amounts of the AuNPs (14–900 ng/mg) than Oryza sativa and Cucurbita mixta (7–59 ng/mg).	Zhu et al. (2012)

(Continued)

TABLE 9.2 (Continued)

The ecotoxicity effects of ENPs based on previously published data

Types of ENPs	ENPs	Organism	Concentration	Effects	References
Carbon ENPs	Single walled CNTs	*Danio rerio*	20–360 mg/L	Showed hatching delay of 52 to 72 h in embryos but with no effect on embryonic development and survival of exposed embryos	Cheng et al. (2007)
		Oncorhynchus mykiss	0.1–0.5 mg/L	A dose-dependent increase in ventilation rate, gill pathologies and mucus secretion was observed	Smith et al. (2007)
		Amphiascus tenuiremis	0.58–10 mg/L	It increased life-cycle mortality, reduced fertilization rates and molting in the 10 mg/mL	Templeton et al. (2006)
	Double walled CNTs	*Danio rerio*	120 and 240 mg/L	It induced a hatching delay at 240 mg/L	Cheng et al. (2007)
	Iron-platinum	*Salmonella typhimurium* strains TA98, TA100, TA1535, TA1537	39.1–5000 µg per plate	Mutagenicity was negative; no growth inhibition was observed with 2-aminoethanethiol capped on iron-platinum ENPs	Maenosono et al. (2009)
		Escherichia coli strain WP2uvrA-	9.1–5000 µg per plate	No mutagenicity	Maenosono et al. (2009)
	Quantum dots	*Elliption complanata*	0–8 mg/L	It exhibited immunotoxicity; decrease in the number and viability of hemocytes; with increase in lipid peroxidation in gills and marginal increase in DNA strand breaks in digestive glands	Gagne et al. (2008)
	Chitosan	*Danio rerio*	5–50 mg/L	Embryo experienced decreased hatching rate and increased mortality. It induced malformations, including a bent spine, pericardial edema, an opaque yolk in the embryos	Hu et al. (2011)

9.5.1 Risk Assessment

The use of ENPs brings forth the need for the warranty of the safe exposure of ENPs to biological systems and environment. The various steps in risk assessment are laid out in Figure 9.4. There is a need to identify the potential hazards due to the exposure of ENPs during the use of ENPs that are in continuous contact with consumers (Grobe and Rissanen 2013), for example, silver ENPs, which are used in textiles and as antibacterial agents, and ZnO and TiO_2 ENPs which are used in powders and creams.

To assess the risks imposed by ENPs on the environment, the following measures are suggested (Crane et al. 2007).

- The toxicity of ENPs should be investigated in standard model organisms for ecotoxicity assessment.
- Standard methods for ecotoxicology studies are to be preferred based on the type of organism selected for the studies.
- The dosimetry of the ENPs should be carefully performed based on the surface-to-volume or number-to-volume ratio.

9.5.2 Risk Management

To determine the considerable effect of ENPs on the environment; it is suggested that the entire life cycle, including production, use, and fate of ENPs, should be examined. It is an urgent requirement to study and understand the mobility, bioavailability, and impacts of ENPs on environment and biological systems. The dissolution of ENPs in a solvent/environmental matrix should also be considered during the assessment of ENP-mediated toxicity in soil and water.

a. *Prevention of ENPs from entering the environment*: ENPs have the potential to enter the environment during use and disposal. It was reported that silver ENPs used in socks to prevent foul smell were found to enter the water supply (FERA 2011); therefore, there is a need to prevent the entry of ENPs into the environment.

b. *Prevention at the production end*: Occupational exposure to ENPs should be controlled by isolating the production of ENPs in separate enclosures. Workers should wear protective suits, gloves, and goggles to prevent ENPs from entering

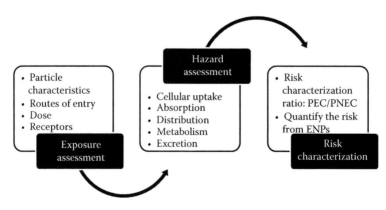

FIGURE 9.4
Steps in risk assessment of ENPs.

the human system by the dermal route. The use of air-purifying respirators should be encouraged.

c. *Prevention at the consumer end*: Consumers should be well aware of the harmful effects of ENPs; the composition of the ENPs should be checked before usage of products. A controlled use of ENP-incorporated products should be ensured.

d. *Responsible production of nanomaterials*: To minimize toxicity due to ENPs, it is recommended that they be used judiciously in commercial goods and environmental applications. The concentration used in consumer goods should be within the permissible limits.

Apart from the above control measures, there are certain issues related to ENPs that need to be resolved.

- ENPs are defined as materials in the conventional size scale of 1–100 nm, which does not account for aggregated and agglomerated ENPs. Therefore, there is a need for a clear description of nanoparticles.
- The scientific community should come up with reference ENPs to harmonize the protocols for toxicity assessment, as the same ENPs have shown strikingly different toxicological results when studied by different research groups in different parts of the world (Europa document—E. 24957).
- When testing the toxicity of ENPs, it is important to characterize the ENPs and examine their purity to ascertain that contamination is not the reason for ENP-induced toxicity.
- The dispersion medium for ENPs should be defined, as different dispersion media generate different levels of toxicity (Europa document—E. 24957). Also, ENPs should be well dispersed in the medium, as lack of dispersion generates false results.

9.6 Conclusion

Before considering any ENP harmful, information should be gained on its production volumes and its lethal concentration. The knowledge gaps in environment technology, including the data for ecotoxicity of ENPs, characterization in different solvent, protocols for the risk assessment, and lack of suitable lab models for hazard identification, among others, are the big hurdles to forming government regulations for the safe use of ENPs. Data for the marine ecosystems should be gathered and analyzed, as the physicochemical properties of nanomaterials in marine habitats differ from those properties in fresh water. The potential ecotoxicities and environmental health effects caused due to ENPs should be well documented. There is a need to understand the biological interactions of ENPs with proteins, nucleic acids, cells, and tissues for the design of safe nanoproducts for environmental applications.

Acknowledgments

Funding received from the Department of Biotechnology, Government of India, under the project "NanoToF: Toxicological evaluation and risk assessment on Nanomaterials in

Food" (grant number BT/PR10414/PFN/20/961/2014) is gratefully acknowledged. Financial assistance by The Gujarat Institute for Chemical Technology (GICT) for the establishment of a facility for environmental risk assessment of chemicals and nanomaterials and Centre for Nanotechnology Research and Applications (CENTRA) is also acknowledged.

References

Adams, L. K., Lyon, D. Y., Alvarez, P. J. J. Comparative eco-toxicity of nanoscale TiO_2, SiO_2, and ZnO water suspensions. *Water Res*, 2006, 40:3527–3532.

Ahn, J. M., Eom, H. J., Yang, X., Meyer, J. N., Choi, J. Comparative toxicity of silver nanoparticles on oxidative stress and DNA damage in the nematode, *Caenorhabditis elegans*. *Chemosphere*, 2014, 108:343–352.

Aiken, J. 1984. On the formation of small clear spaces in dusty air. *Trans Roy Soc Edinburgh*, 30:337–368.

Aitken, R., Creely, C., Tran, C. *Nanoparticles: An occupational hygiene review*. Institute of Occupational Medicine for the Health and Safety Executive, Edinburgh, UK, 2004.

Aitken, R. J., Chaudhry, M. Q., Boxall, A. B. A., Hull, M. Manufacture and use of nanomaterials: Current status in the UK and global trends. *Occup Med*, 2006, 56:300–306.

Arivalagan, K., Ravichandran, S., Rangasamy, K., Karthikeyan, E. Nanomaterials and its potential applications. *Int J ChemTech Res*, 2011, 3:534–538.

Artale, M. A., Augugliaro, V., Drioli, E., Golemme, G., Grande, C., Loddo, V., Molinari, R., Palmisano, L., Schiavello, M. Preparation and characterisation of membranes with entrapped TiO_2 and preliminary photocatalytic tests. *Ann Chim*, 2001, 91:127–136.

Bai, W., Zhang, Z., Tian, W., He, X., Ma, Y., Zhao, Y., Chai, Z. Toxicity of zinc oxide nanoparticles to zebrafish embryo: A physicochemical study of toxicity mechanism. *J Nanoparticle Res*, 2010, 12:1645–1654.

Batley, G. E., Kirby, J. K., McLaughlin, M. J. Fate and risks of nanomaterials in aquatic and terrestrial environments. *Acc Chem Res*, 2013, 46:854–862.

Bilberg, K., Hovgaard, M. B., Besenbacher, F., Baatrup, E. *In vivo* toxicity of silver nanoparticles and silver ions in Zebrafish (Danio rerio). *J Toxicol*, 2012a:293784.

Bilberg, K., Hovgaard, M. B., Besenbacher, F., Baatrup, E. Silver nanoparticles as antimicrobial agent: A case study on *E. coli* as a model for Gram-negative bacteria. *J Toxicol*, 2012b, 1–9, Article ID 293784.

Biswas, P., Wu, C. Y. Nanoparticles and the environment. *J Air Waste Manag Assoc*, 2005, 55:708–746.

Blinova, I., Ivask, A., Heinlaan, M., Mortimer, M., Kahru, A. Ecotoxicity of nanoparticles of CuO and ZnO in natural water. *Environ Pollut*, 2010, 158:41–47.

Boxall, A., Chaudhry, Q., Sinclair, C., Jones, A., Aitken, R., Jefferson, B. *Current and Future Predicted Environmental Exposure to Engineered Nanoparticles*. London: Central Science Laboratory. 2007, 37–68.

Boyle, D., Al-Bairuty, G. A., Ramsden, C. S., Sloman, K. A., Henry, T. B., Handy, R. D. Subtle alterations in swimming speed distributions of rainbow trout exposed to titanium dioxide nanoparticles are associated with gill rather than brain injury. *Aquat Toxicol*, 2013, 126:116–127.

Brayner, R., Ferrari-Iliou, R., Brivois, N., Djediat, S., Benedetti, M. F., Fievet, F. Toxicological impact studies based on Escherichia coli bacteria in ultrafine ZnO nanoparticles colloidal medium. *Nano Lett*, 2006, 6:866–870.

Buzea, C., Pacheco, I. I., Robbie, K. Nanomaterials and nanoparticles: Sources and toxicity. *Biointerphases*, 2007, 2:M.R.17–MR71.

Cai, L., Tong, M., Wang, X., Kim, H. Influence of clay particles on the transport and retention of titanium dioxide nanoparticles in quartz sand. *Environ Sci Technol*, 2014, 48:7323–7332.

Chaturvedi, S., Dave, P. N., Shah, N. K. Applications of nano-catalyst in new era. *J Saudi Chem Soc*, 2012, 16:307–325.

Cheng, J., Flahaut, E., Cheng, S. H. Effect of carbon nanotubes on developing zebrafish (Danio rerio) embryos. *Environ Toxicol Chem*, 2007, 26:708–716.

Clausen, C. A., Green, F., Nami Kartal, S. Weatherability and leach resistance of wood impregnated with nano-zinc oxide. *Nanoscale Res Lett*, 2010, 5:1464–1467.

Colvin, V. L. The potential environmental impact of engineered nanomaterials. *Nat Biotechnol*, 2003, 21:1166–1170.

Crane, M., Handy, R., Garrod, J. *An assessment of regulatory testing strategies and methods for characterizing the ecotoxicological hazards of nanomaterials*. Report for Defra. Department for Environment, Food and Rural Affairs. Report no. CB01097, 2007.

Depledge, M. H., Pleasants, L. J., Lawton, J. H. Nanomaterials and the environment: The views of the royal commission on environmental pollution (UK). *Environ Toxicol Chem*, 2010, 29:1–4.

Dhawan, A., Shanker, R., Das, M., Gupta, K. C. Guidance for safe handling of nanomaterials. *J Biomed Nanotechnol*, 2011, 7:218–224.

Du, J., Wang, S., You, H., Zhao, X. Understanding the toxicity of carbon nanotubes in the environment is crucial to the control of nanomaterials in producing and processing and the assessment of health risk for human: A review. *Environ Toxicol Pharmacol*, 2013, 36:452–462. http://dx.doi.org/10.1016/j.etap.2013.05.007.

Dunphy Guzman, K. A., Finnegan, D. L., Banfield, J. F. Influence of surface potential on aggregation and transport of titania nanoparticles. *Environ Sci Technol*, 2006, 40:7688–7693.

E. 24957. Nanotechnologies: Principles, applications, implications and hands-on activities. https://ec.europa.eu/research/industrial_technologies/pdf/nano-hands-on-activities_en.pdf.

Ebbesen, T. W., Lezec, H. J., Hiura, H., Bennett, J. W., Ghaemi, H. F., Thio, T. Electrical conductivity of individual carbon nanotubes. *Nature*, 1996, 382:54–56.

EPA N. W. Nanotechnology White Paper. 2007, S. P. Council.

FERA. The food and environment research agency. DWI 70/2/246 Review of the risks posed to drinking water by man-made nanoparticles. 2011, http://orbit.dtu.dk/fedora/objects/orbit:110179/datastreams/file_7614568/content

Gagne, F., Auclair, J., Turcotte, P., Fournier, M., Gagnon, C., Sauve, S., Blaise, C. Ecotoxicity of CdTe quantum dots to freshwater mussels: Impacts on immune system, oxidative stress and genotoxicity. *Aquat Toxicol*, 2008, 86:333–340.

Garcia-Alonso, J., Khan, F. R., Misra, S. K., Turmaine, M., Smith, B. D., Rainbow, P. S., Luoma, S. N., Valsami-Jones, E. Cellular internalization of silver nanoparticles in gut epithelia of the estuarine polychaete *Nereis diversicolor*. *Environ Sci Technol*, 2011, 45:4630–4636.

Genc, S., Zadeoglulari, Z., Fuss, S. H., Genc, K. The adverse effects of air pollution on the nervous system. *J Toxicol*, 2012, 2012:782462.

George, S., Gardner, H., Seng, E. K., Chang, H., Wang, C., Yu Fang, C. H., Richards, M., Valiyaveettil, S., Chan, W. K. Differential effect of solar light in increasing the toxicity of silver and titanium dioxide nanoparticles to a fish cell line and zebrafish embryos. *Environ Sci Technol*, 2014, 48:6374–6382.

Gong, N., Shao, K., Feng, W., Lin, Z., Liang, C., Sun, Y. Biotoxicity of nickel oxide nanoparticles and bio-remediation by microalgae *Chlorella vulgaris*. *Chemosphere*, 2011, 83:510–516.

Griffitt, R. J., Weil, R., Hyndman, K. A., Denslow, N. D., Powers, K., Taylor, D., Barber, D. S. Exposure to copper nanoparticles causes gill injury and acute lethality in zebrafish (*Danio rerio*). *Environ Sci Technol*, 2007, 41:8178–8186.

Grobe, A., Rissanen, M. Dialogforum Nano of BASF 2011/2012 Transparency in communication on nanomaterials from the manufacturer to the consumer. Nano (p. 51). 2013, Bonn, Germany. Retrieved from http://www.basf.com/group/corporate/nanotechnology/en/microsites/nanotechnology/dialogue/index.

Gumus, D., Berber, A. A., Ada, K., Aksoy, H. *In vitro* genotoxic effects of ZnO nanomaterials in human peripheral lymphocytes. *Cytotechnology*, 2014, 66:317–325.

Gupta, G. S., Kumar, A., Shanker, R., Dhawan, A. Assessment of agglomeration, co-sedimentation and trophic transfer of titanium dioxide nanoparticles in laboratory scale predator-prey model system. *Sci Rep*, 2016, 6:31422.

Gupta, G. S., Kumar, A., Senapati, V. A., Pandey, A. K., Shanker, R., Dhawan, A. Laboratory scale microbial food chain to study bioaccumulation, biomagnification, and ecotoxicity of cadmium telluride quantum dots. *Environ Sci Technol*, 2017, 51(3):1695–1706.

Hartland, A., Lead, J. R., Slaveykova, V. I., O'Carroll, D., Valsami-Jones, E. The environmental significance of natural nanoparticles. *Nat Educ Knowl*, 2013, 4:7.

He, X., Cai, Y., Zhang, H., Liang, C. Photocatalytic degradation of organic pollutants with Ag decorated free-standing TiO_2 nanotube arrays and interface electrochemical response. *J Mater Chem*, 2011, 21:475–480.

He, R., Hocking, R. K., Tsuzuki, T. Local structure and photocatalytic property of mechanochemical synthesized ZnO doped with transition metal oxides. *J Aust Ceram Soc*, 2013, 49:70–75.

Heinlaan, M., Ivask, A., Blinova, I., Dubourguier, H. C., Kahru, A. Toxicity of nanosized and bulk ZnO, CuO and TiO_2 to bacteria *Vibrio fischeri* and crustaceans *Daphnia magna* and *Thamnocephalus platyurus*. *Chemosphere*, 2008, 71:1308–1316.

Hu, Y. L., Qi, W., Han, F., Shao, J. Z., Gao, J. Q. Toxicity evaluation of biodegradable chitosan nanoparticles using a zebrafish embryo model. *Int J Nanomedicine*, 2011, 6:3351–3359.

Ion, A. C., Ion, I., Culetu, A. Carbon-based nanomaterials. Environmental applications. 2011, http://www.romnet.net/ro/seminar16martie2010/lucrari_extenso/Alina%20CIon_environmental.pdf.

Jarvis, T. A., Miller, R. J., Lenihan, H. S., Bielmyer, G. K. Toxicity of ZnO nanoparticles to the copepod *Acartia tonsa*, exposed through a phytoplankton diet. *Environ Toxicol Chem*, 2013, 32:1264–1269.

Johari, S., Kalbassi, M., Soltani, M., Yu, I. Toxicity comparison of colloidal silver nanoparticles in various life stages of rainbow trout (*Oncorhynchus mykiss*). *IJFS*, 2013, 12:76–95.

Kaegi, R., Ulrich, A., Sinnet, B., Vonbank, R., Wichser, A., Zuleeg, S., Simmler, H., Brunner, S., Vonmont, H., Burkhardt, M., Boller, M. Synthetic TiO_2 nanoparticle emission from exterior facades into the aquatic environment. *Environ Pollut*, 2008, 156:233–239.

Kaegi, R., Sinnet, B., Zuleeg, S., Hagendorfer, H., Mueller, E., Vonbank, R., Boller, M., Burkhardt, M. Release of silver nanoparticles from outdoor facades. *Environ Pollut*, 2010, 158:2900–2905.

Kamat, P. V., Huehn, R., Nicolaescu, R. A "sense and shoot" approach for photocatalytic degradation of organic contaminants in water. *J Phys Chem B*, 2002, 106:788–794.

Karn, B., Kuiken, T., Otto, M. Nanotechnology and *in situ* remediation: A review of the benefits and potential risks. *Environ Health Perspect*, 2009, 117:1813–1831.

Kashiwada, S. Distribution of nanoparticles in the see-through medaka (*Oryzias latipes*). *Environ Health Perspect*, 2006, 114:1697–1702.

Kessler, R. Engineered nanoparticles in consumer products: Understanding a new ingredient. *Environ Health Perspect*, 2011, 119:a120–a125.

Khot, L. R., Sankaran, S., Maja, J. M., Ehsani, R., Schuster, E. W. Applications of nanomaterials in agricultural production and crop protection: A review. *Crop Prot*, 2012, 35:64e70.

Kovriznych, J. A., Sotnikova, R., Zeljenkova, D., Rollerova, E., Szabova, E., Wimmerova, S. Acute toxicity of 31 different nanoparticles to zebrafish (*Danio rerio*) tested in adulthood and in early life stages—Comparative study. *Interdiscip Toxicol*, 2013, 6:67–73.

Kumar, A., Dhawan, A., Shanker, R. The need for novel approaches in ecotoxicity of engineered nanomaterials. *J Biomed Nanotechnol*, 2011a, 7:79–80.

Kumar, A., Pandey, A. K., Shanker, R., Dhawan, A. Microorganisms: A versatile model for toxicity assessment of engineered nanoparticles. *Nano-Antimicrobials*, 2012, 497–524.

Kumar, A., Pandey, A. K., Singh, S. S., Shanker, R., Dhawan, A. Cellular uptake and mutagenic potential of metal oxide nanoparticles in bacterial cells. *Chemosphere*, 2011b, 83:1124–1132.

Kumar, A., Pandey, A. K., Singh, S. S., Shanker, R., Dhawan, A. Engineered ZnO and TiO(2) nanoparticles induce oxidative stress and DNA damage leading to reduced viability of *Escherichia coli*. *Free Radic Biol Med*, 2011c, 51:1872–1881.

Li, L. Z., Zhou, D. M., Peijnenburg, W. J., Van Gestel, C. A., Jin, S. Y., Wang, Y. J., Wang, P. Toxicity of zinc oxide nanoparticles in the earthworm, *Eisenia fetida* and subcellular fractionation of Zn. *Environ Int*, 2011, 37:1098–1104.

Lin, Q. B., Li, B., Song, H., Wu, H. J. Determination of silver in nano-plastic food packaging by microwave digestion coupled with inductively coupled plasma atomic emission spectrometry or inductively coupled plasma mass spectrometry. *Food Addit Contam Part A Chem Anal Control Expo Risk Assess*, 2011, 28:1123–1128.

Lin, S., Reppert, J., Hu, Q., Hudson, J., Reid, M., Ratnikova, T., Rao, A., Luo, H., Ke, P. Uptake, translocation, and transmission of carbon nanomaterials in rice plants. *Small*, 2009, 5:1128–1132.

Lisha, K. P., Anshup, Pradeep, T. Towards a practical solution for removing inorganic mercury from drinking water using gold nanoparticles. *Gold Bull*, 2009, 42:144–152.

Lopez-Serrano Oliver, A., Munoz-Olivas, R., Sanz Landaluze, J., Rainieri, S., Camara, C. Bioaccumulation of ionic titanium and titanium dioxide nanoparticles in zebrafish eleuthero embryos. *Nanotoxicology*, 2015, 9(7):835–842.

Lowry, G. V., Hotze, E. M., Bernhardt, E. S., Dionysiou, D. D., Pedersen, J. A., Wiesner, M. R., Xing, B. Environmental occurrences, behavior, fate, and ecological effects of nanomaterials: An introduction to the special series. *J Environ Qual*, 2010, 39:1867–1874.

Ma, H., Bertsch, P. M., Glenn, T. C., Kabengi, N. J., Williams, P. L. Toxicity of manufactured zinc oxide nanoparticles in the nematode *Caenorhabditis elegans*. *Environ Toxicol Chem*, 2009, 28:1324–1330.

Maenosono, S., Yoshida, R., Saita, S. Evaluation of genotoxicity of amine-terminated water-dispersible FePt nanoparticles in the Ames test and *in vitro* chromosomal aberration test. *J Toxicol Sci.*, 2009, 34:349–354.

Mansoori, G. A., Bastami, T. R., Ahmadpour, A., Eshaghi, Z. Environmental application of nanoscience. *Ann Rev Nano Res*, 2008, 2, Chap.2:1–73.

Markus, A. A., Parsons, J. R., Roex, E. W. M., Voogt, P. D., Laane, R. W. P. M. Modeling aggregation and sedimentation of nanoparticles in the aquatic environment. *Sci Total Environ*, 2015, 506-507:323–329.

Masciangioli, T., Zhang, W. Peer reviewed: Environmental technologies at the nanoscale. *Environ Sci Technol*, 2003, 37 102A–108A, DOI: 110.1021/es0323998.

Miller, R. J., Lenihan, H. S., Muller, E. B., Tseng, N., Hanna, S. K., Keller, A. A. Impacts of metal oxide nanoparticles on marine phytoplankton. *Environ Sci Technol*, 2010, 44:7329–7334.

Morrison, M. Nano and the environment workshop report, Nanoforum, 2006, Brussels. http://nanoforum.org.

Mueller, N. C., Nowack, B. Exposure modeling of engineered nanoparticles in the environment. *Environ Sci Technol*, 2008, 42:4447–4453.

Nazarenko, Y., Han, T. W., Lioy, P. J., Mainelis, G. Potential for exposure to engineered nanoparticles from nanotechnology-based consumer spray products. *J Expo Sci Environ Epidemiol*, 2011, 21:515–528.

NNI (National Nanotechnology Initiative) What is nanotechnology? Available at: https://www.nano.gov, 2004.

Nurmi, J. T., Tratnyek, P. G., Sarathy, V., Baer, D. R., Amonette, J. E., Pecher, K., Wang, C. M., Linehan, J. C., Matson, D. W., Penn, R. L. Characterization and properties of metallic iron nanoparticles: Spectroscopy, electrochemistry, and kinetics. *Environ Sci Technol*, 2005, 39:1221–1230.

Oberdörster, E., McClellan-Green, P., Haasch, M. L. Ecotoxicity of engineered nanomaterials. In *Nanomaterials—Toxicity, Health and Environmental Issues* (ed. Challa S. S. R. Kumar). 2006, Wiley-VCH, Weinheim, 119–148.

Oberdörster, G., Oberdörster, E., Oberdörster, J. Nanotoxicology: An emerging discipline evolving from studies of ultrafine particles. *Environ Health Perspect*, 2005, 113:823–839.

Paknikar, K. M., Nagpal, V., Pethkar, A. V., Rajwade, J. M. Degradation of lindane from aqueous solutions using iron sulfide nanoparticles stabilized by biopolymers. *Sci Technol Adv Mater*, 2005, 6:370–374.

Pal, T., Sau, T. K., Jana, N. R. Reversible formation and dissolution of silver nanoparticles in aqueous surfactant media. *Langmuir*, 1997, 13:1481–1485.

Pan, G., Li, L., Zhao, D., Chen, H. Immobilization of non-point phosphorus using stabilized magnetite nanoparticles with enhanced transportability and reactivity in soils. *Environ Pollut*, 2010, 158:35–40.

Patel, P., Kansara, K., Shah, D., Vallabani, N. V., Shukla, R. K., Singh, S., Dhawan, A., Kumar, A. Cytotoxicity assessment of ZnO nanoparticles on human epidermal cells. *Mol Cytogenet*, 2014, 7:81.

Peralta-Videa, J. R., Zhao, L., Lopez-Moreno, M. L., de la Rosa, G., Hong, J., Gardea-Torresdey, J. L. Nanomaterials and the environment: A review for the biennium 2008–2010. *J Hazard Mater*, 2011, 186:1–15.

Phillips, J. I., Green, F. Y., Davies, J. C., Murray, J. Pulmonary and systemic toxicity following exposure to nickel nanoparticles. *Am J Ind Med*, 2010, 53:763–767.

Pirela, S. V., Sotiriou, G. A., Bello, D., Shafer, M., Bunker, K. L., Castranova, V., Thomas, T., Demokritou, P. Consumer exposures to laser printer-emitted engineered nanoparticles: A case study of life-cycle implications from nano-enabled products. *Nanotoxicology*, 2015, 9(6):760–768.

Ponder, S. M., Darab, J. G., Mallouk, T. E. Remediation of Cr(VI) and Pb(II) aqueous solutions using supported, nanoscale zero-valent iron. *Environ Sci Technol*, 2000, 34:2564–2569.

Powell, B. R., Bloink, R. L., Eickel, C. C. Preparation of cerium dioxide powders for catalyst supports. *J Am Cerurn Soc*, 1988, 71:C.-104–C-I106.

Premanathan, M., Karthikeyan, K., Jeyasubramanian, K., Manivannan, G. Selective toxicity of ZnO nanoparticles toward Gram-positive bacteria and cancer cells by apoptosis through lipid peroxidation. *Nanomedicine*, 2011, 7:184–192.

Project of the Emerging Nanotechnologies (PEN). http://www.nanotechproject.org/inventories/consumer/. Accessed on October 20, 2016. Available at http://www.nanotechproject.org.

Recillas, S., Colon, J., Casals, E., Gonzalez, E., Puntes, V., Sanchez, A., Font, X. Chromium VI adsorption on cerium oxide nanoparticles and morphology changes during the process. *J Hazard Mater*, 2010, 184:425–431.

Reed, M. A. Quantum dots. *Sci Am*, 1993, 268:118–123.

Ren, G., Hu, D., Cheng, E. W., Vargas-Reus, M. A., Reip, P., Allaker, R. P. Characterisation of copper oxide nanoparticles for antimicrobial applications. *Int J Antimicrob Agents*, 2009, 33:587–590.

Rickerby, D. G., Morrison, M. Nanotechnology and the environment: A European perspective. *Sci Technol Adv Mater*, 2007, 8:19–24.

Roh, J., Sim, S. J., Yi, J., Park, K., Chung, K. H., Ryu, D., Choi, J. Ecotoxicity of silver nanoparticles on the soil nematode *Caenorhabditis elegans* using functional ecotoxicogenomics. *Environ Sci Technol*, 2009, 43:3933–3940.

Sánchez, A., Recillas, S., Font, X., Casals, E., Gonzalez, E., Puntes, V. Ecotoxicity of, and remediation with, engineered inorganic nanoparticles in the environment. *Trends Analyt Chem*, 2011, 30:507–516.

Sassolas, A., Prieto-Simón, B., Marty, J. Biosensors for pesticide detection: New trends. *Am J Analyt Chem*, 2012, 3:210–232.

SCENIHR (Scientific Committee on Emerging and Newly Identified Health Risks). The appropriateness of existing methodologies to assess the potential risks associated with engineered and adventitious products of nanotechnologies. 2006.

Schaumann, G. E., Philippe, A., Bundschuh, M., Metreveli, G., Klitzke, S., Rakcheev, D., Grun, A., et al. Understanding the fate and biological effects of Ag- and TiO-nanoparticles in the environment: The quest for advanced analytics and interdisciplinary concepts. *Sci Total Environ*, 2014, 535:3–19.

Scientific Committee on Consumer Products (SCCP). Opinion on safety of nanomaterials in cosmetic products. http://ec.europa.eu/health/ph_ risk/committees/04_sccp/docs/sccp_o_123.pdf, European Commission. SCCP/1147/07, 2007.

Senapati, V. A., Jain, A. K., Gupta, G. S., Pandey, A. K., Dhawan, A. Chromium oxide nanoparticles induced genotoxicity and p53 dependent apoptosis in human lung alveolar cells. *J Appl Toxicol*. 2015a, doi: 10.1002/jat.3174.

Senapati, V. A., Kumar, A., Gupta, G. S., Pandey, A. K., Dhawan, A. ZnO nanoparticles induced inflammatory response and genotoxicity in human blood cells: A mechanistic approach. *Food Chem Toxicol*. 2015b, http://dx.doi.org/10.1016/j.fct.2015.06.018

Sharma, V., Anderson, D., Dhawan, A. Zinc oxide nanoparticles induce oxidative DNA damage and ROS-triggered mitochondria mediated apoptosis in human liver cells (HepG2). *Apoptosis*, 2012a, 17:852–870.

Sharma, V., Kumar, A., Dhawan, A. Nanomaterials: Exposure, effects and toxicity assessment. *Proc. Natl. Acad. Sci., India, Sect. B Biol. Sci.*, 2012b, 88(1):3–11.

Sharma, V., Singh, P., Pandey, A. K., Dhawan, A. Induction of oxidative stress, DNA damage and apoptosis in mouse liver after sub-acute oral exposure to zinc oxide nanoparticles. *Mutat Res*, 2012c, 745:84–91.

Shukla, R. K., Kumar, A., Vallabani, N. V. S., Pandey, A. K., Dhawan, A. Titanium dioxide nanoparticle-induced oxidative stress triggers DNA damage and hepatic injury in mice. *Nanomedicine (Lond)*, 2013, 9:1423–1434.

Singh, S., Kumar, A. Engineered nanomaterials: Safety and health hazard. *Int J Nanotechnol Nanomed*, 2016, 1:1–23.

Sinha, T., Ahmaruzzaman, M. Green synthesis of copper nanoparticles for the efficient removal (degradation) of dye from aqueous phase. *Environ Sci Pollut Res Int*, 2015, 22(24):20092–20100.

Smith, C. J., Shaw, B. J., Handy, R. D. Toxicity of single walled carbon nanotubes to rainbow trout, (*Oncorhynchus mykiss*): Respiratory toxicity, organ pathologies, and other physiological effects. *Aquat Toxicol*, 2007, 82:94–109.

Suman, T. Y., Rajasree, S. R. R., Kirubagaran, R. Evaluation of zinc oxide nanoparticles toxicity on marine algae chlorella vulgaris through flow cytometric,cytotoxicity and oxidative stress analysis. *Ecotoxicol Environ Saf*, 2015, 113:23–30.

Tardif, F. Nanoparticle monitoring. State of the art and development strategies. In HUB meeting of the European Technology Platform for Industrial Safety, Brussels. 2007.

Templeton, R. C., Ferguson, P. L., Washburn, K. M., Scrivens, W. A., Chandler, G. T. Life-cycle effects of single-walled carbon nanotubes (SWNTs) on an estuarine meiobenthic copepod. *Environ Sci Technol*, 2006, 40:7387–7393.

Tungittiplakorn, W., Lion, L. W., Cohen, C., Kim, J. Y. Engineered polymeric nanoparticles for soil remediation. *Environ Sci Technol*, 2004, 38:1605–1610.

USEPA. Nanotechnology white paper. Science Policy Council, United States Environmental Protection Agency. http://www.epa.gov/osa/pdfs/nanotech/epa-nanotechnology-whitepaper-0207.pdf, 2007.

Velzeboer, I., Quik, J. T., Van de Meent, D., Koelmans, A. A. Rapid settling of nanoparticles due to heteroaggregation with suspended sediment. *Environ Toxicol Chem*, 2014, 33:1766–1773.

Vielkind, M., Kampen, I., Kwade, A. Zinc oxide nanoparticles in bacterial growth medium: Optimized dispersion and growth inhibition of *Pseudomonas putida*. *Adv Nanopart*, 2013, 02(04):287–293.

Waalewijn-Kool, P. L., Ortiz, M. D., Lofts, S., Van Gestel, C. A. The effect of pH on the toxicity of zinc oxide nanoparticles to *Folsomia candida* in amended field soil. *Environ Toxicol Chem*, 2013, 32:2349–2355.

Wang, L., Lee, K., Sun, Y. Y., Lucking, M., Chen, Z., Zhao, J. J., Zhang, S. B. Graphene oxide as an ideal substrate for hydrogen storage. *ACS Nano*, 2009, 3:2995–3000.

Wang, C., Liu, H., Qu, Y. TiO_2-based photocatalytic process for purification of polluted water: Bridging fundamentals to applications. *J Nanomaterials*, 2013, 2013:1–14.

Wang, C., Yu, C. Detection of chemical pollutants in water using gold nanoparticles as sensors: A review. *Rev Anal Chem*, 2013, 32:1–14.

Wiesner, M. R., Bottero, J. *Environmental Nanotechnology*. Applications and Impacts of Nanomaterials. The McGraw-Hill Companies, New York, NY, 2007, doi: 10.1036/0071477500.

Wiesner, M. R., Lowry, G. V., Alvarez, P., Dionysiou, D., Biswas, P. Assessing the risks of manufactured nanomaterials. *Environ Sci Technol*, 2006, 40:4336–4345.

Wild, E., Jones, K. C. Novel method for the direct visualization of *in vivo* nanomaterials and chemical interactions in plants. *Environ Sci Technol*, 2009, 43:5290–5294.

Wu, B., Wang, Y., Lee, Y. H., Horst, A., Wang, Z., Chen, D. R., Sureshkumar, R., Tang, Y. J. Comparative eco-toxicities of nano-ZnO particles under aquatic and aerosol exposure modes. *Environ Sci Technol*, 2010, 44:1484–1489.

Yu, B., Zeng, J., Gong, L., Zhang, M., Zhang, L., Chen, X. Investigation of the photocatalytic degradation of organochlorine pesticides on a nano-TiO$_2$ coated film. *Talanta*, 2007, 72:1667–1674.

Zeng, R., Wang, J., Cui, J., Hu, L., Mu, K. Photocatalytic degradation of pesticide residues with RE3þ-doped nano-TiO$_2$. *J Rare Earth*, 2010, 28:253e256.

Zhang, B., Misak, H., Dhanasekaran, P. S., Kalla, D., Asmatulu, R. Environmental Impacts of Nanotechnology and Its Products. *Proceedings of the 2011 Midwest Section Conference of the American Society for Engineering Education*, 2011a, USA.

Zhang, L., Fang, M. Nanomaterials in pollution trace detection and environmental improvement. *Nano Today*, 2010, 5:128–142.

Zhang, R., Khalizov, A., Wang, L., Hu, M., Xu, W. Nucleation and growth of nanoparticles in the atmosphere. *Chem Rev*, 2011b, 112:1957–2011.

Zhang, S., Saebfar, H. *Chemical Information Call-in Candidate: Nano Zinc Oxide*. California Dept Toxic Substances Control, San Francisco, California, 2010, 1–11.

Zhu, X., Zhu, L., Chen, Y., Tian S. Acute toxicities of six manufactured nanomaterial suspensions to *Duphnia magna*. *J Nanoparticle Res*, 2009, 11:67–75.

Zhu, Z. J., Wang, H., Yan, B., Zheng, H., Jiang, Y., Miranda, O. R., Rotello, V. M., Xing B., Vachet, R. W. Effect of surface charge on the uptake and distribution of gold nanoparticles in four plant species. *Environ Sci Technol*, 2012, 46:12391–12398.

10
Iron Nanoparticles for Contaminated Site Remediation and Environmental Preservation

Adam Truskewycz, Sayali Patil, Andrew Ball, and Ravi Shukla

CONTENTS

10.1	What Are Nanoparticles?	324
10.2	Environmental Pollution Due to Anthropogenic Activities	325
10.3	Iron in the Environment	326
10.4	Iron Nanoparticles: Potential Use in Remediation	328
10.5	Different Types of Iron Nanoparticles	329
	10.5.1 Nanoscale Zero-Valent Iron (Fe^0)	329
	10.5.2 Magnetite (Fe_3O_4)	329
	10.5.3 Maghemite (γ-Fe_2O_3)	329
	10.5.4 Hematite (α-Fe_2O_3)	329
	10.5.5 Goethite (α-FeO(OH))	330
	10.5.6 Lepidocrocite (γ-FeO(OH))	330
	10.5.7 Ferrihydrite	330
	10.5.8 Amorphous Mixtures	330
10.6	Parameters for the Application of Iron Nanoparticles to Bioremediation	330
	10.6.1 Size	330
	10.6.2 Agglomeration	331
	10.6.3 Shape and Particle Uniformity	332
	10.6.4 Aging/Storage Times/Stability/Passivation	332
10.7	Synthesis of Iron Nanoparticles	333
	10.7.1 Ball Milling	333
	10.7.2 Co-Precipitation	334
	10.7.3 Reduction of Aqueous Iron Salts	335
	10.7.4 Hydrothermal/Hydrosolvo Synthesis	335
	10.7.5 Thermal Decomposition	336
	10.7.6 Microwave-Assisted Synthesis	336
	10.7.7 Microemulsion	336
	10.7.8 Sonochemical	337
	10.7.9 Optimization	339
10.8	Different Methods of Generating Coatings	339
	10.8.1 Natural Polymers—Starch	340
	10.8.2 Polyelectrolyte-Carboxymethyl Cellulose	341
	10.8.3 Polyvinyl Alcohol	341
	10.8.4 Poly(Methacrylic Acid)-Block-(Methyl Methacrylate)-Block-(Styrenesulfonate) Triblock	341
	10.8.5 Mesoporous Carbon/Silica Coated	342

    10.8.6   Emulsified Nanoscale Zero-Valent Iron 342
    10.8.7   Biological Capping Agents-Plant/Algal-Based Polyphenols 343
    10.8.8   Immobilized Nanoparticles on Solid Supports 343
    10.8.9   Bimetallic Particles 344
10.9   Characterization Techniques 344
    10.9.1   Scanning Electron Microscopy 344
    10.9.2   Transmission Electron Microscopy 345
    10.9.3   X-Ray Diffraction 345
    10.9.4   Energy-Dispersive X-Ray Spectroscopy 346
    10.9.5   Fourier Transform Infrared 346
    10.9.6   Dynamic Light Scattering 347
    10.9.7   X-Ray Absorption Spectroscopy 347
    10.9.8   X-Ray Photoelectron Spectroscopy 347
10.10 Application of Iron Nanoparticles for Pollution Remediation 348
    10.10.1 Comparison of Nanoparticles with Other Technologies 348
10.11 Types of Pollutants Iron Nanoparticles Catabolize 349
    10.11.1 Chlorinated Solvents 349
    10.11.2 Dyes 350
    10.11.3 Pesticides 350
    10.11.4 Nitrogen 351
    10.11.5 Metals 351
    10.11.6 Fenton/Fenton-Like Reactions 352
10.12 *In Situ* Remediation Using Iron Nanoparticles 353
    10.12.1 Groundwater Remediation 353
    10.12.2 Migration and Monitoring 354
    10.12.3 Surface Waters 355
    10.12.4 Terrestrial Remediation 355
    10.12.5 Cost of Iron Nanoparticle Remediation/Feasibility 356
10.13 Fate of Fe NP's in the Environment and Ecotoxicity Concerns 357
    10.13.1 Microorganisms 357
    10.13.2 Plants 359
    10.13.3 Fish 359
    10.13.4 Mammals 360
    10.13.5 Ecotoxicity Extrapolations 361
10.14 Conclusions 361
Acknowledgment 362
References 362

10.1 What Are Nanoparticles?

A nanoparticle (NP) can be defined as any entity that has at least one dimension between 1–100 nm in size. To put their size into perspective, 1 nm is 1 billionth of a meter (approximately half the width of a DNA double helix). Nanoparticles are of great scientific interest, as they exhibit unique properties that are not observed at the atomic and molecular levels or in the bulk form of the materials. At the nanosize scale, the so called *quantum effects* dictate multiple properties of materials that include but are not limited to melting point, fluorescence, electrical conductivity, and magnetic permeability, and chemical reactivity changes as a function of the size of the particles. For instance, in case of metal nanoparticles, particularly gold, silver, and copper, when particle size becomes progressively small at the nano scale, due to optical interactions, a large number of surface electrons leads to the interesting phenomenon of *surface plasmon resonance* (Daniel and Astruc 2004). At the nano scale, environments are quite different from what is experienced at larger dimensions. In bulk materials, tightly packed atoms react with neighboring atoms of the same type and behave as clusters/bulk solids, whereas at the nano scale, the specific behavior of individual atoms has a larger role to play in interactions (Halperin 1986). It is this size range that makes nanoparticles unique for plethora of real-world applications (Daniel and Astruc 2004, Agrawal et al. 2013, Weerathunge et al. 2014, Arakha et al. 2015, Carnovale et al. 2016).

Nanoparticles can be organic (e.g., cholesterol, rhovanil, rhodiarome), inorganic (nanoscale zero-valent iron [nZVI], gold and silver nanoparticles), or a combination of both (e.g., yolk–shell nanoparticle@MOF Petalous Heterostructures) (Liu et al. 2014b, Destrée and Nagy 2006). Their small size and subsequent increased surface area-to-volume ratio allow them to tightly pack together to make smooth, uniform surfaces or, if intermolecular attractive forces can be combated, they can act as individual reactive entities with high surface area-to-volume ratio and elevated reactive properties.

They can exist naturally (e.g., mineral composites, biogenic magnetite) as incidental entities (e.g., combustion products, sand blasting) or as engineered nanoparticles (e.g., nZVI, carbon nanotubes) (Cameotra and Dhanjal 2012). The current use of nanoparticles spans many disparate disciplines from drug delivery in medical applications to conductors of electricity for electronic devices to their use in paint to act as UV light blockers for enhancing the life of paint (Allen et al. 2002) and for enhanced contaminant degradation efficiencies in the environmental remediation field (O'Carroll et al. 2013).

10.2 Environmental Pollution Due to Anthropogenic Activities

Since the industrial revolution, increased population growth and heavy reliance on machinery have led to unprecedented economic growth and development. While previously this has been carried out with little regard to the environment, the concept of environmental preservation has accelerated over the last decade within developing as well as first-world countries. The focus on climate change and human health has led to a plethora of scientific research aimed at tackling the issues related to anthropogenic activities such as design and development of smart nanomaterials/chemicals and their impact on the environment, including their ecotoxicity and the potential health implications they pose for both environmental and human health. Furthermore, technologies that have shown promise in terms of decontaminating the environment are constantly being refined and optimized.

Pollution from human activities often finds its way into the environment in excessive concentrations, and this can cause rapid, significant, and prolonged ecological harm. Persistent organic pollutants and heavy metals are among the most serious environmental polluters (Zhu et al. 2015). Unlike most naturally occurring environmental impacts, these anthropogenic pollutants are particularly recalcitrant to environmental means of detoxification (bioremediation). Remediation, particularly bioremediation (natural attenuation), may be regarded as protracted when it comes to appropriate land restoration for commercial development purposes, animal and human health, and ecosystem regeneration (Prakash et al. 2013).

10.3 Iron in the Environment

Iron is everywhere—inside animal cells (part of the hemoglobin protein responsible for transporting oxygen around the body) (Gutteridge 1986), within plant cells (responsible for processes from respiration to photosynthesis) (Kim and Guerinot 2007), and within microbial cells (required for many processes, including apoptosis and electron shuttling) (Breuer et al. 2014). It is the fourth most abundant element found within the earth's crust (Figure 10.1) and is present in many different forms. It is responsible for the dark red, yellow, and orange sands in the Australian outback and the dark black sands on the south coast of Indonesia's Java Island and Elba island in Italy (Jamiesona et al. 2006, Perfumo et al. 2011).

Despite the huge prevalence of iron in natural ecosystems, metallic iron is seldom found naturally, as it readily reacts with oxygen to form iron oxides. There are a number of iron oxides falling under the categories of oxides, hydroxides, or oxide hydroxides; these may include the following (Génin et al. 2001; Schwertmann 2008) (Figure 10.2).

- Goethite α-FeO(OH)
- Lepidocrocite γ–FeO(OH)
- Akaganeite β–FeO(OH)
- Schwertmannite $Fe_{16}O_{16}(SO_4)_2(OH)_{12} \cdot nH_2O$
- Feroxyhite δ-FeOOH)
- High pressure FeOOH
- Ferrihydrite $5Fe_2O_3 \cdot 9H_2O$
- Bernalite $Fe(OH)_3 \cdot nH_2O$
- Ferrous hydroxide $Fe(OH)_2$
- Green rusts $Fe^{II}_4 Fe^{III}_2 (OH)_{12}]^{2+} \cdot [SO_4^{2-} \cdot nH_2O]^{2-}$ & $Fe^{II}_6 Fe^{III}_2 (OH)_{16}]^{2+} \cdot [C_2O_4^{2-} \cdot 3 H_2O]^{2-}$

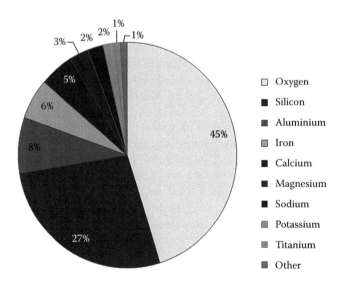

FIGURE 10.1
Abundance of elements in the continental crust. (Adapted from Skinner. B. J. 1979. Earth resources (minerals/metals/ores/geochemistry/mining)., *PNAS, Proceedings of the National Academy of Sciences* 76(9): 4212–4217.)

- Hematite α–Fe_2O_3
- Magnetite Fe_3O_4 ($Fe^{2+}Fe^{3+}O_4$)
- Maghemite γ–Fe_2O_3
- β-Fe_2O_3
- ε-Fe_2O_3
- Wustite $Fe^{2+}O$

Iron oxides naturally exhibit low solubility in water; however, decreasing crystal size, reduction of Fe^{2+}, and the presence of ligands creating Fe complexes can increase their solubility. Furthermore, iron^{III} oxides have surfaces with OH ions attached that adsorb many anions, which makes them encapsulated, thereby limiting their mobility in soils.

The iron oxides routinely used for nanoparticle research for site remediation are mainly Goethite α-FeO(OH), Magnetite Fe_3O_4 ($Fe^{2+}Fe^{3+}O_4$), Hematite α–Fe_2O_3, and Maghemite γ–Fe_2O_3 (Figure 10.2). However, nZVI (metallic iron) nanoparticles tend to have the largest focus in research for environmental remediation (Hua et al. 2012).

The form and stability of iron is reliant on a variety of physicochemical and environmental conditions. Under aerobic conditions where soil is not affected by extreme acidity or alkalinity, Fe^{2+} ions oxidize to Fe^{3+} ions. These Fe^{3+} ions will then try to achieve stability by forming Fe^{III} oxides. The stability of these species is not readily affected by protonation but

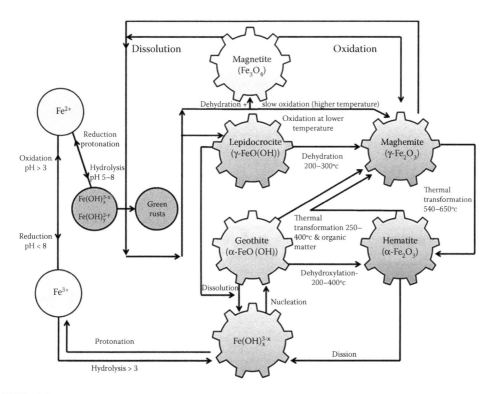

FIGURE 10.2
Dynamics of different iron oxide transformations. (Adapted from Chesworth, W., Perez-Alberti, A., Arnaud, E., Morel-Seytoux, H. J., Schwertmann, U. 2008. *Encyclopedia of Earth Sciences Series*. Chesworth, W. (ed.) Springer, Netherlands, ISBN: "978-1-4020-3995-9", doi: "10.1007/978-1-4020-3995-9_302", pp. 363–369.)

may be dissolved and recycled by biotic reduction or complexation with organic ligands. Under anaerobic conditions, Fe^{3+} in oxides can be reduced to Fe^{2+} via electron transfers resulting from the oxidation of biomass (Schwertmann 2008).

10.4 Iron Nanoparticles: Potential Use in Remediation

Iron addition is a current treatment for the degradation of a wide range of recalcitrant environmental pollutants, with millimetric iron and permeable reactive barriers being routinely used. As reactions between iron and pollutants occur at the surface of iron, it is logical that an increased surface area-to-volume ratio would favor elevated degradation rates. Studies show that nanoparticles exhibit significant increase in degradation rates compared to their millimetric counterparts (Lien and Zhang 2001).

Not only have exceptional pollutant degradation rates been shown in laboratory scale tests, but there are other benefits for particles to enter the nano scale. An example of this lies with groundwater contamination. Contaminated groundwater is locked away underneath complex geographical matrices and is particularly hard to access for remediation purposes (Figure 10.3). Pumping and treating is time intensive and costly, as installation of permeable reactive barriers requires costly excavation, and they can only be installed in shallow aquifers. Chemical treatment methods are often toxic to the environment and can be costly. Emulsified iron nanoparticles and other coated iron nanoparticles have been successfully

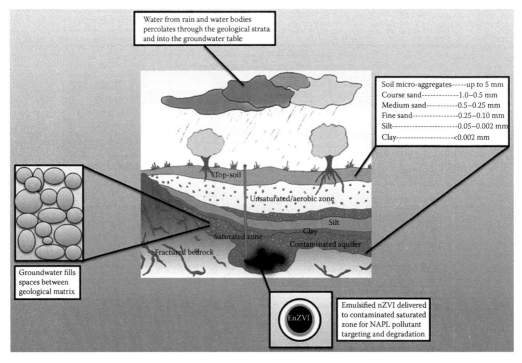

FIGURE 10.3
Groundwater dynamics to consider when applying EnZVI to a contaminated site.

applied to *in vitro* groundwater remediation projects for deep aquifers without the need to excavate or transport to treatment plants (Cook 2009).

Due to iron's high abundance in the earth's crust and nZVI's ability to rapidly form iron oxide following passivation in aerobic environments, its potential environmental compatibility, particularly for groundwater remediation applications, is of great interest.

10.5 Different Types of Iron Nanoparticles

In the field of nanoparticle research, there are a number of different types of iron (Figure 10.2) and iron-based nanoparticles, each with their own inherent properties. Magnetism, high coercivity, low Curie temperature, and reactivity are parameters that may be sought after, and optimizing one or all of these properties is of significant interest for researchers. In the environmental remediation field, reactivity with recalcitrant pollutants and minimization of passivation are of high importance.

10.5.1 Nanoscale Zero-Valent Iron (Fe^0)

Metallic iron, elemental iron, or nanoscaled zero valent iron is ferromagnetic in nature and does not possess a charge. Its structure is made up of only iron atoms and is unstable under aerobic conditions. It will therefore oxidize readily unless coatings are applied. The majority of research conducted with metallic iron nanoparticles uses nZVI with an oxide shell to prevent further rapid oxidation (Fan et al. 2015).

10.5.2 Magnetite (Fe_3O_4)

Magnetite is a ferromagnetic species of iron (superparamagnetic when nanoparticle size is smaller than 15 nm) when it is below its Curie temperature (858K). It has an inverse cubic iron oxide spinel structure and is a mixed valence metal possessing Fe^{2+} and Fe^{3+} states. It exhibits strong magnetism and is black in color but unstable under aerobic conditions and will readily oxidize into maghemite (Laurenzi 2008, Zhang and Olin 2011, Virkutyte and Varma 2013). Therefore, magnetite is synthesized under anaerobic conditions.

10.5.3 Maghemite (γ-Fe_2O_3)

Maghemite is ferrimagnetic below its Curie temperature (645K). Its structure has a meta stable cubic iron oxide spinel structure and possesses Fe^{3+} charged states. It is magnetic and reddish brown in appearance and is often the product of weathered magnetite. It may also be produced by heating other iron oxides (Laurenzi III 2008, Zhang and Olin 2011, Virkutyte and Varma 2013).

10.5.4 Hematite (α-Fe_2O_3)

Hermatite is a weak ferromagnetic or antiferromagnetic iron species. It possesses a hexagonal closed-packed corundum, like maghemite, only with Fe^{3+} charged states. It is very stable under aerobic conditions and has a dark red color when ground or a black/grey color if coarsely crystalline. Hermatite can be formed from magnetite and maghemite upon heating to temperatures above 300°C and between 370–600°C, respectively (Laurenzi III 2008, Virkutyte and Varma 2013, Pereira et al. 2013).

10.5.5 Goethite (α-FeO(OH))

Geothite is an antiferromagnetic iron oxide that is very insoluble, particularly in acidic media. It possesses an orthorhombic unit cell or exists as rhombohedral prisms when nanoparticles. They are thermodynamically stable. When iron^{III} salts undergo hydrolysis, they form ferrihydrite. Upon exposure to higher temperatures (60°C) followed by environmental aging, they can form goethite (Madsen et al. 2009, Guo and Barnard 2011).

10.5.6 Lepidocrocite (γ-FeO(OH))

Lepidocrocite is generally regarded as paramagnetic; however, at extremely low temperatures (52 K), it can become antiferromagnetic (Hirt and Lanci 2011). It has an edge-shared octahedral structure and is relatively unstable under aerobic conditions favoring the formation of Goethite (Chesworth et al. 2008). Lepidocrocite can be formed by mixing solutions of iron salts (Fe^{3+} or Fe^{2+}) with an oxidant through precipitation or hydrolysis reactions (Cui et al. 2012).

10.5.7 Ferrihydrite

Ferrihydrite is an ordered iron^{III} oxyhydroxide whose structure is yet to be determined. Some researchers believe it is made up of various poorly ordered Fe^{III} oxyhydroxides (Carta et al. 2009). It is easily formed upon hydrolysis of iron^{III} salts and is generally considered to be antiferromagnetic; however, a decrease in particle size increases the magnetism of the particles (Carta et al. 2009, Guo and Barnard 2011, Masina et al. 2015).

10.5.8 Amorphous Mixtures

Crystalline structures are easily formed at high temperatures; however, amorphous structures can be generated upon rapid cooling. Furthermore, impurities in nanoparticle structures may also impact regularly arranged atoms/entities, leading to amorphous nature of particles. Boron, oxide, and hydroxides may impact the crystallinity of nanoparticles and reduce their reactivity (Phu et al. 2011). Conversely, Yoo et al. (2007) stated that amorphous iron oxide nanoparticles possess larger catalytic and magnetic attributes.

10.6 Parameters for the Application of Iron Nanoparticles to Bioremediation

10.6.1 Size

In many cases, nanotechnology is just a scaled-down version of current technologies (Hameri 2012); however, by reducing entities to the nanoscale different forces, interactions and properties are observed. Atoms and molecules are nanoscale entities and their interactions within their environment are heavily reliant on other entities of similar size.

In the field of nanoparticle research, interactions between small entities are observed that are not experienced with micro or macro dimensions. Optical properties may differ, that is, gold nanoparticles are red in color (Daniel and Astruc 2004), magnetism may become apparent (Carta et al. 2009, Guo and Barnard 2011, Masina et al. 2015), and so on. These interactions are often termed quantum size effects.

Long et al. (2015) indicated that nanoparticles with sizes of approximately 30 nm have superior quantum effects when used for catalysis, biology, and medical applications; however, Nurmi et al. (2005) reported that a particle size of less than 5 nm was optimal for maximizing the influence of quantum effects on the physical and chemical properties of nanoparticles. However, iron oxide nanoparticles with sizes between 10–150 nm will exhibit similar behavior, as they possess a lower electron density to other nonoxide or nonsemiconductive entities.

In the field of pollutant degradation for environmental preservation/restoration, iron nanoparticle size is particularly important. If the nanoparticle is too large, reactivity rates may be too slow to sufficiently catabolize the anthropogenic entity or may not be deliverable to the pollution source in the case of groundwater remediation where particles need to travel through complex geological matrices (Figure 10.3) to reach the pollutant (Patil et al. 2016). On the other hand, if the particle size is too small, nanoparticles are susceptible to rapid oxidation under aerobic conditions and may also react with entities found naturally in the environment. Due to their increased surface area, rapid oxidation and reactivity with other environmental constituents may exhaust the nanoparticles to the point where their metallic core is reduced in size so that insufficient concentrations of metallic iron remain to take part and complete remediation (Cundy et al. 2008).

Nanoparticles are classified as particles having at least one dimension between 1–100 nm (Jordan et al. 2005). In environmental remediation situations, a nanoparticle is a particle where all three dimensions are below 100 nm in diameter. Although it has been reported that small nanoparticles may react too fast for many site applications (Bardos et al. 2011), there is no accepted "optimal" size for nanoparticle mediated remediation, as particles generated via different synthesis methods will possess variations in porosity and surface area, particle size distribution, and impurities. Furthermore, coatings may be added to stabilize or increase the reactivity of particles (Grieger et al. 2010).

Maleki et al. 2012 optimized parameters for generating nanoparticles with tunable size characteristics. Different ratios of iron salts (Fe^{3+} and Fe^{2+}) were reduced with ammonium hydroxide in conjunction with surfactants (toluene, cetyl trimethylammonium bromide (CTAB), and 1-butanol, respectively). The findings revealed that the reducing agent had little effect on size control; however, tweaking the water, surfactants, and iron salt concentration had an impact on particle size.

When the molar ratio of water to surfactant was increased, the porosity and size of the nanoparticles also increased. When the concentration of iron salts was increased, so was the size of the nanoparticles. This is due to a high nucleation rate and fast nucleus growth rate. Increases in reducing agent concentration have been linked with the formation of smaller nanoparticles. Maleki et al. (2012), however, found only a small correlation, with only 25%–30% size difference experienced.

10.6.2 Agglomeration

Nanoparticle chemistry, when used for water remediation, can be identified as surface science, where physical and chemical interactions/reactions occur at the solid and liquid interphases. Perhaps the greatest benefits of progressing from the macro scale to the micro scale and then to the nano scale are the subsequent increase in particle surface area available for chemical reactions to take place. Increases in surface area lead to faster reaction rates and in turn more rapid pollutant degradation capacities. One of the greatest hurdles to overcome in nanoparticle research is to prevent attractive particle–particle electrostatic and magnetic interactions. These forces cause the particles to form microscaled fractional aggregates that are unfavorable for nanotechnological applications. Interactions between particles can be expressed as the sum of Van der Waals attraction and electrostatic double layer repulsion, as described by the DLVO theory (theory generated by Derjaguin, Landau, Verwey, and Overbeek). These forces are influenced by particle size, surface potential, the Hamaker constant, and the solution's ionic strength (Verwey 1948; Phenrat et al. 2007).

Aggregated nanoparticles do not behave as nanoscale entities, as they have limited mobility in porous media (only up to a few centimeters) (Phenrat et al. 2007), resulting in a significant loss in reactivity due to reduced surface area. Aggregated nanoparticles have a tendency to settle at faster rates in aqueous media, limiting their exposure time with dissolved or floating pollutants, and can block pores in porous media limiting mobility of both nanoparticles and groundwater flows. In order to achieve colloidal stability, researchers have been focusing their efforts on generating coatings and stabilizers.

In 2007, Huang and Ehrman generated bimetallic iron nanoparticles particles with palladium seeds via a chemical reduction approach. To prevent passivation and agglomeration, they incorporated a poly(acrylic acid) (PAA) solution into an iron salt solution (containing palladium chloride), which they reacted with sodium borohydride to form stable nanoparticles with particle–particle repulsive properties. Repulsive properties were greater when iron nanoparticles were generated without palladium salts.

10.6.3 Shape and Particle Uniformity

Nanoparticles of various shapes have been synthesized, that is, spheres, rods, tubes, triangles, pyramids, and cubes, just to name a few (Sajanlal et al. 2011). The shape of nanoparticles may be of importance depending on their application. For instance, nanotubes and wires have been implicated in the conduction of electricity and could benefit the electronics industry (Li et al. 2011a). However, spherical nanoparticles may have fewer edges, which will benefit movement through porous media in groundwater remediation and will have the largest surface area-to-volume ratio for their mass, leading to optimum reactivity (Alam and Haas 2015).

For site remediation, it is important that nanoparticles be delivered to the pollution source quickly and in sufficient concentrations. To ensure this is possible without adding excessive nanoparticles to a specific site, it is important that variations in particle size be minimized. The addition of excess particles to a particular site is unfavorable, as remediation costs are elevated, porous media can be saturated, and the pores can be blocked, preventing particle delivery. In addition, the potential for nanoparticulate-induced toxicity for ecosystems is enhanced. Nanoparticle uniformity allows researchers to better predict reactivity and delivery rates in porous media. Carrying out preliminary simulations is important for assessing site clean-up costs and effectiveness of nanotechnology as a treatment option (Cameselle et al. 2013).

Jana et al. 2004 were able to control the size and shape of iron oxide particles by controlling the reactivity and concentration of the precursors. Simply, iron salts were dissolved in methanol and oleic acid. Sodium hydroxide was then added to form a brown precipitate. The brown precipitate was digested in hydrochloric acid and dissolved in 1-octadecane at 60–70°C. The aforementioned solution was mixed with octadecane and oleac acid and heated up to 300°C in anaerobic conditions. Increasing oleic acid concentrations increased the particle size and extended heating times led to the formation of cubic particles over spherical particles when specific precursor concentrations were used.

10.6.4 Aging/Storage Times/Stability/Passivation

Researchers aim to create nanoparticles for their enhanced reactive properties; however, due to their rapid tendency to react with oxygen and to some extent water, coatings are often generated that reduce their reactivity and extend their storage capacities (Zhang 2003).

1. $2Fe^0 + 4H^+ + O_2 \rightarrow 2Fe^{2+} + 2H_2O$
2. $2Fe^0 + 2H_2O \rightarrow Fe^{2+} + H_2 + 2OH^-$

Coatings have been generated to improve nanoparticle life in aerobic environments; however, nanoparticles have the capacity to catabolize pollutants in anaerobic environments also and may also need to be deliverable to the pollution source before they start to react.

Iron nanoparticles, in particular Fe^0, readily oxidize under aerobic conditions and will form an oxide shell made up of either hydroxides or oxyhydroxides. The core of these particles is shielded from the external environment and can therefore remain as metallic iron (Fe^0). Further weathering utilizes the core to generate more oxides, increasing the thickness of the outer layer, decreasing the percentage of metallic iron, and reducing the reactivity of particles (Grieger et al. 2010).

During contaminant degradation, particle reactivity may also have a role to play in Fe^0 utilization and or oxide layer thickness. Kim et al. (2012) reported that iron nanoparticles generated with the borohydride reduction method (95% Fe^0 content) generated particles where the oxide layer thickness remained the same but the Fe^0 core shrank during pollutant remediation. Liu and Lowry (2006) performed similar tests using iron nanoparticles generated by the reduction of iron oxides by H_2 at high temperatures (48% Fe^0 content). Their observations were different and showed that the thickness of the oxide shell layer increased while the Fe^0 core decreased during pollutant remediation.

Kim et al. (2012) showed that a number of factors were responsible for the utilization of the Fe^0 core of their borohydride generated nanoparticles when used to remediate nitrate in aqueous solutions. The following reactions were responsible for degeneration of the iron core (i) nitrate reduction (39.1%), (ii) initial shell modifications (25%), (iii) hydrogen production (possibly 5%), (iv) oxidation of Fe^0 by dissolved oxygen (1.2%), and (v) acid flushing for nZVI regeneration (possibly 9%). The remaining Fe^0 expected to remain in their experimental vessels was approximately 20.7% of the original concentration. O'Hara et al. 2006 generated emulsified nZVI nanoparticles by adding the particles to a vegetable oil solution mixed with surfactants. These particles have properties that prevent their interaction with atmospheric oxygen and groundwater constituents that may pacify the nanoparticles (Figure 10.3). Emulsified nZVI particles were used to treat trichloroethylene (TCE) contamination in groundwater with degradation rates of between 60%–100% at all depths targeted.

10.7 Synthesis of Iron Nanoparticles

There are many different methods for producing nZVI. A number of the most utilized techniques include co-precipitation, reduction of aqueous iron salts, hydrothermal synthesis, thermal deposition, microwave, microemulsion, ball milling, sonochemical and sol-gel, and forced hydrolysis techniques (Figure 10.4).

10.7.1 Ball Milling

Ball milling is a physical approach to the generation of nanoscaled materials, including crystalline structures, nanoparticles, and nanocomposites using mechanical energy. Starting materials (i.e., iron oxides) are powdered and then put into a milling chamber containing balls (made from silicon, steel, iron, carbide, or tungsten) (Li et al. 2009). The milling chamber rotates, and the powdered iron/iron oxide is crushed against the wall

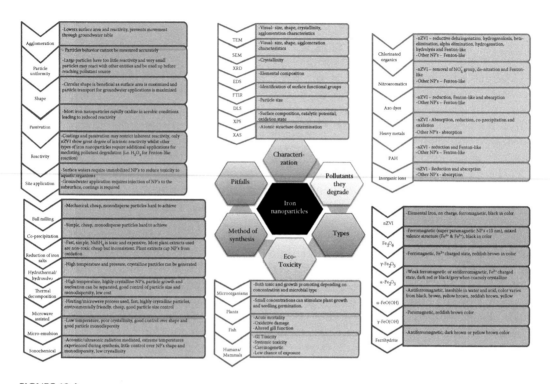

FIGURE 10.4
The complexity involved in the generation of iron nanoparticles for environmental remediation purposes.

of the vessel. The balls aid in crushing the iron particles, leading to smaller particle sizes. Pressures experienced with the crushing of the parent material lead to compositional changes, including hardening, fracturing, and rewelding of powdered particles. Milling mixtures of different materials can lead to chemical reactions and subsequent formation of new materials (Andrews et al. 2010). This method of nanoparticle synthesis is simple and cheap. Surfactants or stabilizers can be added to the freshly ground nanoparticles to prevent agglomeration and passivation (Poudyal 2008). Nanoparticles with monodispersed attributes are difficult to achieve via this method. However, some degree of particle sizing can be controlled by increasing temperatures and milling time (Safarik and Safarikova 2012).

10.7.2 Co-Precipitation

Co-precipitation is perhaps the simplest method of generating iron oxide nanoparticles (mainly used to generate Fe_3O_4, γ-Fe_2O_3, or FeOOH species). Simply, nanoparticles are generated by aging stoichiometric mixture of ferric (Fe^{3+}) and ferrous (Fe^{2+}) salts in a 1:2 M ratio, respectively, under highly aqueous alkaline medium.

As magnetite particles are particularly susceptible to oxidation, this process is usually performed under anaerobic conditions using an inert gas, that is, nitrogen. It is very difficult to generate monodispersed particles using this method. However, in an attempt to control and or alter the size, shape, and monodispersity of the particles, a number of different approaches have been taken.

Factors such as differing iron salts in different ratios, changes in reaction temperatures, optimization of pH conditions, mixing speed of solution, speed of reducing agent addition to iron salt solution, and ionic properties of the media have been all considered and optimized (Wu et al. 2008, Wu et al. 2015). To further maintain and control particle physical attributes, stabilizers and coatings are often incorporated to reduce passivation, agglomeration, size, and uniformity.

10.7.3 Reduction of Aqueous Iron Salts

The most common method for the generation of nZVI for environmental purposes is the reduction of iron salts using sodium borohydride under anaerobic conditions. The borohydride reduction of iron salts can proceed via the following ways:

1. $Fe(H_2O)_6^{3+} + 3BH_4^- + 3H_2O \rightarrow Fe^0 \downarrow + 3B(OH)_3 + 10.5H_2$
2. $4Fe^{3+} (aq) + 3BH^-_4(aq) + 9H_2O(l) \rightarrow 4Fe^0 (s) + 3H_2BO^-_3(aq) + 12H^+ (aq) + 6H_2(g)$
3. $2Fe^{2+} (aq) + BH^-_4(aq) + 3H_2O(l) \rightarrow 2Fe^0 (s) + H_2BO^-_3(aq) + 4H^+ (aq) + 2H_2(g)$

Although this method is fast and simple, sodium borohydride is a toxic compound that is likely to contaminate the environment. Ideally, it should therefore be substituted for more environmental alternatives (Crane and Scott 2012). However, in medical applications where smaller nanoparticle quantities are required, this synthesis technique may be appropriate. Sodium borohydride-generated particles are susceptible to spontaneous combustion in air and accelerated weathering in aerobic solutions. For these reasons, coatings are often generated to prevent these occurrences (Yan et al. 2013).

One such alternative is the use of polyphenols from plant leaves or fruits. Not only are the antioxidative extracts from plant biomass more environmentally friendly than sodium borohydride, but they also provide a protective capping for the nanoparticles. This capping reduces the initial passivation rate by shielding the particles from atmospheric and dissolved oxygen. When using plant-derived polyphenols for nanoparticle synthesis, the end product may not be nZVI due to the incomplete reduction of iron and therefore thorough x-ray diffraction (XRD) analysis should be used to confirm the composition of generated particles (Dorofeev et al. 2012).

Green synthesis is cheap, as costly chemicals are not necessary, and the process does not require increased temperatures, pressures, or additional energy inputs (Stefaniuka et al. 2016). Different plant extracts can lead to differing size, shape, particle uniformity, and agglomeration characteristics. As plant biomass weight is not directly related to its polyphenol-containing concentration, measurements of the reducing capacity of extracts may be required between preparations to ensure reproducible results.

10.7.4 Hydrothermal/Hydrosolvo Synthesis

Hydrothermal synthesis, as the name suggests, utilizes supercritical water to generate nanoparticles. Hydrosolvo synthesis uses the same technology; however, solvents other than water are often used to generate species with specific attributes. Reactants are placed into aqueous reactors or autoclaves and subjected to temperatures that can exceed 300°C and pressures over 2000 psi. During these conditions, the water/solvent hydrolyzes and dehydrates metal salts. In the autoclave, iron salts and stabilizers can be dissolved in concentrations that are not soluble at ambient conditions. There are also regions within the vessel with lower temperatures that allow for seeding and particle growth to occur (Hayashi and Hakuta 2010; Safarik and Safarikova 2012).

Nanoparticles generated via this method possess ordered structures with a high degree of crystallization. Size and shape characteristics can be modified by altering the reaction time temperature, pressure, and solvent (Wu et al. 2015).

10.7.5 Thermal Decomposition

As previously mentioned, in order to obtain nanoparticles with high degrees of crystallinity, they must be subjected to high temperatures in many cases. Thermal decomposition is a method of generating nanoparticles via the thermal decomposition of organometallic precursors, often organometallic complexes. Iron precursors may be thermally decomposed via hot-injection methods, or prepared reaction mixtures can be heated in a reaction vessel at temperatures that may exceed 300°C (Tartaj et al. 2003; Frey et al. 2009). Due to the tendency of iron and certain iron oxides (i.e., Fe^0 and magnetite, respectively) to rapidly oxidize in aerobic conditions, the generation of particles via thermal decomposition is often conducted with coatings/stabilizers including hot organic surfactants. Nucleation/seeding can be separated from growth phases via this synthesis method, unlike co-precipitation methodologies. This aids in the generation of monodispersed particles and the control of size parameters by optimizing temperature, decomposition time, and ferric salts used (Wu et al. 2015). This technology is regarded as simple to use and low in cost, and has the ability to generate high-quality nanoparticles on large scales (Safarik and Safarikova 2012).

10.7.6 Microwave-Assisted Synthesis

It has been shown that nanoparticle synthesis methods incorporating high temperatures can aid in improving or instilling crystallinity in nanoparticles. In microwave-assisted synthesis of nanoparticles, two heating mechanisms are utilized. First, dipolar polarization, in which microwave radiation forces molecules to align their dipoles within the electric field. The oscillating field reorients the molecules, and the dipoles attempt to realign, causing friction and subsequent heat generation and loss.

Ionic conduction is the second heating method, and is a result of charged ions vibrating due to their excitement from the microwave field. This vibration leads to collisions with molecules or atoms, resulting in heat generation (Baghbanzadeh et al. 2011, Wu et al. 2015, Kalyani et al. 2015).

Due to significant and rapid heating, crystalline nanoparticles can be generated in less than half an hour. This NP synthesis method also provides control over particle size distribution, enhanced physiochemical properties, and generation of crystalline particles. It is seen as being faster, more environmentally friendly, and cheaper than conventional NP synthesis methods (Safarik and Safarikova 2012, Singh and Nakate 2013).

10.7.7 Microemulsion

Microemulsion-mediated iron nanoparticle synthesis consists of a nonpolar oil phase, a surfactant phase (containing both hydrophobic and hydrophilic groups), and an aqueous polar phase (Wu et al. 2015). The hydrophobic ends of the surfactant dissolve into the oil phase, and the hydrophilic ends dissolve into the aqueous phase. A cosurfactant may also be used to lower the interfacial tension (Mohapatra and Anand 2010).

The aqueous phase generally contains metal salts, and the oil phase may contain complex mixtures of hydrocarbons, olefins or other nonpolar entities. Small droplets of the immiscible phase become dispersed into the continuous phase (this can be either nanosized aqueous phase droplets dispersed in an oil phase stabilized by surfactants or vice versa) (Mohapatra and Anand 2010). In these nanocavities, self-assembled nanoparticles can generate. The size of the nanoparticles is generally limited to the dimensions of the formed nanocavities and therefore monodispersed nanoparticles with spherical shapes are often generated. Size parameters can be controlled by altering the concentrations of the dispersed and surfactant phases. Due to the low temperatures experienced via this synthesis method, the nanoparticles generated often have poor crystallinity (Hasany et al. 2012).

10.7.8 Sonochemical

In sonochemical synthesis of nanoparticles, water and other secondary organic compounds are subjected to ultrasonic radiation (20 KHz–10 MHz), where acoustic waves create bubbles (cavities) that vibrate and accumulate ultrasonic energy. When a bubble grows too large due to diffusion of the solute vapor, it collapses and releases the stored energy resulting in an implosion with temperatures over 4500°C experienced (temperatures that are rapidly quenched) and pressures up to 1800 atm. This process generates H• and OH• radicals (the addition of organic additives can produce other radical species), which act as reductants. These radical species reduce iron salts to form iron oxide nanoparticles (Bang and Suslick 2010, Mousavi and Ghasemi 2010). Nanoparticle synthesis using this approach has little control over shape and monodispersity of particles. Furthermore, particles with amorphous structures are often formed. This can be beneficial for enhanced reactivity in some cases (Tables 10.1 and 10.2). Furthermore, particles generated by this method have been shown to remain suspended in solution without settling for over 12 months (Teo et al. 2009, Wu et al. 2015).

TABLE 10.1

Types of Nanoparticles Generated from Differing Synthesis Approaches

Method of Synthesis	Nanoparticle Generated	References
Ball milling	nZVI, γ-Fe_2O_3, α-Fe_2O_3	• Li et al. (2009) • Lemine et al. (2009)
Reduction (sodium borohydride)	nZVI, FeOOH, Fe_3O_4, α-Fe_2O_3	• Agarwal et al. (2011) • Farahmandjou and Soflaee 2014
Reduction (plant extracts)	Fe_3O_4, FeOOH, α-Fe_2O_3, amorphous NP's	• Ahmmad et al. (2013) • Herlekar et al. (2014) • Shahwan et al. (2011)
Co-precipitation	Fe_3O_4, γ-Fe_2O_3 or FeOOH species, amorphous NP's	• Mohapatra and Anand (2010) • Wu et al. (2008)
Hydrothermal/hydrosolvo synthesis	α-Fe_2O_3, FeOOH, Fe_3O_4	• Mohapatra and Anand (2010) • Wu et al. (2008)
Thermal decomposition	Fe_3O_4, α-$Fe2O3$, γ-Fe_2O_3	• Wu et al. (2008)
Microwave-assisted synthesis	Fe_3O_4, γ-Fe_2O_3, α-Fe_2O_3, FeOOH, amorphous NP's	• Mohapatra and Anand (2010)
Microemulsion	Fe_3O_4, α-FeOOH, γ-Fe_2O_3, α-Fe_2O_3, amorphous NP's	• Mohapatra and Anand (2010) • Petrova et al. (2006) • Wu et al. (2008) • Avanaki and Hassanzadeh (2013) • Cheng et al. (2011)
Sonochemical	nZVI, Fe_3O_4, Fe_2O_3, amorphous NP's	• Mohapatra and Anand (2010) • Wu et al. (2008)

TABLE 10.2

Surface Modifications for Iron Nanoparticles, Their Properties, and Advantages for Their Use

Surface Coating	Properties	Advantages	References
Natural polymers, that is, starch	• Organic polymer • Various different coatings with different properties (dependent on polymer used)	• Easy to source • Capping agent • Reduce agglomeration • Advantageous for negatively charged metal ion remediation • Nontoxic • Cost effective • Environmentally friendly	• He and Zhao (2009)
Carboxymethyl cellulose (CMC)	• Linear • Anionic • Nongelling polymer • Negatively charged • Polyelectrolyte • High molecular weight	• Negative charge prevents attraction of many groundwater species and reduces agglomeration • Strong binding to nanoparticle • Advantageous for TCE degradation (with increasing chain length) • High density of charged functional groups • Enhances remediation due to absorption	• Liang et al. (2013) • Kim et al. (2009) • Sakulchaicharoen et al. (2010) • He et al. (2009)
Polyvinyl alcohol (PVA)	• Metal-ion-polymer-complex formed • Hydrophilic • Can transform into polymer gel	• Structure can be modified, that is, cross-linked, chemically modified • Reduces agglomeration • Stable dispersions • Nontoxic	• Bepari et al. (2014) • Kundu et al. (2011) • Liu et al. (2013)
PMAA-PMMA-PSS triblock	• Triblock coating • Block 1—poly methacrylic acid (carboxylic acid) • Block 2—poly methyl methacrylate (low polarity) • Block 3—polystyrenesulfonate (strong electrolyte)	• Three coatings, each providing beneficial properties • Block 1—capping agent for protection from oxidation • Block 2—targeting of NAPL pollutants • Block 3—particle–particle electrosteric repulsion	• Sirk et al. (2009) • Phenrat and Lowry (2014) • Krajangpan et al. (2012) • Saleh et al. (2005)
Mesoporous carbon/silica coating	• Inert • May act as an absorbent	• Large surface area • Pore size can be altered • Limited particle–particle attraction • High thermal stability • Potential to absorb and then degrade pollutants	• Petala et al. (2013) • Tang et al. (2014) • Knežević (2014) • Zamani et al. (2009)
Emulsified nZVI	• Viscous • Biodegradable • Hydrophobic	• Interacts with NAPL pollutants by dissolving to expose iron • Protects nanoparticles from oxidation and hydrolysis • May promote bioremediation • Enhances reactivity of nanoparticle	• O'Hara et al. (2006) • Müller and Nowack (2010) • Yanga and Chang (2011) • Rajan (2011)

(Continued)

TABLE 10.2 (*Continued*)
Surface Modifications for Iron Nanoparticles, Their Properties, and Advantages for Their Use

Surface Coating	Properties	Advantages	References
Biological capping using plant- or algal-based polyphenols	• Water soluble • Various different coatings with different properties (dependent on plant extract used)	• Environmental friendly • May promote bioremediation • Easy to apply • Prenanoparticle synthesis capping from oxidation • Cost effective	• Machado et al. (2015) • Shahwan et al. (2011) • Pattanayak et al. (2013) • Mahdavi et al. (2013) • Pattanayak and Nayak (2013) • Machado et al. (2014)
Immobilization	• Fixed nanoparticles • Loaded within matrix or on surfaces of different materials	• Decreases aggregation potential • Removable from system, reducing environmental impact • Can be used for surface water remediation • Optimum colloidal properties • Capacity to incorporate catalysts that would otherwise be toxic to the environment for enhanced remediation • Potential pollutant-absorptive properties	• Petala et al. (2013) • Fu et al. (2013) • Tang et al. (2015) • Luo et al. (2013) • Abbassi et al. (2013) • Arancibia-Miranda et al. (2016)

10.7.9 Optimization

Iron nanoparticles for environmental remediation purposes are a current and successful technology; however, they are far from perfect. Different contaminated environments require different attributes for optimum environmental regeneration. For example, if using nanoparticles for groundwater remediation, particles should possess coatings to facilitate movement through the underground matrices. However, in surface water, nanoparticles may need to be affixed onto solid supports so they can be retrieved and prevent the release of excessive nanoparticles into the environment (Figure 10.5).

There is a vast plethora of different attributes that nanoparticle researchers are aiming to optimize. Some issues that are in focus relate to desired shape, crystallinity, passivation reduction, increased reactivity, increased mobility, agglomeration reduction, coatings that target pollutants, solid support matrix materials, and reduction in synthesis costs.

10.8 Different Methods of Generating Coatings

Currently, there is significant research in the area of generating stabilizers and coatings for nZVI particles. Bare or naked iron nanoparticles may be highly reactive; however, they cannot be used for *in situ* remediation projects due to a number of inherent pitfalls.

Particle agglomeration reduces the surface area available for reactions to occur on nanoparticle surfaces and in turn reduces remediation efficiency. Also, agglomerated nanoparticles are unable to travel significant distances through the underground matrix for

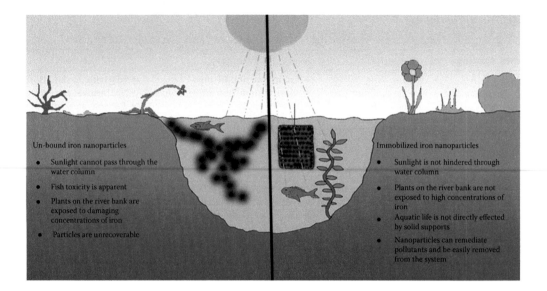

FIGURE 10.5
Comparison between the use of free-iron nanoparticles and immobilized iron nanoparticles for surface water remediation.

groundwater remediation. Coatings may be applied to nanoparticles to prevent the attractive interactions they have with one another (Figure 10.4). Coatings are also added to particles to help them overcome attractive forces of groundwater constituents while targeting pollution sources. Passivation of particles may also be counteracted via surface coatings. These coatings will greatly increase shelf lives and prevent rapid oxidation. Oxidized particles have lower intrinsic reactivity and may be rendered useless for application.

Coatings for nanoparticles are generally applied presynthesis, postsynthesis, or a combination of both. Presynthesis methods usually involve the addition of the coating to the precursors, that is, iron salt solution and reductant addition, while postsynthesis methods, as the name suggests, involve the coating of the nanoparticle bulk material. Generally, iron nanoparticles with post-synthesis coating application show reduced reactivity, whereas presynthesis methods may increase reactivity and aid in controlling particle size, shape, and uniformity.

Although commercially used nanoparticles are currently used for remediation purposes, they still have shortcomings in regards to reactivity, transport in underground matrices, and overall pollution catabolism (Mueller et al. 2012).

Some of the commonplace and emerging coating technologies are described below:

10.8.1 Natural Polymers—Starch

Polysaccharides, including starch and cellulose, are highly abundant and easy-to-source plant biomass components that have been shown to be effective stabilizers for nanoparticles. These entities prevent oxidation and particle–particle attractive forces. Starch is a linear, neutral, and gelling polymer. In some circumstances, that is, the absorption of negatively charged metal ions, the use of starch stabilizers may be beneficial over carboxymethyl cellulose (CMC)-stabilized particles (He and Zhao, 2009).

10.8.2 Polyelectrolyte-Carboxymethyl Cellulose

Carboxymethyl cellulose is a linear, anionic, and nongelling polymer (Liang et al. 2013). Negatively charged polymers are seen as beneficial for *in situ* remediation, as they repel attractions from negatively charged species, that is, soil particles or oxidized humic acids. Polysaccharide coatings for nanoparticles can be applied via both pre- and postsynthesis routes. Kim et al. (2009) showed increases in mobility through porous media with CMC modified with a posttreatment method of nZVIs, and Sakulchaicharoen et al. (2010) showed enhanced stability and dechlorination of bimetallic Fe/Pd nanoparticles using a presynthesis approach. CMC has a stronger affinity to iron nanoparticles compared to that of starch, as it stabilizes particles via electrostatic repulsions as opposed to steric hindrance (Liang et al. 2013). It has been shown that CMCs with increasing chain length have a greater capacity to degrade TCE (He et al. 2009, Fan et al. 2015).

10.8.3 Polyvinyl Alcohol

Polyvinyl alcohol (PVA) generation/coating of iron nanoparticles is a simple procedure whereby PVA aqueous solution is used to precipitate iron salts. PVA's OH groups form a metal-ion-polymer complex, and the resulting nanoparticles are hydrophilic and form electrostatic and steric repulsions, leading to stable dispersions of relatively monodispersed particles (Kundu et al. 2011, Bepari et al. 2014). One drawback of using PAA alone is that it instantly dissolves in water and therefore is short lived for many environmental applications (Liu et al. 2013). Modified PAA coatings (cross-linked PVA or chemically modified structures) have provided additional adsorption onto nanoparticles and reduced solubility in aqueous solutions. However, their use as a host matrix for block addition of further coatings has also been reported (Gao and Seliskar 1998). This polymer is a biodegradable synthetic polymer and therefore its use within *in situ* environments is possible.

10.8.4 Poly(Methacrylic Acid)-Block-(Methyl Methacrylate)-Block-(Styrenesulfonate) Triblock

Poly(methacrylic acid)-block-(methyl methacrylate)-block-(styrenesulfonate), or the PMAA-PMMA-PSS triblock coating, has been specifically engineered to combat a number of pitfalls associated with bare nanoparticles. Poly methacrylic acid is a carboxylic acid that has a strong affinity to adsorb to iron nanoparticle surfaces. This block acts as a capping agent that reduces rapid passivation of the particle while presenting functional groups that will interact with the poly methyl methacrylate coating. The PMMA block not only creates a low polarity region that aids in thermodynamically targeting nonaqueous phase liquids (NAPLs) pollutants, but also prevents water from reaching the particle. This layer prevents passivation and collapses upon interacting with NAPL, allowing the encapsulated nanoparticle to react and catabolize the pollutant. The third block, the polystyrenesulfonate coat, consists of a strong electrolyte that generates electrosteric repulsions between particles so that they remain independent and do not agglomerate. Furthermore, the negative charge on this coat also minimizes particles interactions with negatively charged groundwater species. In summary, PMAA-PMMA-PSS-coated nanoparticles are electrosterically independent particles that target NAPL pollutants and possess extended shelf lives compared to uncoated nanoparticles. Due to electrosteric particle–particle repulsions and subsequent reduction in agglomeration potential, their capacity to travel through groundwater matrices is superior to uncoated nanoparticles. This coating is applied after synthesis of nanoparticles (Saleh et al. 2005, Sirk et al. 2009, Krajangpan et al. 2012, Phenrat and Lowry 2014).

10.8.5 Mesoporous Carbon/Silica Coated

Iron nanoparticles containing mesoporous carbon or silica coatings have had a large focus in the medical nanotechnology field, as drugs or ligands can be added in larger concentrations than bare nanoparticles. This is due to the large number of pores that increase the surface area and the fact that carbon and silica are inert and therefore will not cause secondary reactions within the body (Zamani et al. 2009).

Little research has been conducted with mesoporous silica/carbon nanaoparticles in terms of remediation with NAPL pollutants; however, a significant focus has been directed toward the absorbance of toxic metal ions (Petala et al. 2013). These particles possess large specific surface area, pore size that can be altered, hydrophobicity, high thermal stability, limited particle–particle attractions, and increased magnetism (Tang et al. 2014). Both mesoporous silica and carbon possess absorbent properties that can assist in retrieving metal ions from solutions and in turn speed up the rate of remediation. Addition of these coatings is conducted using postsynthesis techniques. There are a number of mesoporous silica coatings (i.e., SBA-15, KIT-6, MMSN) and mesoporous carbon coatings (i.e., CMK-3, CNC, OMC) being used and researched (Alam et al. 2009, Knežević 2014).

10.8.6 Emulsified Nanoscale Zero-Valent Iron

Iron nanoparticles need to be in contact with water in order to achieve reductive dehalogenation. However, NAPL pollutants are insoluble in water and therefore surface areas available for reaction and/or transport to the oil/water interface may be limited.

Emulsified iron nanoparticles are a technology where nanoparticles are suspended in an aqueous solution, that is, a water droplet, which is then further encapsulated by a surfactant and biodegradable oil (usually a vegetable oil). This technology aids in both transport of particles to the nonedge areas of NAPL pollutants and to thin out NAPL to increase the surface area available for degradation (O'Hara et al. 2006, Müller and Nowack 2010). Reactive nanoparticles are protected against reactive groundwater constituents, passivation is decreased, and breakdown constituents of large NAPL pollution molecules can dissolve in the oil, reducing their prevalence in the groundwater's aqueous phase. Due to pollutants dissolving through the oil layer, a concentration gradient is achieved so that nanoparticles are not overwhelmed by reactants and steady degradation rates can be achieved (Rajan 2011, Yanga and Chang 2011).

Due to the increased viscosity of nanoparticle solution because of the oil, emulsified nanoparticles are only useful in sandy aquifers or for direct injection to the polluted zone, as low permeable soils will not facilitate transport. However, in optimum geographical conditions, their effectiveness against chlorinated solvents has been shown (Yanga and Chang 2011).

The majority of research conducted with emulsified nZVI is directed toward remediation of NAPL pollutants within groundwater scenarios. This is due to the hydrophobic/oil layer being beneficial for dissolving and interacting with these pollutants and its role in facilitating transport through the underground matrix. Quinn et al. (2006) utilized emulsified nZVI for a different approach. Sea water is particularly corrosive toward iron and therefore the use of iron nanoparticles for ocean remediation is not a highly researched area in the field. However, Quinn et al. (2006) were able to demonstrate that emulsified nZVI was able to withstand rapid weathering from sea water salts and also remove over 90% of lead and copper, with initial concentrations of these metal pollutants being at 5 mg/L (Quinn et al. 2006).

10.8.7 Biological Capping Agents-Plant/Algal-Based Polyphenols

Reducing iron salts with plant-based polyphenol solutions is an environmentally friendly and simple method for capping and generating iron nanoparticles. Simply, plant biomass is heated in aqueous solutions below 100°C and then filtered to remove solids. The resulting solution is reacted with iron salts, reducing them into nanoparticles, and a plant-derived coating is generated, preventing them from oxidation.

A number of plant extracts have been used to generate capping properties for iron nanoparticles. Machado et al. (2015) compared extracts from 26 different tree species to determine differences in particle size, shape, reactivity, and agglomeration tendency. Findings from this study showed that different extracts generated particles with different properties; however, the vine and medlar extracts were considered best for site remediation (Machado et al. 2015).

In the vast majority of literature published, plant extracts for nanoparticle synthesis and coating resulted in agglomerated particles (Machado et al. 2015 [various plant extracts], Shahwan et al. 2011 [green tea], Pattanayak et al. 2013 [clove], Mahdavi et al. 2013 [seaweed], Pattanayak and Nayak 2013 [neem tree]). However, exceptions do exist (e.g., Machado et al. 2014 [mandarin]).

10.8.8 Immobilized Nanoparticles on Solid Supports

Iron nanoparticles for the remediation of pollutants in surface waters should be used in conjunction with either permeable reactive barriers or loaded onto immobilized solid supports. Due to their small size, retrieval of free particles is virtually impossible in uncontrolled natural environments. In addition, free iron nanoparticles have been shown to adversely affect fish and their embryos (Chen et al. 2013), which can be detrimental for aquatic health (Figure 10.5).

Considerable research is being conducted in this area with nanoparticles being immobilized onto clays, mesoporous silica, rectorite, resins, alginate, multiwalled carbon nanotubes, and bentonte, and by using two solvent-reduction methods, just to name a few recent examples. Solid-support immobilization of iron nanoparticles allows particles to be added to surface waters and then removed when reactions are complete or nanoparticles are pacified. Nanoparticles may either be synthesized and then added to immobile supports or generated directly onto solid supports. The latter has been shown to reduce large variations in size fluctuations and decrease aggregation potential (Petala et al. 2013). Additionally, some coatings (i.e., zeolites and SBA-15) can increase the catalytic and sorptive properties of iron nanoparticles.

A study conducted by Fu et al. (2013) used ion exchange resins as a solid support matrix for nZVI nanoparticles, as they have been implicated in the removal of either cations or anions from metals in solution. Upon stabilization of nZVI to the resin, the resin/nZVI support was able to remove both anions and cations. At optimal conditions, >84% of Cr(VI) was removed from solution when the initial Cr(VI) concentration was 20mg/L.

Tang et al. (2015) used a dual-solvent method for impregnating SBA-15 (mesoporous silica) with nZVIs. SBA-15 was mixed with n-pentane and then a ferrous solution was added dropwise and allowed to dry. Under anaerobic conditions and in the presence of n-pentane, the supports were exposed to sodium borohydride, forming immobilized nZVI onto SBA-15 supports. These supports were used to degrade p-nitrophenol, a nitro aromatic compound. Following 30 days of aerobic exposure, >80% of p-nitrophenol was degraded.

Immobilized iron nanoparticles on clay supports have been the focus for a number of researchers (Abbassi et al. 2013, Luo et al. 2013, Arancibia-Miranda et al. 2016) due to their pollutant absorption capacity, low cost, and optimal colloidal properties. Luo et al. (2013) generated nanoscale zero valent iron-rectorite (nZVI-R) composites for the removal of orange II dye. Very small nZVI nanoparticles (10 nm) were immobilized between the interlayer regions of the rectorite clay. This ability of this composite to degrade Orange II dye (70mg/L) was compared to commercial nZVI and proved to be superior, achieving complete removal of the parent compound within 10 minutes.

10.8.9 Bimetallic Particles

For contaminant remediation, iron nanoparticles (mainly nZVI) have been considered optimal when compared to other monometallic nanoparticles due to their cost effectiveness, high reactivity, and relatively low toxicity. However, it has been shown that the addition of a metal catalyst can increase reactivity. As iron nanoparticles have a negative redox potential, noble metals with a positive redox potential (i.e., Cu, Ag, Pd, Pt, and Ni) are combined with iron to form bimetallic nanoparticles. With these particles, iron can either be a surface for pollutant absorption/adsorption or act as an electron donor for interface reductive degradation of pollutants (Liu et al. 2014a,b).

To date, there are a number of commercially available iron nanoparticles available for environmental site remediation. Bimetallic particles are less prevalent in the marketplace despite their exceptional degradation efficiencies due to the uncertainties related to their environmental toxicity of the noble catalyst (O'Carroll et al. 2013). Furthermore, bimetallic nanoparticles are subject to elevated corrosion rates and therefore optimization of Fe to catalytic metal and the application of coatings to increase shelf lives may be required (Schrick et al. 2002).

An exception to the rules can be seen with iron/palladium bimetallic particles. They are sold commercially, have shown reduced corrosion compared to their Fe^0 counterparts (Liu et al. 2005), and their usage has, in the degradation of trichloroethane, resulted in a reduction in up to two orders of magnitude of TCE compared to iron nanoparticles alone. The addition of palladium to iron nanoparticles has been shown to be environmentally benign in groundwater conditions, as it is likely to remain insoluble and therefore will not be bioavailable (Cook 2009).

10.9 Characterization Techniques

10.9.1 Scanning Electron Microscopy

Scanning electron microscopy (SEM) is routinely used within nanoparticle research. SEM's primary function is to provide information pertaining to the surfaces of specific entities. Samples to be imaged using SEM must have a surface that is electrically conductive. For dried metallic nanoparticle imaging, no further sample preparation will be required. However, if nanoparticles possess a thick organic surface coating or are immobilized onto a surface without electrical properties, they may require further preparation. Biological samples may first require critical point drying followed by sputter coating of electrically conductive material (often gold or platinum). Nonorganic samples may only require sputter coating before imaging can proceed.

Once a sample is loaded onto the stage, it is irradiated with a high-energy electron beam that, in most circumstances, scans across the sample. There are a number of signals produced when the electron beam hits a sample. These include secondary electrons, backscattered electrons, Auger electrons, characteristic x-rays, and photons of various energies. Although some information relating to composition and crystallinity may be derived from Auger electrons, characteristic x-rays, and photons, the data will not be quantitative. Secondary and backscatter electrons are mainly used for sample imaging. Secondary electrons are mainly responsible for mapping out the surface topography and morphology of samples, while backscattered electrons illustrate the contrast in composition for multiphase samples (Goldstein et al. 2017).

Scanning electron microscopy is a particularly powerful tool for visualizing milli- to nanoparticulate matter; however, in many cases SEM does not have the necessary resolution to clearly image nanoparticles between 1–100 nm in size. If detailed surface imaging is not necessary or possible, the use of a transmission electron microscope (TEM) is used for particle imaging.

10.9.2 Transmission Electron Microscopy

When obtaining visual images for individual particle shape, size, oxide layer, and agglomeration tendency, TEM is the one-stop tool for the job. Like SEM, TEM uses electrons to generate images of miniature-sized samples. In TEM, the electrons interact with the sample as they are transmitted through it (Chekli et al. 2015). The transmitted electrons form an image on a fluorescent screen that can be imaged using a charge-coupled device (CCD) camera. Specimens imaged by TEM need to be thin enough in at least one dimension for electrons to pass through them (<50–100 nm). The interactions between the electrons and the specimen's internal matter aid in generating images that can give information about internal components of entities. For example, in a polymer surface-coated iron oxide nanoparticle, there will be differences in contrast between the surface coating, the oxide layer, and the metallic iron core.

When determining the crystallinity of a sample by means of crystal lattice structure, electron scattering and diffraction are important parameters to test. When the electron beam hits the sample in a TEM, the electrons are not truly transmitted but scattered in a forward direction. With diffraction patterns, the majority of the intensity is found within the direct beam resulting from unscattered electrons. Scattered diffracted waves create diffraction spots. These diffraction waves are then recombined to generate an image. The use of electromagnetic lenses allows diffracted electrons to be focused into a regular arrangement of diffracted spots, which is known as a diffraction pattern (Bendersky and Gayle 2001). These diffraction patterns are characteristic of specific entities depending on the purity and crystallinity of the sample being analyzed.

10.9.3 X-Ray Diffraction

Each crystalline material has a group of atoms ordered with a specific arrangement. This is called a unit cell, and it is repeated continuously in three dimensions to make up a crystal lattice. The number of times the unit cell repeats is unrestricted; however, the repeating unit is characteristic for a specific entity.

X-ray diffraction (XRD) is a nondestructive technique used to determine the crystalline phases and their orientation within a particular solid. In order to do this, an x-ray beam is generated and passed through a substance. The atomic planes of a crystal will cause these x-rays to diffract in different directions. Characteristic scattering patterns are designated to a specific entity (diffraction patterns).

The Joint Committee on Powder Diffraction Standards (JCPDS) has a database of powdered diffraction patterns of more than 50,000 entries. Once a signature diffraction pattern has been generated from XRD analysis, the resulting spectra are compared against species in this database to confirm the crystal makeup of the analyzed sample. (Massa 2004).

Not all iron nanoparticles will possess a XRD pattern, and this is dependent on the degree of order in the nanoparticle structure itself. Different synthesis methods can lead to disordered structures; that is, the sonochemical method for preparing iron nanoparticles involves a step whereby extreme high temperatures are experienced followed by rapid cooling. This process is not beneficial for crystal structure formation.

10.9.4 Energy-Dispersive X-Ray Spectroscopy

Energy dispersive x-ray spectroscopy (EDS) is a technique used in conjunction with SEM or TEM to determine the elemental composition of a sample. An electron beam is focused or scanned across a sample, and this excites electrons within a sample.

Electrons from an inner shell of atoms move to a higher energy level, and there is an unstable gap that needs to be filled. An electron from a higher energy level fills this void, and a signature x-ray emission is created and released. This x-ray is representative of the atomic number of the element it is derived from and is the difference in energy between the high and low energy shells. Elemental composition is measured using an energy-dispersive spectrophotometer (Shindo and Oikawa T. 2002; Rahman et al. 2011).

10.9.5 Fourier Transform Infrared

Fourier transform infrared (FTIR) spectroscopy is a technique used to determine functional groups located on surface of molecules. Infrared absorbance spectroscopy measures the loss of IR radiation transmitted through a sample. An uncoated Fe^0 nanoparticle is expected to have a specific vibrational frequency upon being exposed to IR radiation, as all molecules are made up of the same pure Fe^0 element (Devi and Gayathri 2010). Coatings deposited to the outside of Fe^0 nanoparticles will disrupt this frequency and alter the signature resonance due to absorption of radiation (Bellisola and Sorio 2011).

Simply, IR radiation is directed toward the sample, and this radiation either transmits through or is reflected off its surface. Different frequencies of IR radiation are used, and the variances in transmitted frequency can be due to bond stretching/bond deformations, asymmetric bending, twisting, wagging, rocking, and scissoring motions (Bellisola and Sorio 2011). The aforementioned bond characteristics alter the absorbance potential of the differing IR radiation wavelengths. The absorption bands of specific functional groups are characteristic for each entity, and once a vibrational band assignment is identified, the measured signal is transferred into an IR spectrum. The IR spectrum shows the absorption intensity as percentage transmission vs. wave numbers (cm^{-1}) and is comparable against databases containing spectra for specific entities for species identification (Devi and Gayathri 2010, Baraton 2015).

This tool is used in nanoparticle research when organic polymer coatings need to be identified or plant-based capping agents are used (Heera and Shanmugam 2015).

10.9.6 Dynamic Light Scattering

Dynamic light scattering (DLS) is commonplace for determining the size of nanoparticles among researchers. Unlike SEM and TEM, it is capable of analyzing large numbers of nanoparticle sizes, permitting more representative findings for nonagglomerated monodispersed samples. However, intensities generated from agglomerated particles overlay those of smaller particles, providing skewed results (Lim et al. 2013, Fissan et al. 2014).

DLS measures the Brownian motion of suspended particles and calculates their size based on their velocity or translational diffusion coefficient. Larger particles move more slowly through the liquid, and this alters the frequency of measurable scattered light (Lim et al. 2013, Neoh et al. 2015).

Not only can DLS estimate particle size, it can also determine particle stability due to changes in agglomeration tendency. Furthermore, dispersion characteristics between particles can be determined via zeta potential, where particles are considered individual entities (do not form clusters or have little particle–particle interaction) above +30 mV or below −30 mV (Lewicka and Colvin 2013).

10.9.7 X-Ray Absorption Spectroscopy

X-ray absorption spectroscopy (XAS) is particularly useful for determining the atomic structure of nanoparticles. Unlike XRD, XAS can give structural information (oxidation state, bond length, coordination number, and electronic configuration) of both crystalline and amorphous materials alike (Chekli et al. 2015). Furthermore, samples can be analyzed under varying environmental conditions. For nZVI particles, which have a tendency to rapidly oxidize in aerobic conditions, their properties can be measured in aerobic conditions (Chekli et al. 2015).

Upon a sample being irradiated with synchrotron radiation, an absorption edge is witnessed if the radiation is sufficient to photo-ionize and eject the core electrons from the nanoparticle sample. Higher energies are representative of higher oxidation states, and in turn valences can be identified (Sun et al. 2006, Barcaro et al. 2013a).

10.9.8 X-Ray Photoelectron Spectroscopy

X-ray photoelectron spectroscopy (XPS) is a quantitative method for analyzing surface composition (depth less than 10 nm) and chemical states (Sun et al. 2006). Samples are subjected to monochromatic x-rays of known radiation, and as a result, electrons are ejected from atomic core levels. Kinetic energy is released and recognized by a detector that translates this information to binding energy (Chekli et al. 2015).

This is particularly useful for iron nanoparticles, as surface composition and rapid oxidation are prevalent. X-ray photoelectron spectroscopy can identify all elements involved in nanocomplexes and can also differentiate between each element's oxidation state (Barcaro et al. 2013b). Furthermore, XPS can also be used to determine the catalytic potential of different nanoparticles. If nanoparticles readily oxidize in oxygen environments, the samples must be kept in anoxic conditions until just before use.

10.10 Application of Iron Nanoparticles for Pollution Remediation

Although iron nanoparticles and a limited number of bimetallic nanoparticles are being used commercially for *in situ* site remediation, there are still hesitations and concerns regarding their use due to nanoparticles being a relatively new technology for groundwater research. The first field-scale test of nZVI for groundwater remediation was in Trenton, New Jersey, in 2008 (Zhang and Elliott 2006).

Well-established site remediation methods are likely to be selected over new technologies such as nanoremediation until there is irrefutable evidence the new technology is effective and does not contribute to considerable environmental harm itself. With further research enabling cheaper methods of generating nanoparticles and with more companies competing for a market share, their cost will decrease, and subsequent use is set to increase. With this increase, data on their effectiveness as a remediation strategy will be sufficient for contractors to consider them as a viable technology. Currently companies in the United States and Europe are using iron nanoparticles for remediation of groundwater. However, other countries around the world have not yet adopted them for the aforementioned reasons (Waltlington 2005).

10.10.1 Comparison of Nanoparticles with Other Technologies

Granular zero-valent iron particles (>100 nm in diameter) are currently utilized as a treatment material used to detoxify a wide range of environmental contaminants in groundwater, including viruses, heavy metals, chlorinated hydrocarbons, nutrients (nitrates and phosphates), arsenite (AS [III]), herbicides, phenolic compounds, chelating agents, dyes, and pesticides (Nicole et al. 2011, Wijesekara et al. 2014, Lee et al. 2014). With such a broad spectrum of pollutant detoxification potential, zero-valent iron particles appear to be a "miracle treatment method" at first glance.

However, current approaches require granular zero-valent iron particles to be used in conjunction with permeable reactive barriers (PRBs) due to their size and affinity to agglomerate (particle charges create magnetic attractions). This reduces the site-specific surface area on particles able to react with contaminants and also hampers their ability to be injected directly into the groundwater contaminant zone, as larger and bonded granular iron particles (>100 nm in diameter) cannot follow groundwater flows through the underground matrix (Nicole et al. 2011).

PRBs may be successful in some applications but can only treat pollutants that flow through the barrier. This means pollutants are not treated at the source of contamination. Pollutants that have trouble percolating through the soil matrix may not come into contact with the barriers, and remediation times for polluted areas where groundwater flows are slow may be lengthy. Furthermore, costs associated with excavating land to install and replace PRBs are high, and PRBs can only be installed in shallow unconsolidated aquifers (Grieger et al. 2010).

Since the year 2000, attention has shifted from millimetric zero-valent iron research to nanoscale zero-valent iron research. The key factors favoring nZVIs include large surface area-to-weight ratio, favorable quantum size properties, low standard reduction potential, higher *in situ* reactivity, and potential increase in transport efficiency through groundwater's underground matrix (Kim et al. 2009, 2012, Nurmi et al. 2005).

nZVIs have shown degradation rates of up to four orders of magnitude higher than their millimetric counterpart. However, for groundwater remediation, issues relating to agglomeration and nZVI binding to groundwater matrix constituents (i.e., clay soils, humic acids, and other naturally occurring groundwater elements) still generate problems for particle mobility and targeted contaminant reduction (Grieger et al. 2010, O'Carroll et al. 2013).

nZVIs have shown a great deal of promise for pollution catabolism, but the pitfalls preventing adequate groundwater remediation need to be addressed before they can be utilized and recognized as a viable technology. In order combat these pitfalls, modifications to the nZVI particles are suggested to enable them to become efficient remediation tools for groundwater.

10.11 Types of Pollutants Iron Nanoparticles Catabolize

Iron nanoparticles have shown tremendous capacity to mineralize a number of different pollutant classes either via direct or indirect means (Figure 10.4). Direct avenues may involve the reduction of pollutants, while indirect means can involve their status as a catalyst for the generation of reactive free radical species, that is, Fenton reaction.

Pollutants shown to be degraded in the presence of iron nanoparticles include but are not limited to the following:

- Chlorinated methanes—Dichloromethane (CH_2Cl_2), Chloroform ($CHCl_3$)
- Chlorinated benzenes—Hexachlorobenzene (C_6Cl_6), Trichlorobenzenes ($C_6H_3Cl_3$), Chlorobenzene (C_6H_5Cl)
- Chlorinated ethenes—Vinyl chloride (C_2H_3Cl), Trichloroethene (C_2HCl_3), *trans*-Dichloroethene ($C_2H_2Cl_2$)
- Polychlorinated and nonchlorinated hydrocarbons—PCBs, Pentachlorophenol (C_6HCl_5O)
- Azo and anthroquinone dyes—Orange II ($C_{16}H_{11}N_2NaO_4S$), Remazol Brilliant Blue-R
- Pesticides—Lindane ($C_6H_6Cl_6$), DDT ($C_{14}H_9Cl_5$)
- Nitroaromatics—N-nitrosodimethylamine (NDMA) ($C_4H_{10}N_2O$), RDX $C_3H_6N_6O_6$
- Heavy metals—Mercury (Hg^{2+}), Silver (Ag^+)
- Other organic contaminants—(TNT)
- Inorganic anions—(NO^{3-}, AsO_4^{3-})
- Bromonated compounds—Decabromodiphenyl ether (DBDE) (Zhang 2003)

10.11.1 Chlorinated Solvents

Iron nanoparticles (nZVIs) have been utilized for the remediation of a number of different chlorinated compounds. Reductive dechlorination is the main driving force for their resulting in partial dechlorination of the parent compound or, ideally, complete dechlorination and conversion to ethene and chloride.

A simplified representation of the degradation process can be explained via the following reaction:

$$Fe^0 + RCl + H_2O + H^+ \rightarrow Fe^{2+} + RH + Cl^- + H_2O$$

However, a number of different reductive steps may be at play (particularly for chlorinated methanes, ethanes, and ethenes). These include:

Hydrogenolosis—Substitution of a chlorine atom with hydrogen (reaction requires both an electron and a hydrogen donor to proceed).

Beta Elimination—Release of chlorine atoms and subsequent saturation of a carbon–carbon bond (dominant reductive dechlorination mechanism with nZVI) without the addition of a hydrogen atom.

Alpha Elimination—Removal of two chlorine atoms from a single carbon within the compound, saturating the bond. The reaction proceeds without the addition of hydrogen atoms.

Hydrogenation and hydrolysis—Breakdown of triple bonds to double bonds or double bonds to single carbon–carbon bonds via the addition of hydrogen (in the presence of water for hydrolysis) (O'Carroll et al. 2013).

Rajajayavel and Ghoshal (2015) showed that not only can nZVI degrade TCE, the process can be enhanced up to 40-fold by the addition of sulfide to the surface of the particles. Sulfide is not used in the reaction and therefore degradation is mainly due to β-elimination degradation mechanism from nZVIs.

10.11.2 Dyes

Iron nanoparticles (predominantly nZVI) have been implicated in Azo and anthroquinone dye degradation (Costa et al. 2012). The predominant mechanism for this is reduction. Absorption of dye molecules may also remove dye from solution; however, under passivation, the bound dye molecules are likely to dislodge and re-enter the system (Li et al. 2011b, Truskewycz et al. 2016).

The reduction mechanism for degradation is a consequence of iron nanoparticles reacting with oxygen and water to form unstable transitional compounds (free radicals), which are very reactive toward organic molecules. Of these, the hydroxyl radical is particularly efficient at cleaving Azo bonds. The Fenton reaction (reaction of iron and hydrogen peroxide under acidic conditions) has been implicated in pollutant catabolism with iron nanoparticles (Pereira and Freire 2006, Bokare et al. 2007, Nidheesh et al. 2013) and will be discussed in further detail elsewhere in this chapter (see Fenton/Fenton like reactions).

In acidic media, it has been postulated that iron can donate two electrons to H^+ and transform them to atoms. These atoms can interact with Azo dye molecules, break their structure, and convert them into colorless amines (Rahman et al. 2014).

Nam and Tratnyek (2000) have shown that zero-valent iron was able to directly reduce 9 Azo dyes (each containing SO^{3-} groups) in acidic conditions. One of these reactions was with Orange II, and the resulting products from reductive cleavage of the Azo bond were sulfanilic acid and 1-amino-2-naphthol. This can be seen in the following reaction:

$$C_{16}H_{11}N_2NaO_4S + 5H^+ + Fe^0 \rightarrow C_6H_7NO_3S + C_{10}H_9NO + Fe^{2+}$$

10.11.3 Pesticides

Pesticides are persistent organic pollutants that can be found in surface and groundwater alike. They are made up of diverse structures and may possess specific or nonspecific toxicity to pests and the surrounding environment.

Due to the plethora of different organic pesticide structures, it is impossible to show a specific route for iron nanoparticle degradation of these compounds. In fact, nZVIs may follow degradation routes experienced with a number of different pollutants.

Chlorinated pesticides are likely to be degraded via reductive chlorination (Bezbaruah et al. 2009); nitrogen-containing pesticides can be degraded via reductive denitration. Furthermore, reactions between iron nanoparticles and oxygen/water can lead to reactive oxygen species that can attack complex organic pollutant chemical structures (Choe et al. 2000).

Elliott et al. (2009) used nZVIs to degrade the chlorinated pesticide lindane. Following absorption of lindane to the nZVI particles, two subsequent hydrogenolysis steps occurred, as shown by the following reaction:

$$C_6H_6Cl_6 + Fe^0 \rightarrow C_6H_6Cl_4 + Fe^{2+} + 2Cl^-$$

Following the reduction of lindane to γ −3,4,5,6-tetrachlorocyclohexene, subsequent reduction reactions led to the complete removal of chlorine from degradation intermediates forming benzene as an end product. To further support this finding, Poursaberi et al. (2012) demonstrated that nZVI reductively dechlorinated the pesticide DDT $\rightarrow$ DDD $\rightarrow$ DDMS $\rightarrow$ DPE in a step-by-step process.

The nonchlorinated herbicide molinate was degraded in the presence of nZVI. The mechanism behind this was the generation of free radicals as a result of the interaction between iron and O_2. Disprotonation of the radicals can lead to the formation of hydrogen peroxide, which further reacts with Fe^0 to produce the hydroxyl radical. This reactive species is known to attack the chemical structures of complex organic pollutants (Joo et al. 2004).

10.11.4 Nitrogen

Zero-valent iron has an affinity to remove nitrogen from nitro-aromatic compounds and also interact with nitrate to form stable products, that is, nitrogen gas, nitrite, or ammonium, depending on the reaction conditions.

Najan et al. (2008) demonstrated that the use of nZVI on hexahydro-1,3,5-trinitro-1,3,5-triazine (RDX) led to degradation via two routes. First was the removal of an NO_2^- group to form an unstable intermediate. This species broke down in water to form MEDINA (methylenedinitramine), which was then further transformed to N_2 and NH_4^+.

The second route involved an initial denitration of RDX and the subsequent denitration or reduction of hexahydro-1-nitroso-3,5-dinitro-1,3,5-triazine (MNX) to eventually form hexahydro-1,3,5-trinitroso-1,3,5-triazine (TNX). TNX denitrozation occurs in the presence of iron nanoparticles to eventually break the aromatic structure. Further reactions lead to the formation of N_2 and NH_4^+.

nZVIs can also interact with nitrate to form nitrogen gas, nitrite, and ammonium via the following reactions:

1. $5Fe^0 + 2\,NO_3^- + 6H_2O \rightarrow 5Fe^{2+} + N_2 + 12OH^-$
2. $Fe^0 + NO_3^- + 2H^+ \rightarrow H_2O + NO_2^-$
3. $NO_3^- + 6H_2O + 8e^- \rightarrow NH_3 + 9OH^-$

10.11.5 Metals

Heavy metals are entities that may be naturally occurring; however, due to their high concentrations and wide distributions as a result of industrial, domestic, agricultural, medical, and technological practices, these metals have become serious environmental pollutants. Some heavy metals have high toxicity (e.g., arsenic, cadmium, chromium, lead, and mercury). They are classified as cancer-causing or possible cancer-causing entities that can induce organ damage at low concentrations (Tchounwou et al. 2012).

Iron nanoparticles have a number of mechanisms by which they can be used to treat heavy metal entities in the environment. The process the iron nanoparticle will utilize to remove toxic heavy metals depends on the standard redox potential (E^0) of the metal contaminant compared to that of iron. Iron's standard redox potential is -0.44 V. Metals with an E^0 lower than or similar to this will be absorbed to iron's surface (i.e., Zn, Cd). If the toxic metal species has an E^0 significantly higher, it will in most cases undergo reduction and precipitation (i.e., Cr, As, Cu). If the E^0 is slightly higher than that of Fe^0 (−0.44V), both may be reduced and absorbed. Co-precipitation and oxidation reactions may also occur in some cases, depending on the environmental conditions experienced (O'Carroll et al. 2013).

Absorption of Zn onto nZVIs containing an FeOOH surface layer was observed by Liang et al. (2014). Zinc ion concentrations of 100 mg/L were subjected to nZVI concentrations between 0.1–2.0 g/L. At concentrations of 0.8 g/L or above, the nZVI was able to remove 99% of zinc ions present in solution. Dissolved oxygen increased the removal capacity due to the enhanced absorption efficiency of FeOOH compared to that of nZVI.

10.11.6 Fenton/Fenton-Like Reactions

Many types of iron nanoparticles have a high intrinsic reactivity attributed to them, that is, nZVI; however, this reactivity may also be a downfall due to reactions with oxygen and water leading to passivation. One process that can utilize iron nanoparticles with or without oxidation is the Fenton reaction. The Fenton reaction has the capability to destroy a number of recalcitrant organic compounds structures, including environmental contaminants by the generation of free radicals. This reaction can be coupled with light (photo-Fenton) to further enhance the free radical-generating capacity and lead to elevated pollution degradation rates.

In the Fenton/photo-Fenton reaction, iron itself may only slightly contribute to the overall pollutant degradation outcome (Barbusiński 2009). Iron in the Fenton reaction isn't used up but serves as a catalyst (Nidheesh et al. 2013) for the following cascade of reactions between hydrogen peroxide and iron in acidic environments:

1. $Fe^{2+} + H_2O_2 \rightarrow Fe^{3+} + OH^- + OH^\bullet$
2. $OH^\bullet + H_2O_2 \rightarrow HO_2^\bullet + H_2O$
3. $Fe^{3+} + HO_2^\bullet \rightarrow Fe^{2+} + H^+ + O_2$
4. $Fe^{2+} + HO_2^\bullet \rightarrow Fe^{3+} + HO_2^-$
5. $Fe^{2+} + OH^\bullet \rightarrow Fe^{3+} + OH^-$

There is potential for iron nanoparticles and the $HO_2^\bullet$ radical to participate in pollutant degradation reactions (Chatterjee and Mahata 2004). However, their contribution is likely to be insignificant compared to that of the $OH^\bullet$ radical.

Conclusions for pollutant degradation:

- nZVIs exhibit high reactivity, are able to effectively reduce organic environmental contaminants, and can interact with heavy metals.
- Iron oxide nanoparticles tend to have low reactivity with organic pollutants and have a higher capacity to interact with heavy metals (compared to nZVIs).
- Both iron oxides and nZVI will undergo Fenton reactions when in the presence of H_2O_2.

10.12 *In Situ* Remediation Using Iron Nanoparticles

10.12.1 Groundwater Remediation

Groundwater is an important component of the hydrological cycle, with groundwater and surface waters being an interconnected and interchangeable resource (Evans 2007). Groundwater contamination is often overlooked when compared to that of land-based water systems, as environmental damage is difficult to identify and costly to monitor in the underground matrix.

Groundwater is the source of drinking water for an estimated 2 billion people worldwide and an irrigation resource used to support 40% of the world's agricultural food production. In Australia, Victoria, New South Wales, and South Australia, irrigation is composed of approximately 60% groundwater, while in Western Australia, 72% of groundwater is utilized for urban and industrial processes (Thiruvenkatachari et al. 2007).

Despite groundwater's widespread applications, agricultural, mining, and other industrial processes have been identified as directly and indirectly contaminating these water sources with a vast array of contaminants including chlorinated hydrocarbons, nitrates, phosphates, pesticide, fuels, metals, and radioactive materials (Ball et al. 2001, Khan et al. 2004). In Europe, approximately 20,000 sites are polluted and require remediation, while in the United States, this figure rises to more than 235,000 sites (Nicole et al. 2011). In India's groundwater, severe and widespread heavy metal concentrations are present, with much of the literature reporting on arsenic contamination. Chromium, nickel, cadmium, and mercury are also widespread in India's groundwater, with reported findings in 43 districts of 14 states in India. In addition, fluoride is particularly prevalent, with approximately 20 of the 28 Indian states containing some degree of fluoride in their groundwater. More than 66 million people in India are suspected of suffering from the severe chronic condition fluorosis (Chakraborti et al. 2011, Chakraborti et al. 2016).

Traditional treatments for groundwater remediation have relied heavily on the transportation of groundwater to surface-dwelling plants. These technologies have been deemed to be inefficient within reasonable timeframes and are often costly due to the energy requirements required to pump water to the surface and the cost of infrastructure to facilitate remediation (Bayer and Finkel 2006).

Monitored natural attenuation and bioremediation are other remediation strategies employed to remediate contaminated aquifers. Although these technologies are cheap to implement, lengthy timeframes are required for remediation, and many pollutants may be recalcitrant to break down from these biological entities. Furthermore, introducing microorganisms to deep underground aquifers is troublesome, as the pressures experienced during field-scale groundwater injections may damage microbial cells (O'Carroll et al. 2013).

The complexities relating to groundwater remediation are vast, and each site will have differing characteristics that may be favorable or unfavorable for the use of nanoparticles (mainly nZVIs) as a treatment option. Groundwater flow rates; geological strata; connection with surface waters; use of water for future irrigation or drinking; aerobic/anaerobic conditions; nontargeted species that will react with nanoparticles, that is, clays, humic acids, minerals, and so on; and changes in pH are just a few of the issues one should take into account before commencing site remediation.

In order to combat and bypass some of the aforementioned obstacles in relation to groundwater remediation, researchers have been focusing their efforts on optimizing nanoparticle characteristics and coatings. Ideally, iron nanoparticles should be reactive, uniform in size, not too large to be immobile and not too small to be used up before reaching the contaminant, resistant to passivation, and able to target pollutant sources without interacting with groundwater constituents, and should possess particle–particle repulsive forces. With these factors addressed, nanoparticles that can travel through the underground matrix, target pollutants (mainly NAPL pollutants), and rapidly degrade these pollutants can be generated.

10.12.2 Migration and Monitoring

In vitro proof-of-concept studies assess the migration of nanoparticles through columns packed with sand soil, clays, and other geological entities often found within the underground groundwater matrices. Although these studies are vital for proof of concept, they cannot simulate real-life scenarios (Figure 10.2). Changes in mineral and salt content of groundwater, differences in geological strata, pH variations, differences in oxidation/reduction potential, organic and inorganic microaggregate species that may react with particles, groundwater flows that may spread out or bottleneck in certain areas, and simulating real-life underground injection of particles are just a few parameters that are irreproducible in lab-scale tests. Furthermore, determining the microbial, invertebrate, plant, and animal species likely to be exposed during an on-site application of nanoparticle remediation is difficult, and it is even harder to determine the potential impacts of these particles on their well-being.

Assessing the migration of nanoparticles through the groundwater matrices following site remediation trials is often conducted by indirect measurements such as pH, total iron content, and oxidation-reduction potential by potentiometry. In many cases, when trying to determine the extent of nanoparticle movement, the samples taken are overly dilute with respect to nanoparticle concentration. Aggregation, deposition, reactions with groundwater constituents, and filtration are common causes for this. When samples are taken close to or directly from the injection zone, the presence of iron nanoparticles is visible with the naked eye due to their signature black color. In concentrations as high as these (>10 mg/L), nanoparticle measurements are possible via TEM and inductively coupled plasma mass spectrometry (ICPMS) (Shi et al. 2015).

Recently, work has been conducted to attach fluorescent tags to nanoparticles in order to determine the extent of their movement. Bhomkar et al. (2014) have successfully attached green fluorescent protein with a poly-lysine tag (His-GFP-LYS) to gold, iron oxide, cerium oxide, and zinc oxide nanoparticles, which have then been stabilized with poly-acrylic acid coating. A portable spectrophotometer was capable of distinguishing turbidity changes representative of the presence of nanoparticles. Although this technology is only limited to samples containing low levels of organic matter in surface waters and can only detect particles with a negative charge, its concept is one that should be optimized for groundwater samples.

In the medical field, tagging or labeling nanoparticles for drug delivery is commonplace. Nanoparticles may be labeled with dye flurophores giving amplified optical signals. These particles are unlikely to be useful for environmental remediation purposes, as the dye flurophores are expensive and considered persistent organic pollutants (Miranda et al. 2015). For environmental applications, tagged/labeled nanoparticles should be able to withstand the sheer forces attributed to particle injection and percolation through the porous groundwater matrix. Additional research with respect to adherence of tags/labels during the sheer forces experienced with site application and use of tags that are environmentally benign is important.

10.12.3 Surface Waters

Unlike groundwater, surface water health is easier to monitor, as sampling doesn't require drilling and often visual cues can represent tarnished waterways, that is, visual pollution, deceased animals, and plants. Ecosystem dynamics are completely different in surface waters, also. Higher-order animals are found in rivers and streams, that is, fish, frogs, and freshwater crayfish. Photosynthetic algae and water plants exist within these water bodies, microorganisms are largely anaerobes or facultative anaerobes (unless considering deep sediments), and other animals rely on the water for a direct drinking source.

Beside the living organisms reliant on surface waters, there are also a number of other important differences between surface waters and groundwater that help to paint a picture as to the vastly different dynamics they possess. Some factors impacting surface waters but not groundwaters are as follows; presence of sunlight, flow rate of water, freely flowing water, differing turbidity, differences in gravitational effects, largely fluctuating temperatures, evaporation, recharge, dissolved oxygen, and salt and mineral concentrations. The list goes on ... With such varying dynamics, it can be expected that the treatment of surface waters will vary from that of groundwater.

Delivery of nanoparticles to natural surface waters should not be done without appropriate means of recovering the particles postremediation (Figure 10.5). A number of studies assessing the toxicity of iron nanoparticles on fish have shown considerable health implications indicating that fish and other aquatic organisms may be particularly sensitive to these entities (Chen et al. 2013, Remya et al. 2015, Saravanan et al. 2015).

Due to nanoparticles' properties, they are likely to remain suspended in aquatic solutions for extended periods of time and subsequently will have direct contact with organisms throughout the whole water column. Fish gills may become saturated with particles, and the permeable skin of frogs may accumulate particles, leading to reduced oxygen uptake capacity. In addition, some iron nanoparticles have the capacity to generate reactive radical species that may also harm biological entities (Pereira and Freire 2006, Remya et al. 2015). Additionally, excess nanoparticles within the water column may restrict light from penetrating the water, leading to plant death.

To rectify these problems, iron nanoparticles impregnated into permeable reactive barriers or adsorbed onto solid supports can be used for surface water remediation. Permeable reactive barriers are constructs that allow water or solutions to flow through a reactive matrix. Iron nanoparticles impregnated into these filtration constructs can detoxify contaminants as they pass through.

Solid supports loaded with nanoparticles allow reactions to occur at the surface of the nanoparticles. Nanoparticles are adsorbed onto the support matrices so that the maximum surface area is available for reaction to occur. Furthermore, the addition of metal catalysts or other adsorbent materials can be incorporated into the design to enhance reactivity and absorptive potential of the pollutants (Rajabi et al. 2013).

In these technologies, nanoparticles are not released into the system and in turn are not likely to pose significant environmental harm. Also, metal catalyst-doped structures can often be recharged, leading to their reuse and subsequent reduction in remediation costs (Fan and Gao 2006).

10.12.4 Terrestrial Remediation

Contamination of terrestrial sites has historically been addressed by removing the contaminated soil to landfill (costly), degradation of contaminants via chemical methods (potentially toxic), and bioremediation (slow process). Although the vast majority of research using iron nanoparticles deals with the remediation of water, particularly groundwater, studies have shown successful results for degrading contaminants in soils also.

Naja et al. (2009) showed that CMC coated nZVIs were capable of degrading 98% of Hexahydro-1,3,5-trinitro-1,3,5-triazine (RDX) under anaerobic conditions in soil in less than a day. Furthermore, in soil packed columns, 95% of RDX (60mg/kg^{-1}) was degraded using a CMC-nZVI solution.

Ibrahem et al. (2012) showed that trichloroethylene could be degraded by surfactant modified nZVI particles. Soil was doped with TCE and subjected to nanoparticle treatment. Results showed that this method was 23 orders of magnitude slower than studies conducted in aqueous phases; however, after a week, complete removal was achieved.

Varying reports on effectiveness have been shown above for *in vitro* studies. Upon transferring this technology to site remediation, other factors should also be considered. Nanoparticulate dust has been implicated in a number of health problems via inhalation (Donaldson et al. 2005).

If the technology is to be used around populated areas, proper control measures should be in place to ensure dust is not airborne (nanoparticle slurries are used without chance of drying out). Furthermore, mixed findings for iron nanoparticle impacts on microorganisms and plant germination have been reported. Some microorganisms are killed under iron nanoparticle exposure, while others thrive (Borcherding et al. 2014, Shahzeidi and Amiri 2015). Also, some plant seeds germinate faster in the presence of iron nanoparticles, while at high concentrations, iron nanoparticles cause toxicity (Alam et al. 2015, Chichiriccò and Poma 2015). All of the aforementioned factors should be considered before utilizing this technology for terrestrial systems, and this will be dependent on the site itself.

10.12.5 Cost of Iron Nanoparticle Remediation/Feasibility

With all the build-up about iron nanoparticles' high reactivity, elevated pollution degradation rates, and seemingly environmentally friendly nature, can they realistically deliver what is promised and be a viable technology?

From a landholder's perspective, the cheapest and fastest option for site remediation is preferable. Bioremediation may be the cheapest option; however, the timeframes required to clean up sites are lengthy and limited persistent pollutants are bioavailable for degradation from these organisms. In many cases, contaminated sites are excavated and sent to landfill. This is particularly expensive and environmentally unsound (He et al. 2009). With iron nanoparticle technology still in its infancy in regard to site remediation, many landholders and contractors may prefer to choose well-established technologies over ones with question marks surrounding their efficiency, potential delivery to the source, and ecotoxicity.

Site remediation using iron nanoparticles has been estimated to be competitive in respect to price for a number of priority pollutants, including cadmium, chromium, zinc, lead, arsenic, xylene, toluene, TCE, lindane, chloramphenicol, atenolol, and gemfibrozil. While the price of remediating metals depends on the metal and extent of pollution present, sludge formation is minimized when utilizing nZVI for treatment. In addition, if remediation of pollutants utilizing iron nanoparticles is not capable of completely remediating a site, the remaining pollution may be cost-effectively managed (Adeleye et al. 2016).

It is a common belief in the environmental nanotechnology field that iron nanoparticles can provide comparable or significantly higher pollution degradation rates at a much cheaper price than conventional methods (Grieger et al. 2015). While reactivities and pollutant degradation capabilities are not questioned, the cost associated with producing them is.

Currently, the most popular method for generating nZVI particles is through the sodium borohydride reduction of aqueous salts method (Adeleye et al. 2016). These particles also require capping and/or further coatings to prevent oxidation and enhance remediation. The cost of using sodium borohydride to manufacture particles has been deemed too expensive to give iron nanoparticles the edge they need to be widely implemented in site remediation. Costs of nZVI particles in 2011 were between $50 and $200/kg, and the price has not shifted much to today, when prices are between $50 and $100/kg. It has been forecasted that to be considered viable, nZVI costs need to be reduced by at least 20% (Elliott et al. 2009, Crane and Scott. 2012, Adeleye et al. 2016). Depending on the scale and complexity of the remediation project, the cost to remediate one cubic meter of chlorinated hydrocarbon contaminated soil with nZVI is between $89.50 and $218.66 (Cook 2009).

10.13 Fate of Fe NP's in the Environment and Ecotoxicity Concerns

When a contaminated environment is remediated, the aim is to generate technologies that are low-cost, effective, easy to implement, environmentally benign, and do not have health implications for people who come into contact with them. It is therefore important that when assessing the potential for nanoparticles to be used as a viable technology, we must first determine if the nanoparticles themselves could become environmental contaminants or toxic to humans.

Although iron nanoparticles are currently being used for site remediation (Lowry et al. 2012, Su et al. 2012) there are mixed opinions about their toxicity. Contractors for site remediation have their reservations about using this relatively young technology due to fear of potential environmental damage from the nanoparticles themselves being apparent in the future.

Nanoparticle toxicity is dependent on a number of functional parameters such as particle size, shape, surface charge, chemistry, and composition. The cytotoxicity of nanoparticles is thought to be due to their ability to generate oxidative stresses and activate proinflammatory genes (Yildirimer et al. 2011).

Many studies have been conducted in relation to iron nanoparticle toxicity within microorganisms, human cell lines, mice, fish, and plants (Figure 10.4).

10.13.1 Microorganisms

Microorganisms are very sensitive indicators of change in an environment. Population and diversity changes may be a result of simple natural changes in temperature or may result from more serious events like significant environmental damage caused by large-scale anthropogenic contamination (Dion 2008).

Changes in temperature are unlikely to kill microbes (unless severe increases in temperature are experienced). However, they may instill a competitive advantage for some organisms, reducing the prevalence of those originally present. During a severe contamination event, the majority of microorganisms may have already died as a result of pollution-induced toxicity and inability to access food, nutrient, or water sources trapped by the pollutant.

However, links between iron nanoparticles and microbial toxicity have been shown, although iron nanoparticles are seen as less environmentally damaging compared to others metallic nanoparticles, that is, copper, aluminum, and silver, to name a few. Shahzeidi and Amiri (2015) used nutrient agar plates impregnated with Fe_3O_4 nanoparticles to assess the growth kinetics of *Pseudomonas aeruginosa*, *Escherichia coli*, and *Staphylococcus aureus*. Results indicated that nanoparticle concentrations as low as 1.56 mg/ml were sufficient to slightly inhibit microbial growth. Increasing nanoparticle concentrations did not have a marked effect on increased microbial toxicity except with *P. aeruginosa* at 25 mg/ml nanoparticle concentration, where microbial growth was considerably inhibited.

Conversely, Arakha et al. (2015) showed that iron oxide nanoparticles (n-Fe3O4) only exhibited significant antibacterial activity against *Bacillus subtilis* and *E. coli* at high nanoparticle concentrations (>50 µM). However, the addition of a chitosan coating reversed the surface potential of the nanoparticles from negative to positive (mainly due to –OH groups on the chitosan molecule), and bacterial toxicity was significantly increased. Findings from this study indicated that the positive surface potential of the nanoparticles led to the production of reactive oxygen species (ROS), which is known to inactivate microbial cells (Arakha et al. 2015). With this in mind, regardless of whether the nanoparticles themselves exhibit toxicity, the end product that is to be used should be tested to ensure that changing nanoparticle properties and/or coatings themselves are not toxic.

Borcherding et al. (2014) linked the size of iron oxide nanoparticles with bacterial proliferation. A decrease in nanoparticle size induced growth and increased the viability of *P. aeruginosa* (PA01) and subsequent biofilm generation. In contrast, small iron oxide nanoparticles have been shown to inhibit 5′ adenosine monophosphate-activated protein kinase (AMP) activity and deregulate the host's innate immunity. With a weakened immunity and iron oxide's ability to induce the growth of potential pathogenic bacteria, human health may be at risk from nanoparticles at the smaller side of the nanorange. However, nanoparticles that are too small are not beneficial for environmental restoration, as passivation transforms them into iron oxides before they can reach the pollution source (Bardos et al. 2011).

In cases where the natural microflora have been significantly altered or killed off, the use of nanoparticles for site remediation may be seen as a viable technology. If particles revert to an inert form and/or pose no long-term damage following degradation, there are increases in microfloral diversity and populations are likely to return to the site. Following severe ecosystem-altering events, it is rare for the original microflora to return to the site with the same diversity and populations (Nurulita et al. 2015).

Iron nanoparticles have been hypothesized as being relatively safe for environmental use by many researchers. Iron is the fourth most abundant element found within the earth's crust and exists mainly as metal oxides. Iron nanoparticles initially have a high degree of reactivity, but upon being used in detoxification reactions, their reactivity decreases as passivation increases. The resulting products are iron oxides in most cases, thereby reverting back to natural end products following use. Furthermore, the use of many coatings for iron nanoparticles are composed of organic polymers, which, when broken down, may increase the carbon balance of the soil and in turn stimulate microbial growth (He et al. 2009).

Due to the increase in nanoparticle research and the multitude of different modifications (size, shape, coatings, porosity, crystallinity, stability, functional groups attached, etc.) comparing the degree of toxicity of one type of nanoparticle against another is virtually impossible unless standardized tests are created. These tests should be tailored to the field they are suited for; for example, iron nanoparticles for use in the medical industry may require toxicity testing on animal cells, whereas environmental applications may require testing on microbial, plant, and animal cells (Soenen et al. 2012).

10.13.2 Plants

When it comes to plants, there is such a thing as too much of a good thing. Excessive sunlight can burn leaves, too much water can rot roots, and too much fertilizer also deteriorates plant health. In many cases, this trend can also be seen with the effect of iron oxide nanoparticles on plants.

Sheykhbaglou et al. (2010) showed that small concentrations of iron nanoparticles (0.75 g/L) increased the dry weights of soybean leaf and pod weights. Also, the highest yields of soybean were experienced when soybeans were supplemented with 0.50 g/L iron oxide nanoparticles. Similar increases in tomato, watermelon, and wheat growth have been shown with low concentrations of iron oxide nanoparticles (Alam et al. 2015, Shankramma et al. 2015, Wang et al. 2015). Iron deficiency can lead to a reduction in chlorophyll, reducing photosynthesis, and can induce peroxidation, leading to leaking of intracellular electrolytes. Iron is an essential plant nutrient in the correct quantities. Iron nanoparticles may also increase the water uptake by seeds by penetrating their surface and in turn increase seed germination. Increased seed germination has been seen with peanut plants and wheat (Alam et al. 2015, Li et al. 2015).

When levels of iron are too high (including from the presence of iron nanoparticles), plant growth is adversely affected. Wang et al. (2015) showed elevated watermelon growth with up to 20 mg/L iron nanoparticles; however, at 100 mg/L, protein denaturation in plant tissue was apparent, leading to cellular aging.

Small ferrofluid concentrations (10–50 μL/L) may initially increase nucleic acid and chlorophyll *a* concentrations; however, at higher concentrations (100–250 μL/L) chlorophyll *a* and *b* ratios are significantly disrupted (decreased by approximately 35% each), significantly affecting photosynthesis (Chichiriccò and Poma. 2015).

In addition, if nanoparticle concentrations are excessively high around the rhizosphere of plants, they may interfere with the microbial populations that colonize the area. Changes in microbial diversity and population dynamics around the plant's root/soil interface may hamper stimulatory plant growth interactions.

With this in mind, iron nanoparticles for environmental restoration, particularly in groundwater systems, are unlikely to impact plant growth dynamics dramatically. The use of iron nanoparticles for soil restoration is not commonplace, and the dynamics of the site, including the potential damage to flora, should be taken into consideration before attempting such a venture.

10.13.3 Fish

As there is potential for nanoparticles to enter surface water upon remediation, their toxicity on aquatic life forms is important to determine. The predominant routes of exposure for fish are likely to be through absorption through eggs, via the gills, or through ingestion.

Chen et al. (2013) studied the effect of CMC-stabilized-nZVI, nFe_3O_4 and Fe^{2+}(aq). Their findings showed that CMC-stabilized nZVI and Fe^{2+} (aq) induced acute mortality, developmental toxicity, and oxidative stress responses in *Oryzias latipes* embryos and hatchlings. Fe_3O_4 showed minimal acute mortality on embryos, as the embryo-containing solutions were void of Fe^{2+} and ROS.

Saravanan et al. (2015) showed that at concentrations of 1 and 25 mg/L, Fe_2O_3 nanoparticles showed toxicity responses in *Labeo rohita* over a short exposure period of 96 hours. Increases in plasma glucose concentrations were observed, a stress response that may be a result of respiratory disturbances. Plasma proteins were significantly decreased for both 1 and

25 mg/L nanoparticle exposure concentrations. Decreases in plasma protein can be an indicator of direct nanoparticle toxicity to cells or the production of ROS, which can damage cells. Additionally, gill Na^+/K^+-ATPase concentrations were significantly reduced, indicating that nanoparticles are directly impacting the function of the fish's gills.

Fish and their embryos tend to show an elevated susceptibility to iron nanoparticles compared to mammalian counterparts. This may be due a number of factors, including permeable eggs that allow transfer of nanoparticles to developing embryos; ease of nanoparticles to be in contact with the gills, leading to a depletion in oxygen exchange and/or allowing NPs to enter the bloodstream; and direct and nonselective ingestion of nanoparticles (Remya et al. 2015).

It is suggested that the use of nanoparticles within surface waters should be in conjunction with permeable reactive barriers or loading into immobilized supports. In this way, the release of nanoparticles that can impact aquatic organisms can be minimized, particles can be retrieved, and environmental harm can be minimized (Figure 10.5). Furthermore, application of nanoparticles to soils or injection of nanoparticles into subsurface environments should include risk assessments as to the potential of iron nanoparticles to reach surface waters.

10.13.4 Mammals

Although iron is an essential trace element for mammals, excessive concentrations have resulted in a number of different disorders in humans. Iron concentrations between 20 to 60 mg/kg body weight have been considered potentially serious, inducing GI toxicity, and concentrations over 60 mg/kg body weight have been considered potentially lethal, inducing systemic toxicity. High concentrations of iron entering the body via ingestion can lead to pathological changes in the GI tract, liver, and cardiovascular system (Singh et al. 2013). People with hemochromatosis would also be considered at higher risk from high iron concentrations in drinking water.

The size of nanoparticles is one of their key attributes, instilling increased surface area for faster reactions to occur. However, it may be regarded as a detrimental feature with regards to toxicology. Nanoparticles that are smaller than 100 nm are absorbed by intestinal cells and lymphatic tissues and can enter the circulatory system. The particles' surface and surface chemistry also impact their toxicity. Small particles will absorb into the lining layer of the lungs due to their low surface tension. Smaller particles have been shown to be more damaging to lungs due to increased surface area, greater ability to conjugate, and ability to sustain energy (Ai et al. 2011). Furthermore, particles with sizes of less than 70 nm are not phagocyted by macrophages in the alveolar region efficiently, and this could lead to accumulation (Bergeron and Archambault 2005).

The major concern in relation to nanoparticle toxicity is their ability to generate ROS, which can lead to DNA damage and cancer. Apopa et al. (2009) showed that the presence of iron oxide nanoparticles (Fe_2O_3) led to the increased permeability of microvascular endothelial cells through their generation of ROS. These cells are largely implicated in carcinogenesis and tumor growth.

Inhalation studies of Fe_3O_4 dust in Wistar rats were conducted by Pauluhn (2011). Rats were exposed to nanoparticle concentrations up to 52.1 mg m^{-3} for 6 hours per day, 5 days a week, for 13 consecutive weeks. Results indicated that no mortality, consistent changes in body weight, changes in eating habits, or systemic toxicity were observed. Only some subchronic symptoms were experienced in rats with poorly soluble particles. Furthermore, no cytotoxicity was observed when healthy Sprague Dawley rats were made to inhale up to 90 μg m^3 Fe_2O_3 iron nanoparticles (average size = 72 nm) for 6 hrs/day for 3 days (Zhou et al. 2003).

With such extensive exposures leading to relatively minor health implications, rats and potentially other mammalian systems have great capacity to withstand iron nanoparticle-induced toxicity. For environmental nanoparticle applications, aerosol nanoparticle concentrations are estimated to be low and minimal inhalation exposure is expected. The majority of nanoparticle-based remediation is conducted with nanoparticle slurries, preventing the occurrence of volatile and aerosol nanoparticles.

10.13.5 Ecotoxicity Extrapolations

Although research into the ecotoxicity dynamics of iron nanoparticles is still in its infancy, trends can be extrapolated from the current scientific research. In environments where organic and metallic contaminants pose a risk to environmental harm (Ball and Truskewycz 2013), a number of considerations should be addressed before selecting the use of iron nanoparticles as a treatment option. Briefly, nanoparticles themselves should not be tested for toxicity alone, as surface coatings can change the charge of nanoparticles themselves, or coatings may have inherent toxicity. Iron nanoparticles have shown both positive and negative growth dynamics for microorganisms and plants. In subsurface groundwater injections of iron nanoparticles, the impact on the environment is not likely to be excessively damaging. This is due to the particles pacifying over time, and their transport through the underground matrices is likely to be limited. Organic coatings on particles may also stimulate microbial growth due to the increased carbon balance of the environment.

In terrestrial environments or in surface waters, the introduction of iron nanoparticles may be troublesome. High concentrations of nanoparticles induced protein denaturation and cellular aging in plants and also altered chlorophyll *a* and chlorophyll *b* ratio dynamics by lowering them affecting usual photosynthesis functioning. Plant/microbe interactions may also be altered in the rhizosphere. Fish and their embryos have been shown to be particularly sensitive to iron nanoparticles, inducing mortality, oxidative damage, and altered gill functions. Nanoparticles are therefore recommended to be used in conjunction with PRBs or loaded onto surface supports for surface water remediation.

Additionally, standardized ecotoxicological tests should be in place so that nanoparticle toxicity can be compared. With this testing, bare nanoparticles and coated nanoparticles alike should be tested to determine the end product toxicity. Ecotoxicity on a range of model organisms should be conducted to ensure the impact on all environments can be assessed. Parameters to be assessed are likely to include but should not be limited to surface characteristics and coatings, shape, size, physical composition and chemical reactivity, and passivation potential.

10.14 Conclusions

The use of iron nanoparticles for environmental remediation offers great promise for fast, effective, and targeted pollution remediation. However, each specific site will require research as to which particles to use.

- Surface waters are likely to require particles to be loaded onto solid matrices or within PRBs to prevent toxicity to aquatic life forms and to ensure nanoparticles are removable from the system.

- Application of iron nanoparticles to terrestrial soils may be site specific depending on the type and abundance of threatened/endangered flora and fauna in the region. Furthermore, minimization of airborne nanoparticles is important, as their effects on humans are not completely understood.
- Groundwater remediation needs to factor in the geological strata, flow rate of groundwater, aquifer characteristics, pollutant type, closeness to surface waters, and future use of groundwater, just to name a few.

Like any new and emerging remediation technology, there are concerns regarding safety and effectiveness. One of the major drawbacks with providing community peace of mind with regards to this is that there are no widely accepted ecotoxilogical standards to compare particle toxicity. Furthermore, every new alteration made to nanoparticles, whether it is a new coating, different size characteristic, addition of a catalyst for enhanced remediation, or another change, has the potential to significantly alter particle behavior. Without ecotoxilogical standards, concerns and hesitations are likely to restrict their usage until sufficient *in situ* pilot-scale tests have returned positive outcomes without adverse effects.

Great advances have been made in regards to iron nanoparticle properties with surface functionalization strategies used to combat agglomeration; reduce rapid passivation; instill pollutant targeting; and control size, shape and particle uniformity. There is still plenty of scope to enhance nanoparticle properties. Increased reaction rates, reduction in production costs, isolation of environmentally friendly coatings, catalysts, and generation of novel solid support matrices and further optimization of particle size, shape agglomeration, and pollutant targeting are of importance.

Despite the benefits of iron nanoparticles over competing technologies (increased pollutant degradation speeds, ability to treat groundwater pollutants on site, little persistent environmental toxicity for groundwater remediation, etc.), their cost is too high for many contractors to consider them as an option. Millimetric iron particles cost between \$0.36 and \$1.08/kg, while iron nanoparticles are on the order of \$50–\$100.00/kg (Elliott et al. 2009, Bardos et al. 2011, Crane and Scott. 2012, Adeleye et al. 2016). The use of sodium borohydride for generation of nZVIs is expensive, and waste products from this process are toxic. Alternative routes for the generation of nZVIs may hold significant commercial promise if iron nanoparticle production costs can be reduced.

Acknowledgment

The authors acknowledge funding support of Australia India Council grant (AIC086-2015) for this work.

References

Abbassi, R., Yadav A. K., Kumar, N., Huang, S., Jaffe, P.R. 2013. Modeling and optimization of dye removal using "green" clay supported iron nano-particles. *Ecological Engineering* 61: 366–370.

Adeleye, A. S., Conway, J. R., Garner, K., Huang, Y., Su, Y., Kelle, A. A. 2016. Engineered nanomaterials for water treatment and remediation: Costs, benefits, and applicability. *Chemical Engineering Journal* 286: 640–662.

Agarwal, A., Joshia, H., Kumar, A. 2011. Synthesis, characterization and application of nano lepidocrocite and magnetite in the degradation of carbon tetrachloride. *South African Journal of Chemistry* 64: 218–224.

Agrawal, G. K., Timperio, A. M., Zolla, L., Bansal, V., Shukla, R., Rakwal, R. 2013. Biomarker discovery and applications for foods and beverages: Proteomics to nanoproteomics. *Journal of Proteomics* 93: 74–92.

Ahmmad, B., Leonard, K., Islam, M. S., Kurawaki, J., Muruganandham, M., Ohkubo, T., Kuroda, Y. 2013. Green synthesis of mesoporous hematite (a-Fe_2O_3) nanoparticles and their photocatalytic activity. *Advanced Powder Technology* 24: 160–167.

Ai, J., Biazar, E., Jafarpour, M., Montazeri, M., Majdi, A., Aminifard, S., Zafari, M., Akbari, H. R., Rad, H. G. 2011. Nanotoxicology and nanoparticle safety in biomedical designs. *International Journal of Nanomedicine* 6: 1117–1127.

Alam, J., Sultana, F., Iqbal, T. 2015. Potential of iron nanoparticles to increase germination and growth of wheat seedling. *Journal of Nanoscience with Advanced Technology* 1(3): 14–20.

Alam, S., Anand, C., Logudurai, R., Balasubramanian, V. V., Ariga, K., Bose, A. C., Mori, T., Srinivasu, P., Vinu, A. 2009. Comparative study on the magnetic properties of iron oxide nanoparticles loaded on mesoporous silica and carbon materials with different structure. *Microporous and Mesoporous Materials* 121(1–3): 178–184.

Alam, S. M. N., Haas, Z. J. 2015. Coverage and connectivity in three-dimensional networks with random node deployment. *Ad Hoc Networks* 34: 157–169.

Allen, N. S., Edge, M., Sandoval, G., Ortega A., Liauw, C. M., Stratton, J., McIntyre R. B. 2002. Interrelationship of spectroscopic properties with the thermal and photochemical behaviour of titanium dioxide pigments in metallocene polyethylene and alkyd based paint films: Micron versus nanoparticles. *Polymer Degradation and Stability* 76(2): 305–319.

Andrews, D., Scholes, G., Wiederrecht, G. 2010. *Comprehensive Nanoscience and Technology, Five-Volume set., Theories of Superparamagnetic Relaxation*. Academic Press, London. ISBN 0123743966, 9780123743961.

Apopa, P. L., Qian, Y., Shao, R., Guo, N. L., Schwegler-Berry, D., Pacurari, M., Porter, D., Shi, X., Vallyathan, V., Castranova, V., Flynn, D. C. 2009. Iron oxide nanoparticles induce human microvascular endothelial cell permeability through reactive oxygen species production and microtubule remodeling. *Particle and Fibre Toxicology* 6: 1, doi: 10.1186/1743-8977-6-1.

Arakha, M., Pal, S., Samantarrai, D., Panigrahi, T. K., Mallick, B. C., Pramanik, K., Mallick, B., Jha, S. 2015. Antimicrobial activity of iron oxide nanoparticle upon modulation of nanoparticle-bacteria interface. *Nature; Scientific Reports* 5: 14813, doi: 10.1038/srep14813.

Arancibia-Miranda, N., Baltazar, S. E., García, A., Munoz-Lira, D., Sepúlveda, P., Rubioa, M. A., Altbir, D. 2016. Nanoscale zero valent supported by Zeolite and Montmorillonite: Template effect of the removal of lead ion from an aqueous solution. *Journal of Hazardous Materials* 301: 371–380.

Avanaki, Z. A., Hassanzadeh, A. 2013. Modified Brillouin function to explain the ferromagnetic behavior of surfactant-aided synthesized α-Fe_2O_3 nanostructures. *Journal of Theoretical and Applied Physics* 7: 19.

Baghbanzadeh, M., Carbone, L., Cozzoli, P. D., Kappe, C. O. 2011. Microwave-assisted synthesis of colloidal inorganic nanocrystals. *Angewandte Chemie International Edition* 50(48): 11312–11359.

Ball, A., Truskewycz, A. 2013. Polyaromatic hydrocarbon exposure: an ecological impact ambiguity. *Environmental Science and Pollution Research* 20(7): 4311–4326.

Ball, J., Donnelley, L., Erlanger, P., Evans, R., Kollmorgen, A., Neal, B., Shirley, M. 2001. Inland Waters, Australia State of the Environment Report 2001 (Theme Report), CSIRO Publishing on behalf of the Department of the Environment and Heritage, Canberra.

Bang, J. H., Suslick, K. S. 2010. Applications of ultrasound to the synthesis of nanostructured materials. *Adv. Mater* 22: 1039–1059.

Baraton, M. 2015. Life cycle analysis of nanoparticles: Reducing risk and liability. In: *Chapter 10: Surface Functionalization of Semiconductor*. Vaseashta, A. (ed.) DEStech Publications, Norwich University Applied Research Institute, Northfield, Vermont, USA, ISBN 1605950238, 9781605950235, pp. 255–259.

Barcaro, G., Caro, A., Fortunelli, A. 2013a. Springer handbook of nanomaterials. *Chapter 11: Alloys on the Nanoscale.* Vajtai, R. (ed.), Springer-Verlag, Berlin, Heidelberg, pp. 409–458, doi: 10.1007/978-3-642-20595-8_11.

Barcaro, G., Caro, A., Fortunelli, A. 2013b. Springer handbook of nanomaterials. *Chapter 11: Alloys on the Nanoscale.* Vajtai, R. (ed.) Springer Science & Business Media, ISBN 364220595X, 9783642205958, pp. 417–420.

Bardos, P., Bone, B., Elliott, D., Hartog, N., Henstock, J., Nathanail, P. 2011. A Risk/Benefit Approach to the Application of Iron Nanoparticles for the Remediation of Contaminated Sites in the Environment. Defra Research Project Final Report, Defra Project Code: CB0440.

Bayer, P., Finkel, M. 2006. Life cycle assessment of active and passive groundwater remediation technologies. *Journal of Contaminant Hydrology* 83: 171–199.

Barbusiński, K. 2009. Fenton reaction—Controversy concerning the chemistry. *Ecological Chemistry And Engineering S* 16(3): 347–358.

Bellisola, G., Sorio, C. 2011. Review article: Infrared spectroscopy and microscopy in cancer research and diagnosis. *American Journal of Cancer Research* 2(1): 1–21.

Bendersky, L. A., Gayle, F. W. 2001. Electron diffraction using transmission electron microscopy. *Journal of research of the National Institute of Standards and Technology* 106: 997–1012.

Bepari, R. A., Bharali, P., Das, B. K. 2014. Controlled synthesis of a- and c-Fe_2O_3 nanoparticles via thermolysis of PVA gels and studies on a-Fe_2O_3 catalyzed styrene epoxidation. *Journal of Saudi Chemical Society* 21: 170–178.

Bergeron, S., Archambault, É. 2005. Canadian Stewardship Practices for Environmental Nanotechnology. *Science-Metrix, Montréal,* 65. Environment Canada, http://www.ethique.gouv.qc.ca/index.php?option=comdocman&task=catview&gid=18&Itemid=14&phpMyAdmin=bb1c4a2d0bactf758937. Accessed on March 7, 2016.

Bezbaruah, A. N., Thompson, J. M., Chisholm, B. J. 2009. Remediation of alachlor and atrazine contaminated water with zero-valent iron nanoparticles. *Journal of Environmental Science and Health Part B* 44: 518–524.

Bhomkar, P., Goss, G., Wishart, D. S. 2014. A simple and sensitive biosensor for rapid detection of nanoparticles in water. *Journal of Nanoparticle Research* 16: 2253.

Bokare, A., Chikate, T., Rode, C. V., Paknikar, K. 2007. Effect of surface chemistry of Fe-Ni nanoparticles on mechanistic pathways of Azo dye degradation. *Environmental Science & Technology* 41: 7437–7443.

Borcherding, J., Baltrusaitis, J., Chen, H., Stebounova, L., Wu, C. -M., Rubasinghege, G., Mudunkotuwa, I. A., Caraballo, J. C., Zabner, J., Grassian, V. H., Comellas, A. P. 2014. Iron oxide nanoparticles induce *Pseudomonas aeruginosa* growth, induce biofilm formation, and inhibit antimicrobial peptide function. *Environmental Science: Nano* 1: 123.

Breuer, M., Rosso, K. M., Blumberger, J., Butt, J. N. 2014. Multi-haem cytochromes in *Shewanella oneidensis* MR-1: structures, functions and opportunities. *Journal of The Royal Society Interface* 12(102): 20141117.

Cameotra, S. S., Dhanjal, S. 2012. Chapter 13; Environmental Nanotechnology: Nanoparticles for Bioremediation of Toxic Pollutants. *Bioremediation Technology: Recent Advances,* 1st edition. Fulekar, H. M. (ed.) ISBN 9048136784, 9789048136780, Springer, Netherlands, pp. 348–369.

Cameselle, C., Reddy, K. R., Darko-Kagya, K., Khodadoust, A. 2013. Effect of dispersant on transport of nanoscale iron particles in soils: Zeta potential measurements and column experiments. *Journal of Environmental Engineering* 139(1): 23–33.

Carnovale, C., Bryant, G., Shukla, R., Bansal, V. 2016. Size, shape and surface chemistry of nano-gold dictate its cellular interactions, uptake and toxicity. *Progress in Materials Science* 83: 152–190.

Carta, D., Casula, M. F., Corrias, A., Falqui, A., Navarra, G., Pinna, G. 2009. Structural and magnetic characterization of synthetic ferrihydrite nanoparticles. *Materials Chemistry and Physics* 113(1): 349–355.

Chakraborti, C., Das, B., Murrill, M. T. 2011. Examining India's groundwater quality management. *Environmental Science & Technology* 45: 27–33.

Chakraborti, D., Rahman, M. M., Chatterjee, A., Das, D., Das, B., Nayak, B., Pal et al. 2016. Fate of over 480 million inhabitants living in arsenic and fluoride endemic Indian districts: Magnitude, health, socio-economic effects and mitigation approaches. *Journal of Trace Elements in Medicine and Biology* 38: 33–45.

Chatterjee, D., Mahata, A. 2004. Evidence of superoxide radical formation in the photodegradation of pesticide on the dye modified TiO_2 surface using visible light. *Journal of Photochemistry and Photobiology A: Chemistry* 165: 19–23.

Chekli, L., Bayatsarmadi, B., Sekine, R., Sarkar, B., Shen, A. M., Scheckel, K. G., Skinner, W., Naidu, R., Shon, H. K., Lombi, E., Donner, E. 2015. Analytical characterisation of nanoscale zero-valent iron: A methodological review. *Analytica Chimica Acta* 903: 13–35.

Chen, P., Wu, W., Wu, K. C. 2013. The zero valent iron nanoparticle causes higher developmental toxicity than its oxidation products in early life stages of medaka fish. *Water Research* 47(12): 3899–3909.

Cheng, X., Wu, B., Yang, Y., Li, Y. 2011. Synthesis of iron nanoparticles in water-in-oil microemulsions for liquid-phase Fischer–Tropsch synthesis in polyethylene glycol. *Catalysis Communications* 12: 431–435.

Chesworth, W., Perez-Alberti, A., Arnaud, E., Morel-Seytoux, H. J., Schwertmann, U. 2008. Iron Oxides, Encyclopedia of Soil Science. In: *Encyclopedia of Earth Sciences Series*. Chesworth, W. (ed.) Springer, Netherlands, ISBN: "978-1-4020-3995-9", doi: "10.1007/978-1-4020-3995-9_302", pp. 363–369.

Chichiriccò, G., Poma, A. 2015. Penetration and toxicity of nanomaterials in higher plants. *Nanomaterials* 5: 851–873.

Choe, S., Chang, Y., Hwang, K., Khim, J. 2000. Kinetics of reductive denitrifcation by nanoscale zero-valent iron. *Chemosphere* 41: 1307–1311.

Cook, S. M. 2009. *Assessing the Use and Application of Zero-Valent Iron Nanoparticle Technology for Remediation at Contaminated Sites*. U.S. Environmental Protection Agency., Office of Solid Waste and Emergency Response & Office of Superfund Remediation and Technology Innovation, Washington, DC.

Costa, M. C, Mota, F. S, Santos, A. B. D., Mendonça, G. L. F., Nascimento, R. F. 2012. Effect of dye structure and redox mediators on anaerobic azo and anthraquinone dye reduction, *Quim. Nova* 35(3): 482–486.

Crane, R. A., Scott, T. B. 2012. Nanoscale zero-valent iron: Future prospects for an emerging water treatment technology. *Journal of Hazardous Materials* 211–212: 112–125.

Cui, H., Ren, W., Lin, P., Liu, Y. 2012. Structure control synthesis of iron oxide polymorph nanoparticles through an epoxide precipitation route. *Journal of Experimental Nanoscience* 8(7-8): 869–875.

Cundy, A. B., Hopkinson, L., Whitby, R. L. D. 2008. Use of iron-based technologies in contaminated land and groundwater remediation: A review. *Science of The Total Environment* 400(1–3): 42–51.

Daniel, M., Astruc, D. 2004. Gold nanoparticles: Assembly, supramolecular chemistry, quantum-size-related properties, and applications toward biology, catalysis, and nanotechnology. *Chemical Reviews* 104(1): 293–346.

Destrée, C., Nagy, J. B. 2006. Mechanism of formation of inorganic and organic nanoparticles from microemulsions. *Advances in Colloid and Interface Science* 123–126: 353–367.

Devi, T. S. R., Gayathri, S. 2010. FTIR and Ft-Raman spectral analysis of paclitaxel drugs. *International Journal of Pharmaceutical Sciences Review and Research* 2(2)-Article 019.

Dion, P. 2008. The microbiological promises of extreme soils. In *Microbiology of Extreme Soils*. Soil Biology 13, Dion, P., Nautiyal, C. S. (eds), Springer-Verlag, Berlin, Heidelberg, p. 35.

Donaldson, K., Tran, L., Jimenez, L. A., Duffin, R., Newby, D. E., Mills, N., MacNee, W., Stone, V. 2005. Combustion-derived nanoparticles: A review of their toxicology following inhalation exposure. *Particle and Fibre Toxicology* 2: 10.

Dorofeev, G. A., Streletskii, A. N., Povstugar, I. V., Protasov, A. V., Elsukov, E. P. 2012. Determination of nanoparticle sizes by xray diffraction. *Colloid Journal* 74(6): 675–685.

Elliott, D., Lien, H., Zhang, W. 2009. Degradation of lindane by zero-valent iron nanoparticles. *Journal of Environmental Engineering* 135: 5(317), 317–324, doi: 10.1061/(ASCE)0733-9372.

Evans, R. 2007. *The impact of groundwater use on Australian rivers*, Technical report, Australian Government, Land and Water Australia, www.lwa.gov.au

Fan, J., Gao, Y. 2006. Nanoparticle-supported catalysts and catalytic reactions—A mini-review. *Journal of Experimental Nanoscience* 1(4): 457–475.

Fan, J, Guo, Y, Wang, J, Fan, M. 2009. Rapid decolorization of Azo dye methyl orange in aqueous solution by nanoscale zerovalent iron particles. *Journal of Hazardous Materials* 166: 904–910.

Fan, W., Cheng, Y., Yu, S., Fan, X., Deng, Y. 2015. Preparation of wrapped nZVI particles and their application for the degradation of trichloroethylene (TCE) in aqueous solution. *Journal of Water Reuse and Desalination* 5(3): 335–343.

Farahmandjou, M., Soflaee, F. 2014. Synthesis of iron oxide nanoparticles using borohydride reduction. *International Journal of Bio-Inorganic Hybrid Nanomaterials* 3(4): 203–206.

Fissan, H., Ristig, S., Kaminski, H., Asbach, C., Epple, M. 2014. Comparison of different characterization methods for nanoparticle dispersions before and after aerosolization. *Analytical Methods* 6: 7324–7334.

Frey, N. A., Peng, S., Cheng, K., Sun, S. 2009. Magnetic nanoparticles: synthesis, functionalization, and applications in bioimaging and magnetic energy storage. *Chemical Society Reviews* 38(9): 2532–2542.

Fu, F., Ma J., Xie, L., Tang, B., Han, W., Lin, S. 2013. Chromium removal using resin supported nanoscale zero-valent iron. *Journal of Environmental Management* 15(128): 822–827.

Gao, L., Seliskar, C. J. 1998. Formulation, characterization, and sensing applications of transparent Poly(vinyl alcohol)-Polyelectrolyte blends. *Chemistry of Materials* 10(9): 2481–2489.

Génin, J. R., Refait, P., Bourrie, G., Abdelmoula, M., Trolard, F. 2001. Structure and stability of the Fe(II)-Fe(III) green rust "Fougerite" mineral and its potential for reducing pollutants in soil solutions. *Applied Geochemistry* 16: 559–570.

Goldstein, J. I., Newbury, D. E., Michael, J. R., Ritchie, N. W. M., Scott, J. H. J., Joy, D. C. 2017. Backscattered electrons. In: *Scanning Electron Microscopy and X-Ray Microanalysis*. Springer (online). ISBN 1493966766, 9781493966769, pp. 15–28.

Grieger, K. D., Fjordbøge, A., Hartmann, N. B., Eriksson, E., Bjerg, P. L., Baun, A. 2010. Environmental benefits and risks of zero-valent iron nanoparticles (nZVI) for *in situ* remediation: Risk mitigation or trade-off? *Journal of Contaminant Hydrology* 118: 165–183.

Grieger, K. D., Hjorth, R., Rice, J., Kumar, N., Bang, J. 2015. Nano-remediation: tiny particles cleaning up big environmental problems. *Blog Entry For IUCN*. http://cmsdata.iucn.org/downloads/nanoremediation.pdf. Accessed on September 21, 2016.

Guo, H., Barnard, A. S. 2011. Thermodynamic modeling of nanomorphologies of hematite and goethite. *Journal of Materials Chemistry* 21: 11566–11577.

Gutteridge, J. M. C. 1986. Iron promoters of the Fenton reaction and lipid peroxidation can be released from haemoglobin by peroxides. *FEBS Letters* 201: 1873–3468.

Halperin, W. P. 1986. Quantum size effects in metal particles. *Reviews of Modern Physics*. 58(3): 533–606.

Hameri, K. 2012. Nanotechnology Commercialization for Managers and Scientists. *Chapter 10: Environment, Health and Safety within the Nanotechnology Industry*. Helwegen, W., Escoffier, L. (eds), Pan Stanford Publishing Pte. Ltd., Singapore, ISBN 9814316229, 9789814316224, pp. 340–342.

Hasany, S. F., Ahmed, I., Rehman, R. J. A. 2012. Systematic review of the preparation techniques of iron oxide magnetic nanoparticles. *Nanoscience and Nanotechnology* 2(6): 148–158.

Hayashi, H., Hakuta, Y. 2010. Hydrothermal synthesis of metal oxide nanoparticles in supercritical water. *Materials*. 3: 3794–3817.

He, F., Zhao, D., Roberts, C. 2009. Nanotechnology Applications for Clean Water: Solutions for Improving Water Quality. In: *Stabilization of Zero-Valent Iron Nanoparticles for Enhanced In Situ Destruction of Chlorinated Solvents in Soils and Groundwater*. Diallo, M., Street, A., Sustich, R., Duncan, J., Savage, N., William, A. (eds), Elsevier, Norwich, USA, ISBN 0815519737, 9780815519737, pp. 282–284.

Heera, P., Shanmugam, S. 2015. Nanoparticle characterization and application: An overview. *International Journal of Current Microbiology and Applied Sciences* 4(8): 379–386.

Herlekar, M., Barve, S., Kumar, R. 2014. Plant-mediated green synthesis of iron nanoparticles. *Journal of Nanoparticles* 2014: 9 pages. doi: 10.1155/2014/140614

Hirt, A. M., Lanci, L. 2011. Low-temperature magnetic properties of lepidocrocite. *Journal of Geophysical Research* 107(B1), doi: 10.1029/2001jb000242

Hua, M., Zhang, S., Pan, B., Zhang, W., Lv, L., Zhang, Q. 2012. Heavy metal removal from water/wastewater by nanosized metal oxides: A review. *Journal of Hazardous Materials* 211–212: 317–331.

Huang, K., Ehrman, S. H. 2007. Synthesis of iron nanoparticles via chemical reduction with palladium ion seeds. *Langmuir.* 23: 1419–1426.

Ibrahem, A. K., Moghny, T. A., Mustafa, Y. M., Maysour, N. E., Dars, F. M. S. E. D. E., Hassan, R. F. 2012. Degradation of trichloroethylene contaminated soil by zero-valent iron nanoparticles. *ISRN Soil Science.* 2012, article ID 270830, 9 pages, doi: 10.5402/2012/270830

Jamiesona, E., Jonesa, A., Coolinga, D., Stockton, N. 2006. Magnetic separation of red sand to produce value. *Minerals Engineering* 19(15): 1603–1605.

Jana, N. R., Chen, Y., Peng, X. 2004. Size- and shape-controlled magnetic (Cr, Mn, Fe, Co, Ni) oxide nanocrystals via a simple and general approach. *Chemistry of Materials* 16: 3931–3935.

Joo, S., Feitz, A. J., Waite, T. D. 2004. Oxidative degradation of the carbothioate herbicide, molinate, using nanoscale zero-valent iron. *Environmental Science & Technology* 38: 2242–2247.

Jordan, J., Jacob, K. I., Tannenbaum, R., Sharaf, M. A., Jasiuk, I. 2005. Experimental trends in polymer nanocomposites—a review. *Material Science and Engineering: A* 393(1–2): 1–11.

Kalyani, S., Sangeetha, J., Philip, J. 2015. Microwave assisted synthesis of ferrite nanoparticles: Effect of reaction temperature on particle size and magnetic properties. *Journal of Nanoscience and Nanotechnology* 15, 1–7.

Khan, I., Husain, T., Hejazi, R. 2004. An overview and analysis of site remediation technologies. *Journal of Environmental Management* 71: 95–122.

Kim, H., Kim, T., Ahn, J., Hwang, K., Park, J., Lim, T., Hwang, I. 2012. Aging characteristics and reactivity of two types of nanoscale zero-valent iron particles (FeBH and FeH$_2$) in nitrate reduction. *Chemical Engineering Journal* 197: 16–23.

Kim, H., Phenrat, T., Tilton, R. D., Lowry, G. V. 2009. Fe$_0$ nanoparticles remain mobile in porous media after aging due to slow desorption of polymeric surface modifiers. *Environmental Science & Technology* 43(10): 3824–3830.

Kim, S. A., Guerinot, M. L. 2007. Mining iron: Iron uptake and transport in plants. *FEBS Letters* 581: 2273–2280.

Knežević, N. Z. 2014. Core/shell magnetic mesoporous silica nanoparticles with radially oriented wide mesopores. *Processing and Application of Ceramics* 8(2): 109–112.

Krajangpan, S., Kalita, H., Chisholm, B. J., Bezbaruah, A. N. 2012. Iron nanoparticles coated with amphiphilic polysiloxane graft copolymers: Dispersibility and contaminant treatability. *Environmental Science & Technology* 46: 10130–10136.

Kundu, T. K., Karak, N., Barik, P., Saha, S. 2011. Optical properties of ZnO nanoparticles prepared by chemical method using poly (VinylAlcohol) (PVA) as capping agent. *International Journal of Soft Computing and Engineering (IJSCE).* ISSN: 2231–2307, 1: 19–24.

Laurenzi, M. A. 2008. Transformation kinetics & magnetism of magnetite nanoparticles., Chapter IV-Nanostructure, *ProQuest.* Dissertations and Theses; Thesis (Ph.D.)—The Catholic University of America, 2008, ISBN 9780549650621, Publication Number: AAI3311966, Dissertation Abstracts International, vol. 69(05), pp. 44–46.

Lee, H., Lee, H., Kima, H., Kweon, J., Lee, B., Lee, C. 2014. Oxidant production from corrosion of nano- and micro-particulate zero-valent iron in the presence of oxygen: A comparative study. *Journal of Hazardous Materials* 265: 201–207.

Lemine, O. M., Alyamani, A., Sajieddine, M., Bououdina, M. 2009. Characterization of α-Fe$_2$O$_3$ Nanoparticles Produced by High Energy Ball Milling. *Proceedings of the 1st WSEAS International Conference on Nanotechnology*, Cambridge, UK. ISBN: 978-960-474-059-8.

Lewicka, Z. A., Colvin, V. L. 2013. Springer Handbook of Nanomaterials. In: *Chapter 32.4: Nanoparticle Physiochemical Characteristics of Relevance for Toxicology*. Vajtai, R. (ed.) Springer-Verlag, Berlin, Heidelberg. ISBN 364220595X, 9783642205958, pp. 1127–1129.

Li, C., Gu, L., Tong, J., Maier, J. 2011a. Carbon nanotube wiring of electrodes for high-rate lithium batteries using an imidazolium-based ionic liquid precursor as dispersant and binder: A case study on iron fluoride nanoparticles. *ACS Nano* 5(4): 2930–2938.

Li, F., XiaoPeng, G., Yi, L., DongSheng, W., HongXiao, T. 2011b. Dechlorination of 2,4-dichlorophenol by nickel nanoparticles under the acidic conditions. *Chinese Science Bulletin* 56(21): 2258–2266.

Li, S., Yan, W., Zhang, W. 2009. Solvent-free production of nanoscale zero-valent iron (nZVI) with precision milling. *Green Chemistry* 11: 1618–1626.

Li, X., Yang, Y., Gao, B., Zhang, M. 2015. Stimulation of peanut seedling development and growth by zero-valent iron nanoparticles at low concentrations. *PLoS ONE* 10(4): e0122884. doi: 10.1371/journal.pone.0122884.

Liang, Q., An, B., Zhao, D. 2013. Monitoring Water Quality: Pollution Assessment, Analysis, and Remediation. In: *Chapter 12: Removal and Immobilization of Arsenic in Water and Soil Using Polysaccharide Modified Magnetite Nanoparticles*. Ahuja, S. (ed.) Elsevier, Kidlington, Oxford and Waltham, USA, ISBN: 0444594043, 9780444594044, pp. 287–291.

Liang, W., Dai, C., Zhou, C., Zhang, Y. 2014. Application of zero-valent iron nanoparticles for the removal of aqueous zinc ions under various experimental conditions. *PLoS ONE*. 9(1): e85686. doi: 10.1371/journal.pone.0085686.

Lien, H., Zhang, W. 2001. Nanoscale iron particles for complete reduction of chlorinated ethenes. *Colloids and Surfaces A: Physicochemical and Engineering Aspects* 191: 97–105.

Lim, J., Yeap, S. P., Che, H. X., Low, S. C. 2013. Characterization of magnetic nanoparticle by dynamic light scattering. *Nanoscale Research Letters* 8(1): 381.

Liu, F., Ni, Q., Murakami, Y. 2013. Preparation of magnetic polyvinyl alcohol composite nanofibers with homogenously dispersed nanoparticles and high water resistance. *Textile Research Journal* 83(5): 510–518.

Liu, W., Qian, T., Jiang, H. 2014a. Bimetallic Fe nanoparticles: Recent advances in synthesis and application in catalytic elimination of environmental pollutants. *Chemical Engineering Journal* 236: 448–463.

Liu, Y., Lowry, G. V. 2006. Effect of particle age (Fe_0 contents) and solution pH on NZVI reactivity: H2 evolution and TCE dechlorination. *Environmental Science & Technology* 40: 6085–6090.

Liu, Y., Majetich, S.A., Tilton, R. D., Sholl, D. S., Lowry, G. V. 2005. TCE dechlorination rates, pathways, and efficiency of nanoscale iron particles with different properties. *Environmental Science & Technology* 39: 1338–1345.

Liu, Y., Zhang, W., Li, S., Cui, C., Wu, J., Chen, H., Huo, F. 2014b. Designable yolk-shell nanoparticle@ MOF petalous heterostructures. *Chemistry of Materials* 26: 1119–1125.

Long, N. V., Teranishi, T., Yang, Y., Thi, C. M., Cao, Y., Nogami, M. 2015. Iron oxide nanoparticles for next generation gas sensors. *International Journal of Metallurgical & Materials Engineering* 1: 119.

Lowry, G., Phenrat, T., Fagerlund, F. 2012. *Fundamental Study of the Delivery of Nanoiron to DNAPL Source Zones in Naturally Heterogeneous Field Systems*. SERDP Project ER-1485., Final Report.

Luo, S., Qin, P., Shao, J., Peng, L., Zeng, Q., Gu, J. 2013. Synthesis of reactive nanoscale zero valent iron using rectorite supports and its application for Orange II removal. *Chemical Engineering Journal* 223: 1–7.

Machado, S., Grosso, J. P., Nouws, H. P. A., Albergaria, J. T., Delerue-Matos, C. 2014. Utilization of food industry wastes for the production of zero-valent iron nanoparticles. *Science of the Total Environment* 496: 233–240.

Machado, S., Pacheco, J. G., Nouws, H. P. A., Albergaria, J. T., Delerue-Matos, C. 2015. Characterization of green zero-valent iron nanoparticles produced with tree leaf extracts. *Science of the Total Environment* 533: 76–81.

Madsen, D. E., Cervera-Gontard, L., Kasama, T., Dunin-Borkowski, R. E., Koch, C. B., Hansen, M. F., Frandsen, C., Mørup, S. 2009. Magnetic fluctuations in nano-sized goethite (α-FeOOH) grains. *Journal of Physics: Condensed Matter*. 21:016007 (11pp).

Mahdavi, M., Namvar, F., Ahmad, M. B., Mohamad, R. 2013. Green biosynthesis and characterization of magnetic iron oxide (Fe_3O_4) Nanoparticles using seaweed (*Sargassum muticum*) aqueous extract. *Molecules* 18: 5954–5964.

Maleki, H., Simchia, A., Imanic, M., Costa, B. F. O. 2012. Size-controlled synthesis of superparamagnetic iron oxide nanoparticles and their surface coating by gold for biomedical applications. *Journal of Magnetism and Magnetic Materials* 324(23): 3997–4005.

Masina, C. J., Neethling, J. H., Olivier, E. J., Manzini, S., Lodya, L., Srotc, V., Van Akenc, P. A. 2015. Structural and magnetic properties of ferrihydrite nanoparticles. *RSC Advances* 5: 39643–39650.

Massa, W. 2004. *Crystal Structure Determination, Crystal Lattices*, Springer-Verlag, Berlin, Heidelberg, pp. 3–5, ISBN: 978-3-662-06431-3.

Miranda, M. M., Fernández, J. M. C., Encinar, J. R., Paraka, W. J., Carrion, C. C. 2015. Determination of the ratio of fluorophore/nanoparticle for fluorescence-labelled nanoparticles. *Analyst* 141: 1266–1272.

Mohapatra, M., Anand, S. 2010. Synthesis and applications of nano-structured iron oxides/hydroxides—A review. *International Journal of Engineering, Science and Technology* 2(8):127–146.

Mousavi, M. F., Ghasemi, S. 2010. Sonochemistry: Theory, reactions, syntheses, and applications. In: *Chapter 1: Sonochemistry: A Suitable Method for Synthesis of NanoStructured Materials*. Nowak, F. M. (eds) Nova Science Publishers, Inc., ISBN 1617286524, 9781617286520, pp. 1–16.

Mueller, N. C., Braun, J., Bruns, J., Černík, M., Rissing, P., Rickerby, D., Nowack, B. 2012. Application of nanoscale zero valent iron (NZVI) for groundwater remediation in Europe. *Environmental Science and Pollution Research* 19: 550–558.

Müller, N. C., Nowack, B. 2010. *Nano zero valent iron—The solution for water and soil remediation?* Report of the ObservatoryNANO, www.observatorynano.eu

Naja, G., Apiratikul, R., Pavasant, P., Volesky, B., Hawari, J. 2009. Dynamic and equilibrium studies of the RDX removal from soil using CMC-coated zerovalent iron nanoparticles. *Environmental Pollution* 157: 2405–2412.

Najan, G., Halasz, A., Thiboutot, S., Ampleman, G., Hawari, J. 2008. Degradation of hexahydro-1,3,5-trinitro-1,3,5-triazine (RDX) using zerovalent iron nanoparticles. *Environmental Science & Technology* 42: 4364–4370.

Nam, S., Tratnyek, P. G. 2000. Reduction of Azo dyes with zero-valent iron. *Water Research* 34(6): 1837–1845.

Neoh, K. G., Li, M., Kang, E. 2015. Nanotechnology in Endodontics: Current and Potential Clinical Applications. In: *Chapter 3: Characterization of Nanomaterials/Nanoparticles*. Kishen, A. (ed.) Springer, Switzerland, ISBN: 3319135759, 9783319135755, pp. 25–29.

Nicole, C., Braun, M., Braun, J., Bruns, J., Černík, M., Rissing, P., Rickerby, D., Nowack, B. 2011. Application of nanoscale zero valent iron (NZVI) for groundwater remediation in Europe. *Environmental Science and Pollution Research* 19: 550–558.

Nidheesh, P. V., Gandhimathi, R., Ramesh, S. T. 2013. Degradation of dyes from aqueous solution by Fenton processes: a review. *Environmental Science and Pollution Research* 20: 2099–2132.

Nurmi, J. T., Tratnyek, P. G., Sarathy, V., Baer, D. R., Amonette, J. E., Pecher, K., Wang, C., Linehan, J. C., Matson, D. W., Penn, R. L., Driessen, M. D. 2005. Characterization and properties of metallic iron nanoparticles: Spectroscopy, electrochemistry, and kinetics. *Environmental Science and Pollution Research* 39: 1221–1230.

Nurulita, Y., Adetutu, E. M., Kadali, K. K., Zul, D., Mansur, A. A., Ball, A. S. 2015. The assessment of the impact of oil palm and rubber plantations on the biotic and abiotic properties of tropical peat swamp soil in Indonesia. *International Journal of Agricultural Sustainability* 13(2): 150–166.

O'Carroll, D., Sleep, B., Krol, M., Boparai, H., Kocur, C. 2013. Nanoscale zero valent iron and bimetallic particles for contaminated site remediation. *Advances in Water Resources* 51: 104–122.

O'Hara, S., Krug, T., Quinn, J., Clausen, C., Geiger, C. 2006. Field and laboratory evaluation of the treatment of DNAPL source zones using emulsified zero-valent iron. *Remediation* doi: 10.1002.rem

Patil, S. S., Shedbalkar, U. U., Truskewycz, A., Chopade, B. A., Ball, A. S. 2016. Nanoparticles for environmental clean-up: A review of potential risks and emerging solutions. *Environmental Technology & Innovation* 5: 10–21.

Pattanayak, M., Mohapatra, D., Nayak, P. L. 2013. Green synthesis and characterization of zero valent iron nanoparticles from the leaf extract of *Syzygium aromaticum* (clove). *Middle-East Journal of Scientific Research* 18(5): 623–626.

Pattanayak, M., Nayak, P. L. 2013. Green synthesis and characterization of zero valent iron nanoparticles from the leaf extract of *Azadirachta indica* (neem). *World Journal of Nano Science & Technology* 2(1): 06–09.

Pauluhn, J. 2011. Subchronic inhalation toxicity of iron oxide (magnetite, Fe_3O_4) in rats: Pulmonary toxicity is determined by the particle kinetics typical of poorly soluble particles. *Journal of Applied Toxicology* 32: 488–504.

Pereira, N., Mujika, M., Arana, S., Correia, T., Silva, A. M. T., Gomes, H. T., Rodrigues, P. J., Lima, R. 2013. Visualization and Simulation of Complex Flows in Biomedical Engineering. *The Effect of Static Magnetic Field on the Flow of Iron Oxide Magnetic Nanoparticles Through Glass Capillaries.* Lima, R., Imai, Y., Ishikawa, T., Oliveira, V. (eds) Springer, Netherlands, ISBN 9400777698, 9789400777699. pp. 181–182.

Pereira, W. S., Freire, R. S. 2006. Azo dye degradation by recycled waste zero-valent iron powder. *Journal of the Brazilian Chemical Society* 17(5): 832–838.

Perfumo, A., Cockell, C., Elsaesser, A., Marchant, R. 2011. Microbial diversity in Calamita ferromagnetic sand. *Environmental Microbiology Reports* 3(4): 483–490.

Petala, E., Dimos, K., Douvalis, A., Bakas, T., Tucek, J., Zboril, R., Karakassides, M. A. 2013. Nanoscale zero-valent iron supported on mesoporous silica: Characterization and reactivity for Cr(VI) removal from aqueous solution. *Journal of Hazardous Materials* 26: 295–306.

Petrova, O. S., Gudilin, E. A., Chekanova, A. E., Knot'ko, A. V., Murav'eva, G. P., Maksimov, Y. V., Imshennik, V. K., Suzdalev, I. P., Tret'yakov, Y. D. 2006. Microemulsion synthesis of mesoporous γ-Fe_2O_3 nanoparticles. *Doklady Akademii Nauk* 410(5): 629–632.

Phenrat, T., Lowry, G. V. 2014. Nanotechnology applications for clean water, solutions for improving water quality. In: *Chapter 30.4: Contaminant Targeting of Polymeric Functionalized Nanoparticles*. 2nd edition, Street, A., Sustich, R., Duncan, J., Savage, N., William, A. (eds) Elsevier, Kidlington, Oxford and Waltham, USA, ISBN: 1455731854, 9781455731855, pp. 484–485.

Phenrat, T., Saleh, N., Sirk, K., Tilton, R. D., Lowry, G. V. 2007. Aggregation and sedimentation of aqueous nanoscale zerovalent iron dispersions. *Environmental Science & Technology* 41: 284–290.

Phu, N. D., Ngo, D. T., Hoang, L. H., Luong, N. H., Ha, N. H. 2011. Crystallization process and magnetic properties of amorphous iron oxide nanoparticles. *Journal of Physics D: Applied Physics* 44: 345002.

Poudyal, N. 2008. Fabrication of superparamagnetic and ferromagnetic nanoparticles., Chapter 2: Small particle magnetism and fabrication. *ProQuest*. Dissertation Abstracts International, ISBN 054994740X, 9780549947400, 25–27.

Poursaberi, T., Konoz, E., Sarrafib, A. H. M., Hassanisadi, M., Hajifathli, F. 2012. Application of nanoscale zero-valent iron in the remediation of DDT from contaminated water. *Chemical Science Transactions* 1(3): 658–668.

Prakash, D., Gabani, P., Chandel, A. K., Ronen, Z., Singh, O. V. 2013. Bioremediation: A genuine technology to remediate radionuclides from the environment. *Microbial Biotechnology* 6: 349–360.

Quinn, J. W., Brooks, K. B., Geiger, C. L., Clausen, C. A., Milum, K. M. 2006. *Application of Emulsified Zero-Valent Iron to Marine Environments*. NASA Technical Reports. Document ID: 20120003625., Report/Patent Number: KSC-2006-079.

Rahman, M. M., Khan, S. B., Jamal, A., Faisal, M., Aisiri, A. M. 2011. Nanomaterials Chapter 3. *Iron Oxide Nanoparticles*. Rahman, M. M. (ed.), InTech, Shangai, China, ISBN: 978-953-307-913-4, pp. 43–64.

Rahman, N., Abedin, Z., Hossain, M. A. 2014. Rapid degradation of Azo dyes using nano-scale zero valent iron. *American Journal of Environmental Science* 10(2): 157–163.

Rajabi, F., Kakeshpour, T., Saidi, M. R. 2013. Supported iron oxide nanoparticles: Recoverable and efficient catalyst for oxidative S-S coupling of thiols to disulfides. *Catalysis Communications* 40: 13–17.

Rajajayavel, S. R. C., Ghoshal, S. 2015. Enhanced reductive dechlorination of trichloroethylene by sulfidated nanoscale zerovalent iron. *Water Research* 78: 144–153.

Rajan, C. S. 2011. Nanotechnology in groundwater remediation. *International Journal of Environmental Science and Development* 2(3): 182–187.

Remya, A. S., Ramesh, M., Saravanan, M., Poopal, R. K., Bharathi, S., Nataraj, D. 2015. Iron oxide nanoparticles to an Indian major carp, *Labeo rohita*: Impacts on hematology, iono regulation and gill Na+/K+ ATPase activity. *Journal of King Saud University – Science* 27(2): 151–160.

Safarik, I., Safarikova, M. 2012. Magnetic nanoparticles: from fabrication to clinical applications. *Chapter 8: Magnetic Nanoparticles for In Vitro Biological and Medical Applications, An Overview.* Thanh, N. (ed.) CRC Press, ISBN: 1439869332, 9781439869338, pp. 218–219.

Sajanlal, P. R., Sreeprasad, T. S., Samal, A. K., Pradeep, T. 2011. Anisotropic nanomaterials: structure, growth, assembly, and functions. *Nano Reviews* 2: 5883.

Sakulchaicharoen, N., O'Carroll, D. M., Herrera, J. E. 2010. Enhanced stability and dechlorination activity of pre-synthesis stabilized nanoscale FePd particles. *Journal of Contaminant Hydrology* 118(3–4): 117–27.

Saleh, N., Phenrat, T., Sirk, K., Dufour, B., Ok, J., Sarbu, T., Matyjaszewski, K., Tilton, R. D., Lowry, G. V. 2005. Adsorbed triblock copolymers deliverreactive iron nanoparticles to theoil/water interface. *Nano Letters* 5(12).

Saravanan, M., Suganya, R., Ramesh, M., Poopal, R. K., Gopalan, N., Ponpandian, N. 2015. Iron oxide nanoparticles induced alterations in haematological, biochemical and ionoregulatory responses of an Indian major carp *Labeo rohita*. *Journal of Nanoparticle Research* 17: 274.

Schrick, B., Blough, J. L., Jones, A. D., Mallouk, T. E. 2002. Hydrodechlorination of trichloroethylene to hydrocarbons using bimetallic nickel-iron nanoparticles. *Chemistry of Materials* 14: 5140–5147.

Schwertmann, U. 2008. Encyclopaedia of Soil Science. *Chapter: Iron Oxides*. Chesworth, W. (ed.) Springer, Netherlands. doi: 10.1007/978-1-4020-3995-9_302, pp. 363–369.

Shahwan, T., Sirriah, S. A., Nairat, M., Boyacı, E., Eroglu, A. E., Scott, T. B., Hallam, K. R. 2011. Green synthesis of iron nanoparticles and their application as a Fenton-like catalyst for the degradation of aqueous cationic and anionic dyes. *Chemical Engineering Journal* 172: 258–266.

Shahzeidi, Z. S., Amiri, G. 2015. Antibacterial activity of Fe_3O_4 nanoparticles. *International Journal of Bio-Inorganic Hybrid Nanomaterials* 4(3): 135–140.

Shankramma, K., Yallappa, S., Shivanna, M. B., Manjanna, J. 2015. Fe_2O_3 magnetic nanoparticles to enhance *S. lycopersicum* (tomato) plant growth and their biomineralization. *Applied Nanoscience* 6(7): 893–990.

Sheykhbaglou, R., Sedghi, M., Shishevan, M. T., Sharifi, R. S. 2010. Effects of nano-iron oxide particles on agronomic traits of soybean. *Notulae Scientia Biologicae* 2(2): 112–113.

Shi, Z., Fan, D., Johnson, R. L., Tratnyek, P. G., Nurmi, J. T., Wu, Y., Williams, K. H. 2015. Methods for characterizing the fate and effects of nano zerovalent iron during groundwater remediation. *Journal of Contaminant Hydrology* 181: 17–35.

Shindo, D., Oikawa, T. 2002. *Analytical Electron Microscopy for Materials Science, Chapter 4: Energy Dispersive X-ray Spectroscopy*. Springer, Japan. ISBN 978-4-431-66988-3, pp. 81–100.

Singh, A. K., Nakate, U. T. 2013. Microwave synthesis, characterization and photocatalytic properties of SnO_2 nanoparticles. *Advances in Nanoparticles* 2: 66–70.

Singh, U. K., Layland, F. C., Prasad, R., Singh, S. 2013. *Poisoning in Children, Chapter 5, Metals and Non-Metals*. Jaypee Brothers Publishers, ISBN: 9789350257739, pp. 41–46.

Sirk, K. M., Saleh, N. B., Phenrat, T., Kim, H., Dufour, B., Ok, J., Golas, P. L., Matyjaszewski, K., Lowry, G.V., Tilton, R. D. 2009. Effect of adsorbed polyelectrolytes on nanoscale zero valent iron particle attachment to soil surface models. *Environmental Science & Technology* 43: 3803–3808.

Skinner, B. J. 1979. Earth resources (minerals/metals/ores/geochemistry/mining). *PNAS, Proceedings of the National Academy of Sciences* 76(9): 4212–4217.

Soenen, S. J., De Cuyper, M., De Smedt, S. C., Braeckmans, K. 2012. Chapter ten—Investigating the toxic effects of iron oxide nanoparticles. *Methods in Enzymology* 509: 195–224.

Stefaniuka, M., Oleszczuka, P., Okb, Y. S. 2016. Review on nano zerovalent iron (nZVI): From synthesis to environmental applications. *Chemical Engineering Journal* 287(1): 618–632.

Su, C., Puls, R. W., Krug, T. A., Watling, M. T., O'Hara, S. K., Quinn, J. W., Ruiz, N. E. 2012. A two and half-year-performance evaluation of a field test on treatment of source zone tetrachloroethene and its chlorinated daughter products using emulsified zero valent iron nanoparticles. *Water Research* 46: 5071–5084.

Sun, Y., Li, X., Cao, J., Zhang, W. 2006. Characterization of zero-valent iron nanoparticles. *Advances in Colloid and Interface Science* 120(1–3): 47–56.

Tang, L., Tang, J., Zeng, G., Yang, G., Xie, X., Zhou, Y., Pang, Y., Fang, Y., Wang, J., Xiong, W. 2015. Rapid reductive degradation of aqueous p-nitrophenol usingnanoscale zero-valent iron particles immobilized on mesoporoussilica with enhanced antioxidation effect. *Applied Surface Science* 333: 220–228.

Tang, L., Yang, G., Zeng, G., Cai, Y., Li, S., Zhou, Y., Pang, Y., Liu, Y., Zhang, Y., Luna, B. 2014. Synergistic effect of iron doped ordered mesoporous carbon on adsorption-coupled reduction of hexavalent chromium and the relative mechanism study. *Chemical Engineering Journal* 239: 114–122.

Tartaj, P., Morales, M. P., Veintemillas-Verdaguer, S., Gonzalez-Carreno, T., Serna, C. J. 2003. The preparation of magnetic nanoparticles for applications in biomedicine. *Journal of Physics D: Applied Physics* 36: R182–R197.

Tchounwou, P. B., Yedjou, C. G., Patlolla, A. K., Sutton, D. J. 2012. Heavy metals toxicity and the environment. *Experientia Supplementum* 101: 133–164.

Teo, B. M., Chen, F., Hatton, T. A., Grieser, F., Ashokkumar, M. 2009. Novel One-pot synthesis of magnetite latex nanoparticles by ultrasound irradiation. *Langmuir* 25(5): 2593–2595.

Thiruvenkatachari, R., Vigneswaran, S., Naidu, R. 2007. Permeable reactive barrier for groundwater remediation. *Journal of Industrial and Engineering Chemistry* 14: 145–156.

Truskewycz, A., Shukla, R., Ball, A. S. 2016. Iron nanoparticles synthesized using green tea extracts for the fenton-like degradation of concentrated dye mixtures at elevated temperatures. *Journal of Environmental Chemical Engineering* 4(4): 4409–4417.

Verwey, E. J. W. 1948. Theory of the stability of lyophobic colloids. *The Journal of Physical Chemistry A* 51(3): 631–636.

Virkutyte, J., Varma, R. S. 2013. Nanotechnology for Water and Wastewater Treatment. In: *Chapter 17-Green Synthesis of Nanoparticles and Nanomaterials*. Lens, P., Virkutyte, J., Jegatheesan, V., Kim, S., Al-Abed, S., IWA Publishing, London. ISBN 1780404581, 9781780404585, pp. 380–381.

Waltlington, K. 2005. *Emerging Nanotechnologies for Site Remediation and Wastewater Treatment., U.S. Environmental Protection Agency., Office of Solid Waste and Emergency Response & Office of Superfund Remediation and Technology Innovation.* Technology Innovation and Field Services Division, Washington, DC.

Wang, M., Liu, X., Hu, J., Li, J., Huang, J. 2015. Nano-ferric oxide promotes watermelon growth. *Journal of Biomaterials and Nanobiotechnology* 6: 160–167.

Weerathunge, P., Ramanathan, R., Shukla, R., Sharma, T., Bansal, V. 2014. Aptamer-controlled reversible inhibition of gold nanozyme activity for pesticide sensing. *Analytical Chemistry* 86: 11937–11941.

Wijesekara, S. S. R. M. D. H. R., Basnayake, B. F. A., Vithanage, M. 2014. Organic-coated nanoparticulate zero valent iron for remediation of chemical oxygen demand (COD) and dissolved metals from tropical landfill leachate. *Environmental Science and Pollution Research* 21(11): 7075–7087.

Wu, W., He, Q., Jiang, C. 2008. Magnetic iron oxide nanoparticles: Synthesis and surface functionalization strategies. *Nanoscale Research Letters* 3: 397–415.

Wu, W., Wu, Z., Yu, T., Jiang, C., Kim, W. 2015. Recent progress on magnetic iron oxide nanoparticles: synthesis, surface functional strategies and biomedical applications. *Science and Technology of Advanced Materials* 16: 023501 (43pp).

Yan, W., Lien, H., Koel, B. E., Zhang, W. 2013. Iron nanoparticles for environmental clean-up: recent developments and future outlook. *Environmental Science: Processes & Impacts* 15: 63.

Yanga, G. C. C., Chang, Y. 2011. Integration of emulsified nanoiron injection with the electrokinetic process for remediation of trichloroethylene in saturated soil. *Separation and Purification Technology* 79(2): 278–284.

Yildirimer, L., Thanh, N. T. K., Loizidoua, M., Seifalian, A. M. 2011. Toxicology and clinical potential of nanoparticles. *Nanotoday* 6(6): 585–607.

Yoo, B. Y., Hernandez, S. C., Koo, B., Rheem, Y., Myung, N. V. 2007. Electrochemically fabricated zero-valent iron, iron-nickel, and iron-palladium nanowires for environmental remediation applications. *Water Science & Technology* 55(1-2): 149–56.

Zamani, C., Illa, X., Abdollahzadeh-Ghom, S., Morante, J., Romano Rodríguez, A. 2009. Mesoporous silica: A suitable adsorbent for amines. *Nanoscale Research Letters* 4(11): 1303–1308.

Zhang, R., Olin, H. 2011. *OMICS: Biomedical Perspectives and Applications, Chapter 7. Magnetic Nanoparticles in Biomedical Applications.* CRC Press. ISBN 1439850089, 9781439850084., pp. 119–121.

Zhang, W. 2003. Nanoscale iron particles for environmental remediation: An overview. *Journal of Nanoparticle Research* 5: 323–332.

Zhang, W., Elliott, D. W. 2006. Applications of iron nanoparticles for groundwater remediation. *Remediation Journal* 16(2): 7–21.

Zhou, Y. M., Zhong, C. Y., Kennedy, I. M., Pinkerton, K. E. 2003. Pulmonary responses of acute exposure to ultrafine iron particles in healthy adult rats. *Environmental Toxicology* 18(4): 227–35.

Zhu, Y., Yang, Y., Liu, M., Zhang, M., Wang, J. 2015. Concentration, distribution, source, and risk assessment of PAHs and heavy metals in surface water from the Three Gorges Reservoir, China. *Human and Ecological Risk Assessment* 21(6): 1593–1607.

11

Solubility of Nanoparticles and Their Relevance in Nanotoxicity Studies

Archini Paruthi and Superb K. Misra

CONTENTS

11.1 Dissolution of Nanoparticles 375
11.2 Factors Affecting Dissolution 376
11.3 Measurement of Nanoparticle Dissolution 377
11.4 Biological Implications of Dissolution 378
11.5 Case Study for Silver Nanoparticles 381
References 385

11.1 Dissolution of Nanoparticles

Nanomaterials in their direct/embedded form have become contributors toward enhancement in functionalities for a range of commodities. As per the report furnished by the United States government accountability office in May 2010, it was predicted that the world market for products containing nanomaterials was expected to reach $2.6 trillion by 2015. Keeping in mind the huge influx of nano-embedded products available in the market, it becomes imperative that we understand the true biological impact of nanomaterials. Nanosafety studies have mainly been conducted through nanotoxicological studies wherein the toxicological impact of nanomaterial was examined on a range of biological systems (microbial, cellular, *in vivo*, and ecotoxicological). Unlike other physicochemical properties of nanomaterials, which have been studied extensively as a contributing factor for toxicity, dissolution is rather complex because nanoparticle solubility can be looked upon as one of the physicochemical parameter stand-alone, or as a property that is affected by changing other physicochemical parameters (e.g., size, shape, surface chemistry, porosity crystallinity, etc.). Irrespective of how dissolution is looked upon, there is no doubt that dissolution of nanoparticles in the relevant exposure media is often the key to understanding the true biological response of nanoparticles. Dissolution of nanoparticles, being an important parameter, influences the mode of action for nanoparticles, the uptake pathway, and the toxicity mechanisms. A nanoparticle exposed to an aqueous medium may dissolve to release its ionic components until the concentrations of these components in its free ionic form in solution reach a maximum concentration controlled by solubility. In practice, and in relevance to nanosafety, the degree of dissolution of a nanomaterial may show a huge variation as a function of its structure, chemistry, size, and surface functionalization. In addition to this, all other components in the media, organic (e.g., proteins) or inorganic (dissolved salts), will interact both with the particle surface and the released ionic components in the media to further affect dissolution. Generally, tetravalent metal oxides such as titanium or silicon are much less soluble than divalent metal oxides such as copper or zinc, which are less soluble than pure metal nanoparticles or metal sulfide nanoparticles. The dissolution of nanoparticles can affect their biopersistence in the environment, the exposure route, and the nature of the toxicant (ions/nanoparticles). Dissolution of nanoparticles must be addressed to correctly interpret the biological response of nanoparticles, establish the risk to nanoparticle exposure in different biological/environmental compartments, and postulate their cellular uptake mechanisms. Therefore, from a regulatory perspective, it becomes important to find out whether the toxicity observed is due to the nanoparticles, caused by the released of ions, or a combination of both. Through this chapter, we demonstrate some of the complexities involved in assessing solubility of nanoparticles and also look into a case study involving silver nanoparticles.

11.2 Factors Affecting Dissolution

A common assumption of solubility increasing as particle size decreases has been shown to be theoretically true, through the Ostwald-Freundlich equation (Kaptay 2012).

$$x_{A(\beta)} = x^o_{A(\beta)} \times \exp\left(\frac{2 \times V^o_{A(\alpha)} \times \sigma^o_{\alpha/\beta}}{R \times T \times r_\alpha}\right)$$

where $x_{A(\beta)}$ is the solubility component A (mole fraction) in the form of a spherical, pure phase α of radius r_α (m) in a given solution β at temperature T (K) and at a fixed pressure P (Pa); $x^o_{A(\beta)}$ is the same of an infinitely large phase α; $\sigma^o_{\alpha/\beta}$ is the interfacial energy (J/m²) between the two phases (supposed to be size independent); $V^o_{A(\alpha)}$ is the molar volume of the pure phase $A(\alpha)$; and $R = 8.3145$ J/mol K. A numerical analysis performed by Kaptay (2012) shows that the solubility of particles increases with decreasing size not due to the increased curvature, but rather due to the increased specific surface area of the phase.

Although the above-mentioned models are able to predict solubility theoretically, capturing solubility data experimentally is challenging. The challenge stems from the fact that there are several factors that can affect dissolution, and controlling one often has a cascading effect on other parameters. For example, size-dependent solubility is difficult to explain, because often, size control is achieved through surface modification. Published reports suggest nanoparticles have a wide spectrum of size dependent solubility. For example, there are reports that demonstrate size dependent solubility (Derfus et al. 2004, Karlsson et al. 2015), no differences in dissolution between nano and macroscopic particles (Zhang et al. 2012), and even inhibition in dissolution by size reduction (Hahn et al. 2012). Figure 11.1 depicts the complexity in terms of the various physicochemical factors (surface morphology, surface chemistry, crystallinity, crystal structure, the presence of impurities) that may influence nanoparticle dissolution. Capping of nanoparticles is performed to create controlled-size nanoparticles, and, interestingly, the surface modifications themselves may affect the solubility of the nanoparticle. Therefore, two nanoparticles of identical size and composition but of different surface modification may dissolve very differently, in a way that nanoparticle suspensions can contain different forms, that is, as nanoparticles, free/complexed ions, and adsorbed ions on nanoparticles (Kittler et al. 2010, Liu et al. 2010). Silver nanoparticles are a good example to illustrate this scenario, wherein all three forms of silver can be present in the suspension. In addition to the intrinsic factors of the nanoparticles,

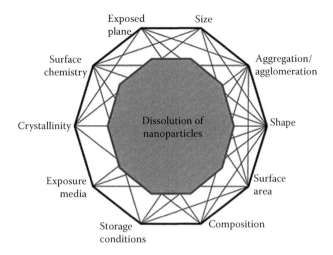

FIGURE 11.1
Some of the physicochemical factors affecting solubility of nanoparticles and also the interconnectivity of these factors. (Adapted from Misra, S.K. et al. 2012a. *Environ Sci Technol* 46: 1216; Misra, S.K. et al. 2012b. *Sci Total Environ* 438: 225–232.)

attributes of the surrounding media (pH, ionic strength, presence of organic matter/proteins) can also affect nanoparticle suspension stability. This further leads to agglomeration of nanoparticles, eventually affecting the dissolution (Tejamaya et al. 2012). The solubility parameter, that is, the kinetics of dissolution and the saturation concentration, may vary for the same nanoparticles when suspended in different environmental/physiological media. An example of where a combination of intrinsic and extrinsic factors affect nanoparticle dissolution is for electrostatically stabilized nanoparticles, which are more sensitive to changes in ionic strength and pH of the media, and this in turn affects their behavior in solution with respect to aggregation. Temperature, on the other hand, has more impact on sterically stabilized suspensions.

11.3 Measurement of Nanoparticle Dissolution

Apart from the complexities of physicochemical and external factors that affect dissolution, a lack of coherent dissolution protocols also makes assessment of dissolution for nanoparticles a complicated task. This has resulted in producing research results that are conflicting. As an example, two nanoparticles with similar size and composition (but different surface chemistry) can have different solubility. Experimental dissolution protocols for macroscopic materials have long been established, with techniques such as ultracentrifugation and membrane filtration commonly used. However, assessment of dissolution for nanoparticles is rather tricky from a separation (between particles and ions) point of view. This is because there are size thresholding challenges in using existing techniques for measuring dissolution of bulk materials to nanoparticles. Conventionally, the dissolved fraction is separated from the suspension by using filtration through a filter, centrifugation, or dialysis methods. Clearly, the size threshold of the filters is unsuitable in the case of nanoparticles and for methods involving ultracentrifugation and equilibrium dialysis. Some of the techniques that are used to measure dissolution are (i) ion-selective electrodes, (ii) dialysis membranes, (iii) centrifugation, (iv) diffusion, (v) complexation/speciation, (vi) filtration/ultrafiltration, and (vii) centrifugal ultrafiltration. Once the nanoparticles and dissolved species are separated, a range of detection methods (spectroscopy, inductively coupled plasma-mass spectrometry, atomic force microscopy, etc.) can be coupled to measure the particle solubility. Table 11.1 illustrates some of the common methods that are used in nanotoxicological studies to assess dissolution of nanoparticles. As stated by Misra et al. (2012a, 2012b), a robust method to measure the solubility of nanoparticles must be efficient in separating nanoparticles from their respective ionic species and also account for the interaction of nanoparticles with the exposure media and its effect on dissolution.

11.4 Biological Implications of Dissolution

The state of solubility of nanoparticles within the surrounding media will play a significant role in their bioavailability, uptake rates, and toxicity (Figure 11.2). For example, aggregated nanoparticles can sediment in the media and become available to sediment-dwelling organisms, thus opening up the dietary uptake route. Similarly, if the nanoparticles solubilize in the exposure media, speciation of ions can be prominent, and the uptake path and uptake mechanism of these complex species in organisms will be different from that of the nanoparticles themselves (Li et al. 2012). Although it is envisaged that dissolution can help in understanding the observed nanotoxicological response, recent studies seem to favor a complex relationship between nanoparticles and dissolved species (Navarro et al. 2008). Studies with selected nanoparticles (silver, copper oxide, zinc oxide) highlight the complexity of nanoparticle solubility. Nanoparticle suspension can have different forms of the compositional species (nanoparticles, free/complexed ions, and adsorbed ions on nanoparticles). These scenarios can significantly affect the biological response of the nanoparticles. In addition, the presence of metal ions can also interfere with some common *in vitro* cytotoxicity assays (Kroll et al. 2009). The unique behavior (agglomeration/aggregation and dissolution) of nanoparticles, along with several possible uptake mechanisms, poses a challenge to the current equilibrium models employed for calculating free metal ion concentrations in biouptake/bioaccumulation studies. This is highlighted in Figure 11.2, which shows how the extent of nanoparticle dissolution can dictate whether free metal ion concentration, endocytosis-mediated nanoparticle response, or a combination of both mechanisms will govern toxicity. For example, if nanoparticles solubilize in the medium, then ion transporters/ion channels will become the preferred route for cellular entry. However, if nanoparticles are not soluble, then a combination of possible routes (viz. endocytosis, ion transportation, or a combination of both) may occur. Nanoparticle solubility is crucial for correct interpretation of nanotoxicological data and eventually for "safety by design of nanoparticles." For example, modifying the surface of carbon nanotubes can improve their solubility (Lin et al. 2004) and reduce their biopersistence.

As much as nanoparticle solubility will govern the biological response, the biological milieu also has a significant role to play in controlling the dissolution of nanoparticles. One such biological factor is protein–nanoparticle interaction, which is a highly dynamic and protein-dependent process. This protein–nanoparticle interaction takes place as soon as the nanoparticle system enters a biological system and is facilitated by the exchange between

TABLE 11.1
Selected Results Showing Nanoparticle Solubility in Various Ecotoxicological/Biological Media

Size (nm)	Observation	References
Copper Oxide Nanoparticles		
30	2% dissolution in Osterhout's medium for CuO NPs compared to 0.12% for bulk CuO particles	Mortimer et al. (2010)
25	1.2% of particles dissolved immediately within 5 min in gastric fluid	Lei et al. (2008)
20	95% dissolution at pH 5.5 in ultrapure water 0.3% dissolution at pH 7 and <0.1% dissolution at pH 7.4	Studer et al. (2010)
14	Dissolution is enhanced by the presence of the organic tryptone and yeast extract. (80%–95% dissolution) Deionized water or NaCl has little effect on the dissolution CuO NPs dissolved much more in the LB broth medium compared to CuO microparticles	Gunawan et al. (2011)
7 spheres 7 × 40 rods	Spherical 7 nm NPs showed a higher dissolution compared to rod-shaped NPs and micron-sized CuO particles in 1 mM $NaNO_3$, up to 3.5% and 1% dissolution for spherical and rod-shaped NPs	Misra et al. (2012a,b)
Silver Nanoparticles		
39 ± 9	2% dissolution over a 1-month period in 10 mM $NaNO_3$, sulfidation of the Ag-NPs caused reduction in dissolution rates	Levard et al. (2011)
19 ± 13 8 ± 2	Less than 10% dissolution of AG NPs in the absence of cysteine. Presence of cysteine increased the amount of dissolved silver from the NPs. Up to 36% for citrate-Ag NPs and 47% for PVP-Ag NPs	Gondikas et al. (2012)
82 88	Ag dissolution of up to 3% after just 6 h in the tween-Ag NPs and similar level in 15 days for citrate and bare Ag NPs in natural river water	Li and Lenhart (2012)
6, 25 5–38	Solubility was not affected by the synthesis method and coating as much as by their size in 1 mM $NaHCO_3$	Ma et al. (2012)
10–200	0.7%–0.8% of dissolution over 3-h period in MOPS media No increases of dissolved Ag in presence of cysteine (10–500 nM)	Navarro et al. (2008)
30–50	Dissolved Ag^+ decreased with increased EDTA loading in 1 or 100 mM $NaNO_3$ (varying amount of EDTA, Brij-35, and alginic acid) Brij-35 and alginate enhanced Ag solubility	Chappell et al. (2011)
20 80	After 3 h 30% dissolution for 20 nm and 5% dissolution for 80 nm Ag NPs in 2.5 mM $(NH_4)_2SO_4$, 30 mM HEPES buffer	Radniecki et al. (2011)
50 ± 20	Citrate-Ag NPs: Up to 14% and 37% dissolution at 25°C and 37°C, respectively, in ultrapure water PVP-Ag NPs: Up to 50% and 90% dissolution at 25°C and 37°C, respectively Starting concentration affects the apparent equilibrium concentration	Kittler et al. (2010)
4.8 ± 1.6	Ag dissolution is dependent on pH and dissolved oxygen in deionized water Three distinct forms of silver exist: Ag NPs, dissolved Ag^+, and adsorbed Ag^+	Liu and Hurt (2010)
Zinc Oxide Nanoparticles		
50–70	Dissolution rate reduced with increasing NP concentration in LB broth (3.2% dissolution)	Baek and An (2011)
50–70	No difference in dissolution between nano and bulk ZnO particles, with both showing up to 80% dissolution in Osterhout's medium	Mortimer et al. (2010)
30	No difference in dissolution rate and equilibrium Zn concentration for both bulk and nano ZnO in Algal test medium (19% dissolution)	Franklin et al. (2007)

(Continued)

TABLE 11.1 (Continued)
Selected Results Showing Nanoparticle Solubility in Various Ecotoxicological/Biological Media

Size (nm)	Observation	References
13	Dissolution followed the order: DMEM (90 μM) > BEGM (80 μM) > H_2O (60 μM)	Xia et al. (2008)
	Organic components enhance dissolution, (~80% within 3 h)	
19	Equilibrium solubility followed the order: LB (68 mg/L) > DM (38 mg/L) > NaCl (14 mg/L) > H_2O (6.9 mg/L) > PBS (0.57 mg/L)	Li et al. (2012)
	Nano and bulk ZnO dissolution similar in ultrapure water.	
	No difference reported in the measurement of dissolved Zn by ultracentrifugation and centrifugal filtration	
Spheres rods	Dissolution kinetics did not differ significantly between particle types in artificial seawater	Peng et al. (2011)
	4.1%–4.9% dissolution, with highest equilibrium solubility for smallest spherical particles and lowest for largest rods	

Source: Adapted from Misra, S.K. et al. 2012a. *Environ Sci Technol* 46: 1216; Misra, S.K. et al. 2012b. *Sci Total Environ* 438: 225–232.

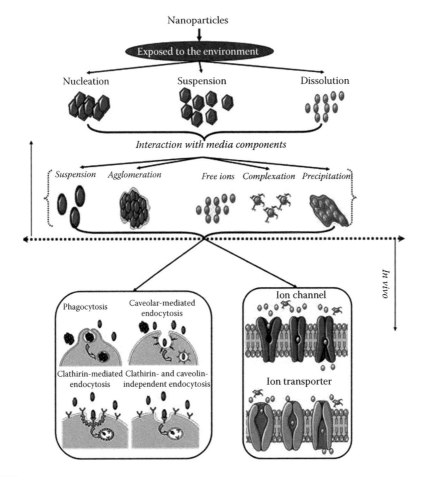

FIGURE 11.2
Schematic illustrations showing how nanoparticles once released in medium can form various species and hence affect the biological response. (Adapted from Misra, S.K. et al. 2012a. *Environ Sci Technol* 46: 1216; Misra, S.K. et al. 2012b. *Sci Total Environ* 438: 225–232.)

surface-adhered proteins and bulk proteins. With time, the proteins with greater affinity to the nanoparticle surface form a "hard corona," whereas proteins that are loosely bound to the surface constitute a "soft corona." It is expected that the physicochemical properties of nanoparticles will affect the protein adsorption. However, Wen et al. (2013) showed that the presence of cytoskeletal proteins (actin and tubulin) prevented the release of silver ions from citrate coated-silver nanoparticles (36 nm). Similarly, citrate capped silver nanoparticles, when associated with human serum albumin, bovine serum albumin, and high density lipoprotein, showed varying dissolution traits (Shannahan et al. 2015). The addition of serum albumin (human and bovine) reduced the dissolution of silver nanoparticles, but the presence of lipoprotein increased the dissolution of the nanoparticles.

11.5 Case Study for Silver Nanoparticles

The increase in the number of applications and usage of silver nanoparticles (Ag NPs) also brings increased risk to the environment and health due to potential toxicity caused by silver (particulate and dissolved species) to organisms. Taking exposure into consideration the mechanisms of action causing toxicity depends on the dose–response relationships and on the physicochemical form of Ag NPs in the complex media. This makes it imperative to understand how Ag NPs dissolve and quantify this release in a range of environments. Dissolution of Ag NPs depends on a large number of intrinsic and extrinsic factors. Intrinsic factors include the physicochemical features of the nanoparticles and surface coating on the particles, whereas extrinsic factors include water chemistry, pH, ionic strength of the media, dissolved oxygen content, the presence of natural organic matter (NOM), and complexing agents in the media (Dobias and Bernier-Latmani 2013). Therefore, these factors controlling the state of the nanoparticles also inadvertently influence the toxicity of the nanoparticles. As explored by Tejamaya et al. (2012), due to reactivity (aggregation/dissolution), further changes in the surface chemistry and the morphology mediate changes in the nature of the toxicant in the media or can change the exposure dose significantly when subjected to ecotoxicological media, thereby affecting the bioavailability. It was primarily hypothesized that the major contributor to toxicity caused by Ag NPs is the release of ionic silver (Li and Lenhart 2012). The concentration of these free ions targets the ion regulatory mechanisms in a cell and can lead to dysregulation of normal biological machinery, resulting in necrosis or apoptosis (Devi et al. 2015). The particles were found to oxidize in presence of dissolved oxygen, which is facilitated at low pH (high proton content in the media), leading to dissolution and etching of NPs (Tejamaya et al. 2012). After the release of Ag ions from the nanoparticles, it can be readsorbed onto the available particle surfaces in the media or else can link to other reactive surfaces. Ag ions also have the potency to react with complexing species (S^{2-}, NO_3^-, Cl^-, SO_4^{2-}, PO_4^{3-}, EDTA, etc.) in order to form secondary precipitates, especially in the aquatic environment. Quantification of these complexes (e.g., AgCl) is very important, as they precipitate out at higher concentrations.

There have been various *in vivo* studies in addition to *in vitro* assessments performed for toxicity assessment using animal models like embryonic staged zebrafishes, microbes, Daphnia, and Oncorhynchus (Unrine et al. 2012, Poda et al. 2013, Devi et al. 2015). A recent study reflected on how nature combats toxicity by the process of "sulfidation." Ag_2S formation leads to a decrease in ion mobility, especially in the case of sewage sludge products, mediating the reduction in the dissolution rate (Devi et al. 2015). There is also a possibility that silver ions undergo reduction to form Ag^0 in presence of agents like fulvic acid and citric acid. The negative surface charge on the particles when screened by the electrolytes present in the media induces a rapid aggregation process. This aggregation is usually controlled by the use of capping agents on the surface of the particles (Li et al. 2012). The ionic strength or salinity of the media plays an important role, as Ag NPs are usually found to aggregate and eventually precipitate at high salt concentrations or when the concentration of Ag NPs is increased in the given media. Capping agents usually prevent aggregation either by electrostatic and/or steric repulsion. When citrate is used as a stabilizer, these high ionic salts interact with citrate and change the electrorepulsive capacity. As the concentration lowers, the repulsion capacity reduces, mediating the decrease in aggregation (Park et al. 2013). Reports have shown that citrate coating on Ag NPs reduces their dissolution because Ag ions bind to the carboxylic groups of the acid, which further leads to decrease in their solubility, and citrate is also a very good reducing agent (reducing Ag^+ to Ag^0 at the surface) (Tejamaya et al. 2012, Dobias and Bernier-Latmani 2013). Citrate is also photoreactive in nature and tends to alter the particle morphology in presence of sunlight, which is discussed subsequently (Li and Lenhart, 2012). In addition to citrate, there are various other capping agents used for making Ag NPs (e.g., Tween-80, tannic acid, PVP, PEG, hydrocarbons, and polysaccharides). The choice of capping agent governs whether the particles would stay well dispersed or will aggregate when subjected to a range of media (e.g., environmental or biological media). The agglomeration and dissolution kinetics are the ones that govern the cellular consequences based on the rate of internalization and inflammatory responses generated in response to these particles. When hydrocarbon-coated Ag NPs were subjected to lysosomal and alveolar fluids, they lost their distinctive morphology and extensively aggregated. The case of discontinuous and partial coating led to an increase in reactivity, leading to effectively higher dissolution. Polysaccharide coating on Ag NPs was found to minimize agglomeration and effectively increase the dissolution (35%), leading to enhancement in particle reactivity (Braydich-Stolle et al. 2014). It is suggested that oxidative dissolution has a minor contribution to Ag ion release in the initial phase as soon as the Ag NPs are deployed into the system. Recent (Dobias and Bernier-Latmani 2013) studies have proposed that there is always a rapid initial release of Ag^+ ions due to desorption of Ag from nanoparticle surfaces in addition to a slow release rate of silver ions due to oxidative dissolution. Another angle on these kind of studies is that the dissolution that happens under oxic conditions is a mass-based phenomenon reflecting the faster and complete dissolution of smaller nanoparticles compared to larger particles. Larger particles tend to be a part of the system for a longer period of time and serve as a constant source of silver ions. These silver ions are further sorbed by the rest of the particles in the system and mediate the delivery of high doses of ions to the cell, which is popularly known as the "Trojan Horse Effect." Even after dissolution, the fate of Ag ions in the given media and environmental conditions turns the final decision either in favor of toxicity or safety.

Li and Lenhart (2012) collected natural fresh water (Olentangy River, Ohio) in which the dissolution and stability of Ag NPs was studied for 15 days in the presence and absence of sunlight. Additionally, this study also emphasized the role of capping agents. Figure 11.3 shows the aggregation behavior of various types of Ag NPs (tween, citrate, and uncoated) with respect to time and sunlight conditions. Within 24 hrs, citrate-coated and bare Ag NPs started to show aggregation (82 nm to 500–800 nm), whereas sterically dispersed tween-coated Ag NPs did not show significant aggregation (82–88 nm) during the same period. The total content of silver release within 6 hours accounted to 3% (40 μg/L) of the total mass. On Day 3, the hydrodynamic diameter of tween-coated particles began to decrease under both light (88–81 nm) and dark (88–57 nm) conditions. This indicates the

Solubility of Nanoparticles and Their Relevance in Nanotoxicity Studies

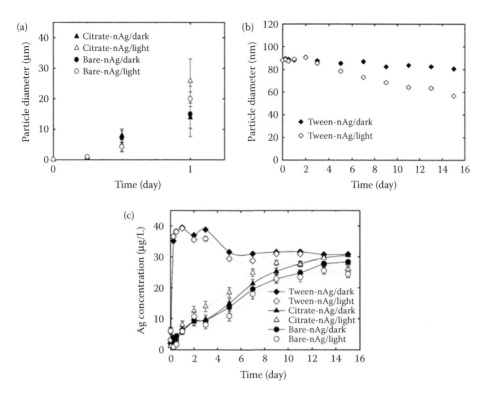

FIGURE 11.3
Variation in the hydrodynamic diameter as a function of time in presence and absence of light with respect to different capping agents (a) citrate, (b) tween, and (c) dissolution profile of the capped nanoparticles under dark and light conditions. (Reprinted with permission from Li, X. and Lenhart, J.J. Aggregation and dissolution of silver nanoparticles in natural surface water. *Environ Sci Technol* 46: 5378–86. Copyright 2012, American Chemical Society.)

absence of aggregation and degradation of the tween layer, leading to the best dissolution of particles under light conditions. The particle surface was oxidized to Ag+, and Ag was eventually released in the solution. Under light conditions, the particles are also expected to undergo photofragmentation. As shown in Figure 11.3c, the release of Ag ions for all the capped systems was similar for 3 days. For tween-capped Ag NPs, the significant decline in concentration of dissolved Ag also confirms the readsorption of the ions on the surface of the nanoparticles. It was also found in the study that citrate-capped Ag NPs (28.0 ± 0.9 μg/L) released more Ag as opposed to bare Ag NPs (24.5 ± 0.6 μg/L), and this is due to the photoreactive nature of citrate (Li and Lenhart 2012).

$$\text{Citrate} \xrightarrow{h\nu} \text{acetone-1,3-dicarboxylate} + CO_2 + e^-$$

The equation above signifies the action of citrate in the presence of sunlight. The released electrons in the process are eventually stored by Ag NPs, and this further increases the chemical reactivity at the surface. Hence, silver ions readily form secondary complexes after the release of ions on oxidation. It was also observed that Ag NP dissolution is dependent on the pH of the media. Hence, in a natural water system, the controlling factor behind the release of Ag ions is the aggregation state of the particles. The equation above signifies the action of citrate in the presence of sunlight. The released electrons in the process are eventually stored by Ag NPs, and this further increases the chemical reactivity at the surface. Hence, silver ions

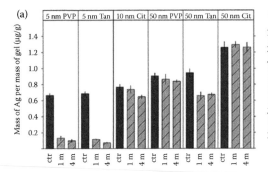 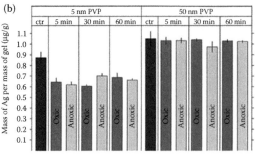

FIGURE 11.4
(a) Mass of Ag remnant in the gel pucks deployed on site. This shows the variation in the release of Ag ions from the gel pucks with size and different coatings (tannic acid, PVP, and citrate coating) with respect to time. (b) Mass of Ag in the gel pucks subjected to oxic and anoxic conditions in the filtered water obtained from the lake. (Reprinted with permission from Dobias, J. and Bernier-Latmani R. Silver release from silver nanoparticles in natural waters. *Environ Sci Techno* 47: 4140–6. Copyright 2013, American Chemical Society.)

readily form secondary complexes after the release of ions on oxidation. It was also observed that Ag NP dissolution is dependent on the pH of the media. Hence, in a natural water system, the controlling factor behind the release of Ag ions was the aggregation state of the particles.

A similar line of research was performed by Dobias and Bernier-Latmani (2013), wherein they tested dissolution of Ag NPs with varying sizes and surface chemistry under oxic or anoxic conditions. The media chosen in this study were water samples from Lake Geneva. Figure 11.4 demonstrates that release of Ag ions is more from the smaller particles (5 nm) as opposed to larger particles (50 nm). As observed in Figure 11.4, there was no significant difference in the release of silver ions observed from the control over the initial phase of deployment of the gel pucks, leading to the conclusion that oxidation dissolution does not majorly contribute to the initial release of ions into the system. There is a significant difference in the release of Ag ions when citrate, polyvinyl pyrrolidone (PVP), and tannic acid are used as the capping agents; this is hypothesized to be due to the release of chemisorbed Ag ions in addition to oxidative dissolution. In the course of study, it was also observed that the release of Ag from Ag NPs significantly decreases as the pH values go above 8.0, and dissolution is slower when in the range of 7.9–8.4.

Nanomedicine is a stream of nanotechnology, which is employed for repair, monitoring, construction, and control of biological systems. Hence, safety issues like toxicity are of major concern in this highly regulated field to regulatory authorities like FDA. Nanomedicine's impact, especially for the development of drug delivery vehicles, has been posed because of its ability to cross biological barriers (e.g., blood-brain barrier ~35 nm). In addition, a high surface area-to-volume ratio favors a large quantity of drug loading in carriers. Nanoparticles are of great significance for medical imaging techniques like iron oxide nanoparticles (NPs) (<20 nm) and CdSe quantum dots (<10 nm), and, further, they can be conjugated with specific antibodies or coatings of peptides to target particles at a specific site (Lim et al. 2015). *In vitro* imaging usually includes positron emission tomography (PET) and single-photon emission tomography (SPECT); these also involve fluorescent trackers and contrast agents. Karlsson et al. (2008) made a detailed study pertaining to the toxicity of metal oxide nanoparticles (ZnO, TiO2, CuO, Fe_2O_3, and Fe_3O_4) and their exposure to epithelial cell lines, and these particles caused varied degrees of damage to the cells due to oxidative stress and dissolution of nanoparticles leading to formation of ions. Derfus et al. (2004) studied the toxicity caused due to CdSe quantum dots in primary rat hepatocytes, and the major reason behind toxicity turned out to be the release of Cd. Degradability of nanocarriers turns out to be the major matter of concern because if the nanomaterial is nondegradable, toxicity is caused due to accumulation (nonclearance from the body), and in cases when it is degradable, then toxicity is usually caused by the degraded components or ions or the complexes formed due to interaction with the surrounding milieu (intercellular fluids and plasma with varied salt concentrations) (Zhang et al. 2012). Cardiovascular devices undergo corrosion by the mechanism of metal ion release. The severity of the toxicity is dependent on the rate of release of ions or wear debris. Nitinol (alloy with 55% Ni and 45% Ti) is majorly used for devices like stents, patches, and so on used for atrial correction or stenting. The scope of release of Ni either due to faulty manufacturing or the patient's physiological condition is quite high. Usually, patients implanted with Nitinol have elevated Ni levels for a month in their serum. Later, a reduction in the levels occurs as alloy passivation happens by deposition of calcium phosphate or endothelialization. The complexity of toxicity mechanisms is also dependent on the bionanointerface (the bonding characteristics of the cell, physicochemical characteristics of the nanoparticles released from the implant) (Hahn et al. 2012). Nickel titanium or cobalt-chromium alloys are also used as cardiovascular implants, but they tend to release alloy nanoparticles due to production failure at times or abrasion. In the case of total hip or total knee replacements, hip resurfacing with Co–Cr alloys is usually practiced. Aseptic loosening of metal on metal causes adverse tissue responses and heavy generation of wear particles. Larger surface area further increases the metal ions from the debris (which is usually reported to be in nanometer ranges), which are mostly Co and Cr ions. Macrophages phagocytose these particles, then activation of biochemical cascades is initiated by the release of cytokines, growth factors, and chemokines. These induce inflammatory responses, which lead to osteolysis. Under physiological conditions, Co differentially corrodes faster than Cr and stays mobile (high concentrations characterized in the blood), which mediates its entry into distant organs. On the other hand, Cr ions mostly tend to bind with the local tissue. Cr also forms stable complexes with the media and deposits in the implant's periphery. As Co is mobile, consistently high levels can lead to hearing loss, tremors, cognitive losses, and hormonal imbalances. Fibrous capsule formation or deposition around the implant's periphery *in vivo* might also deteriorate the diffusion of metal ions to the surrounding tissue. Under certain conditions due to acute and chronic inflammation, local acidification occurs and the pH of the region drops to 5.4 (tissue inflammation region) due to increased concentration of H^+, and such an environment accelerates the corrosion, leading to an increase in the amount of ion release. CoCr has potentially shown cytotoxic effects on numerous cell lines, preferentially MG-63 (osteoblast cell lines) and U937 (human monocyte cell lines) (Posada et al. 2015). Similarly, stainless steel is also widely used as an orthopedic implant material in addition to its food applications. The biological environment leads to surface adsorption of proteins, which affects the passive oxide film layer and stimulates the release of ions in case there is an increased proportion of Cr at the surface.

References

Baek, Y.K. and An, Y.J. 2011. Microbial toxicity of metal oxide nanoparticles (CuO, NiO, ZnO, and Sb_2O_3) to *Escherichia coli*, Bacillus subtilis, and *Streptococcus aureus*. Sci Total Environ 409: 1603.

Braydich-Stolle, L.K., Breitner, E.K., Comfort, K.K., Schlager, J.J. and Hussain, S.M. 2014. Dynamic characteristics of silver nanoparticles in physiological fluids: Toxicological implications. *Langmuir* 30: 15309–16.

Chappell, M.A., Miller, L.F., George, A.J., Pettway, B.A., Price, C.L., Porter, B.E., Bednar, A.J., Seiter, J.M., Kennedy, A.J. and Steevens, J.A. 2011. Simultaneous dispersion-dissolution behavior of concentrated silver nanoparticle suspensions in the presence of model organic solutes. *Chemosphere* 84: 1108.

Derfus, A.M., Chan, W.C.W. and Bhatia, S.N. 2004. Probing the cytotoxicity of semiconductor quantum dots. *Nano Lett.* 4: 11–18.

Devi, G.P., Ahmed, K.B., Varsha, M.K., Shrijha, B.S., Lal, K.K., Anbazhagan, V. and Thiagarajan, R. 2015. Sulfidation of silver nanoparticle reduces its toxicity in zebrafish. *Aquat Toxicol* 158: 149–156.

Dobias, J. and Bernier-Latmani R. 2013. Silver release from silver nanoparticles in natural waters. *Environ Sci Technol* 47: 4140–6.

Franklin, N.M., Rogers, N.J., Apte, S.C., Batley, G.E., Gadd, G.E. and Casey, P.S. 2007. Comparative toxicity of nanoparticulate ZnO, bulk ZnO, and $ZnCl_2$ to a freshwater microalga (*Pseudokirchneriella subcapitata*): The importance of particle solubility. *Environ Sci Technol* 41: 8484.

Gondikas, A.P., Morris, A., Reinsch, B.C., Marinakos, S.M., Lowry, G.V. and Hsu-Kim, H. 2012. Cysteine-induced modifications of zero-valent silver nanomaterials: Implications for particle surface chemistry, aggregation, dissolution, and silver speciation. *Environ Sci Technol* 46: 7037.

Gunawan, C., Teoh, W.Y., Marquis, C.P. and Amal, R. 2011. Cytotoxic origin of copper(ii) oxide nanoparticles: Comparative studies with micron-sized particles, leachate, and metal salts. *ACS Nano* 5: 7214.

Hahn, A., Fuhlrott, J., Loos, A. and Barcikowski, S. 2012. Cytotoxicity and ion release of alloy nanoparticles. *J Nanoparticle Res* 14: 1–10.

Kaptay, G. 2012. On the size and shape dependence of the solubility of nano-particles in solutions. *Int J Pharm* 430: 715–22.

Karlsson, H.L., Toprak, M.S. and Fadeel, B. 2015. Toxicity of metal and metal oxide nanoparticles. In *Handbook on the Toxicology of Metals*, Fourth eds., Sweden: Elsevier (Academic Press). Chapter 4, Volume I, 75–112.

Karlsson, H.L., Cronholm, P., Gustafsson, J. and Moller, L. 2008. Copper oxide nanoparticles are highly toxic: A comparison between metal oxide nanoparticles and carbon nanotubes. *Chem Res in Toxicol* 21: 1726–1732.

Kittler, S., Greulich, C., Diendorf, J., Koller, M. and Epple, M. 2010. Toxicity of silver nanoparticles increases during storage because of slow dissolution under release of silver ions. *Chem Mater* 22: 4548.

Kroll, A., Pillukat, M.H., Hahn, D. and Schnekenburger, J. 2009. Current *in vitro* methods in nanoparticle risk assessment: Limitations and challenges. *Eur J Pharm Biopharm* 72: 370.

Lei, R., Wu, C., Yang, B., Ma, H., Shi, C., Wang, Q., Yuan, Y. and Liao, M. 2008. Integrated metabolomic analysis of the nano-sized copper particle-induced hepatotoxicity and nephrotoxicity in rats: A rapid *in vivo* screening method for nanotoxicity. *Toxicol Appl Pharmacol* 232: 292.

Levard, C., Reinsch, B.C., Michel, F.M., Oumahi, C., Lowry, G.V. and Brown, Jr. G.E. 2011. Sulfidation processes of PVP-coated silver nanoparticles in aqueous solution: Impact on dissolution rate. *Environ Sci Technol* 45: 5260.

Li, X. and Lenhart, J.J. 2012. Aggregation and dissolution of silver nanoparticles in natural surface water. *Environ Sci Technol* 46: 5378–86.

Li, X., Lenhart, J.J. and Walker, H.W. 2012. Aggregation kinetics and dissolution of coated silver nanoparticles. *Langmuir* 28: 1095–104.

Lim, E.K., Kim, T., Paik, S., Haam, S., Huh, Y.M. and Lee, K. 2015. Nanomaterials for theranostics: Recent advances and future challenges. *Chem Rev* 115: 327–394.

Lin, Y., Taylor, S., Li, H., Fernando, K.S., Qu, L., Wang, W., Gu, L., Zhou, B. and Lin, Y.P. 2004. Advances towards bioapplications of carbon nanotubes. *J Mater Chem* 14: 527.

Liu, J. and Hurt, R.H. 2010. Ion release kinetics and particle persistence in aqueous nano-silver colloids. *Environ Sci Technol* 44: 2169.

Liu, J., Sonshine, D.A., Shervani, S. and Hurt, R.H. 2010. Controlled release of biologically active silver from nanosilver surfaces. *ACS Nano* 4: 6903.

Ma, R., Levard, C., Marinakos, S.M., Cheng, Y., Liu, J., Michel, F.M., Brown Jr, G.E. and Lowry, G.V. 2012. Size-controlled dissolution of organic-coated silver nanoparticles. *Environ Sci Technol* 46: 752.

Misra, S.K, Dybowska, A., Berhanu, D., Croteau, M.N., Luoma, S.N., Boccaccini, A.R. and Valsami-Jones, E. 2012a. Isotopically modified nanoparticles for enhanced detection in bioaccumulation studies. *Environ Sci Technol* 46: 1216.

Misra, S.K., Dybowska, A., Berhanu, D., Luoma, S.N. and Valsami-Jones, E. 2012b. The complexity of nanoparticle dissolution and its importance in nanotoxicological studies. *Sci Total Environ* 438: 225–232.

Mortimer, M., Kasemets, K. and Kahru, A. 2010. Toxicity of ZnO and CuO nanoparticles to ciliated protozoa Tetrahymena thermophila. *Toxicology* 269: 182.

Navarro, E., Piccapietra, F., Wagner, B., Marconi, F., Kaegi, R., Odzak, N., Sigg, L. and Behra, R. 2008.Toxicity of silver nanoparticles to Chlamydomonas reinhardtii. *Environ Sci Technol* 42: 8959.

Park, K., Tuttle, G., Sinche, F. and Harper, S.L. 2013. Stability of citrate-capped silver nanoparticles in exposure media and their effects on the development of embryonic zebrafish (*Danio rerio*). *Arch Pharm Res* 36: 125–133.

Peng, X., Palma, S., Fisher, N. S. and Wong, S.S. 2011. Effect of morphology of ZnO nanostructures on their toxicity to marine algae. *Aquat Toxicol* 102: 186.

Poda, A.R., Kennedy, A.J., Cuddy, M.F. and Bednar, A.J. 2013. Investigations of UV photolysis of PVP-capped silver nanoparticles in the presence and absence of dissolved organic carbon. *J Nanoparticle Res* 15: 1–10.

Posada, O.M., Tate, R.J. and Grant, M.H. 2015. Toxicology *in vitro* effects of CoCr metal wear debris generated from metal-on-metal hip implants and Co ions on human monocyte-like U937 cells. *Toxicol In Vitr* 29: 271–280.

Radniecki, T.S., Stankus, D.P., Neigh, A., Nason, J.A. and Semprini, L. 2011. Influence of liberated silver from silver nanoparticles on nitrification inhibition of *Nitrosomonas europaea*. *Chemosphere* 85: 43.

Shannahan J.H., Podila R., Aldossari A.A., Emerson H., Powell B.A., Ke P.C., Rao A.M. and Brown J.M. 2015. Formation of a protein corona on silver nanoparticles mediates cellular toxicity via scavenger receptors. *Toxicol Sci* 143: 136–146.

Studer, A.M., Limbach, L.K., Van Duc, L., Krumeich, F., Athanassiou, E.K., Gerber, L.C., Moch, H. and Stark, W.J. 2010. Nanoparticle cytotoxicity depends on intracellular solubility: Comparison of stabilized copper metal and degradable copper oxide nanoparticles. *Toxicol Lett* 197: 169.

Tejamaya, M., Merrifield, R.C. and Lead, J.R. 2012. Stability of citrate, PVP, and PEG coated silver nanoparticles in ecotoxicology media. *Environ Sci Technol.* 46: 7011–7017.

Unrine, J.M., Colman, B.P., Bone, A.J., Gondikas, A.P. and Matson, C.W. 2012. Biotic and abiotic interactions in aquatic microcosms determine fate and toxicity of Ag nanoparticles. Part 1. Aggregation and dissolution. *Environ Sci Technol* 46: 6915–6924.

Wen Y., Geitner N.K., Chen R., Ding F., Chen P., Andorfer R.E., Govindan P.N. and Ke P.C. 2013. Binding of cytoskeletal proteins with silver nanoparticles. *RSC Adv* 3: 22002–22007.

Xia, T., Kovochich, M., Liong, M., Mädler, L., Gilbert, B., Shi, H., Yeh, J.I., Zink, J.I. and Nel, A.E. 2008. Comparison of the mechanism of toxicity of zinc oxide and cerium oxide nanoparticles based on dissolution and oxidative stress properties. *ACS Nano* 2: 2121.

Zhang, X.Q., Xu, X., Bertrand, N., Pridgen, E., Swami, A. and Farokhzad, O.C. 2012. Interactions of nanomaterials and biological systems: Implications to personalized nanomedicine. *Adv Drug Deliv Rev* 64: 1363–1384.

12

Nanotechnology in Functional Foods and Their Packaging

Satnam Singh, Shivendu Ranjan, Nandita Dasgupta, and Chidambaram Ramalingam

CONTENTS

12.1 Introduction 389
12.2 Nanotechnology in Development of Functional Foods 390
    12.2.1 Nutrition and Food Processing 390
    12.2.2 Food Nanoadditives and Nanoingredients 393
    12.2.3 Nanoencapsulation of Nutrients and Their Delivery Mechanisms 394
    12.2.4 Antimicrobial Activity 397
12.3 Nanotechnology in Food Packaging Research Trends 398
    12.3.1 Nanoreinforcements in Food Packaging Materials 398
    12.3.2 Nanocomposite Active Food Packaging 399
        12.3.2.1 Antimicrobial Systems 399
        12.3.2.2 Oxygen Scavengers 400
        12.3.2.3 Enzyme Immobilization Systems 401
    12.3.3 Nanocomposite Smart Packaging Systems 403
        12.3.3.1 Time Temperature Integrators and Moisture Indicators 403
        12.3.3.2 Freshness Indicators 403
        12.3.3.3 Detection of Gases 404
        12.3.3.4 Oxygen Sensors 404
        12.3.3.5 Detection of Microorganisms 405
12.4 Conclusion 405
Acknowledgments 405
References 406

12.1 Introduction

Nanotechnology is one of the prime needs of the modern world, and food demands have rapidly increased in recent years owing to rapid advancements in industrialization and population explosion. Conventional technologies for food development are becoming less efficient and cannot meet consumer demand. To meet the need of the modern world's advancement, food technologies are required. The emerging technology of food, such as nanotechnology, is an ideal option for fulfilling ever-increasing consumer demands. Researchers all over the world are working in the nanofood sector, and this has led to drastic improvements in food research. However, a technology gap between research and industrial application still exists. Exploring the current trends and future prospects for nanofood technology, the information presented here will be valuable to the international industrial community for identifying new options to circumvent problems that exist in the food sector (Momin et al. 2013, Giri et al. 2016).

Nanotechnology has many applications tissue engineering, drug delivery, biomedical engineering, and so on (Kingsley et al. 2013). Nanotechnology is also administered in the food sector, which includes nanoencapsulation of nutrients that regulate their delivery and protect them from unfavorable conditions, nanosensors, nanosized food additives and ingredients, targeted delivery of required components, food safety, new product developments, processing, active and smart packaging, and so on (McClements et al. 2009, Huang et al. 2010). The unique physiochemical properties of nanostructure open windows of opportunities in food industries like food manufacturing, packaging, and storage.

In this review, we discuss the current research trends, opportunities, challenges, and future aspects and also discuss a few ideas to nourish nanofood research. We critically discuss the overall basic research as well as the application of nanotechnology in functional foods and nanopackaging involving novel products and marketed products.

12.2 Nanotechnology in Development of Functional Foods

The 21st century is witnessing a replacement scientific and industrial revolution as a consequence of the manipulation of matter at the nanometric level. This rising discipline (i.e., nanotechnology) has expanded techniques for planning, fabricating, measuring, and manipulating matter at the nm scale, and is growing at a thoughtless pace, with unique phenomena enabling novel applications. Nanomaterials exhibit different chemical and physical properties, such as size, size distribution, surface area, surface properties, shape, chemical composition, and agglomeration state, that are not apparent in bulk materials. Hence, it is not surprising that applications of nanotechnology in several industrial areas, such as cosmetics, medicine, food, construction materials, and so on, have greatly grown in the last decade (Pérez-Esteve et al. 2013).

Nanotechnology applications in the food industry include encapsulating and targeting delivery of particles, enhancing flavor and sensory properties, adding antibacterial nanoparticles to food, increasing the shelf life of food, detecting contamination, and improving food storage. Nanotechnology in food also results in changes to the nutritional functionality and removal of chemicals or pathogens from food (Chellaram et al. 2014). Nanotechnology has been revolutionizing the approach to food engineering systems from production to processing, storage, and creation of innovative materials, products, and applications. Currently, the market for nanotechnological food industry products has reached US$1 billion. The application of nanotechnology in the areas of food and food packaging is emerging rapidly (Ezhilarasi et al. 2013). Nanoparticles exhibit many features like high surface/volume ratio, reassembling and self-reassembling capability, capacity of creating porous structure, and so on, giving them potential for use in the food industry (Pérez-Esteve et al. 2013, Bernardes et al. 2014).

12.2.1 Nutrition and Food Processing

Nanotechnology in food development gives specific characteristics to food by altering the physiochemical properties and microscopic characteristics like texture, taste, other

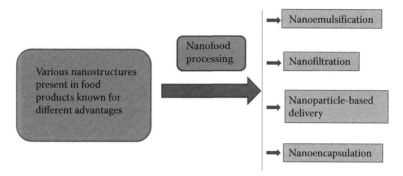

FIGURE 12.1
Different forms of processing methods used in development of functional nanofoods for specific functions.

sensory attributes, color, strength, processability, water solubility, stability during shelf life, thermal stability, and oral bioavailability of functional compounds of natural food (Silva et al. 2012). Nanotechnology in food cultivation, production, processing, and packaging or nanoadditives in original food results in nanofood (Sekhon 2010). The main objectives of nanofood are to improve the quality, safety, and nutritional value of food and at the same time reduce cost. Nanofood processing mainly focuses on the development of nanosized food ingredients and additives and delivery systems for nutrients and supplements in the form of nutraceuticals. Various types of processing methods have been introduced, such as nanoemulsions, surfactant micelles, emulsion layers, reverse micelles, and functionally designed nanocapsules (Figure 12.1) (Handford et al. 2014).

Functional ingredients added for food fortification (including vitamins, antimicrobials, antioxidants, probiotics, prebiotics, peptides and proteins, carotenoids, omega fatty acids, flavorings, colorants, and preservatives) are not administered directly in their pure form, but are sometimes incorporated into the delivery system (e.g., nanostructures) (Elliott 2002). However, these active molecules are often degraded in the processing steps and offer poor bioavailability. Nanoencapsulation and nanoemulsion formulations can be utilized so that functional food ingredients can be incorporated, absorbed, or dispersed in nanostructures. Nanostructures of inorganic materials have also been studied as a coating material to provide a moisture or oxygen barrier (e.g., silicon dioxide [E551], magnesium oxide [E530], titanium dioxide [E171], and antibacterial "active" coatings, especially silver)(Chaudhry and Groves 2010). Different products have been developed based on the Nanoclusters™ system, such as SlimShake Chocolate, which incorporates silica nanoparticles that are coated with cocoa to enhance the chocolate flavor through an increase in surface area that targets the taste buds. NanoCluster™, from RBC Life Sciences® Inc. (Irving, TX, USA), is another such delivery system for food products (Ranjan et al. 2014).

New food products containing nanostructures have been introduced or are currently being developed for different purposes (Chaudhry et al. 2008, Luykx et al. 2008, Weiss et al. 2008): (i) to protect nutraceuticals against degradation during manufacturing, distribution, and storage, improving their stability; (ii) to enhance the bioavailability of poorly soluble functional food ingredients (e.g., hydrophobic vitamins), thus improving their nutritional value; (iii) to increase food shelf life by protecting it from oxygen and water (Leclercq et al. 2009); (iv) to produce low-fat, low-carbohydrate, or low-calorie products (e.g., mayonnaise, spreads, and ice creams); (v) to optimize and modify the sensory characteristics of food products, creating new consumer sensations (e.g., texture, consistency, development of new taste or taste masking, flavor enhancement, color alteration); and (vi) to control functional food ingredient delivery (e.g., flavor, nutrients). Additionally, encapsulation also aids in controlled release of active ingredients wherein gastrointestinal retention time can be prolonged and the site of release of active ingredients can be adjusted according to the need (Krasaekoopt et al. 2003, Medina et al. 2007). Nanotechnology can also be employed to solubilize lipophilic ingredients such as carotenoids, phytosterols, and antioxidants in water so as to disperse in water or fruit drinks (Chen et al. 2006a,b). Various compounds such as lycopene, beta-carotenes, and phytosterols are integrated with carriers and are used in the production of healthy foods, especially to prevent the accumulation of cholesterol (Gruère 2012, Ranjan et al. 2014). The added health benefits arising from this application of nanotechnology are therefore particularly beneficial for consumers with health concerns. Lycopene nanostructures (particle size of 100 nm, US5968251) have been developed and accepted as a generally regarded as safe (GRAS) substance by the Food and Drug Administration (FDA). Water-dispersible lycopene nanostructures can be added to soft drinks to provide color and health benefits (Limpens et al. 2006). Lycopene has been included in other products such as baking mixtures and blancmanges (Chaudhry and Groves 2010). Conclusively, it can be stated that the recent trends of nanofood focus on encapsulation techniques, but at the same time, they should have proper quality checks and toxicity analyses before launching in the market.

Certain ingredients with high nutritional value are prone to degradation due to processing steps. For example, ω-3 polyunsaturated fatty acids are easily oxidized, so they require both stabilization procedures and protection against deterioration factors (Ruxton et al. 2004, Lavie et al. 2009). Zimet et al. (2011) developed casein nanostructures and reformed casein micelles that showed a remarkable protective effect against docosahexaenoic acid oxidation, up to 37 days at 4°C (Zimet et al. 2011). BioDelivery Sciences International has developed the Bioral™ nanocochleate nutrient delivery system, which is a phosphatidylserine carrier (~50 nm) derived from soya beans (GRAS status). This system has been used for protecting micronutrients and antioxidants from degradation during manufacture and storage (Chaudhry and Groves 2010). Recently, natural dipeptide antioxidants (e.g., L-carnosine) have been used as bi preservatives in food technology. However, their direct application in food is not possible, as they are unstable and may lead to proteolytic degradation and a potential interaction of peptide with food components. Maherani et al. (2012) have successfully encapsulated these natural antioxidant peptides by nanoliposomes to overcome the limitations related to the direct application in food. Coenzyme Q10 (CoQ10) has low oral bioavailability, which does not allow it to reach its therapeutic concentration easily (Maherani et al. 2012). Ankola et al. (2007) demonstrated the potential of nanotechnology in improving the therapeutic value of molecules like CoQ10, facilitating its usage as a first-line therapeutic agent for prophylaxis of consumers for better health. Biodegradable polymeric nanostructures based on poly (lactide-co-glycolide) (PLGA), using quaternary ammonium salt didodecyl dimethyl ammonium bromide as a stabilizer, are also developed for this purpose (Ankola et al. 2007). To deliver safe nanofoods, nanoadditives or stabilizers used in nanofoods should be of natural grade—which will ultimately support the delivery of safe nanofoods.

Wen and co-workers (2006) described the applications of liposomes in oral delivery of functional food ingredients like proteins, enzymes, flavors, and antimicrobial compounds (Wen et al. 2006). The entrapment of proteolytic enzymes in liposomes for cheese production (Mozafari et al. 2006) can reduce the production time to half without compromising flavor and texture. Similarly, Wu et al. (2012) have recently demonstrated that the entrapment of essential oils within the zein nanostructure allows their dispersion in water, enhancing their potential for use as antioxidant and antimicrobial agents in food preservation (Wu et al. 2012). Tang et al. (2013) developed self-assembled nanostructures composed of chitosan and an edible polypeptide, poly(γ-glutamic acid), for oral delivery of tea catechins, which can be used as food additives for drinks, foods, and dietary supplements (Tang et al. 2013). Likewise, food additives can be entrapped into nanostructures to mask the taste and odor of fish oil, which may be further added to bread for health benefits, and live probiotic microbes can be added to promote gut functions (Chaudhry et al. 2008). With the future aspect of facilitating the rational development of zein-based nanoscale delivery systems for the transportation of hydrophobic bioactive compounds with ideal surface morphology and functionalities for food and pharmaceutical applications, many works have been undertaken to synthesize zein-, casein-, and gelatine-based nanoparticles (Chen et al. 2013, Ye et al. 2014, Li et al. 2015). Similarly, different natural product-based active compounds should be identified to be nanoencapsulated. Alternatively, the same active compounds should be researched in different oils for efficient functioning.

Nanoemulsions reduce the need for stabilizers, as they will protect against breakdown and separation of food (Cushen et al. 2012) and can thus significantly reduce the quantity of fat needed. The products of such nanoemulsions are as "creamy" as conventional food products, with no compromise on the mouth feel and flavor. The use of zein nanostructures, a major protein found in corn, has also been explored as a vehicle of flavor compounds and essential oils (e.g., thymol and carvacrol) because they have the potential to form a tubular network resistant to microorganisms (Sozer and Kokini 2009, Wu et al. 2012). Another nanostructure proposed to encapsulate nutraceuticals (e.g., vitamins) or to mask disagreeable flavor/odor compounds is the α-lactalbumin nanotube, which can be obtained from milk protein (Srinivas et al. 2010). Based on the origin of these nanostructures, they can be regarded as food grade, which makes their common application in the entrapment of functional food ingredients easier (Cushen et al. 2012).

12.2.2 Food Nanoadditives and Nanoingredients

The role of nanotechnology in food industries is very much important and enables them to develop food products with some value-added qualities. Nanotechnology in food development gives specific characteristics to food by altering the physiochemical properties and microscopic characteristics like texture, taste, other sensory attributes, color, strength, processability, water solubility, stability during shelf life, thermal stability, and oral bioavailability of functional compounds of natural food (Figure 12.2) (Silva et al. 2012).

The application of nanotechnology in the development of bioactive compounds with good bioavailability increases their stability (by preventing degradation during processing and storage) and solubility. For instance, the absorption and bioavailability of lycopene

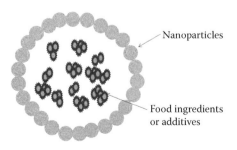

FIGURE 12.2
Food additives/ingredients encapsulated in nanosized droplets or lipid nanoparticles, which increase their bioavailability in the body and protect the from the unfavorable environment.

and reservatrol in the body (Hsieh 2009) can be enhanced by using nanocrystals of these compounds. Very high absorption has been shown in the case of nanoparticulated nutrients that contain calcium nanoparticles, which show an excellent absorption rate (Jeon and Lee 2009). Nanoemulsions can be used to encapsulate functional food components, have great potential, and have received a great deal of attention due to their ability to solubilize lipophilic molecules (e.g., coenzyme Q10, an oil-soluble vitamin-like substance) and enhance their stability and bioavailability (Yu et al. 2009a,b). A comparative study of authorized additives titanium dioxide (TiO_2) or E-171 showed high visual transparency with high shielding against UV light when used as nanoscale crystals compared to microsize crystals (Latva-Nirva et al. 2009).

12.2.3 Nanoencapsulation of Nutrients and Their Delivery Mechanisms

Nanoencapsulation packaging techniques for substances in miniature provide many benefits, such as higher bioavailability, high shelf stability, controlled release, and prolonged residence time of active compounds. This same nanoencapsulation can reduce the quantity of active ingredients required at particular sites in the human body due to efficient delivery. The protection of bioactive compounds, such as vitamins, antioxidants, proteins, and lipids as well as carbohydrates from unfavorable conditions, can be achieved using this technique for the production of various functional foods with improved functionality and stability. Various compounds (biogenic and synthetic) that are used for delivering additional nutrients within food are being developed. Table 12.1 shows the most promising of these. Generally, compounds used in nanoencapsulation are not good for health; they should be nontoxic and biodegradable in the human body. There must be some toxicological safety evaluation that declares whether a specific product is safe for the human body. The testing of nanoparticles (NPs) is currently not completely regulated, and common toxicity tests may not be appropriate for NPs; hence, there is a need to develop other techniques to test NPs. The effectiveness of nanoencapsulated compounds must also be tested to ensure

TABLE 12.1

List of Nanostructures with Suitable Shell Materials and Applications for Nutrient Delivery

Compound	Detail	Application
Liposome	Phospholipid capsule consisting of lipid bilayer outside and aqueous inside.	Can deliver both hydrophobic and hydrophilic material.
Nanoencapsulation	Made up of lipid or other polymer, stable and of nanosize approx. 100 nm.	Can deliver both hydrophobic and hydrophilic material simultaneously.
Micelles	Droplets of surfactants (lipids or biopolymers) in a liquid.	Generally, deliver hydrophobic materials.
Solid lipid nanoparticles	Stable system of crystalline or semicrystalline, which is stabilized by a surfactant coating. Formulated by emulsion technologies.	Deliver hydrophobic materials.
Casein	Micellar structure made of milk protein by self-assembly.	Delivery of vitamins, minerals, and proteins.
Whey protein	Largely α-lactoglobulin and β-lactalbumin that form fibrils, nanoparticles, and hydrogels, dependent on processing conditions.	Can deliver different hydrophilic compounds to the intestinal mucosa.
Silica	Nanoporous, biocompatible along with a degradability property in the stomach environment	Can deliver various hydrophilic compounds to the stomach.

Source: Rao, P. J. and M. M. Naidu. 2016. *Nanoscience in Food and Agriculture* 2: 129–56. Springer.

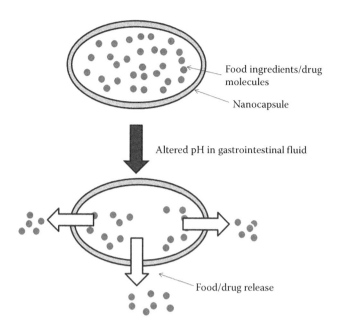

FIGURE 12.3
Release mechanism of active compounds from nanocapsule.

that the process is actually improving the characteristics of nonencapsulated compounds, but few methods have been developed (Figure 12.3) (Gutiérrez et al. 2013). One patented technique for nanoencapsulation has the ability to encapsulate bioactive and active compounds in nutraceutical products. These nanocapsules can easily degrade at targeted tissues and deliver the active ingredients. Octenyl succinic anhydride-ε-polylysine has two roles, as either surfactants or emulsifiers in the encapsulation of nutraceuticals or drugs, and antimicrobial activity (Yu et al. 2009a).

Lipid-based nanoencapsulates used in food/supplements include nanoliposomes, archaeosomes (at an experimental stage), and nanocochleates (Mozafari et al. 2006, Ranjan et al. 2014). Single- and bilayer arrangements of liposomes have hydrophobic/hydrophilic interactions among lipid/lipid and lipid/water interfaces. They can easily entrap water and lipid-soluble materials and regulate their delivery and release due to their unique properties like small size and hydrophilicity/hydrophobicity in a single compound (Mozafari et al. 2006, 2008). A very stable and precise delivery system is represented by "cochleates"—a phospholipid-divalent cation precipitate composed of naturally occurring materials, developed and patented by BioDelivery Sciences International, Inc., Newark, NJ (Ranjan et al. 2014).

Another system—nanocochleates—has a multilayered structure consisting of a large, continuous, solid lipid bilayer sheet rolled up in a spiral fashion with little or no internal aqueous space (Mozafari et al. 2006). Seventy-five percent weight of soy-based phospholipid of nanocochleates is composed of lipids that can be phosphotidyl serine, dioleoyl phosphatidylserine, phosphatidic acid, phosphatidylinositol, phosphatidyl glycerol, and/or a mixture of one or more of these lipids with other lipids. The nonocoiled nature of nanocochleates stabilizes and protects various micronutrients by wrapping around them and improving the nutritional value of processed foods (Sekhon 2010). They deliver their contents to target cells through the fusion of the outer layer of the cochleate to the cell membrane. They can be used for the encapsulation and delivery of many bioactive materials, including compounds with poor water solubility, protein and peptide drugs, and large hydrophilic molecules (Moraru et al. 2003). Liposomes can also deliver functional components such as nutraceuticals, antimicrobials, and flavors to food (Were et al. 2003, Sastry et al. 2013, Ranjan et al. 2014), but ultimately it is decided by the preparation methods, which may involve non-food-grade solvents and detergents that might leave residues in the delivery systems and make them toxic (Mozafari et al. 2006). Lipid nanocarriers can entrap lipid-soluble and water-soluble vitamins, which results in stability in different media. The nanoencapsulation of vitamin E into nanoliposomes with tea polyphenol (water soluble) has seen good bioavailability (Ma et al. 2009). The combined nanoencapsulation of vitamin E with vitamin C has also been done (Marsanasco et al. 2011). The major factors that influence liposome structure are incubation in buffer solution and stomach pH. The higher absorption of the bioactive compound is attributed to the greater bioavailability of vitamin E (Katouzian and Jafari 2016). Dietary phytoconstituent capsaicin-loaded nanoliposome has been developed and evaluated against liver oxidative stress. It has been shown that liposome-encapsulating capsaicin acts as a promising therapeutic agent in reducing liver oxidative stress produced by different stress factors (Figure 12.4) (Giri et al. 2016).

The naturally occurring bioactive compounds in certain foods have physiological properties that can reduce the risk of various diseases, including cancer. Compounds such as omega-3 and omega-6, probiotics, b-carotene, prebiotics, vitamins, and minerals have been encapsulated by various methods to enhance the efficiency of these compounds; their good bioavailability reflects gastrointestinal retention time being extended due to better bioadhesive capacity in the mucus that covers the intestinal epithelium (Gutiérrez et al. 2013, Bernardes et al. 2014). Nanocapsules have been developed to mask the unpleasant tuna fish oil taste and odor and to deliver it to the stomach (Neethirajan and Jayas

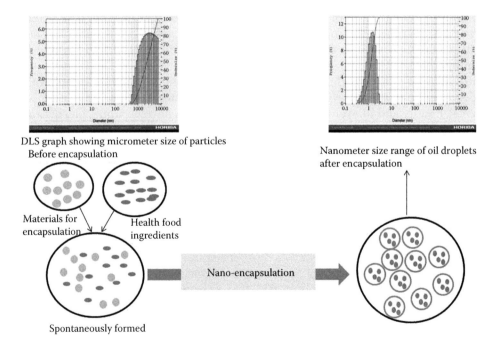

FIGURE 12.4
Basic mechanism and steps to be followed in nanoemulsion or nanoencapsulation fabrication.

2011). Various methods of encapsulating bioactive compounds include nanoemulsion, liposomes, nanocapsules, and so on. Nanoemulsion systems can be multiphase, which corresponds to preparation of an oil-in-water (O/W) emulsion; that is, emulsions consist of oil droplets within an aqueous phase and improve solubility and bioavailability of bioactive compounds due to the reduction of incomplete dissolution of lipids and decreased droplet size (Yin et al. 2009), while liposomes are formulated with lipids and phospholipids (such as phosphatidylcholine), which can be in the form of single or multiple bilayers, providing protection to the compounds (such as vitamins) encapsulated in their cores (Soppimath et al. 2001). However, they are highly unstable in acidic pH conditions and therefore degrade rapidly in the stomach, whereas liposomes are thermodynamically unstable, unlike nanoemulsions (Observatory Nano and Commission 2009).

Molecules composed of regions (typically one hydrophobic and one hydrophilic monomer) with opposite affinities for an aqueous solvent can form nanoencapsulated polymers. Some natural biopolymers like gelatin (protein), albumin (protein) (Zwiorek et al. 2004), alginate (saccharide), chitosan (saccharide), collagen (protein), and the milk protein α-lactalbumin (Graveland-Bikker and De Kruif 2006) are examples of such nanoencapsulated polymer delivery systems. Protein-based nanoencapsulates are particularly interesting because they are relatively easy to prepare and can form complexes with polysaccharides, lipids, or other biopolymers. A wide variety of nutrients can also be incorporated (Chen et al. 2006a,b). In addition, numerous copolymers have been synthesized to date, leading to the formation of micelles, nanospheres, polymersomes, and nanocapsules (Letchford and Burt 2007). But a fundamental understanding of polymer–polymer and polymer–nutraceutical interactions at the molecular level and their impact on functional properties of the delivery systems are required to ensure the design of ideal nutraceutical carriers for use in the food industry.

Probiotics are generally referred to as a mixture of bacterial species such as *Lactobacillus acidophilus*, *Lactobacillus casei*, *Lactobacillus rhamnosus*, and *Bifidobacterium* spp. and are present in dairy foods like yogurts, yogurt-type fermented milk, cheese, puddings, fruit-based drinks, and so on. Their viability in food products can be increased by nanoencapsulation. Nanoencapsulation is desirable to develop designer probiotic bacterial formulations that trigger their delivery to certain parts of the gastrointestinal tract, where they interact with specific receptors (Kailasapathy and Rybka 1997, Vidhyalakshmi et al. 2009). An enhanced shelf life of probiotic organisms has been seen when nanoencapsulated with calcium alginate (Kailasapathy and Rybka 1997). Curcumin, a natural pigment present in turmeric and responsible for its yellow color, has health benefits that can be enhanced by encapsulation in nanoemulsions (Wang et al. 2009). The bioavailability of lycopene can be enhanced by fortifying nanoparticles of lycopene in tomato juice, pasta sauce, and jam (Auweter et al. 1999). The milk protein casein can act as a neutral nanocarrier and be employed as a vehicle for delivering mineral nutrients such as vitamin D2 (Semo et al. 2007).

12.2.4 Antimicrobial Activity

Research studies have well established the use of bioactive compounds (essential oils) as natural antimicrobial agents that can be used to increase the shelf life of food. Interestingly, the same bioactive compounds in nanoemulsion form exhibit enhanced antimicrobial activity compared to the conventional forms due to their minute droplet size (Ranjan et al. 2016). Nanoemulsions of other various essential oils such as carvacrol cinnamaldehyde and limonene have shown antimicrobial activity against three different microorganisms, *Escherichia coli*, *Saccharomyces cerevisiae*, and *Lactobacillus delbrueckii* (Ranjan et al. 2016). On other hand, silver has shown antimicrobial activity. Moreover, its effect can be increased by combination with other antimicrobial substances such as extracts of grapefruit, purslane, or essential oils (Kim 2006).

Silver acts differently from other antimicrobials and shows broad-spectrum toxicity to numerous strains of bacteria, fungi, and algae, and possibly to some extent for some viruses. The mechanism of action of nanosilver on microbes is such that it brings structural changes in bacterial cell walls and nuclear cell walls due to its binding with tissue proteins and causes cell disruption and finally death. The binding of nanosilver to DNA and RNA can also cause death by inhibiting bacterial replication (Rai et al. 2009). The use of silver nanoparticles as sterilizer of freeze-dried foods has been seen in food companies. Another function of nanosilver is antiripening activity (Malshe and Malshe 2007).

12.3 Nanotechnology in Food Packaging Research Trends

Nanotechnology in food packaging materials may increase food life due to high-barrier packaging with improved mechanical and thermal properties, while nanosensors may be incorporated in the packaging systems to alert consumers if a food product is no longer safe to eat, improve food safety, alert consumers that food is contaminated or spoiled, repair tears in packaging, and even release preservatives to extend the life of the food in the package. Nanosupplements can be easily incorporated by encapsulation techniques for nutritional and drug targeted delivery systems. And, as health plays a major role in food, the disadvantages of the technology should be considered (Chellaram et al. 2014). Food industries are always searching for new, cheaper methods to produce and preserve food, and with this need, we enter into the realm of nanotechnology. Recent trends in food packaging are related to nanoreinforcement, nanocomposite active packaging, and nanocomposite smart packaging. Nanoreinforcement is mainly used to give extra tensile strength to food packets by different reinforcement methods using nanoclays, cellulose, and so on. Nanocomposite active packaging is the integration of many useful systems along with food packets, for example, antimicrobial, oxygen scavenging, and enzyme immobilization systems. Similarly, nanocomposite smart packaging mainly involves sensors, for example, time–temperature integrators (TTIs), gas detectors, and other nanosensors (Ranjan et al. 2014).

12.3.1 Nanoreinforcements in Food Packaging Materials

Polymers have replaced conventional materials (glass, ceramics, metals, paper, and board) in food packaging owing to their functionality, light weight, low cost, and ease of processability. However, their strength and capability to resist deformation are lower as compared to metals and ceramics (Jordan et al. 2005). Furthermore, their inherent permeability to gases and vapors is another limiting property in food packaging (Arora and Padua 2010). Nanoreinforcement techniques are used to increase viability and tensile strength of packaging materials by filling the gaps of them. Polymer nanocomposites usually have much better polymer/filler interactions than the usual composites (Luduena et al. 2007). Nanofillers have a vital role in enhancing composite performance by improving their properties such as mechanical strength, thermal stability, and barrier properties. The characteristic parameters that contribute greatly to modifying the properties of various composites are filler loading, their size and shape, and their affinity toward the matrix material. Nanoclay-filled polymer matrix-based nanocomposites have generated a significant amount of attention within the materials industry for their enhanced performance. This area emerged with the recognition that exfoliated clays could exhibit superior strength, modulus, and higher barrier properties in comparison to virgin polymer matrix (Goettler et al. 2007). With the recent advancements in nanotechnology, the correlation of material properties with filler size has become a point of great interest. NPs have attracted a great deal of interest owing to their extraordinary potential to exhibit novel characteristics that cannot be achieved with their traditional microscale counter parts. There have been various studies on the incorporation of different NPs into the matrix (Kriegel et al. 2008); however, the packaging industry has recently focused its attention mainly on nanoclay particulates due to their easy availability, low cost, easy processability, and good performance. Carbon-based materials like carbon nanotubes and graphene nanosheets are also being developed (Arora and Padua 2010). A uniform distribution of nanofillers into a polymer matrix results in a very large matrix/filler interfacial area, which restricts the mechanical mobility of the matrix and improves its mechanical, thermal (especially glass transition temperature—Tg), and barrier properties. Fillers with higher aspect ratios have higher specific surface area, providing better reinforcing effects (Dalmas et al. 2007). The aspect ratio is defined by the ratio of the largest to the smallest dimension of filler. Further, an interphase region of decreased mobility surrounding each nanofiller results in a percolating interphase network in the composite that plays an important role in improving nanocomposite properties (Qiao and Brinson 2009). For constant filler content, a reduction in particle size increases the number of filler particles, bringing them closer to one another; thus, the interface layers from adjacent particles overlap, altering the bulk properties more significantly (Jordan et al. 2005).

12.3.2 Nanocomposite Active Food Packaging

Active food packaging, unlike conventional food packaging, may be defined as a system that not only acts as a passive barrier but also interacts with the food in some desirable way, for example, by releasing desirable compounds (antimicrobial or antioxidant agents, for instance) or by removing some detrimental factor (such as oxygen or water vapor). The consequences of such interactions are usually related to improvements in food stability. Some of the main examples of food packaging systems are given in the following sections (Ranjan et al. 2014).

12.3.2.1 Antimicrobial Systems

Nanoparticles have been patented as antimicrobial agents. Several particles have been used for antimicrobial activity, such as nanoparticulated silver, gold, zinc, copper, cerium (Ismail et al. 2008), aluminum, iron, cadmium, palladium, rhodium, or chromium (Park and Kim 2003). These nanometallic materials act by releasing metal ions in microorganisms that react with proteins and thereby inactive the protein. The inactive protein prevents cellular function, disrupts membranes, and hinders normal activity and DNA replication, finally killing the microorganism (Ismail et al. 2008). Antimicrobial systems are fast emerging as viable solutions due to their help in controlling the growth of pathogenic and spoilage-causing organisms on food surfaces where microbial growth predominantly takes place. Further interest is drawn by the use of nanocomposites in these systems due to naturally different properties at the nanoscale range when compared to the microscale range. This can be highlighted by the higher surface-to-volume ratio of nanoscale materials in relation to microscale products. Such properties enable nanomaterials to attach more copies of microbial molecules and cells. Nanoscale materials have been investigated for antimicrobial activity as growth inhibitors (Cioffi et al. 2005), killing agents (Huang et al. 2005, Kumar and Münstedt 2005), or antibiotic carriers (Gu et al. 2003).

Silver, apart from its known toxicity toward a variety of microorganisms (He and Hwang 2016), also offers some other advantages, such as higher temperature stability and low volatility (Kumar and Münstedt 2005). A larger surface area allows greater interaction with microbial cells (Kvitek et al. 2008), and this is shown by the greater effectiveness of silver NPs when compared to the proven antimicrobial activity (Yu et al. 2007) of larger silver particles. Silver is found as a commonly used basis for nanocomposite formation due to described mechanisms of its antimicrobial activity, namely: (i) adhesion to the cell surface, degradation of lipopolysaccharides, and formation of "pits" in the membranes, largely increasing permeability (Sondi and Salopek-Sondi 2004); (ii) penetration inside bacterial cell, damaging DNA (Li et al. 2009b); and (iii) releasing antimicrobial Ag^+ ions by dissolution of silver nanoparticles (Morones et al. 2005). The work of Morones et al. (2005) is consistent with the findings of Kumar and Münstedt (2005), and this leads us to believe that released Ag^+ ions bind to electron donor groups in biological groups containing sulfur, oxygen, or nitrogen (Kumar and Münstedt 2005, Morones et al. 2005). Silver NPs have also been reported to absorb and decompose ethylene, which may contribute to their effects on extending the shelf life of fruits and vegetables (Li et al. 2009a).

Nanosilver has been widely used for the development of contact materials for preservation of foods for a longer time by preventing growth or killing microorganisms. For example, BlueMoonGoods LLC has developed fresh-box super-airtight containers using silver NPs for food storage that can reduce bacteria by up to 99.9%. Foods can easily be stored for a longer time than in conventional food containers. The naturally antifungal, antibacterial, and antimicrobial properties of the finely dispersed nanosilver particles permanently imbedded in the container offer the consumer benefits of fresher, good-quality food for a longer period of time, and, subsequently, reduced food wastage (The Project on Emerging Nanotechnology 2013). Costa et al. (2011) demonstrated that the shelf life of fresh fruit salad can be increased by adding Ag-MMT nanoparticles into the bottom of polypropylene boxes (Costa et al. 2011). The shelf life of a carrot was also tested by the same research group, and it can be increased by more than 2 months when stored under $4 \pm 1°C$ (Costa et al. 2012).

Two other antimicrobial mechanisms were proposed by Rabea et al. (2003), namely chelation of trace metals by chitosan, inhibiting microbial enzyme activities, and (in fungal cells) penetration through the cell wall and membranes to bind DNA and inhibit RNA synthesis (Rabea et al. 2003). Carbon nanotubes have also been reported to have antibacterial properties. Direct contact with aggregates of carbon nanotubes has been demonstrated to kill *E. coli*, possibly because the long and thin nanotubes puncture microbial cells, causing irreversible damage and leakage of intracellular material. On the other hand, there are studies suggesting that carbon nanotubes may also be cytotoxic to human cells, at least when in contact with skin (Monteiro-Riviere et al. 2005) or lungs (Warheit et al. 2004). Once present in food packaging material, the nanotubes might eventually migrate into food. Thus, it is mandatory to know any eventual health effects of ingested carbon nanotubes.

12.3.2.2 Oxygen Scavengers

Oxygen (O_2) participates in several forms of food deterioration. Direct oxidation reactions result in browning reactions and rancid flavors, to name only two examples. Food deterioration by indirect action of O_2 includes food spoilage by aerobic microorganisms. The incorporation of O_2 scavengers into food packaging systems can maintain very low O_2 levels, which is useful for several applications. Oxygen scavenger films were successfully developed by Xiao-e et al. (2004) by adding TiO_2 NPs to different polymers (Xiao-e et al. 2004). Nanocomposite materials could be used as packaging films for a variety of oxygen-sensitive food products. Since TiO_2 acts by a photocatalytic mechanism, its major drawback would be the requirement of UVA light (Mills et al. 2006). The commonly used polymer polythene is known for moisture and barrier applications, but it shows inadequate results in oxygen barrier properties for the packaging of oxygen-sensitive food products. So, to overcome this problem, high-density polyethylene (HDPE) films have been modified by incorporating them with iron containing kaolinite to create oxygen-scavenging packaging films (Busolo et al. 2010).

A nanocomposite called Imperm is made by Voridian collaboration with Nanocor and is used in plastic beer bottles. It makes the bottles as hard as glass but stronger and lighter. Further, the nanoparticles are designed in such a way that they minimize the loss of carbon dioxide trying to escape from the beer and keep oxygen out of the bottle, therefore retaining the freshness of the beer and extending its shelf life (The Project on Emerging Nanotechnology 2013). Similarly, Honeywell (USA) also have designed similar product named Aegis®, OX. They developed multilayer polyethylene terephthalate bottles that integrate nanocomposites (nylon resin) to enhance the barrier properties and extend shelf life (The Project on Emerging Nanotechnology 2013). Another hybrid plastic, Durethan KU2-2601 (Bayer AG), was developed by using numerous silicate nanoparticles. The plastic incorporates Nanocor clay to produce a film that is lighter, stronger, and more heat resistant than traditional packaging materials. The film is designed to prevent the entry of oxygen and other gases and the exit of moisture, thus preventing food spoilage (Han et al. 2011, Gruère 2012). For more examples related to marketed nanotechnology packaging-related products, see Table 12.2.

12.3.2.3 Enzyme Immobilization Systems

Enzymes have various applications in the food industry. The specificity of enzyme catalysts ensures improvements in many applications, like sugar and corn syrups, dairy, baking products, alcohol drinks, meat tenderness, or cheese ripening, but due to their short lifetime, it limits their usefulness in many applications (Pérez-Esteve et al. 2013). However, their sensitivity to manufacturing conditions and/or to enzyme inhibitors can sometimes limit the applicability of the direct enzyme addition to foods. Immobilization is usually an effective way to improve enzyme stability to pH and temperature and resistance to proteases and other denaturing compounds, as well as to provide an adequate environment for their repeated use or controlled release (Lopez-Rubio et al. 2006). Enzyme immobilization has been considered for packaging applications (Soares and Hotchkiss 1998). The incorporation of enzymes like lactase or cholesterol reductase to packaging materials could increase the value of food products and answer the needs of consumers with enzyme deficiencies (Fernández et al. 2008). Nanoscale enzyme immobilization systems would have enhanced performance when compared to conventional ones because of their much higher surface contact area and mass transfer rate, which are probably the most important factors affecting the effectiveness of such systems (Fernández et al. 2008, Sun-Waterhouse and Waterhouse 2016).

Approaches might be expected to deal with enzyme adsorption into nanoclays incorporated into polymers (Rhim and Ng 2007), since NCs have a high affinity for protein

TABLE 12.2
Other Nanotechnology Packaging-Related Products Present on the Market

Product	Company	Details	Applications
FresherLonger™ Miracle Food Storage	Sharper Image® USA	Food storage containers infused with naturally antibacterial silver nanoparticles (25 nm)	It's easy to keep foods like fruits, vegetables, herbs, breads, cheeses, and soups fresher three or even four times longer.
FresherLonger™ Plastic Storage Bags	Sharper Image® USA	Resealable zip-top food storage bags infused silver nanoparticles	To reduce the growth of bacteria, mold, and fungus and thereby increase shelf life.
Nano Silver Baby Mug Cup	Baby Dream® Co., Ltd. Korea	Silver nanoparticles infused with containers	To reduce the growth of bacteria, mold, and fungus by 99.9% and thereby maintain freshness.
Samsung® Refrigerator	Samsung® Korea	Used silver nanoparticles	Interior coating of refrigerators for effective sterilization deodorization and anti-bacterial effects.
Nano Silver NS-315 Water Bottle	A-DO Global Korea	Used silver nanoparticles	Antibacterial, antibiotic effect by nanosilver. Silver has no smell and is harmless to food.

Source: The Project on Emerging Nanotechnology. 2013. The Project on Emerging Nanotechnologies (PEN). Nanotechnology Consumer Products Inventory. [Cited 2013 29th October]; Available from: http://www.nanotechproject.org/cpi/browse/categories/food-and-Beverage/

adsorption and have been reported to be efficient enzyme carriers (Gopinath and Sugunan 2007). Conductive polymers may also be used as immobilizing matrices for biomolecules (Ahuja et al. 2007), as reported by Sharma et al. (2004), who immobilized glucose oxidase onto films of poly (aniline-co-fluoroaniline) (Sharma et al. 2004). SiO_2 nanoparticles have been modified to immobilize glutamate dehydrogenase and lactate dehydrogenase, which have shown excellent enzyme activity upon immobilization (Qhobosheane et al. 2001).

Graphene and its derivatives are known for advantages over other nanomaterials, consisting of uniform nanosheets of single carbon atoms, which provide an enormous and uniform area for biomolecule (enzyme) attachment (Shan et al. 2009, Zhang et al. 2010) and are used as suitable carriers for enzyme immobilization (Agyei et al. 2015). Nanosheets of graphene derivatives are used as immobilization of α- and β-galactosidase enzymes using cysteamine as the spacer arm and glutaraldehyde as the cross-linker (Singh et al. 2014). Separation of these nanobiocatalysts from reaction can be done by low-speed centrifugation or sedimentation methods with good recovery. Immobilization of α-galactosidase using graphene showed enhanced thermal stability from 50°C to 80°C. The optimum temperatures for soluble and immobilized Cicer α-galactosidase are 50°C and 80°C, respectively. The soluble enzyme retained more than 90% activity when kept at 50°C for 15 min, but lost 90% activity within 2 min when kept at 70°C, whereas immobilized Cicer α-galactosidase retained almost 90% activity at 80°C for 60 min and also retained 70% residual activity after 12 repeated uses. The immobilized enzyme has been used for the hydrolysis of the raffinose family of oligosaccharides from soybean-derived food products, reducing gas problems in the alimentary canal caused by consumption of soybean products (Singh et al. 2014). Similarly, immobilization of β-galactosidase on nanosheets of graphene was used for the hydrolysis of whey lactose, a byproduct of the cheese industry. Immobilization of β-galactosidase had shown 92% activity after 10 successive reuses (Kishore et al. 2012).

Nanotechnology in Functional Foods and Their Packaging 403

12.3.3 Nanocomposite Smart Packaging Systems

Nanotechnology can be used to establish a system (nanosensor system) that can detect minor changes in the environment, such as temperature changes, changes in humidity or oxygen exposure during storage, chemical changes due to degradation of food components, or contamination by microorganisms (Bouwmeester et al. 2009). A "smart" food packaging system may be defined as a system that "perceives" some property of the packaged food and uses a variety of mechanisms to register and transmit information about the current quality or status of the food with regard to its safety and digestibility. Nanosensors incorporated into food packaging systems can be used to detect spoilage-related changes; contaminations by pathogens; and chemical contaminations such as veterinary drugs, pesticides, phytotoxins, and marinetoxins resulting from processing techniques and spoiled food, therefore eliminating the need for inaccurate expiration dates providing real-time status of food freshness (Liao et al. 2005).

12.3.3.1 Time Temperature Integrators and Moisture Indicators

For determining the kinetics of physical, chemical, and microbial spoilage in food products, the most concerning environment factor is temperature. Time–temperature indicators or integrators are designed to monitor, translate and record if a certain food product is safe to be consumed after checking the temperature history, which includes whether the threshold temperature has been exceeded over time and/or estimating the minimum amount of time a product has spent above the threshold temperature. This is particularly important when food is stored in conditions other than the optimal ones. For instance, if a product is supposed to be frozen, a TTI can indicate whether it had been inadequately exposed to higher temperatures and the time of exposure. TTIs are categorized into three basic types, namely abuse indicators, partial temperature history indicators, and full temperature history indicators. Abuse indicators and critical temperature indicators merely indicate whether a reference temperature has been achieved. Partial temperature history indicators integrate the time–temperature history only when the temperature exceeds a critical predetermined value.

Finally, full temperature history indicators provide a continuous register of temperature changes with time (Singh 1994). The communication is usually manifested by a color development (related to a temperature-dependent migration of a dye through a porous material) or a color change (using a temperature-dependent chemical reaction or physical change). Timestrip has developed a system (iStrip) for chilled foods based on gold nanoparticles, which is red at temperatures above freezing. Accidental freezing leads to irreversible agglomeration of the gold nanoparticles, resulting in loss of the red color (Robinson and Morrison 2010).

12.3.3.2 Freshness Indicators

The real-time information of products that includes actual product quality during storage and distribution can be provided by freshness indicators to producers, retailers, and/or consumers. Food products can be tested to check whether they are spoiled by detecting the marker spoilage compounds or microbial metabolites like volatile sulfides, amine, and the metabolites or compounds present in food products when a food product is spoiled. A technique involving deposition of a transition metal (silver and/or copper) coating 1–10 nm thick on plastic film or paper packaging structures that can check the quality of meat was invented by Smolander et al. (2004). The distinctive dark color appears when this thin coating reacts with sulfide volatile components present in spoiled food products (Smolander et al. 2004). A peptide receptor-based portable "bioelectronic nose" is created to check the freshness of food by detecting trimethylamine, which is present only when raw seafood is spoiled (Lim et al. 2016).

12.3.3.3 Detection of Gases

Food spoilage is caused by microorganisms whose metabolisms produce gases that may be detected by several types of gas sensors that have been developed to translate chemical interactions between particles on a surface into response signals. Nanosensors to detect gases are usually based on metal oxides or, more recently, conducting polymer nanocomposites, which are able to quantify and/or identify microorganisms based on their gas emissions. Sensors based on conducting polymers (or electroactive conjugated polymers) consist of conducting particles embedded into an insulating polymer matrix. The change in resistance of the sensors produces a pattern corresponding to the gas under investigation (Arshak et al. 2007). Conducting polymers are very important because of their electrical, electronic, magnetic, and optical properties, which are related to their conjugate π electron backbones (Ahuja et al. 2007). Polyene and polyaromatic-conducting polymers such as polyaniline, polyacetylene, and polypyrrole have been widely studied (Ahuja et al. 2007). Electrochemically polymerized "conducting polymers" have a remarkable ability to switch between conducting oxidized (doped) and insulating reduced (undoped) states, which is the basis for several applications (Rajesh et al. 2004). Nanosensors containing carbon black and polyaniline developed by Arshak et al. (2007) have been demonstrated to be able to detect and identify three food-borne pathogens by producing a specific response pattern for each microorganism (Arshak et al. 2007).

12.3.3.4 Oxygen Sensors

There has been an increasing consensus on ensuring a complete absence of O_2 in oxygen-free food packaging systems, and to this effect, there is an interest in developing nontoxic and irreversible O_2 sensors with packaging done under vacuum or nitrogen. Lee et al. (2005) developed a UV-activated colorimetric O_2 indictor that uses TiO_2 NPs to photosensitize the reduction of methylene blue (MB) by triethanolamine in a polymer encapsulation medium using UV-A light. Upon UV irradiation, the sensor bleaches and remains colorless until it is exposed to oxygen, when its original blue color is restored (Lee and Sheridan 2005). The rate of color change is proportional to the level of O_2 exposure (Gutiérrez-Tauste et al. 2007). Mills and Hazafy (2009) used nanocrystalline SnO_2 as a photosensitizer in a colorimetric O_2 indicator comprising a sacrificial electron donor (glycerol), a redox dye (MB), and an encapsulating polymer (hydroxyethyl cellulose). Exposure to UV-B light led to photobleaching of the indicator and photoreduction of MB by the SnO_2 NPs. The color of the films varied according to O_2 exposure: bleached when not exposed and blue upon exposure. Mihindukulasuriya and Lim (2013) adopted a technique to develop a UV-activated oxygen indicator membrane using the electrospinning method. By encapsulating the active components (TiO2 nanoparticles, glycerol, and methylene blue) within electrospun poly(ethylene oxide) fibers that are submicron in diameter, the sensitivity of the membrane was higher as compared to the cast membrane carrier.

12.3.3.5 Detection of Microorganisms

Another role of nanosenors is the detection of pathogens and toxins in food products when incorporated into food packaging (Bhattacharya et al. 2007). The fluorescent nanoparticles method is the one of the methods to detect pathogens and toxins in food (Burris and Stewart 2012). Foodborne pathogenic bacteria species (such as *Salmonella typhimurium*, *Shigella flexneri*, and *E. coli* O157:H7) can be detected by quantum dots coupled with immunomagnetic (Fe_2O_3 magnetic nanoparticles) separation in milk and apple juice (Zhao et al. 2008). Another technique that is used in detecting of *E. coli* in food samples is based on light scattering by cell mitochondria; measurement of the light scattering confirms the presence of *E. coli*. The basic principle behind this technique is that a protein of a known and characterized bacterium set on a silicon chip can bind with any other *E. coli* bacteria present in the food sample and result in nanosized light scattering that will be detected by analyzing the digital image (Horner et al. 2006). Fu et al. (2008) developed a technique for detection of *Salmonella* present in food products. Silicon/gold nanorods are incorporated with anti-*Salmonella* antibodies and fluorescent dye, which become visible when they attach to the bacteria (Fu et al. 2008).

12.4 Conclusion

Nanotechnology is fast emerging in the field of food science and technology and its industries. In recent decades, most research has been done on nanofunctional foods and nonopackaging, and more patents and marketed products have been launched on the market. The major factors that enable nanomaterial applicability in the food sector are their smaller size, greater size-volume ratio, higher penetration, and higher efficiency. Overall, nanotechnology has a major role in nutraceutical and active compound delivery systems, which ultimately is beneficial for human health. Recently, researchers have identified nanosized additives and nanoingredients that may have positive and negative roles in nanofood industries and products. Nanotechnology has major and emerging applications in food packaging, which ultimately affects the strength and function of food packages; for example, nanoreinforcement materials increases strength, while active and smart packaging is helpful in providing different roles to packages. However, deep research needs to be done based on nanomaterials' applications in food sectors, which should identify the mechanism of action of nanomaterials at the molecular level, and its role on targeted and nontargeted delivery systems should be explored. Additionally, applications of nanofoods should be researched, which will ultimately be beneficial for defense or astronauts in unfavorable conditions. The toxic effects of all the researched nanofoods should not be avoided, and the foods should pass quality checks.

Acknowledgments

SS acknowledges Mr. Sukhdev Singh and Mrs. Gurmeet Kaur, Palia Kalan, Uttar Pradesh, India, for all their technical support throughout the work. The authors acknowledge the Department of Biotechnology (DBT, India) for the funding with grant number—BT/PR10414/PFN/20/961/2014. SR acknowledges Veer Kunwar Singh Memorial Trust, Chapra, Bihar, India for partial support—VKSMT/SN/NFNA/0011.

References

Agyei, D., B. K. Shanbhag, and L. He. 2015. Enzyme engineering (immobilization) for food applications. *Improving and Tailoring Enzymes for Food Quality and Functionality* 1984: 213–35. doi: 10.1016/B978-1-78242-285-3.00011-9.

Ahuja, T., I. A. Mir, D. Kumar, and Rajesh. 2007. Biomolecular immobilization on conducting polymers for biosensing applications. *Biomaterials* 28 (5): 791–805. doi: 10.1016/j.biomaterials.2006.09.046.

Ankola, D. D., B. Viswanad, V. Bhardwaj, P. Ramarao, and M. N. V. Ravi Kumar. 2007. Development of potent oral nanoparticulate formulation of coenzyme Q10 for treatment of hypertension: Can the simple nutritional supplements be used as first line therapeutic agents for prophylaxis/therapy? *European Journal of Pharmaceutics and Biopharmaceutics* 67 (2): 361–69. Elsevier.

Arora, A., and G. W. Padua. 2010. Review: Nanocomposites in food packaging. *Journal of Food Science* 75 (1): 43–49. doi: 10.1111/j.1750-3841.2009.01456.x.

Arshak, K., C. Adley, E. Moore, C. Cunniffe, M. Campion, and J. Harris. 2007. Characterisation of polymer nanocomposite sensors for quantification of bacterial cultures. *Sensors and Actuators, B: Chemical* 126 (1): 226–31. doi: 10.1016/j.snb.2006.12.006.

Auweter, H., H. Bohn, H. Haberkorn, D. Horn, E. Luddecke, and V. Rauschenberger. 1999. Production of Carotenoid Preparations in the Form of Coldwater-Dispersible Powders, and the Use of the Novel Carotenoid Preparations. Google Patents.

Bernardes, P. C., N. J. De Andrade, and N. De Fátima Ferreira Soares. 2014. Nanotechnology in the food industry = Nanotecnologia na indústria de alimentos. *Bioscience Journal* 30 (6): 1919–32. http://www.seer.ufu.br/index.php/biosciencejournal/article/view/22931

Bhattacharya, S., J. Jang, L. Yang, D. Akin, and R. Bashir. 2007. BioMEMS and nanotechnology-based approaches for rapid detection of biological entities. *Journal of Rapid Methods & Automation in Microbiology* 15 (1): 1–32. Wiley Online Library.

Bouwmeester, H., S. Dekkers, M. Y. Noordam, W. I. Hagens, A. S. Bulder, C. de Heer, S. E. C. G. ten Voorde, S. W. P. Wijnhoven, H. J. P. Marvin, and A. J. A. M. Sips. 2009. Review of health safety aspects of nanotechnologies in food production. *Regulatory Toxicology and Pharmacology* 53 (1): 52–62. doi: 10.1016/j.yrtph.2008.10.008.

Burris, K. P., and C. Neal Stewart. 2012. Fluorescent nanoparticles: Sensing pathogens and toxins in foods and crops. *Trends in Food Science & Technology* 28 (2): 143–52. Elsevier.

Busolo, M. A., P. Fernandez, M. J. Ocio, and J. M. Lagaron. 2010. Novel silver-based nanoclay as an antimicrobial in polylactic acid food packaging coatings. *Food Additives & Contaminants: Part A* 27 (11): 1617–26. doi: 10.1080/19440049.2010.506601.

Chaudhry, Q., and K. Groves. 2010. *Nanotechnology Applications for Food Ingredients, Additives and Supplements*. RSC Publishing: Cambridge, UK.

Chaudhry, Q., M. Scotter, J. Blackburn, B. Ross, A. Boxall, L. Castle, R. Aitken, and R. Watkins. 2008. Applications and implications of nanotechnologies for the food sector. *Food Additives & Contaminants: Part A* 25 (3): 241–58. doi: 10.1080/02652030701744538.

Chellaram, C., G. Murugaboopathi, A. A. John, R. Sivakumar, S. Ganesan, S. Krithika, and G. Priya. 2014. Significance of nanotechnology in food industry. *APCBEE Procedia* 8 (Caas 2013): 109–13. Elsevier B.V. doi: 10.1016/j.apcbee.2014.03.010.

Chen, H., J. Weiss, and F. Shahidi. 2006a. Nanotechnology in nutraceuticals and functional foods. *Food Technology*. http://agris.fao.org/agris-search/search.do?recordID=US201301069370

Chen, L., G. E. Remondetto, and M. Subirade. 2006b. Food protein-based materials as nutraceutical delivery systems. *Trends in Food Science & Technology* 17 (5): 272–83. Elsevier.

Chen, Y., R. Ye, and J. Liu. 2013. Understanding of dispersion and aggregation of suspensions of Zein nanoparticles in aqueous alcohol solutions after thermal treatment. *Industrial Crops and Products* 50: 764–70. Elsevier.

Cioffi, N., L. Torsi, N. Ditaranto, G. Tantillo, L. Ghibelli, L. Sabbatini, T. Bleve-Zacheo, M. D'Alessio, P. Giorgio Zambonin, and E. Traversa. 2005. Copper nanoparticle/polymer composites with antifungal and bacteriostatic properties. *Chemistry of Materials* 17 (21): 5255–62. ACS Publications.

Costa, C., A. Conte, G. G. Buonocore, and M. A. Del Nobile. 2011. Antimicrobial silver-montmorillonite nanoparticles to prolong the shelf life of fresh fruit salad. *International Journal of Food Microbiology* 148 (3): 164–67. doi: 10.1016/j.ijfoodmicro.2011.05.018.

Costa, C., A. Conte, G. G. Buonocore, M. Lavorgna, and M. A. Del Nobile. 2012. Calcium alginate coating loaded with silver-montmorillonite nanoparticles to prolong the shelf-life of fresh-cut carrots. *Food Research International*, 48, 164–169.

Cushen, M., J. Kerry, M. Morris, M. Cruz-Romero, and E. Cummins. 2012. Nanotechnologies in the food industry—Recent developments, risks and regulation. *Trends in Food Science & Technology* 24 (1): 30–46. Elsevier.

Dalmas, F., J.-Y. Cavaillé, C. Gauthier, L. Chazeau, and R. Dendievel. 2007. Viscoelastic behavior and electrical properties of flexible nanofiber filled polymer nanocomposites. Influence of processing conditions. *Composites Science and Technology* 67 (5): 829–39. Elsevier.

Elliott, R. 2002. Science, medicine, and the future: Nutritional genomics. *BMJ* 324 (7351): 1438–42. doi: 10.1136/bmj.324.7351.1438.

Ezhilarasi, P. N., P. Karthik, N. Chhanwal, and C. Anandharamakrishnan. 2013. Nanoencapsulation techniques for food bioactive components: A review. *Food and Bioprocess Technology* 6 (3): 628–647. doi: 10.1007/s11947-012-0944-0.

Fernández, A., D. Cava, M. J. Ocio, and J. M. Lagarón. 2008. Perspectives for biocatalysts in food packaging. *Trends in Food Science & Technology* 19 (4): 198–206. Elsevier.

Fu, J., B. Park, G. Siragusa, L. Jones, R. Tripp, Y. Zhao, and Y.-J. Cho. 2008. An Au/Si hetero-nanorod-based biosensor for *Salmonella* detection. *Nanotechnology* 19 (15): 155502. IOP Publishing.

Giri, T. K., P. Mukherjee, T. K. Barman, and S. Maity. 2016. Nano-encapsulation of capsaicin on lipid vesicle and evaluation of their hepatocellular protective effect. *International Journal of Biological Macromolecules* 88: 236–43. Elsevier B.V. doi: 10.1016/j.ijbiomac.2016.03.056.

Goettler, L. A., K. Y. Lee, and H. Thakkar. 2007. Layered silicate reinforced polymer nanocomposites: Development and applications. *Polymer Reviews* 47 (2): 291–317. Taylor & Francis.

Gopinath, S., and S. Sugunan. 2007. Enzymes immobilized on Montmorillonite K 10: Effect of adsorption and grafting on the surface properties and the enzyme activity. *Applied Clay Science* 35 (1): 67–75. Elsevier.

Graveland-Bikker, J. F., and C. G. De Kruif. 2006. Unique milk protein based nanotubes: Food and nanotechnology meet. *Trends in Food Science & Technology* 17 (5): 196–203. Elsevier.

Gruère, G. P. 2012. Implications of nanotechnology growth in food and agriculture in OECD countries. *Food Policy* 37 (2): 191–98. doi: 10.1016/j.foodpol.2012.01.001.

Gu, H., P. L. Ho, E. Tong, L. Wang, and B. Xu. 2003. Presenting vancomycin on nanoparticles to enhance antimicrobial activities. *Nano Letters* 3 (9): 1261–63. ACS Publications.

Gutiérrez, F. J., S. M. Albillos, E. Casas-Sanz, Z. Cruz, C. Garcia-Estrada, A. Garcia-Guerra, J. Garcia-Reverter et al. 2013. Methods for the nanoencapsulation of β-carotene in the food sector. *Trends in Food Science & Technology* 32 (2): 73–83. Elsevier.

Gutiérrez-Tauste D., X. Domènech, N. Casañ-Pastor, and J.A. Ayllón. 2007. Characterization of methylene blue/TiO_2 hybrid thin films prepared by the liquid phase deposition (LPD) method: Application for fabrication of light-activated colorimetric oxygen indicators. *Journal of Photochemistry and Photobiology A* 187 (1): 45–52.

Han W., Y. Yu, N. Li, and L. Wang. 2011. Application and safety assessment for nano-composite materials in food packaging. *Chinese Science Bulletin* 56 (12):1216–1225.

Handford, C. E., M. Dean, M. Henchion, M. Spence, C. T. Elliott, and K. Campbell. 2014. Implications of nanotechnology for the agri-food industry: Opportunities, benefits and risks. *Trends in Food Science and Technology* 40 (2): 226–41. Elsevier Ltd. doi: 10.1016/j.tifs.2014.09.007.

He, X., and H.-M. Hwang. 2016. Nanotechnology in food science: Functionality, applicability, and safety assessment. *Journal of Food and Drug Analysis* 24 (4): 671–81. Elsevier.

Horner, S. R., C. R. Mace, L. J. Rothberg, and B. L. Miller. 2006. A proteomic biosensor for enteropathogenic *E. coli*. *Biosensors and Bioelectronics* 21 (8): 1659–63. Elsevier.

Hsieh, K. L. 2009. Lycopene and Resveratrol Dietary Supplement. Google Patents.

Huang, L., D.-Q. Li, Y.-J. Lin, M. Wei, D. G. Evans, and X. Duan. 2005. Controllable preparation of nano-MgO and investigation of its bactericidal properties. *Journal of Inorganic Biochemistry* 99 (5): 986–93. Elsevier.

Huang, Q., H. Yu, and Q. Ru. 2010. Bioavailability and delivery of nutraceuticals using nanotechnology. *Journal of Food Science* 75 (1): R50–R57. Wiley Online Library.

Ismail, A. A., L. Pinchuk, O. R. Pinchuk, and D. Pinchuk. 2008. Polymerbased Antimicrobial Agents, Methods of Making Said Agents, and Products and Applications Using Said Agents. WO2008058272.

Jeon, J. S., and H. S. Lee. 2009. Nano-Particles Containing Calcium and Method for Preparing the Same. WO2009011520.

Jordan, J., K. I. Jacob, R. Tannenbaum, M. A. Sharaf, and I. Jasiuk. 2005. Experimental trends in polymer nanocomposites—A review. *Materials Science and Engineering: A* 393 (1): 1–11. Elsevier.

Kailasapathy, K., and S. Rybka. 1997. *L. Acidophilus* and *Bifidobacterium* spp.—Their therapeutic potential and survival in yogurt. *Australian Journal of Dairy Technology* 52 (1): 28. Dairy Industry Association of Australia.

Katouzian, I., and S. M. Jafari. 2016. Nano-encapsulation as a promising approach for targeted delivery and controlled release of vitamins. *Trends in Food Science & Technology* 53: 34–48. Elsevier Ltd. doi:10.1016/j.tifs.2016.05.002.

Kim, T. 2006. Antimicrobial Composition Containing Natural Extract, Nano Silver and Natural Essential Oil. WO2006109898.

Kingsley, J. D., S. Ranjan, N. Dasgupta, and P. Saha. 2013. Nanotechnology for tissue engineering: Need, techniques and applications. *Journal of Pharmacy Research* 7 (2): 200–204. Elsevier.

Kishore, D., M. Talat, O. N. Srivastava, and A. M. Kayastha. 2012. Immobilization of β-Galactosidase onto functionalized graphene nano-sheets using response surface methodology and its analytical applications. *PloS One* 7: e40708. doi: 10.1371/journal.pone.0040708.

Krasaekoopt, W., B. Bhandari, and H. Deeth. 2003. Evaluation of encapsulation techniques of probiotics for yoghurt. *International Dairy Journal* 13 (1): 3–13. Elsevier.

Kriegel, C., A. Arrechi, K. Kit, D. J. McClements, and J. Weiss. 2008. Fabrication, functionalization, and application of electrospun biopolymer nanofibers. *Critical Reviews in Food Science and Nutrition* 48 (8): 775–97. Taylor & Francis.

Kumar, R., and H. Münstedt. 2005. Silver ion release from antimicrobial polyamide/silver composites. *Biomaterials* 26 (14): 2081–88. Elsevier.

Kvitek, L., A. Panáček, J. Soukupova, M. Kolar, R. Vecerova, R. Prucek, M. Holecova, and R. Zboril. 2008. Effect of surfactants and polymers on stability and antibacterial activity of silver nanoparticles (NPs). *The Journal of Physical Chemistry C* 112 (15): 5825–34. ACS Publications.

Latva-Nirva, E., G. Dahms, A. Jung, and B. Fiebrig. 2009. Ultrafine Titanium Dioxide Nanoparticles and Dispersions Thereof. WO2009141499.

Lavie, C. J., R. V. Milani, M. R. Mehra, and H. O. Ventura. 2009. Omega-3 polyunsaturated fatty acids and cardiovascular diseases. *Journal of the American College of Cardiology* 54 (7): 585–94. Am Coll Cardio Found.

Leclercq, S., K. R. Harlander, and G. A. Reineccius. 2009. Formation and characterization of microcapsules by complex coacervation with liquid or solid aroma cores. *Flavour and Fragrance Journal* 24 (1): 17–24. doi: 10.1002/ffj.1911.

Lee S. K., M. Sheridan, and A. Mills. 2005. Novel UV-activated colorimetric oxygen indicator. *Chem Mater* 17 (10): 2744–2751.

Letchford, K., and H. Burt. 2007. A review of the formation and classification of amphiphilic block copolymer nanoparticulate structures: Micelles, nanospheres, nanocapsules and polymersomes. *European Journal of Pharmaceutics and Biopharmaceutics* 65 (3): 259–69. Elsevier.

Li, H., F. Li, L. Wang, J. Sheng, Z. Xin, L. Zhao, H. Xiao, Y. Zheng, and Q. Hu. 2009a. Effect of nanopacking on preservation quality of Chinese Jujube (*Ziziphus Jujuba Mill. Var. Inermis (Bunge) Rehd*). *Food Chemistry* 114 (2): 547–52. Elsevier.

Li, H. L., X. B. Zhao, Y. K. Ma, G. X. Zhai, L. B. Li, and H. X. Lou. 2009b. Enhancement of gastrointestinal absorption of Quercetin by solid lipid nanoparticles. *Journal of Controlled Release* 133 (3): 238–44. Elsevier.

Li, X., A. Liu, R. Ye, Y. Wang, and W. Wang. 2015. Fabrication of gelatin—Laponite composite films: Effect of the concentration of laponite on physical properties and the freshness of meat during storage. *Food Hydrocolloids* 44: 390–98. Elsevier.

Liao, F., C. Chen, and V. Subramanian. 2005. Organic TFTs as gas sensors for electronic nose applications. *Sensors and Actuators, B: Chemical* 107 (2): 849–55. doi: 10.1016/j.snb.2004.12.026.

Lim J. H., J. Park, J. H. Ahn, H. J. Jin, S. Hong, and T. H. Park. 2016. A peptide receptorbased bioelectronic nose for the realtime determination of seafood quality. *PubMed Commons* 39 (1): 22901715. doi: 10.1016/j.bios.2012.07.054.

Limpens, J., F. H. Schröder, C. M. A. de Ridder, C. A. Bolder, M. F. Wildhagen, U. C. Obermüller-Jevic, K. Krämer, and W. M. van Weerden. 2006. Combined lycopene and Vitamin E treatment suppresses the growth of PC-346C human prostate cancer cells in nude mice. *The Journal of Nutrition* 136 (5): 1287–93. Am Soc Nutrition.

Lopez-Rubio, A., R. Gavara, and J. M. Lagaron. 2006. Bioactive packaging: Turning foods into healthier foods through biomaterials. *Trends in Food Science and Technology* 17 (10): 567–75. doi: 10.1016/j.tifs.2006.04.012.

Luduena, L. N., V. A. Alvarez, and A. Vazquez. 2007. Processing and microstructure of PCL/clay nanocomposites. *Materials Science and Engineering: A* 460: 121–29. Elsevier.

Luykx, D. M. A. M., R. J. B. Peters, S. M. Van Ruth, and H. Bouwmeester. 2008. A review of analytical methods for the identification and characterization of nano delivery systems in food. *Journal of Agricultural and Food Chemistry*. doi: 10.1021/jf8013926.

Ma, Q. H., Y. Z. Kuang, X. Z. Hao, and N. Gu. 2009. Preparation and characterization of tea polyphenols and Vitamin E loaded nanoscale complex liposome. *Journal of Nanoscience and Nanotechnology* 9 (2): 1379–83. American Scientific Publishers.

Maherani, B., E. Arab-Tehrany, A. Kheirolomoom, F. Cleymand, and M. Linder. 2012. Influence of lipid composition on physicochemical properties of nanoliposomes encapsulating natural dipeptide antioxidant L-carnosine. *Food Chemistry* 134 (2): 632–40. Elsevier.

Malshe, V. C., and A. P. Malshe. 2007. Non-Metallic Nano/Micro Particles Coated with Metal, Process and Applications Thereof. Google Patents.

Marsanasco, M., A. L. Márquez, J. R. Wagner, S. del V. Alonso, and N. S. Chiaramoni. 2011. Liposomes as vehicles for Vitamins E and C: An alternative to fortify orange juice and offer Vitamin C protection after heat treatment. *Food Research International* 44 (9): 3039–46. Elsevier.

McClements, D. J., E. A. Decker, Y. Park, and J. Weiss. 2009. Structural design principles for delivery of bioactive components in nutraceuticals and functional foods. *Critical Reviews in Food Science and Nutrition* 49 (6): 577–606. Taylor & Francis.

Medina, C., M. J. Santos-Martinez, A. Radomski, O. I. Corrigan, and M. W. Radomski. 2007. Nanoparticles: Pharmacological and toxicological significance. *British Journal of Pharmacology* 150 (5): 552–58. doi: 10.1038/sj.bjp.0707130.

Mihindukulasuriya, S. D. F., and Lim, L.-T. 2013. Oxygen detection using UV-activated 835 Electrospun poly (ethylene oxide) fibers encapsulated with TiO_2 nanoparticles. *Journal of 836 Materials Science*, 48, 5489–5498.

Mills, A., G. Doyle, A. M. Peiro, and J. Durrant. 2006. Demonstration of a novel, flexible, photocatalytic oxygen-scavenging polymer film. *Journal of Photochemistry and Photobiology A: Chemistry* 177 (2–3): 328–31. doi: 10.1016/j.jphotochem.2005.06.001.

Mills A., and D. Hazafy. 2009. Nanocrystalline SnO_2-based, UVBactivated, colourimetric oxygen indicator. *Sens Actuators B* 136 (2): 344–349.

Momin, J. K., C. Jayakumar, and J. B. Prajapati. 2013. Potential of nanotechnology in functional foods. *Emirates Journal of Food and Agriculture* 25 (1): 10–19. doi: 10.9755/ejfa.v25i1.9368.

Monteiro-Riviere, N. A., R. J. Nemanich, A. O. Inman, Y. Y. Wang, and J. E. Riviere. 2005. Multi-walled carbon nanotube interactions with human epidermal keratinocytes. *Toxicology Letters* 155 (3): 377–84. Elsevier.

Moraru, C. I., C. P. Panchapakesan, Q. Huang, P. Takhistov, L. I. U. Sean, and J. L. Kokini. 2003. Nanotechnology: A new frontier in food science. *Food Technology* 57 (12): 24–29. Institute of Food Technologists.

Morones, J. R., J. L. Elechiguerra, A. Camacho, K. Holt, J. B. Kouri, J. T. Ramirez, and M. J. Yacaman. 2005. The bactericidal effect of silver nanoparticles. *Nanotechnology* 16 (10): 2346. IOP Publishing.

Mozafari, M. R., J. Flanagan, L. Matia-Merino, A. Awati, A. Omri, Z. E. Suntres, and H. Singh. 2006. Recent trends in the lipid-based nanoencapsulation of antioxidants and their role in foods. *Journal of the Science of Food and Agriculture* 86 (13): 2038–45. Wiley Online Library.

Mozafari, M. R., C. Johnson, S. Hatziantoniou, and C. Demetzos. 2008. Nanoliposomes and their applications in food nanotechnology. *Journal of Liposome Research* 18 (4): 309–27. Taylor & Francis.

Neethirajan, S., and D. S. Jayas. 2011. Nanotechnology for the food and bioprocessing industries. *Food and Bioprocess Technology* 4 (1): 39–47. Springer.

Observatory Nano, and European Commission. 2009. Report on Nanotechnology in Agrifood, no. May: 46.

Park, C.-J., and Y.-S. Kim. 2003. Method of Providing Antibacterial Activity on a Surface of a Body Using Nano-Sized Metal Particles. Google Patents.

Pérez-Esteve, E., A. Bernardos, R. Martínez-Máñez, and J. M. Barat. 2013. Nanotechnology in the development of novel functional foods or their package. An overview based in patent analysis. *Recent Patents on Food, Nutrition & Agriculture* 5 (1): 35–43. doi: 10.2174/2212798411305010006.

Qhobosheane, M., S. Santra, P. Zhang, and W. Tan. 2001. Biochemically functionalized silica nanoparticles. *Analyst* 126 (8): 1274–78. Royal Society of Chemistry.

Qiao, R., and L. Catherine Brinson. 2009. Simulation of interphase percolation and gradients in polymer nanocomposites. *Composites Science and Technology* 69 (3): 491–99. Elsevier.

Rabea, E. I., M. E.-T. Badawy, C. V. Stevens, G. Smagghe, and W. Steurbaut. 2003. Chitosan as antimicrobial agent: Applications and mode of action. *Biomacromolecules* 4 (6): 1457–65. ACS Publications.

Rai, M., A. Yadav, and A. Gade. 2009. Silver nanoparticles as a new generation of antimicrobials. *Biotechnology Advances* 27 (1): 76–83. Elsevier.

Rajesh, W. Takashima, and K. Kaneto. 2004. Amperometric phenol biosensor based on covalent immobilization of tyrosinase onto an electrochemically prepared novel copolymer poly (N-3-Aminopropyl Pyrrole-Co-Pyrrole) film. *Sensors and Actuators, B: Chemical* 102 (2): 271–77. doi: 10.1016/j.snb.2004.04.028.

Ranjan, S., N. Dasgupta, A. R. Chakraborty, S. Melvin Samuel, C. Ramalingam, R. Shanker, and A. Kumar. 2014. Nanoscience and nanotechnologies in food industries: Opportunities and research trends. *Journal of Nanoparticle Research* 16 (6): 1–23. doi: 10.1007/s11051-014-2464-5.

Ranjan, S., N. Dasgupta, C. Ramalingam, and A. Kumar. 2016. Nanoemulsions in food science and nutrition. In *Nanotechnology in Nutraceuticals: Production to Consumption*, edited by Sen S. and Pathak Y., pp. 135–64. Boca Raton, Florida: CRC Press.

Rao, P. J., and M. M. Naidu. 2016. Nanoencapsulation of bioactive compounds for nutraceutical food. *Nanoscience in Food and Agriculture* 2: 129–56. Springer.

Rhim, J.-W., and P. K. W. Ng. 2007. Natural biopolymer-based nanocomposite films for packaging applications. *Critical Reviews in Food Science and Nutrition* 47 (4): 411–33. Taylor & Francis.

Robinson, D. K. R., and M. J. Morrison. 2010. *Nanotechnologies for Food Packaging: Reporting the Science and Technology Research Trends: Report for the Observatory NANO*. http://www.observatorynano.eu/project/filesystem/files/Food%20Packaging%20Report%202010%20DKR%20Robinson.pdf (accessed September 15, 2017).

Ruxton, C. H. S., S. C. Reed, M. J. A. Simpson, and K. J. Millington. 2004. The health benefits of Omega-3 polyunsaturated fatty acids: A review of the evidence. *Journal of Human Nutrition and Dietetics* 17 (5): 449–59. Wiley Online Library.

Sastry, R. K., S. Anshul, and N. H. Rao. 2013. Nanotechnology in food processing sector—An assessment of emerging trends. *Journal of Food Science and Technology* 50 (5): 831–41. Springer.

Sekhon, B. S. 2010. Food nanotechnology—An overview. *Nanotechnology, Science and Applications*. doi: 10.2147/NSA.S8677.

Semo, E., E. Kesselman, D. Danino, and Y. D. Livney. 2007. Casein micelle as a natural nano-capsular vehicle for nutraceuticals. *Food Hydrocolloids* 21 (5): 936–42. Elsevier.

Shan, C., H. Yang, J. Song, D. Han, A. Ivaska, and L. Niu. 2009. Direct electrochemistry of glucose oxidase and biosensing for glucose based on graphene. *Analytical Chemistry* 81 (6): 2378–82. doi: 10.1021/ac802193c.

Sharma, A. L., R. Singhal, A. Kumar, K. K. Pande, and B. D. Malhotra. 2004. Immobilization of glucose oxidase onto electrochemically prepared poly (Aniline-Co-Fluoroaniline) films. *Journal of Applied Polymer Science* 91 (6): 3999–4006. Wiley Online Library.

Silva, H. D., M. Â. Cerqueira, and A. A. Vicente. 2012. Nanoemulsions for food applications: Development and characterization. *Food and Bioprocess Technology*. 5 (3): 854–67. doi: 10.1007/s11947-011-0683-7.

Singh, N., G. Srivastava, M. Talat, H. Raghubanshi, O. N. Srivastava, and A. M. Kayastha. 2014. Cicer alpha-galactosidase immobilization onto functionalized graphene nanosheets using response surface method and its applications. *Food Chem* 142: 430–38. doi: 10.1016/j.foodchem.2013.07.079.

Singh, R. P. 1994. Scientific principles of shelf-life evaluation. In *Shelf Life Evaluation of Food*, edited by Man C. M. D. and Jones A. A., 1st ed., pp. 3–26. US: Springer US. doi: 10.1007/978-1-4615-2095-5_1.

Smolander, M., Hurme, E., Koivisto, M., and Kivinen, S. 2004. Vol. International Patent WO2004/102185 A1.

Soares, N. F. F., and J. H. Hotchkiss. 1998. Naringinase immobilization in packaging films for reducing Naringin concentration in grapefruit juice. *Journal of Food Science* 63 (1): 61–65. Wiley Online Library.

Sondi, I., and B. Salopek-Sondi. 2004. Silver nanoparticles as antimicrobial agent: A case study on *E. coli* as a model for gram-negative bacteria. *Journal of Colloid and Interface Science* 275 (1): 177–82. Elsevier.

Soppimath, K. S., T. M. Aminabhavi, A. R. Kulkarni, and W. E. Rudzinski. 2001. Biodegradable polymeric nanoparticles as drug delivery devices. *Journal of Controlled Release* 70 (1): 1–20. Elsevier.

Sozer, N., and J. L. Kokini. 2009. Nanotechnology and its applications in the food sector. *Trends in Biotechnology* 27 (2): 82–89. Elsevier.

Srinivas, P. R., M. Philbert, T. Q. Vu, Q. Huang, J. L. Kokini, E. Saos, H. Chen et al. 2010. Nanotechnology research: Applications in nutritional sciences. *The Journal of Nutrition* 140 (1): 119–24. Am Soc Nutrition.

Sun-Waterhouse, D., and G. I. N. Waterhouse. 2016. 2—Recent advances in the application of nanomaterials and nanotechnology in food research. *Novel Approaches of Nanotechnology in Food*. Elsevier Inc. doi: 10.1016/B978-0-12-804308-0.00002-9.

Tang, D.-W., S.-H. Yu, Y.-C. Ho, B.-Q. Huang, G.-J. Tsai, H.-Y. Hsieh, H.-W. Sung, and F.-L. Mi. 2013. Characterization of tea catechins-loaded nanoparticles prepared from chitosan and an edible polypeptide. *Food Hydrocolloids* 30 (1): 33–41. Elsevier.

The Project on Emerging Nanotechnology. 2013. The Project on Emerging Nanotechnologies (PEN). *Nanotechnology Consumer Products Inventory*. [Cited 2013 29th October]; Available from: http://www.nanotechproject.org/cpi/browse/categories/food-and-Beverage/

Vidhyalakshmi, R., R. Bhakyaraj, and R. S. Subhasree. 2009. Encapsulation "The future of probiotics"—A review. *Advances in Biological Research* 3 (3–4): 96–103.

Wang, X., Y.-W. Wang, and Q. Huang. 2009. Enhancing stability and oral bioavailability of polyphenols using nanoemulsions. *ACS Symposium Series* 1007: 198–212.

Warheit, D. B., B. R. Laurence, K. L. Reed, D. H. Roach, G. A. M. Reynolds, and T. R. Webb. 2004. Comparative pulmonary toxicity assessment of single-wall carbon nanotubes in rats. *Toxicological Sciences* 77 (1): 117–25. Soc Toxicology.

Weiss, J., E. A. Decker, D. Julian McClements, K. Kristbergsson, T. Helgason, and T. Awad. 2008. Solid lipid nanoparticles as delivery systems for bioactive food components. *Food Biophysics* 3 (2): 146–54. doi: 10.1007/s11483-008-9065-8.

Wen, H.-W., T. R. DeCory, W. Borejsza-Wysocki, and R. A. Durst. 2006. Investigation of NeutrAvidin-tagged liposomal nanovesicles as universal detection reagents for bioanalytical assays. *Talanta* 68 (4): 1264–72. Elsevier.

Were, L. M., B. D. Bruce, P. Michael Davidson, and J. Weiss. 2003. Size, stability, and entrapment efficiency of phospholipid nanocapsules containing polypeptide antimicrobials. *Journal of Agricultural and Food Chemistry* 51 (27): 8073–79. ACS Publications.

Wu, Y., Y. Luo, and Q. Wang. 2012. Antioxidant and antimicrobial properties of essential oils encapsulated in zein nanoparticles prepared by liquid–liquid dispersion method. *LWT-Food Science and Technology* 48 (2): 283–90. Elsevier.

Xiao-e, L., A. N. M. Green, S. A. Haque, A. Mills, and J. R. Durrant, 2004. Light-driven oxygen scavenging by titania/polymer nanocomposite films. *J Photochem Photobiol A* 162:253–259.

Ye, C., R. Ye, and J. Liu. 2014. Effects of different concentrations of ethanol and isopropanol on physicochemical properties of Zein-based films. *Industrial Crops and Products* 53: 140–47. Elsevier.

Yin, L.-J., B.-S. Chu, I. Kobayashi, and M. Nakajima. 2009. Performance of selected emulsifiers and their combinations in the preparation of β-Carotene nanodispersions. *Food Hydrocolloids* 23 (6): 1617–22. Elsevier.

Yu, H., Y. Huang, and Q. Huang. 2009a. Synthesis and characterization of novel antimicrobial emulsifiers from ε-polylysine. *Journal of Agricultural and Food Chemistry* 58 (2): 1290–95. ACS Publications.

Yu, H., X. Xu, X. Chen, T. Lu, P. Zhang, and X. Jing. 2007. Preparation and antibacterial effects of PVA-PVP hydrogels containing silver nanoparticles. *Journal of Applied Polymer Science* 103 (1): 125–33. Wiley Online Library.

Yu, H. G., H. G. Ji, S. D. Kim, and H. K. Woo, 2009b. Nano-Emulsion Composition of Coenzyme Q10. WO2009005215.

Zhang, J., J. Zhang, F. Zhang, H. Yang, X. Huang, H. Liu, and S. Guo. 2010. Graphene oxide as a matrix for enzyme immobilization. *Langmuir* 26 (9): 6083–85. doi: 10.1021/la904014z.

Zhao, Y., M. Ye, Q. Chao, N. Jia, Y. Ge, and H. Shen. 2008. Simultaneous detection of multifood-borne pathogenic bacteria based on functionalized quantum dots coupled with immunomagnetic separation in food samples. *Journal of Agricultural and Food Chemistry* 57 (2): 517–24. ACS Publications.

Zimet, P., D. Rosenberg, and Y. D. Livney. 2011. Re-assembled casein micelles and casein nanoparticles as nano-vehicles for ω-3 polyunsaturated fatty acids. *Food Hydrocolloids* 25 (5): 1270–76. Elsevier.

Zwiorek, K., J. Kloeckner, E. Wagner, and C. Coester. 2004. Gelatin nanoparticles as a new and simple gene delivery system. *Journal of Pharmacy and Pharmaceutical Sciences* 7 (4): 22–28.

13

Use of Nanotechnology as an Antimicrobial Tool in the Food Sector

María Ruiz-Rico, Édgar Pérez-Esteve, and José M. Barat

CONTENTS

13.1 Introduction 413
13.2 Metallic Nanoparticles 415
    13.2.1 Silver and Gold Nanoparticles 416
    13.2.2 Iron Oxide Nanoparticles 416
    13.2.3 Other Metal Oxide Nanoparticles 417
    13.2.4 Application of Metallic Nanoparticles as Components of Antimicrobial Food Packaging 420
13.3 Nanoencapsulation Systems 420
    13.3.1 Organic Nanoencapsulation Systems 424
        13.3.1.1 Nanoemulsions 424
        13.3.1.2 Microemulsions 426
        13.3.1.3 Liposomes 426
        13.3.1.4 Solid Lipid Nanoparticles 427
        13.3.1.5 Nanofibers 428
        13.3.1.6 Application of Food-Grade Nanoparticles in Food Packaging 428
    13.3.2 Inorganic Nanoencapsulation Systems 429
        13.3.2.1 Antimicrobial-Loaded Mesoporous Silica Particles 431
        13.3.2.2 Antimicrobial Functionalized MSPs 434
        13.3.2.3 Antimicrobial Loaded and Functionalized Mesoporous Silica Particles 435
13.4 Limitations of Nanomaterials for Food Applications 436
    13.4.1 Nanomaterial Stability 436
    13.4.2 Nanomaterials Biocompatibility 436
    13.4.3 Nanomaterial Toxicity 437
13.5 Final Remarks 439
Acknowledgments 441
References 441

13.1 Introduction

Nanotechnology is becoming increasingly important for the agrifood sector. The potential applications of nanotechnology in the food sector are already being developed to improve shelf life, food quality and safety, control delivery, and biosensors for contaminated food or food packaging (Ranjan et al. 2014).

Nanomaterials are particles with at least one dimension that ranges from about 1 to 100 nm. Although a general consensus on using 100 nm as a limit has been reached, suggestions have been made to also consider nanoparticles to be materials with properties or phenomena related to their dimensions, even if they fall beyond the nanoscale range (Rossi et al. 2014). Despite the possibility of using nanoparticles (NPs) in the food industry in all food chain stages (Figure 13.1), the most relevant applications of nanomaterials in the food sector are the following (Weiss et al. 2006, Durán and Marcato 2013, Pérez-Esteve et al. 2013):

- Targeted delivery systems of nutrients or bioactive compounds through bioactive nanoencapsulation
- Sensory improvements and stabilization of active ingredients
- Packaging innovation
- Development of novel nanostructured processing aids, such as filters, membranes, and reactors
- Biosensors to detect pathogens, allergens, contaminants, and quality degradation indicators in foods
- Antimicrobial agents against pathogen and food-borne microorganisms

The increased presence of nanotechnology it is not at all surprising if we consider all the possibilities of using nanotechnology in the food sector. As observed in Figure 13.1, nanotechnology as an antimicrobial tool offers possibilities in all food production stages.

Nanomaterials have emerged in the microbiological area because of their unique physicochemical properties (chemical stability, size, surface area, and the possibility of functionalization) that allow antimicrobial effects to be improved.

New technologies to prevent spoilage and to guarantee the safety of food products have been developed because resistant microorganisms and inadequate traditional antimicrobial methods have been extended. Inappropriate use of chemical preservatives can lead to resistant microorganisms and can provide them with favorable conditions to emerge, spread, and persist (Capeletti et al. 2014). In this context, nanomaterials with antimicrobial activity are seen as a promising solution. Antimicrobial nanomaterials affect multiple

Food production
and processing
Nanopesticides
Nanosensors
Animal feed fortification
Nanoencapsulation systems
Antimicrobial nanoparticles

Nutrition
Nutraceuticals
Nanoencapsulation systems
Controlled delivery supports

Food safety
Nanosensors
Antimicrobial nanoparticles

Packaging
Nanosensors
Antimicrobial nanoparticles
Smart/active packaging

FIGURE 13.1
Schematic representation of the use of nanotechnology in the agrifood sector.

biological pathways of the target microorganism. Thus, in order to generate resistance, microorganisms should be able to generate simultaneous mutations, which is very unlikely (Weir et al. 2008, Huh and Kwon 2011).

Current methods used in the food industry to inhibit microbial growth affect the quality properties of food (thermal treatment), or are related to toxicological problems (traditional additives) (Zengin et al. 2011). As a result, consumers demand alternative methods to inhibit microorganisms in food while maintaining the quality and safety of the food product. A novel approach for the inhibition of microbial growth could be the application of nanomaterials, which could be responsible for antimicrobial activity because of their composition, structure, or size, such as metallic nanoparticles; or can be used as antimicrobial carriers that allow the encapsulation or attachment of antimicrobial molecules to protect them from degradation, such as food-grade nanoencapsulation systems (nanoemulsions, microemulsions, liposomes, solid lipid nanoparticles, or nanofibers) or inorganic encapsulation systems (zeolites, clays, or mesoporous silica particles). Antimicrobial nanoparticles help reduce toxicity by overcoming resistance and have lower costs compared to conventional antimicrobial agents. Nanosized carriers are also able to improve bioavailability and the inhibitory effect of antimicrobial biomolecules (Huh and Kwon 2011).

This chapter is an overview of recently developed nanoparticles that have been used to prevent the growth of microorganisms for food applications. Diverse approaches to creating antimicrobial nanoparticles are explained. On the one hand, metallic nanoparticles are introduced and examples of their application in food packaging are provided. On the other hand, nanoencapsulation systems for food antimicrobials, including food-grade encapsulation systems and mesoporous silica particles, are explained and some examples of their potential applications are offered. Finally, the current limitations of using nanotechnology in the food sector are presented.

13.2 Metallic Nanoparticles

The antimicrobial activity of metals, such as silver (Ag), copper (Cu), or gold (Au), has been known and applied for centuries. Preparing engineered NPs with these elements improves the optical, magnetic, electrical, and catalytic properties of metals, and increases their antimicrobial effectiveness thanks to the higher surface:volume ratio of nanoparticles (Dizaj et al. 2014), and also to certain nanoproperties that are attributable to their dimensions, such as the ability to penetrate prokaryote and eukaryote cells (Rossi et al. 2014).

There are different metallic NPs, of which Ag or Au, and metal oxide NPs, such as iron oxide (Fe_2O_3 or Fe_3O_4), zinc oxide (ZnO), cerium oxide (CeO_2), titanium dioxide (TiO_2), magnesium oxide (MgO), calcium oxide (CaO), and copper oxide (CuO), are the most reported ones.

The exact mechanisms for the antibacterial effect of metallic NPs are still being investigated, although their main possibilities include: free metal ion toxicity arising from the dissolution of metals from the surface of nanoparticles and oxidative stress via the generation of reactive oxygen species (ROS) on the surfaces of nanoparticles (Dizaj et al. 2014). Furthermore, the antimicrobial activity of engineered NPs depends on diverse factors such as composition, size, shape, concentration, and type of microorganism (Morones et al. 2005, Pal et al. 2007, He et al. 2011, Kanmani and Lim 2013).

13.2.1 Silver and Gold Nanoparticles

Silver nanoparticles are the most popular engineered NPs used as antimicrobial agents. In the last few years, nanosilver has been included in different patents of detergents, food processing devices, clothes, toys, personal care products, and antimicrobial films (Pérez-Esteve et al. 2013, Gaillet and Rouanet 2015).

The increased toxicity of Ag NPs compared to silver is due to the size and shape of NPs, which favor them easily passing into a cell and releasing silver ions (Gilmartin and O'Kennedy 2012). The antibacterial action of Ag NPs results from the damage of the bacterial outer membrane. These particles can induce pits and gaps in the bacterial membrane, and can then fragment the cell. There are also reports that Ag ions interact with disulfide or sulfhydryl groups of enzymes, which leads to the disruption of metabolic processes and, in turn, causes cell death (Dizaj et al. 2014). The antibacterial activity of gold nanoparticles (Au NPs) has been attributed to the attachment of NPs to the bacterial membrane, followed by membrane potential modification, a lowering ATP level and the inhibition of tRNA binding to the ribosome (Dizaj et al. 2014). The antimicrobial activity of these nanoparticles against food-related microorganisms has been extensively studied (Petrus et al. 2011, Zawrah and El-Moez 2011, Kanmani and Lim 2013, Okafor et al. 2013, Zarei et al. 2014, Velmurugan et al. 2014), as has their inclusion in different encapsulation systems and packaging applications (An et al. 2008, Fernández et al. 2010a, Fortunati et al. 2012, Kanmani and Rhim 2014).

13.2.2 Iron Oxide Nanoparticles

Iron oxide NPs are magnetic nanoparticles (MNPs) that can be used to capture microorganisms and inhibit their growth. MNPs are usually functionalized with biorecognition molecules, such as carbohydrates (El-Boubbou et al. 2007), antibiotics (Gu et al. 2003b), antibodies (Shim et al. 2014), amines (Huang et al. 2010), and amino acids (Jin et al. 2014). After molecular functionalization, biofunctional MNPs show high selectivity and sensitivity for many biological applications because of the large number of active sites, and they also improve the aqueous stability of NPs. MNPs bind with the target microorganism; then the application of a magnetic field allows the microorganism to be removed from media or the food system (Honda et al. 1998). Otherwise, some MNPs have the ability to cluster microorganisms and to inhibit them. One of the most interesting examples of MNPs is NPs functionalized with amines because of the simplicity of the design and their ability to capture and kill microorganisms (Zhan et al. 2014). Bacterial removal is influenced by surface charge, hydrophobicity, surface properties of pathogens, and the nature of the matrix. Bacterial cell surfaces possess net negative electrostatic charge by virtue of an ionized phosphoryl and carboxylate substituent on outer cell envelope macromolecules, which are exposed to the extracellular environment. Otherwise, the reported amine-functionalized MNPs possess positive zeta potential; thus, the removal of bacteria is driven by attractive electrostatic interaction between negatively charged cellular membranes and the positively charged surface of NPs (Huang et al. 2010, Zhan et al. 2014). Huang et al. (2010) reported the use of amine-functionalized magnetic nanoparticles (AF-MNPs) for the capture and removal of bacteria pathogens from water samples. Transmission electron microscopy (TEM) images showed that bacteria indeed aggregated with AF-MNPs, and AF-MNPs covered most regions of the cell surface and were nonuniformly distributed. The work of Singh et al. (2011) studied the efficiency of MNPs functionalized with carboxyl (succinic acid), amine (ethylenediamine), and thiol (2,3-dimercaptosuccinic acid) for the removal of toxic metal ions and bacterial pathogens from water. TEM images revealed that bacterial pathogens were successfully trapped by MNPs. Moreover, the bacterial cell wall trapped by MNPs was almost damaged, and it has been clearly observed that MNPs penetrated into the lipid bilayer membrane and disrupted its structural integrity.

13.2.3 Other Metal Oxide Nanoparticles

Similar to silver, gold, and iron oxide nanoparticles, the application of other metal oxide NPs has a stronger antimicrobial effect compared with conventional metal oxide powders. One of the metal oxide particles most widely used as an antimicrobial agent is ZnO NPs. They produce cellular internalization and cell wall disorganization, which result in the leakage of intracellular contents and the death of bacterial cells (Brayner et al. 2006, Huang et al. 2008, Liu et al. 2009). Tayel et al. (2011) reported the antibacterial activity of ZnO NPs against several food-borne pathogens and showed their complete inhibition after a few hours of exposure. The antibacterial activity of ZnO NPs was stronger than that of ZnO powder because of the higher surface-to-volume ratio of NPs. Microscopic studies have displayed the disruptive effect of ZnO NPs on cell membranes, which produced the lysis of bacterial cells.

The antimicrobial property of TiO_2 NPs has been related to its crystal structure, shape, and size (Dizaj et al. 2014). Upon irradiation by UV light, excited electrons are emitted from TiO_2 to create positive cavities, which can react with surface oxygen and water to yield active oxygen and hydroxyl free radicals (Huang et al. 2005). The antimicrobial activity of CaO NPs is due not only to the generation of superoxide on the surface of these particles, but also to an increase in the pH value by the hydration of CaO with water (Dizaj et al. 2014). CeO_2 NPs are absorbed on the outer membrane of the microorganism due to electrostatic attraction. Speciation of particles is modified after absorption (oxidoreduction) to produce toxicity to the microorganism because such adsorption onto the surface is linked to oxidative stress for bacteria (Thill et al. 2006).

In the same way, Mg ions on the surface of MgO NPs damage the cell membrane and produce a leakage of intracellular contents which, in turn, leads to the death of bacterial cells (Huang et al. 2005, Dizaj et al. 2014). Jin and He (2011) studied the antibacterial activity of MgO NPs alone or in combination with ZnO NPs and nisin. Their results revealed strong bactericidal MgO NPs activity against *Escherichia coli* O157:H7 and *Salmonella*, and their synergistic effect in combination with nisin. Microscopic assays have demonstrated that the presence of MgO NPs alone or in combination with nisin brings about significant changes in cell morphology and membrane integrity, which results in cell growth inhibition and death. Indeed, the toxic effect of CuO NPs can be caused by solubilized Cu-ions, which produce the disruption of the cell wall and membrane by ROS production and then damage the vital enzymes of bacteria (Kasemets et al. 2009, Dizaj et al. 2014).

After taking these results into account, the direct application of metal oxide NPs in food products has been recently described (Mirhosseini and Firouzabadi 2013, Firouzabadi et al. 2014). Firouzabadi et al. (2014) investigated the antibacterial activity of ZnO suspensions that contained 0.3% citric acid against different food-borne bacteria in mango juice, and showed a significant inhibitory effect on microbial growth during 24 h of incubation. In the same way as Ag NPs, the incorporation of metal oxide NPs into food packaging applications has been reported in recent years (Chawengkijwanich and Hayata 2008, Cardenas et al. 2009, Jin et al. 2009, Akbar and Anal 2014). A summary of the different metallic nanoparticles and their activity against food-related microorganisms is shown in Table 13.1.

TABLE 13.1
Overview of Metallic Nanoparticles and Their Mechanisms of Action

Metallic NP	Microorganism	Mechanism of Antimicrobial Action	References
Ag NPs	*Pseudomonas aeruginosa, Vibrio cholerae, E. coli, Salmonella typhi*	Attachment to cell membrane, which produces changes in the membrane morphology resulting in an increase in its permeability and internalization. Reaction with sulfur-containing proteins and phosphorus-containing compounds such as DNA in the interior of the cell that affect functional processes.	Morones et al. (2005) Pal et al. (2007)
Ag NPs	*E. coli, Listeria monocytogenes, V. cholerae, Salmonella enterica, Vibrio parahaemolyticus, Bacillus cereus, Staphylococcus aureus*	Internalization and inhibition of intracellular enzyme activity due to Ag^+ ions.	Petrus et al. (2011)
Ag NPs	*E. coli, B. cereus, P. aeruginosa, L. monocytogenes, Klebsiella pneumonia, Aspergillus* spp., *Penicillium* spp.	Pore formation, disruption of the cell membrane, and leakage of intracellular materials. Inhibition of biofilms due to the induction of bacterial cell detachment.	Kanmani and Lim (2013)
Vancomycin-capped Au NP	Vancomycin-resistant enterococci and *E. coli*	Polyvalent inhibitor due to combined activity of vancomycin and Au NPs.	Gu et al. (2003a)
Au NPs	*Salmonella typhimurium*	Attachment to the surface of cell membrane which disturbed cell functions.	Zawrah and El-Moez (2011)
Au NPs	*E. coli* and *S. typhi*	Variation of the electric field in the plasmatic membrane because of Au NPs stabilized on zeolites.	Lima et al. (2013)
Chitosan-functionalized MNPs	*S. aureus, Citrobacter freundii, Enterobacter cloacae, Flavobacterium ferrugineum*	Capture by attractive electrostatic forces.	Honda et al. (1998)
Vancomycin-functionalized MNPs	*S. aureus, Staphylococcus epidermidis,* coagulase negative staphylococci (CNS)	Capture due to vancomycin binds to D-Ala-D-Ala receptors of the cell wall.	Gu et al. (2003b)
AF-MNPs	*Sarcina lutea, E. coli, P. aeruginosa, B. cereus, Proteus vulgaris, Bacillus subtilis*	Capture by attractive electrostatic forces.	Huang et al. (2010), Singh et al. (2011), Zhan et al. (2014)
Amino acids functionalized MNPs	*B. subtilis* and *E. coli*	Capture by attractive electrostatic forces.	Jin et al., (2014)

(Continued)

TABLE 13.1 (Continued)
Overview of Metallic Nanoparticles and Their Mechanisms of Action

Metallic NP	Microorganism	Mechanism of Antimicrobial Action	References
Antibody-functionalized MNPs	S. typhimurium	Cluster and selective filtration	Shim et al. (2014)
ZnO NPs	E. coli, L. monocytogenes, B. cereus, S. aureus	Accumulation in bacterial membrane and cellular internalization resulting in damage, leakage of intracellular contents and death.	Brayner et al. (2006), Zhang et al. (2007), Liu et al. (2009), Mirhosseini and Firouzabadi (2013), Firouzabadi et al. (2014)
ZnO NPs	Streptococcus agalactiae and S. aureus	Surface modification and cellular internalization with increase in membrane permeability and membrane disorganization.	Huang et al. (2008)
ZnO NPs	B. cereus, Enterobacter cloacae, E. coli, P. aeruginosa, S. aureus, Pseudomonas fluorescens, S. enteriditis, S. typhimurium	Production of ROS resulting in damage of cell membranes.	Applerot et al. (2009) Tayel et al. (2011)
ZnO NPs	Botrytis cinerea and Penicillium expansum	Effect on cellular functions which led to the death of fungal hyphae.	He et al. (2011)
Chitosan-capped ZnO NPs	E. coli	Combined activity of chitosan (attachment) and ZnO NPs.	Bhadra et al. (2011)
ZnO and TiO$_2$ NPs	E. coli and S. typhimurium	Internalization and oxidative stress resulting in genotoxicity and cytotoxicity.	Kumar et al. (2011a) Kumar et al. (2011b)
TiO$_2$ NPs	Candida albicans	Production of ROS induce destructive effects inside the microbial cells and cause death cell.	Haghighi et al. (2013)
CeO$_2$ NPs	E. coli	Adsorption into cell membrane, oxidoreduction, and cytotoxicity due to oxidative stress.	Thill et al. (2006)
MgO NPs	E. coli and Salmonella	Damage of cell membrane resulting in a leakage of intracellular contents and death.	Jin and He (2011)
MgO NPs	E. coli	Physical damage to the cell membrane which results in downregulation of outer membrane components and then mechanical damage of the membrane.	Leung et al. (2014)
CuO NPs	E. coli	Release of Cu^{2+} ions, generating hydroxyl radicals in the cytoplasm, which damage DNA, proteins, and other molecules leading to cell death.	Rispoli et al. (2010)
CuO NPs	Campylobacter jejuni	Disruption of the cell membrane and oxidative stress.	Xie et al. (2011)
CuO NPs	E. coli, P. aeruginosa, B. subtilis, S. aureus	Release of Cu^{2+} ions, resulting in damage of biochemical processes.	Azam et al. (2012)

13.2.4 Application of Metallic Nanoparticles as Components of Antimicrobial Food Packaging

The reported properties of metallic NPs, besides allowing the creation of antimicrobial agents to be added directly to food matrices, open the possibility to create a new generation of antimicrobial food packaging materials. It is well known that the primary functions of packaging are to isolate food from external environment and to protect food against deterioration by the action of microorganisms, moisture, gases, dusts, and odors, as well as mechanical forces (Sung et al. 2013). Hence, there is current trend of food packaging systems comprising the use of active packaging (packaging systems used to extend shelf life or improve safety among other functions). In this context, antimicrobial packaging containing metallic nanoparticles as antimicrobial agents is gaining weight in the packaging industry.

The most popular metallic nanoparticles incorporated in packaging matrixes are Ag NPs. It is probably due to their reported antimicrobial activity, as well as the possibility of synthetizing Ag NPs from $AgNO_3$ inside the film via green chemistry (Rhim et al. 2006, Shankar and Rhim 2016). The commonest biological resources used, as both reducing and stabilizing agents, are polysaccharides such as chitosan or agar and proteins like gelatine (Kanmani and Rhim 2014). These compounds include hydroxyl groups that allow the reduction of Ag^+ ions to Ag NPs. The produced NPs become attached to the composite, providing the stabilization of Ag NPs. Hence, the synthesis and stabilization of Ag NPs are produced in a single-step procedure, simplifying the preparation method (Sanpui et al. 2008). The obtained composites show good antimicrobial activity against food-borne microorganisms (Rhim et al. 2006, Ghosh et al. 2010, Kanmani and Rhim 2014).

One of the examples involving the use of films containing metallic NPs for food packaging applications reports the properties of Ag NP-polyvinylpyrrolidone (PVP) coatings to protect asparagus (An et al. 2008). These authors reported the effect of Ag NP-PVP coatings on the physicochemical and microbial qualities of asparagus spears during 25 days of storage. The coated asparagus showed lower microbial growth and better quality properties compared with the control samples among storage time. In another example, Fernández et al. (2010a) reported the antimicrobial activity of cellulose-Ag NP hybrid materials during storage of minimally processed melon. These hybrid materials released Ag^+ ions after melon juice impregnated the cellulose pad, which allowed the reduction of spoilage-related microorganisms. Furthermore, the presence of Ag-loaded absorbent pads retarded the senescence of the fruit after 10 days of storage. More examples of food packaging that incorporates metallic nanoparticles can be found in Table 13.2.

13.3 Nanoencapsulation Systems

Microbial inhibition has traditionally been achieved by adding chemical antimicrobial agents or preservatives. Despite their effectiveness, the continuous use of antimicrobials may result in resistant strains appearing (EFSA 2013) and in some toxicity problems (Zengin et al. 2011). Therefore, consumers demand alternative antimicrobials, such as natural products and/or new administration forms, which help lower concentrations of doses (Gyawali and Ibrahim 2014). Although naturally occurring antimicrobial compounds have particularly attracted much attention, the number of compounds allowed by regulations to be used in foods is very limited. So, making better use of available approved antimicrobial compounds is necessary and, for this reason, research has focused on developing technologies, such

TABLE 13.2

Summary of Metallic Nanoparticles for Food Packaging Applications

Metallic NP	Food Packaging/Surface	Microorganism	Food	References
Ag NPs	Polyethylene film	*Alicyclobacillus acidoterrestris*	Apple juice	Del Nobile et al. (2004)
Ag NPs	Chitosan composite	*E. coli* O157:H7, *L. monocytogenes*, *S. aureus*, and *S. typhimurium*	–	Rhim et al. (2006)
Ag NPs	Cellulose acetate nanofibers	*E. coli*, *S. aureus*, *Klebsiella pneumoniae*, and *P. aeruginosa*	–	Son et al. (2006)
Ag NPs	PVP	Psychrotrophic bacteria	Asparagus	An et al. (2008)
Ag NPs	Chitosan composite	*E. coli*	–	Sanpui et al. (2008)
Ag NPs	Cellulose	*B. subtilis*, *S. aureus*, and *K. pneumoniae*	–	Pinto et al. (2009)
Ag NPs	Cellulose composite	*E. coli* and *S. aureus*	–	Gopiraman et al. (2016)
Ag NPs	Cellulose pad	Mesophilic bacteria, psychrotrophic microorganisms, fungi, and yeast	Melon	Fernández et al. (2010a)
Ag NPs	Cellulose pad	Aerobic bacteria, lactic acid bacteria, *Pseudomonas* spp., and *Enterobacteriaceae*	Beef meat	Fernández et al. (2010b)
Ag NPs	Agar film	*E. coli*, *S. aureus*, and *C. albicans*	–	Ghosh et al. (2010)
Ag NPs	Chitosan film	*E. coli*, *Bacillus* spp., and *K. pneumoniae*	–	Vimala et al. (2010)
Ag and TiO$_2$ NPs	Low density polyethylene (LDPE) film	*Penicillium citrinum*, yeast, and molds	Bayberries	Wang et al. (2010)
Ag NPs	Hydroxypropyl methylcellulose film	*E. coli* and *S. aureus*	–	De Moura et al. (2012)
Ag NPs	Polylactic acid (PLA)	*E. coli* and *S. aureus*	–	Fortunati et al. (2012)
Ag NPs	LDPE film	Aerobic mesophilic and coliform bacteria, *E. coli*, *S. aureus*, yeast, and molds	Barberries	Motlagh et al. (2012)
Ag NPs	Gelatin nanocomposite film	*S. typhimurium*, *L. monocytogenes*, *E. coli*, *S. aureus*, and *B. cereus*	–	Kanmani and Rhim (2014)
Ag NPs	Poly(butylene adipate-co-terephthalate) film	*E. coli* and *L. monocytogenes*	–	Shankar and Rhim (2016)
Ag NPs	Agar/banana powder films	*E. coli* and *L. monocytogenes*	–	Orsuwan et al. (2016)
Au NPs	Quinoa film	*E. coli* and *S. aureus*	–	Pagno et al. (2015)
ZnO NPs	Paper	*E. coli*	–	Ghule et al. (2006)
ZnO NPs	PVP coating	*L. monocytogenes*, *Salmonella* Enteritidis, and *E. coli* O157:H7	Liquid egg white	Jin et al. (2009)
ZnO NPs	Polyurethane coating	*E. coli* and *B. subtilis*	–	Li et al. (2009a)

(*Continued*)

TABLE 13.2 (*Continued*)
Summary of Metallic Nanoparticles for Food Packaging Applications

Metallic NP	Food Packaging/Surface	Microorganism	Food	References
ZnO NPs	Poly(vinyl chloride) (PVC)	*E. coli* and *S. aureus*	—	Li et al. (2009b)
ZnO NPs	Glass	*C. albicans*	—	Eskandari et al. (2011)
ZnO NPs	Cellulose film	*E. coli* and *S. aureus*	—	Jebel and Almasi (2016)
ZnO NPs	PLA coating	*S. enterica*	Liquid egg white	Jin and Gurtler (2011)
ZnO NPs	PVC film	Psychrophilic bacteria, yeast, and molds	Apple	Li et al. (2011)
ZnO NPs	Calcium alginate film	*S. typhimurium* and *S. aureus*	Poultry meat	Akbar and Anal (2014)
Ag and ZnO NPs	LDPE film	Aerobic mesophilic bacteria, yeast, and molds	Orange juice	Emamifar et al. (2010)
Ag and ZnO NPs	LDPE film	*Lactobacillus plantarum*	Orange juice	Emamifar et al. (2011)
Ag, CuO and ZnO NPs	LDPE film	Coliform bacteria	Cheese	Beigmohammadi et al. (2016)
TiO$_2$ NPs	Ethylene-vinyl alcohol copolymer composite	*Bacillus* spp., *E. coli*, *Erwinia caratovora*, *L. plantarum*, *P. fluorescens*, *S. aureus*, *Zygosaccharomices rouxii*, and *Pichia jadirii*	—	Cerrada et al. (2008)
TiO$_2$ NPs	Polypropylene film	*E. coli*	Lettuce	Chawengkijwanich and Hayata (2008)
TiO$_2$ NPs	Chitosan composite	*E. coli*, *S. aureus*, and *S. enterica*	—	Daz-Visurraga et al. (2010)
TiO$_2$ NPs	Stainless steel	*E. coli*	Meat exudates	Verran et al. (2010)
TiO$_2$ NPs	Stainless steel and glass	*L. monocytogenes*	—	Chorianopoulos et al. (2011)
TiO$_2$ NPs	LDPE film	Mesophilic bacteria and yeast	Pears	Bodaghi et al. (2013)
TiO$_2$ NPs	Chitosan/poly(vinylalcohol) composite	*E. coli*, *S. aureus*, *P. aeruginosa*, and *C. albicans*	Cheese	Youssef et al. (2015)
Cu NPs	Cellulose	*E. coli*	—	Mary et al. (2009)
Cu NPs	Chitosan composite	*S. enterica* and *S. aureus*	—	Cardenas et al. (2009)
Ag and Cu NPs	Fish skin gelatin	*L. monocytogenes* and *S. enterica*	—	Arfat et al. (2017)

as nanoencapsulation systems, to improve the functionality of food antimicrobials (Weiss et al. 2009).

Encapsulation systems have emerged in the food industry in order to entrap bioactive phytochemicals, antimicrobials, nutraceuticals, vitamins, enzymes, or flavors (Liang et al. 2012). These systems must fulfill certain requirements to be considered adequate encapsulation supports: (i) the system should efficiently entrap the bioactive compound and allow its release at the targeted site at a controlled rate and/or in response to an external stimulus, (ii) the carrier should protect the bioactive compound from chemical degradation or environmental agents in all the supply chain stages and ingredient interactions that may negatively impact the beneficial effect of the bioactive substance, and (iii) the support has to be incorporated into food without interfering with its properties (McClements et al. 2009). Nanoencapsulation of food ingredients has provided several benefits, including improvement in their bioavailability due to the increased surface area per unit of mass, improved solubility of poorly water-soluble ingredients, greater activity and better retention during food processing, and improved physical properties that allow their incorporation into different food systems (Khare and Vasisht 2014).

Naturally occurring antimicrobial compounds include plant metabolites such as essential oils, free fatty acids or phytochemicals (i.e., curcuminoids, organosulfurs), animal enzymes or proteins, and microorganism-based bacteriocins. Among existing antimicrobial compounds, encapsulation of essential oils (EOs) to protect bioactive substances from environmental stress, to mask undesirable sensory properties of EOs, and to achieve the controlled release of the antimicrobial compounds at the site of action has been broadly reported. These lipophilic extracts of bioactive compounds have displayed antimicrobial activity against several pathogen and food-borne microorganisms. Inhibitory activity has been attributed to their phenolic compounds and their interaction with microbial cell membranes, which cause leakage of ions and cytoplasmatic content, and thus lead to cellular breakdown (Burt 2004). The direct application of EOs in food products has several limitations due to their strong sensory properties (odor and flavor), high volatility, poor solubility, and potential toxicity. The concentration of an essential oil needed to inhibit microbial growth in a food system is higher than in *in vitro* studies because of not only interactions with food components (Jo et al. 2015), but also difficulties in their dispersion in the food water phase (Weiss et al. 2009). Therefore, nanoencapsulation systems help lower the concentration of EOs and reduce their incorporation into food with a different polarity, which, in turn, helps improve the balance between food sensory properties and antimicrobial efficacy, enhance stability and protection from environmental stress, and prevent interactions with food components (Gaysinsky et al. 2005, Liang et al. 2012).

Other biomolecules with proven antimicrobial activity are proteins and enzymes. These hydrophilic compounds are fragile, and even small conformational changes may reduce their activity. Therefore, stabilization of peptides or enzymes through entrapment in nanoencapsulation systems may be a potential alternative to protect antimicrobials from denaturation by proteolysis and dilution effects and to enhance their efficacy and stability for food applications. As a result of entrapments in the confined environment of nanopores, the molecular motions of water molecules are modified. Consequently, biochemical reaction rates lower and the stabilization of biological materials during storage reduces (Balcão et al. 2013). Encapsulation of antimicrobial proteins in organic vesicles may offer several advantages, such as reducing the affinity between the antimicrobial compound and nontarget components, which could imply undesired interactions; protecting protein from inhibitors or unfavorable conditions; or cutting the risk of emergence of resistant strains (Mozafari et al. 2008). However, the main limitation of the entrapment of antibacterial peptides in lipid carriers is the ability of these compounds to interact with liposomal membranes and to disrupt them. So, the entrapment process and the lipid composition of matrices should be optimized to minimize this disadvantage (Laridi et al. 2003).

When selecting supports for encapsulating antimicrobial compounds, two principal approaches can be chosen: organic nanoencapsulation systems based on food-grade nanoparticles and inorganic nanoencapsulation systems such as zeolites, nanoclays, or mesoporous silica particles.

13.3.1 Organic Nanoencapsulation Systems

Organic nanoencapsulation systems are based on food-grade NPs, which are particles that exist naturally or have been manufactured within the nanosize range from food-compatible sources like chitosan, lecithin, alginate, arabic and xanthan gum, proteins, and so on (Hannon et al. 2015). These supports have different structures and properties that allow the entrapment of varied bioactive compounds and offer considerable advantages of usage. Liquids substances can be handled since solids, smell, or taste can be masked in a food product; sensitive compounds can be protected; and delivery can be controlled and targeted (Gomes et al. 2011). Figure 13.2 shows current nanostructured encapsulation systems, which include liposomes, solid lipid nanoparticles (SLNs), nanoemulsions, microemulsions, and nanofibers (Weiss et al. 2009). Among these nanoencapsulation systems, there are particles that exceed the size range for NPs (see Introduction). Notwithstanding, the scientific community now considers these systems to be nanoencapsulation systems due to the encapsulation concept of small molecules. In our opinion, this nomenclature may be confusing, and standard definitions are needed.

One of the most important factors that should be considered to select a suitable organic nanoencapsulation system for antimicrobial application is the polarity of the bioactive compound. Many antimicrobial substances are lipophilic and are only partially soluble in the aqueous phase. Therefore, nanoemulsions and SLNs may be adequate systems for highly hydrophobic compounds, while microemulsions, nanoemulsions, and SLNs may be suitable carriers for amphiphilic compounds and liposomes for hydrophilic antimicrobials (Weiss et al. 2009).

13.3.1.1 Nanoemulsions

Nanoemulsions are a class of emulsions (liquid-liquid dispersions of two completely or partially immiscible liquids with one phase being dispersed in the second phase) with droplet sizes on the nanometric scale, typically between 50 and 500 nm, although the scientific community has reached no clear agreement on this point. Reduction in droplet size displays some advantages compared to classical emulsions, such as the improved

Nanoemulsions   Microemulsions   Liposomes   Solid lipid nanoparticles   Nanofibers

FIGURE 13.2
Schematic representation of different food-grade encapsulation systems.

dispersibility of lipophilic compounds in water and bioavailability during gastrointestinal passage. The obtained emulsions are kinetically stable, which means better stability to particle aggregation and gravitation separation than conventional emulsions, and also being optically transparent (Ahmed et al. 2012, Jo et al. 2015). Nanoemulsions also have a larger surface area with a decreasing particle size and thus a larger amount of the antimicrobial compound at the droplet interface, which could be beneficial for interactions with microbial surfaces and enhanced delivery (Weiss et al. 2009).

Nanoemulsions are thermodynamically unstable systems that typically consist of oil, surfactant, and water. Thus, they are predisposed to breakdown with time and to separate into constituent phases with time (Gutiérrez et al. 2008). They are prepared by solubilizing the lipophilic bioactive components in the oil phase, then homogenizing this phase with an aqueous phase that contains a water-soluble emulsifier. The size of the produced droplets depends on the composition of the system and the homogenization method used (Ahmed et al. 2012).

The use of nanoemulsions to encapsulate different EOs (Chang et al. 2012, Liang et al. 2012, Bhargava et al. 2015, Salvia-Trujillo et al. 2015a) and bioactive compounds from EOs, such as carvacrol (Landry et al. 2014); eugenol (Gomes et al. 2011); isoeugenol (Nielsen et al. 2016); thymol (Chen et al. 2015); trans-cinnamaldehyde (Jo et al. 2015); D-limonene (Zahi et al. 2015); and amphiphilic substances like curcumin, nisin, or lactoferrin (Ahmed et al. 2012, Balcão et al. 2013, Sadiq et al. 2016), has been widely reported in recent years.

Encapsulation of EOs in nanoemulsions based on food-grade ingredients has shown good stability during long-term storage and prolonged antimicrobial activity against food-borne pathogens, such as *Listeria monocytogenes*, *E. coli*, *Salmonella typhimurium* and *Staphylococcus aureus* (Liang et al. 2012, Bhargava et al. 2015, Jo et al. 2015), or against spoilage microorganisms, such as *Lactobacillus delbrueckii* and *Zygosaccharomyces bailii* (Donsi et al. 2011, Chang et al. 2012). Such encapsulation has been reported to enhance the antimicrobial potential of an entrapped substance compared to coarse emulsion. The study of Salvia-Trujillo et al. (2015a), with microfluidized nanoemulsions that contained lemongrass or clove EOs, showed a faster release of antimicrobial compounds in comparison with coarse emulsions and presented enhanced inactivation kinetics in an *E. coli* population. This enhancement was probably produced by the solubilization of lipophilic antimicrobials within nanoemulsions that can increase the binding sites able to interact with porins of microbial membranes. Nielsen et al. (2016, 2017) reported the improved antibacterial activity of isoeugenol due to emulsion encapsulation using β-lactoglobulin and starch as emulsifiers. The nanoencapsulated isoeugenol efficiently reduced the bacterial population in growth media and carrot juice. In addition, the antibacterial properties of the entrapped isoeugenol against biofilms of different pathogens and spoilage bacteria were studied, resulting in good bactericidal activity in media but not in carrot juice due to interactions with food components.

Another advantage of such nanoencapsulation is the increased bioavailability of the entrapped compound. The encapsulation of curcumin, a natural polyphenolic antimicrobial with low water solubility and low bioavailability, in nanoemulsions has been recently reported (Ahmed et al. 2012). In this study, curcumin-loaded nanoemulsions were prepared using different lipid phases. The results revealed that the bioaccessibility of curcumin depended on lipid type and appeared to be slightly higher in conventional emulsions than in nanoemulsions, but the physical stability of nanoemulsions was much better.

In a final step, the encapsulation of bioactive compounds from EOs in nanoemulsion matrices has enabled their incorporation into different food products. Donsi et al. (2011) studied the encapsulation of a terpenes mixture and D-limonene in nanoemulsions and their introduction into fruit juices in order to enhance their antimicrobial activity while minimizing the impact on juice quality attributes. In this study, antimicrobial activity was firstly tested *in vitro* against *E. coli, L. delbrueckii*, and *Saccharomyces cerevisiae*, and resulted in lower values of the minimum inhibitory concentrations. Later, the application of the most efficient antimicrobial nanocapsules was tested in pear and orange juices inoculated with *L. delbrueckii*. The addition of low concentrations of the nanoencapsulated bioactive compounds showed the inhibition of the microorganism with minimal alteration to the organoleptic properties of the fruit juices. Along the same lines, Jo et al. (2015) reported the application of trans-cinnamaldehyde nanoemulsions in watermelon juice, and showed good encapsulation efficiency and inhibition of *S. typhimurium* and *S. aureus*. Recently, Bhargava et al. (2015) tested the antimicrobial activity of oregano oil nanoemulsions against food-borne bacteria on fresh lettuce. Dipping the contaminated lettuce into oregano oil nanoemulsions reduced approximately three logarithmic cycles of *L. monocytogenes, S. typhimurium*, and *E. coli* O157:H7. Landry et al. (2014, 2015) examined the efficacy of a carvacrol nanoemulsion against *Salmonella enterica* and *E. coli* O157:H7-contaminated sprouting seeds. Seeds were soaked in the nanoemulsion for different times and at various concentrations. The results indicated that this treatment successfully inactivated microorganisms.

13.3.1.2 Microemulsions

Microemulsions are self-assembled oil-in-water systems that contain water and oil, such as micelles. In contrast to nanoemulsions, they are thermodynamically stable, transparent, low-viscous, and isotropic dispersions with diameters that range from 5 to 100 nm (Jo et al. 2015). As previously mentioned, the size range for microemulsions is narrower than the range for nanoemulsions, which can be misleading because this is not the usual nomenclature. Clear definitions in size terms, among other features, are required to standardize methodologies and results.

Some examples of the entrapment of lipophilic antimicrobial compounds in microemulsions have been described in the last few years. Gaysinsky et al. (2007) studied the antimicrobial activity of eugenol microemulsions in pasteurized milk against *L. monocytogenes* and *E. coli* O157:H7. Microemulsions were able to inhibit the growth of strains similarly to unencapsulated eugenol due to the diffusional mass transport of eugenol from microemulsions to the macroemulsion (milk). The encapsulation of fatty acids like monolaurin microemulsions has demonstrated enhanced antimicrobial activities against *Bacillus subtilis, E. coli, Aspergillus niger*, and *Penicillium digitatum* than unencapsulated fatty acid, and has also improved the shelf life of fresh noodles (Fu et al. 2008). A similar microemulsion matrix with glycerol monolaurate has led to the complete inhibition of *E. coli* and *S. aureus* in *in vitro* assays (Zhang et al. 2009).

13.3.1.3 Liposomes

Liposomes are spherical vesicles in which an aqueous volume is entirely enclosed by a membranous lipid bilayer, composed mainly of natural or synthetic phospholipids (Rawat et al. 2008). Depending on the manufacturing method, the size of liposomes can range from tens of nanometers to tens of micrometers (Weiss et al. 2009). Liposomes can serve as a versatile carrier for lipophilic molecules, such as EO compounds and hydrophilic molecules, like proteins and enzymes, by offering certain advantages like an amphiphilic character, biocompatibility, low toxicity, and surface modification ease (Rawat et al. 2008). For lipophilic compounds, the bilayer membrane acts as the host environment to compounds, which are solubilized inside the bilayer. Hydrophilic compounds are encapsulated inside liposomes, and the liposomal membrane must provide a barrier to diffuse compounds from the inside to the outside (Weiss et al. 2009).

One example of the encapsulation of EOs in this type of carrier is found in the study of Liolios et al. (2009). These authors reported the encapsulation of *Origanum dictamnus* EO and bioactive compounds like carvacrol, thymol, and c-terpinene in phosphatidyl choline-based liposomes, and they tested their antimicrobial activity against pathogenic bacteria and fungi. The results showed that all the tested compounds exhibited enhanced antimicrobial activities after encapsulation given the improvement of the cellular transport and delivery of the active component inside the cell.

There are numerous references to the encapsulation of antimicrobial proteins (bacteriocins) and enzymes in liposomes, such as pediocin (Degnan and Luchansky 1992), nisin (Benech et al. 2002, Colas et al. 2007, Taylor et al. 2008), or lysozyme (Were et al. 2004). Some of these encapsulated compounds have been included and tested in food systems. Nisin is an antimicrobial peptide produced by strains of *Lactococus lactis*, and has been recognized as being safe for food applications by the Joint Food and Agriculture Organization/World Health Organization (FAO/WHO) Expert Committee on Food Additives (da Silva Malheiros et al. 2010). Using nisin in its free form is expensive and is associated with loss of activity due to the degradation or deactivation, and emergence, of nisin-resistant microbial strains. Benech et al. (2002) added encapsulated and free nisin to cheese-milk solution and observed that *Listeria innocua* diminished during the ripening process of cheese. Laridi et al. (2003) incorporated nisin-loaded liposomes during the fermentation process of cheddar cheese with long-term stability of vesicles during storage time. More recently, da Silva Malheiros et al. (2010) tested the antimicrobial activity of nisin-loaded liposomes in combination with refrigeration in milk inoculated with *L. monocytogenes*. The inhibitory rate for liposomes and free nisin was similar, but encapsulation seemed important to help overcome stability issues and interaction with food components during long-term storage.

13.3.1.4 Solid Lipid Nanoparticles

Solid lipid nanoparticles are submicron colloidal carriers with a mean diameter of 50–1000 nm. They consist in a solid hydrophobic core with a monolayer of phospholipid coating (Rawat et al. 2008). These supports combine the advantages of nanoemulsions and liposomes, such as high dissolution velocities associated with good permeability of the bioactive compound through cell membranes, and they simultaneously avoid some of their disadvantages, such as stability and toxicity problems (Lai et al. 2006). SLNs have been sought as vehicles for functional lipids and bioactive peptides (Almeida et al. 1997).

Lai et al. (2006) studied the incorporation into SLNs of essential fatty acids from *Artemisia arborescens*, which have shown antiviral activity. Their results indicated that SLNs maintained better activity of fatty acids because of the reduced volatility of the encapsulated compounds in comparison with other carriers, which tend to lose activity through evaporation. Other active compounds that have been encapsulated in this carrier type are curcuminoids. Tiyaboonchai et al. (2007) reported the development of curcuminoid-loaded SLN and showed a prolonged release of the phytochemical compound and physical and chemical stability of curcuminoids in lyophilized SLN after several months of storage.

The first study into the incorporation of lysozyme into SLNs was by Almeida et al. (1997). The results showed that lysozyme remained intact throughout the process and lost none of its activity, as shown by electrophoresis and the lysis rate of *Micrococcus lysolideikticus*. Encapsulation of another antimicrobial peptide, nisin, in SLNs has also been reported (Prombutara et al. 2012). Nisin-loaded SLNs were synthesized to provide protection from the environment and to prolong the biological activity of the peptide. Slow-releasing nisin was achieved over several days according to pH and the salt concentration. The antibacterial activity of nisin-loaded SLN against *L. monocytogenes* and *Lactobacillus plantarum* was prolonged for 15–20 days, compared to only 1–3 days for free nisin (Prombutara et al. 2012).

13.3.1.5 Nanofibers

Nanofibers are fibers whose average diameters range between 10–100 nm and have unique properties (superior mechanical performance, large surface-to-mass ratio, high porosity) that may be used as food-packaging materials, ingredients, sensors, and processing aids. Electrospinning is an emerging technology that is applied to produce these ultrafine polymer fibers by applying a high-voltage electrical field to a (bio)polymer solution or melt, which is pumped through a small capillary orifice. The applied electrical field forces polymers to be ejected from the tip of the orifice in the form of a fiber jet. The solvent is rapidly evaporated due to the high velocity and large surface area of the jet, which leads to the deposition of a solidified nanofiber on a grounded target. Fiber size, and thus the surface-to-mass ratio and fiber porosity, can be controlled precisely by the varying solution composition, solution properties, and electrospinning conditions (Kriegel et al. 2009).

The antimicrobial activity of this type of nanofibers has been reported for their polymer composition, such as chitosan nanofibers (Ignatova et al. 2006; Torres-Giner et al. 2008). The mechanism proposed for these materials is related to the presence or release of positively charged amino groups from the nanofibers when they are dissolved.

Nanofibers can also be used as carriers of entrapped antimicrobial compounds. Kriegel et al. (2010) studied the encapsulation of eugenol in microemulsions and their incorporation into nanofibers. Release studies have suggested a burst release of encapsulated eugenol due to the high porosity of the fibers and the hydrophilicity of the polymeric carrier, which results in the rapid dissolution of the carrier matrix. The antimicrobial activity of the loaded nanofibers has been evaluated against *S. typhimurium* and *L. monocytogenes*, which inhibits the growth of food-borne pathogens. Another applied strategy is encapsulation of bioactive compounds in cyclodextrins, followed by their entrapment in nanofibers. Vega-Lugo and Lim (2009) reported the incorporation of allyl isothiocyanate into nanofibers encapsulated in β-cyclodextrin or added directly to fiber-forming solutions. Entrapment in cyclodextrins protected allyl isothiocyanate from evaporation loss during the electrospinning process and allowed the controlled release of the antimicrobial compound under different relative humidity conditions. Kayaci et al. (2013) investigated the incorporation of eugenol-cyclodextrin complexes into nanofibers, and achieved good thermal stability and the slow release of the EO component. The results revealed the thermal evaporation of eugenol at higher temperatures compared to the free compound, and a slow release due to its encapsulation in cyclodextrin complexes.

13.3.1.6 Application of Food-Grade Nanoparticles in Food Packaging

Besides the direct application of organic nanoparticles in food products, the incorporation of these nanoencapsulation systems in food packaging systems is a remarkable application that has been developed in recent years to create films.

Edible films have been proposed as an alternative to food packaging to help improve the quality and safety of food products. This bioactive food packaging technology protects foods from dehydration; acts as a gas barrier with surrounding media; lowers respiration and oxidative reaction rates; reduces or even suppresses physiological disorders; and may serve as carriers of active compounds, such as antimicrobials, antioxidants, and texture enhancers, among others (Acevedo-Fani et al. 2015).

The commonest bioactive compounds entrapped in edible films are EOs. Thanks to their volatile nature, EOs that are added to films in a free form can evaporate from film to form dispersions during drying. Therefore, the effectiveness of bioactive compounds diminishes. Nanoencapsulation of EOs before incorporating films has been reported as a solution to maintain their usefulness for a longer time by the controlled release of compounds (Jiménez et al. 2014). Edible coatings that incorporate lemongrass EO nanoemulsions and their antimicrobial activity against food-borne pathogens in fruits have been newly reported (Kim et al. 2013, 2014, Salvia-Trujillo et al. 2015b). Kim et al. (2013, 2014) described the effect of nanoemulsion coatings on the microbial safety and physicochemical qualities of plums and grape berries during storage. These coatings inhibited the growth of the studied bacteria and did not significantly alter the quality parameters of fruits during storage. More recently, Salvia-Trujillo et al. (2015b) stated the contribution of lemongrass EO nanoemulsion-based coatings to extend the shelf life of fresh-cut Fuji apples in a faster and greater way than coatings obtained from conventional emulsions.

Regarding encapsulation of antimicrobial peptides, Imran et al. (2012) studied the nanoencapsulation of nisin to improve its bioavailability and the immobilization of vesicles to formulate biodegradable films embedded with the active compound. Nisin migrated through the liposome by pore formation, then diffused from the network of film to reach the food in contact. This strategy guaranteed the availability of nisin due to protection from inhibitors of the food system. The results showed that nisin-loaded films effectively reduced the growth of *L. monocytogenes* in *in vitro* assays.

13.3.2 Inorganic Nanoencapsulation Systems

Inorganic nanoencapsulation systems are characterized by a well-organized regular system of channels or pores that allow the entrapment of bioactive molecules. Different inorganic containers have been described to encapsulate and control the release of bioactive compounds, such as zeolites, nanoclays, or mesoporous silica particles. Zeolites and clays are aluminosilicate solids organized into different three-dimensional frameworks with adequate features (simple manufacturing, low cost, and biocompatibility) that act as delivery systems. However, according to the literature, mesoporous silica particles are the most important carriers for entrapping and delivering bioactive compounds in a sustained manner. Therefore, this section focuses on these mesoporous materials.

Mesoporous silica particles (MSPs) are silicon dioxide (SiO_2) porous materials with pore diameters between 2 and 50 nm, according to the International Union of Pure and Applied Chemistry (IUPAC). These materials have tunable sizes from the nano- to the microscale, different shapes from spheres to rods, and are easily functionalizable surfaces (Asefa and Tao 2012). Compared to organic encapsulation systems, MSPs are more rigid (higher thermal and chemical stability and more resistant to changes in pH, ion concentrations, or pressure) supports that can protect cargoes from chemical and biological (enzymatic or microbial) degradation (Tao 2014, Song and Yang 2015). Different MSPs can be obtained by making minor changes in synthesis routes, such as hexagonal or cubic Mobil Corporation Matter materials (MCM-n) (i.e., MCM-41 and MCM-48), Santa Barbara Amorphous Silica (i.e., SBA-15), Universidad Valencia Material (i.e., UVM-7), and hollow silica spheres (Beck et al. 1992, Zhao et al. 1998, el Haskouri et al. 2002, Cao et al. 2013). Microscopic images of some mesoporous silica materials with different shapes and sizes, among other features, are presented in Figure 13.3.

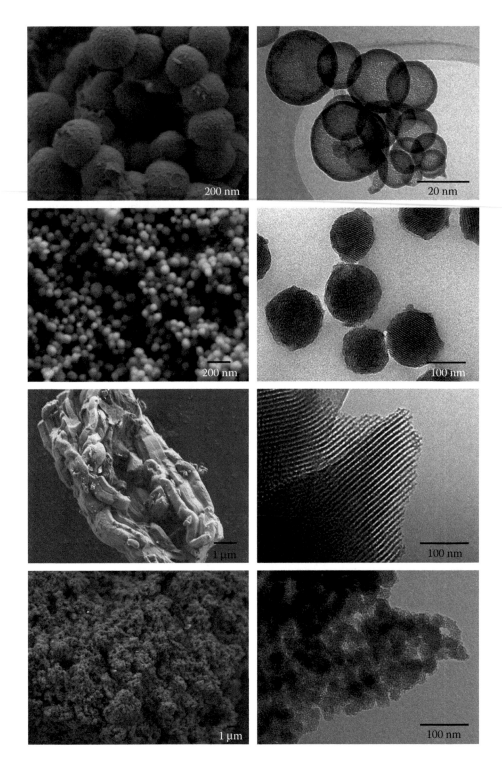

FIGURE 13.3
Field emission scanning electron microscopy (FESEM) (left) and TEM (right) images of different MSPs. H: Hollow silica spheres, M: nanoparticulated MCM-41, S: SBA-15, U: UVM-7.

MSPs display excellent features as delivery supports, such as high stability, good biocompatibility, optical transparency, large pore volume (~1 cm^3/g), large surface area (~1000 m^2/g), and the possibility of surface functionalization (Trewyn et al. 2007, Slowing et al. 2008, Al Shamsi et al. 2010, Tang et al. 2012, Song and Yang 2015).

MSPs are considered biocompatible materials for microbial populations in the literature. For instance, a study by Wehling et al. (2013) investigated the effect of silica particles with sizes between 15 and 500 nm on bacterial viability. The results determined that particles displayed no inhibitory properties independently of their particle size. Other research works, which have assessed the antibacterial effect of loaded nanoparticles, have shown that unloaded supports display no inhibitory activity against the studied microorganisms (Izquierdo-Barba et al. 2009, Mas et al. 2013, Wang et al. 2014, Ruiz-Rico et al. 2015). Nevertheless, when MSPs are loaded and/or functionalized with antimicrobial compounds, they can be used as antimicrobial materials in different applications.

13.3.2.1 Antimicrobial-Loaded Mesoporous Silica Particles

The incorporation of antimicrobial compounds into the pores of MSPs has been reported in recent years as an efficient strategy to create antimicrobial agents. A study by Izquierdo-Barba et al. (2009) encapsulated antimicrobial peptide LL-37 and chlorhexidine in a mesoporous silica monolith and achieved controlled release by attaching thiol groups to the silica surface. Both loaded particles showed bactericidal activity against Gram-positive and Gram-negative bacteria, but the particles loaded with chlorhexidine displayed high toxicity according to the hemolysis, lactate dehydrogenase release, and cell proliferation assays. In another study, nitroimidazole PA-824, with high antitubercular activity, was entrapped in MSPs and its solubility and thus its bioavailability improved. However, antibacterial activity was no greater than the effect of the free compound. Lack of improved activity could be due to difficulties in releasing nitroimidazole from the MSPs produced by molecules, which could interfere with the cellular environment (Xia et al. 2014).

The encapsulation of natural antimicrobial compounds in mesoporous supports has also been recently described. Allyl isothiocyanate is a natural antibacterial compound that can be added to food, but problems with its volatility, pungency, and poor water solubility exist (Siahaan et al. 2013). Encapsulation of this compound in MSPs and its controlled release has been achieved according to pore size distribution, and antibacterial properties remained after adsorption and desorption processes (Park et al. 2011, 2012, Park and Pendleton 2012). Siahaan et al. (2013) compared the suitability of MCM-41 and dried algae *Laminaria japonica* as carriers of allyl isothiocyanate. Their results revealed that loading was achieved in both delivery vectors, despite the MCM-41 support allowing greater higher adsorption and desorption and consequently greater bacteriostatic activity.

Along the same lines, Bernardos et al. (2015) reported the entrapment of some EOs components (carvacrol, cinnamaldehyde, eugenol, and thymol) in MCM-41 and β-cyclodextrin in order to control the volatility of the bioactive compounds. The *in-vitro* antifungal activity of the free and encapsulated EOs components against *A. niger* was established. The results revealed that carvacrol and thymol in MCM-41 displayed remarkably enhanced antifungal properties compared to free or β-cyclodextrin-encapsulated compounds. The antifungal activity of silica-entrapped carvacrol and thymol maintained antifungal activity and inhibited fungal growth after 30 days of incubation. This clearly contrasts with corresponding compounds when alone or encapsulated in β-cyclodextrin, which were unable to display antifungal properties to such an extent.

Other examples of encapsulation of a naturally occurring antimicrobial compounds are described for lysozyme entrapment (Yu et al. 2015, Wang et al. 2016). Yu et al. (2015) reported the antimicrobial activity of core-shell (magnetic Fe_3O_4 core and MSNs shell) nanoparticles loaded with lysozyme and functionalized with the thermo-responsive polymer poly(N-isopropylacrylamide). The developed material showed minimal lysozyme release at room temperature (25°C) but abrupt release of the enzyme at physiological temperature (37°C). Moreover, the core-shell NPs reduced the bacterial growth of *Bacillus cereus* and *Micrococcus luteus* at physiological temperature due to the release of the antimicrobial enzyme.

The antimicrobial activity of caprylic acid, entrapped in MCM-41 nanoparticles against different food-borne pathogens, has been studied and compared with the bactericidal effect of free fatty acid (Ruiz-Rico et al. 2015). Caprylic acid is a medium chain-length saturated fatty acid with reported antimicrobial activity. However, the application or supplementation of caprylic acid in food products has its disadvantages given its sensorial properties (i.e., unpleasant rancid-like smell and taste) and diminished antimicrobial activity through interactions with food constituents. Entrapment of caprylic acid in silica nanoparticles maintained the antimicrobial activity of fatty acid against *E. coli*, and slightly reduced its antibacterial effect for *Salmonella enterica*, *S. aureus*, and *L. monocytogenes*. Despite these slight variances, MSPs were suitable supports for caprylic acid encapsulation and allowed its controlled delivery to inhibit microbial growth in nutrient broth. These authors also attempted to elucidate the mechanism of action of entrapped caprylic acid against *L. monocytogenes* with microscopic studies. As seen in Figure 13.4, the TEM images confirmed the inhibitory effect of the entrapped fatty acid, which had severe effects on the morphology

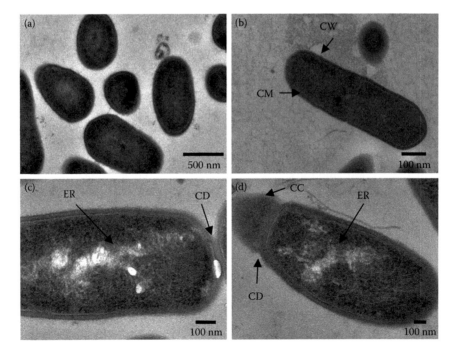

FIGURE 13.4
TEM micrographs by ultrathin sectioning of *Listeria monocytogenes*. Images (a–b) represent untreated cells, and images (c–d) show cells in the presence of caprylic-loaded nanoparticles. CW: cell wall; CM: cell membrane; ER: empty regions; CC: cytoplasmic content; CD: cell wall and membrane damage.

of the bacterium given the disruption of the cell membrane and the leakage of intracellular contents (Ruiz-Rico et al. 2015).

Besides natural organic molecules such as EO components, organosulfur compounds and fatty acids, encapsulation of antibiotics, such as amoxicillin, gentamicin, erythromycin, vancomycin, rifampicin, linezolid, and tetracycline in the MSPs also allows the preparation of novel biopesticides (Doadrio et al. 2004, Vallet-Regí et al. 2004, Li et al. 2010, Molina-Manso et al. 2012, Capeletti et al. 2014). Amoxicillin has been entrapped in bare MSPs (Vallet-Regí et al. 2004) and 3-aminopropyl trimethoxysilane and 3-chloropropyltriethoxysilane functionalized MSPs (Li et al. 2010). It was also possible to modulate the delivery of the antibiotic by functionalization, while release was based on the diffusion mechanism. These studies have revealed that an *in vitro* controlled release of amoxicillin in mesoporous material has advantages over traditional administration forms (capsules, tablets), where antibiotic release is faster and not controlled. Xue and Shi (2004) studied gentamicin encapsulation in an SBA-15 solid functionalized with a poly (DL-lactide-co-glycolide) (PLGA). This hybrid support produced an initial burst release for 1 day, followed by a sustained release over a 5-week period due to polymer degradation. Another functionalization strategy for controlled delivery has been evaluated by Doadrio et al. (2006), who studied the functionalization of SBA-15 with hydrophobic groups (octyl and octadecyl groups) to control erythromycin release. This functionalization produced reduced loading and induced slower delivery kinetics because of the support's diminished wettability. More recently, entrapment of vancomycin, rifampicin, and linezolid in SBA-15 has been evaluated against a strain of *S. aureus*, which produced biofilms. This support was able to release high antibiotic concentrations in the initial stage of experiments, with concentrations well over the breakpoints (Molina-Manso et al. 2012). Capeletti et al. (2014) reported the use of silica nanospheres to entrap tetracycline and found that antibacterial activity was produced in the presence of the antibiotic in the support's pores.

As mentioned in the section on metallic nanoparticles, Ag NPs have inhibitory properties against bacteria and are used as antimicrobial agents. Thus, their incorporation into MSPs, such as hollow spheres (Lin et al. 2013, Shen et al. 2014), MCM-41 materials (Li et al. 2013), or the SBA-15 matrix (Min et al. 2010, Naik et al. 2011, Wang et al. 2014), has been investigated. Ag NPs can be attached to the matrix surface or supported on the interior wall of MSPs. Loading Ag NPs in pores can protect the inner Ag NP core from the outer environment and can increase its chemical stability, thus avoiding aggregation before acting as antimicrobial agents. The incorporation of Ag NPs into the interior of MSPs allows Ag^+ ions to be released slowly from the interior of pores, which results in high antibacterial durability. Long-term antibacterial activity can remain strong for long periods (Shen et al. 2014). The inhibitory effect of these loaded MSPs is due to the formation of extracellular ROS and the toxicity of Ag^+ ions, which induce the production of intracellular ROS and cell death (Wang et al. 2014). The studies of Liong et al. (2009) and Yang et al. (2012) fabricated a mesoporous silica capsule, which was assembled on previously prepared Ag nanocrystals or spheres. This different approach allowed Ag to be obtained with similar properties to the previously described nanoparticles NPs, such as good antibacterial activity, slow Ag^+ ion release over long times, and low aggregation tendency.

Zinc oxide NPs have aggregation problems in the same way as Ag NPs do. ZnO NPs are hydrophobic compounds; thus, their application in aqueous media is usually restricted due to an aggregation process. Controlling the size and morphology of NPs by incorporating them into an inert nanoporous inorganic matrix like MSPs can avoid the aggregation effect, which facilitates NPs handling and reduces health risks. The channels of nanoporous materials with a diameter between 0.3 and 10 nm limit the growth of ZnO NPs and diminish their agglomeration. In the study of Copcia et al. (2012), a zeolite and SBA-15 were loaded with ZnO NPs. The antibacterial activity of ZnO NPs was dependent on the size and concentration of NPs. The results showed that SBA-15 allowed smaller-sized ZnO NPs to be obtained. Therefore, this support had greater antibacterial activity than ZnO-zeolite.

13.3.2.2 Antimicrobial Functionalized MSPs

As mentioned above, MSPs not only have voids capable of entrapping active compounds, but also present a large surface capable of reacting with organic molecules. This process is commonly known as functionalization. Antimicrobial functionalized MSPs are particles that display antimicrobial properties because of the effect of molecules that are attached to the surface of supports. This approach to create antimicrobial devices is really new and in accordance there are very few references of this type of antimicrobial MSPs found in the literature.

MSPs coated with didodecyldimethylammonium bromide (DDAB) showed strong antimicrobial activity against bacteria, yeast, molds, and viruses. The nanoparticles coated with quaternary ammonium cationic surfactant had a lower minimum inhibitory concentration than that obtained for free DDAB. This improvement may be due to differences in the size, charge density and the stiffness of both formulations, which may have an impact on their interaction with biological systems. It has also been reported that coated MSPs can be reused without losing their antimicrobial activity, and that DDAB did not show leaching. Thus the antimicrobial effect is mediated mainly by contact (Botequim et al. 2012).

In a later study, Li and Wang (2013) reported lysozyme-coated mesoporous silica nanoparticles (Lys-MSNs) to be antibacterial agents that exhibit efficient antibacterial activity against *E. coli* both *in vitro* and *in vivo,* and had low cytotoxicity and negligible hemolytic side effects. Bacterial growth was 5-fold lower for Lys-MSNs than that of free lysozyme (Lys) *in vitro*. These particles were able to reduce the microorganism by two orders of magnitude using an intestine-infected mouse model compared to the untreated group. This enhanced antibacterial efficacy originated from the Lys corona, and presented a multivalent interaction between the Lys of the functionalized nanoparticles and peptidoglucans of the bacterial walls. This multivalent interaction increased the local Lys concentration on the surface of cell walls, which produced the hydrolysis of peptidoglycans and damaged the bacterial cell wall.

The use of particles functionalized with an antibiotic has been recently studied by Qi et al. (2013). They investigated the development of vancomycin-modified mesoporous silica nanoparticles (Van-MSNs) for targeting and killing Gram-positive bacteria on macrophage-like cells. Vancomycin shows specific hydrogen bonding interactions towards the terminal D-alanyl-D-alanine moieties of Gram-positive bacteria, as well as inhibited bacterial cell wall synthesis capability. The vancomycin corona on Van-MSNs provided multivalent interactions and enhanced specific recognition towards Gram-positive bacteria. The high local vancomycin concentration on surfaces enabled NPs to strongly immobilize on the cell walls of *S. aureus,* which resulted in a significant inhibition effect on cell wall synthesis. Van-MSNs also showed a weak recognition with Gram-negative bacteria, which was attributed to the non-specific electrostatic interactions between Van-MSNs and the negatively charged bacterial membranes. Detectable non-specific binding of Van-MSNs with macrophage-like cells was not observed.

Pędziwiatr-Werbicka et al. (2014) synthesized novel silicates using silylated natural fatty acids as co-condensing reagents to produce functionalized MSPs. Different mono- and bis-silylated fatty acid precursors were used to modify the SBA-15 structure. The obtained functionalized materials displayed pronounced hydrophobicity owing to the presence of fatty length chains. The antimicrobial activity of fatty acid tethered silicates at different concentrations was evaluated. The result revealed that the inhibitory effect was sensitive to the nature of the fatty acid precursor, which reached a maximum inhibition of 50% compared to the control samples.

13.3.2.3 Antimicrobial Loaded and Functionalized Mesoporous Silica Particles

The combination of both previously described mechanisms, loading and functionalizing MSPs, could well be another effective system against microbial growth. This section presents reported studies that have used an antimicrobial compound entrapped in supports functionalized on their external surface with antimicrobial molecules. The presence and activity of both compounds enhance them synergistically. A schematic representation of antimicrobial-loaded and functionalized MSPs is offered in Figure 13.5. As this figure illustrates, the interaction between the coating of MSPs and bacteria allowed cargo release and thus cell damage and microorganism death due to the combined effect of attached and entrapped molecules.

One of the examples found in the literature is the development of MSPs loaded with peracetic acid (PAA) and silver NPs that decorate the structure of particles and have a strong synergistic bactericidal effect on antibiotic-resistant and biofilm-forming *S. aureus* (Carmona et al. 2012). Minimum reduction was observed for PAA- or Ag-loaded materials, which indicates limited antimicrobial action of the released compounds when used alone. Conversely, when the Ag-PAA-SBA-15 samples were used in bactericidal tests, a clear synergistic effect was observed with total bacteria inhibition. The synergic effect was probably produced by a deaggregation effect on the biofilm caused by PAA increasing the exposed surface of the bacteria susceptible to being affected by Ag^+ ions, and the amplified Ag^+ production was likely due to the accelerated dissolution of Ag NPs.

Another example is a study by Mas et al. (2013), in which MCM-41 nanoparticles were loaded with vancomycin and capped with ε-poly-L-lysine (ε-PL). The developed NPs showed enhanced ε-PL toxicity against Gram-negative bacteria. ε-PL is a cationic polymer with antimicrobial activity against Gram-negative bacteria, while vancomycin is an antibiotic that has an inhibitory effect against Gram-positive bacteria, as mentioned above. ε-PL attached to NPs, which were positively charged, and was bound to the negatively charged cell wall to induce bacterial wall damage, and allowed entrapped vancomycin to gain access to the cell. The achieved antibacterial activity indicated a strong synergistic effect when the nanoformulation was used.

More recently, MSNs, with different aspect ratios, loaded with silver ions and coated with chitosan, were designed as new antimicrobial nanodevices (Karaman et al. 2016). The antimicrobial activity of the materials was studied against different Gram-positive and Gram-negative bacteria, resulting in an enhancement of the antimicrobial effect for

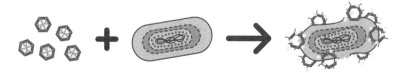

FIGURE 13.5
Representation of the synergistic effect of loaded and functionalized MSPs.

E. coli. The synergistic effect resulted from the combination of aspect ratio, Ag ion release, and electrostatic interaction of chitosan molecules and microbial cells. The highest-aspect-ratio MSNs showed the maximal antimicrobial activity against the different tested bacteria because they aligned their long axes parallel to the cell membrane, resulting in increased attachment and thus higher microbial membrane disruption. The release of Ag ion produced the previously described antimicrobial effects (see the section on metallic nanoparticles) and increased the positive charge density of chitosan due to chelating process, which improves the electrostatic interaction between chitosan and membrane cells.

13.4 Limitations of Nanomaterials for Food Applications

Strategies for the development of new antimicrobial agents based on nanostructured particles have been presented in previous sections. As has been shown, these systems improve the antimicrobial effect of non-nanostructured equivalent systems as well as increasing the number of benefits (improved bioavailability and stability, controlled release, masking unpleasant properties, etc.).

Other uses of nanomaterials in the food industry include food ingredients, food additives, food packaging, supplements, and food contact surfaces. Engineered nanomaterials are used as food additives with different applications. TiO_2 (E-171) NPs are used to preserve white color, to prevent agglomeration, and to improve handling in food products. Calcium, magnesium, and iron-based NPs are used as health supplements (Smolkova et al. 2015). Ag NPs are the most frequently used metallic NP as antimicrobial protection in water filtration systems, packaging, and dietary supplements (Vance et al. 2015). Organic nanoencapsulation systems are used as carriers for food ingredients, flavoring or supplements, or are incorporated in food packaging. Regarding inorganic encapsulation systems, Si-based materials are not considered dangerous and have largely been used in the agrifood sector. For instance, SiO_2 (E-551) is used as a carrier and a clarifying and anticaking agent (Jaganathan and Godin 2012), and nanoclays are applied in food packaging to improve the barrier properties of the films (Smolkova et al. 2015).

However, despite the large amount of research showing the advantages of antimicrobial systems based on nanomaterials, few examples of nanostructured antimicrobial agents can be found on the market. Factors that should be considered before incorporating nanomaterials into food are the stability of supports, the biocompatibility of systems with food matrices, and their potential toxicity.

13.4.1 Nanomaterial Stability

As previously mentioned, inorganic nanoparticles exhibit good stability, which would allow them to be incorporated into different food chain stages (Trewyn et al. 2007, Dizaj et al. 2014). However, some food-grade nanoencapsulation carriers, like nanoemulsions or liposomes, are unstable systems that tend to break down over time (Weiss et al. 2009).

13.4.2 Nanomaterials Biocompatibility

Another factor that should be considered for food applications is the biocompatibility of nanomaterials with the food matrix. Introducing these nanomaterials into food may affect the properties of the food product (physicochemical and sensory characteristics). Nevertheless, there are practically no data about the influence of metallic NPs on the features of food products. Some studies into food-grade NPs have indicated the impact of nanoencapsulation carriers on some chemical and physical characteristics of food matrices (Benech et al. 2002, Donsi et al. 2011). More references to the indirect effect of food packaging that contains metallic NPs or organic nanoencapsulation supports on some shelf life parameters of food products can be found (An et al. 2008, Emamifar et al. 2011, Li et al. 2011, Kim et al. 2014, Salvia-Trujillo et al. 2015b). As far as we know, only one study has established the influence of different MSPs on the physical properties of a food model system (Pérez-Esteve et al. 2014). Its results have shown that functionalized particles improve the filler-matrix interaction and modify some physical properties of gelatine gels. Nevertheless, the amount of MSPs required for food applications is very small due to their high load and functionalization capacity. So, it has been assumed that the concentrations used should not affect the properties of food products.

13.4.3 Nanomaterial Toxicity

As introducing NP-based products in the last few years has grown, a better understanding of the potential toxicity of NPs in diverse biological systems is required before applying them in the food industry.

NPs could enter the food chain directly as antimicrobial agents or as delivery systems of bioactive compounds, such as nutrient or antimicrobial compounds (Dasgupta et al. 2015). Another factor that should be considered is the potential indirect contamination of food when NPs are incorporated into packaging materials for the purpose of prolonging shelf life or detecting some components of food products due to potential migrations (Bouwmeester et al. 2009).

Some studies have assessed the potential migration of antimicrobial metallic NP-coated surfaces into food products or food simulants, but results are inconsistent (Metak et al. 2015, Hannon et al. 2016). Metak et al. (2015) tested commercial nanocomposites with Ag NPs in real food matrices, obtaining nonsignificant released Ag ions but the presence of Ag NPs in some food products. In contrast, Hannon et al. (2016) determined ionic silver migration levels above the regulatory limits and absence of Ag NPs in the food simulant suggesting that the main mechanism for migration is dissolution of NPs from the packaging rather than the migration of NPs.

The safety of NPs for human health and the environment is still uncertain, but some leakage, accumulation, and toxicity problems have been described (Neal 2008, Werlin et al. 2011, Kumar and Dhawan 2013, Hannon et al. 2015). Furthermore, the accurate characterization of NPs *in vitro* and in complex food matrices is necessary because certain features (i.e., particle size, size distribution, potential agglomeration, and surface charge) can change in different biological matrices. However, in general, it has been observed that the toxicity of nanoparticles is selective. While prokaryotic cells are really sensitive to small concentrations of nanoparticles, eukaryotic cells have a larger tolerance range than bacteria before toxicity is observed (Reddy et al. 2007).

NPs included in food products undergo different processes and transformations after food has been ingested. Firstly, nanomaterials can certainly be released from the food matrix. In the gastrointestinal tract, NPs can undergo a variety of transformations, such as agglomeration, aggregation, adsorption, or binding with other food components, and might react with acid and digestive enzymes. Afterward, they may pass through the gastrointestinal epithelium and end up in the systemic blood circulation during a multistep process. In a first step, diffusion of NPs through the first barrier takes place, this being the mucus that lines the gut wall, which is followed by contact with enterocytes or M-cells of the gastrointestinal epithelium by cellular or paracellular transport, and finally by post-translocation events. Absorption, metabolism, distribution, and excretion (ADME) profiles depend on the properties of NPs, such as size (smaller particles are absorbed faster than larger ones), surface charge, or functionalized groups (Rossi et al. 2014). Exposure to NPs may lead to oxidative damage and inflammatory reactions of the gastrointestinal tract (Handford et al. 2014). Having entered the bloodstream, NPs are quickly distributed throughout the body and produce internalization in tissues and retention in cells and organelles. Scientific evidence has shown a widespread distribution of NPs inside animal bodies (e.g., brain, bone marrow, spleen, kidney, and liver). The potential passage of natural barriers, like cellular barriers, the blood-brain barrier, the placental barrier, and the blood-milk barrier, is crucial for risk assessments (Bouwmeester et al. 2009). In general, long-term exposure to nanomaterials has been related to kidney and liver lesions, different cancer types, and cytotoxicity and genotoxic effects (Rossi et al. 2014).

Some properties of engineered NPs, such as the ability to form suspensions and high diffusion rates, allow their dispersion in the environment, which increases the risk of environmental impact (Kumar and Dhawan 2013). The surface properties of metallic NPs are another feature that determines the potential toxicological effect of nanomaterials (Reddy et al. 2012). The surface properties of MNPs influence their interactions with the biological system. Neutral and negatively charged MNPs interact less with plasma proteins, which produces nonspecific binding that can cause cell lysis. Positively charged MNPs interact strongly with blood components, which causes rapid clearance from systemic circulation and leads to nonspecific tissue uptake (Reddy et al. 2012). Once NPs access the body, both internalization in cells and interaction with cellular organelles and macromolecules are quite likely because of their small size. These interactions can damage genetic material and cellular organelles by physical injury and by modulating biochemical pathways (Kumar and Dhawan 2013). Moreover, indirect damage can be caused by the generation of ROS and oxidative stress, which induce lipid peroxidation and also damage mitochondria, macromolecules, and cell membranes, and can then induce inflammatory responses, tissue degradation, carcinogenesis, ageing, and other diseases (Johnston et al. 2010; Kumar and Dhawan 2013). When NPs are ingested, they are transported and accumulate in different organs, of which the liver and intestinal tract are the primary deposition site and target organs (Gaillet and Rouanet 2015). Although the toxicity mechanism of metallic NPs is unclear, the main hypotheses suggest that toxicity can derive from small-sized NPs or through the release of ions from particles, or perhaps be due to a combination of both concepts since smaller particles are likely to allow the better release of ions (Johnston et al. 2010). In-vitro and in-vivo studies have been performed, but discrepancies between both methods do not provide a clear opinion of the potential toxicity of these nanomaterials (Sharma et al. 2012, Tsoi et al. 2012, Shukla et al. 2013, Hannon et al. 2016).

Despite there now being little evidence for the potential toxicity of food-grade nanoemulsions, some physicochemical and physiological mechanisms associated with their small particle size may potentially cause toxicity (McClements and Rhao 2011). The cytotoxicity of food-grade nanoemulsions has been studied by Yu and Huang (2013) with Caco-2 cell monolayers and HepG2 cells to mimic the small intestine epithelium and liver hepatocytes, respectively, as the most important tissues exposed. These authors found that nanoemulsions revealed no significant difference from their micron-sized emulsions (used as a control), which suggests no apparent toxicity of nanoemulsions to the small intestine.

However, their results also showed that nanoemulsions affected the cell viability of HepG2 cells more than micron-sized emulsions did. Moreover, certain components used to prepare nanoemulsions, such as emulsifiers or solvents, are toxic when consumed at sufficiently high levels. Another factor to bear in mind is the enhancement of the bioavailability of some bioactive compounds that normally possess low bioavailability. This improvement in bioaccesibility could produce toxic effects, which were not predicted from the data obtained on the nonencapsulated compound (McClements and Rhao 2011).

Regarding MSPs, the modification of some features of these materials, such as particle size, may produce changes in their biocompatibility and toxicological properties. Therefore, different *in-vitro* and *in-vivo* studies have been performed in the last few years to elucidate the potential toxicity of silica supports. The results are conflicting and it is necessary to establish standard protocols to accurately extrapolate the results (Xu et al. 2012). The potential toxicological effects of MSPs depends on some characteristics, such as particle size, shape, mesoporosity, aggregation rate, surface modifications, and dosage (Asefa and Tao 2012). The most important factor to determine if MSPs result in harmful effects is particle size. Microsized MSPs (>100 nm) are absorbed and distributed in different organs, but can be easily decomposed and eliminated after a few days of oral exposure (Fu et al. 2013; Lee et al. 2014). The relationship between nanomaterial size and dosage is quite important for determining the possible biological response. The amount of MSPs that can cause, or not, a toxicological effect can vary for a given MSP depending on its size (Asefa and Tao 2012). Surface modifications of MSPs also seem significant given their potential toxicology. Surface functionalization with organic moieties is quite important for enhancing biocompatibility and reducing collisions of particles with some undesired organelles (Tang et al. 2012). Another feature that should be taken into account is mesoporosity. This structural character is also substantially responsible for the lower rigidity of MSPs, which buffer their interactions with biological entities (Asefa and Tao 2012).

A clear understanding of the possible health effects of nanoparticles is still unavailable, and this is evidently a limitation of the widespread use of these clearly extraordinary nanobiotechnologies. If the problems associated with the widespread use of nanomaterials, such as robust functionalization techniques and simple cost-effective fabrication, can be overcome, the potential of these nanotechnologies will be vast (Gilmartin and O'Kennedy 2012).

13.5 Final Remarks

Nanostructured materials exhibit unique physicochemical properties that offer the opportunity for creating new technologies for manufacturing, packaging, and storage of food products (Dasgupta et al. 2015). Metallic nanoparticles display better antimicrobial effectiveness than conventional preservatives because of the higher surface area-to-volume ratio of NPs, among other features. Nanoscalar delivery systems allow the encapsulation or attachment of antimicrobial compounds with high volatility or poor solubility, which protect them from environmental degradation. Nanoencapsulation systems also have a good loading capacity that allow antimicrobial inhibitory concentrations to lower, as well as the controlled or stimuli-responsive delivery of the antimicrobial compound at the site of action. Figure 13.6 summarizes the most relevant features of each nanoscalar delivery system.

Carrier		Remarkable features
	Nanoemulsions	Slightly turbid Increased bioavailability and controlled delivery Tendency to break down Susceptible to chemical degradation
	Microemulsions	Transparent High stability Increased bioavailability
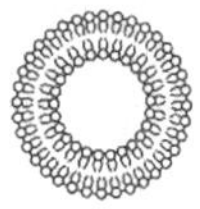	Liposomes	Biocompatibility Easy functionalization Increased bioavailability
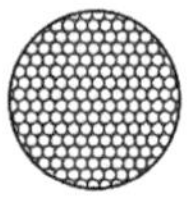	Solid lipid nanoparticles	High stability Increased bioavailability
	Nanofibers	Mechanical performance Entrapment of other encapsulation systems
	Mesoporous silica particles	High stability Controlled relaease Easy functionalization Delivery in the site of action

FIGURE 13.6
Overview of the features of different nanoencapsulation systems.

Despite these advantages, application of nanotechnology in the agrifood sector is still a relatively new concept given issues relating to product labeling, potential health risks, and lack of unifying regulations (Handford et al. 2014). Currently, there are major gaps in knowledge on nanomaterials in the food sector that need further research. First, a clear strict definition of nanoparticles is needed. Validated methods for *in situ* detection and characterization of nanomaterials in complex food matrices should be developed by, ideally, using relatively easily procedures with equipment that is currently available in most laboratories. Precise toxicology and ADME profile studies should be conducted. The long-term health consequences of ingesting nanoparticles via food have to be investigated. Accurate risk assessment methodologies should be provided by scientific entities. Finally, adequate regulations on nanotechnology applications for food and related products should be developed by the regulatory authorities. Once these limitations have been solved, approved nanomaterials might be used as safe technologies in the food industry in order to make the most of their excellent antimicrobial properties.

Acknowledgments

The authors gratefully acknowledge the financial support from the Ministerio de Economía y Competitividad and FEDER-EU (Project AGL2015-70235-C2-1-R, [MINECO/FEDER]). M.R.R. is grateful to the Ministerio de Educación, Cultura y Deporte for her grant (AP2010-4369).

References

Acevedo-Fani, A., Salvia-Trujillo, L., Rojas-Graü, and Martín-Belloso, O. 2015. Edible films from essential-oil-loaded nanoemulsions: Physicochemical characterization and antimicrobial properties. *Food Hydrocolloids* 47:168–177.

Ahmed, K., Li, Y., McClements, D. J., and Xiao, H. 2012. Nanoemulsion- and emulsion-based delivery systems for curcumin: Encapsulation and release properties. *Food Chemistry* 132:799–807.

Akbar, A. and Anal, A. K. (2014). Zinc oxide nanoparticles loaded active packaging, a challenge study against *Salmonella typhimurium* and *Staphylococcus aureus* in ready-to-eat poultry meat. *Food Control* 38:88–95.

Almeida, A. J., Runge, S., and Müller, R. H. 1997. Peptide-loaded solid lipid nanoparticles (SLN): Influence of production parameters. *International Journal of Pharmaceutics*, 149:255–265.

Al Shamsi, M., Al Samri, M. T., Al-Salam, S. et al. 2010. Biocompatibility of calcined mesoporous silica particles with cellular bioenergetics in murine tissues. *Chemical Research in Toxicology* 23:1796–1805.

An, J., Zhang, M., Wang, S., and Tang, J. 2008. Physical, chemical and microbiological changes in stored green asparagus spears as affected by coating of silver nanoparticles-PVP. *LWT-Food Science and Technology* 41(6):1100–1107.

Applerot, G., Lipovsky, A., Dror, R., Perkas, N., Nitzan, Y., Lubart, R., and Gedanken, A. 2009. Enhanced antibacterial activity of nanocrystalline ZnO due to increased ROS-mediated cell injury. *Advanced Functional Materials*, 19(6), 842–852.

Arfat, Y. A., Ahmed, J., Hiremath, N., Auras, R., and Joseph, A. 2017. Thermo-mechanical, rheological, structural and antimicrobial properties of bionanocomposite films based on fish skin gelatin and silver-copper nanoparticles. *Food Hydrocolloids* 62:191–202.

Asefa, T. and Tao, Z. 2012. Biocompatibility of mesoporous silica nanoparticles. *Chemical Research in Toxicology* 25(11):2265–2284.

Azam, A., Ahmed, A. S., Oves, M., Khan, M. S., Habib, S. S., and Memic, A. 2012. Antimicrobial activity of metal oxide nanoparticles against Gram-positive and Gram-negative bacteria: A comparative study. *International Journal of Nanomedicine* 7:6003.

Balcão, V. M., Costa, C. I., Matos, C. M., Moutinho, C. G., Amorim, M., Pintado, M. E., Gomes, A. P., Vila, M. M., and Teixeira, J. A. 2013. Nanoencapsulation of bovine lactoferrin for food and biopharmaceutical applications. *Food Hydrocolloids* 32:425–431.

Beck, J. S., Vartuli, J. C., Roth, W. J. et al. 1992. A new family of mesoporous molecular sieves prepared with liquid crystal templates. *Journal of the American Chemical Society* 114(27):10834–10843.

Beigmohammadi, F., Peighambardoust, S. H., Hesari, J., Azadmard-Damirchi, S., Peighambardoust, S. J., and Khosrowshahi, N. K. 2016. Antibacterial properties of LDPE nanocomposite films in packaging of UF cheese. *LWT-Food Science and Technology* 65:106–111.

Benech, R. O., Kheadr, E. E., Laridi, R., Lacroix, C., and Fliss, I. 2002. Inhibition of *Listeria innocua* in cheddar cheese by addition of nisin Z in liposomes or by *in situ* production in mixed culture. *Applied and Environmental Microbiology* 68(8):3683–3690.

Bernardos, A., Marina, T., Žáček, P., Pérez-Esteve, É., Martínez-Máñez, R., Lhotka, M., Kouřimská, L., Pulkrábek, J., and Klouček, P. 2015. Antifungal effect of essential oil components against *Aspergillus niger* when loaded into silica mesoporous supports. *Journal of the Science of Food and Agriculture* 95:2824–2831.

Bhadra, P., Mitra, M. K., Das, G. C., Dey, R., and Mukherjee, S. 2011. Interaction of chitosan capped ZnO nanorods with *Escherichia coli*. *Materials Science and Engineering: C* 31(5):929–937.

Bhargava, K., Conti, D. S., da Rocha, S. R. P., and Zhang, Y. 2015. Application of an oregano oil nanoemulsion to the control of foodborne bacteria on fresh lettuce. *Food Microbiology* 47:69–73.

Bodaghi, H., Mostofi, Y., Oromiehie, A., Zamani, Z., Ghanbarzadeh, B., Costa, C., Conte, A., and Del Nobile, M. A. 2013. Evaluation of the photocatalytic antimicrobial effects of a TiO_2 nanocomposite food packaging film by *in vitro* and *in vivo* tests. *LWT-Food Science and Technology* 50(2):702–706.

Botequim, D., Maia, J., Lino, M. M. F., Lopes, L. M. F., Simoes, P. N., Ilharco, L. M., and Ferreira, L. 2012. Nanoparticles and surfaces presenting antifungal, antibacterial and antiviral properties. *Langmuir* 28(20):7646–7656.

Bouwmeester, H., Dekkers, S., Noordam, M. Y., Hagens, W. I., Bulder, A. S., de Heer, C., ten Voorde, S. E. C. G., Wijnhoven, S. W. P., Marvin, H. J. P., and Sips, A. J. A. M. 2009. Review of health safety aspects of nanotechnologies in food production. *Regulatory Toxicology and Pharmacology* 53:52–62.

Brayner, R., Ferrari-Iliou, R., Brivois, N., Djediat, S., Benedetti, M. F., and Fiévet, F. 2006. Toxicological impact studies based on *Escherichia coli* bacteria in ultrafine ZnO nanoparticles colloidal medium. *Nano Letters* 6(4):866–870.

Burt, S. 2004. Essential oils: Their antimicrobial properties and potential applications in food—A review. *International Journal of Food Microbiology* 94:223–253.

Cao, Z., Yang, L., Yan, Y., Shang, Y., Ye, Q., Qi, D., Ziener, U., Shan, G., and Landfester, K. 2013. Fabrication of nanogel core-silica shell and hollow silica nanoparticles via an interfacial sol-gel process triggered by transition-metal salt in inverse systems. *Journal of Colloid and Interface Science* 406:139–147.

Capeletti, L. B., de Oliveira, L. F., Gonçalves, K. D. A., de Oliveira, J. F. A., Saito, A., Kobarg, J., dos Santos, J. J. Z., and Cardoso, M. B. 2014. Tailored silica-antibiotic nanoparticles: Overcoming bacterial resistance with low cytotoxicity. *Langmuir* 30:7456–7464.

Cardenas, G., Diaz, J., Melendrez, M. F., Cruzat, C., and Garcia Cancino, A. 2009. Colloidal Cu nanoparticles/chitosan composite film obtained by microwave heating for food package applications. *Polymer Bulletin* 62:511–524.

Carmona, D., Lalueza, P., Balas, F., Arruebo, M., and Santamaría, J. 2012. Mesoporous silica loaded with peracetic acid and silver nanoparticles as a dual-effect, highly efficient bactericidal agent. *Microporous and Mesoporous Materials* 161:84–90.

Cerrada, M. L., Serrano, C., Sánchez-Chaves, M., Fernández-García, M., Fernández-Martín, F., de Andres, A., Jiménez-Riobóo, R. J., Kubacka, A., Ferrer, M., and Fernández-García, M. 2008. Self-sterilized EVOH-TiO_2 nanocomposites: Interface effects on biocidal properties. *Advanced Functional Materials* 18:1949–1960.

Chang, Y., McLandsborough, L., and McClements, D. J. 2012. Physical properties and antimicrobial efficacy of thyme oil nanoemulsions: Influence of ripening inhibitors. *Journal of Agricultural and Food Chemistry* 60:12056–12063.

Chawengkijwanich, C. and Hayata, Y. 2008. Development of TiO_2 powder-coated food packaging film and its ability to inactivate *Escherichia coli in vitro* and in actual tests. *International Journal of Food Microbiology* 123:288–292.

Chen, H., Zhang, Y., and Zhong, Q. 2015. Physical and antimicrobial properties of spray-dried zein–casein nanocapsules with co-encapsulated eugenol and thymol. *Journal of Food Engineering* 144:93–102.

Chorianopoulos, N. G., Tsoukleris, D. S., Panagou, E. Z., Falaras, P., and Nychas, G. J. E. 2011. Use of titanium dioxide (TiO_2) photocatalysts as alternative means for *Listeria monocytogenes* biofilm disinfection in food processing. *Food Microbiology* 28:164–170.

Colas, J. C., Shi, W. L., Rao, V. S. N. M., Omri, A., Mozafari, M. R., and Singh, H. 2007. Microscopical investigations of nisin-loaded nanoliposomes prepared by Mozafari method and their bacterial targeting. *Micron* 38:841–847.

Copcia, V. E., Gradinaru, R., Mihai, G. D., Bilba, and Sandu, I. 2012. Antibacterial activity of nanosized ZnO hosted in microporous clinoptilolite and mesoporous silica SBA-15 matrices. *Revista De Chimie* 63(11):1124–1131.

Dasgupta, N., Ranjan, S., Mundekkad, D., Ramalingam, C., Shanker, R., and Kumar, A. 2015. Nanotechnology in agro-food: From field to plate. *Food Research International* 69:381–400.

da Silva Malheiros, P., Daroit, D. J., da Silveira, N. P., and Brandelli, A. 2010. Effect of nanovesicle-encapsulated nisin on growth of *Listeria monocytogenes* in milk. *Food Microbiology* 27(1):175–178.

Daz-Visurraga, J., Melendrez, M. F., Garcia, A., Paulraj, M., and Cardenas, G. 2010. Semitransparent chitosan-TiO_2 nanotubes composite film for food package applications. *Journal of Applied Polymer Science* 116:3503–3515.

De Moura, M. R., Mattoso, L. H., and Zucolotto, V. 2012. Development of cellulose-based bactericidal nanocomposites containing silver nanoparticles and their use as active food packaging. *Journal of Food Engineering* 109(3):520–524.

Degnan, A. J. and Luchansky, J. B. 1992. Influence of beef tallow and muscle on the antilisterial activity of pediocin AcH and liposome-encapsulated pediocin AcH. *Journal of Food Protection* 55:552–554.

Del Nobile, M. A., Cannarsi, M., Altieri, C., Sinigaglia, M., Favia, P., Iacoviello, G., and D'agostino, R. 2004. Effect of Ag-containing nanocomposite active packaging system on survival of *Alicyclobacillus acidoterrestris*. *Journal of Food Science* 6:379–383.

Dizaj, S. M., Lotfipour, F., Barzegar-Jalali, M., Zarrintan, M. H., and Adibkia, K. 2014. Antimicrobial activity of the metals and metal oxides nanoparticles. *Materials Science and Engineering: C* 44:278–284.

Doadrio, A. L., Sousa, E. M. B., Doadrio, J. C., Pérez Pariente, J., Izquierdo-Barba, I., and Vallet-Regí, M. 2004. Mesoporous SBA-15 HPLC evaluation for controlled gentamicin drug delivery. *Journal of Controlled Release* 97(1):125–132.

Doadrio, J. C., Sousa, E. M., Izquierdo-Barba, I., Doadrio, A. L., Perez-Pariente, J., and Vallet-Regí, M. 2006. Functionalization of mesoporous materials with long alkyl chains as a strategy for controlling drug delivery pattern. *Journal of Materials Chemistry* 16(5):462–466.

Donsi, F., Annunziata, M., Sessa, M., and Ferrrari, G. 2011. Nanoencapsulation of essential oils to enhance their antimicrobial activity in foods. *LWT—Food Science and Technology* 44:1908–1914.

Durán, N. and Marcato, P. D. 2013. Nanobiotechnology perspectives. Role of nanotechnology in the food industry: A review. *International Journal of Food Science and Technology* 48:1127–1134.

EFSA. 2013. The European Union Summary Report on antimicrobial resistance in zoonotic and indicator bacteria from humans, animals and food in 2011. *EFSA Journal* 11(5):3196–3359. (http://www.efsa.europa.eu/en/efsajournal/pub/3196.htm). [30/07/2015]

El-Boubbou, K., Gruden, C., and Huang, X. 2007. Magnetic glyco-nanoparticles: A unique tool for rapid pathogen detection, decontamination, and strain differentiation. *Journal of the American Chemical Society* 129(44):13392–13393.

el Haskouri, J., Ortiz de Zárate, D., Guillem, C. et al. 2002. Silica-based powders and monoliths with bimodal pore systems. *Chemical Communications* 21(4):330–331.

Emamifar, A., Kadivar, M., Shahedi, M., and Soleimanian-Zad, S. 2010. Evaluation of nanocomposites packaging containing Ag and ZnO on shelf-life of fresh orange juice. *Innovative Food Science and Emerging Technologies* 11:742–748.

Emamifar, A., Kadivar, M., Shahedi, M., and Soleimanian-Zad, S. 2011. Effect of nanocomposite packaging containing Ag and ZnO on inactivation of *Lactobacillus plantarum* in orange juice. *Food Control* 22:408–413.

Eskandari, M., Haghighi, N., Ahmadi, V., Haghighi, F., and Mohammadi, S. R. 2011. Growth and investigation of antifungal properties of ZnO nanorod arrays on the glass. *Physica B: Condensed Matter* 406(1):112–114.

Fernández, A., Picouet, P., and Lloret, E. 2010a. Cellulose-silver nanoparticle hybrid materials to control spoilage-related microflora in absorbent pads located in trays of fresh-cut melon. *International Journal of Food Microbiology* 142(1):222–228.

Fernandez, A., Picouet, P., and Lloret, E. 2010b. Reduction of the spoilage-related microflora in absorbent pads by silver nanotechnology during modified atmosphere packaging of beef meat. *Journal of Food Protection* 73(12):2263–2269.

Firouzabadi, F. B., Noori, M., Edaltpanah, Y., and Mirhosseini, M. 2014. ZnO nanoparticle suspensions containing citric acid as antimicrobial to control *Listeria monocytogenes*, *Escherichia coli*, *Staphylococcus aureus* and *Bacillus cereus* in mango juice. *Food Control* 42:310–314.

Fortunati, E., Armentano, I., Zhou, Iannoni, A., Saino, E., Visai, L., Berglund, L. A., and Kenny, J. M. 2012. Multifunctional bionanocomposite films of poly (lactic acid), cellulose nanocrystals and silver nanoparticles. *Carbohydrate Polymers* 87(2):1596–1605.

Fu, C., Liu, T., Li, L., Liu, H., Chen, D., and Tang, F. 2013. The absorption, distribution, excretion and toxicity of mesoporous silica nanoparticles in mice following different exposure routes. *Biomaterials* 34(10):2565–2575.

Fu, X., Huang, B., and Feng, F. 2008. Shelf life of fresh noodles as affected by the food grade monolaurin microemulsion system. *Journal of Food Process Engineering* 31:619–627.

Gaillet, S. and Rouanet, J. M. 2015. Silver nanoparticles: Their potential toxic effects after oral exposure and underlying mechanisms–a review. *Food and Chemical Toxicology* 77:58–63.

Gaysinsky, S., Davidson, P. M., Bruce, B. D., and Weiss, J. 2005. Growth inhibition of *Escherichia coli* O157:H7 and *Listeria monocytogenes* by carvacrol and eugenol encapsulated in surfactant micelles. *Journal of Food Protection* 68(12):2559–2566.

Gaysinsky, S., Taylor, T. M., Davidson, P. M., Bruce, B. D., and Weiss, J. 2007. Antimicrobial efficacy of eugenol microemulsions in milk against *Listeria monocytogenes* and *Escherichia coli* O157:H7. *Journal of Food Protection* 70(11), 2631–2637.

Ghosh, S., Kaushik, R., Nagalakshmi, K., Hoti, S. L., Menezes, G. A., Harish, B. N., and Vasan, H. N. 2010. Antimicrobial activity of highly stable silver nanoparticles embedded in agar–agar matrix as a thin film. *Carbohydrate Research* 345(15):2220–2227.

Ghule, K., Ghule, A. V., Chen, B. J., and Ling, Y. C. 2006. Preparation and characterization of ZnO nanoparticles coated paper and its antibacterial activity study. *Green Chemistry* 8(12):1034–1041.

Gilmartin, N. and O'Kennedy, R. 2012. Nanobiotechnologies for the detection and reduction of pathogens. *Enzyme and Microbial Technology* 50(2):87–95.

Gomes, C., Moreira, R. G., and Castell-Perez, E. 2011. Poly (DL-lactide-co-glycolide) (PLGA) nanoparticles with entrapped trans-cinnamaldehyde and eugenol for antimicrobial delivery applications. *Journal of Food Science* 76(2):16–24.

Gopiraman, M., Jatoi, A. W., Hiromichi, S., Yamaguchi, K., Jeon, H. Y., Chung, I. M., and Soo, K. I. 2016. Silver coated anionic cellulose nanofiber composites for an efficient antimicrobial activity. *Carbohydrate Polymers* 149:51–59.

Gu, H., Ho, P. L., Tong, E., Wang, L., and Xu, B. 2003a. Presenting vancomycin on nanoparticles to enhance antimicrobial activities. *Nano Letters* 3(9):1261–1263.

Gu, H., Ho, P. L., Tsang, K. W. T., Wang, L., and Xu, B. 2003b. Using biofunctional magnetic nanoparticles to capture vancomycin-resistant enterococci and other gram-positive bacteria at ultralow concentration. *Journal of the American Chemical Society* 125:15702–15703.

Gutiérrez, J. M., González, C., Maestro, A., Solè, I., Pey, C. M., and Nolla, J. 2008. Nano-emulsions: New applications and optimization of their preparation. *Current Opinion in Colloid & Interface Science* 13:245–251.

Gyawali, R. and Ibrahim, S. A. 2014. Natural products as antimicrobial agents. *Food Control* 46:412–429.

Haghighi, F., Roudbar Mohammadi, S., Mohammadi, P., Hosseinkhani, S., and Shipour, R. 2013. Antifungal activity of TiO_2 nanoparticles and EDTA on *Candida albicans* biofilms. *Infection, Epidemiology and Medicine* 1(1):33–38.

Handford, C. E., Dean, M., Henchion, M., Spence, M., Elliott, C. T., and Campbell, K. 2014. Implications of nanotechnology for the agri-food industry: Opportunities, benefits and risks. *Trends in Food Science & Technology* 40:226–241.

Hannon, J. C., Kerry, J. P., Cruz-Romero, M., Azlin-Hasim, S., Morris, M., and Cummins, E. 2016. Human exposure assessment of silver and copper migrating from an antimicrobial nanocoated packaging material into an acidic food simulant. *Food and Chemical Toxicology* 95:128–136.

Hannon, J. C., Kerry, J. P., Cruz-Romero, M., Morris, M., and Cummins, E. 2015. Advances and challenges for the use of engineered nanoparticles in food contact materials. *Trends in Food Science & Technology* 43(1):43–62.

He, L., Liu, Y., Mustapha, A., and Lin, M. 2011. Antifungal activity of zinc oxide nanoparticles against *Botrytis cinerea* and *Penicillium expansum*. *Microbiological Research* 166(3):207–215.

Honda, H., Kawabe, A., Shinkai, M., and Kobayashi, T. 1998. Development of chitosan-conjugated magnetite for magnetic cell separation. *Journal of Fermentation and Bioengineering* 86(2):191–196.

Huang, L., Li, D. Q., Lin, Y. J., Wei, M., Evans, D. G., and Duan, X. 2005. Controllable preparation of Nano-MgO and investigation of its bactericidal properties. *Journal of Inorganic Biochemistry* 99(5):986–993.

Huang, Y. F., Wang, Y. F., and Yan, X. P. 2010. Amine-functionalized magnetic nanoparticles for rapid capture and removal of bacterial pathogens. *Environmental Science & Technology* 44(20):7908–7913.

Huang, Z., Zheng, X., Yan, D. et al. 2008. Toxicological effect of ZnO nanoparticles based on bacteria. *Langmuir* 24(8):4140–4144.

Huh, A. J. and Kwon, Y. J. 2011. "Nanoantibiotics": A new paradigm for treating infectious diseases using nanomaterials in the antibiotics resistant era. *Journal of Controlled Release* 156(2):128–145.

Ignatova, M., Starbova, K., Markova, N., Manolova, N., and Rashkov, I. 2006. Electrospun nanofibre mats with antibacterial properties from quaternised chitosan and poly (vinyl alcohol). *Carbohydrate Research* 341(12):2098–2107.

Imran, M., Revol-Junelles, A. M., René, N., Jamshidian, M., Akhtar, M. J., Arab-Tehrany, E., Jacquot, M., and Desobry, S. 2012. Microstructure and physico-chemical evaluation of nano-emulsion-based antimicrobial peptides embedded in bioactive packaging films. *Food Hydrocolloids*, 29:407–419.

Izquierdo-Barba, I., Vallet-Regí, M., Kupferschmidt, N., Terasaki, O., Schmidtchen, A., and Malmsten, M. 2009. Incorporation of antimicrobial compounds in mesoporous silica film monolith. *Biomaterials* 30(29):5729–5736.

Jaganathan, H. and Godin, B. 2012. Biocompatibility assessment of Si-based nano- and microparticles. *Advanced Drug Delivery Reviews* 64:1800–1819.

Jebel, F. S. and Almasi, H. 2016. Morphological, physical, antimicrobial and release properties of ZnO nanoparticles-loaded bacterial cellulose films. *Carbohydrate Polymers* 149:8–19.

Jiménez, A., Sánchez-González, L., Desobry, S., Chiralt, A., and Tehrany, E. A. 2014. Influence of nanoliposomes incorporation on properties of film forming dispersions and films based on corn starch and sodium caseinate. *Food Hydrocolloids* 35:159–169.

Jin, T. and Gurtler, J. B.. 2011. Inactivation of *Salmonella* in liquid egg albumen by antimicrobial bottle coatings infused with allyl isothiocyanate, nisin and zinc oxide nanoparticles. *Journal of Applied Microbiology* 110:704–712.

Jin, T. and He, Y. 2011. Antibacterial activities of magnesium oxide (MgO) nanoparticles against foodborne pathogens. *Journal of Nanoparticle Research* 13:6877–6885.

Jin, T., Sun, D., Su, J. Y., Zhang, H., and Sue, H. 2009. Antimicrobial efficacy of zinc oxide quantum dots against *Listeria monocytogenes*, *Salmonella* Enteritidis, and *Escherichia coli* O157:H7. *Journal of Food Science* 74(1):M46–52.

Jin, Y., Liu, F., Shan, C., Tong, M., and Hou, Y. 2014. Efficient bacterial capture with amino acid modified magnetic nanoparticles. *Water Research* 50:124–134.

Jo, Y. J., Chun, J. Y., Kwon, Y. J., Min, S. G., Hong, G. P., and Choi, M. J. 2015. Physical and antimicrobial properties of trans-cinnamaldehyde nanoemulsions in water melon juice. *LWT—Food Science and Technology* 60:444–451.

Johnston, H. J., Hutchison, G., Christensen, F. M., Peters, S., Hankin, S., and Stone, V. 2010. A review of the *in vivo* and *in vitro* toxicity of silver and gold particulates: Particle attributes and biological mechanisms responsible for the observed toxicity. *Critical Reviews in Toxicology* 40(4):328–346.

Kanmani, P. and Lim, S. T. 2013. Synthesis and characterization of pullulan-mediated silver nanoparticles and its antimicrobial activities. *Carbohydrate Polymers* 97(2):421–428.

Kanmani, P. and Rhim, J. W. 2014. Physicochemical properties of gelatin/silver nanoparticle antimicrobial composite films. *Food Chemistry* 148:162–169.

Karaman, D. Ş., Sarwar, S., Desai, D., Björk, E. M., Odén, M., Chakrabarti, P., Rosenholm, J. M., and Chakraborti, S. 2016. Shape engineering boosts antibacterial activity of chitosan coated mesoporous silica nanoparticle doped with silver: A mechanistic investigation. *Journal of Materials Chemistry B* 4(19):3292–3304.

Kasemets, K., Ivask, A., Dubourguier, H. C., and Kahru, A. 2009. Toxicity of nanoparticles of ZnO, CuO and TiO_2 to yeast *Saccharomyces cerevisiae*. *Toxicology in Vitro* 23(6):1116–1122.

Kayaci, F., Ertas, Y., and Uyar, T. 2013. Enhanced thermal stability of eugenol by cyclodextrin inclusion complex encapsulated in electrospun polymeric nanofibers. *Journal of Agricultural and Food Chemistry* 61:8156–8165.

Khare, A. R. and Vasisht, N. 2014. Nanoencapsulation in the food industry: Technology of the future. *Microencapsulation in the Food Industry*, ed. A.G.Gaonkar, N.Vasisht, and A.R.Khare, R.Sobel, pp. 151–155. Elsevier Inc.

Kim, I. H., Lee, H., Kim, J. E., Song, K. B., Lee, Y. S., Chung, D. S., and Min, S. C. 2013. Plum coatings of lemongrass oil-incorporating carnauba wax-based nanoemulsion. *Journal of Food Science* 78(10):1551–1559.

Kim, I. H., Oh, Y. A., Lee, H., Song, K. B., and Min, S. C. 2014. Grape berry coatings of lemongrass oil-incorporating nanoemulsion. *LWT—Food Science and Technology* 58:1–10.

Kriegel, C., Kit, K. M., McClements, D. J., and Weiss, J. 2009. Nanofibers as carrier systems for antimicrobial microemulsions. Part I: Fabrication and characterization. *Langmuir* 25:1154–1161.

Kriegel, C., Kit, K. M., McClements, D. J., and Weiss, J. 2010. Nanofibers as carrier systems for antimicrobial microemulsions. II. Release characteristics and antimicrobial activity. *Journal of Applied Polymer Science* 118:2859–2868.

Kumar, A. and Dhawan, A. 2013. Genotoxic and carcinogenic potential of engineered nanoparticles: An update. *Archives of Toxicology* 87(11):1883–1900.

Kumar, A., Pandey, A. K., Singh, S. S., Shanker, R., and Dhawan, A. 2011a. Engineered ZnO and TiO_2 nanoparticles induce oxidative stress and DNA damage leading to reduced viability of *Escherichia coli*. *Free Radical Biology and Medicine* 51(10):1872–1881.

Kumar, A., Pandey, A. K., Singh, S. S., Shanker, R., and Dhawan, A. 2011b. Cellular uptake and mutagenic potential of metal oxide nanoparticles in bacterial cells. *Chemosphere* 83(8):1124–1132.

Lai, F., Wissing, S. A., Müller, R. H., and Fadda, A. M. 2006. *Artemisia arborescens* L essential oil–loaded solid lipid nanoparticles for potential agricultural application: Preparation and characterization. *AAPS PharmSciTech* 7(1):E10–E18.

Landry, K. S., Chang, Y., McClements, D. J., and McLandsborough, L. 2014. Effectiveness of a novel spontaneous carvacrol nanoemulsion against *Salmonella enterica* Enteritidis and *Escherichia coli* O157:H7 on contaminated mung bean and alfalfa seeds. *International Journal of Food Microbiology* 187:15–21.

Landry, K. S., Micheli, S., McClements, D. J., and McLandsborough, L. 2015. Effectiveness of a spontaneous carvacrol nanoemulsion against *Salmonella enterica* Enteritidis and *Escherichia coli* O157:H7 on contaminated broccoli and radish seeds. *Food Microbiology* 51:10–17.

Laridi, R., Kheadr, E. E., Benech, R. O., Vuillemard, J. C., Lacroix, C., and Fliss, I. 2003. Liposome encapsulated nisin Z: Optimization, stability and release during milk fermentation. *International Dairy Journal* 13:325–336.

Lee, J. A., Kim, M. K., Paek, H. J., Kim, Y. R., Kim, M. K., Lee, J. K., Jeong, J., and Choi, S. J. 2014. Tissue distribution and excretion kinetics of orally administered silica nanoparticles in rats. *International Journal of Nanomedicine* 9:251–260.

Leung, Y. H., Ng, A., Xu, X., Shen, Z., Gethings, L. A., Wong, M. T., Chan, C. M. N., Guo, M. Y., Ng, Y. H., Djurišić, A. B., Lee, P. K. H., Chan, W. K., Yu, L. H., Phillips, D. L., Ma, A. P. Y., and Leung, F. C. C. 2014. Mechanisms of antibacterial activity of MgO: Non-ROS mediated toxicity of MgO nanoparticles towards *Escherichia coli*. *Small* 10(6):1171–1183.

Li, J. H., Hong, R. Y., Li, M. Y., Li, H. Z., Zheng, Y., and Ding, J. 2009a. Effects of ZnO nanoparticles on the mechanical and antibacterial properties of polyurethane coatings. *Progress in Organic Coatings* 64(4):504–509.

Li, L. L. and Wang, H. 2013. Enzyme-coated mesoporous silica nanoparticles as efficient antibacterial agents *in vivo*. *Advanced Healthcare Materials* 2(10):1351–1360.

Li, X., Li, W., Xing, Y., Jiang, Y., Ding, Y., and Zhang, P. 2011. Effects of nano-ZnO power-coated PVC film on the physiological properties and microbiological changes of fresh-cut "Fuji" apple. *Advanced Materials Research* 152–153:450–453.

Li, X., Xing, Y., Jiang, Y., Ding, Y., and Li, W. 2009b. Antimicrobial activities of ZnO powder-coated PVC film to inactivate food pathogens. *International Journal of Food Science and Technology* 44:2161–2168.

Li, X. L., Zuo, W. W., Luo, M., Shi, Z. S., Cui, Z. C., and Zhu, S. 2013. Silver chloride loaded mesoporous silica particles and their application in the antibacterial coatings on denture base. *Chemical Research in Chinese Universities* 29(6):1214–1218.

Li, Z., Su, K., Cheng, B., and Deng, Y. 2010. Organically modified MCM-type material preparation and its usage in controlled amoxicillin delivery. *Journal of Colloid and Iinterface Science* 342(2):607–613.

Liang, R., Xu, S., Shoemaker, C. F., Li, Y., Zhong, F., and Huang, Q. 2012. Physical and antimicrobial properties of peppermint oil nanoemulsions. *Journal of Agricultural and Food Chemistry* 60:7548–7555.

Lima, E., Guerra, R., Lara, V., and Guzmán, A. 2013. Gold nanoparticles as efficient antimicrobial agents for *Escherichia coli* and *Salmonella typhi*. *Chemistry Central Journal* 7(11):1.

Lin, L., Zhang, H., Cui, H., Xu, M., Cao, S., Zheng, G., and Dong, M. 2013. Preparation and antibacterial activities of hollow silica–Ag spheres. *Colloids and Surfaces B: Biointerfaces* 101:97–100.

Liolios, C. C., Gortzi, O., Lalas, S., Tsaknis, J., and Chinou, I. 2009. Liposomal incorporation of carvacrol and thymol isolated from the essential oil of *Origanum dictamnus* L. and *in vitro* antimicrobial activity. *Food Chemistry* 112(1):77–83.

Liong, M., France, B., Bradley, K. A., and Zink, J. I. 2009. Antimicrobial activity of silver nanocrystals encapsulated in mesoporous silica nanoparticles. *Advanced Materials* 21(17):1684–1689.

Liu, Y., He, L., Mustapha, A., Li, H., Hu, Z. Q., and Lin, M. 2009. Antibacterial activities of zinc oxide nanoparticles against *Escherichia coli* O157: H7. *Journal of Applied Microbiology* 107(4):1193–1201.

Mary, G., Bajpai, S. K., and Chand, N. 2009. Copper (II) ions and copper nanoparticles-loaded chemically modified cotton cellulose fibers with fair antibacterial properties. *Journal of Applied Polymer Science* 113:757–766.

Mas, N., Galiana, I., Mondragón, L. et al. 2013. Enhanced efficacy and broadening of antibacterial action of drugs via the use of capped mesoporous nanoparticles. *Chemistry-A European Journal* 19(34):11167–11171.

McClements, D. J., Decker, E. A., Park, Y., and Weiss, J. 2009. Structural design principles for delivery of bioactive components in nutraceuticals and functional foods. *Critical Reviews in Food Science and Nutrition* 49(6):577–606.

McClements, D. J. and Rhao, J. 2011. Food-grade nanoemulsions: Formulation, fabrication, properties, performance, biological fate, and potential toxicity. *Critical Reviews in Food Science and Nutrition* 51:285–330.

Metak, A. M., Nabhani, F., and Connolly, S. N. 2015. Migration of engineered nanoparticles from packaging into food products. *LWT-Food Science and Technology* 64(2):781–787.

Min, S. H., Yang, J. H., Kim, J. Y., and Kwon, Y. U. 2010. Development of white antibacterial pigment based on silver chloride nanoparticles and mesoporous silica and its polymer composite. *Microporous and Mesoporous Materials*, 128(1):19–25.

Mirhosseini, M. and Firouzabadi, F. B. 2013. Antibacterial activity of zinc oxide nanoparticle suspensions on food-borne pathogens. *International Journal of Dairy Technology* 66:291–295.

Molina-Manso, D., Manzano, M., Doadrio, J. C., Del Prado, G., Ortiz-Pérez, A., Vallet-Regí, M., Gómez-Barrena, E., and Esteban, J. 2012. Usefulness of SBA-15 mesoporous ceramics as a delivery system for vancomycin, rifampicin and linezolid: A preliminary report. *International Journal of Antimicrobial Agents* 40(3):252–256.

Morones, J. R., Elechiguerra, J. L., Camacho, A., Holt, K., Kouri, J. B., Ramírez, J. T., and Yacaman, M. J. 2005. The bactericidal effect of silver nanoparticles. *Nanotechnology* 16(10):2346.

Motlagh, N. V., Mosavian, M. T. H., Mortazavi, S. A., and Tamizi, A. 2012. Beneficial effects of polyethylene packages containing micrometer-sized silver particles on the quality and shelf life of dried barberry (*Berberis vulgaris*). *Journal of Food Science* 71(1):E2–E9.

Mozafari, M. R., Johnson, C., Hatziantoniou, S., and Demetzos, C. 2008. Nanoliposomes and their applications in food nanotechnology. *Journal of Liposome Research* 18:309–327.

Naik, B., Desai, V., Kowshik, M., Prasad, V. S., Fernando, G. F., and Ghosh, N. N. (2011). Synthesis of Ag/AgCl–mesoporous silica nanocomposites using a simple aqueous solution-based chemical method and a study of their antibacterial activity on *E. coli*. *Particuology* 9(3):243–247.

Neal, A. L. 2008. What can be inferred from bacterium–nanoparticle interactions about the potential consequences of environmental exposure to nanoparticles? *Ecotoxicology* 17:362–371.

Nielsen, C. K., Kjems, J., Mygind, T., Snabe, T., Schwarz, K., Serfert, Y., and Meyer, R. L. 2016. Enhancing the antibacterial efficacy of isoeugenol by emulsion encapsulation. *International Journal of Food Microbiology* 229:7–14.

Nielsen, C. K., Kjems, J., Mygind, T., Snabe, T., Schwarz, K., Serfert, Y., and Meyer, R. L. 2017. Antimicrobial effect of emulsion-encapsulated isoeugenol against biofilms of food pathogens and spoilage bacteria. *International Journal of Food Microbiology* 242:7–12.

Okafor, F., Janen, A., Kukhtareva, T., Edwards, V., and Curley, M. 2013. Green synthesis of silver nanoparticles, their characterization, application and antibacterial activity. *International Journal of Environmental Research and Public Health* 10(10):5221–5238.

Orsuwan, A., Shankar, S., Wang, L. F., Sothornvit, R., and Rhim, J. W. 2016. Preparation of antimicrobial agar/banana powder blend films reinforced with silver nanoparticles. *Food Hydrocolloids* 60:476–485.

Pagno, C. H., Costa, T. M., de Menezes, E. W. et al. 2015. Development of active biofilms of quinoa (*Chenopodium quinoa* W.) starch containing gold nanoparticles and evaluation of antimicrobial activity. *Food Chemistry* 173:755–762.

Pal, S., Tak, Y. K., and Song, J. M. 2007. Does the antibacterial activity of silver nanoparticles depend on the shape of the nanoparticle? A study of the gram-negative bacterium *Escherichia coli*. *Applied and Environmental Microbiology* 73(6):1712–1720.

Park, S. Y., Barton, M., and Pendleton, P. 2011. Mesoporous silica as a natural antimicrobial carrier. *Colloids and Surfaces A: Physicochemical and Engineering Aspects* 385(1):256–261.

Park, S. Y., Barton, M., and Pendleton, P. 2012. Controlled release of allyl isothiocyanate for bacteria growth management. *Food Control* 23(2):478–484.

Park, S. Y. and Pendleton, P. 2012. Mesoporous silica SBA-15 for natural antimicrobial delivery. *Powder Technology* 223:77–82.

Pędziwiatr-Werbicka, E., Miłowska, K., Podlas, M., Ubong, A., Tunung, R., Elexson, N., Chai, L. F., and Son, R. 2014. Oleochemical-tethered SBA-15-type silicates with tunable nanoscopic order, carboxylic surface, and hydrophobic framework: Cellular toxicity, hemolysis, and antibacterial activity. *Chemistry-A European Journal* 20(31):9596–9606.

Pérez-Esteve, E., Bernardos, A., Martínez-Máñez, R., and Barat, J. M. (2013). Nanotechnology in the development of novel functional foods or their package. An overview based in patent analysis. *Recent Patents on Food, Nutrition & Agriculture* 5(1):35–43.

Pérez-Esteve, E., Oliver, L., García, L., Nieuwland, M., de Jongh, H. H., Martínez-Máñez, R., and Barat, J. M. 2014. Incorporation of mesoporous silica particles in gelatine gels: Effect of particle type and surface modification on physical properties. *Langmuir* 30:6970–6979.

Petrus, E. M., Tinakumari, S., Chai, L.C., Ubong, A., Tunung, R., Elexson, N., Chai, L. F., and Son, R. 2011. A study on the minimum inhibitory concentration and minimum bactericidal concentration of nano colloidal silver on food-borne pathogens. *International Food Research Journal* 18(1):55–66.

Pinto, R. J. B., Marques, P. A. A. P., Neto, C. P., Trindade, T., Daina, S., and Sadocco, P. 2009. Antibacterial activity of nanocomposites of silver and bacterial or vegetable cellulosic fibers. *Acta Biomaterialia* 5:2279–2289.

Prombutara, P., Kulwatthanasal, Y., Supaka, N., Sramala, I., and Chareonpornwattana, S. 2012. Production of nisin-loaded solid lipid nanoparticles for sustained antimicrobial activity. *Food Control* 24:184–190.

Qi, G., Li, L., Yu, F., and Wang, H. 2013. Vancomycin-modified mesoporous silica nanoparticles for selective recognition and killing of pathogenic gram-positive bacteria over macrophage-like cells. *ACS Applied Materials & Interfaces* 5(21):10874–10881.

Ranjan, S., Dasgupta, N., Chakraborty, A. R., Samuel, S. M., Ramalingam, C., Shanker, R., and Kumar, A. 2014. Nanoscience and nanotechnologies in food industries: Opportunities and research trends. *Journal of Nanoparticle Research* 16(6):1–23.

Rawat, M., Singh, D., Saraf, S., and Saraf S. 2008. Lipid carriers: A versatile delivery for proteins and peptides. *Yakugaku Zasshi* 128(2):269–280.

Reddy, K. M., Feris, K., Bell, J., Wingett, D. G., Hanley, C., and Punnoose, A. 2007. Selective toxicity of zinc oxide nanoparticles to prokaryotic and eukaryotic systems. *Applied Physics Letters* 90(21):213902.

Reddy, L. H., Arias, J. L., Nicolas, J., and Couvreur, P. 2012. Magnetic nanoparticles: Design and characterization, toxicity and biocompatibility, pharmaceutical and biomedical applications. *Chemical Reviews* 112:5818–5878.

Rhim, J. W., Hong, S. I., Park, H. M., and Ng, P. K. W. 2006. Preparation and characterization of chitosan-based nanocomposite films with antimicrobial activity. *Journal of Agricultural and Food Chemistry* 54:5814–5822.

Rispoli, F., Angelov, A., Badia, D., Kumar, A., Seal, S., and Shah, V. 2010. Understanding the toxicity of aggregated zero valent copper nanoparticles against *Escherichia coli*. *Journal of Hazardous Materials* 180(1):212–216.

Rossi, M., Cubadda, F., Dini, L., Terranova, M. L., Aureli, F., Sorbo, A., and Passeri, D. 2014. Scientific basis of nanotechnology, implications for the food sector and future trends. *Trends in Food Science & Technology* 40:127–148.

Ruiz-Rico, M., Fuentes, C., Pérez-Esteve, E., Jiménez-Belenguer, A. I., Quiles, A., Marcos, M. D., Martínez-Máñez, R., and Barat, J. M. 2015. Bactericidal activity of caprylic acid entrapped in mesoporous silica nanoparticles. *Food Control* 56:77–85.

Sadiq, S., Imran, M., Habib, H., Shabbir, S., Ihsan, A., Zafar, Y., and Hafeez, F. Y. 2016. Potential of monolaurin based food-grade nano-micelles loaded with nisin Z for synergistic antimicrobial action against *Staphylococcus aureus*. *LWT-Food Science and Technology* 71:227–233.

Salvia-Trujillo, L., Rojas-Graü, A., Soliva-Fortuny, R., and Martín-Belloso, O. 2015a. Physicochemical characterization and antimicrobial activity of foodgrade emulsions and nanoemulsions incorporating essential oils. *Food Hydrocolloids* 43:547–556.

Salvia-Trujillo, L., Rojas-Graü, A., Soliva-Fortuny, R., and Martín-Belloso, O. 2015b. Use of antimicrobial nanoemulsions as edible coatings: Impact on safety and quality attributes of fresh-cut Fuji apples. *Postharvest Biology and Technology* 105:8–16.

Sanpui, P., Murugadoss, A., Prasad, P. D., Ghosh, S. S., and Chattopadhyay, A. 2008. The antibacterial properties of a novel chitosan–Ag-nanoparticle composite. *International Journal of Food Microbiology* 124(2):142–146.

Shankar, S. and Rhim, J. W. 2016. Tocopherol-mediated synthesis of silver nanoparticles and preparation of antimicrobial PBAT/silver nanoparticles composite films. *LWT-Food Science and Technology* 72:149–156.

Sharma, V., Singh, P., Pandey, A. K., and Dhawan, A. 2012. Induction of oxidative stress, DNA damage and apoptosis in mouse liver after sub-acute oral exposure to zinc oxide nanoparticles. *Mutation Research* 745:84–91.

Shen, Q., Wang, J., Yang, H., Ding, X., Luo, Z., Wang, H., Pan, C., Sheng, J., and Cheng, D. 2014. Controllable preparation and properties of mesoporous silica hollow microspheres inside-loaded Ag nanoparticles. *Journal of Non-Crystalline Solids* 391:112–116.

Shim, W. B., Song, J. E., Mun, H., Chung, D. H., and Kim, M. G. 2014. Rapid colorimetric detection of *Salmonella typhimurium* using a selective filtration technique combined with antibody–magnetic nanoparticle nanocomposites. *Analytical and Bioanalytical Chemistry* 406:859–866.

Shukla, R. K., Kumar, A., Gurbani, D., Pandey, A. K., Singh, S., and Dhawan, A. 2013. TiO_2 nanoparticles induce oxidative DNA damage and apoptosis in human liver cells. *Nanotoxicology* 7(1):48–60.

Siahaan, E. A., Meillisa, A., Woo, H. C., Lee, C. W., Han, J. H., and Chun, B. S. 2013. Controlled release of allyl isothiocyanate from brown algae *Laminaria japonica* and mesoporous silica MCM-41 for inhibiting food-borne bacteria. *Food Science and Biotechnology* 22(1):19–24.

Singh, S., Barick, K. C., and Bahadur, D. 2011. Surface engineered magnetic nanoparticles for removal of toxic metal ions and bacterial pathogens. *Journal of Hazardous Materials* 192(3):1539–1547.

Slowing, I. I., Vivero-Escoto, J. L., Wu, C. W., and Lin, V. S. Y. 2008. Mesoporous silica nanoparticles as controlled release drug delivery and gene transfection carriers. *Advanced Drug Delivery Reviews* 60:1278–1288.

Smolkova, B., El Yamani, N., Collins, A. R., Gutleb, A. C., and Dusinska, M. 2015. Nanoparticles in food. Epigenetic changes induced by nanomaterials and possible impact on health. *Food and Chemical Toxicology* 77:64–73.

Son, W. K., Youk, J. H., and Park, W. H. 2006. Antimicrobial cellulose acetate nanofibers containing silver nanoparticles. *Carbohydrate Polymers* 65:430–434.

Song, N. and Yang, Y. W. 2015. Molecular and supramolecular switches on mesoporous silica nanoparticles. *Chemical Society Reviews* 44(11):3474–3504.

Sung, S. Y., Sin, L. T., Tee, T. T., Bee, S. T., Rahmat, A. R., Rahman, W. A. W. A., Tan, A. C., and Vikhraman, M. 2013. Antimicrobial agents for food packaging applications. *Trends in Food Science & Technology* 33(2):110–123.

Tang F., Li L. and Chen D. 2012. Mesoporous silica nanoparticles: Synthesis, biocompatibility and drug delivery. *Advanced Materials* 24(12):1504–1534.

Tao, Z. 2014. Mesoporous silica-based nanodevices for biological applications. *RSC Advances* 4(36):18961–18980.

Tayel, A. A., El-Tras, W. F., Moussa, S., El-Baz, A. F., Mahrous, H., Salem, M. F., and Brimer, L. 2011. Antibacterial action of zinc oxide nanoparticles against foodborne pathogens. *Journal of Food Safety* 31:211–218.

Taylor, T. M., Bruce, B. D., Weiss, J., and Davidson, P. M. 2008. *Listeria monocytogenes* and *Escherichia coli* O157:H7 inhibition *in vitro* by liposome-encapsulated nisin and ethylene diaminetetraacetic acid. *Journal of Food Safety* 28:183–197.

Thill, A., Zeyons, O., Spalla, O., Chauvat, F., Rose, J., Auffan, M., and Flank, A. M. 2006. Cytotoxicity of CeO_2 nanoparticles for *Escherichia coli*. Physico-chemical insight of the cytotoxicity mechanism. *Environmental Science & Technology* 40(19):6151–6156.

Tiyaboonchai, W., Tungpradit, W., and Plianbangchang, P. 2007. Formulation and characterization of curcuminoids loaded solid lipid nanoparticles. *International Journal of Pharmaceutics* 337:299–306.

Torres-Giner, S., Ocio, M. J., and Lagaron, J. M. 2008. Development of active antimicrobial fiber based chitosan polysaccharide nanostructures using electrospinning. *Engineering in Life Sciences* 8(3):303–314.

Trewyn, B. G., Slowing, I. I., Giri, S., Chen, H. T., and Lin, V. S. 2007. Synthesis and functionalization of a mesoporous silica nanoparticle based on the sol-gel process and applications in controlled release. *Accounts of Chemical Research* 40(9):846–853.

Tsoi, K. M., Dai, Q., Alman, B. A., and Chan, W. C. W. 2012 Are quantum dots toxic? Exploring the discrepancy between cell culture and animal studies. *Accounts of Chemical Research* 46(3):662–671.

Vallet-Regí, M., Doadrio, J. C., Doadrio, A. L., Izquierdo-Barba, I., and Pérez-Pariente, J. 2004. Hexagonal ordered mesoporous material as a matrix for the controlled release of amoxicillin. *Solid State Ionics* 172(1):435–439.

Vance, M. E., Kuiken, T., Vejerano, E. P., McGinnis, S. P., Hochella Jr, M. F., Rejeski, D., and Hull, M. S. 2015. Nanotechnology in the real world: Redeveloping the nanomaterial consumer products inventory. *Beilstein Journal of Nanotechnology* 6(1):1769–1780.

Vega-Lugo, A. C. and Lim, L. T. 2009. Controlled release of allyl isothiocyanate using soy protein and poly(lactic acid) electrospun fibers. *Food Research International* 42:933–940.

Velmurugan, P., Anbalagan, K., Manosathyadevan, M., Lee, K. J., Cho, M., Lee, S. M., Park, J. H., Oh, S. G., Bang, K. S., and Oh, B. T. 2014. Green synthesis of silver and gold nanoparticles using *Zingiber officinale* root extract and antibacterial activity of silver nanoparticles against food pathogens. *Bioprocess and Biosystems Engineering* 37(10):1935–1943.

Verran, J., Packer, A., Kelly, P., and Whitehead, K. A. 2010. Titanium coating of stainless steel as an aid to improved cleanability. *International Journal of Food Microbiology* 141:S134–S139.

Vimala, K., Mohan, Y. M., Sivudu, K. S., Varaprasad, K., Ravindra, S., Reddy, N. N., Padma, Y., Sreedhar, B., and MohanaRaju, K. 2010. Fabrication of porous chitosan films impregnated with silver nanoparticles: A facile approach for superior antibacterial application. *Colloids and Surfaces B: Biointerfaces* 76:248–258.

Wang, K., Jin, P., Shang, H., Li, H., Xu, F., Hu, Q., and Zheng, Y. 2010. A combination of hot air treatment and nano-packing reduces fruit decay and maintains quality in postharvest Chinese bayberries. *Journal of the Science of Food and Agriculture* 90:2427–2432.

Wang, L., He, H., Zhang, C., Sun, L., Liu, S., and Yue, R. 2014. Excellent antimicrobial properties of silver-loaded mesoporous silica SBA-15. *Journal of Applied Microbiology* 116(5):1106–1118.

Wang, Y., Nor, Y. A., Song, H., Yang, Y., Xu, C., Yu, M., and Yu, C. 2016. Small-sized and large-pore dendritic mesoporous silica nanoparticles enhance antimicrobial enzyme delivery. *Journal of Materials Chemistry B* 4(15):2646–2653.

Wehling, J., Volkmann, E., Grieb, T., Rosenauer, A., Maas, M., Treccani, L., and Rezwan, K. 2013. A critical study: Assessment of the effect of silica particles from 15 to 500 nm on bacterial viability. *Environmental Pollution* 176:292–299.

Weir, E., Lawlor, A., Whelan, A., and Regan, F. 2008. The use of nanoparticles in anti-microbial materials and their characterization. *Analyst* 133:835–845.

Weiss, J., Gaysinsky, S., Davidson, M., and McClements, D. J. 2009. Nanostructured encapsulation systems: Food antimicrobials. *Global Issues in Food Science and Technology* 1:425–479.

Weiss, J., Takhistov, P., and Weiss, J. 2006. Functional materials in food nanotechnology. *Journal of Food Science* 71(9):R107–R116.

Were, L. M., Bruce, B., Davidson, P. M., and Weiss, J. 2004. Encapsulation of nisin and lysozyme in liposomes enhances efficacy against *Listeria monocytogenes*. *Journal of Food Protection* 67(5):922–927.

Werlin, R., Priester, J. H., Mielke, R. E., Krämer, S., Jackson, S., Stoimenov, P. K., Stucky, G. D., Cherr, G. N., Orias, E., and Holden, P. A. 2011. Biomagnification of cadmium selenide quantum dots in a simple experimental microbial food chain. *Nature Nanotechnology* 6(1):65–71.

Xia, X., Pethe, K., Kim, R., Ballell, L., Barros, D., Cechetto, J., Jeon, H., Kim, K., and Garcia-Bennett, A. E. 2014. Encapsulation of anti-tuberculosis drugs within mesoporous silica and intracellular antibacterial activities. *Nanomaterials* 4:813–826.

Xie, Y., He, Y., Irwin, P. L., Jin, T., and Shi, X. 2011. Antibacterial activity and mechanism of action of zinc oxide nanoparticles against *Campylobacter jejuni*. *Applied and Environmental Microbiology* 77(7):2325–2331.

Xu, W., Riikonen, J., and Lehto, V. P. 2012. Mesoporous systems for poorly soluble drugs. *International Journal of Pharmaceutics* 453(1):181–197.

Xue, J. M. and Shi, M. 2004. PLGA/mesoporous silica hybrid structure for controlled drug release. *Journal of Controlled Release* 98(2):209–217.

Yang, H., Liu, Y., Shen, Q., Chen, L., You, W., Wang, X., and Sheng, J. 2012. Mesoporous silica microcapsule-supported Ag nanoparticles fabricated via nano-assembly and its antibacterial properties. *Journal of Materials Chemistry* 22(45):24132–24138.

Youssef, A. M., El-Sayed, S. M., Salama, H. H., El-Sayed, H. S., and Dufresne, A. 2015. Evaluation of bionanocomposites as packaging material on properties of soft white cheese during storage period. *Carbohydrate Polymers* 132:274–285.

Yu, H. and Huang, Q. 2013. Investigation of the cytotoxicity of food-grade nanoemulsions in Caco-2 cell monolayers and HepG2 cells. *Food Chemistry* 141:29–33.

Yu, E., Galiana, I., Martínez-Máñez, R., Stroeve, P., Marcos, M. D., Aznar, E., Sancenón, F., Murguía, J. R., and Amorós, P. 2015. Poly (N-isopropylacrylamide)-gated Fe_3O_4/SiO_2 core shell nanoparticles with expanded mesoporous structures for the temperature triggered release of lysozyme. *Colloids and Surfaces B: Biointerfaces* 135:652–660.

Zahi, M. R., Liang, H., and Yuan, Q. 2015. Improving the antimicrobial activity of D-limonene using a novel organogel-based nanoemulsion. *Food Control* 50:554–559.

Zarei, M., Jamnejad, A., and Khajehali E., 2014. Antibacterial effect of silver nanoparticles against four foodborne pathogens. *Jundishapur Journal of Microbiology* 7(1):8720.

Zawrah, M. F. and El-Moez, S. I. A. 2011. Antimicrobial activities of gold nanoparticles against major foodborne pathogens. *Life Science Journal* 8(4):37–44.

Zengin, N., Yüzbaşıoğlu, D., Ünal, F., Yılmaz, S., and Aksoy, H. 2011. The evaluation of the genotoxicity of two food preservatives: Sodium benzoate and potassium benzoate. *Food and Chemical Toxicology* 49(4):763–769.

Zhan, S., Yang, Y., Shen, Z., Shan, J., Li, Y., Yang, S., and Zhu, D. 2014. Efficient removal of pathogenic bacteria and viruses by multifunctional amine-modified magnetic nanoparticles. *Journal of Hazardous Materials* 274:115–123.

Zhang, H., Shen, Y., Weng, P., Zhao, G., Feng, F., and Zheng, X. 2009. Antimicrobial activity of a food-grade fully dilutable microemulsion against *Escherichia coli* and *Staphylococcus aureus*. *International Journal of Food Microbiology* 135:211–215.

Zhang, L., Jiang, Y., Ding, Y., Povey, M., and York, D. 2007. Investigation into the antibacterial behaviour of suspensions of ZnO nanoparticles (ZnO nanofluids). *Journal of Nanoparticle Research* 9(3):479–489.

Zhao, D., Feng, J., Huo, Q., Melosh, N., Fredrickson, G. H., Chmelka, B. F., and Stucky, G. D. 1998. Triblock copolymer syntheses of mesoporous silica with periodic 50 to 300 angstrom pores. *Science* 279(5350):548–52.

14
Interface Considerations in the Modeling of Hierarchical Biological Structures

Parvez Alam

CONTENTS

14.1 Introduction 453
14.2 Atom–Atom Interfaces 453
14.3 Molecular Interfaces 454
14.4 The Nano Effect 456
14.5 Biomechanical Interfaces at Higher Length Scales 459
14.6 Final Remarks 460
References 460

14.1 Introduction

This chapter concerns property transitions between length scales in the hierarchical modeling of biological structures (biostructures). Biological materials are inherently complex, and mechanical properties such as toughness and fracture toughness are very often directly associated with the hierarchy of structure. At each length scale, there is an interface. Inclusion of the subtle effects each interface may have on both local and global properties into a model may give rise to more reliable model predictions. In this chapter, we elucidate the importance in considering these subtle interface variations.

14.2 Atom–Atom Interfaces

Atom–atom interface interactions and the way in which they are defined are essentially the cornerstones for atomistic modeling. Two predominating philosophies are routinely incorporated into atomistic modeling software. These include empirical as well as quantum mechanical representations. Both representations are currently indispensable in the more prominent atomistic modeling methods, including Monte Carlo, molecular dynamics, and the lattice energy methods.

Empirical representations are born, essentially, from currently accepted empirical equations that have been derived to describe atomic separation. A benefit of this approach to the atom–atom interface is that it can be easily modified to the parameters of a specific equation. A classical example of an empirical model that is still used in simulations concerning certain ionic materials is the long-range Coulomb interaction model with short-range repulsion; see Equation 14.1 (Born and Mayer 1932). Here, E is energy, r is the atom–atom spacing, and L and η are correction factors that correspond to the empirical data set. An alternative, simpler model has greater functionality in that, unlike the Coulomb interaction model, harmonic functions such as torque and bond angle can be easily incorporated; see Equation 14.2.

$$E(r) = Le^{(-r/\eta)} \tag{14.1}$$

$$E = Lx^{\eta} \tag{14.2}$$

Quantum mechanical models are different and are essentially probabilistic models. They have no reliance on empirical data and have no need for *correction factors* or simulation-specific modifications to constitutive interatomic relations. Quantum mechanical models yield reliable simulation results and have a particular benefit in *ab initio* simulations where electrostatic optimizations of complex molecules may be necessary. Being a probabilistic approach, quantum mechanical models also have the added benefit of predicting the charge-sharing effect of any deviation of a biomolecule, thus allowing for subtle variations in chirality and isomerism to be modeled and compared with relative ease. In the quantum mechanical approach, electrons travel within electrical fields created by nuclei. The coinciding wave functions, Ψ, are shared between existing nuclear, vibrational, rotational, and electronic wave functions.

$$\Psi = \Psi_{electronic} + \Psi_{vibrational} + \Psi_{rotational} + \Psi_{nuclear} \tag{14.3}$$

$$E = E_{electronic} + E_{vibrational} + E_{rotational} + E_{nuclear} \tag{14.4}$$

An alternative to using wave functions in the quantum mechanical methods is density functional theory (DFT) (Payne et al. 1992). The fundamental difference between the two methods here is that DFT employs electron densities in the determination of interatomic activity. DFT uses the approximation shown in Equation 14.5, in which energy is a summation of both kinetic and potential energies. Here, U_k is the kinetic energy, U_{p-e} is the potential energy of electrons arising through repulsion, and U_{n-e} is the potential energy of atomic electrons arising through attraction to the nucleus, which is positively charged.

$$E = U_k + U_{p-e} + U_{n-e} \tag{14.5}$$

14.3 Molecular Interfaces

Biological macromolecules are typically modeled from the atomistic principles previously described. When considering *molecular interfaces*, there is a need to contextualize the specific interface under scrutiny. Molecular modeling methods such as MD and MC can be useful for modeling, for example, primary protein folding mechanisms and secondary structure interactions. However, when modeling large sets of tertiary and quaternary structure interactions, the need for structural and interfacial approximations becomes apparent. There are a few important articles that can be cited as useful templates to dealing with these inherently complex systems.

Interface Considerations in the Modeling of Hierarchical Biological Structures

One such article (Bradford et al. 2006) uses probabilistic functions in the form of Bayesian networks to distinguish interfacial characteristics between interacting proteins. The Bayesian network is very useful since it is essentially a steered probabilistic method representing variables and their *conditional dependencies* through a directed noncyclic graph. The strength of the method is in its simplicity. Dependent variable outputs are either *true* or *false*, thus allowing for the incorporation of numerous variables with relative ease. The Bayes hierarchical Bayesian network can most simplistically be written,

$$p(a, b | x) \propto p(x|a)p(a|b)p(b) \qquad (14.6)$$

Here, a is a prior originating from the prior probability $p(a)$ of the fundamental Bayesian form $p(a|x) \propto p(x|a)p(a)$. The prior a depends on parameter b, all of which relate to x data. A likelihood, $p(a|b)$, and a new prior, $p(b)$, replaces the prior probability, $p(a)$, for the hierarchical form in Equation 14.6. Bradford and co-workers distinguish between interacting and noninteracting patches at protein interfaces by five variables. These are shown in Figure 14.1.

Jones and Thornton (1997) considered a different approach to studying protein–protein interaction sites. They worked from the basis that protein–protein interfaces are predominantly hydrophobic (Korn and Burnett 1991, Young et al. 1994). They studied essentially the same variables shown in Figure 14.1, but in discrete patches defined by a finite number of residues. A similar patch-by-patch method used by Preissner et al. (1998) suggests that approximations based on residue-type and structure can be made for larger-scale simulations. Larger-scale simulations of higher-order structures are currently indispensable for drug discovery, and such larger-scale simulations are possible through interfacial (residue–residue) approximations. Various established mathematical methods have already been implemented into docking software to deal with the interfacial approximation. This might include Monte Carlo-based approximations, Equation 14.7, such as in GlamDock (Tietze and Apostolakis 2007) and ICM (Cardozo et al. 1995). This is a probabilistic (statistical) approach to approximating interface properties and is relatively slow. In the equation, we have N configuration of a system with $\xi_1, \xi_2, \xi_3, \xi_4 \ldots \xi_P$, where $P(\xi)$ is a given probability distribution.

$$\lim_{N \to \infty} \frac{N_\xi}{N} = P(\xi) \qquad (14.7)$$

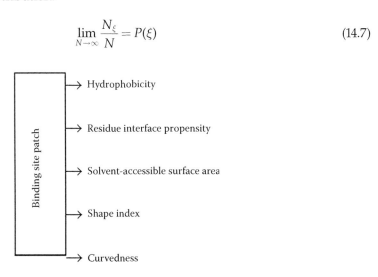

FIGURE 14.1
Bayesian parameters employed by Bradford et al. (2006) for protein–protein interactions.

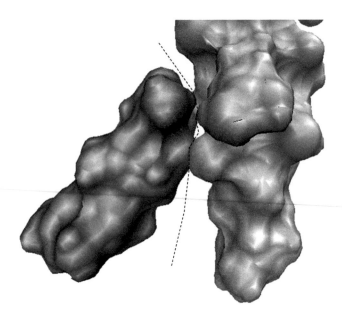

FIGURE 14.2
Simplified pictographical representation of a bisecting line passing through the interacting points of two higher order proteins.

Genetic algorithms have also been used to approximate docking interfaces (Ördög and Grolmusz 2008) and are used in software such as Autodock and MolDock. Incremental methods of optimization of ligand pose, made faster through GPU (Heinzerling et al. 2012), are used in software such as DOCK, and geometrical surface approximations for faster solving of ligand-type interactions have been made in LigandFit and Glide (Venkatachalam et al. 2003, Friesner et al. 2004). A combined approach of genetic algorithms with multilevel local optimizations has been suggested by Novikov and Chilov (2009). Recently, a bisector method developed by Seong et al. (2011) is perhaps one of the more unconventional methods by which interface interactions can be approximated with greater speed. This method can take, essentially, large clusters of atoms (typically higher-order protein structures) and bisect between them to form a plane traversing through Euclidean space yet solely connecting the intersecting points. The bisector, Figure 14.2, exists in a real-time environment, and as a consequence, a great variety of computations can easily be applied with very little cost in memory.

14.4 The Nano Effect

Many biological structures are nanoscale, in that they will exhibit dimensions below 100 nm in at least one direction. There are a plethora of examples of such nanostructures in nature, including nanocages (diatoms), nanospatulae (geckos), nanoflakes (butterflies), nanoparticles (corals and shells), nanofilms (water-repellent leaves), and nanospikes (insect wings) (Jackson et al. 1988, Menig et al. 2000, Wang et al. 2001, Sitti 2003, Wee et al. 2005, Przenioslo et al. 2008, Malvadkar et al. 2010, Alam et al. 2013, Kumar et al. 2013, Pogodin et al. 2013, Stillfried et al. 2013). More recent work has shown that nanocrystals in contact with soft amorphous phases give rise to semicrystalline metastable states that can take both amorphous and crystalline forms depending on the loading conditions (Sintya and Alam 2016a,b). In hierarchical modeling, there may be a need to transition from the molecular to nano, or even micron scale. Typically higher-scale models will lack atomistic detail. Yet, details in low to high scale transitions, and, in the better-case scenario, specific properties should be incorporated for more apposite model predictions. One example is the nano effect. The nano effect can best be described as relating to the balance between bulk and free surface properties. At the molecular level, surface/molecular energies dominate, whereas at the macro scale, the properties of a body of matter are determined by the energy of the bulk. Surface and bulk properties are considerably different, and at the nano scale, as a function of atomic volume, the surfaces contribute considerably to the overall properties of the nanomaterial and cannot be neglected.

We may begin by considering a nano-object as consisting of two phases, a bulk, α, and a surface, β (Alam 2015). Energetically, these can be considered to exist in a state of continuum with an interface separating each phase. The elastic surface energy (Gibbs 1961) between the two phases, U_s, can be represented by Equation 14.8.

$$U_s = U - (U_\alpha + U_\beta) \tag{14.8}$$

Here, U is the internal energy of the system and $(U_\alpha + U_\beta)$ is the total uniform energy in both parts. Notably, $(U_\alpha + U_\beta)$ can be modified to include any variations in energy at the surface. It should be noted that the surface is in a continual state of (molecular) motion. When the nanostructure surface dominates the bulk, this motion can be considerable, and work energy at the surface needs to be accounted for. This can be defined according to Equation 14.9.

$$W_s = \gamma \, dA \tag{14.9}$$

Here, A is the area of the surface and γ is the surface stress. Surface deformation of a compressible material induces a volumetric, V, change in each phase such that:

$$dV_\alpha = -dV_\beta \tag{14.10}$$

The general surface deformation can be expressed as a strain, ε_s:

$$\varepsilon_s = dA/A \tag{14.11}$$

and the surface stress is approximated from Shuttleworth (1950):

$$\gamma = E_s + \partial E_s/\partial \varepsilon_s \tag{14.12}$$

where E_s is the free energy of the surface. Different approximations for the surface stress can be made following Couchman and Davidson (1977) as:

$$\gamma = E_s + \partial E_s/\partial \varepsilon_e \tag{14.13}$$

which can alternatively be expressed:

$$-\partial E_s/\partial E = q + (E_s - \gamma) \, \partial \varepsilon_e/\partial E \tag{14.14}$$

where ε_e is the elastic surface deformation, E is the electrode potential, and q is the surface charge density. With respect to the above equations, it is well worth noting that to date, none have yet been proven experimentally (Marichev 2011). Numerous flaws can be brought to light in both the Shuttleworth and the Couchman forms. Marichev suggests that since the first and second Gokhstein equations have been confirmed experimentally (see Gokhshtein 1970, 1975), they should rather be used as determinants of surface tension for solid materials.

$$(\partial \gamma / \partial q)_\varepsilon = (\partial E / \partial \varepsilon_e)_Q \text{—first Gokhstein eq.} \tag{14.15}$$

where Q is the surface charge.

$$-\partial \gamma / \partial q = q + \partial q / \partial \varepsilon_e \text{—second Gokhstein eq.} \tag{14.16}$$

In a nanomaterial, both phases, α and β, can be considered solids. It follows then, that when the surface of α strains, work is done at the interface between α and β. During deformation, the total work for a solid–solid interface, W_{s-s}, combines three separate work terms.

$$W_{s-s} = E_{s-s}(dA_{s-s}) + \gamma_{s-s}(dA_\alpha) + \gamma_{s-s}(dA_\beta) \tag{14.17}$$

where the term $E_{s-s}(dA_{s-s})$ is the work required to create a new surface, $\gamma_{s-s}(dA_\alpha)$ is the work required to deform phase α about the interface, and $\gamma_{s-s}(dA_\beta)$ is the work required to deform phase β about the interface to the same extent as α. E_{s-s} is the interface energy, γ_{s-s} is the interface stress, A_{s-s} is the interface area, and A_α and A_β are the areas related to deformation in α and β solids, respectively. Though total work also exists in micro- and macroscale materials with interfaces, because the bulk phase dominates dimensionally and with regards to properties, the total work term at a nanointerface can effectively be reduced to:

$$W_{s-s} = E_{s-s}(dA_{s-s}) \tag{14.18}$$

This simplification is possible because $\{\gamma_{s-s}(dA_\alpha) + \gamma_{s-s}(dA_\beta)\}$ is negligible and the magnitude of $\{\gamma_{s-s}(dA_\alpha) + \gamma_{s-s}(dA_\beta)\}$ relative to $\{E_{s-s}(dA_{s-s})\}$ increases as the dimensions of each phase come closer to the nano scale. Thus, both surface and interfacial effects described here cannot be ignored at the nano scale. Indeed, surface-dominating characteristics at the nano scale may well be a cause for mechanically beneficial phenomena such as molecular pinning (Touaiti et al. 2010).

The total mechanical work in a composite material, W_c, is a summation of the work done in the bulk material of phases α and β, W_α and W_β, respectively, with the work done at the solid–solid interfaces. In much larger-than-nanoscale materials,

$$(W_\alpha + W_\beta) \gg W_{s-s} \tag{14.19}$$

and therefore

$$W_c \to (W_\alpha + W_\beta) \tag{14.20}$$

while in nanostructured materials,

$$(W_\alpha + W_\beta) \ll W_{s-s} \tag{14.21}$$

thus

$$W_c \to W_{s-s} \tag{14.22}$$

14.5 Biomechanical Interfaces at Higher Length Scales

At the nano scale, tractive forces may still be modeled as molecular sliding. However, as we increase the length scale, molecular models turn out to be impractical in many ways. There is an increased (perhaps unnecessarily so) CPU/GPU demand alongside memory consumption. There are also more data that require considerable effort to analyze. Finally, there is the unfortunate but oft-experienced scenario of long run-time models that turn out to have been developed incorrectly. The predominating question would be: Is there any added benefit of scaling an atomistic model up with more atoms, or can we reach the same conclusions and interpretations from our model by making use of more appropriate modeling methods for higher-scale materials? The author of this chapter considers that observed incoherency between length scales may in fact be a function of inadequately or indeed inaccurately described interfaces. In this section, we describe some simple means by which larger-scale structural models can be made more representative of hierarchical biostructures.

Energy transfer and interfacial motion have already to some extent been covered in the previous section on the nano effect. Nevertheless above the nano scale, work energy at the interface takes the full form of Equation 14.17 with no further simplification. This energy is holistic in that the deformation that gives rise to it may be one or a combination of a multitude of forms. These may include, typically, friction, viscoelastic or viscoplastic resistance, crack resistance, and topographical interlocking. It is worth mentioning that each characteristic is essentially a function of molecular-level behavior. Thus, appropriate approximations must be made when applying molecular mechanical computations to macroscale interface properties. There are a number of ground-level relationships that are already in use that can be manipulated to semiempirical forms or from basis molecular computations.

We can start simply by considering mechanical energy transfer across a boundary. Assuming molecular-level computations are correct, that a boundary system is in continuum, and that the energy scalar follows from the previous boundary, we can easily return to Equation 14.17, which accounts for the three phases of α, β, and the interface. One important issue to address in regards to Equation 14.17 is that the effects from specific interface characteristics need to be included for the form to function more realistically in transition from molecular modeling to macroscale (e.g., finite-element) modeling. When friction is the dominant mode of irrecoverable interfacial deformation, changes to this general understanding of deformational work are to be modified. Energy will be lost when molecules slide, and the term $E_{s-s}(dA_{s-s})$ thus requires modification to account for nonlinear reductions in work energy as a function of the initial transition from stored work energy to molecular sliding, f_s, and subsequent kinetic friction, f_d, where molecules are in continuous sliding motion. In such a case, therefore,

$$E_{s-s}(dA_{s-s}) = f(f_s N, f_d N) \tag{14.23}$$

where N is the normal force acting between phases α and β. This approximation has numerous limitations, including an assumption that the normal force and frictional force are linearly proportional, that the contacting area is not in a state of atomic saturation (i.e., atom–atom contact is not in continuum), and that theoretical materials can follow what is essentially an empirical construction. The great versatility of this model, however, is its simplicity and it is thanks to the simplicity that it can be modified via semiempirical or theoretical corrections.

Such corrections can be tuned to the material properties of the interface. Keeping the assumption that normal and frictional forces are proportional, alternative force approximations related to, for example, viscoelasticity can be made. The important point here is that the variable N can be manipulated relatively painlessly in series form to establish (in the case of viscoelasticity) the force per unit time as a function of the strain rate. The series may be applied from empirically determined (or, indeed, computationally determined) data sets, or from a direct mathematical function, such as in a Prony series.

Defining an interface with material properties *and* fractal topographical irregularity is somewhat more challenging and requires consideration of the local deformational characteristics of individual protuberances, the way by which the protuberance affects force transfer and interface motion, and the subsequent nonlinear variations as a function of protuberance shape. Discretization is, of course, the normal route to dealing with complex geometry. That said, fractal surfaces are problematic in that they will reach a dimension that can no longer be rendered, and interpolation functions should be rather used to define fractal-induced behavior. Good derivations of such functions can be found in Warren and Krajcinovic (1996) and Mistakidis and Panagouli (2003). In short, the fractal effect increases friction at lower orders, but as the fractal order increases, its effects on friction becomes less obvious than its preceding orders.

14.6 Final Remarks

In this chapter, we briefly highlight the importance of interface considerations in the hierarchical modeling of biological structures. We consider the fundamentals of interface modeling from the atomistic level, through the molecular and nano levels, and finish at the macro level of modeling. The understanding and application of justifiable interface conditions at every length scale is critical to the development of stable, fast-working, effective, yet accurate biomodels.

References

Alam P. 2015. Mechanical properties of bio-nanostructured materials. In: *Handbook of Mechanical Nanostructuring* (ed. M. Aliofkhazraei), Wiley-VCH Verlag GmbH & Co. KGaA, Weinheim, Germany. pp. 211–231. ISBN 978-3-527-33506-0

Alam P, Stillfried DG, Celli J, Toivakka M. 2013. Effects of fibre-surface morphology on the mechanical properties of Porifera-inspired rubber-matrix composites. *Applied Physics A: Materials Science and Processing*, 111(4):1031–1036.

Born M, Mayer JE. 1932. Zur Gittertheorie der Ionenkristalle. *Zeitschrift für Physik*, 75:1–18.

Bradford JR, Needham CJ, Bulpitt AJ, Westhead DR. 2006. Insights into protein–protein interfaces using a Bayesian network prediction method. *Journal of Molecular Biology*, 362:365–386.

Cardozo T, Totrov MM, Abagyan RA. 1995. Homology modeling by the ICM method. *Proteins*, 23:403–414.

Couchman PR, Davidson CR. 1977. The Lippmann relation and surface thermodynamics. *Journal of Electroanalytical Chemistry*, 85(1–2):407–409.

Friesner RA, Banks JL, Murphy RB, Halgren TA, Klicic JJ, Mainz DT, Repasky MP et al. 2004. Glide: A new approach for rapid accurate docking and scoring. 1. Method and assessment of docking accuracy. *Journal of Medicinal Chemistry*, 47:1739–1749.

Gibbs JW. 1961. On the equilibrium of heterogeneous substances. In: *The Scientific Papers of J. Willard Gibbs, Volume 1: Thermodynamics*, New York, Dover, pp. 55–353.

Gokhshtein, AY. 1970. Investigation of surface tension of solid electrodes at several frequencies simultaneously. *Electrochimica Acta*, 15(1):219–223.

Gokhshtein AY. 1975. The estance method. *Russian Chemical Reviews*, 44(11):921–932.

Heinzerling L, Klein R, Rarey M. 2012. Fast force-field based optimization of protein-ligand complexes with graphical processor. *Journal of Computational Chemistry*, 33:2554–2565.

Jackson AP, Vincent JFV, Turner RM. 1988. The mechanical design of nacre. *Proceeding s of the Royal Society of London Series B-Biological Sciences*, 234:415–440.

Jones S, Thornton JM. 1997. Analysis of protein–protein interaction sites using surface patches. *Journal of Molecular Biology*, 272:121–132.

Korn AP, Burnett RM. 1991. Distribution and complimentarity of hydropathy in multisubunit proteins. *Proteins: Structure, Function, Genetics*, 9:37–55.

Kumar R, Smith S, McNeilan J, Keeton M, Sanders J, Talamo A, Bowman C, Xie Y. 2013. Butterfly-wing inspired nanotechnology. In: *The Nanobiotechnology Handbook*, Ed. Xie Y. CRC Press, Taylor and Francis Group, Boca Raton.

Malvadkar NA, Hancock MJ, Sekeroglu K, Dressick WJ, Demirel MC. 2010. An engineered anisotropic nanofilm with unidirectional wetting properties. *Nature Materials*, 9(12):1023–1028.

Marichev VA. 2011. The Shuttleworth equation: Is modifications and current state. *Protection of Metals and Physical Chemistry of Surfaces*, 47(1):25–30.

Menig R, Meyers MH, Meyers MA, Vecchio KS. 2000. Quasi-static and dynamic mechanical response of *haliotis rufescens* (abalone) shells. *Acta Materialia*, 48:2383–2398.

Mistakidis ES, Panagouli OK. 2003. Friction evolution as a result of roughness in fractal interfaces. *Emgineering Computations*, 20:40–57.

Novikov FN, Chilov GG. 2009. Molecular docking: Theoretical background, practical applications and perspectives. *Mendeleev Communications*, 19:237–242.

Ördög R, Grolmusz V. 2008. Evaluating genetic algorithms in protein-ligand docking. *Bioinformations Research and Applications Lecture Notes in Computer Science*, 4983:402–413.

Payne MC, Teter MP, Allan DC, Arias TA, Joannopoulos JD. 1992. Iterative minimization techniques for ab initio total-energy calculations—Molecular dynamics and conjugate gradients. *Reviews of Modern Physics*, 64:1045–1097.

Pogodin S, Hasan J, Baulin VA, Webb HK, Truong VK, Nguyen THP, Boshkovikj V et al. 2013. Biophysical model of bacterial cell interactions with nanopatterned Cicada wing surfaces. *Biophysical Journal*, 104(4):835–840.

Preissner R, Goede A, Frömmel C. 1998. Dictionary of interfaces in proteins (DIP) data bank of complementary molecular surfaces patches. *Journal of Molecular Biology*, 280:535–550.

Przenioslo R, Storlaski J, Mazur M, Brunelli M. 2008. Hierrchically structured scleractinian coral biocrystals. *Journal of Structural Biology*, 161:74–82.

Seong JK, Baek N, Kim KJ. 2011. Real-time approximation of molecular interaction interfaces based on hierarchical space decomposition. *Computer-Aided Design*, 43:1598–1605.

Shuttleworth R. 1950. The surface tension of solids. *Proceedings of the Physical Society. Section A*, 63(5):444–457.

Sintya E, Alam P. 2016a. Self-assembled semi-crystallinity at parallel-beta sheet nanocrystal interfaces in clustered MaSp1 (spider silk) proteins. *Materials Science and Engineering C*, 58:366–371.

Sintya E, Alam P. 2016b. Localised semi-crystalline phases of MaSp1 proteins show high sensitivity to overshearing in beta-sheet nanocrystals. *International Journal of Biological Macromolecules*, 92:1006–1011.

Sitti M. 2003. Synthetic gecko foot-hair micro/nano-structures as dry adhesives. *Journal of Adhesion Science and Technology*, 18(7):1055–1074.

Stillfried DG, Toivakka M, Alam P. 2013. Crispatotrochus-mimicking coatings improve the flexural properties of organic fibres. *Journal of Materials Science*, 48(24):8449–8453.

Tietze S, Apostolakis J. 2007. Glamdock: Development and validation of a new docking tool on several thousand protein-ligand complexes. *Journal of Chemical Information and Modeling*, 47:1657–1672.

Touaiti F, Alam P, Toivakka M, Bousfield DW. 2010. Polymer chain pinning at interfaces in $CaCO_3$-SBR latex composites, *Materials Science and Engineering A*, 527: 2363–2369.

Venkatachalam CM, Jiang X, Oldfield T, Waldman M. 2003. LigandFit: A novel method for the shape-directed rapid docking of ligands to protein active sites. *Journal of Molecular Graphics and Modelling*, 21:289–307.

Wang RZ, Suo Z, Evans AG, Yao N, Aksay IA. 2001. Deformation mechanisms in nacre. *Journal of Materials Research*, 16:2485–2493.

Warren, T. L. and Krajcinovic, D. 1996. A fractal model for the static coefficient of friction at the fiber-matrix interface. *Composites: Part B*, 27B:421–430.

Wee KM, Rogers TN, Altan BS, Hackney SA, Hamm C. 2005. Engineering and medical applications of diatoms. *Journal of Nanoscience and Nanotechnology*, 5(1):88–91.

Young L, Jernigan RL, Covell DG. 1994. A role for surface hydrophobicity in protein–protein recognition. *Protein Science*, 3:77–729.

Index

A

A549 cell line, 125, 134
AA, *see* Ascorbic acid
Abatement of pollution, 304
Abaxis Piccolo Xpress Company, 94
Ab initio simulations, 454
Abraxane, 116, 149–153
Absorption, metabolism, distribution, and excretion profiles (ADME profiles), 438
Abuse indicators, 403
Acetone, 124
Acidic media, 350
ACS, *see* Applied compressive stress
ACSs, *see* Acute coronary syndromes
Active food packaging, 399
Active packaging, 420
Active targeting for lung cancer, 120
    lung cancer cells, 120
    nanocarriers, 121
    tumoral endothelium, 120–122
Acute coronary syndromes (ACSs), 81
Acute myocardial infarction (AMI), 70
ADCs, *see* Antibody-drug conjugates
Adenosine 5′-monophosphate (AMP), 181
5′ Adenosine monophosphate-activated protein kinase (AMP), 358
Adenosine triphosphate (ATP), 261
Adipose derived stromal cells (ASCs), 269
Adjuvants, 114
Adjuvant therapy, 110
ADME profiles, *see* Absorption, metabolism, distribution, and excretion profiles
Adoptive cell therapy, 115
Adsorbents, engineered nanoparticles as, 303
Adsorption mechanism, 303
AEE, *see* 2-(2-Aminoethoxy) ethanol
Aerobic microorganisms, 401
Aerosol behavior of ENPs, 309
Aerosol nanoparticle, 361
AF-MNPs, *see* Amine-functionalized magnetic nanoparticles
Affordability of quality care, 97
Affordable, Sensitive, Specific, User-friendly, Rapid and Robust, Equipment-free, and Deliverable to end users criteria (ASSURED criteria), 93

Ag-PAA-SBA-15, 435
Agar, 420
Agglomeration, 331–332
Aggregated nanoparticles, 332
Aging/storage times/stability/passivation, 332–333
AgNPs, *see* Silver nanoparticles
Agrifood sector, nanotechnology in, 414
Ag/TiO$_2$-nanotube-composed nanoelectrodes, 302
AHA, *see* American Heart Association
a-IgG, *see* Antihuman γ-globulin
Air, 309
    pollution, 73
Akt1 protein, 126
AKT signaling, 292
5-ALA, *see* 5-Aminolevulinic acid
AL-AuNR-DTX, *see* Docetaxel-loaded anionic lipid-coated gold nanorods
Alginate, 39, 224
    nanomaterials, 12
Alkaline phosphatase (ALP), 83, 94
Allyl isothiocyanate, 431
Alopecia, 152
ALP, *see* Alkaline phosphatase
Alpha elimination, 350
α-acetylene-poly (tert-butyl acrylate) (ptBA), 173
α-lactalbumin nanotube, 393
αvβ6-dependent H2009 lung cancer cells, 125
American Heart Association (AHA), 81
AMI, *see* Acute myocardial infarction
Amine-functionalized magnetic nanoparticles (AF-MNPs), 416
Amine (NH$_2$), 127, 416
    functionalized CNT, 87
Amino acids, 416
2-(2-Aminoethoxy) ethanol (AEE), 204
5-Aminolevulinic acid (5-ALA), 269
3-Aminopropyltriethoxysilane (APTES), 87
Ammonia-functionalized MWNT vector (MWNT-NH$_3^+$ vector), 137
Amorphous mixtures, 330
Amoxicillin, 433
AMP, *see* 5′ Adenosine 5′-monophosphate; Adenosine monophosphate-activated protein kinase

Amphiphilic polymer coating, 186
  carbon nanotubes, 188–189
  gold NPs, 186–187
  iron oxide NPs, 187–188
  quantum dots, 189–190
Amphiphilic polyurethane ENPs, 303
Amphiphilic substances, 425
Amplified surface-to-volume ratio, 15
Analytes, 141
Angina, 70
Angiogenesis, 252, 291
Anthropogenic activities, environmental pollution to, 325
Anti-EGFR antibody-conjugated nanorods, 134
Anti-immunoglobulin (Anti-IgG), 185
Antiangiogenesis agents, 124
Antibiotics, 416, 433
Antibodies, 79, 80, 113, 416
  antibody-conjugated pH-triggered drug-releasing magnetic NPs, 47
  antibody-sandwich immunoassays, 80
Antibody-drug conjugates (ADCs), 114
Anticancer
  drugs, 41
  vaccines, 114
Antiferromagnetic iron oxide, 330
Antigens, 82–83, 113
Antihuman γ-globulin (a-IgG), 88
Antimicrobial
  activity, 397–398
  agents, 399, 414
  antimicrobial-loaded mesoporous silica particles, 431–434
  food packaging, metallic NPs application as components of, 420
  functionalized MSPs, 434–435
  loaded and functionalized MSPs, 435–436
  mechanisms, 400
  molecules, 415
  nanomaterials, 414–415
  nanoparticles, 415
  packaging, 420
  peptide LL-37, 431
  proteins, 427
  substances, 398, 424
  systems, 399–400
Antioxidant agents, 399
Antiripening activity, 398
Aorta, 70–71
Apoferritin-CeO$_2$ nanotruffle, 145
Apoferritin-encapsulated Pt nanoparticles, 145
Apoferritin (apoFt), 145
Apoptosis process, 326

Applied compressive stress (ACS), 262
Aptamer, 96–97
APTES, *see* 3-Aminopropyltriethoxysilane
APTES: H$_2$O amination solution, 97
Aquatic environment, nanoparticles and
  ecotoxicological impacts of ENPs, 305–306
  environmental applications of ENPs, 301–305
  fate and accumulation of ENPs, 307–309
  parameters determining fate of ENPs in environment, 302
  safety of nanotechnology, 309–315
Aquatic organisms, 306
Aqueous iron salts reduction, 335
Aqueous solution, 342
Archaeosomes, 395
Architectural nanoparticles for lung cancer, 127–128
Area under the curve (AUC), 152
Arg-Gly-Asp (RGD)-HPMA-Indium-111 (In-111)-conjugated nanotheranostics, 40
Arginine-glycin-aspartate (RGD), 145, 265, 288–289
Array technique, 89
Artemisia arborescens (*A. arborescens*), 427
Arteries, walls of, 72
Ascorbic acid (AA), 189
ASCs, *see* Adipose derived stromal cells
Aspect ratio, 399
Aspergillus niger (*A. niger*), 426
ASSURED criteria, *see* Affordable, Sensitive, Specific, User-friendly, Rapid and Robust, Equipment-free, and Deliverable to end users criteria
ASSURED guidelines, 97
Atezolizumab, 114
Atherosclerosis, 76, 80
  diseases, 74
  plaque, 81
Atmospheric biomes, 300
Atom–atom interfaces, 453–454
Atomic saturation, 459
Atomistic modeling methods, 453
Atom transfer radical polymerization (ATRP), 184
ATP, *see* Adenosine triphosphate
ATRP, *see* Atom transfer radical polymerization
Au-based nanomaterials, *see* Gold-based nanomaterials
Au-S bond, *see* Gold-sulfur bond
Au@SiO$_2$, *see* Silica-coated AuNPs
AUC, *see* Area under the curve

Auger electrons, 345
AuNPs, see Gold nanoparticles
Autoanalyzers, 80
Autodock software, 456
Autologous DRibbles vaccine, 115
Avastin®, see Bevacizumab
Azide-terminated poly(N-isopropylacrylamide) (N3-PNIPAM), 185

B

Bacillus cereus (*B. cereus*), 432
Bacillus subtilis (*B. subtilis*), 358, 426
Backscattered electrons, 345
BACPT, see 7-Butyl-10-amino camptothecin
Bacteria, 12
Bacterial removal, 416
Bacteriocins, 427
Ball milling approach, 333–334
Basic fibroblast growth factor (bFGF), 265
Bavituximab, 113
Bayes hierarchical Bayesian network, 455
Bayesian network, 455
Bayesian parameters, 455
BBBs, see Blood-brain barriers
BCC research report, see Business Communications Company research report
B-cell lymphoma 2 gene (BCL2 gene), 127
B-cells, 79
BCL2 gene, see B-cell lymphoma 2 gene
bcp-GQDs, see Block copolymer-integrated graphene quantum dots
Beta-carotenes, 392
Beta elimination, 350
Beta fibroblast growth factor (bFGF), 289
3β-[N-(N,N-dimethylaminoethane)-carbamoyl] cholesterol (DC-chol), 149, 202
β-lactoglobulin, 425
Bevacizumab, 114, 120
bFGF, see Basic fibroblast growth factor; Beta fibroblast growth factor
B/F separation, see Bound/free separation
Bi-phospholipid-coated PLGA nanoparticles, 124
Bifidobacterium spp, 397
Bimetallic Fe/Pd nanoparticles, 341
Bimetallic iron nanoparticles particles, 332
Bimetallic particles, 344
Bio-based nanocarriers, 117
Bio-based nanoparticles, 141, 150–151; see also Polymer-based nanoparticles for lung cancer
    apoFt, 145
    liposomes, 147–148
    protein-based nanoparticles, 146–147
    SLNs, 148–149
    viral NPs, 145–146
Bioactive compounds, 394, 397, 427, 429, 437
Bioactive food packaging technology, 428
Bioactivity, 280
Biochemical oxygen demand (BOD), 305
Biocompatibility, 280, 284
    biocompatible polyester dendrimer, 127
    nanomaterials, 436–437
Biodegradable polymers, 37
    synthetic polymer, 341
Biodiesel production, 305
"Bioelectronic nose", 404
Biofunctional MNPs, 416
Biological/biology, 17
    capping agents-plant/algal-based polyphenols, 343
    implications of dissolution, 378–381
    macromolecules, 454
    materials, 453
    water window, 7
Biological identity
    to bare nanoparticle, 18
    of particle, 21
Biologic therapy, see Immunotherapy
Biomacromolecules, 189
Biomarkers, 80–82
Biomaterials, 279
    for tissue engineering scaffolds, 280
Biomechanical interfaces at higher length scales, 459–460
Biomimetism, 280
Biomolecules
    biomolecule-coated NPs, 194
    carbon nanotubes, 196–197
    gold NPs, 194–195
    iron oxide NPs, 195–196
    quantum dots, 197
    surface modification with, 193–194
BIONs, see Bismuth iron oxide nanoparticles
Bioral™ nanocochleate nutrient delivery system, 392
Biorecognition molecules, 416
Bioremediation, 353
    agglomeration, 331–332
    aging/storage times/stability/passivation, 332–333
    shape and particle uniformity, 332
    size, 330

Bioscaffolds, 283
   custom scaffold, 284
   injectables, 283–284
Biosensor, 81, 82, 414
   platforms, 82
Biosphere, 305
Biosynthetic route, 135
Biotemplating, 282
Biotherapy, *see* Immunotherapy
Biotin-conjugated cTnI detection antibodies, 85
Biotin–streptavidin capture technology, 84
Biotinylation, 87
Bisector method, 456
Bismuth, 222, 227
Bismuth iron oxide nanoparticles (BIONs), 227
Block copolymer-integrated graphene quantum dots (bcp-GQDs), 201
Block copolymers, 197
   block copolymer-modified NPs, 198
   carbon nanotubes, 200
   gold NPs, 199
   iron oxide NPs, 199–200
   quantum dots, 200–201
Blood-brain barriers (BBBs), 5
BMPs, *see* Bone morphogenetic proteins
BMSCs, *see* Bone marrow derived mesenchymal stem cells
BNP, *see* B-type natriuretic peptide; Brain natriuretic peptide
BOD, *see* Biochemical oxygen demand
Bone marrow derived mesenchymal stem cells (BMSCs), 289
Bone morphogenetic proteins (BMPs), 292
Bone tissue engineering, 282
Bound/free separation (B/F separation), 84
Bovine serum albumin (BSA), 196
Brachytherapy, *see* Internal beam radiation therapy
Brain natriuretic peptide (BNP), 81
Breast adenocarcinoma (MCF-7), 127
BSA, *see* Bovine serum albumin
B-type natriuretic peptide (BNP), 81
Business Communications Company research report (BCC research report), 3
7-Butyl-10-amino camptothecin (BACPT), 127

C

c-Jun N-terminal kinases (JNK), 188
C-reactive protein (CRP), 81
c-terpinene, 427
CAD, *see* Computer-aided design
Cadmium selenide (CdSe), 10, 138
Cadmium sulfide QDs (CdS QDs), 189
Cadmium telluride (CdTe), 136
Caelyx®, 116
CAG, *see* Coronary angiography
Calcium, 436
Calcium fluoride/fluorinated hydroxyapatite nanocrystals (CF/FHAp nanocrystals), 12–13
Calcium nanoparticles, 394
Calcium oxide (CaO), 415
   NPs, 417
Cancer, 106, 116, 251
   partner with, 72
Cancer antigens (CAs), 138
Cancer cells, 114
   behavior regulation, extracellular matrix role in, 256–257
   migration, 260
Cancer stem cells (CSCs), 265
Capping agents, 382
Caprylic acid, 432
Capsaicin-loaded nanoliposome, 396
Capsids, 146, 224
Carbamates, 304
Carbodithiolates, 176
Carbohydrates, 416
Carbon
   dots, 11
   nanodots, 234
   nanofibers, 136
   nanostructures, 234–235
Carbon-based materials, 10–13
Carbon-based nanoparticles (CNPs), 136
   CNTs, 136–137
   ND, 137–138
Carbon-based nanostructures (CBNs), 55–57
Carbon nanohorn (CNH), 136
Carbon nanotubes (CNTs), 10, 56, 87, 136–137, 399; *see also* Gold nanoparticles (AuNPs); Iron oxide nanoparticles (IONPs); Quantum dots (QDs)
   amphiphilic polymer coating, 189–190
   block copolymers, 200
   click chemistry, 184–185
   CNT-based drug delivery system, 137
   ligand exchange reactions, 177–178
   lipid coating over NP surfaces, 203
   PEG, 192–193
   silanization of NP surface, 180
   surface modification with biomolecules, 196–197
Carboplatin, 112, 116
Carboxylated grapheme, 88
Carboxyl derivatized polymer (COOHP), 190

Carboxymethyl cellulose (CMC), 340, 341
Carcinoembryonic antigen (CEA), 138
Cardiac biomarker, 80
    chemiluminescence-based immunoassays, 83–84
    colorimetric detection-based assay, 90
    detection, 82–90
    electrical detection-based assay, 85–88
    ELISA, 82–83
    fluorescence-based immunoassays, 84–85
    sandwich immunoassay, 85
    silver enhancement based colorimetric detection of cardiac troponin I, 90
    summary of primary clinically utilizing, 81
    surface plasmon resonance-based detection, 88–89
Cardiac biomarkers, 98
Cardiac catheterization, 78
Cardiac CT, 78
Cardiac troponin I (cTnI), 81, 98
    detection and quantification, 82
Cardiovascular disease (CVD), 70, 76, 80
    biomarkers and role in cardiovascular risk, 80–82
    CAG, 78
    cardiac biomarkers detection, 82–90
    causes of death, 74
    challenges, 97–98
    comparison of CVD deaths in developed *vs.* developing countries, 74
    current tools for diagnosis of, 76–80
    in developing world, 73–76
    ECG, 77
    echocardiography, 78
    human body indicating particular body area, 71
    imaging techniques, 77–78
    immunoassay technique, 78–80
    POCT, 90–97
    proportion of global deaths and number of deaths, 76
    rate of CVD increasing in developing world, 75–76
    risk factors for, 72–73
    types, 70–72
Cardiovascular risk (CVR), 81
Cardiovascular risk, role in, 80–82
Carrying acrylate groups (HA-AC), 267
Cartilage bioengineering, 291
Carvacrol, 425, 427, 431
    cinnamaldehyde, 397
    nanoemulsion, 426
CAs, *see* Cancer antigens

Catalysis, 305
Catheter, 78
Cationic lipid-coated gold nanoparticle (CL-AuNP-DTX), 202
Cat-MDBC, *see* Multidentate block copolymer with catechol groups
Caveolae, 152
CBNs, *see* Carbon-based nanostructures
CCD, *see* Charge-coupled device
CD, *see* Compact disc
CD24 mRNA expression, 267
CDC, *see* Centers for Disease Control and Prevention
CDP-Star/Sapphire-II, 84
CdSe, *see* Cadmium selenide
CdS QDs, *see* Cadmium sulfide QDs
CdTe, *see* Cadmium telluride
CdTe/CdSe QDs, 189
CdTe QD-loaded antibody-conjugated pluronic triblock copolymer-based micelles, 53
CEA, *see* Carcinoembryonic antigen
Cell migration studies, 260
Cells, biomaterial interactions with, 284
    biocompatibility, 284
    mechanosensing, 291
    mitogen-activated protein kinase pathway, 292
    nanoscale composition, 288–289
    nanoscale features, 286–288
    PI3 kinase/Akt, 292
    porosity, 289–291
    responses of cells and signaling pathways, 291–292
    rho-associated protein kinases, 292
    surface of nanomaterial, 285–286
    transforming growth factor-β pathways, 292
    Wnt/β-catenin, 292–293
Cellulose, 340, 398
CE-MRA, *see* Contrast-enhanced magnetic resonance angiography
Centers for Disease Control and Prevention (CDC), 81
Central laboratory test, 91
Ceramic nanoparticles, 282
Cerebrovascular disease, 70
Cerenkov luminescence imaging (CLI), 235
Ceria, 13
Cerium oxide (CeO_2), 415
    NPs, 13, 417
CEST agents, *see* Chemical exchange saturation transfer agents
Cetuximab (Erbitux), 113

Cetyltrimethylammonium bromide (CTAB), 20, 331
CF/FHAp nanocrystals, *see* Calcium fluoride/fluorinated hydroxyapatite nanocrystals
Ch-g-PEI, *see* Polyethylenimine grafted chitosan
Characteristic x-rays, 345
Charge-coupled device (CCD), 345
    camera, 83
CHD, *see* Coronary heart disease
Chelating process, 436
Chemical composition in tumor ECM, 265
Chemical exchange saturation transfer agents (CEST agents), 230–231
Chemical oxygen demand (COD), 305
Chemical preservatives, 414
Chemical treatment methods, 328–329
Chemiluminescence (CL), 83
    CL-based immunoassays, 83–84
Chemiluminescence immunoassays (CLIAs), 83
Chemiluminescent enzyme immunoassay (CLEIA), 84
Chemiluminometric ELISA-on-chip biosensor, 83
Chemotherapeutic agents, 112, 124
Chemotherapeutic drugs, 112, 121, 122
Chemotherapy, 108, 112, 252
Chip-based devices, 96
Chitosan, 39, 420
    chitosan-based controlled release system, 126
    nanofibers, 428
    nanoparticles for lung cancer, 126
Chlorhexidine, 431
Chlorinated pesticides, 351
Chlorinated phenols, 303
Chlorinated solvents, 349–350
Cholesterol, 202
Chondroitin sulfate (CS), 253
Chronic obstructive pulmonary disease (COPD), 73
Cigarette smoking, 107
Cinnamaldehyde, 431
Circulating tumor cells (CTCs), 136
Circulatory system, 70
Cisplatin, 112, 136, 154
    cisplatin-coated sodium alginate-functionalized NDs, 138
    cisplatin-DPPG reverse micelles, 154
Citrate, 382
CK-MB, 86, 87, 98
CL-AuNP-DTX, *see* Cationic lipid-coated gold nanoparticle
CL, *see* Chemiluminescence
Classical emulsions, 424
Clays, 429
Cleavable MSNs-cMSNs, 140
CLEIA, *see* Chemiluminescent enzyme immunoassay
CLI, *see* Cerenkov luminescence imaging
CLIAs, *see* Chemiluminescence immunoassays
Click chemistry, 181
    carbon nanotubes, 184–185
    gold NPs, 181–183
    iron oxide NPs, 183–184
    quantum dots, 185–186
Climate change, 325
CMC, *see* Carboxymethyl cellulose; Critical micelle concentration
CMOS, *see* Complementary metal oxide semiconductor
CNH, *see* Carbon nanohorn
CNPs, *see* Carbon-based nanoparticles
CNTs, *see* Carbon nanotubes
Co-precipitation method, 334–335
Coating generation methods, 339
    bimetallic particles, 344
    biological capping agents-plant/algal-based polyphenols, 343
    emulsified nanoscale zero-valent iron, 342–343
    immobilized nanoparticles on solid supports, 343–344
    mesoporous carbon/silica coated, 342
    natural polymers, 340
    PMAA-PMMA-PSS triblock, 341–342
    polyelectrolyte-carboxymethyl cellulose, 341
    PVA, 341
Cobalt-chromium alloys, 385
Cochleates, 395
COD, *see* Chemical oxygen demand
Coefficient of variation (CV), 83
Coenzyme Q10 (CoQ10), 392
Coinciding wave functions, 454
Collagen, 287
    fibers, 288
Colloidal gold, 133
Colorimetric detection-based assay, 90
Colorimetric sensing system, 90
Comberstatin, 124
Combination therapy, 112
Comet assay, 23
Compact disc (CD), 94
Competitive immunoassays, 79
Complementary metal oxide semiconductor (CMOS), 86
Computed tomography (CT), 36, 77, 224, 225
    CT scanner, 226

gold nanoparticles, 227
heavy metal nanoparticles, 227–228
iodine-containing nanoparticles, 226–227
UCNPs, 227
Computer-aided design (CAD), 283
Conditional dependencies, 455
Congenital heart disease, 71
Conjugated monoclonal antibodies (Conjugated mAbs), 113–114
Consumers, 315
demand alternative antimicrobials, 420
Contact guidance, 287
Contrast agents, 220, 224–225
carriers, 221
Contrast-enhanced magnetic resonance angiography (CE-MRA), 77
Contrast-generating components, 221, 224
Contrast-induced nephropathy, 226
Contrast media, see Contrast agents
Conventional treatment modalities, 108
chemotherapy, 112
immunotherapy, 112–115
radiation therapy, 110–112
surgery, 110
Converging technology, 299
COOHP, see Carboxyl derivatized polymer
COPD, see Chronic obstructive pulmonary disease
Copper (Cu), 375, 415
Copper nanoparticles (CuNPs), 194, 302
Copper oxide (CuO), 415
CoQ10, see Coenzyme Q10
Core-forming PDLLA, 155
Coronary angiography (CAG), 78
Coronary artery disease, see Coronary heart disease (CHD)
Coronary heart disease (CHD), 70, 72, 75, 76, 80
Correction factors, 454
Cost of iron NP remediation/feasibility, 356–357
Cremophor®, 124
Cremophor® EL, 149, 152, 155, 156
cRGD, see Cyclic Arg-Gly-Asp
Critical micelle concentration (CMC), 223
CRMM-LC, see Lung Cancer module of the Cancer Risk Management Model
Cross-flow chromatogram, 83–84
CRP, see C-reactive protein
Crystalline nanoparticles, 336
Crystalline structures, 330
Crystalline structures, 333–334
CS, see Chondroitin sulfate
CSCs, see Cancer stem cells

CT, see Computed tomography
CTAB, see Cetyltrimethylammonium bromide
CTCs, see Circulating tumor cells
cTnI, see Cardiac troponin I
cTnT, 98
Roche Elecsys hs-cTnT POC assay, 91
CT scan, see Computed tomography
Cu(I)-catalyzed azide-alkyne cycloaddition, 182
CuNPs, see Copper nanoparticles
Curcumin-loaded nanoemulsions, 425
Curcumin, 425
Curcuminoids, 423
Custom scaffold, 284
CV, see Coefficient of variation
CVD, see Cardiovascular disease
CVR, see Cardiovascular risk
Cy3 dye, 146
Cyclic Arg-Gly-Asp-D-Phe-Lys [c(RGDfK)], 40
Cyclic Arg-Gly-Asp (cRGD), 49, 204, 233
Cys-Pro-Leu-Gly-Leu-Ala-Gly-Gly peptide linker, 44–45
Cytokines, 188
Cytotoxic effect of Ag NPs, 134

D

D-biotin, 87
DCC, see N, N'-dicyclohexyl carbodiimide
DC-chol, see 3β-[N-(N,N-dimethylaminoethane)-carbamoyl] cholesterol
DCS, see Drug carrier system
DDAB, see Didodecyldimethylammonium bromide
DEAPA, see 3-(Diethyl amino) propyl amine-PVA-g-PLGA
Deep vein thrombosis (DVT), 71–72
Dendrimers, 41–43, 127–128, 224, 226–227
dendrimer-mediated encapsulation method, 127
Density functional theory (DFT), 454
Deoxyglucose labeled with ^{18}F (^{18}FDG), 236
Derjaguin, Landau, Verwey, and Overbeek theory (DLVO theory), 332
Dermatan sulfate (DS), 253
Detectors, 225
Dextran, 39, 224, 229–230
dextran-coated iron oxides, 227
Df-Bz-NCS, see p-isothiocyanatobenzyl-desferrioxamine
DFT, see Density functional theory
DHLA, see Dihydrolipoic acid

Diabetes, 73
Diblock copolymers, 125
Didodecyldimethylammonium bromide
    (DDAB), 434
Diet, 73
3-(Diethyl amino) propyl amine-PVA-g-PLGA
    (DEAPA), 124
Diethylene triamine pentaacetic acid
    (DTPA), 236
Diffraction pattern, 345
Digital microfluidic device, 96
Dihydrolipoic acid (DHLA), 176–177
DiIC@^{89}Zr-SCL, see Long-circulating bimodal
    ^{89}Zr-labeled liposomes
DiI dye, 39
Dimensionality, 258–260
3-(4,5-Dimethylthiazol-2-Yl)-2,5-
    diphenyltetrazolium bromide assay
    (MTT assay), 23
Dioleoylphosphatidyl ethanolamine
    (DOPE), 149
Dipalmitoyl-phosphatidyl glycerol (DPPG),
    154, 202
1,2 Dipalmitoylphosphatidylcholine
    (DPPC), 202
Dipeptide antioxidants, 392
Dipolar polarization, 336
Direct immunoassay, 79
Discretization, 460
Dispersion
    characteristics, 347
    medium for ENPs, 315
Dissolution
    biological implications, 378–381
    factors affecting, 376–377
    of nanoparticles, 375–376
Dithiothreitol, 176
Divalent metal oxides, 375
Diverse approaches, 415
D-limonene, 425
DLS, see Dynamic light scattering
DLTs, see Dose-limiting toxicities
DLVO theory, see Derjaguin, Landau, Verwey,
    and Overbeek theory
DMA, see N,N-dimethylacrylamide
Docetaxel, 113
Docetaxel-loaded anionic lipid-coated gold
    nanorods (AL-AuNR-DTX), 202
DOCK software, 456
Docosahexaenoic acid oxidation, 392
DOPE, see Dioleoylphosphatidyl
    ethanolamine; L-α-dioleoyl
    phosphatidylethanolamine

Dose-limiting toxicities (DLTs), 152
DOTA, see 1,4,7,10-Tetraazacyclododecane-
    1,4,7,10-tetraacetic acid (DOTA)
Dox-nanoparticles (Dox-NPs), 267
Dox-NPs, see Dox-nanoparticles
Dox, see Doxorubicin
Doxil®, 116
Doxorubicin (Dox), 38, 39, 52, 124, 267
    doxorubicin-HPMA-Gly-Phe-Leu-Gly
        conjugates, 40
    doxorubicin-loaded magnetic iron oxide, 125
DPPC, see 1,2 Dipalmitoylphosphatidylcholine
DPPG, see Dipalmitoyl-phosphatidyl glycerol
DRibbles (DPV-001), 115
Drug carrier system (DCS), 148
Drug docetaxel (DTX), 202
Drug-loaded NPs, 124
Drug release from theranostic nanoemulsions, 54
Drug screening, mimicking approaches by 3D
    platforms for, 265–270
DS, see Dermatan sulfate
DTPA, see Diethylene triamine pentaacetic acid
DTX, see Drug docetaxel
Dual-color QDs, 138, 139
Durvalumab, 114
DVT, see Deep vein thrombosis
Dyes, 350
Dye-tagged T4 VNPs, 146
Dynamic light scattering (DLS), 347

E

E-171, 394, 436
EBRT, see External beam radiation therapy
ECG, see Electrocardiography
Echocardiography (Echo), 78, 82
ECM, see Extracellular matrix
Ecotoxicity
    effects of ENPs, 310–313
    extrapolations, 361
Ecotoxicological/biological media, nanoparticle
    solubility in, 379–380
EDAC, see N-ethyl-N-(3-dimethyl-aminopropyl)
    carbodiimide hydrochloride
EDC, see 1-Ethyl-3-(3-dimrthylaminopropyl)
    carbodiimide hydrochloride
Edible films, 428
EDS, see Electron-dispersive x-ray analysis
EELS, see Electron energy loss spectroscopy
EGF, see Epidermal growth factor
EGFP, see Enhanced green fluorescence protein
EGFR-targeted SPIONs, 135
EGFR, see Epidermal growth factor receptor

Index 471

EIA system, *see* Enzyme immunosorbent assay system
Elecsyss Troponin T third-generation assay, 83
Electrical detection-based assay, 85–88
Electro-addressed protein A-aryl diazonium adducts, 84
Electroactive conjugated polymers, 404
Electrocardiography (ECG), 77, 82
Electrocardium, 77
Electrochemical-based Abbott i-STAT system, 94
Electron-dispersive x-ray analysis (EDS), 22, 346
Electron beam CT, *see* Ultrafast computer tomography (UFCT)
Electron energy loss spectroscopy (EELS), 22
Electrons, 302–303, 346
  shuttling process, 326
Electrospinning, 129, 282, 404, 428
Electrosteric stabilization, 307
Electrowetting-on-dielectric platforms (EWOD platforms), 96
Elemental iron, 329
ELISA-on-chip (EOC), 83
ELISAs, *see* Enzyme-linked immunosorbent assays
EMT, *see* Epithelial mesenchymal transition
Emulsified iron nanoparticles, 328–329, 342
Emulsified nanoscale zero-valent iron, 342–343
Emulsifiers, 395
Emulsions, 224, 229
"Enabling technology", 299
Encapsulation, 137, 392
  of curcumin, 425
  of EOs, 425
  systems, 423
Endocytosis, 120
Energy dispersive x-ray spectroscopy, *see* Electron-dispersive x-ray analysis (EDS)
Energy transfer, 459
Engineered nanomaterials, 436
Engineered nanoparticles (ENPs), 300–301, 306, 438
  behaviour in air, 301
  ecotoxicity effects, 310–313
  ecotoxicological impacts of, 305–306, 309
  ENP-incorporated products, 305
  environmental applications of, 301–305
  fate and accumulation, 307–309
  FeS, 303
  parameters determining fate of, 302
  zinc oxide, 300, 302
Engineered T cells, 115
Enhanced green fluorescence protein (EGFP), 126
Enhanced permeability and retention (EPR), 40, 118, 220
ENPs, *see* Engineered nanoparticles
Environmental applications of ENPs, 301
  abatement of pollution, 304
  catalysis, 305
  construction, 304
  degradation of organic pollutants, 302–303
  ENPs as adsorbents, 303
  nanomembranes, 304–305
  nanosensors, 304
  treatment and remediation, 303
Environmental nanoparticle applications, 361
Environmental pollution to anthropogenic activities, 325
Environmental remediation, 331
Environment, iron in, 326–328
Enzymatic hydrolysis, 137
Enzyme-linked immunosorbent assays (ELISAs), 78, 79, 82–83, 184
Enzyme immunosorbent assay system (EIA system), 95
Enzymes, 401
  catalysts, 401
  immobilization systems, 401–402
EOC, *see* ELISA-on-chip
EOs, *see* Essential oils
Epidermal growth factor (EGF), 122
Epidermal growth factor receptor (EGFR), 113
Epithelial mesenchymal transition (EMT), 258
EPR, *see* Enhanced permeability and retention
ε-poly-L-lysine (ε-PL), 435
ERK, *see* Extracellular signal-regulated kinase
Erlotinib, 114, 115, 120
Erythromycin, 433
Escherichia coli (*E. coli*), 358, 397, 405, 425, 426
  cells, 10–11
  O157:H7, 426
ESNPs, *see* Extremely small iron oxide nanoparticles
Essential oils (EOs), 423, 431
1-Ethyl-3-(3-dimrthylaminopropyl) carbodiimide hydrochloride (EDC), 196
Ethylene, 400
Ethylene glycol, 304
Eugenol, 425, 431
  microemulsions, 426
Euphorbia nivulia (*E. nivulia*), 135
European Respiratory Society's Annual Congress, 73
EWOD platforms, *see* Electrowetting-on-dielectric platforms
Excessive alcohol consumption, 73

Exercise testing, 77
Exfoliation, 196
Exogenous "synthetic biomarkers", 96
External beam radiation therapy (EBRT), 111
Extracellular matrix (ECM), 251, 286
    components, 254–255
    fibers, 128
    role in regulation of cancer cell behavior, 256–257
Extracellular signal-regulated kinase (ERK), 261–262
Extremely small iron oxide nanoparticles (ESNPs), 199
Ex vivo 3D lung tissue cultures, 120

F

[^{18}F]fluorodipalmitin ([^{18}F]FDP), 236
FA, *see* Folic acid
Fabrication of nanoscaffolds, 281–283
FAO/WHO, *see* Joint Food and Agriculture Organization/World Health Organization
Fate and accumulation of engineered nanoparticles, 307
    air, 309
    measuring parameters for fate and transport of ENPs, 308
    soil, 309
    water, 307–308
FDA-approved superparamagnetic iron oxide nanoparticle, 229
FDA, *see* Food and Drug Administration; US Food and Drug Administration
^{18}FDG, *see* Deoxyglucose labeled with ^{18}F
Fe^{III} oxyhydroxides, 330
Fenton/fenton-like reactions, 352
Fenton/photo-Fenton reaction, 352
Ferrihydrite, 330
Ferritin (Ft), 145, 224
FESEM, *see* Field emission scanning electron microscopy
FETs, *see* Field-effect transistors
FHBP, *see* Fibronectin-derived heparin binding peptide
FI, *see* Fluorescence imaging
Fibronectin-derived heparin binding peptide (FHBP), 265
Field-effect transistors (FETs), 86
Field emission scanning electron microscopy (FESEM), 430
Filler interfacial area, 399
Filtration membranes, 304–305

First Gokhstein equation, 458
FIs, *see* Fluorescence immunoassays
FISH, *see* Fluorescence *in situ* hybridization
Fish, 359–360
    oil, 393
FITC, *see* Fluorescein isothiocyanate
Flavor compounds, 393
Flow cytometry analysis, 267
Fluorescein isothiocyanate (FITC), 96, 127
Fluorescence imaging (FI), 141, 224–225, 232; *see also* Magnetic resonance imaging (MRI); Nuclear imaging
    carbon nanostructures, 234–235
    CLI, 235
    fluorescently labeled liposomes, 235
    gold nanoparticles, 233
    quantum dots, 232–233
Fluorescence immunoassays (FIs), 84
Fluorescence *in situ* hybridization (FISH), 10, 139
Fluorescent/fluorescence
    assay-based Triage system, 94
    dye, 405
    fluorescence-based immunoassays, 84–85
    Fluorescence tags, 84
    labeled liposomes, 235
    metal nanoshells, 139
    nanoparticles method, 405
    properties of quantum dots, 304
Fluorinated graphene sheets, 286
Fluorine
    contrast agents, 231–232
    MRI, 231
Fluoro-microbead guiding chip-based sandwich immunoassay (FMGC-based sandwich immunoassay), 85
Fluorophores, 238–239
fNDs, *see* Functionalized NDs
Folate receptor (FR), 120
Folic acid (FA), 135
Food
    fortification, 391
    grade, 393
    industry, 390
    nanoadditives, 393–394
    processing, 390–393
    sectors, 390, 405
    spoilage, 404
Food and Drug Administration (FDA), 224, 392
Food-borne pathogens, 425
Food-grade
    encapsulation systems, 424
    nanoencapsulation systems, 415

Index

nanoparticles application in food packaging, 428–429
Food packaging
    food-grade nanoparticles application in, 428–429
    nanocomposite active food packaging, 399–402
    nanocomposite smart packaging systems, 403–405
    nanoreinforcements in food packaging materials, 398–399
    nanoreinforcements in materials, 398–399
    nanotechnology in research trends, 398
Fourier transform infrared spectroscopy (FTIR spectroscopy), 346–347
Fourier transform mass spectrometry (FTMS), 24
FR, see Folate receptor
Fractal effect, 460
Fractal topographical irregularity, 460
Framingham study, 77
Free fatty acids, 423
Freshness indicators, 403–404
Friction, 459
FTIR spectroscopy, see Fourier transform infrared spectroscopy
FTMS, see Fourier transform mass spectrometry
Full temperature history indicators, 403
Functional components, 396
Functional food ingredients, 391
Functional ingredients, 391
Functionalization, 434
Functionalized NDs (fNDs), 138

G

Gadolinium (Gd), 40, 77, 229
GAGs, see Glycosaminoglycans
γ-globulin, 88
γ-methacryloxypropyltrimethoxysilane (MAPS), 183
Gamma rays, 235
Gas chromatography coupled to mass spectrometry (GC/MS), 24, 133
Gasoline, 304
Gastrointestinal tract (GI tract), 220
GCE, see Glassy carbon electrode
GC/MS, see Gas chromatography coupled to mass spectrometry
Gefitinib, 115
Gemcitabine (Gem), 154, 267
Generally regarded as safe (GRAS), 392
Genetic algorithms, 456

Genexol®, 116
Genexol®-PM, 155–157
Gentamicin, 433
Getein FIA 8000 Quantitative immunoassay analyzer, 94
GF, see Growth factor
GFP, see Green fluorescent protein
GI tract, see Gastrointestinal tract
GlamDock, 455
Glassy carbon electrode (GCE), 97
Glioblastoma cells (SF-268), 127
Glioma U373-MG cells, 265
Glucose oxidase (GOX), 200
Glutamate dehydrogenase, 402
Glycosaminoglycans (GAGs), 253
GO, see Graphene oxide
Goethite (α-FeO(OH)), 330
Gold-based nanomaterials (Au-based nanomaterials), 7–9
    biomedical applications, 9
Gold-sulfur bond (Au-S bond), 43, 174
Gold (Au), 133, 415
    Au/Ag hollow nanospheres, 133
    ENPs, 303
    nanorods, 133, 233
    nanoshells, 237
    nanostars, 233
Gold nanoparticles (AuNPs), 43–45, 90, 94, 133–134, 222, 227, 233, 269, 403; see also Carbon nanotubes (CNTs); Iron nanoparticles (FeNPs); Iron oxide nanoparticles (IONPs); Quantum dots (QDs)
    amphiphilic polymer coating, 186–187
    block copolymers, 199
    click chemistry, 181–183
    ligand exchange reactions, 174–175
    lipid coating over NP surfaces, 202
    PEG, 190–191
    silanization of NP surface, 178–179
    surface modification with biomolecules, 194–195
Gold nanoparticles (GNPs), see Gold nanoparticles (AuNPs)
GOX, see Glucose oxidase
Gp60, 152
Grade IV granulocytopenia, 152
Gradients, 228–229
Gram-positive bacteria, 434
Granular zero-valent iron particles, 348
Graphene, 11, 56, 87–88
    nanosheets, 399
Graphene oxide (GO), 11

GRAS, *see* Generally regarded as safe
Green fluorescent protein (GFP), 124, 145
Green synthesis, 335
Groundwater, 355
    remediation, 331, 353–354
Growth factor (GF), 261–262
Guanosine triphosphate (GTP), 261
Gyrolab xPLORE Company, 94

H

H446 lung cancer cell lines, 135–136
HA-AC, *see* Carrying acrylate groups
HA, *see* Hyaluronan; Hyaluronic acid; Hydroxyapatite
Habitual smoking, 72
"Half-cylinder" model, 203
"Hard corona", 381
Hardening of arteries, 76
Harmonic functions, 454
10HCPT, *see* 10-Hydroxycamptothecin
HDA, *see* Hexadecyl amine
HDPE films, *see* High-density polyethylene films
HDR, *see* High dose rate
Healthcare data, 98
Heart-type fatty acid binding protein (H-FABP), 81–82, 98
Heart attack, 70
Heating oil, 304
Heavy metal nanoparticles, 227–228
Hematite (α-Fe_2O_3), 329
Hemoglobin, 238
Heparin sulfate (HS), 253
Hepatocyte growth factor (HGF), 12
HER3, *see* Human epidermal growth factor receptor-3
Hermatite, 329
Heterogeneous photocatalysis method, 305
Hexadecyl amine (HDA), 173
Hexahydro-1-nitroso-3,5-dinitro-1,3,5-triazine (MNX), 351
Hexahydro-1,3,5-trinitro-1,3,5-triazine (RDX), 351, 356
Hexahydro-1,3,5-trinitroso-1,3,5-triazine (TNX), 351
H-FABP, *see* Heart-type fatty acid binding protein
HGF, *see* Hepatocyte growth factor
Hierarchical modeling, 457
HIF-1α, *see* Hypoxia inducible factor
High-density polyethylene films (HDPE films), 401
High-sensitivity CRP (hs-CRP), 95
High blood cholesterol, *see* Hyperlipidemia
High blood pressure, *see* Hypertension
High dose rate (HDR), 111
Higher-sensitivity POC assays, 91
hNIS gene, *see* Sodiumiodide symporter gene
HOEt-MWNTs, *see* Hydroxyethylated MWNTs
Hollow silica spheres, 429
Hollow spheres, 433
Horseradish peroxidase (HRP), 83, 94, 185
Hounsfield units (HU), 226
HPMA, *see* N-(2-hydroxypropyl) methacrylamide
HRP, *see* Horseradish peroxidase
hs-CRP, *see* High-sensitivity CRP
HS, *see* Heparin sulfate
HSA, *see* Human serum albumin
HU, *see* Hounsfield units
Human breast cancer cells (MCF-7), 269
Human bronchial epithelial cell line (16HBE14o-), 147
Human colorectal adenocarcinoma (HT-29), 127
Human epidermal growth factor receptor-3 (HER3), 114
Human H-chain ferritin (rHF), 145
Human health, 325
Human hepatoma 7402 cell lines, 135–136
Human serum albumin (HSA), 147, 196
Human umbilical vein endothelial cells (HUVEC), 267
HUVEC, *see* Human umbilical vein endothelial cells
Hyaluronan (HA), 39, 253
    hydrogel-based 3D tumor model, 267
Hyaluronic acid, *see* Hyaluronan (HA)
Hydrogel, 266–268
Hydrogenation, 350
Hydrogenolosis, 350
Hydrolysis, 350
Hydrophilic/hydrophilicity, 16, 43, 137, 264, 395, 428
    compounds, 423, 426
    interactions, 395
    therapeutic molecules, 41
Hydrophobic/hydrophobicity, 16, 19, 147, 264, 395, 416, 435
    bioactive compounds, 393
    core, 223
    groups, 433
    hydrophobic/oil layer, 342–343
    interactions, 301, 395
    vitamins, 391
Hydrothermal/hydrosolvo synthesis, 335–336

Hydroxyapatite (HA), 282
10-Hydroxycamptothecin (10HCPT), 127
Hydroxyethylated MWNTs (HOEt-MWNTs), 189
Hydroxyethyl cellulose, 404
Hydroxyl (OH), 127
    groups, 420
    radical, 302–303
HyperAcute®, see Tergenpumatucel-L
Hyperlipidemia, 72
Hypertension, 72
Hypertensive diseases, 74
Hypoxia, 120, 256
Hypoxia inducible factor (HIF-1α), 263–264

I

ICM, 455
IC-MS, see Inductively coupled mass spectroscopy
ICP-AES, see Inductively coupled plasma atomic emission spectroscopy
ICPMS, see Inductively coupled plasma mass spectrometry
IFN-γ, see Interferon gamma
Igs, see Immunoglobulins
IGF, see Insulin-like growth factor
ILs, see Interleukins
Imaging
    modalities, 224–225
    techniques, 54, 77–78
Immobilized capture antibodies, 84
Immobilized nanoparticles on solid supports, 343–344
Immune checkpoint inhibitors, 114
Immune system, 114
Immunoassay
    detection, 80
    technique, 78–80
Immunoglobulins (Igs), 12, 79
Immunotherapy, 112
    adoptive cell therapy, 115
    immune checkpoint inhibitors, 114
    mAbs, 113–114
    therapeutic vaccines, 114–115
Imperm nanocomposite, 401
In-vitro
    antifungal activity, 431
    cytotoxicity of conjugate, 137
    imaging, 139
    screening process, 96
    studies, 147, 438, 439
In-vivo
    imaging, 139
    studies, 438, 439

Indirect immunoassay, 79
Induced pluripotent stem cells (iPSCs), 281
Inductively coupled mass spectroscopy (IC-MS), 23
Inductively coupled plasma atomic emission spectroscopy (ICP-AES), 22
Inductively coupled plasma mass spectrometry (ICPMS), 354
Infrared absorbance spectroscopy, 346
Injectable scaffold, 283–284
Inorganic encapsulation systems, 415
Inorganic nanoencapsulation systems, 424, 429; see also Organic nanoencapsulation systems
    antimicrobial-loaded mesoporous silica particles, 431–434
    antimicrobial functionalized MSPs, 434–435
    antimicrobial loaded and functionalized MSPs, 435–436
    FESEM and TEM images of different MSPs, 430
    MSPs, 431
Inorganic nanoparticles, 325, 436
In situ remediation, 341
    cost of iron NP remediation/feasibility, 356–357
    groundwater remediation, 353–354
    using iron NPs, 353
    migration and monitoring, 354
    surface waters, 355
    terrestrial remediation, 355–356
    water remediation, 303
Insulin-like growth factor (IGF), 292
    IGF-1, 12
Integrated diagnostic devices, 82
Interface in hierarchical biological structures modeling
    atom–atom interfaces, 453–454
    biomechanical interfaces at higher length scales, 459–460
    molecular interfaces, 454–456
    nano effect, 456–458
Interfacial deposition method, 124
Interfacial motion, 459
Interferon gamma (IFN-γ), 188
Interleukins (ILs)
    IL-1β, 188
    IL-2, 146
    IL-6, 188
    IL-8, 147, 188, 265
    IL-10, 188
Internal beam radiation therapy, 111
International Union of Pure and Applied Chemistry (IUPAC), 429

Intravascular nanoparticle-conjugated peptide, 96
Intravenous injection (IV injection), 12
In vitro knockdown of GFP, 124
In vivo
    imaging, 234
    residence time, 125
    testing, 124
Iodine-containing nanoparticles, 226–227
Ionic conduction, 336
Ionic materials, 453–454
Ionizable amino groups, 126
IONPs, *see* Iron oxide nanoparticles
iPSCs, *see* Induced pluripotent stem cells
Iressa, *see* Gefitinib
Iron (Fe), 133, 304
    addition, 328
    in environment, 326–328
    iron-based NPs, 436
    iron/palladium bimetallic particles, 344
Iron nanoparticles (FeNPs); *see also* Gold nanoparticles (AuNPs)
    amorphous mixtures, 330
    application of iron NPs for pollution remediation, 348–349
    ball milling, 333–334
    characterization techniques, 344
    complexity in generation of iron NPs for environmental remediation, 334
    co-precipitation, 334–335
    different methods of generating coatings, 339–344
    DLS, 347
    ecotoxicity extrapolations, 361
    EDS, 346
    environmental pollution to anthropogenic activities, 325
    in environment and ecotoxicity concern, 357
    ferrihydrite, 330
    fish, 359–360
    FTIR spectroscopy, 346–347
    goethite, 330
    hydrothermal/hydrosolvo synthesis, 335–336
    in situ remediation using iron NPs, 353–357
    iron in environment, 326–328
    lepidocrocite, 330
    maghemite, 329
    magnetite, 329
    mammals, 360–361
    microemulsion, 336–337
    microorganisms, 357–358
    microwave-assisted synthesis, 336
    nanoscale zero-valent iron, 329
    optimization, 339
    parameters for application of iron NPs to bioremediation, 330–333
    plants, 359
    potential use in remediation, 328–329
    reduction of aqueous iron salts, 335
    SEM, 344–345
    sonochemical, 337–339
    synthesis of iron NPs, 333
    TEM, 345
    thermal decomposition, 336
    types, 329
    types of pollutants iron NPs catabolize, 349–352
    XAS, 347
    XPS, 347
    XRD, 345–346
Iron oxide (Fe_2O_3), 415
Iron oxide nanoparticles (IONPs), 9, 125, 136, 222, 416–417; *see also* Carbon nanotubes (CNTs); Gold nanoparticles (AuNPs); Quantum dots (QDs)
    amphiphilic polymer coating, 187–188
    block copolymers, 199–200
    click chemistry, 183–184
    ligand exchange reactions, 175–176
    lipid coating over NP surfaces, 202–203
    PEG, 191–192
    silanization of NP surface, 179
    surface modification with biomolecules, 195–196
Ischemic heart disease, *see* Coronary heart disease
Isoeugenol, 425
IUPAC, *see* International Union of Pure and Applied Chemistry
IV injection, *see* Intravenous injection

J

JNK, *see* c-Jun N-terminal kinases
Joint Committee on Powder Diffraction Standards (JCPDS), 346
Joint Food and Agriculture Organization/World Health Organization (FAO/WHO), 427

K

Keratan sulfate (KS), 253
Keytruda, *see* Pembrolizumab

Index

Kinetic model of organic degradation rate, 303
K*ras* mutant, 140

L

LA, *see* Linoleic acid
Labeled antibodies, *see* Conjugated monoclonal antibodies (Conjugated mAbs)
Lactate dehydrogenase, 402
Lactobacillus acidophilus (*L. acidophilus*), 397
Lactobacillus casei (*L. casei*), 397
Lactobacillus delbrueckii (*L. delbrueckii*), 397, 425, 426
Lactobacillus rhamnosus (*L. rhamnosus*), 397
Lactococus lactis (*L. lactis*), 427
Lactoferrin, 425
Lactose dehydrogenase enzyme, 125
LAg NPs, *see* Latex-capped silver nanoparticles
L-α-dioleoyl phosphatidylethanolamine (DOPE), 202
Laminaria japonica (*L. japonica*), 431
Laminin-related cellular pathways, 286
Lancet, The, 73
Langmuir adsorption isotherm, 303
Lanthanide nanoparticles (LNPs), 140–141
Lanthanides, 140
Lateral flow devices (LFDs), 78
Lateral flow immunoassays (LFIAs), 93–94
Latex-capped silver nanoparticles (LAg NPs), 135
Lattice energy methods, 453
LChNPs, *see* Lomustine-loaded chitosan nanoparticles
LC/MS, *see* Liquid chromatography/mass spectrometry
LCST, *see* Lower critical solution temperature
LCTP, *see* Lung cancer targeting peptide
LDCT, *see* Low-dose computerized tomography
LDR, *see* Low dose rate
Le Chatellier's principle, 174
Lepidocrocite (γ-FeO(OH)), 330
LEPU Quant-Gold-1, 94
Level of detection (LOD), 94, 95
Lewis lung carcinoma (LLC1), 264
LFDs, *see* Lateral flow devices
LFIAs, *see* Lateral flow immunoassays
LHRH peptide, *see* Luteinizing hormone-releasing hormone peptide
Ligand(s), 120
    carbon nanotubes, 177–178
    exchange reactions, 173–174, 176
    gold NPs, 174–175
    iron oxide nanoparticles, 175–176
    quantum dots, 176–177
Limonene, 397
Linezolid, 433
Linoleic acid (LA), 190
Lipid(s), 397
    carbon nanotubes, 203
    coating over NP surfaces, 201
    gold NPs, 202
    iron oxide NPs, 202–203
    lipid-based nanocarriers, 121, 122
    lipid-based nanoencapsulates, 395
    lipid-soluble vitamins, 396
    nanoparticles, 393
    quantum dots, 204
Lipid oligonucleotide conjugates (LON-QDs), 204
Lipofectin®, 149
Lipophilic
    compounds, 426
    ingredients, 392
    molecules, 394, 426
Lipoplatin™, 116, 118, 147, 148, 154–155
Lipoproteins, 224, 229
Liposomal cisplatin, *see* Lipoplatin™
Liposomal doxorubicin, 116
Liposome-polycation-hyaluronic acid (LPH), 122
Liposome(s), 49–51, 147–148, 223, 226, 229, 235, 236, 424, 426–427
    liposome-encapsulating capsaicin, 396
    in oral delivery, 392
    structure, 396
Liquid chromatography/mass spectrometry (LC/MS), 24
Liquids substances, 424
Listeria innocua (*L. innocua*), 427
Listeria monocytogenes (*L. monocytogenes*), 425, 426, 432
LLC1, *see* Lewis lung carcinoma
LMR, *see* Lossy mode resonance
LNPs, *see* Lanthanide nanoparticles
LNs, *see* Lymph nodes
Loaded antibodies, *see* Conjugated monoclonal antibodies (Conjugated mAbs)
Lobectomy, 110
Localized surface plasmon resonance effect (LSPRE), 7
Local surface plasmon resonance (LSPR), 187
LOD, *see* Level of detection
Lomustine-loaded chitosan nanoparticles (LChNPs), 126
LON-QDs, *see* Lipid oligonucleotide conjugates
Long-circulating bimodal ^{89}Zr-labeled liposomes (DiIC@^{89}Zr-SCL), 236

Lossy mode resonance (LMR), 187
Low-dose computerized tomography (LDCT), 108
Low-energy photons, 141
Low-resource settings, 97
Low dose rate (LDR), 111
Lower critical solution temperature (LCST), 199
LPH, see Liposome-polycation-hyaluronic acid
LSPR, see Local surface plasmon resonance
LSPRE, see Localized surface plasmon resonance effect
Lung
   reduced lung function, 73
   tumors, 110
Lung cancer, 106
   Abraxane® (ABI-007), 149–153
   active targeting for, 120–122
   causes, 107
   cells, 120
   clinical studies of nanosystems for, 149–157
   conventional treatment modalities, 108–115
   diagnosis, 108
   drawbacks associated with conventional therapies for, 111
   Genexol®-PM, 155–157
   Lipoplatin™, 154–155
   nanocarrier-based drugs for, 156
   need for nanotechnology in diagnosis and therapy of, 116–122
   passive targeting for, 118–119
   screening techniques for, 109
   small molecule inhibitors, 115
   stages, 107–108
   types, 107
Lung Cancer module of the Cancer Risk Management Model (CRMM-LC), 108
Lung cancer targeting peptide (LCTP), 125
Lung-specific X protein (LUNX protein), 136
Luteinizing hormone-releasing hormone peptide (LHRH peptide), 127
LVs, see Lymph vessels
Lycopene, 392
   absorption and bioavailability, 393–394
   nanostructures, 392
Lycurgus Cup, fourth-century, 1–2
Lymph nodes (LNs), 179
Lymph vessels (LVs), 179
Lys-MSNs, see Lysozyme-coated mesoporous silica nanoparticles
Lys, see Lysozyme
Lysozyme-coated mesoporous silica nanoparticles (Lys-MSNs), 434
Lysozyme (Lys), 427, 434
   entrapment, 432

M

MAA, see Mercaptoacetic acid
mAbs, see Monoclonal antibodies
Maghemite (γ-Fe_2O_3), 9
Magnesium, 305, 436
Magnesium oxide (MgO), 415
Magnetic hyperthermia, 135
Magnetic liposomes, 51
Magnetic nanoparticles (MNPs), 46–48, 135–136, 141, 416; see also Carbon-based nanoparticles (CNPs)
Magnetic resonance imaging (MRI), 9, 36, 77, 128, 175, 220, 228; see also Fluorescence imaging; Nuclear imaging
   CEST agents, 230–231
   fluorine contrast agents, 231–232
   T_1 contrast agents, 229
   T_2 contrast agents, 229–230
Magnetite (Fe_3O_4), 9, 329, 334, 415
MALDI/MS, see Matrix-assisted laser desorption/ionization mass spectrometry
Malignant tumor, 251
Mammals, 360–361
Mammary epithelial cells (MECs), 260
MAP kinase kinase (MAP2K), 292
MAP kinase kinase kinase (MAP3K), 292
MAPK pathway, see Mitogen-activated protein kinase pathway
Marinetoxins, 403
Mass transfer rate, 401
Material safety data sheets (MSDSs), 16–17
Matrix-assisted laser desorption/ionization mass spectrometry (MALDI/MS), 24
Matrix metalloproteinase (MMPs), 45, 120, 253
   MMP-based novel probes, 45
   MMP2, 140
   MMP9, 120, 140
Maximum tolerated dose (MTD), 152
MBC, see Metastatic breast cancer
MB dye, see Methylene blue dye
MCLW sensor, see Metal clad leaky waveguide sensor
MCM-41
   materials, 433
   nanoparticles, 435
MCM-n, see Mobil Corporation Matter materials
Mechanosensing, 291

Index

MECs, see Mammary epithelial cells
Medical imaging
    CT, 225–228
    dendrimers, 224
    emulsions, 224
    fluorescence imaging, 232–235
    imaging modalities and contrast agents, 224–225
    liposomes, 223
    micelles, 223
    MRI, 228–232
    nanocrystals, 222–223
    nanomaterials in medical imaging and tagging, 7, 8
    nanoparticle contrast agents for, 220, 221
    natural nanoparticles, 224
    nuclear imaging, 235–237
    PA imaging, 237–239
    polymeric nanoparticles, 224
    QDs, 221–222
    surface-enhanced Raman spectroscopy, 239–241
MEDINA, see Methylenedinitramine
Membrane-based immunoassays, 78
Menstruation, age of first, 73
3-Mercapto-1-propanesulfonate (MPS), 175
Mercaptoacetic acid (MAA), 176
Mercaptohexadecanoic acid (MHDA), 89
3-Mercaptopropionic acid (MPA), 193
Mesenchymal stem cells (MSCs), 267, 286
Mesoporous carbon/silica coated, 342
Mesoporous silica nanoparticles (MSNs), 120, 140
Mesoporous silica particles (MSPs), 424, 429, 431
    antimicrobial loaded and functionalized, 435–436
Metal-based and miscellaneous nanoparticles for lung cancer, 133, 142–144; see also Polymer-based nanoparticles for lung cancer
    Ag NPs, 134–135
    Au NPs, 133–134
    CNPs, 136–138
    LNPs, 140–141
    MNPs, 135–136
    MSN, 140
    nanoshells, 139–140
    QD, 138–139
Metal clad leaky waveguide sensor (MCLW sensor), 95
Metallic iron, 326, 329, 333
Metallic nanoparticles, 415
    application as components of antimicrobial food packaging, 420
    for food packaging applications, 421–422
    iron oxide nanoparticles, 416–417
    and mechanisms of action, 418–419
    other metal oxide nanoparticles, 417
    silver and gold nanoparticles, 416
Metal oxide nanoparticles, 415, 417
Metal(s), 351–352, 415
    elements and oxides, 222
    metal-based nanocarriers, 117
    metal-ion-polymer complex, 341
    nanoparticles, 12, 238
Metastatic breast cancer (MBC), 153
Metastatic tumors, 106
Methoxy-PEG (mPEG), 192
Methylene blue dye (MB dye), 188, 404
Methylenedinitramine (MEDINA), 351
Methyl tertiary butyl ether (MTBE), 304
MFM, see Multifunctional micelle
MgO NPs, 417, 419
MHDA, see Mercaptohexadecanoic acid
MI, see Myocardial infarction
Micelles, 223, 226, 229, 426
Microbial cells, 326, 353, 400, 436
Microbial growth inhibition, 415
Microbial inhibition, 420
Micrococcus luteus (*M. luteus*), 432
Micrococcus lysolideikticus (*M. lysolideikticus*), 427
Microemulsion, 336–337, 424, 426
Microfluidic(s), 269
    based-approaches, 91, 92
    chip-based capillary systems, 96
    chip-based immunoassays, 94
    microfluidic-based immunoassay, 96
    models, 269–270
    technology, 269
    in vitro 3D cancer model, 270
Micron-sized emulsions, 438
Microorganisms, 357–358
    detection, 405
microRNA (miRNA), 122, 139
    miRNA-486 molecule, 139
Micro-scale topography, 287
Microscopic
    assays, 417
    characteristics, 390–391
Microsized MSPs, 439
Microwave-assisted synthesis, 336
Mimicking
    approaches by 3D platforms for drug screening, 265
    biochemical properties, 263–265
    dimensionality, 258–260

Mimicking (*Continued*)
    geometrical properties, 257–261
    hydrogel, 266–268
    mechanical properties, 261–263
    microfluidics models, 269–270
    nanofibers, 268
    porosity, 260–261
    properties of tumor extracellular matrix, 257
    topography, 258
Miniaturized immunoassays, 96
miRNA, *see* microRNA
Mitogen-activated protein kinase pathway (MAPK pathway), 291, 292
Mitsubishi PATHFASTs, 84
MMPs, *see* Matrix metalloproteinase
Mn-doped ZnS quantum dots, 304
MNPs, *see* Magnetic nanoparticles
MNX, *see* Hexahydro-1-nitroso-3,5-dinitro-1,3,5-triazine
Mobil Corporation Matter materials (MCM-n), 429
Modern epidemiological transition, 75
Modifiable risk factors, 72–73
Modified T cells, 115
Moisture indicators, 403
MolDock software, 456
Molecular-level computations, 459
Molecular dynamics method, 453
Molecular interfaces, 454–456
Molecular modeling methods, 454
Molecular pinning, 458
Monoclonal antibodies (mAbs), 79, 83, 88, 113, 120
    conjugated mAbs, 113–114
    naked mAbs, 113
Monolaurin microemulsions, 426
Monte Carlo
    method, 453
    Monte Carlo-based approximations, 455
MPA, *see* 3-Mercaptopropionic acid
mPEG-poly(ethylene glycol) monomethyl ether (PEGylated chitosan derivatives), 193
mPEG, *see* Methoxy-PEG
MPR, *see* MRI, photoacoustic imaging, and Raman imaging
MPS, *see* 3-Mercapto-1-propanesulfonate
MRI, *see* Magnetic resonance imaging
MRI, photoacoustic imaging, and Raman imaging (MPR), 240
MRP1, *see* Multidrug resistance-associated protein 1
MSCs, *see* Mesenchymal stem cells
MSDSs, *see* Material safety data sheets
MSNs, *see* Mesoporous silica nanoparticles
MSPs, *see* Mesoporous silica particles
MTBE, *see* Methyl tertiary butyl ether
MTD, *see* Maximum tolerated dose
MTT assay, *see* 3-(4,5-Dimethylthiazol-2-Yl)-2,5-diphenyltetrazolium bromide assay
Multidentate block copolymer with catechol groups (Cat-MDBC), 199
Multidisciplinary working group, 90
Multidrug resistance-associated protein 1 (MRP1), 148
Multifunctional mesoporous nanocomposites, 135
Multifunctional micelle (MFM), 125
Multifunctional PEG-b-PDLLA micellar carrier, 125
Multipotent mesenchymal stem cells, 287
Multiwalled carbon nanotube (MWNT), 10–11, 136, 137, 188–189, 305
    MWNT-g-PCA-PTX conjugate, 137
    MWNT-PNIPAM, 185
Murine B16F10 melanoma cells, 122
MWNT, *see* Multiwalled carbon nanotube
MYO, 98
Myocardial damage, 82
Myocardial infarction (MI), 12
Myoglobin, 81, 82, 96

N

N-acetyl-L-cysteine (NAC), 134
Naked mAbs, *see* Naked Monoclonal Antibodies
Naked Monoclonal Antibodies (Naked mAbs), 113
Nano-based product market, 4
Nanobelt field effect transistor, 87
Nanobiomaterials, 3–4
Nanobiotechnology, 3, 6
Nanocages, 456
Nanocarriers, 35, 118
    active targeting for lung cancer, 120–122
    advantages, 118
    marketed nanocarriers used for passive targeting of lung cancer, 119
    passive targeting for lung cancer, 118–119
    targeting approaches for, 118–122
Nanoclays, 398, 424
Nanoclusters™ system, 391
Nanocochleates, 395
Nanocomposite(s), 333–334
    active food packaging, 398, 399
    antimicrobial systems, 399–400

detection of gases, 404
detection of microorganisms, 405
enzyme immobilization systems, 401–402
freshness indicators, 403–404
oxygen scavengers, 400–401
oxygen sensors, 404
properties, 399
smart packaging systems, 403
time temperature integrators and moisture indicators, 403
Nanocrystals, 222–223
Nanodiamond (ND), 136, 137–138
Nano effect, 456–458
Nano-electro-mechanical-systems (NEMS), 3
Nanoemulsion(s), 53–55, 226, 393, 424–426, 439
    formulations, 391
    systems, 397
Nano-enabled products, 4
Nanoencapsulation, 391, 425
    antimicrobial compounds, 420–423
    biomolecules, 423–424
    inorganic nanoencapsulation systems, 429–436
    of nutrients and delivery mechanisms, 394–397
    organic nanoencapsulation systems, 424–429
    packaging techniques, 394
    systems, 415, 420, 423, 440
Nanofabrication of materials or scaffolds, 281
Nanofibers, 128, 129, 268, 282, 424, 428
Nanofibrous scaffolds, 268, 282
Nanofillers, 398
Nanofilms, 456
Nanoflakes, 456
Nanofood processing, 391
Nanoformulations, 116, 121
Nanoingredients, 393–394
Nanoliposomes, 395
Nanomaterial(s), 1, 11, 172, 300, 315, 414, 439
    assessing nanotoxicity, 21–24
    biocompatibility, 436–437
    carbon-based materials, 10–13
    categories of bioscaffolds, 283–284
    in dental practice, 12
    in development of protein arrays, 3
    exposure state and routes of uptake, 16–17
    fabrication of nanoscaffolds, 281–283
    gold (Au)-based nanomaterials, 7–9
    interactions with cells, 284–293
    interaction with biological systems, cellular uptake, and distribution, 17–20
    interaction with living systems, 279
    iron-oxide nanoparticles, 9

limitations for food applications, 436
nanotoxicity and molecular mechanisms, 20
physicochemical properties, 13–16
quantum dots, 9–10
role in traditional medical formulations, 2–3
scaffold types, 283
sources contributing to growing environmental amounts, 17
stability, 436
surface, 285–286
tissue engineering, 280
toxicity, 437–439
types, 5–6
Nanomedicine, 116, 384
    clinical translation, 152
    imaging-guided, 35
    for lung cancer therapy, 116
    nanomedicine-based dendrimers, 41
    tunable, 48
Nanomembranes, 304–305
Nanomicelles, 125
Nano-overdose, 16–17
Nanoparticles (NPs), 6, 172–173, 281, 324–325, 333–334, 384, 390, 394, 399, 414, 433, 437, 456
    and aquatic environment, 299
    CEST agents, 230
    contrast agents for medical imaging, 220, 221
    CT, 225–228
    dendrimers, 224
    dissolution, 375–376
    dissolution measurement, 377–378
    ecotoxicological impacts of engineered nanoparticles, 305–306
    emulsions, 224
    environmental applications of engineered nanoparticles, 301–305
    fate and accumulation of engineered nanoparticles, 307–309
    fluorescence imaging, 232–235
    imaging modalities and contrast agents, 224–225
    lipid exchange process, 174
    liposomes, 223
    micelles, 223
    modifications, 173
    MRI, 228–232
    nanocrystals, 222–223
    natural nanoparticles, 224
    nuclear imaging, 235–237
    PA imaging, 237–239
    parameters determining fate of ENPs in environment, 302

Nanoparticles (*Continued*)
  PEGylation, 191
  polymeric nanoparticles, 224
  QDs, 221–222
  safety of nanotechnology, 309–315
  surface-enhanced Raman spectroscopy, 239–241
  synthesis, 337
  with technologies, 348–349
  toxicity, 357
Nanoparticle solubility
  biological implications of dissolution, 378–381
  case study for silver nanoparticles, 381–385
  dissolution of nanoparticles, 375–376
  factors affecting dissolution, 376–377
  measurement of nanoparticle dissolution, 377–378
  in various ecotoxicological/biological media, 379–380
Nanoplatforms for lung cancer, 122; *see also* Lung cancer
  bio-based nanoparticles, 141–149
  metal-based and miscellaneous nanoparticles for lung cancer, 133–141
  nanoplatforms used for lung cancer therapy and diagnosis, 123
  polymer-based nanoparticles for lung cancer, 122–132
Nanoporous material, 283
Nanoreinforcement techniques, 398
Nanoremediation, 348
Nanoscaffolds fabrication, 281–283
Nanoscale
  composition, 288–289
  enzyme immobilization systems, 401
  features, 286–288
  materials, 300
  topography, 287
Nanoscale zero-valent iron (Fe^0), 303, 329
Nanoscale zero valent iron-rectorite (nZVI-R), 344
Nanosciences, 17, 299–300
Nanosensors, 304, 398, 403, 405
Nanoshells, 139–140
Nanosilver, 400
Nanosized carriers, 415
Nanosized droplets, 393
Nanosized particulate carriers, 115
Nanospatulae, 456
Nanospikes, 456
Nanostructured encapsulation systems, 424
Nanostructured films, 11

Nanostructures, 391, 393
  carbon, 234–235
  of inorganic materials, 391
  with shell materials and applications for nutrient delivery, 394
Nanotechnology, 1, 4, 299, 389–390, 413
  in agrifood sector, 414
  antimicrobial activity, 397–398
  antimicrobial nanomaterials, 414–415
  as antimicrobial tool in food sector, 413
  applications, 390
  in development of functional foods, 390–398
  different forms of processing methods, 391
  ecotoxicity effects of ENPs, 310–313
  features of different nanoencapsulation systems, 440
  food nanoadditives and nanoingredients, 393–394
  in food packaging research trends, 398–405
  metallic nanoparticles, 415–420
  nanoencapsulation of nutrients and their delivery mechanisms, 394–397
  nanoencapsulation systems, 420–436
  nanomaterials limitations for food applications, 436–439
  nanotechnology packaging-related products present on market, 402
  need in diagnosis and therapy of lung cancer, 116–122
  nutrition and food processing, 390–393
  platforms in lung cancer, 117
  risk assessment, 314
  risk management, 314–315
  safety, 309
  targeting mechanism in lung cancer cells by, 119
Nanotheranostics, 35–36
  molecular imaging techniques, 36
  theranostic nanomedicine types, 37–57
Nanotoxicity
  assessment, 21–24
  biological implications of dissolution, 378–381
  case study for silver nanoparticles, 381–385
  dissolution of nanoparticles, 375–376
  factors affecting dissolution, 376–377
  measurement of nanoparticle dissolution, 377–378
  and molecular mechanisms, 20
  of nanomaterials, 20
  nanoparticle solubility in various ecotoxicological/biological media, 379–380

in organisms, 4
  solubility of nanoparticles and relevance in nanotoxicity studies, 375
Nanowires (NWs), 85, 300
NAPLs, see Nonaqueous phase liquids
National Lung Screening Trial (NLST), 108
National Nanotechnology Initiative (NNI), 1
Naturally occurring antimicrobial compounds, 423, 432
Natural nanoparticles, 224
Natural organic matters (NOMs), 301, 381
Natural organic molecules, 433
Natural polymers, 38, 39, 340
NC, see Nitrocellulose
NCI-H460 NSCLC cells, 127
ND, see Nanodiamond
NE, see Norepinephrine
Near-infrared (NIR), 7, 134
  fluorescence dyes, 38
  fluorophores, 238
  window, 232
Near-infrared-fluorescence (NIRF), 233
NEMS, see Nano-electro-mechanical-systems
N-ethyl-N-(3-dimethyl-aminopropyl) carbodiimide hydrochloride (EDAC), 196–197
Neuron-specific enolase (NSE), 138
Neurotrophins (NGF), 292
Next-generation theranostics nanomedicine, 36
NGF, see Neurotrophins
N-(2-hydroxypropyl) methacrylamide (HPMA), 40
Nickel titanium alloys, 385
N-isopropylacrylamide (NIPAAM), 46
NIPAAM, see N-isopropylacrylamide
NIR, see Near-infrared
NIRF, see Near-infrared-fluorescence
Nisin, 417, 425
  nanoencapsulation, 429
  nisin-loaded SLNs, 427–428
  nisin-resistant microbial strains, 427
Nitrocellulose (NC), 95
Nitrogen, 234, 351
Nitroimidazole PA-824, 431
Nitrosonium tetrafluoroborate (NOBF$_4$), 175
Nivolumab, 114
NLST, see National Lung Screening Trial
N-methyl-2-pyrrolidone (NMP), 202
NMP, see N-methyl-2-pyrrolidone
NMR spectroscopy, see Nuclear magnetic resonance spectroscopy
N,N′-dicyclohexyl carbodiimide (DCC), 197
N,N-dimethylacrylamide (DMA), 183

NNI, see National Nanotechnology Initiative
NOBF$_4$, see Nitrosonium tetrafluoroborate
Noble metal NPs, 133
N-octyl-N-mPEG-chitosan (OPEGC), 193
NOEL, see No observable effect level
NOMs, see Natural organic matters
Non-lipid-based nanocarriers, 121
Non-RTKs, see Nonreceptor tyrosine kinases
Nonaqueous phase liquids (NAPLs), 341
Nonchlorinated herbicide molinate, 351
Noncleavable MSNs-ncMSNs, 140
Nonencapsulated compounds, 395
"Nonmodifiable" risk factors, 72
Nonpolymeric materials, 47
Nonreceptor tyrosine kinases (Non-RTKs), 115
Non–small cell lung cancer (NSCLC), 110, 115, 116, 117, 127
Nonsulfated GAGs, 253
No observable effect level (NOEL), 17
Norepinephrine (NE), 189
NOTA, see 1,4,7-Triazacyclononane-1,4,7-triacetic acid
N3-PNIPAM, see Azide-terminated poly(N-isopropylacrylamide)
NPs, see Nanoparticles
NSCLC, see Non–small cell lung cancer
NSE, see Neuron-specific enolase
Nuclear imaging, 235; see also Fluorescence imaging; Magnetic resonance imaging (MRI)
  liposomes, 236
  polymeric nanoparticles, 236
  radiolabeled inorganic nanoparticles, 236–237
  ultrasound imaging, 36
Nuclear magnetic resonance spectroscopy (NMR spectroscopy), 24
Nucleation, 309
Nucleic acids, 97
Nutrition, 390–393
NWs, see Nanowires
nZVI-R, see Nanoscale zero valent iron-rectorite

O

OA, see Oleic acid
Objective response rate (ORR), 113
OEG-terminated alkanethiolate, see Oligo(ethylene glycol)-terminated alkanethiolate
Oil-in-water emulsion (O/W emulsion), 397
Oleic acid (OA), 173

Oligo(ethylene glycol)-terminated alkanethiolate (OEG-terminated alkanethiolate), 89
OMICS technology, 24
Oncolytic viruses (OVs), 146
"One-size-fits-all" treatment strategies, 80
One-step sandwich assay principle, 83
Opdivo, see Nivolumab
OPEGC, see N-octyl-N-mPEG-chitosan
Opinion on Safety of Nanomaterials in Cosmetic Products, 301
Optical imaging, 36
Oral bioavailability, 148, 391, 392
Oral squamous cell carcinoma cells (OSCC-3), 265
Organic nanoencapsulation systems, 424, 436; see also Inorganic nanoencapsulation systems
    food-grade nanoparticles application in food packaging, 428–429
    liposomes, 426–427
    microemulsions, 426
    nanoemulsions, 424–426
    nanofibers, 428
    solid lipid nanoparticles, 427–428
Organic NPs, 129, 325
Organic pollutants, degradation of, 302–303
Organic polymers, 358
Organochlorines, 304
Organophosphates, 304
Organosulfurs, 423
Origanum dictamnus EO, 427
ORR, see Objective response rate
OS, see Overall survival
OSCC-3, see Oral squamous cell carcinoma cells
Ostwald-Freundlich equation, 376
Ostwald ripening, 54
Overall survival (OS), 113
OVs, see Oncolytic viruses
O/W emulsion, see Oil-in-water emulsion
Oxidative stress, 438
Oxygen (O_2), 399–401
    oxygen-free food packaging systems, 404
    scavengers, 400–401
    sensors, 404

P

p38 mitogen-activated protein kinases (p38 MAPK), 188
p53 tumor suppressor gene, 115, 149
PAA, see Peracetic acid; Poly(acrylic acid)
PAA-coated gold NPs (PAA-GNPs), 186, 187
PAA-coated magnetite nanoparticles (PAA-MNPs), 187–188
PAA-GNPs, see PAA-coated gold NPs
PAA-MNPs, see PAA-coated magnetite nanoparticles
PAA-OA, see PAA-octylamine
PAA-octylamine (PAA-OA), 190
Paclitaxel, 53, 124, 148, 156
    paclitaxel-encapsulated chitosan nanoparticles, 38
    paclitaxel-incorporated perfluoropentane nanoemulsion, 54
PAH, see Poly (allyl amine hydrochloride)
PA imaging, see Photoacoustic imaging
Palladium chloride, 332
PAMAM, see Poly (amidoamine)
pan-cytokeratin (pan-ck), 136
Pancreatic ductal adenocarcinoma (PDAC), 267
PANI, see Polyaniline
Paper-based devices, 96
Paraplatin, 116
Parent cell, 113
Partial temperature history indicators, 403
Particle
    agglomeration, 339–340
    particle–particle attractive forces, 340
    reactivity, 333
Passive targeting for lung cancer, 118–119
Patch-by-patch method, 455
Pathogens, 79
Patritumab, 114, 120
PBS-BSA, see PBS with bovine serum albumin
PBS-DCM, see PBS with dichloromethane
PBS with bovine serum albumin (PBS-BSA), 190
PBS with dichloromethane (PBS-DCM), 190
PC, see Phosphocholine
PCA, see Poly (citric acid)
PC-AuNPs, see Phosphatidylcholine-coated AuNPs
PCL, see Poly(ε-caprolactone)
PD-1 proteins, 114
PD-L1 proteins, 114
PDAC, see Pancreatic ductal adenocarcinoma
PDMAEMA, see Polydimethylaminoethyl methacrylate
PDMS, see Polydimethylsiloxane
PDT, see Photodynamic therapy
PE, see Phosphoethanolamine; Polyethylene; Pulmonary embolism
PECs, see Predicted environmental concentration
Pediocin, 427
PEG, see Polyethylene glycol

PEG-AuNPs, *see* PEG coating over AuNPs
PEG-block-poly-ε-caprolactone (PEG-PCL), 53
PEG-CNTs, *see* PEG coating on carbon nanotubes
PEG-coated IONPs (PEG-IONPs), 191
PEG coating on carbon nanotubes (PEG-CNTs), 192
PEG coating over AuNPs (PEG-AuNPs), 190
PEGDA, *see* Polyethylene glycol diacrylate
PEG-IONPs, *see* PEG-coated IONPs
PEGP, *see* Polyethylene glycol functionalized polymer
PEG-PCL, *see* PEG-block-poly-ε-caprolactone
PEG-PLA, *see* Polyethylene glycol-poly(lactic acid)
PEGylated chitosan derivatives, *see* mPEG-poly(ethylene glycol) monomethyl ether
PEGylation, 20
    nanocarrier, 125
    of nanographene, 56–57
    of NPs, 191
    process, 221
Pembrolizumab, 114
PEN, *see* Project on Engineering Nanotechnologies
Penicillium digitatum (*P. digitatum*), 426
PEO-b-PDMAEMA, *see* Poly-(N,N-dimethylaminoethyl methacrylate)
PEO-b-PPDSM, *see* Poly(ethylene oxide)-block-poly(pyridyldisulfide ethylmethacrylate)
PEO, *see* Poly(ethylene oxide)
Peracetic acid (PAA), 435
Perfluoro-15-crown-5 ether, 231
Perfluorocarbons, 54
Peripheral arterial disease, 70–71
Permeable reactive barriers (PRBs), 248
Pesticides, 350–351, 403
PET, *see* Positron emission tomography
PFS, *see* Progression-free survival
PGA, *see* Polyglutamic acid
PGLSA, *see* Poly (glycerol succinic acid)
Phase-separation techniques, 282
Phosphatidylcholine-coated AuNPs (PC-AuNPs), 202
Phosphatidylcholine, 397
Phosphocholine (PC), 203
Phosphoethanolamine (PE), 203
Phospholipid(s), 397
    coating, 148
    encapsulation, 204
    PEGylated, 56

phospholipid-coated AuNPs, 202
phospholipid-coated MWNTs, 203
Photoacoustic imaging (PA imaging), 237; *see also* Fluorescence imaging; Magnetic resonance imaging (MRI); Nuclear imaging
    fluorophores, 238–239
    metal nanoparticles, 238
Photoacoustic imaging, 36
Photocatalytic
    mechanism, 401
    property of ZnO ENPs, 302–303
Photo-degradation cycles, 305
Photodynamic therapy (PDT), 269
    efficacy, 269, 270
    photodynamic therapy-assisted cancer treatment, 39
Photoluminescence (PL), 56, 175, 189
Photons, 345
Photosensitive molecule (PS molecule), 269
Photosensitizer-encapsulated micellar delivery, 52–53
Photosensitizer-encapsulated polymeric nanoparticles, 38
Photo-stable QDs, 185
Photosynthesis, 359, 361
Photothermal therapy (PTT), 56
Physical inactivity, 73
Physicochemical characteristics, 301
Physiochemical properties, 390–391
Phytochemicals, 423
Phytosterols, 392
Phytotoxins, 403
PI3K/Akt, 291–292
p-isothiocyanatobenzyl-desferrioxamine (Df-Bz-NCS), 236
PK1, 40, 116
PL, *see* Photoluminescence
PLA, *see* Poly lactic acid
PLA-PEG, *see* Poly(lactide)-poly(ethylene glycol)
Plant(s), 305, 359
    metabolites, 423
    plant-based polyphenol solutions, 343
PLAPCL-TPGS, *see* Poly lactic acid-polycaprolactone-d-α-tocopheryl polyethylene glycol 1000 succinate
Plaque, 76, 77, 80
Plasmid DNA-encoding p53-EGFP (pp53-EGFP), 149
PLA-TPGS, *see* Poly(lactic acid)-D-α-tocopheryl polyethylene glycol 1000 succinate
PLA-TPGS–based nanoparticles, 37

Platinum ENP-modified glassy carbon electrodes, 304
PLGA-PEG, *see* Poly(D,L-lactide-co-glycolide)-polyethylene glycol
PLGA, *see* Poly(D,L-lactide-co-glycolide) (PLGA)
PLK1 gene, *see* Polo-like Kinase gene
PMA, *see* Polymethacrylic acid
PMAA-PMMA-PSS triblock, *see* Poly(methacrylic acid)-block-(methyl methacrylate)-block-(styrenesulfonate) triblock
PMMA, *see* Poly(methylmethacrylate)
PNECs, *see* Predicted no-effect concentrations
Pneumonectomy, 110
PNIPAM, *see* Poly(N-isopropylacrylamide)
POCT, *see* Point-of-care assay and technology
Point-of-care assay and technology (POCT), 90
    analytical characteristics of current cardiac troponin I and T assays, 92
    aptamer, 96–97
    classification of current and highly sensitive cTn assays, 93
    comparison of performances of commercially available techniques, 95
    for cTnI and cTnT, 91
    in market, 90, 93–94
    microfluidic-based immunoassay and chip-based devices, 96
    paper-based devices, 96
    point of care under development in laboratory, 95
    roadmap in, 90
Pollutants iron NPs catabolization, 349
    chlorinated solvents, 349–350
    dyes, 350
    Fenton/Fenton-like reactions, 352
    metals, 351–352
    nitrogen, 351
    pesticides, 350–351
Pollution, 325
    abatement, 304
    iron NPs application for pollution remediation, 348
    NPs with technologies, 348–349
Polo-like Kinase gene (PLK1 gene), 137
Poloxamer 188 solution, 124
Polyacetylene, 404
Poly(7-(4-(acryloyloxy)butoxy)coumarin)-*b*-poly(N-isopropylacrylamide) (P7AC-*b*-PNIPAAm), 201
Poly(acrylic acid) (PAA), 46, 148, 173, 332
Poly (allyl amine hydrochloride) (PAH), 148

Poly (amidoamine) (PAMAM), 41, 127
    PAMAM-Ac-FITC-LCTP conjugate, 127
    PAMAM-and PPI-based dendrimers, 41–42
Polyaniline (PANI), 86, 404
Polyaromatic-conducting polymers, 404
Polycaprolactone, *see* Poly(ε-caprolactone) (PCL)
Poly (citric acid) (PCA), 137
Polyclonal antibody, 79
Polydimethylaminoethyl methacrylate (PDMAEMA), 177
Polydimethylsiloxane (PDMS), 90, 260–261, 287
    cancer cell migration in, 260
    microcontact-printed, 287
    PDMS-based microfluidic device, 269
    PDMS–AuNP composite film, 90
Poly(D,L-lactide-co-glycolide)-polyethylene glycol (PLGA-PEG), 37
Polyelectrolyte-carboxymethyl cellulose, 341
Polyene, 404
Poly(ε-caprolactone) (PCL), 199, 224
Polyethylene (PE), 125
Polyethylene glycol-poly(lactic acid) (PEG-PLA), 38
Polyethylene glycol (PEG), 10, 155, 175, 190, 220–221
    carbon nanotubes, 192–193
    gold NPs, 190–191
    iron oxide NPs, 191–192
    PEG-modified PLGA, 125
    PEG-SWNTs, 192–193
    poly ethylene glycol–coated nanoparticles for lung cancer, 125
    quantum dots, 193
Polyethylene glycol diacrylate (PEGDA), 265
Polyethylene glycol functionalized polymer (PEGP), 190
Polyethylenimine grafted chitosan (Ch-g-PEI), 126
Poly(ethylene oxide)-block-poly(pyridyldisulfide ethylmethacrylate) (PEO-b-PPDSM), 199
Poly (ethylene oxide)-poly (β-benzyl-L-aspartate), 128
Poly(ethylene oxide) (PEO), 199, 200
Polyglutamic acid (PGA), 40
Poly(glycerol succinic acid) (PGLSA), 127
Poly(hydroyethyl acrylate-b-N-isopropylacrylamide) block copolymer-coated iron oxide NPs [Fe_3O_4@P(HEA-b-NIPAm)], 199
Poly lactic acid (PLA), 289
Poly(lactic acid)-D-α-tocopheryl polyethylene glycol 1000 succinate (PLA-TPGS), 37

Poly lactic acid-polycaprolactone-d-α-
        tocopheryl polyethylene glycol 1000
        succinate (PLAPCL-TPGS), 121
Poly lactide co-glycolic acid (PLGA), see
        Poly(D,L-lactide-co-glycolide)
        (PLGA)
Poly(lactide)-poly(ethylene glycol)
        (PLA-PEG), 200
Polymer-based nanoparticles for lung cancer,
        122; see also Bio-based nanoparticles
    chitosan nanoparticles for lung cancer, 126
    dendrimers, 127–128
    PLGA nanoparticles for lung administration,
        123–125
    polymer-based nanocarriers, 130–132
    polymeric micelles, 128
    polymeric nanofibers, 128–129
    stealth polymer, 125
Polymeric colloidal particles, 122
Polymeric micelles, 51–53, 128
Polymeric nanocarriers, 122
Polymeric nanofibers, 128–129
Polymeric nanoparticles, 37–39, 224, 236
Polymer(s), 224, 236, 398
    drug conjugates, 39–41
    nanocomposites, 398, 404
    in NPs coating, 187
    polymer-based nanocarriers, 117
    polymer–nutraceutical interactions, 397
    polymer–polymer interactions, 397
Polymethacrylic acid (PMA), 135, 341
Poly(methacrylic acid)-block-(methyl
        methacrylate)-block-(styrenesulfonate)
        triblock (PMAA-PMMA-PSS
        triblock), 341
Poly(methylmethacrylate) (PMMA), 201
Poly (N-isopropylacrylamide)-polystyrene, 128
Poly(N-isopropylacrylamide) (PNIPAM), 185
Poly(N-vinylimidazole-co-N-vinylpyrrolidone)-
        g-poly(d-, l-lactide), 38
Poly(pentafluorophenylmethacrylate)
        (PPFPMA), 201
Polypropylenimine (PPI), 41, 121, 127
Poly(propylene oxide) (PPO), 200
Polypyrrole, 404
Polysaccharides, 340, 420
Polyvinyl alcohol (PVA), 124, 341
Polyvinyl pyrrolidone (PVP), 134, 173, 384, 420
Poly-(4-vinyl)phenol nanoparticles, 236
Poly-(N,N-dimethylaminoethyl methacrylate)
        (PEO-b-PDMAEMA), 199
Poly(D,L-lactide-co-glycolide) (PLGA), 37, 38,
        122, 224, 258, 392, 433

nanoparticles for lung administration,
        123–125
PoP, see Porphyrin-phospholipid
Pore size, 260, 261, 267, 290, 291, 342
Porosity, 260–261, 289–291
Porphyrin-phospholipid (PoP), 239
Portable readers, 78
Positron emission tomography (PET), 36, 40,
        235, 384
pp53-EGFP, see Plasmid DNA-encoding
        p53-EGFP
PPFPMA, see Poly(pentafluorophenylmethacr
        ylate)
PPI, see Polypropylenimine
PPO, see Poly(propylene oxide)
PRBs, see Permeable reactive barriers
Predicted environmental concentration (PECs),
        305–306
Predicted no-effect concentrations (PNECs),
        305–306
Pristine carbon nanotubes, 137
Probabilistic functions, 455
Probiotics, 391, 396, 397
PROCAM Study, see Prospective
        Cardiovascular Münster Study
Processing methods
    different forms, 391
    types, 391
Progression-free survival (PFS), 113
Project on Engineering Nanotechnologies
        (PEN), 300
Prokaryotic cells, 437
Prospective Cardiovascular Münster Study
        (PROCAM Study), 77
Prostate-specific membrane antigen
        (PSMA), 234
Protected PMA, see 3-Trimethylsilyl-prop-2-ynyl
        methacrylate
Protein(s), 420
    corona, 18–19
    microarray assays, 11–12
    protein-based nanoparticles, 146–147
    protein–nanoparticle interaction, 378
    protein–protein interaction, 11
    protein–protein interfaces, 455
Proteoglycans, 256
Prussian blue, 238
PSMA, see Prostate-specific membrane antigen
PS molecule, see Photosensitive molecule
ptBA, see α-acetylene-poly (tert-butyl acrylate)
PTT, see Photothermal therapy
Pulmonary delivery, 148
Pulmonary embolism (PE), 71–72

Pulmonary route, 121, 122
PVA, see Polyvinyl alcohol
PVP, see Polyvinyl pyrrolidone

Q

QD@SiO$_2$, see Silica-coated QDs
QDs, see Quantum dots
Quantum dots (QDs), 6, 9–10, 37, 136, 138–139, 221–222, 232–233, 236–237; see also Carbon nanotubes (CNTs); Gold nanoparticles (AuNPs); Iron oxide nanoparticles (IONPs)
    amphiphilic polymer coating, 188–189
    block copolymers, 200–201
    click chemistry, 185–186
    ligand exchange reactions, 176–177
    lipid coating over NP surfaces, 204
    PEG, 193
    routes for QD biofunctionalization, 198
    silanization of NP surface, 180–181
    surface modification with biomolecules, 197
Quantum effects, 4, 301, 324–325, 331
Quantum mechanical models, 454
Quantum size effects, 330
Quantum yield (QY), 189, 225
Quercetin, 125
QY, see Quantum yield

R

"Race Against Time", 74
Radiation therapy, 72, 110–112
Radiofrequency ablation, 110
Radioimmunoassay, 79
Radioimmunotherapy (RIT), 113
Radiolabeled inorganic nanoparticles, 236–237
Radiotracers, 235, 236
RAFT, see Reversible addition fragmentation chain-transfer
Raman spectroscopy, 239
Rare earth metals, see Lanthanides
Rate of relaxation, 229
rBM, see Reconstituted basement membrane
RDX, see Hexahydro-1,3,5-trinitro-1,3,5-triazine
Reactive oxygen species (ROS), 21, 22, 134, 269, 358, 415, 438
    ROS-induced genotoxicity, 135
Recapitulating tumor ECM
    mimicking approaches by 3D platforms for drug screening, 265–270
    mimicking properties of tumor extracellular matrix, 257–265
    tumor microenvironment, 252–257
Receptor-mediated endocytosis (RME), 19
Receptor tyrosine kinases (RTKs), 115
Recombinant human cTnT, 83
Reconstituted basement membrane (rBM), 260
Reduction mechanism for degradation, 350
Reference quantity, 17
Refractive index, 304
Remediation, 325
    application of iron nanoparticles for pollution, 348–349
    cost of iron nanoparticle, 356–357
    groundwater, 353–354
    potential use in, 328–329
    terrestrial, 355–356
    treatment and, 303
RES, see Reticuloendothelial system
Response rate (RR), 113, 156
Resting testing, 77
Reticuloendothelial system (RES), 10, 190, 220–221
Reversible addition fragmentation chain-transfer (RAFT), 185
RGD, see Arginine-glycin-aspartate
RHD, see Rheumatic heart disease
Rhenium-doped nano TiO$_2$, 302–303
Rheumatic fever, 71
Rheumatic heart disease (RHD), 71
rHF, see Human H-chain ferritin
Rho-associated protein kinase (RhoA/ROCK), 291, 292
RhoA/ROCK, see Rho-associated protein kinase
Rho GTPases, 261
Rhombohedral prisms, 330
Rifampicin, 433
Right-sized nanomaterials, 13–14
Rigidity, see Stiffness
Risk
    assessment, 314
    management, 314–315
RIT, see Radioimmunotherapy
RME, see Receptor-mediated endocytosis
RMS roughness, see Root-mean-squared roughness
Robust functionalization techniques, 439
Roche Cobas h 232 POC system, 94
Root-mean-squared roughness (RMS roughness), 258
ROS, see Reactive oxygen species
RR, see Response rate
RTKs, see Receptor tyrosine kinases

S

Saccharomyces cerevisiae (*S. cerevisiae*), 397, 426
Safety of nanotechnology, 309–315
Saliva-based nano-biochip immunoassay, 96
Salmonella typhimurium (*S. typhimurium*), 405, 425
SAM, *see* Self-assembled monolayer
Sandwich immunoassays, 79
Santa Barbara Amorphous Silica, 429
SBA 15, 343, 429
    functionalization, 433
    matrix, 433
Scanning electron microscopy (SEM), 344–345
Scanning tunneling microscope, 3
SCCP, *see* Scientific Committee on Consumer Products
SCENIHR, *see* Scientific Committee on Emerging and Newly Identified Health Risks
scFv, *see* single chain antibody fragment
SCID mice, *see* Severe combined immune-deficient mice
Scientific Committee on Consumer Products (SCCP), 301
Scientific Committee on Emerging and Newly Identified Health Risks (SCENIHR), 301
Screen-printed carbon electrode surface (SPCE surface), 189
Screen-printed electrode (SPE), 87
Screen-printed microarray (SP microarray), 84
SDS, *see* Sodium dodecyl sulfate
Second-harmonic light scattering (SHS), 175
Secondary electrons, 345
Second Gokhstein equation, 458
Segmentectomy, 110
SELEX, *see* Systematic evolution of ligand by exponential enrichment
Self-assembled monolayer (SAM), 88, 184
SEM, *see* Scanning electron microscopy
SERS, *see* Surface-enhanced Raman spectroscopy
Severe combined immune-deficient mice (SCID mice), 137
S–H groups, *see* Sulfhydryl groups
Shigella flexneri (*S. flexneri*), 405
SHS, *see* Second-harmonic light scattering
SI-ATRP, *see* Surface-initiated atom transfer radical polymerization
Signaling pathways, responses of cells and, 291–292
Silanization of NP surface, 178
    carbon nanotubes, 180
    gold NPs, 178–179
    iron oxide NPs, 179
    quantum dots, 180–181
Silica-coated AuNPs (Au@SiO$_2$), 178, 179
Silica-coated QDs (QD@SiO$_2$), 181
Silica@Au nanoshells, 133
Silica nanoparticles (SNs), 48
Silica nanospheres, 433
Silicon, 333, 375
Silicon dioxide (SiO$_2$), 429
    SiO$_2$ (E-551), 436
Silicon nanowire (SiNW), 86
Silicon quantum dots (SiQDs), 185
Silver (Ag), 133, 400, 415
    ENPs, 306, 308, 314
Silver nanoparticles (AgNPs), 12, 134–135, 381, 416, 433, 435, 436
    case study for, 381–385
    AgNP-PVP, 420
Silylated natural fatty acids, 434
Simple cost-effective fabrication, 439
Simplex detection, 94
Single-photon emission-computed tomography (SPECT), 36, 235, 384
Single-walled carbon nanotube (SWNT), 10–11, 136, 234, 307
Single carbon atoms, 402
single chain antibody fragment (scFv), 122
Single PANI nanowire, 86, 87
SiNW, *see* Silicon nanowire
SiQDs, *see* Silicon quantum dots
siRNA, *see* small-interfering RNA
SlimShake Chocolate, 391
SLNs, *see* Solid lipid nanoparticles
small-interfering RNA (siRNA), 122
Small molecule inhibitors, 115
"Smart" food packaging system, 403
Smartphone based-approaches, 91, 92
Smoking, 72, 75
SNs, *see* Silica nanoparticles
SOD, *see* Superoxide dismutase
Sodium borohydride, 335, 344
    for generation of nZVIs, 362
    reduction of aqueous salts method, 357
    sodium borohydride-generated particles, 335
Sodium dodecyl sulfate (SDS), 89
Sodium hydroxide, 332
Sodiumiodide symporter gene (hNIS gene), 42
"Soft corona", 381
Soil, 309, 356
Sol-gel process, 180, 283
Solid-base KF/CaO nanocatalyst, 305

Solid-support immobilization of iron nanoparticles, 343
Solid lipid nanoparticles (SLNs), 48–49, 148–149, 424, 427–428
Solid supports, immobilized nanoparticles on, 343–344
Solid tumors, 257
Solvents, 304, 439
    chlorinated, 342, 349–350
    dispersive, 197
    non-food-grade, 396
    organic or aqueous, 137
    polar and nonpolar, 175
Sonochemical, 337–339
SP-A, *see* Surfactant protein-A
SPCE surface, *see* Screen-printed carbon electrode surface
SPE, *see* Screen-printed electrode
SPECT, *see* Single-photon emission-computed tomography
Spheroids, 266
SPI-077, 147
SPIONs, *see* Superparamagnetic iron oxide nanoparticles
SP microarray, *see* Screen-printed microarray
SPMNBs, *see* Super-paramagnetic nanobeads
Spoilage microorganisms, 425
SPR, *see* Surface plasmon resonance
SPR AUTOLAB SPIRITs, 89
Staphylococcus aureus (*S. aureus*), 425
Starch, 340, 425
Stealth polymer, 125
STEMI patients, *see* ST-segment elevation myocardial infarction patients
Stiffness, 261–263
Stress, 73, 292
Stress testing, *see* Exercise testing
Stroke, 70, 72
ST-segment elevation myocardial infarction patients (STEMI patients), 91
Succinimide 4-(N-maleimidomethyl)cyclohexane-1-carboxylate (Sulfo-SMCC), 197
Sulfation pattern, 256
Sulfhydryl groups (S–H groups), 133
Sulfo-SMCC, *see* Succinimide 4-(N-maleimidomethyl)cyclohexane-1-carboxylate
Sulfuric acid, 309
Super-paramagnetic nanobeads (SPMNBs), 95
Superoxide dismutase (SOD), 145
Superoxide radical, 302–303
Superparamagnetic iron oxide nanoparticles (SPIONs), 125, 135, 183–184

SPIONs-P(GMA-co-PEGMA)-FA, 184
SPIONs-P(GMA-co-PEGMA)-N_3-FA functionalized NPs, 184
Surface-activated poly(L-lactic acid), 285
Surface-engineered targeted dendrimers, 127
Surface-enhanced Raman spectroscopy (SERS), 8, 239–241
Surface-initiated atom transfer radical polymerization (SI-ATRP), 199
Surface area of zerovalent iron, 303
Surface chemistry, 264–265, 285
Surface chemistry of tumor ECM, 264–265
Surface functionalization, 7, 15, 439
Surface modification, 283
    amphiphilic polymer coating, 186–190
    click chemistry use, 181–186
    ligand exchange reactions, 173–178
    lipid coating over NP surfaces, 201–204
    of nanomaterials, 172–173
    polyethylene glycol use, 190–193
    silanization of NP surface, 178–181
    surface modification with biomolecules, 193–197
    use of block copolymers, 197–201
Surface of nanomaterial, 18, 204, 285–286
Surface plasmon, 7, 8
Surface plasmon resonance (SPR), 43, 88, 238, 325
    angle, 88
    changes in refractive index, 89
    immunoassay technique, 88
    immunosensor, 88
    pattern, 178–179
    surface plasmon resonance-based detection, 88–89
Surface waters, 355
Surfactant protein-A (SP-A), 136
Surfactants, 53, 331, 395
    charged, 204
    hot organic, 336
    nonionic, 179
Surgery, 108, 110
Sustainable nanotechnology, 4
SWNT, *see* Single-walled carbon nanotube
Synchrotron radiation, 347
Synthetic polymers, 224, 266, 268, 290
Systematic evolution of ligand by exponential enrichment (SELEX), 97
Systemic radiotherapy, 108

T

Tagged antibodies, *see* Conjugated monoclonal antibodies (Conjugated mAbs)

TAK1, see TGF-β-activated kinase
Tarceva, see Erlotinib
Targeted antigens, 114–115
Taxol®, 124, 152
T4 bacteriophage, 146
TBL approach, see Triple-bottom-line approach
TC, see Tricaprin
TCE, see Trichloroethylene
T_1 contrast agents, 229
T_2 contrast agents, 229–230
TDPA, see Tetradecyl phosphoric acid
Tea catechins, 393
Tecentriq, see Atezolizumab
TEM images, see Transmission electron microscopy images
Temperature history, 403
TEOS, see Tetraethoxysilane
Tergenpumatucel-L, 115
Terminal deoxynucleotidyl transferase dUTP-mediated nick end labeling assay (TUNEL assay), 23
Terrestrial remediation, 355–356
1,4,7,10-Tetraazacyclododecane-1,4,7,10-tetraacetic acid (DOTA), 236
Tetracycline, 433
Tetradecyl phosphoric acid (TDPA), 173
Tetraethoxysilane (TEOS), 178
Tetrahydrofuran (THF), 181
Tetra octylammonium bromide (TOAB), 174
Tetravalent metal oxides, 375
TF-SCMNPs, see Thiol-functionalized silica-coated magnetite NPs
TG4010, 146
TGF-β, see Transforming growth factor beta
TGF-β-activated kinase (TAK1), 188
TGF, see Transforming growth factor
Theranostic approaches, 116, 157
Theranostic nanomedicine; see also Nanomedicine
  carbon-based nanostructures, 55–57
  dendrimers, 41–43
  gold nanoparticles, 43–45
  liposomes, 49–51
  magnetic nanoparticles, 46–48
  nanoemulsion, 53–55
  polymer drug conjugates, 39–41
  polymeric micelles, 51–53
  polymeric nanoparticles, 37–39
  silica nanoparticles, 48
  solid lipid nanoparticles, 48–49
Therapeutic agents, 38, 43
Therapeutic/diagnostic nanocarriers, 117–118
Therapeutic substances, 35–36
Therapeutic vaccines, 114–115
Thermal decomposition method, 336
Thermal stability, 192, 391, 393, 398, 428
THF, see Tetrahydrofuran
Thin-layer chromatography coupled to mass spectrometry (TLC/MS), 24
Thiol-functionalized silica-coated magnetite NPs (TF-SCMNPs), 179
Thiolated chitosan, 121
Thiols, 176, 240, 267, 416
Three-dimension (3D)
  cultures, 252
  printing techniques, 283
  scaffolds, 289
Thymol, 425, 427, 431
TILs, see Tumor-infiltrating lymphocytes
Timestrip, 403
Time–temperature integrators (TTIs), 398, 403
Tin oxide (SnO_2), 87
Tissue(s), 221
  engineering, 280
  engineering/regenerative medicine strategies, 293
  morphogenesis, 252
  stiffness, 291
  targets, 120
Titanium, 128, 375
Titanium dioxide (TiO_2), 394, 415
  nanoparticles, 404
  nanotube arrays, 304
  NPs, 417
  TiO_2 (E-171) NPs, 436
TLC/MS, see Thin-layer chromatography coupled to mass spectrometry
TLR, see Toll-like receptor
TNF, see Tumor necrosis factor
TnT, see Troponin T
TNX, see Hexahydro-1,3,5-trinitroso-1,3,5-triazine
TOAB, see Tetra octylammonium bromide
Tobacco, 75, 107
Tocopheryl polyethylene glycol succinate (TPGS), 50
  TPGS-based theranostic micelles, 52
Tocopheryl polyethylene glycol succinic acid (TPGS-COOH), 37
Toll-like receptor (TLR), 115
TOPO, see Trioctylphosphine oxide
Topography, 258
  local nanotopography, 287
  surface, 268, 345
Topology of gold nanoparticles, 7–8

Toxicity
    of ENPs, 305–306
    mechanism of metallic NPs, 438
    nanomaterial, 437–439
TPGS, see Tocopheryl polyethylene glycol succinate
TPGS-COOH, see Tocopheryl polyethylene glycol succinic acid
Tractive forces, 459
TRAIL protein, see Tumor necrosis factor-related apoptosis-inducing ligand protein
Trans-cinnamaldehyde nanoemulsions, 425, 426
Transcytosis, 152
Transforming growth factor (TGF), 291–292
Transforming growth factor beta (TGF-β), 188
    pathways, 292
    TGF-β1, 184
Transmission electron microscopy images (TEM images), 22, 345, 416
1,4,7-Triazacyclononane-1,4,7-triacetic acid (NOTA), 236
Triblock polymer (PLGA-PEG-PLGA), 125
Tricaprin (TC), 149
Trichloroethylene (TCE), 333, 356
Trimethylamine, 404
3-Trimethylsilyl-prop-2-ynyl methacrylate (Protected PMA), 183
Trioctylphosphine oxide (TOPO), 181
Triple-bottom-line approach (TBL approach), 4
"Trojan Horse Effect", 382
Troponin T (TnT), 81
    primary antibody, 96
    TnT 1 and 2, 83
TTIs, see Time–temperature integrators
Tumoral endothelium, 120–122
Tumor extracellular matrix; see also Extracellular matrix (ECM)
    biochemical properties, 263–265
    components, 253–256
    geometrical properties, 257–261
    mechanical properties, 261–263
    mimicking properties, 257
Tumor-infiltrating lymphocytes (TILs), 115
Tumor necrosis factor-related apoptosis-inducing ligand protein (TRAIL protein), 147
Tumor necrosis factor (TNF), 147
    TNF-α, 188
Tumor(s), 261
    cell invasion and migration, 260
    components of tumor extracellular matrix, 253–256
    extracellular matrix role in regulation of cancer cell behavior, 256–257
    microenvironment, 252
    microvasculature, 119
    tumor-selective paclitaxel accumulation, 42
TUNEL assay, see Terminal deoxynucleotidyl transferase dUTP-mediated nick end labeling assay
Tungsten x-ray targets, 78
Type 2 diabetes, 73

U

UA, see Uric acid
UCNPs, see Upconverting nanoparticles
UCST, see Upper critical solution temperature
Ultrafast computer tomography (UFCT), 78
Ultrasound imaging, 36
Ultraviolet (UV), 185
    irradiation, 404
    radiation, 139
    UV-activated oxygen indicator membrane, 404
    UV-A light, 404
Uncoated QDs, 139
UNGA, see United Nations General Assembly
Unhealthy eating, 73
Unit cell, 345
United Nations General Assembly (UNGA), 73–74
United States Environmental Protection Agency, 300–301
Universidad Valencia Material–7 (UVM-7), 429
Upconverting nanoparticles (UCNPs), 141, 227
Upper critical solution temperature (UCST), 199
Urban modernization, 75
Uric acid (UA), 189
US Environmental Protection Agency (USEPA), 300
USEPA, see US Environmental Protection Agency
US Food and Drug Administration (FDA), 114, 224, 392
US National Nanotechnology, 300
UV, see Ultraviolet
UVM-7, see Universidad Valencia Material–7

V

Vaccine, 114
    anticancer, 114
    autologous Dribbles, 115
    therapeutic, 114–115

Van-MSNs, *see* Vancomycin-modified mesoporous silica nanoparticles
Vancomycin-modified mesoporous silica nanoparticles (Van-MSNs), 434
Vancomycin, 433, 435
Vascular cell adhesion molecule-1 (VCAM-1), 120
Vascular endothelial growth factor (VEGF), 12, 120, 256, 284
   VEGF-165, 12
   VEGFR2, 115
VATS, *see* Video-assisted thoracoscopic surgery
VCAM-1, *see* Vascular cell adhesion molecule-1
V-chip, 96
VEGF, *see* Vascular endothelial growth factor
Vertical flow immunoassay (VFIA), 95
Vessel wall MRI (VWMRI), 77
VFIA, *see* Vertical flow immunoassay
Video-assisted thoracoscopic surgery (VATS), 110
Viral nanoparticles (Viral NPs), 145–146
Viruses, 229, 398, 434
Virus shells, 224
Volatile organic compounds (VOCs), 133
VWMRI, *see* Vessel wall MRI

W

Water, 307–308
   droplet, 342
   remediation, 331–332
   vapour, 399
   water-dispersible lycopene nanostructures, 392
   water-soluble MWCNTs, 189
   water-soluble vitamins, 396
Wedge resection, 110
WHF, *see* World Heart Federation
Wild-type p53, 149
Wnt/β-catenin, 291–293
World Health Organization, 73
World Heart Federation (WHF), 75
Wrapping-time effect, 20

X

X-ray absorption spectroscopy (XAS), 347
X-ray diffraction (XRD), 335, 345–346
X-ray photoelectron spectroscopy (XPS), 347
X-rays, 78, 346
   monochromatic, 347
   source, 225
   special, 78
Xyotax®, 116

Z

Zeolites, 304, 415, 424, 429
Zero-valent iron, 350, 351
Zinc, 375–376
Zinc oxide (ZnO), 415
   ENPs, 300
   nanomaterials, 21
Zinc oxide nanoparticles (ZnO NPs), 354, 417, 433
Zinc sulfide (ZnS), 10
Zygosaccharomyces bailii (*Z. bailii*), 425

PGSTL 04/05/2018